# MATHEMATICAL AND PHYSICAL PAPERS

IN TWO VOLUMES
VOLUME II

# MATHEMATICAL AND PHYSICAL PAPERS

BY

## SIR JOSEPH LARMOR, Sc.D., F.R.S.

HON. F.R.S. EDIN.: HON. MEM. R. IRISH ACAD.: HON. MEM. ASIATIC
SOC. OF BENGAL, MANCHESTER LIT. AND PHIL. SOC.: HON. FOR. MEM.
U.S. NATIONAL ACAD. OF SCIENCES, AMERICAN ACAD. OF SCIENCE
AND ARTS, BOSTON, AMERICAN PHILOSOPHICAL SOCIETY, PHILA-
DELPHIA, WASHINGTON ACAD., R. ACCADEMIA DEI LINCEI, ROME,
ISTITUTO DI BOLOGNA: CORRESPONDANT OF THE INSTITUTE OF
FRANCE: LUCASIAN PROFESSOR IN THE UNIVERSITY OF CAMBRIDGE,
AND FELLOW OF ST JOHN'S COLLEGE

VOLUME II

CAMBRIDGE
AT THE UNIVERSITY PRESS
1929

## CAMBRIDGE
### UNIVERSITY PRESS

University Printing House, Cambridge CB2 8BS, United Kingdom

Cambridge University Press is part of the University of Cambridge.

It furthers the University's mission by disseminating knowledge in the pursuit of
education, learning and research at the highest international levels of excellence.

www.cambridge.org
Information on this title: www.cambridge.org/9781107536401

© Cambridge University Press 1929

First published 1929
First paperback edition 2015

A catalogue record for this publication is available from the British Library

ISBN 978-1-107-53640-1 Paperback

# PREFACE TO VOLUME II

EXTENSIVE groups of omissions have been found to be necessary in the present volume. This has arisen mainly from unforeseen insertion of new material in both volumes during the last two years. The fascination of the origins of thermodynamics, the science which essays to make the connections between practical mechanical principles and the underlying atomism, has again asserted itself: and further attempt has been made in Appendices and Notes to disentangle the romantic history of the evolution of foundations in that domain, with the mainly statistical outlook which its generality imposes. Here the names of Carnot, Kelvin, Clausius, Maxwell, Boltzmann, Rayleigh, Willard Gibbs are predominant. Also the temptation to probe closer into the nature of natural radiation, its specification and its mode of dispersal, was not to be resisted. But there it has been confined mainly to verifying just how far, in various respects, the original well-worn ideas of electrical radiation, based on the limited analogy of transmission of sound, yet perhaps in many ways as will appear holding the kernels of recent abstractions, can carry. Apology for such restriction is hardly necessary: the modern constructs in the problems of quantified spectroscopy, with their atomic storage of energy, many of them highly successful in their special fields, have hardly been entered upon, because the vast and tentative literature could not be justly appreciated except by a critic closely cognizant of the diverse evolutions of the last fifteen years in this field of knowledge. That is the modern penalty against expansion of interests: yet complete knowledge up to date is not indispensable to profitable consideration. Even in the astrophysical domain history can repeat itself (cf. p. 586) on wider foundation of facts: while by aid of analysis, of transcendental types as introduced originally through Hamilton's Quaternions, constructive thought had long ago essayed to pioneer beyond the range of formal proof, bringing back to physical reality its harvest of results. Meantime the secured territory behind the ragged frontiers of knowledge repays intensive cultivation.

In these subjects that are for the future to consolidate, as in all other departments of physical theory, the absence of Professor Lorentz, recently removed from science, leaves a blank that cannot be filled: which adjoins to other recent severe losses, including Arrhenius, Dewar, K. Onnes, Wien.

As before, in addition to various Postscripts and eight Appendices, the extensive new footnotes are again indicated by marks such as asterisks instead of numbers: other substantial additions or modifications are put in square brackets. The index, in addition to being a record, may help to consolidate the correlations of the variety of subjects that come under discussion.

Corrections are not infrequent. It has been judged to be informing, as well as historically incumbent, to amend without obliterating the original mistakes, except in trivial cases. This course may tend to interpose delays on a rapid survey of the conclusions. But clarity and finality are reached largely through the clash of alternatives, all kept well in view: there is authority for the opinion that systematic effort towards due articulation of the historical course of progress in an intricate domain, not always improved by division into chapters, may instruct the writer himself and the critic, at least as much as it is calculated to inform the general reader in quest of final results.

The additions are thus largely of the nature of general survey, not encumbered by details of partial developments. Especially in topics of widespread physical concatenation, a condensed phrasing, even though elliptical, in ordinary language, may be a time-saving compromise: exploring students, requiring rapid recovery at will of a field of view as a single whole, may prefer it to ramifying general description or to algebraic formulation with its apparatus of special schedules, however ill it may be adapted for more cursory reading. Algebraic analysis, outside the realm of computation, has indeed to run in blinkers, though they may be ultramundane as in the multiplex geometry: only in combination with a general physical intuition does it become the source and the expression of expanded outlook.

Among omissions that need not be specified, including a set of Articles around 1902 for a supplement of the *Encyclopaedia Britannica*, the writer cannot help regretting the absence of a series of scientific biographical Notices prepared at various times: especially an intensive account of the activities of Lord Kelvin, drawn up with considerable care and research for the Royal Society (*Proceedings*, 1908, pp. i–lxxvi), and intended to assist in exhibiting a synopsis of his career in relation to the scientific progress of his age, from an evolutionary point of view hardly perhaps now much in sight.

The other main omission is a series of papers aiming towards an understanding of the scope and significance, as distinct from the mathematical development, of the modern doctrines of relativity, identified here with invariance of physical aspect as regards all frames of reference belonging to the proper group,—those that are permissible in conformity with the slowness of the messenger rays of light,—for the external world in its electric and its astronomical features. On the astronomical side this problem may claim to have been set once for all by Bradley, the great founder of modern practical astronomy of precision, as far back as Newton's day: he provided the practical solution which has remained adequate ever since. The exploration towards an ideal solution which would be mathematically exact, without any need for approximation, has been stimulated afresh in recent times by the null results of very close physical experimentation starting with Michelson, ultimately evolving towards a formal mathematical calculus through novel avenues of approach to the subject opened out mainly by Einstein. Perhaps in its present stage the theory may be held to lay alongside the actual world which we know, as in recent times most closely explored and formulated in the marvellously precise records of the astronomers, a self-contained artificial map of reference, made up solely of vectors and tensors after the manner of line-geometry, thus subject to the rules of the intricate special discipline known as the Mathematical Theory of Relativity. In this auxiliary cosmos space and time and motion do not occur: yet it can be of great value in a geographical sense, after the manner of a spherical map of the Earth, for unravelling the intricacies of relations connecting regions in our actual world, which astronomers are permitted to formulate only in terms of the complications of delay in the messages of the informing rays of light. The parallelism of relations between the world of actual perception and memory and this particular conglomerated fourfold assembly of permanent ray-models, so to say, extends over a prominent yet necessarily limited range whose boundaries have hardly yet been very closely examined. There may be, however, people who aim at transferring their whole life into this new cosmos. The papers now omitted are concerned largely with various general aspects of this correlation: naturally much of their contents has now become transcended, or modified into improved presentations. A list of these papers has been inserted after the Table of Contents. .

In such ideal constructions one may readily continue to go astray: yet the writer has ventured now, partly for his own clarification, on a sketch of what appears to him at present to be the effective mode of formulation in this field, describable perhaps as complete coherence between recipient minds and a unique external world. It is postulated as a basis for systematic exploration that there may be potential astronomers everywhere in the universe, each group of them active only in their own local environment, except so far as information from outside arrives by the delayed rays of light and in no other way. It is postulated also that such astronomers can be all in instantaneous understanding with one another as regards results thus acquired—a state of affairs at first sight prohibited*, which yet is approached indirectly through the mental process of memory in relation to the adjusted records of the observations of actual astronomers made in all positions along their extensive terrestrial orbit round the Sun, this being in fact the method of formulation tacitly adopted for the construction of the actual science. The question is whether the worlds that these groups of astronomers separately analyse out for themselves could be recognized by them, in joint ideal conference across space and time, as the same extraneous world for all, whether in fact there is an unambiguous external world. Absolutely, such recognition is presumably always possible, if there is a world, for signals can be conceived as instantaneous; but the complexity of an actual identification, mathematically exact, based on the delayed light signals, might be overwhelming. Thus practically the procedure of optical relativity— much more involved than the original instantaneous inertial relativity of Newton—would be to explore this field, with a view to verify that no overt discrepancies appear that are definitely outside feasible modes of adjustment. To this limited extent may astronomy and general physics perhaps make contact with psychology. A summary on these lines has been included in the final Appendix: the subject is our mode of apprehension of the Universe, as valid for all observers in space and time, and some degree of repetition may be pardoned.

The main feature that is involved practically is, as would be expected, a closer formulation as regards the specification of times. Here also a different and more direct problem has been ripening for the last half-century, to which attention is paid in various papers in this collection; namely, how to provide a scale of Newtonian time for dynamical astronomy that shall include correction for the ascertained irregularity in the Earth's axial rotation (cf. p. 479). The scale of time thus sought for would on these ideas (p. 769) be the one belonging to astronomers attached to the Sun, on the presumption that the Solar motion in space has negligible acceleration.

The survey, undertaken systematically after a long interval, of procedure in the early fundamental constructs, has involved some reconsideration of the resulting classical scheme of electrical and optical theory, a delicate undertaking as now conducted in the high light of its modern incapacities. Important and surprising as these are, and fruitful as are the mobile tentative syntheses in spectroscopy and thermionics

---

* This however actually expresses the situation in the cognate problem of ranging for positions in modern artillery practice: the observers acquire direct knowledge only through the delayed sound-waves, but they can confer with one another instantaneously (though not with the sources) by telephone.

and other domains internal to the atoms that they have promoted, the impression may be allowed to persist that, neither in extent as compared with the coherent territory of our inherited science—with its lacunae concerned with defect of convergence to a limit in atomic radiative phenomena, which were never out of sight—nor in inevitableness as regards a comprehensive understanding of the new phenomena that are involved, do they justify a general negative attitude such as it is easy to acquire. The question indeed rather forces itself, whether the load of duly verified discrepancy, especially in regions involving statistical thermodynamic theory, has not been somewhat exaggerated.

In another aspect, the view that modern physical science had erred by basing itself on a self-sufficient narrow materialism is historically one-sided, as has been recognized for example long ago (1899) by the psychologist James Ward in the Preface to his treatise *Naturalism and Agnosticism*. In recent years the magnificent and in many ways startling output in constructive astrophysical results, mainly from the great observatories of the Pacific Coast, Mount Hamilton and Mount Wilson and now Victoria, following the early leads of Herschel, Rosse and I. Roberts, has in its turn promoted the new exploring trend in dynamical molecular studies, thus stimulated by the spectacle of vastness interacting with primordial simplicity of law. Brief discussions in this direction have here been adventured, for example as regards transmission through nebulae and also the Solar magnetic fields.

Even the working hypothesis of the earlier papers of this collection, involving a rotationally elastic aether with its essential innate fluidity, which towards the end of last century had been eliminated largely as too materialist and naïve as a representation of the electro-optical field, can now find more to be advanced on its behalf than would have been at first anticipated. It implied naturally an intrinsic isotropic pressure in the fluent aether. The problems of physical optics, then the most prominent, did not however require its intervention: the dynamical laws of refraction after Fresnel involved no excitation of pressure in the optical medium, a fact that became familiar in the physical school of Stokes and Kelvin. That was on the postulate that the optical elastic aether was to be taken as virtually incompressible, so that any initiation of pressure however arising became adjusted throughout with speed relatively infinite: finite pressural waves propagated in time had to be excluded in optics. The standard Maxwellian equations of the electric field, supporting themselves in this theory on its electronic singularities, were naturally restricted to this implied condition. But the wider possibilities that became obvious in the rotational model could not of course entirely be put aside. If the elastic aether were almost but not quite incompressible, pressure originating anywhere, atomically if not optically, would be promptly equalized universally, not indeed to a static distribution after the manner of an instantaneous Laplacian potential, but by elastic wave-adjustment with velocity very great yet definite, and therefore proportionately difficult to excite or sustain except in intimate processes within the very intense field of the atom itself. The electric vector potential is thus to have divergence, which is elastically propagated. And questions then presented themselves obviously, such as whether this hydrodynamic pressure, an essential feature of the aether model, might not even involve a clue to the universal gravitation, so potent in phenomena on a large scale yet locally so slight. Cf. Vol. i, p. 500.

The Maxwellian electric scheme permitted a principle of relativity to be imposed upon it, as interpreted here into the postulate that there is an external world, of electrons and atoms composed of them, which is identifiable as the same world by all observers who are informed by the messages of radiation, however rapidly these observers may be travelling. The origin of this theory was the ascertained null effects, in various directions, of the convection of the observer's field, up to the second order. In contrast to this permissive relativity, is the fact that in the rotationally elastic model such optical relativity is compelled, being essential to the coherence of the scheme: the necessary aether pressure there demands the convective shrinkage of the system. Moreover this intrinsic pressure imparts work to the shrinking electron, putting into its system energy of amount proportional to $v^2$ in addition to that required for its electrodynamic field, making up in all $\frac{1}{2}e^2/C$, where $C$ is the electrostatic capacity of the shrunken system, with inertia in proportion; thus interpreting dynamically Vol. i, p. 672.

When the detection of free travelling electrons thirty years ago, and their enormous speeds as $\beta$-rays, led to a closer scrutiny of the original working model of an electron (cf. *supra*, Vol. i, p. 520, in relation to Vol. ii, p. 226) as a small spherical region on which lines of the elastic twist representing an electric field abutted, at right angles on account of complete loss of elasticity inside the sphere, the rotational aether scheme still remained effective for that extension of the picture. Under the influence of rapid convection through the fluid aether this spherical core still managed to subsist, deforming itself into a steady ellipsoid, thus becoming in fact, and of its own inner necessity without possibility of any alternative, the Lorentz model of a relativity electron (cf. p. 276). But it has been emphasized at various times, among others by Poincaré and by Lorentz, that even the mere existence of a mechanical model of this kind should be regarded as debarred, because forces would have to be introduced from outside in order to sustain it which are foreign to the electric field. The forces that are needed prove, as they ought, to be none other than an excess in the model,—necessarily constant all over each surface of an electron,—of this hydrostatic aethereal pressure on the outside of the ellipsoidal boundary over that on the inside. So far then from a dynamical scheme failing to include electrons on account of such necessity for extraneous force, this rotational model of the aether, from which the concept of an electron was originally derived, proved to be precisely adapted to the new emergency, the sole type of adjusting internal stress which it could supply in addition to the Maxwellian electric forces being just what was required. Any initial inclination to restrict the core of the electron by constraint to a constant volume had, on the initiation of Lorentz, to be definitely forgone, optical relativity forbidding: yet here again, this feature conforms, for the excess of fluid pressure can then remain constant as regards the same electron however convected. Cf. p. 810: also the more general statement in *Aether and Matter*, § 96. For different electrons, once the charge is somehow determined, this difference of pressures fixes the radii and the masses. If then the cores are regarded as vacuous spaces devoid of pressure the conditions are thus satisfied for identical electrons simply by a constitutive pressure unchanging throughout the aether. The drawback, however, which has always been in mind, is the circumstance that this hollow electron model would have to be made secure against obvious instability as regards change of form: though this can be effected by modification relatively very slight provided the disturbances are not too great,—*e.g.* conceivably by putting a limit to the intensities for which the linear equations of the field hold good, which would be exceeded in the enormously intense fields close to an atom, a procedure which amounts to falling back upon local dispersive quality such as the more recent developments in atomic dynamics have strikingly demanded. But the problem of the positive electron remains.

Coming down now to recent atomic schemes, this old idea of hydrodynamic pressure in the aether, as a natural supplement to the Maxwellian field of electrodynamic force, is again forcibly brought to mind. The most arresting of the recent constructions, in interpretation of the *corpus* of empirical numerical *quantum* relations built up on the spectroscopic side, has been the wave equation set up by E. Schrö-dinger, involving only one scalar independent variable, emerging from earlier more geometrical constructs by L. de Broglie, as on p. 552; he discovers that one suitably adapted differential equation could prescribe the complete *corpus* of precise *quantum* numbers for a vibrating atomic system in very various tractable cases, much as the single wave equation of aereal pressure determines frequency numbers for the various types of sound-vibration in an organ-pipe or other bounded aereal region. This equation regulates the distribution of one scalar quantity, at first sight very foreign to the vectors which describe the electric field: and being successful it does not contradict the formal presentations by matrices. Yet let us revert again to the problem of modes of vibration of a set of atomic singularities imbedded in an isotropic elastic medium, resisting deformation, or rotation its field equivalent in the electric aether, now in combination with strong but not infinite resistance to compression. The procedure for the treatment of such problems for an isotropic medium, familiar since the early days of Stokes and Lamé, manages to separate out by itself the pressural vibration: the scalar field pressure satisfies its own differential equation, recalling therefore this esoteric equation of Schrödinger. For, pursuing the analogy, if the electric singular points of the aethereal vibrating system of the atom are not to be also singularities, after the manner of the actual vesicles filled with compressed gas that occur in crystals, from or to which intense hydrostatic pressure radiates other than the intrinsic electrically sustained constant universal pressure aforesaid—and any such added source of pressure would probably spoil the type of model of a convected electron—we are invited to seek out forms of solution for disturbance superposed on this intrinsic aethereal pressure, adapted to the local field of the atom, of types which indeed converge radially on the positions of its electrons or other nuclei in a way prescribed by their electric field, but without tendency towards values there increasing without limit; for stable modes of local vibration of the ambient aether could only occur around the definite configurations of the material atom permitting of such solutions of the pressure-equation, which alone could subsist. The selective radiation which escapes from an atom would be presumed to arise, in the manner now frequently envisaged on the basis of the combinatory law of spectral series, from temporary interference excited between pairs of states of pure vibration, after the manner of the Helmholtz theory of difference tones in acoustics except that these states are themselves now self-contained and non-radiating. This train of ideas hardly amounts to more than illustration, possibly far removed in its simplicity from the necessities of the picture: though the mental analysis of Nature has never yet turned out to be other than simple in the long run. Half the Schrödinger procedure is left still in obscurity. Yet it may reasonably give some support to the position that the original working electronic model, as directly compelled by a rotational aether, and now recognizing explicitly and adapting pressural propagation in conjunction with electric, may still—crude perhaps but within its range wholly concatenated—have claims to be retained in mind alongside other clues to consistent theoretical formulation.

There is another impressive, because less special yet wholly coherent, analytic procedure for linking up the sources with the field through the universal Action formula, generalized if need be so as to include aethereal pressure—a method which originated indirectly in Maxwell's stress-formulations for the transfer into precise analytic form of the general idea of Faraday (dating from the earliest beginnings of electrodynamics in 1821, cf. p. 763) of a connecting continuous medium in internal tension. If this elastic ultimate medium is imagined mathematically to be severed along an interface, such stress would be determinate as the type required to be applied over the severed face to replace the previous support of the medium beyond, itself already kinetically equilibrated internally by conforming to the equations of the field: the stress, which is quadratic, is relative to a frame of reference though it presents a fourfold invariance, because the variational virtual displacement from which it arises (p. 800) is a movement in that frame. As thus inherent directly in the ultimate formulation by the Action, this quadratic stress, which became the essential foundation of the relativist fourfold dynamical construct by tensors, does now acquire significance other than merely algebraic, by appearing as a necessary element in the interlaced systematization, based on the single local Action function, of the physical manifestations of the field which are otherwise mainly of undulatory type.

With avenues such as these as yet not fully explored, it must be premature to conclude that the classical foundations of electric and optical science are to be regarded as undermined. It has never of course been suggested that any mechanical model is more than an aid to insight, by providing a vivid picture of the particular group of interacting relations of the phenomena that is in hand: the oft-repeated demur, that if there could be one model there would be an infinite number, can only be a different mode of expression of the familiar common ground that, after the groups of relations relevant to the model have been reduced by its aid to coherent and condensed algebraic statement, the scheme of equations that survives may be interpreted as the manifestation of a unique abstract dynamical Action of local structure, independent of models altogether.

Recently there has been a revival of emphasis on the limitations which must affect the representation of a molecular world by the machinery of a continuous mathematical analysis. The foundations here have been an affair of the Fourier theorem: for natural radiation the ultimate analysis is into groups of waves, each presenting itself as a differential element in the Fourier integral, and the restrictions that arise have come into view in various connections (cf. p. 549). As physical knowledge has developed in recent years, the analysis has trended empirically towards arithmetization, under statistical treatment concerned with elements or *quanta* of correlated magnitudes grouped into cells, dynamically consistent and of definite extents (cf. p. 401). It would thus seem that any prospect of assistance from further improvement of the abstract Fourier analysis has vanished, just because the transition from atomism to continuous fields is possible only in the rough. But why the continuous presentation of the physical world has to stop at a definite scale of coarseness remains an unfathomed problem.

The principles of Hydrodynamics, in the hands of Stokes, Helmholtz and Kelvin, and later of an active school headed by Osborne Reynolds and Rayleigh, developed at first largely in contact with the practical problems of propulsion of ships. Besides exerting a conspicuous reaction with general physical theory, these ascertained laws with their empirical limitations have now come to an apotheosis of their own, as the rational foundation, notwithstanding some disparagement, of all improvement in the principles of aviation. In later papers attempt is made to bring this historical mode of discussion, resting on general principles, under concise review in that domain, so far as the requisite technical information has been found to be readily accessible in systematic form.

The writer is again greatly indebted to the skill and attention of his collaborators at the Cambridge University Press.

J. L.

CAMBRIDGE
*September* 1928

# CONTENTS

## PAPERS RELATING TO PHYSICAL RELATIVITY
## NOT HERE REPRINTED· (*See* p. vii.)

On the Essence of Physical Relativity.
   [*Proc. U.S. National Academy of Sciences*, Vol. IV, Nov. 1918, pp. 334–337.]

On generalized Relativity in connexion with Mr W. J. Johnston's
   Calculus.
                [*Proc. Roy. Soc.* Vol. XCVI A, Aug. 1919, pp. 334–363.]

The Relativity of the Forces of Nature.
   [*Monthly Notices of the R. Astron. Soc.* Vol. LXXX, Dec. 1919, pp. 109–138.]

Gravitation and Light.
                [*Proc. Camb. Phil. Soc.* Vol. XIX, Jan. 1920, pp. 324–344.]

Questions in Physical Inter-Determination.
   [Lecture to International Congress of Mathematicians, Strasbourg,
          Sept. 1920, *Comptes rendus du Congrès*, pp. 3–40.]

On the nature and amount of the Gravitational Deflection of Light.
          [*Philosophical Magazine*, Vol. XLV, Jan. 1923, pp. 243–256.]

Can Gravitation really be absorbed into the Frame of Space and Time?
                [*Nature*, Feb. 10, 1923, p. 200.]

Newtonian Time essential to Astronomy.
          [Supplement to *Nature*, April 9, 1927, pp. 1–12.]

The Grasp of Mind on Nature.
   [James Scott Lecture, *Proc. Roy. Soc. Edin.* Vol. XLVII, Aug. 1927,
                pp. 307–325.]

# Addenda *and* Corrigenda (*Jan.* 1929).

**Vol. I,** p. 81. The adaptation of this screw-theory to aerodynamic practice would involve inclusion of the action of the propeller in steady transfer of angular momentum from the machine to the surrounding air, thus originating the instabilities of spin. *(Spin of aircraft.)*

p. 106, footnote. The criterion for validity is that the free paths of the ions should be a multiple of the molecular interval large enough, say $10^3$, to enable the Poisson averaging for potentials to be independent of the statistical averaging for densities. *(Electrolytes.)*

p. 130, line 4 from foot, *for* $r^2$ *read* r *twice.*

p. 293. If the two sheets of a wave-surface are the outer and inner boundaries of one expanding pulse, it would perhaps be anomalous for them to intersect. The general elastic wave-surface has however three sheets. *(The wave-surface in actuality.)*

p. 430. Kelvin's elastic paradox, correcting Green, that a medium of negative modulus of compression may be stable as regards internal disturbance of density, is still intriguing when thus expressed. *(Elastic paradox.)*

p. 641. A list of dates is significant as regards progress in scientific formulations. The General Dynamics of Lagrange, as also the Calculus of Variations, arose from the idea of Least Action in 1758: the path-breaking activities of Hamilton in regard to Systems of Rays and General Dynamics ended with the two great dynamical memoirs, *Phil. Trans.* 1834–5: Jacobi's Lectures on Dynamics were delivered at Königsberg in 1842–3, and published by Clebsch from Borchardt's script as a hearer in 1866. From another side, the Dioptrics of Gauss, following the less formal initiatives of Huygens, Cotes, Möbius, appeared in 1853. Thomson and Tait's *Nat. Phil.*, which placed fresh emphasis on the fundamental character of the Action principle, appeared in 1867, and was soon improved in formal rigour as regards the hydrodynamic analysis by Kirchhoff and by Boltzmann: use had been made, in more tentative and exploratory manner, of the cognate generalized dynamical ideas relating to momentum in Maxwell's final constructive electric memoir of *Phil. Trans.* 1863. *(History of Generalized Dynamics.)*

p. 641, footnote. The key to the argument (p. 189) is that if $(p - s)^2 + (q + r)^2$ were invariant for the rotations of axes round the central ray at the two ends, then if these rotations were chosen so as to make $p = s$, $q = 0$, the further condition $r = 0$ would be satisfied so that $U$ would reduce to the form (21). But this invariance does not hold, for all that has been there done is to express the central terms in bipolar coordinates, the two angles being any whatever.

Let us then make another attempt. The two relations $p = s$, $q = -r$ can be attained by two different rotations at the two ends: and then the middle terms in p. 188 take on the form $\rho_1 \rho_2$ multiplied by $p \cos (\theta_1 - \theta_2) + q \sin (\theta_1 - \theta_2)$, that is by $A \cos (\theta_1 - \theta_2 + a)$: thus if there is a rotation imposed through $a$ at the second end so that $\theta_2 - a$ replaces $\theta_2$, they assume the form $A (x_1 x_2 + y_1 y_2)$ appropriate to an axial optical system. Indeed they were already of that type, in the extended sense that such a system is one that (when straightened out) is not altered by rotation round its axis. And this wider sense appears to be the proper optical one, for if two successive reflexions occurring at the second terminal are part of the system they are equivalent to a rotation round the axis parallel to both their planes. With this extension of the definition of an axial system, and simple verbal modifications, the discussions referred to in the footnote appear to stand. *(Analysis of an asymmetric optical system.)*

p. 643. Cf. Cotes' "noble and beautiful theorem," his "last invention," that there is an apparent distance of any pair of points across an optical system as a reciprocal property of the pair, with wealth of consequent general relations *(Cotes' theorem.)*

for image points, in his cousin R. Smith's *Opticks*, Vol. II, notes, p. 78, where reference is made to Huygens' much earlier but more special propositions.

p. 652, line 11 from foot, *read* $r$ $(1 - v'/c)$.

p. 664, line 17, *for* $\frac{1}{45}$ *read* $\frac{1}{27}$

p. 667, line 13 from foot, corrected in Vol. II, p. 752.

p. 679, line 4, *read* $\frac{1}{2}\rho c \varpi v$

**Vol. II,** p. 39, line 7. The procedure for the Fresnel convection term in phase velocity of wave-trains, here indicated, was followed out in *Aether and Matter* (1900), § 113, under the heading "The Correlation between a stationary and a moving Medium as regards trains of Radiation." The result was that the Fresnel expression was found to require correction by a term of the second order $- (\mu^{-1} - \mu^{-3})\, v^2/c$. The relativity formula gives the same correction to the second order: but as was to be expected on present ideas (p. 40) where matter is concerned, the two disagree widely when $v/c$ is considerable. As however the Fresnel result without any change has more recently been widely appealed to, as one of the most arresting of the exact illustrations of an unrestricted relativity, it seems worth while to probe that conclusion further, more directly. Unless the discrepancy can be adjusted, it will illustrate the precariousness of the fourfold as a substitute for direct comparison of frames of reference (p. 803). The Fresnel term is an addition to the group velocity, as Lorentz pointed out, thus removing experimental discrepancies.

*The Fresnel convection contrasted with relativity.*

The problem before us is simply to transfer a wave-train

$$A \cos \frac{2\pi}{\lambda} \left( \frac{c}{\mu} t - lx - my - nz \right), \quad l^2 + m^2 + n^2 = 1,$$

referred to the frame $(t, x, y, z)$ of the material medium which carries it, into the form

$$A' \cos \frac{2\pi}{\lambda'} \left( \frac{c}{\mu} t' - l'x' - m'y' - n'z' \right), \quad l'^2 + m'^2 + n'^2 = 1$$

appropriate to the observer's frame $(t', x', y', z')$ with regard to which the carrying medium is in motion with velocity $(v, 0, 0)$. The formulae of transformation (p. 39) are

$$t = \epsilon^{-\frac{1}{2}} \left( t' - \frac{v}{c^2} x' \right), \quad x = \epsilon^{-\frac{1}{2}} (x' - vt'), \quad y = y', \quad z = z',$$

with a transformation from $A$ to $A'$ which is not here required. The wave-train referred to the observer's frame is thus expressed by

$$A' \cos \frac{2\pi}{\lambda} \left\{ \epsilon^{-\frac{1}{2}} \frac{c}{\mu} \left( t' - \frac{v}{c^2} x' \right) - \epsilon^{-\frac{1}{2}} l\, (x' - vt') - my' - nz' \right\},$$

which is $\quad A' \cos \frac{2\pi}{\lambda} \left\{ \epsilon^{-\frac{1}{2}} \left( \frac{c}{\mu} + lv \right) t' - \epsilon^{-\frac{1}{2}} \left( l + \frac{v}{\mu c} \right) x' - my' - nz' \right\}.$

We are at present concerned only with the phase velocity in this frame, with regard to which the material system is convected with velocity $v$: for a wave-train travelling in the direction of $v$ for which $(l, m, n) = (1, 0, 0)$ it is

$$\left( \frac{c}{\mu} + v \right) \Big/ \left( \mu + \frac{v}{c} \right), \quad \text{which is} \quad \frac{c}{\mu} + (1 - \mu^{-2})\, v \left( 1 + \frac{v}{\mu c} \right)^{-1},$$

the addition to $c/\mu$ thus differing from the Fresnel expression by the last factor. The value deduced as an example of composition of velocities in relativity is

$$\left( \frac{c}{v} + v \right) \Big/ \left( 1 + \frac{v}{\mu c} \right).$$

When the wave-train is oblique to the velocity of convection, there is change of direction as well as change of velocity when the frame is thus changed. The former, for the change from the observer's to the Solar frame, is the Bradley aberration of the rays; for a ray at inclination $\theta$ with $v$, changed to $\theta'$ by aberration, simplifying by taking $n$ null, we have

$$\cot\theta = l/m, \quad \cot\theta' = \epsilon^{-\frac{1}{2}}\left(l + \frac{v}{\mu C}\right)\Big/ m;$$

so that the aberration $\theta - \theta'$ is given by the law

$$\epsilon^{\frac{1}{2}}\cot\theta' - \cot\theta = v/\mu C . \sin\theta,$$

or approximately    $\sin(\theta - \theta') = \mu^{-1}\dfrac{v}{C}\sin\theta' - \frac{1}{2}\dfrac{v^2}{C^2}\cot\theta'.$

This is the aberration for an observing system in an atmosphere of index $\mu$: its value to the first order appears to involve the factor $\mu^{-1}$, not $\mu$ as considerations of velocity would suggest, which can be neglected for air. But the index of the aqueous humour of the eye, or of the water in Airy's water telescope, is not relevant, because it belongs to the observing system. *(margin: Bradley aberration: atmospheric influence.)*

The Fresnel problem refers to a system in motion without change of configuration: the formula has been finally verified by Zeeman, with the necessary intricate adjustments, for water running through a tube. The difference for transmission across a slab of glass at rest and set in motion has also been verified by Zeeman. Here the theory is simpler, though an incident ray gives rise to a reflected ray and a transmitted ray each made up of contributions from successive to-and-fro reflexions in the slab: for each of these resultants is a simple train of waves, and we have only to determine, by change to a moving frame, how its phase is modified from the value that holds for a slab at rest. Cf. Lorentz, *Pasadena Lectures*, § 35. *(margin: Effect of convection of a slab of glass.)*

p. 45. It is to be observed that the inverse square law which is essential to molecular refractions, as here analysed in terms of equivalent cavitations, derives ultimately from the general ideas of elastic propagation: after the manner of the cognate, but more deep-seated, relation of regular propagation to internal molecular scattering considered on p. 753. *(margin: Continuous optical analysis implies inverse square laws.)*

p. 50. For a medium whose atoms admit of storage of part of the incident radiant energy, as in modern theories inspired by spectroscopic relations, the direct way of arriving at the formula of dispersion would be, following Rayleigh's remark (p. 759), to calculate the vibration dispersed or emitted by the atoms in the direction of the beam, coherently but in quadrantal phase, and deduce the velocity of the compound transmitted beam by combining it with the original incident beam. *(margin: The slowing of light waves by matter.)*

p. 119. Dust or gas released from the head of a comet, even by explosion for the velocity is relatively slight, will accompany the comet in its gravitational orbit, while it is also exposed to the full blast of repulsion by the Solar radiation: thus whatever be the size of the particles, or the velocity of the head, the comet and dust would be expected to form a system with one tail pointing at first directly away from the Sun, however it be broken up later. A massive planet would hold back the dust, so could have no tail. *(margin: Comet's tail at first away from Sun: depends on its small mass.)*

pp. 251, 261, footnote. For propagation in $n$ dimensions the solution of the final equation in $r$ can be derived, when $n$ is odd (after Gaskin), by differentiation from the result for $n = 3$: the pulses can therefore travel without trail in any medium of odd dimensions, in agreement with general theory, after Volterra, *Acta Math.* XVI, as quoted, p. 261. *(margin: Pulses can travel only in spaces of odd dimensions.)*

p. 321. Cf. "Note on...Theory of the Action of Magnetism on Light," *American J. of Math.* Vol. XIX, pp. 371–376 (1897), not here reprinted.

Constants of
Nature as of
variational
origin.

p. 401. In one of the alternative modes of connecting up the Planck formula for natural radiation with light *quanta*, Einstein (1917) has asserted a threefold balance between incident and scattered radiation and spontaneous atomic emission. This introduces by the variational method undetermined constants in excess, each also presumably absolute for all types of atoms: the number appears to be reduced to the one that is representative of temperature by equating two of them.

Momentum
of waves.

p. 434. The relation of travelling momentum to energy is instructively different if the cord illustrates a dispersive medium, with its group transmission. Cf. Rayleigh, *Papers*, Vol. VI, p. 236 (1914).

p. 448, last line, *for* double *read* is another aspect of *and delete* twice

Variation of
latitude.

p. 485. Annual features are found to emerge on comparing the international Kimura term in the latitude with that derived from Greenwich observations alone: see the Greenwich reduction (1928), p. 29, for the records of the zenith telescope.

Critical range
for water:
usual scheme
too simple.

p. 511. According to H. L. Callendar's preliminary account of experimental exploration for water substance, claimed to be very coherent over a wide field, in the correlative scheme of curves connecting total heat, instead of volume, with pressure at the various temperatures (*Engineering*, Nov. 30, 1928, from *Roy. Soc. Proc.* Aug. p. 466) the unstable region separating liquid from gas ends in a sharp cusp, instead of bending round gradually as the usual illustrative equations of state connecting volume with pressure have required. The system was found to be very sensitive to dissolved air. The properties of the critical transition are thereby upset fundamentally: *e.g.* just above the critical point the elastic reaction to compression would no longer tend to vanish, with resulting instability of density. No single equation, such as that of van der Waals, seemed possible to even illustrate both the phases, which were treated separately with satisfactory results on a hypothesis of gradual coaggregation of molecules of the vapour.

An anticipa-
tion con-
firmed.

But more than this is involved. The collapse of sharp transition appears to conform to what was predicted, with reluctance, *supra*, p. 511: even extremely rapid transition in density would involve abolition of surface energy, and so the disappearance of a curved meniscus which with Callendar is the criterion.

Stellar
distances.

p. 587. It appears that this principle had been used also by Newton himself, after Huygens, with Jupiter as the intermediary, to estimate the distance of the stars, in a posthumous tract on the "System of the World."

p. 615, line 17 *seq.* By confusion of thought these estimates have been attempted for free electrons at rest, and $l$ is free path of the electron not of the

Sun's general
magnetic
field:

radiation. For actual circumstances, when the electrons are in motion like a gas, the electronization required to produce the Solar magnetic fields proves to be much smaller, as follows.

From a stream of radiation of intensity $\epsilon$ it appears (Vol. I, p. 663) that a free electron disperses $\frac{8}{3}\pi\,(e/m)^2 e^2 \epsilon$, being the same for all wave-lengths. The momentum in a ray is energy divided by c, and momentum dispersed by the electron per unit time measures its repulsion by the incident radiation. If $l$ is

as due to
rotation and
radiation:

mean free path and $v_0$ velocity of the electron, this repulsion $F$ may be regarded as operating for the free time of order $l/v_0$ (cf. p. 338), thus generating velocity $v$ along its direction, of rough mean value for purpose of reckoning steady electric flux, equal to half its final amount $Fl/mv_0$, which is not held up by reaction but ends by constantly degrading into the fortuitous motions of temperature by the collisions with obstructing atoms. The aberration factor of the emergent Solar radiation, at its maximum at the Sun's equator, is $\frac{7}{10}10^{-5}$: this gives for a numerical density $N$ of electrons an electric flux of intensity $\frac{7}{10}10^{-5}Nev$ in the direction of the motion of the electrons with the rotating Sun: this is at the Sun's equator

$$\tfrac{8}{3}\pi \left(\frac{e}{m}\right)^{3} e \,\frac{\epsilon}{c}\,\tfrac{1}{2}\,\frac{l}{v_0}\,\tfrac{7}{10}10^{-5}Ne.$$

If $v_0$ has the temperature value it is $6.10^8$ at six thousand degrees, and only 30 times greater at six million: for the former surface value this electric flux appears to figure out to intensity $\tfrac{1}{2}10^{-31}Nl$, for the latter to $\tfrac{1}{2}10^{-32}Nl$. It changes with position in the Sun as indicated *loc. cit.* To produce a magnetic field $H$ at the equator of the Sun, of radius $a$, would require a total magnetic moment of order $\tfrac{1}{2}Ha^3$, involving axial electric flux within the Sun of averaged order of intensity $Ha^{-1}$: attending only to very rough averages, to produce a general field of order 30, as observed, $Nl$ in the Sun would have to range around $10^{12}$ for this lower temperature and around $\tfrac{1}{2}10^{14}$ for the higher, values which are both abundantly within the facts for electronization of gases. The value of $l$ would be determined mainly by collisions with the atoms; but if the electronization were nearly complete $Nl$, and therefore the electric flux, would tend to be the same at all densities, a rarefied atmosphere being thus very potent magnetically. the electronization required.

It appears (p. 616) that for a large sunspot of radius $10^9$ with velocity of whirl as much as $10^6$ a magnetic field as high as $10^4$ would require $N$ to be increased from this lower estimate appropriate to surface temperature about 10 times: thus the vortical velocity in sunspots seems to be adequate without any call for abnormally intense electronization. Also the intense fields of sunspots.

This formulation of the effects of aberration of the emergent radiation implies the view of the relations of propagation and scattering in atomic media, advanced on p. 759. The outward radiation in the Sun is to be regarded as on the whole coherent, within bounds as regards wave-length, however great the opacity of the medium may be. It is not as if the reinforcement from each atom it passes over were sporadic, relative to that atom and its motion alone: then there could be no regularity in the propagation and no velocity such as local thermal equilibrium demands. On the other hand, viscosity is an affair of the atoms alone, affected by their relative motion in pairs, and outside any influence of a stream of radiation passing across the medium. The contributions from the elements and ions conspire in phase in a certain direction, that of the ray along which energy travels, while their quadrantal difference is the cause of the diminution of velocity from c to $c/\mu$. A convection of the medium affects this reduced velocity in free space by adding the Fresnel term $(1 - \mu^{-2})\,v$, modifying the paths of the rays by Fermat's principle to an extent which for the Solar gas is negligible. Coherence of stream of radiation is involved.

p. 752. Subject to unimportant corrections, Poynting's conclusions and suggestions in two lectures of dates 1904–6 (*Scientific Papers*, pp. 629–720) on the astronomical effects of pressure of radiation, deserve closer attention. For example, he points out how, in a mass of cometary dust, the incident Solar radiation would sift away the smaller particles into a trail which, he suggests, might be the replenisher of the haze within the Solar system which reveals itself by its illumination as the zodiacal light, as it is constantly being drawn into the Sun by his radiation. On modern ideas the radiation from an adjacent bright star might even be imagined to sift out to some degree the different kinds of gas in a nebula (cf. p. 755) if time were long enough. Sifting effect of radiation.

pp. 615, 811. One is tempted to speculate further from the appearances presented by sunspots. In a vortex underneath the surface the pressure and density of the Solar gas would diminish inward on account of the centrifugal force: at depths where the density is considerable the stream of radiation near the vortex will thus be bent by refraction, after the manner of *mirage* in the Earth's atmosphere. Although this outward stream is everywhere being intensely absorbed or scattered, and renewed by fresh contributions from the atoms over which it passes, the Fermat principle of least time of transit may still be expected to be applicable, at any rate in a general way, for its rays; on the foundation of the Rayleigh synthesis (p. 757) of reduced velocity of propagation in material media, and consequent refraction, as itself a result of molecular scattering in direction along the ray where it is completely coherent in phase and thus far more potent (p. 758) than the sideway scattering which weakens the beam. In usual circumstances, bending can thus be important while scattering is negligible. For a vortex in the radial direction, thus roughly parallel to the emergent stream of radiation, the rays will therefore (pp. 486, 645) curve outwards from the vortex core, because the time of transit between two points on the path will be shortened by approach of the middle part to the core where the velocity of radiation is greater. The upward stream would thus be diverted sideways by the vortex, so that there would be a deficiency of radiation emerging centrally that would show as a dark spot. For a perfect radiator the temperature at the surface is determined solely by the intensity of emission into free space; likewise the strata at the dark spot, though not perfect radiators, must be of lower temperature, which will annul to some degree, near the surface, the centrifugal diminution of density. The effort to adjustment of temperatures by internal radiation is thus an essential part of the complex scheme of interactions. When the vortex is radial, so that all the features are of cylindrical type, the reduced density towards its core will be in adjustment vertically with Solar gravity; the mathematical analysis of this, if it were feasible, might require to some degree a local depression or dimple at the surface as well as reduced density along a column. The penumbra is a conspicuous feature in the appearance of sunspots, with its radial structure. If we may regard the outward radiation as made up of component parallel streams in definite directions each nearly radial, they will be deviated differently in travelling along the vortex: the dark regions will overlap for them in the central region of the spot, but only partially on the outside where striation may conceivably be established, on the lines of an analogy that follows.

For a strikingly vivid illustration of this train of ideas, from which in fact they in part arose, can readily be arranged. A bath filled with water is illuminated by a lamp some distance above it, and the usual turbulent fluid motion is set up: as the motion subsides the more permanent residual vortical features become localized in time into regular whirls: these are projected on to the matt white surface of the bottom of the bath as round dark spots which may be surrounded by spiral dark converging streaks, the whole simulating rather closely the patterns revealed by Hale in the rarefied hydrogen gas high over sunspots. Here the dark spots arise from refraction of the downward illumination at the sides of the vortical dimples on the surface of the water, the analogue of the axial change of density in the Solar gas: while the patterns converging spirally around them have been impressed on the beam of light by refraction at the spirally ribbed pattern on the surface, being able to persist in it until the bottom is reached. Granular patterns recalling small Solar faculae may also be noted, due possibly to consolidation of residual undulatory disturbance between the vortices. In the Solar pattern the spirality is due to emission from the hydrogen itself: thus its dark ribs might arise either from diminished aggregate column of the gas in uniform illumination or from a uniform distribution of gas scattering a ribbed illumination from below, after the example of the bath: and further speculation becomes too precarious. Especially the phenomena of sunspots near the edge of the disc may present

Fermat's principle applicable in the Sun.

A radial vortex diverts the stream of radiation sideways,

producing a dark spot: of lower temperature.

The penumbra.

Spot formation simulated by refraction at water whirls,

especially the hydrogen spirals.

difficulties. But the general combination of vortical effects with shadows due to refraction may invite further scrutiny by astronomers in this connection.

As another instance, the dark bands along a spiral nebula need not perhaps always be due to opaque matter in front of it: they might be indications of clefts of reduced density along the line of sight, not extremely deep, though smallness of refraction* has to be compensated by length of path so that the deviation, equal to transverse gradient of index of refraction multiplied by length over which it is present along the ray, ought to be of the order of the angular breadth of the observed dark band. If we imagine reversal of the light, drawing a sheaf of rays *from* the observer, which are refracted away from the cleft, a boundary would be obtained of the region from which alone radiation whether of stars or nebular matter could reach him along this direction, all the depths beyond this sheaf of rays being shut off by refraction†. If this order of ideas were tolerable some nebulae that are apparently flat spirals might come under suspicion of being elongated spindles. Here study is essential of the wealth of material and discussions for judgment on such questions, which constitutes Vol. xiii (1918) of *Lick Observatory Publications*.

*Dark bands on nebulae need not be opaque.*

p. 667. *For* line 13 from foot *read* "yet capable of dispersal by a grating into independent beams," and cf. p. 758.

p. 692, footnote. This remark may profitably be expanded. The exact dynamical equations of vorticity $(\xi, \eta, \zeta)$ in fluid are

$$(D/dt - \nu\nabla^2)\,(\xi, \eta, \zeta) = \left(\xi\frac{\partial}{\partial x} + \eta\frac{\partial}{\partial y} + \zeta\frac{\partial}{\partial z}\right)(u, v, w),$$

where $\nu$ is viscosity divided by density: they are cognate, where gradient of the motion is small, to diffusion of temperature $\theta$ from sources, $(d/dt - k\nabla^2)\,\theta = 0$, where $k$ is conductivity divided by specific heat and density. If $\nu$ were null, vorticity would be simply convected, after the manner explored by Helmholtz: but actually while it is being convected it diffuses into the surrounding fluid according to a kinematic modulus $\nu$, thus comparable with thermal diffusion.

*Diffusion of vorticity,*

Some orders of magnitude will vivify this mode of representation. It appears that for metals such as copper $k$ is of the order $15 \cdot 10^{-1}$, for rock such as granite it is of the order $2 \cdot 10^{-2}$, for wood of the order $10^{-3}$: while $\nu$ for water is of the order $10^{-2}$ at 20° C. and for air is of the order $15 \cdot 10^{-2}$. Thus, for example, heat diffuses in granite on about the same scale as vorticity diffuses in water.

*as compared with heat.*

Consider a sphere anchored in a stream of water. If the surface conditions could adapt themselves to perfect slip, viscosity would not come into play at all and the stream lines would be of the familiar frictionless type. This type

* On this Rayleigh order of ideas the particles to which the refraction is due need not be gaseous, provided they are densely distributed in the cubic wave-length (p. 758): which is corroborated by the molecular refraction persisting nearly constant into the liquid and solid states.

*Optical refraction possible by fine dust.*

† The generalization to cones of vision, of the Cotes theorem of apparent distance (p. xxi) which is of the very essence of practical relativity, involves that in a refracting cosmic field where apparent extent is increased the brightness of each of the objects in it is increased in the same proportion, save for a factor the square of the index. This appears to afford a sort of probe, with whatever results, for the rays bending slowly over vast distances within nebulae. From a different aspect, if the Cotes principle did not hold good natural radiation in an isolated region could never attain to an equilibrium. This aspect involves that the illumination per unit area at each position is not altered by the visual distortion though its direction is altered, which concurs only partially with the surmise in the text. Yet the total brightness of each luminous object appears to be increased in proportion to transverse areal shrinkage; which involves partial obliteration of the individual stars in lanes or other regions where the nebula is seen expanded. (Cf. p. 519.)

*Distances in nebulae as apparent.*

of flow can be established throughout instantaneously by the interaction of fluid pressure and inertia, for waves of compression travel with the speed of sound which in this connection may be regarded as infinite. But actually such establishment is inhibited locally by the surface of the sphere gripping the

**Vortical bore over surface of obstacle,** fluid, with the result that the adjacent layer of fluid tumbles over upon itself after the manner of a tidal bore advancing up a river, giving rise ultimately to a cushion of vortices which ease off, like friction wheels, the slip between solid

**carried away as a wake:** and fluid. These vortices are, by Helmholtz's principle, carried away with the stream, and form the wake behind the obstacle: at the same time they are modified by diffusing of the vorticity into the adjacent parts of the fluid with which they travel, for the equations of vorticity are unaltered by uniform convection. Cf. various relevant investigations of surface flow by Lord Rayleigh.

**while flow in front is irrotational.** But if the stream is moderately fast the water, except close to the obstacle, has flowed past before vorticity can have diffused to any sensible degree out to it from the surface layer in which it originates. The fluid motion around the obstacle is thus very nearly the irrotational type as originally established instantly by propagated pressure, except in this easing vortical layer close to the surface of the obstacle: but behind it the vorticity may gradually spread. This description of the character of the motion near the solid accords with what is seen on looking down over the side of a ship travelling through smooth water: the resistance, other than that due to surface waves, arises from energy expended in the continual creation of the wake. Cf. p. 675.

**Two different problems.** It seems reasonable that the small deviation from irrotational flow will be established differently, according as it is a sphere being set in motion in resting fluid or a spherical portion of a stream suddenly solidified and reduced to rest.

Cf. a systematic discussion, with references, by Rayleigh, in *Papers*, Vol. VI (1911), pp. 28–40: also various later papers on stability of flow.

p. 741. These ideas may receive apt illustration in the theory of the magnetism of gases (p. 729). If the magnetic molecules are endowed with intrinsic momentum of spin round their axes, it must stiffen the reaction against rotation

**Influence of spin in gas theory.** in the magnetic field. The modified Action density with this rotational velocity eliminated contains terms of the first order in the remaining component velocities (cf. Vol. I, pp. 47, 67). It may be reduced, as in this gravity problem, to a quadratic in terms of local momentoids, together with a modified form for the positional part. This will alter the Langevin expression for induced magnetization, in a way that may be readily worked out.

Note also that if one, or several, of the internal coordinates of a system is locked by pure constraint, thus giving rise to a new system, that system also has its Liouville differential invariant. The original invariance thus involves invariance as regards variation of every partial set of its coordinates and their momenta.

**Scattering of radiation.** pp. 753, 760. Cf., and contrast, as regards theory of scattering in general, the ideas of a discussion by C. V. Raman, "A Classical Derivation of the Compton Effect," *Indian Journal of Physics*, Dec. 15, 1928.

**Nebulae show no colour.** p. 759. The stoppage of the light from streaks in a nebula, whether gaseous or composed of dust, by bending of the rays after the manner of terrestrial *mirage*, would be non-selective, being deviation rather than absorption. The former would in any case be potent before the latter becomes effective.

p. 761. *For* $vd/dt - read - vd/dt$.

**Thresholds.** p. 766. Cf. the memoir of B. van der Pol and J. van der Maek on "The Heartbeat considered as a Relaxation Oscillation...," *Phil. Mag.* Nov. 1928, where the mode of establishment of a regular beat is illustrated from special types of equations.

p. 795. The value of $|\,\eta_{s}\,|$ has for the purpose of the argument to be independent of the distance, a condition which appears on examination to be satisfied. — *Retarded potentials.*

p. 807. To elucidate in another direction the essential distinction at the beginning of the next note, between the field of algebraic formulae for continuous electric density superposed on aethereal structure in space, and the physical theory of one complex medium of the type of aether associated with and controlled by atomic matter, it seems profitable to revert to the difficulties that were encountered in the discussions relating to the stress-energy tensor, as made fundamental with the aim of its absorbing into itself all mechanical science. It was noted by Hilbert and also Lorentz, followed by Einstein and by F. Klein, that this mechanical tensor theory could be adapted so as to arise out of an Action procedure, after the electrodynamic model, for the continuous single extension, with its appropriate continuous densities as above, provided various restrictions and adjustments, perhaps essentially empirical, held good *[A stress tensor for a continuous medium:]* as regards the character of the permitted variations: the perplexity that showed itself in this regard, amidst the variety of tentative processes of abstract analysis, with converging agreement, may be illustrated from the important exchange of Notes between Klein and Hilbert within the Göttingen Academy (*loc. cit.*, F. Klein, *Abhandl.* I, pp. 553–612). The annulling of the various types of variation in the Action, after the manner perhaps suggested by the type of result thus aimed at, leads to a large scheme of equations of control for this continuous field, of which four are proved algebraically to be superfluous as being involved in the others, doubtless ultimately on account in fact of the invariance of the form of the Action: a result which can be expressed in the form that the divergence of a certain quadratic tensor is identically null. *[identically balanced:]* The science of mechanics thus appears to present itself in the vacuous guise of a set of algebraic identities involving vectors and tensors belonging to a geometric continuum: and this point of view has even been widely stressed.

The discussions above referred to leave the impression that this scheme, like *[the Action procedure then empirical.]* various *quantum* constructs, works by virtue of special algebraic devices introduced solely with a view to that end, in fact taking the place of physical imagery with its natural limitations. For example, the restriction required on the variation as regards densities and their field is to make it (with Klein, p. 570) a tangential transformation subject to the equations of that field: a postulate there tentative, but foreshadowing some more coherent procedure which the physically duplex structure of the medium can reveal.

The dynamic of free aether, as it has evolved historically, is in line with the *[Evolution of elastic theory by Action:]* primitive dynamic, after Lagrange and Green, of a continuous elastic vibrating medium: the variation of the distribution of Action is first to be annulled with regard to the most general variation of the strain, with no regard to whether this variation is a connected geometric displacement in the frame of reference or has a value at each place entirely arbitrary. In fact a purely geometric displacement for ordinary space ought not to alter the essential physical strain, when no electric field is involved: so if the Action function has been constructed subject to the proper invariances, the result ought for this particular type of variation to be a tensorial identity, like the one here in question: as is not *[identities:]* unfamiliar in elastic theory: it is not however connected with the Poynting formulation for flux of energy. But the variation of the Action has to vanish as above for unrestricted variation, not however now as an identity but as *[equilibration of medium:]* providing the aggregate of conditions determining local equilibration in the free aether. The free aether is then taken to have settled to equilibrium in the wider kinetic sense, as regards its field, which abuts on and is sustained by its electrons and other atomic nuclei; and their own equilibration, again in the *[interaction of sources.]* full sense of kinetic balance, remains to be secured by annulling the further variation of the Action, thus internally determined when their positions are given, arising from mutual displacement of these positions. The variation thus produced in the field is not now uncontrolled and so arbitrary at each point in free aether, but is continuous, subject to the equilibrium conditions in the field as already secured (cf. Klein, p. 570) as illustrated here on p. 801: the procedure there shown is that each variation of the field is split into one arising by way of gradient from an unconditioned displacement together with terms involving the gradient of the equilibrated displacement. But for both types of displacement, of which they are the difference, the variation of the Action is already annulled: therefore for them alone it is null identically, as on p. 801.

Emergence of the stress tensor:

it gives the interaction with each source:

which conversely can be absorbed into the tensor.

For any finite region in free aether it can be expressed by integration by parts of this gradient of displacement, as the equilibration of a stress tensor over its boundary. But each electron or other nucleus is a singular locality of stress locked into the aether: if any such are within the region each of them must be cut out, after the manner introduced by Green, by inserting a boundary closely surrounding it, and a stress acting over that boundary then takes part in the equilibration. Expressed conversely, the nucleus makes its local contribution to the balancing of the stress tensor over the outer boundary as regards that place, its reversed resultant being the reaction on the nucleus in the form of a set of local forces acting there and arising from its structure as a centre of intrinsic stress. In the usual circumstances, contemplating an interlocked dense atomic system forming a material body of ordinary experience, these atomic forces themselves consolidate into the expression of a material stress, as in the standard theory of an elastic atomic medium; and this stress can be added to the stress tensor making the customary aggregate which has now null divergence everywhere. It is in this way that the internal equilibrium of the free aether involves its own stress which is linear and is for example opera-

The Action analysis involves more than propagation by contact.

tive in the transmission of radiation, while interaction of the electrons or other sources is also involved in the Action and can be appropriately expressed by a formal quadratic field stress between them, equilibrating over every boundary which does not contain such sources, in identical manner, on account of the equilibrium of the aether previously secured by its own linear equations. But the essential here is that the variation is to be made as that of a straining

The stress tensor belongs to the sources.

frame along with which the sources are displaced, so that they have themselves the same virtual displacements as the frame in general, the equilibration of the free aether between the sources, as already secured, involving that the displacements throughout it are not arbitrary, but are determined by the virtual displacements of the sources, thus having at each point definite gradients which, as above, particularize the result.

This would explain how the Faraday-Maxwell stress theory, promoted by Einstein to be the concentrated expression of all dynamics, arises on an atomic theory: if the universe is not made up of aether and matter, it can appear only as a plausible algebraic process, with accidentally resulting identities, avowedly tentative, and liable to internal misfits of the type that can arise from undue narrowness of the foundation scheme.

Historical evolution of the method of variations:

now generalized.

The method of virtual displacements or variations, for a mechanical system, originated in systematic manner with Newton in the *Scholium* to his Third Law of Motion, which laid down the general principle of interaction. It required a century for its full development and consolidation, largely in the hands of d'Alembert and Lagrange. If, to gain a coherent view of the dynamics of electric and radiational science, it has to be envisaged as an affair of a duplex medium, aether and its atomic matter, it is only to be expected that the application of the variational calculus will require a closer formulation, whether on the foundation of a physical scheme in touch with the other branches of natural science, or of empirical mathematical adaptations which may alternatively be too narrow for consistency or too wide for experience. Actually the two methods, physical intuition in touch with experience and new modes in algebraic geometries, may assist each other; yet the vitalizing ideas still range back to Faraday, Kelvin, and Maxwell.

The Maxwellian stress tensor as here determined is consistent with the vanishing of a special type of variation in the free aether, whereas a free stress of the radiational field has to be consistent with the vanishing of all types: thus this quadratic tensor need not represent an actual distribution of free stress, being merely a compendious scheme of resultant interactions of the sources, perhaps even describable as direct actions, that are involved in the specification of the Action. But such interaction, based on the scheme of Action, is not as if there were no aether, for in it the tensor is expressed in terms of the field of the sources.

*Abstract character of the stress tensor.*

pp. 811, xii. These primitive considerations* appear remote from the vast manifolds of recent *quantum* developments. The control by an aether-pressure, on the foundation of Action, here required for a synthesis of perhaps the simplest conceivable type (pp. 810, xi), may be compared with the following sketch (if here rightly summarized) of an application of the method of Action, in the appropriate tensorial analysis, to a consolidation of the wave mechanics as dissected by Dirac, which has been carried through by J. M. Whittaker, in concise but necessarily intricate analysis, following on more special problems treated in cognate manner by C. G. Darwin, in *Proc. Roy. Soc.* Dec. 1928. The scheme relates to distributions located in terms of fourfold relativist *quasi-*geometry, thus is distinct from the alternative effort towards more generalized interpretation (p. 808) in terms of a Hamiltonian dynamic of an atomic system.

*Action for triple vector field in the fourfold.*

In the Lorentz equations of the electrodynamic field there were no structural atoms or electrons, involving internal boundaries or interfaces, but only continuous volume densities of electricity† which in unknown way happened to condense themselves into *quasi*-atomic accumulations: that was doubtless why the electrodynamic force had to be adjoined to the scheme as an additional hypothesis. When the field is consolidated into one coherent model, such as the present rotationally elastic aether, there has to be also a distribution of stress of the nature of a scalar pressure, usually however remaining latent in the overt electrodynamic phenomena. In the atomic wave presentations of Schrödinger, as modified into linear forms by Dirac, these two scalar variables, density and pressure, appear to be expanded into a pair of fourfold vectors, each analogous in some degree to the single fourfold vector which characterizes an electrodynamic field. An invariant function expressing scalar Action density would appear to be constructable from these three field vectors, on the usual lines of a general quadratic form; the special type adopted for systematic development by Whittaker, perhaps the appropriate generalization of the scheme aforesaid, appears to be made up of an electrodynamic part arising in the usual form from its own vector alone, conjoined to a mutual part of product type arising from the other pair jointly. This Action, when minimized, over the field now devoid of singularities, with regard to all its independent variables, gives the differential equations sufficient to prescribe the course of behaviour of all three vectors therein. A stress tensor does not appear as yet to arise.

*The Lorentz scheme a continuous one:*

*in contrast with the wider rotational aether scheme,*

*which generalizes into the Dirac triple vector formulation.*

---

* The Bohr atom can be saved by assuming that somehow the extra pressure at the proton is structurally counteracted.

† Maxwell's own rather careless introduction of convected charge through its vector potential, as essential to make his total current circuital (*Treatise*, §§ 618, 631, and indeed *Phil. Trans.* 1864, §§ 68, 73, 74, 99), had been corrected and systematized by FitzGerald in a brief note so long ago as *Brit. Assoc. Report* 1883, reprinted in his *Scientific Writings*, p. 127.

*Maxwell's field theory for moving charge.*

But the new part of this Action density has, like the Schrödinger function, to involve imaginary quantities, so that the eight components of the two new vectors constitute 16 independent variables, in addition to the four electric variables, also complex. Even the simple Lorentz transformation has to take on the imaginary forms (p. 803) when transformed into the fourfold. The geometrical connections in the background can also be non-uniform and so permit a gravitational term of arbitrary intensity in the Action, in the Einstein manner. This 16-fold scheme of a universe is of unfathomable extent: what would be our concern with it practically is confined to certain very exceptional circumstances, which it is the task of the special problems in wave mechanics to search out and isolate, in which these vectors, thus controlled in 24-fold way, attract notice by concentration of the electric one towards nuclear regions which, being not very diffuse and yet somehow permanent and all identical, are identified as the world tracks of atoms and electrons in the imaginary 12-fold. When there is no disturbing electrodynamic field, the governing equations in these problems prove all to be of one form, that of the original Schrödinger equation; and the possible states, simulating the various atomic systems of experience, are thus restricted to conform to special modes and periods on the analogy of the familiar vibrating systems of continuous dynamics. This distribution of two interlocked complex vectors is thus to constitute the hypercosmos, which is presumed to take on, on occasion but very exceptionally, a localized atomic character which alone is recognizable. Instead of the earlier 4-fold world-history with two distributions of complex scalars, density and pressure, the former taken as cohering into atomic aggregations, there is to be this 12-fold, thus inconceivably wider than any experience, evolving itself in unfathomable manner which on occasion comes within the range of observation by concentration into features resembling atomic nuclei, with the usual direct electric and radiational interpretations in terms of their electric vector: while the laws of spectroscopic transition which gave rise to it all perhaps remain a riddle.

*A 16-fold universe,*

*but almost completely latent.*

Vol. II, p. 2. As the Earth is nearly spherical the axis of rotation $OI$ nearly coincides with $OL$ which is fixed in space.

*The Pole follows the Moon.*

If the Lunar tidal accumulation were fixed as a part of the solid Earth it would merely modify the Lunar nutation. But in addition there is its motion as pulled round with only $\frac{1}{28}$ of the diurnal rotation. This involves a precessional torque of motional type; it is at right angles to its gravitational torque so that there must be a balancing torque on the solid Earth, bending the Pole towards the Moon when she is above the horizon and away when she is below it. It appears (roughly as on p. 323) that this type of short-period tidal reaction is of substantial amount. An analysis of the records, recently announced from Harvard (Stenson, *Science*, Jan. 1929), seems to countenance it in a general way.

*Action has local structure yet involves direct activity at a distance.*

p. 802. It seems proper to emphasize finally the result that in the duplex medium, Aether and Matter, the Action, on Hamilton's principles, reintroduces stress relations at a distance which cannot be traced across. Moreover, if this synthesis is allowed, the gravitational stress tensor derived from the Action after the surface integral terms have been abstracted and thrown away, is not only not invariant, but can hardly be right as an expression of the interaction between the sources. Further it is only in the fourfold limited map of the cosmos, as made absolute by gravitation, that energy and matter are identical, neither of them however being presumed to have any existence at all except as one of the components of a continuous stress tensor which represents the reality in that auxiliary construct. The analysis for the duplex medium as conducted in space and time is as on p. 166 *supra*, or *A ther and Matter*, Ch. VI: and as regards the new interactions with light, as on p. 789.

*Summary.*

## Additions to Corrigenda in Vol. II.

p. xxii, line 14, *insert* not.

p. xxii, line 21, *add* approximately. But if the velocity of phase ($p/n$ for a wave-type involving $pt - nx$) thus conforms to relativity, the velocity of energy ($dp/dn$) could hardly do so. The exact velocity of the energy groups appears to be, from what here follows,

$$\frac{c}{\mu}\left(1 + \frac{\mu v}{c} + \frac{\mu}{\lambda}\kappa\right) \Big/ \left(1 + \frac{v}{\mu c} + \frac{v}{\lambda c}\kappa\right),$$

where $\kappa$ is $d\mu^{-1}/d\lambda^{-1}$ expressing the effect of dispersion.

p. xxii, lines 3 and 6 from foot, *read* $\left(\frac{c}{\mu} + v\right) \Big/ \left(1 + \frac{v}{\mu c}\right)$ *twice*.

p. xxvii, footnote †, line 4, *for* increased *read* diminished.

p. xxxii, third paragraph, line 3, *for* were fixed as a part *read* did not partake of the rotation.

p. 805, footnote †. *Add*, The relation of L. de Broglie, thus confirmed by experiment, is expressible in the form that each electron, in addition to its charge $e$ and its intrinsic inertia $m_0$, has an intrinsic frequency $v_0$ of field-pulsation in its own frame, equal to $m_0 c^2/h$ which thus implies a meaning for the *quantum* constant $h$.

# 48

## ON THE PERIOD OF THE EARTH'S
## FREE EULERIAN PRECESSION*.

[*Proc. Camb. Phil. Soc.* Vol. IX (1896), pp. 183–193.]

General
theorem.

1. The main object of this note is to state a principle which allows us to estimate the effect of elastic yielding of a rotating solid on the period and character of the free precession of its axis of rotation. It has been suggested by the recent papers of Professor Newcomb and Mr Hough[1], in which the discussion is chiefly confined to the case of an incompressible homogeneous spheroid.

2. In order to obtain a clear notion of the action of the internal forcive which causes the free precession, let us examine in a geometrical manner the well-known case of a perfectly rigid solid rotating round its axis $OC$ of greatest moment of inertia. When this motion is disturbed, let $OI$ represent the instantaneous axis of the rotation $\omega$ of the solid, and $OL$ the axis of resultant angular momentum which is fixed in direction in space†. In the case of a symmetrical solid $OC$, $OI$, $OL$ lie in one plane. With reference to the solid itself, $OI$ and

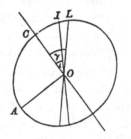

The free
precession
of a rigid
body:

$OL$ will describe right cones round $OC$ with common angular velocity $\Omega$, that of the free precession in the solid body of the instantaneous axis of rotation. The line $OL$ being fixed in space, the point $L$ of the body will thus move away from the point $L$ fixed in space with velocity $-\Omega . OL \sin \gamma$, where $\gamma$ represents the angle $COL$; but as the body is rotating at the instant round the axis $OI$ with angular velocity $\omega$, the velocity of its point $L$ must also be $\omega . OL \sin (\gamma - \iota)$, where $\gamma - \iota$ represents the angle $IOL$. Thus we have the geometrical relation

$$- \Omega \sin \gamma = \omega \sin (\gamma - \iota). \qquad (1)$$

The angular momentum of the rotating solid with respect to its centre of gravity $O$ is at time $t$ made up of

$$C\omega \cos \iota \text{ round } OC \text{ and } - A\omega \sin \iota \text{ round } OA.$$

---

* See further papers on geodynamics, *infra*, of dates 1906 onwards.

[1] S. Newcomb, *Monthly Notices R.A.S.* 1892; S. S. Hough, "The Rotation of an Elastic Spheroid," *Phil. Trans.* 1896.

† The momental ellipsoid of the diagram is oblate; hence the order of the points $CIL$ is wrong, $OL$ being normal to the tangent plane at $I$ on which the ellipsoid rolls, after Poinsot.

After an infinitesimal lapse of time $\delta t$, $OC$ and $OA$ have turned round the axis $OL$ fixed in space through an angle $\omega'\delta t$ say; thus $OC$ has moved through an angle $\omega' \sin \gamma . \delta t$, and $OA$ through an angle

**balance of angular momentum:** $\omega' \cos \gamma . \delta t$. These changes in the axes of the constant component angular momenta will introduce two reacting couples, acting round the same axis perpendicular to the plane $COI$ and equal to

$$C\omega \cos \iota . \omega' \sin \gamma \text{ and } - A\omega \sin \iota . \omega' \cos \gamma:$$

and in the steady precessional motion these couples must equilibrate each other, so that

$$C \cos \iota \sin \gamma + A \sin \iota \cos \gamma = 0;$$

that is,
$$C \tan \gamma = A \tan \iota. \tag{2}$$

Combining equations (1) and (2),

$$\Omega = \omega \, (\cot \gamma \sin \iota - \cos \iota)$$

**free period.**
$$= \omega \frac{C - A}{A} \cos \iota. \tag{3}$$

Thus the angular velocity of free precession, as seen by an observer partaking of the motion of the solid[1], bears the ratio $(C - A)/A \sec \iota$ to the angular velocity of the rotation: when the amplitude of the precession is small, the ratio reduces to $(C - A)/A$.

**Influence of elastic yielding:** 3. We have now to examine in what respects this argument is modified when there is elastic yielding to the centrifugal force of rotation, instead of absolute rigidity. As the period of relaxation of a strain in the Earth is a fraction of a day, while that of the Eulerian precession is measured in hundreds of days, it follows that

**statical.** the distortion due to centrifugal force will follow with great exactness the movements of the instantaneous axis of rotation: and the same will usually hold good in other problems of slow and small free

**The strain follows the axis of rotation and so is ineffective:** precessional motion. Thus if $C'$ and $A'$ represent the moments of inertia, round $OC$ and $OA$, of the configuration which the body would assume when the strain due to centrifugal force is removed,—or, what is the same thing, of the configuration it would assume under the action of an applied bodily forcive equal and opposite to the centrifugal force of the rotation—the angular momentum will now be made up of the components,

$$C'\omega \cos \iota \text{ round } OC,$$
$$- A'\omega \sin \iota \text{ round } OA,$$
$$I\omega \qquad \text{round } OI,$$

**so is removed:** where $I$ is the increase of moment of inertia round $OI$ due to the centrifugal bulging.

[1] This is different from the precession of $OC$ or $OI$ in space, which is round the fixed axis $OL$, and of angular velocity $\Omega'$, where $\Omega' \sin \gamma = \omega \sin \iota$.

Thus the reacting couples, in the plane $COI$, are

$$C'\omega \cos \iota \, . \, \omega' \sin \gamma,$$

$$- A'\omega \sin \iota \, . \, \omega' \cos \gamma,$$

$$I\omega \, . \, \omega' \sin (\gamma - \iota);$$

and these must balance. When the axis $OI$ does not deviate far from $OC$, the moment of inertia round it is equal, to the second order of small quantities, to that round $OC$; thus $I = C - C'$, and the condition for steady precession is therefore *(giving altered period.)*

$$C' \tan \gamma + (C - C') (\tan \gamma - \tan \iota) = A' \tan \iota,$$

that is,          $C \tan \gamma = \{C - (C' - A')\} \tan \iota;$

so that, by the geometrical relation (1)

$$\Omega = \frac{C' - A'}{C - (C' - A')} \cos \iota. \qquad (4)$$

The validity of this result is also, as above stated, confined to the case in which $2\pi/\Omega \sin \iota$ is large compared with the time of relaxation of elastic strain in the solid of revolution to which it belongs.

Since $C - C'$ is small compared with $C$ and $A$, as the elastic deformation only slightly alters the shape of the body, the third of these balancing reacting couples can be neglected compared with the other two. Thus *the free precession, under the influence of elastic yielding, will be the same as would belong to an absolutely rigid body, whose configuration is the one that the actual solid would assume were the centrifugal force removed, its elasticity being supposed unimpaired and the same as actually exists for smaller distortions.* *(Approximate general principle.)*

4. In order that the principle stated in this form may hold good, it is not necessary that the solid should be symmetrical round the axis of rotation. The couple due to the elastic deformation by the centrifugal force can be similarly neglected when the moments of inertia of the solid are all unequal, provided, as above, the oscillation of its axis of rotation remains small, as will be the case if it is spinning round the axis of greatest moment and is but slightly disturbed. Thus, in accordance with Poinsot's theorem, the motion may be represented by the rolling of the modified momental ellipsoid of the solid on a fixed plane; so that the axis of instantaneous rotation will trace out a small ellipse on the surface of the rotating body and the precession will therefore be exactly periodic[1]. It follows easily from the Eulerian equations of motion of a solid free from external forcive, *(Asymmetric solid, precession small. Exact repetition of the precessional motion:)*

[1] It follows from Poinsot's kinematic representation of the Eulerian motion, that for any solid rotating with no couples acting on it, the motion of the axis of rotation in the body itself is in all cases strictly periodic. It is not difficult to extend this result to the more general case in which there are flywheels or other sources of gyrostatic momentum attached to the solid. *(Generalized principle.)*

that the period of the free precessional motion of the axis of rotation
its free<br>period. round this ellipse is to that of the rotation of the body in the ratio
$\left(\dfrac{A'B'}{(C'-A')\,(C'-B')}\right)^{\frac{1}{2}}$, where $A'$, $B'$, $C'$ are the effective moments
of inertia, viz. those that would exist when the strain due to the
centrifugal force is supposed removed as above; agreeing with the
result already given where $A'$ and $B'$ are equal.

Inference as<br>to the Earth's<br>original<br>fluidity. 5. If when the centrifugal force is thus taken off, the rotating
solid became dynamically symmetrical like a homogeneous sphere,
there would be no free precession at all: if it became effectively pro-
late, the precession would be in the negative direction. In the actual
case of the Earth, the precession is in the positive direction; which is
in agreement with the accepted doctrine that the Earth assumed
originally a form of fluid equilibrium, in which it has been gradually
solidifying, so that were the centrifugal force _now_ removed it would
still retain an oblate form.

6. The practical outcome, as regards Astronomy, is that, as the
precessional constant $(C-A)/C$ of the _actual Earth_ is determined
precisely by the periods of the ordinary astronomical precessions and
nutations, so also the modified precessional constant $(C'-A')/A'$
_of the Earth as it would be_ were the centrifugal force of the axial
rotation removed, can be estimated from the period of the free
Eulerian precession,—in so far, that is, as this period can be disentangled from the actual observations of changes of latitude, which
are also affected by large irregular variations due to meteorological
causes, and more or less of an annual character. As the value of
$C-A$ is determined by a knowledge of the variation of gravity at
New _datum_<br>for geo-<br>physics. the Earth's surface, the values of $C$ and $A$ are separately known; and
therefore the value of $C'-A'$ is approximately known in terms of
the Eulerian period, thus affording an additional _datum_ of somewhat
precise character for investigations relating to the physical condition
of the Earth's interior.

On the other hand, the forced nutations due to extraneous astronomical forcives, even those of short period, are of course not sensibly
affected by the circumstance that the centrifugal part of the Earth's
ellipticity follows the movements of the axis of rotation.

7. The conclusions above stated are restricted to the case of a
solid body, that is, one in which parts which were initially near
together cannot wander far apart during the motion. If a large
portion of the interior is fluid, other considerations will influence
the result[1]. If the interior of the Earth were all fluid except an

[1] Cf. Kelvin, _Math. and Phys. Papers_, Vol. III, p. 132; Hough, _Phil. Trans._ 1895.

enclosing shell of moderate thickness, say a few hundred miles, there could be no sensible free steady precession at all, unless the rigidity of this shell were far beyond any that is actually known.

Influence of internal fluidity.

The slight change of configuration due to transfer of the centrifugal force to a new axis of rotation, which is here contemplated, need not be purely elastic, in order to make the conclusions valid. It may be in part viscous: but the wide discrepancy between the actual period of the Earth's free precession and the one that would apply to a perfectly rigid Earth is evidence that this deformation is really in large part elastic. For if elasticity proportional to, or a function of, strain were entirely absent, then viscosity proportional to, or a function of, rate of alteration of strain would (unless its coefficient were excessively large) be ineffective in preventing the assumption of the configuration of fluid equilibrium in the slow motions here considered, and free precession of periodic character would therefore be non-existent.

Free precession inhibited by internal viscosity.

8. The influence of the mobility of the surface waters on the period of the free precession has however yet to be considered. For them also, if they are sufficiently deep, the time of subsidence of a disturbance will be small compared with the period of the precession: so that the principle already employed will still have an application. The simplest case to consider first is that in which the waters cover the whole surface. When the axis of rotation changes, the centrifugal strain will be shifted to the new axis; thus the ellipticity of the underlying solid Earth will be altered owing to its elastic yielding, while that of the ocean surface will be altered by a different amount easily calculated from the conditions of fluid equilibrium.

Influence of the oceanic tides:

If, on the other hand, the ocean consisted only of a thin layer, its law of depth would be changed considerably, and this may involve sensible currents in it. Taking the angular momentum arising from the velocity of these currents (which will diminish the correction due to the surface waters) to be negligible in the actual case, the precessional couple will be as before that of the configuration the Earth would assume if the centrifugal force were removed; it will be that of a spheroid of water with the free surface that would exist on the removal of that force, together with that of the underlying solid Earth with its centrifugal force removed, but with its specific gravity reduced by unity.

induced currents.

If the underlying Earth were absolutely rigid, the effect of the layer of water on the precessional couple would be to diminish it by the couple arising from a spheroid of the same density as water, with ellipticity equal to that part of the ellipticity of the ocean surface which is due to centrifugal force alone: if the underlying Earth is yielding, the ellipticity of this effective water spheroid must be that of the ocean surface less that of the solid Earth, both due to centrifugal force. In any case, it is clear that under these circumstances the free ocean will cause an increase in the precessional period. This, as above stated, is on the supposition that the ocean is on the average deep enough for the angular momentum of the currents created in the alteration of its free surface to be neglected: if for instance it were very shallow it could obviously have no appreciable effect, and that would be in part because the momentum of the currents would compensate the effect of altered ellipticity, in part because its surface would not then have time to attain the equilibrium form.

9. The effect of the transference of masses of water by ordinary ocean currents and other causes has been considered long ago by Lord Kelvin and Professor G. H. Darwin, and more recently by Professor Newcomb, with the result that such disturbances are amply sufficient to originate displacements of the Earth's axis comparable with the amplitudes of the observed changes of latitude.

10. If the Earth were absolutely rigid but covered throughout by surface waters, the momental difference $C' - A'$ which gives the free precessional velocity would thus be less than the one $C - A$ which gives the forced astronomical precession, by that of a spheroid of water of the ellipticity $\epsilon_1 (= \frac{1}{517})$[1] which is due to centrifugal force alone. This latter spheroid, of semi-axes $a (1 - \frac{2}{3}\epsilon_1)$ and $a (1 + \frac{1}{3}\epsilon_1)$, has the moments of inertia

$$C_1 = \tfrac{2}{5}a^2 E_1 (1 + \tfrac{2}{3}\epsilon_1), \quad A_1 = \tfrac{2}{5}a^2 E_1 (1 - \tfrac{1}{3}\epsilon_1);$$

so that for it $C_1 - A_1 = \tfrac{2}{5}a^2 E_1 \epsilon_1$, where $E_1$ is the mass of a sphere of water of the dimensions of the Earth.

11. The potential of the attraction of the actual Earth at distant points is, by Laplace's formula,

$$V = \gamma \left( \frac{E}{r} + \frac{A + B + C - 3I}{2r^3} + \dots \right),$$

where $E$ is the Earth's mass, $a$ its radius, and $\gamma E / a^2 = g$. In the case of symmetry when $A = B$, we have thus

$$\frac{V}{\gamma} = \frac{E}{r} - \frac{C - A}{2r^3} (3 \cos^2 \theta - 1) + \dots,$$

$\theta$ being the co-latitude.

The potential of the centrifugal force of rotation is

$$V_1 = \tfrac{1}{2}\omega^2 r^2 \sin^2 \theta.$$

[1] Thomson and Tait, *Nat. Phil.* § 821.

Over the surface of the ocean of ellipticity $\epsilon$, given by

$$r = a\,(\mathrm{I} + \epsilon \sin^2 \theta),$$

the total potential $V + V_1$ must be constant: thus

$$\frac{E}{a}\,(\mathrm{I} - \epsilon \sin^2 \theta) - \frac{C - A}{2a^3}\,(3 \cos^2 \theta - \mathrm{I}) + \tfrac{1}{2}\omega^2 a^2 \gamma \sin^2 \theta = \text{const.},$$

so that

$$\frac{E}{a}\,\epsilon - \frac{3}{2}\frac{C - A}{a^3} - \tfrac{1}{2}\omega^2 a^2 \gamma = 0,$$

which gives

$$C - A = \tfrac{2}{3}a^2 E\,(\epsilon - \tfrac{1}{2}m);$$

where $m$, $= \omega^2 a/g$, is the ratio of centrifugal force to gravity at the equator, viz. is $\frac{1}{289}$. The value of $\epsilon$, the ellipticity of the sea-level, is about $\frac{1}{294}$[1]: so that the value of $C - A$ is just one half of that which would belong to the actual free surface in the absence of centrifugal force.

Relation to the Earth's ellipticity.

(We observe, incidentally, that

$$\frac{V}{\gamma} = \frac{E}{r} - \tfrac{2}{3}a^2 E\,(\epsilon - \tfrac{1}{2}m)\frac{3 \cos^2 \theta - \mathrm{I}}{2r^3};$$

and $g = -\dfrac{dV}{dr} - \omega^2 r \sin^2 \theta$

$$= \gamma\frac{E}{a^2}\,(\mathrm{I} - 2\epsilon \sin^2 \theta) - \gamma\frac{E}{a^2}\,(\epsilon - \tfrac{1}{2}m)\,(3 \cos^2 \theta - \mathrm{I}) - \omega^2 a \sin^2 \theta$$

$$= g_0\,\{\mathrm{I} - (\tfrac{5}{2}m - \epsilon) \sin^2 \theta\},$$

where $g_0$ is the value of $g$ at the Pole; which is Clairaut's formula for gravity.

The Clairaut gravity formula.

Again, the value of $(C - A)/C$, derived from the precession of the equinoxes, is $\cdot00327$[2]: thus

$$C = \tfrac{2}{3}a^2 E\,(\epsilon - \tfrac{1}{2}m)/\cdot00327 = \cdot35a^2 E.$$

If the Earth were homogeneous of density $5\tfrac{1}{2}$, we should have $C = 0\cdot40a^2 E$: thus the increase of density towards the centre diminishes the axial moment of inertia by 12 per cent.[3])

Effect of increase in internal density.

Thus for the actual Earth, $C - A = \tfrac{1}{3}Ea^2 \cdot \frac{1}{289}$; from which the value for the centrifugal water spheroid, that is

$$C_1 - A_1 = \tfrac{2}{3}\frac{E}{5\tfrac{1}{2}}a^2 \cdot \tfrac{1}{517},$$

is to be deducted, leaving the effective moment reduced in the ratio of 72 to 82.

Increase in free period due to a covering ocean:

In this case therefore of a rigid Earth covered by a fluid ocean, the period of the free precession would be increased in the ratio of 82 to 72 or $1\cdot14$ to unity by the presence of the surface waters.

---

[1] Col. Clarke's *Geodesy*, p. 305.

[2] As quoted from Leverrier in *Thomson and Tait*, § 828.

[3] As to how this fits with Laplace's hypothetical law of internal density, cf. G. H. Darwin, in *Thomson and Tait*, § 828.

In the actual case the surface waters do not cover the whole Earth, so that this is only a superior limit. Moreover the distribution of surface waters is not symmetrical, so that the result of deducting the centrifugal spheroid of water will be to leave an effective solid with moments of inertia all unequal, and the path of the instantaneous axis of rotation on the Earth will be an ellipse, not a circle. But one axis of the ellipse cannot possibly exceed the other from this cause by more than 7 per cent., even if the Earth were rigid.

*reduced for actual surface waters.*

Actually, the diminution of precessional moment due to the surface waters will be much reduced by the elastic yielding of the core, for when the centrifugal force is removed not only the surface but also the bed of the ocean will diminish in ellipticity, so that the differential effect will be smaller. It follows therefore that 6 to 8 per cent. is an outside limit to the actual increase of period of the precession that can be due to mobility of the surface waters.

*Final estimate.*

12. On the equilibrium theory here advanced, the portion of the ellipticity of the surface waters which arises from centrifugal force alone, shifts in company with the axis of rotation, thus causing a tide of the same period as the free precession, as was first pointed out by Lord Kelvin[1]. The equation of this ellipticity is

$$r = a \left(1 + \epsilon_1 \sin^2 \theta\right),$$

where $\epsilon_1 = \frac{1}{517}$. Thus a shift $\delta\theta$ in the axis of rotation involves a change of level $dr/d\theta \,.\, \delta\theta$; which is at the rate of $0 \cdot 2 \sin 2\theta$ of a foot for each second of arc. The reductions of tidal observations in Holland by Bakhuysen, and on the Pacific and Atlantic coasts of North America by Christie, made with this object in view, actually give a tidal component, of the period of the Eulerian precession, with amplitudes respectively about $\cdot 03$, $\cdot 04$, and $\cdot 05$ of a foot, thus corresponding to a precessional radius of the order of a quarter of a second of arc, which is the observed order of magnitude.

*The observed free precessional tide:*

The somewhat close agreement which has been found between the phase of this tide and the phase of the precession tends towards verifying both that the equilibrium theory here worked out is sufficient and that the free precession of period 427 days is fairly constant in amplitude. The amount of the tide as here computed will however be diminished by the centrifugal yielding of the solid Earth; though, this diminution being regular, we should still find that if the precession were quite regular in amplitude, the tide would at each station be proportional to the amplitude of the precession multiplied by the sine of twice the latitude. But it would appear from the observed latitude variations that the amplitude of the precession is altered

*its magnitude and phase:*

[1] *Collected Papers*, Vol. III, p. 332.

so irregularly that the harmonic analysis of a long series of observations can hardly bring out anything more definite than a sort of mean amplitude of tide. The actual tide thus produced follows, on the equilibrium theory, the movement of the Pole exactly; but it can only be disentangled from the observations of aggregate tide by an assumption that it is of a periodic harmonic character. Were it possible to pick out this tide with fair approximation, we should be able to derive information as to how much the surface of the solid Earth yields to the centrifugal force.

*not determinable with any precision.*

13. In all the foregoing, the relaxation of strain in the Earth that would follow the removal of the centrifugal force means such change of strain as would occur if the materials retained their existing moduli of elasticity and were of unlimited strength. As a matter of fact, a removal of the centrifugal force of the Earth's rotation would result in strains so great that the solid constitution of the interior would break down, and the materials would flow into new configurations in which the strains would be eased. In so far in fact as the configuration actually existing is due to consolidation from a fluid form, there can be in it no internal stresses other than simple hydrostatic pressure. In all the previous argument, what has been really considered has been the very slight change of configuration arising from a very slight shift of the axis of rotation: and we can treat of that as the difference of the centrifugal strains that would belong to the two positions of that axis, when we assume the elastic constants to be the actual ones that belong to small strains, and any limitation of elastic strength is ignored. The question of imperfect elasticity thus hardly comes up at all in such a case.

*Elastic limits not concerned in free precession:*

14. The question as to whether the interior of the Earth is in the main solid or fluid is, when precisely expressed, really a question as to what intensity of deforming stress will make the materials in its interior begin to flow. Every actual solid substance has its own elastic limit: and when subject to deforming forces sufficiently beyond this limit, the substance will flow like oil. The result here arrived at is that for the small stresses involved in the slight changes of the axis of rotation that accompany the free precession,—stresses that are of the order of a moderate number of grammes per square centimetre —the materials of the Earth's interior are solid and elastic, and as will appear below (§ 15) of very high rigidity. Beyond a certain limit they must give way and flow: the stresses due to the astronomical precessional forces are not however in the main beyond this limit.

*smallness of stress produced.*

As the Earth slowly alters its form owing to change in the distribution of surface materials and loss of heat, strains will arise in

Accumulated large stresses ultimately relieved in the Earth,

its interior and slowly increase until the material gives way somewhere, with earthquake-like effects. At such a weak place there may be a great deal of strain energy degraded into heat, sufficient it may be to melt the material. The region under a range of volcanic mountains may have a plane of weakness of this kind, up which the molten

to a state of hydrostatic pressure.

material is forced to the surface; which finally heals up after the surrounding stress is reduced in the main to hydrostatic pressure*.

15. It has been estimated by Mr Hough[1] that, were the Earth a homogeneous solid with the actual surface ellipticity, the Chandler

The Earth as rigid as steel:

period for the free precession would require that it should have the rigidity of steel. The order of magnitude of this result will not be entirely altered by actual heterogeneity: so that, if we can assume the correctness of this period, we may conclude that, as regards the slight stresses here involved, the Earth is an elastic solid of about the same order of rigidity as steel. This is in fair accord with the observations of seismologists[2], who find that earthquake shocks are propagated to

confirmation from earthquakes.

distant parts of the Earth not only by the ordinary surface waves, but also by minute tremors which enormously outrun the earthquake proper, arriving in a time that would correspond to propagation in a straight line across the interior of the Earth†, provided it possessed an average rigidity about $\frac{2}{3}$ of that of steel.

* For application of this principle to the internal constitution of celestial solid bodies under cosmic forces see Lord Rayleigh, *Roy. Soc. Proc.* LXXVII (1906) pp. 486–499: *Sci. Papers*, Vol. v, p. 300.

[1] *Loc. cit., Phil. Trans.* 1896.

[2] Cf. Professor Milne, *Brit. Assoc. Report*, 1896, "On Earthquake Phenomena."

† Recent earthquake records have been held to indicate the presence of a central region that is impervious to waves. Cf. H. Jeffreys' treatise on *The Earth*.

# A DYNAMICAL THEORY OF THE ELECTRIC AND LUMINIFEROUS MEDIUM.—PART III: RELATIONS WITH MATERIAL MEDIA.

[*Philosophical Transactions of the Royal Society.* Received April 21, 1897, read May 13, 1897. Published Feb. 9, 1898.]

## CONTENTS.

*(The numbers refer to sections.)*

**The rotational aether scheme.**  1. In two previous memoirs[1] it has been explained that the various hypotheses involved in the theory of electric and optical phenomena, which has been developed by Faraday and Maxwell, can be systematized by assuming the aether to be a continuous, homogeneous, and incompressible medium, endowed with inertia and with elasticity purely rotational. In this medium unitary electric charges, or electrons, exist as point singularities, or centres of intrinsic strain, which can move about under their mutual actions; while atoms of matter are in whole or in part aggregations of electrons in stable orbital motion. In particular, this scheme provides a consistent foundation for the electrodynamic laws, and agrees with the actual relations between radiation and moving matter.

An adequate theory of material phenomena is necessarily ultimately atomic. The older mathematical type of atomic theory which regards the atoms of matter as acting on each other from a distance by means of forces whose laws and relations are gradually evolved by observation and experiment, is in the present method expanded and elucidated by the introduction of a medium through whose intervention these actions between the material atoms take place. It is interesting to recall the circumstance that Gauss in his electrodynamic speculations, which remained unpublished during his lifetime, arrives substantially at this point of view; after examining a law of attraction, **Direct attractions inadequate.** of the Weberian type, between the "electric particles," he finally discards it and expresses his conviction, in a most remarkable letter

[1] *Phil. Trans.* 1894, A, pp. 719–822; 1895, A, pp. 695–743; referred to subsequently as Part I and Part II. In the abstract of the present Memoir, *Roy. Soc. Proc.* LXI, on p. 281, line 6, read $2\pi n'^2 + \int i'dF$ for $2\pi n'^2$; line 35 read $\frac{2}{3} . \frac{1}{3}\pi i'^2$ for $\frac{1}{3}\pi i'^2$; and on p. 284, line 18, read $m/2c . E (1 - m^2)$ for $E (1 - m^2)$ [here corrected].

to Weber,[1] that the keystone of electrodynamics will be found in an
action propagated in time from one "electric particle" to another.
The abstract philosophical distinction between actions at a distance
and contact actions, which dates for modern science from Gilbert's
adoption of the scholastic axiom[2] *Nulla actio fieri potest nisi per
contactum*, can have on an atomic theory of matter no meaning other
than in the present sense. The question is simply whether a wider and
more consistent view of the actions between the molecules of matter
is obtained when we picture them as transmitted by the elasticity
and inertia of a medium by which the molecules are environed, or
when we merely describe them as forces obeying definite laws. But
this medium itself, as being entirely supersensual, we must refrain
from attempting to analyse further. It would be possible (cf. § 6)

*Aether combined with atomic matter a necessity:*

*Historical.*

---

[1] Gauss, *Werke*, v. p. 629, letter to Weber of date 1845; quoted by Maxwell,
*Treatise*, II, § 861. After the present memoir had been practically completed,
my attention was again directed, through a reference by Zeeman, to H. A.
Lorentz's Memoir *La Théorie Electromagnetique de Maxwell et son application
aux corps mouvants*, Archives Néerlandaises 1892, in which (pp. 70 *seqq.*) ideas
similar to the above are developed. The electrodynamic scheme at which he
arrives is formulated differently from that given in § 13 *infra*, the chief
difference being that in the expression for the electric force $(P, Q, R)$ the
term $- d/dt\,(F, G, H)$ is eliminated by introducing the aethereal displacement
$(f, g, h)$. This applies also to the later *Versuch einer Theorie...in bewegten
Körpern*, 1895. The author remarks on the indirect manner in which dynamical
equations had to be obtained, mainly on account of the absence of any notion
as to the nature of the connection between the stagnant aether and the molecules that are moving through it. "Dans le chemin qui nous a conduit à ces
équations nous avons rencontré plus d'une difficulté sérieuse, et on sera
probablement peu satisfait d'une théorie qui, loin de dévoiler le mécanisme
des phénomènes, nous laisse tout au plus l'espoir de le découvrir un jour"
(§ 91). In the following year (1893) similar general ideas were introduced by
von Helmholtz, in his now well-known memoir on the electrical theory of
optical dispersion, in which currents of conduction are included: but his
argument is very difficult, and the results are in discrepancy with those of
Lorentz and the present writer in various respects in which the latter agree;
moreover they are not consistent with the optical properties of moving material
media. Both these discussions, of Lorentz and of von Helmholtz, are in the
main confined to electromotive phenomena: the treatment of the mechanical
forces acting on matter in bulk would require for basis a theory of the
mechanical relations of molecular media such as is developed in this paper.
[See also Appendix added at the end of this memoir.] The results in the paper
by Zeeman, above referred to, "On the Influence of Magnetism on the Light
emitted by a substance," *Verslagen Akad. Amsterdam*, Nov. 28, 1896, have an
important bearing on the view of the dynamical constitution of a molecule
that has been advanced in these papers, and illustrated by calculation in an
ideal simple case in Part I, §§ 114–18; cf. *Roy. Soc. Proc.* LX, 1897, p. 514. (Also
*Phil. Mag.* Dec. 1897: where the loss of energy by radiation from the moving
ions is also examined.)

[2] Gilbert, *de Magnete*, 1600.

or else some analytic formal equivalent.

even to ignore the existence of an aether altogether*, and simply hold that actions are propagated in time and space from one molecule of matter to the surrounding ones in accordance with the system of mathematical equations which are usually associated with that medium; in strictness nothing could be urged against such a procedure, though, in the light of our familiarity with the transmission of stress and motion by elastic continuous material media such as the atmosphere, the idea of an aethereal medium supplies so overwhelmingly natural and powerful an analogy as for purposes of practical reason

The limited practical aim.

to demonstrate the existence of the aether. The aim of a theory of the aether is not the impossible one of setting down a system of properties in which everything that may hereafter be discovered in physics shall be virtually included, but rather the practical one of simplifying and grouping relations and of reconciling apparent discrepancies in existing knowledge.

2. It would be an unwarranted restriction to assume that the properties of the aether must be the same as belong to material media. The modes of transmission of stress by media sensibly con-

Aether is not like matter.

tinuous were however originally formulated in connection with the observed properties of elastic matter; and the growth of general theories of stress action was throughout checked and vivified by comparison with those properties. It was thus natural in the first instance to examine whether a restriction to the material type of elastic medium forms an obstacle in framing a theory of the aether; but when that restriction has been found to offer insuperable difficulties it seems to be equally natural to discard it. Especially is this the case when the scheme of properties which specifies an available medium turns out to be intrinsically simpler than the one which specifies ordinary isotropic elastic matter treated as continuous.

A medium, in order to be available at all, must transmit actions across it in time; therefore there must be postulated for it the property of inertia,—of the same kind as ordinary matter possesses, for there can hardly be a more general kind,—and also the property of elasticity or statical resistance to change either of position or of form. In ordinary matter the elasticity has reference solely to deformation; while in the constitution here assumed for the aether there is perfect fluidity as regards form, but elastic resistance to rotational displacement[1]. This latter is in various ways formally the simpler scheme;

* Cf. however Vol. I, *supra*, Appendix v.

[1] I find that the rotational aether of MacCullagh, which was advanced by him in the form of an abstract dynamical system (for reasons similar to those that prompted Maxwell finally to place his mechanism of the electric field on an abstract basis), was adopted by Rankine in 1850, and expounded with full and clear realization of the elastic peculiarities of a rotational medium: by

elasticity depending on rotation is geometrically simpler and more absolute than elasticity depending on change of shape; and moreover [it is an independent type, and] no phenomenon has been discovered which would allow us to assume that the property of elasticity of volume, which necessarily exists in any molecularly constituted medium such as matter, is present in the aether at all. The objection that rotational elasticity postulates absolute directions in space need hardly have weight when it is considered that a definite space, or spatial framework fixed or moving, to which motion is referred, is a necessary part of any dynamical theory. The other fundamental query, whether such a scheme as the one here sketched could be consistent with itself, has perhaps been most convincingly removed by Lord Kelvin's actual specification of a gyrostatic cellular structure constituted of ordinary matter, which has to a large extent these very properties; although the deduction of the whole scheme of relations from the single formula of Least Action, in its ordinary form in which the number of independent variables is not unnaturally increased, includes its ultimate logical justification in this respect.

*Its scheme: rotational elasticity combined with fluidity.*

*Absoluteness of directions.*

*Kelvin gyrostatic model.*

*Least Action, the final test, can dispense with the models.*

### On Material Models and Illustrations of the Aether and its associated Electrons.

3. Although the Gaussian aspect of the subject, which would simply assert that the primary atoms of matter exert actions on each other which are transmitted in time across space in accordance with Maxwell's equations, is a formally sufficient basis on which to construct physical theory, yet the question whether we can form a valid conception of a medium which is the seat of this transmission is of fundamental philosophical interest, quite independently of the fact that in default of the analogy at any rate of such a medium this theory would be too difficult for development. With a view to further assisting a judgment on this question, it is here proposed to describe a process by which a dynamical model of this medium can be theoretically built up out of ordinary matter,—not indeed a permanent model, but one which can be made to continue to represent the aether for any assignable finite time, though it must ultimately decay. The

*Simplicity promoted by idea of a medium.*

*Rotational fluid aether.*

him also the important advantage for physical explanation, which arises from its fluid character, was first emphasized. Cf. *Miscellaneous Scientific Papers*, pp. 63, 160. In Rankine's special and peculiar imagery, the aether was however a polar *medium* or *system* (as contrasted with a *body*) made up of polarized nuclei (cf. Part I, §§ 37–8) whose vortical atmospheres, where such exist, constitute material atoms. The supposed necessity of having the vibration at right angles to the plane of polarization also misled him to the introduction of complications into the optical theory, such as aeolotropic inertia, and to deviations from MacCullagh's rigorous scheme.

*Rankine's exposition of a rotationally elastic medium.*

aether is a perfect fluid endowed with rotational elasticity; so in the first place we have—and this is the most difficult part of our undertaking—to construct a material model of a perfect fluid, which is a type of medium nowhere existing in the material world. Its characteristics are continuity of motion and absence of viscosity: on the other hand, in an ordinary fluid, continuity of motion is secured by diffusion of momentum by the moving molecules, which is itself viscosity, so that it is only in motions such as vibrations and slight undulations where the other finite effects of viscosity are negligible, that we can treat an ordinary fluid as a perfect one. If we imagine an aggregation of frictionless solid spheres, each studded over symmetrically with a small number of frictionless spikes (say four) of length considerably less than the radius[1], so that there are a very large number of spheres in the differential element of volume, we shall have a possible though very crude means of representation of an ideal perfect fluid. There is next to be imparted to each of these spheres the elastic property of resisting absolute rotation; and in this we follow the lines of Lord Kelvin's gyrostatic vibratory aether. Consider a gyrostat consisting of a flywheel spinning with angular momentum $\mu$, with its axis $AB$ pivoted as a diameter on a ring whose perpendicular diameter $CD$ is itself pivoted on the sphere, which may for example be a hollow shell with the flywheel pivoted in its interior; and examine the effect of imparting a small rotational displacement to the sphere. The direction of the axis of the gyrostat will be displaced only by that component of the rotation which is in the plane of the ring; an angular velocity $d\theta/dt$ in this plane will produce a torque measured by the rate of change of the angular momentum, and therefore by the parallelogram law equal to $\mu d\theta/dt$ turning the ring round the perpendicular axis $CD$, thus involving a rotation of the ring round that axis with angular acceleration $\mu/i \cdot d\theta/dt$, that is with velocity $\mu/i \cdot \theta$, where $i$ is the aggregate moment of inertia of the ring and the flywheel about a diameter of the wheel. Thus when the sphere has turned through a small angle $\theta$, the axis of the gyrostat will be turning out of the plane of $\theta$ with an angular velocity

*The ideal physical perfect fluid.*

*Kelvin gyrostatic model:*

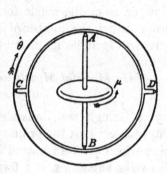

---

[1] The use of these studs is to maintain continuity of motion of the medium without the aid of viscosity; and also (§ 4) to compel each sphere to participate in the rotation of the element of volume of the medium, so that the latter shall be controlled by the gyrostatic torques of the spheres.

$\mu/i$ . $\theta$, which will persist uniform so long as the displacement of the sphere is maintained. This angular velocity again involves, by the law of vector composition, a decrease of gyrostatic angular momentum round the axis of the ring at the rate $\mu^2/i$ . $\theta$; accordingly the displacement $\theta$ imparted to the sphere originates a gyrostatic opposing torque, equal to $\mu^2/i$ . $\theta$ so long as $\mu/i$ . $\int \theta dt$ remains small, and therefore of purely elastic type. If then there are mounted on the sphere three such rings in mutually perpendicular planes, having equal free angular momenta associated with them, the sphere will resist absolute rotation in all directions with isotropic elasticity. But this result holds only so long as the total displacement of the axes of the flywheels is small: it suffices however to confer rotatory elasticity, as far as is required for the purpose of the transmission of vibrations of small displacement through a medium constituted of a flexible framework with such gyrostatic spheres attached to its links, which is Lord Kelvin's gyrostatic model[1] of the luminiferous working of the aether. For the present purpose we require this quality of perfect rotational elasticity to be permanently maintained, whether the disturbance is vibratory or continuous. Now observe that if the above associated free angular momentum $\mu$ is taken to be very great, it will require a proportionately long time for a given torque to produce an assigned small angular displacement, and this time we can thus suppose prolonged as much as we please: observe further that the motion of our rotational aether in the previous papers is irrotational except where electric force exists which produces rotation proportional to its intensity, and that we have been compelled to assume a high coefficient of inertia of the medium, and therefore an extremely high elasticity in order to conserve the ascertained velocity of radiation, so that the very strongest electric forces correspond to only very slight rotational displacements of the medium: and it follows that the arrangement here described, though it cannot serve as a model of a field of steady electric force lasting for ever, can yet theoretically represent such a field lasting without sensible decay for any length of time that may be assigned.

*its complex working.*

*But elastic only for oscillations,*

*unless the inertia density is great.*

4. It remains to attempt a model (cf. Part I. § 116) of the constitution of an electron, that is of one of the point singularities in the uniform aether which are taken to be the basis of matter, and at any rate are the basis of its electrical phenomena. Consider the medium composed of studded gyrostatic spheres as above: although the

*Aether model of an electron:*

---

[1] Lord Kelvin, *Comptes Rendus*, Sept. 1889; *Math. and Phys. Papers*, Vol. III, p. 466.

motions of the aether, as distinct from the matter which flits across it, are so excessively slow on account of its great inertia that viscosity might possibly in any case be neglected, yet it will not do to omit the studs and thus make the model like a model of a gas, for we require rotation of an individual sphere to be associated with rotation of the whole element of volume of the medium in which it occurs. Let then in the rotationally elastic medium a narrow tubular channel be formed, say for simplicity a straight channel $AB$ of uniform section: suppose the walls of this channel to be grasped, and rotated round the axis of the tube, the rotation at each point being proportional for the straight tube to $AP^{-2} + PB^{-2}$: this rotation will be distributed through the medium, and as the result there will be lines of rotational displacement all starting from $A$ and terminating at $B$: and so long as the walls of the channel are held in this position by extraneous force, $A$ will be a positive electron in the medium, and $B$ will be the complementary negative one. They will both disappear together when the walls of the channel are released. But now suppose that before this release the channel is filled up (except small vacuous nuclei at $A$ and $B$ which will assume the spherical form) with studded gyrostatic spheres so as to be continuous with the surrounding medium; the effort of release in this surrounding medium will rotate these spheres slightly until they attain the state of equilibrium in which the rotational elasticity of the new part of the medium formed by their aggregate provides a balancing torque, and the conditions all round $A$ or $B$ will finally be symmetrical. We shall thus have created two permanent conjugate electrons $A$ and $B$; each of them can be moved about through the medium, but they will both persist until they are destroyed by an extraneous process the reverse of that by which they are formed. Such constraints as may be necessary to prevent division of their vacuous nuclei are outside our present scope; and mutual destruction of two complementary electrons by direct impact is an occurrence of infintely small probability. The model of an electron thus formed will persist for any finite assignable time if the distribution of gyrostatic momentum in the medium is sufficiently intense: but the constitution of our model of the medium itself of course prevents, in this respect also, absolute permanence. It is not by any means here suggested that this circumstance forms any basis for speculation as to whether matter is permanent, or will gradually fade away. The position that we are concerned in supporting is that the cosmical theory which is used in the present memoirs as a descriptive basis for ultimate physical discussions is a consistent and thinkable scheme; one of the most convincing ways of testing the possibility of the existence of any hypothetical type of mechanism

*permanent because created only by supernatural processes:*

*and in conjugate pairs:*

*but the model is not eternal.*

being the scrutiny of a specification for the actual construction of a model of it*.

5. An idea of the nature and possibility of a self-locked intrinsic strain, such as that here described, may be facilitated by reference to the cognate example of a *material* wire welded into a ring after twist has been put into it. We can also have a closer parallel, as well as a contrast; if breach of continuity is produced across an element of interface in the midst of an incompressible medium endowed with *ordinary material rigidity*, for example by the creation of a lens-shaped cavity, and the material on one side of the breach is twisted round in its plane, and continuity is then restored by cementing the two sides together, a model of an electric doublet or polar molecule will be produced, the twist in the medium representing the electric displacement and being at a distance expressible as due to two conjugate poles in the ordinary manner. Such a doublet is permanent, as above; it can be displaced into a different position, at any distance, as a strain form, without the medium moving along with it; such displacement is accompanied by an additional strain at each point in the medium, namely, that due to the doublet in its new position together with a negative doublet in the old one. A series of such doublets arranged transversely round a linear circuit will represent the integrated effect of an electric polarization current in that circuit; they will imply irrotational linear displacement of the medium round the circuit after the manner of vortex motion, but this will now involve elastic stress on account of the rigidity†. Thus with an ordinary elastic solid medium, the phenomena of dielectrics, including wave

*A material ring with intrinsic twist inserted.*

*Nuclei of essential strain in a solid.*

* An exposition of this model of the electron is also given in *Aether and Matter* (1900), Appendix. In Maxwell's original mechanical model of his early days, electricity was introduced crudely as a system of idle wheels necessary to make successive elements of the magnetic field revolve the same way. Although a main object of the *Treatise* was to transcend the model, it may well have been this that kept him away from the electron.

† Displacement of an electron is equivalent to gradual creation of doublet moment along its path: that involves putting rotational displacement into the aether around its path: that implies in the model an accompanying rotation of the electron, after the manner of a screw, as it grips the aether, which resumes its continuity behind as the electron passes on. Thus is its travelling magnetic field maintained: and, as the spin it produces in the aether is irrotational and so inelastic, it is maintained without expenditure of the energy of the system. Thus the screwing progress of the electron through the aether is, on our model, not so unlike the previous typical illustration of the revolving progress of a loose intrinsic knot along a rope. Cf. Vol. 1, Appendix IX.

But this inelastic character of the surrounding motion remains adjusted only so long as the velocity of the electron is uniform: when it is accelerated, the field of motion becomes elastic and eases itself by radiation. This motion is superposed on the electron, whose own field persists throughout.

Torsional and
rotational
elasticities
differ in
effect only
on account
of boundaries.

propagation, may be kinematically illustrated; but we can thereby obtain no representation of a single isolated electric charge or of a current of conduction, and the laws of optical reflexion would be different from the actual ones*. This material illustration will clearly extend to the dynamical laws of induction and electromagnetic attraction between alternating currents, but only in so far as they are derived from the kinetic energy; the law of static attraction between doublets of this kind would [as boundaries are involved] be different from the actual electric law.

### [*Aether Stress and Material Stress.*]

Mechanical
forces derived
from energy:
how trans-
mitted.

6. According to the present scheme the ponderomotive forces acting on matter arise from the forces acting on the electrons which it involves; the application of the principle of virtual work to the expression for the strain energy shows that, for each electron at rest, this force is equal to its charge multiplied by the intensity of the electric field where it is situated. It has been urged that a model of the aethereal electric field cannot be complete, and so must be rejected, unless it exhibits a direct mechanism by which the pondero-motive normal traction $F^2/8\pi$ is transmitted across the aether from the surface of one conducting region to that of another: but the position can be maintained that such a representation would tran-scend the limitations belonging to a mechanical model of a process which is in part mechanical and in part ultra-mechanical. Indeed if this force were transmitted in the ordinary elastic sense, the trans-mitting stress would have to be of the nature of a self-balancing Faraday-Maxwell stress involving the square of the aether strain instead of its first power, and thus not directly related to elastic propagation. The model above described is so to speak made of aether, and ought to represent all the tractions that exist in aether, vanishing as they do over the surface of a conducting region: but the model does not in the ordinary sense represent matter at all, except in so far as the aethereal strain form which constitutes the electron is associated with matter†. It therefore cannot represent directly, after the manner of a stress across a medium, a force acting on matter, for that would from this ultimate standpoint be a force acting on a

---

* This is much overstated. A single electron can be constructed in an elastic solid medium, just as well as in the rotational aether on p. 18: it will not unwind itself, because its conical elements are anchored to the remoter part of the medium. The structure for an atomic medium thus produced with polar doublets alone, and without introducing rotational elasticity, satisfies all the optical and alternating electric requirements of an aether, including optical relativity. As regards mobility and inertia, see Appendix.

† On constitutive origin of mechanical elasticity in atomic media, see §§ 41 *seq., infra.*

strain form spreading from its nucleus all through the medium, not a traction on a definite surface bounding the matter.

The fact is that transmission of force by a medium, or by contact action so-called, remains merely a vague phrase until the strain properties of that medium are described; the scientific method of describing them is to assign the mathematical function which represents its energy of strain, and thence derive its relations of stress by the principle of virtual work; a real explanation of the transmission of a force by contact action must be taken to mean this process. Now in an elastic medium permeated by centres of permanent intrinsic strain, whether it be the rotational aether with its contained electrons, or an ordinary elastic solid permeated by polar strain nuclei as described above, the specification of the strain energy of the medium involves a mathematical function, not only of the displacement at each material point of the medium, but also of the positions of these intrinsic strain centres which can move independently through it. To derive the play of internal force, this energy function must be varied with respect to all these independent quantities; the result is elastic tractional stress in the medium across every ideal interface, *together with* forcive tending to displace each strain centre, which we can consider either as resisted by extraneous constraint preventing displacement of the strain centre, or as compensated by the reaction of the inertia of the strain form against acceleration[1]. Consider, for example, the analogy of the elastic solid medium, and suppose a portion of it to be slowly strained by extraneous force; two strains are thereby set up in it, namely that strain which would be thus originated if the solid were initially devoid of intrinsic strain, and that strain which has to be superposed in order to attain the new configuration of the intrinsic strain arising from the displacement of its nuclei. The latter part is conditioned by the displacement of these strain centres, and in its production forces acting on them must be considered to assist, whose intensities may be determined as has been already done in the aethereal problem.

The attractions between material bodies are therefore not transmitted by the aether in the way that mechanical tractions are transmitted by an ordinary solid, for it is electric force that is so transmitted: but neither are they direct actions at a distance. The point of view has been enlarged: the ordinary notion of the transmission of force, as framed mathematically by Lagrange and Green for a

*A stress function analyzed out of the discrete molecular data:*

*expressing the material system in the manner of Lagrange and Green:*

---

[1] Thus when the medium is in equilibrium, there is in it only the static intrinsic strain diverging from these centres, which gives rise to the forces between them; but when it is disturbed by radiation or otherwise, there is also the strain thence arising.

[On this general question of systematic material stress formulation cf. Appendix now added at end of the memoir.]

not wide enough for optics and electrics:

simple elastic medium without singularities, is not wide enough to cover the phenomena of a medium containing intrinsic strain centres which can move about independently of the substance of the medium. But the same mathematical principles lead to the necessary extension of the theory, when the energy function thus involves the positions of the strain centres as well as the elastic displacement in the medium; and the theory which in the simpler case answers fairly to the description of transmission by contact action has features in the wider case to which that name does not so suitably apply[1]. The strain centres (that is, the matter) have, in the strict sense of the term, *energy of position*, or *potential energy*, due to their mutual configuration in the aether, which can come out as work done by mutual forces between them when that configuration is altered, which work may be used up either in accumulating other potential energy elsewhere, or in increasing the kinetic energy of the matter, which is itself, in whole or in part, energy in the aether arising from the movement of the strain forms across it. Discussions as to transmission by contact are not the fundamental ones, as the above actual material

but formulation by an Action function is sufficiently comprehensive.

illustration shows: the single comprehensive basis of dynamics into which all such partial modes of explanation and representation must fit and be coordinated is the formula of Stationary Action, including, as the particular case which covers all the domain of steady systems, the law that the mutual forces of such a system are derived from a single analytical function which is its available potential energy.

The circumstance that no mode of transmission of the mechanical forces, of the type of ordinary stress across the aether, can be put in evidence, thus does not derogate from the sufficiency of the present standpoint. The transmission of material traction by an ordinary solid, which is now often taken as the type to which all physical action must conform, is merely an undeveloped notion arising from experience, which must itself be analysed before it becomes of scientific value: the explanation thereof is the quantitative development of the notion from the energy function by the method of virtual work

Contact action and fields of stress.

in the manner indicated in § 10, *infra*. This orderly development of the laws of activity across a distance, from an analytical specification of a distribution of energy pervading the surrounding space, is the essence of the so-called principle of contact action. It is precisely what the present procedure carries out, with such generalization as

---

[1] An analogous principle applies in the vortex theory illustration of matter. If we consider rigid cores round which the fluid circulates, they are moved

Vortex atom analogue.

about by the fluid pressure: but if we consider vortex rings, say with vacuous cores, these are mere forms of motion that move across the fluid, and if we take them to represent atoms, the interactions between aggregations of atoms cannot be traced by means of fluid pressures, but can only be derived from the analytical character of the function which expresses the energy.

the scope of the problem demands; besides attaining a correlation of the whole range of the phenomena, it avoids the antinomies of partial theories which accumulate on the aether contradictory and unrelated properties, and sometimes even save appearances by passing on to the simple fundamental medium those complex properties of viscous matter whose real origin is to be found in its molecular discreteness.

### Aether contrasted with Matter.

7. The order of development here followed is thus avowedly based on the hypothesis that the aether is a very simple uniform medium, about which it may be possible to know all that concerns us; and the present state of the theories of optics and electricity does much to encourage that idea. This procedure is of course at variance with the extreme application of the inductive canon, which would not allow the introduction of any hypothesis not based on direct observation and experiment. But though that philosophy has abundantly vindicated itself as regards the secondary properties of matter, which are amenable to direct examination, its rigid application would debar us from any theory of the aether at all, as we can only learn about it from circumstantial evidence. We could then merely go on heaping up properties on the aether, on the analogy of what is known of matter, as circumstances necessitated; and this medium would be a sort of sink to dispose of relations that could not be otherwise explained. Whereas matter, with which we are familiar, is the really complicated thing on which all the maze of physical phenomena depends, so that it is doubtful whether much can ever be known definitely as to its ultimate dynamical constitution; our best chance is to try to approach it through the presumably simple and homogeneous aether in which it subsists. *The aether substratum not like matter.* *Atomic matter the source of physical complexity.*

For example, it is found that the transmission of electrostatic force is affected by the constitution of the material dielectric through which it passes, and this is explained by a perfectly valid theory of polarization of the molecules of the matter: to press the analogy and ascribe the possibility of transmission through a vacuum to polarization of the aether may be convenient for some purposes of description, but in the majority of cases the impression is left that the so-called polarization of the aether is thereby explained. Whereas the processes being, almost certainly, of totally different character in the two cases, it will conduce to accurate thought to altogether avoid using the same term in the two senses, and to speak of the *displacement* of the aether which transmits electric force across a vacuum as producing *polarization* in the molecules of a material dielectric which exists in its path, which latter in turn affects the transmission of the electric *Material polarization:* *its contrast with aethereal strain displacement:*

force by reaction. In trying to pass beyond this stage, we may accumulate descriptive schemes of equations, which express, it may be with continually increasing accuracy, the empirical relations between their inter-action: these two phenomena; but we can never reach very far below the surface without the aid of simple dynamical working hypotheses, more or less *à priori*, as to how this interaction between continuous aether and molecular matter takes place.

8. On the present view, physical theory divides itself into two Radiation theory, regions, but with a wide borderland common to both: the theory of radiation or the kinetic relations of this ultimate medium; and the in contrast with the special electro-dynamics. theory of the forces of matter which deals for the most part with molecular movements so slow that the surrounding aether is at each instant practically in an equilibrium condition, so that the material atoms practically act on each other from a distance with forcives obeying definite laws, derivable from the formula for the energy. It is only in electromagnetic phenomena and molecular theory that non-vibrational movements of the aether are involved. The aether not being matter, it need not obey the laws of the dynamics of matter, provided it obey another scheme of dynamical laws consistent among themselves; these laws must however be such that we can construct in the aether an atomic system of matter which itself obeys the actual material laws. The sole spatial relations of the aether itself, on which Aether a dynamical generalization of the spatial frame: its dynamics depend, those namely of incompressibility and rotational elasticity, are thus to be classed along with the existing Euclidean relations of measurements in space (which also might *à priori* be different from what they are) as part of the ultimate scheme of mental representation of the actual physical world. The elastic and other characteristics of ordinary matter, including its viscous relations, are on the other hand a direct consequence of its molecular constitution, is the seat of material energies. in combination with the law of material energy which is itself a consequence of the fact that the energies of the atoms are wholly located in the surrounding simple continuous aether and are thus functions of their mutual configurations. In this way we come round again to an order of procedure similar to that by which Cauchy and Poisson originally based the elastic relations of material bodies on the mutual actions of their constituent molecules.

Consider any two portions of matter which have a potential energy function depending, as above explained, on their mutual configuration alone, the material movements being thus comparatively slow compared with the velocity of radiation; any displacement of them as a single rigid system, whether translational or rotational, can involve no expenditure of work; hence the resultant forcive exerted by the Stresses are mutual: first system on the second must statically equilibrate that exerted by the second system on the first, these forcives must in fact be equal and opposite wrenches on a common axis; and the energy principle thus involves the principle of the balance of action and reaction, in its most general form. This stress, between two molecules, is usually sensible only at molecular range; hence the action of the surrounding parts on a portion of a solid body is practically made up of tractions exerted over the interface between them. Further, since rotation of the body without deformation cannot alter the potential energy of mutual configuration of the molecules, it follows that for a rectangular element of ordinary solid matter the tangential components of these tractions must be self-conjugate, as they are taken to be in the ordinary theory of elasticity. On the other hand, for a medium not molecularly constituted we can hardly treat at all of mutual configuration of parts, and the self-conjugate stress relation will not be a necessary one.

*specified as interfacial force:*

*in terms of strain in the discrete matter.*

A certain similarity may be traced with the view of Faraday, who was disinclined to allow that ray vibrations are transmitted by any medium of the molecular character of ordinary matter, but considered them rather as affections of the lines which represent electric force, the propagation being influenced by the material nuclei which in ponderable media disturb these lines. This propagation in time requires inertia and elasticity for its mathematical expression, and the problem of the free aether is to find what kind of each is requisite.

*Faraday's notion of rays of force:*

*they require inertia.*

9. A theory which, like the present one, explains atoms of matter as made up of singularities of strain and motion in the aether, is bound to look for an explanation of gravitation by means of the properties of that medium; it cannot avail itself of Cotes' dogma that gravitation at a distance is itself as fundamental and intelligible as any explanation thereof could be. In further development of the illustrative possibilities of the pulsatory theory of gravitation, mentioned in the previous papers, we can (ideally) imagine the pulsation to have been applied initially over the outside boundary of the aethereal universe, and thence instantaneously communicated throughout the incompressible medium to the only places that can respond to it, the vacuous nuclei of the electrons; and we can even imagine the pulsations thus established as spontaneously keeping time and phase ever after, when the exciting cause which established this harmony has been discontinued.

*Gravitation not ultimate:*

*pulsatory analogue:*

It has been noticed in Part I, § 103, that gravitation cannot be transmitted by any action of the nature of statical stress; for then the approach of two atoms would increase the strain, and therefore also the stress, and therefore also in a higher ratio the energy of strain which depends on their product, and hence the mutual forces

*if dynamical, must be kinetic:*

of the atoms would resist approach. As gravitation must belong to the ultimate constituents of matter, that is on this theory to the electrons, and must be isotropic all round each of them, it would appear that no mediate aethereal representation of it is possible except the one here considered. The radially vibrating field might be described formally as the magnetic field of the electron considered as a unipolar magnet, necessarily of very rapidly alternating type because otherwise a field of gravitation would be an ordinary magnetic field. The bare groundwork of this hypothesis may thus be formally expressed in Maxwell's language and developed along his pulsating lines, by postulating that the electron is not only a centre of steady electrons, intrinsic electric force, but also a centre of alternating intrinsic magnetic force, instantaneously transmitted because it would otherwise involve condensation, each force being necessarily radial[1]. The ununsatisfactory. satisfactory feature is that this radial *quasi*-magnetic field is introduced for the sake of gravitation alone, which does not present itself as in any direct correlation with other physical agencies.

The following sections are occupied chiefly with an attempt to logically systematize, and in various respects extend, the electric aspect of molecular theory. The preceding paper dealt mainly with the molecular side of directly aethereal phenomena, such as electric and radiative fields; of the present one the earlier part follows up the same subject, and the remainder relates to the actions of the molecules of polarized material bodies on one another, and the Formal material stresses and physical changes thereby produced. As in the theory of preceding papers, the quantitative results are to a large extent indepolarized pendent of any special theory of the constitution of matter, such as material is here employed to bind together and harmonize the separate groups media: of phenomena, and to form a mental picture of their mutual relations; so far as they are electric they may be based directly on Maxwell's equations of the electric field in *free space*, which form a sufficient description of the free aether, and have been verified by experiment. In the Faraday-Maxwell theory, however, as usually expounded, an explanation of these equations is found, explicitly or tacitly, in an assumption that the aether is itself polarizable in the same manner as a material medium, and aether is in fact virtually considered to involves a be matter; on the present theory the equations for free space are an compound analytical statement of the ultimate dynamical definition of the scheme. continuous aethereal medium, and the polarization of material bodies

Pulsatory      [1] Two steady magnetic poles of like sign would repel each other: but in the
force is   case of two poles pulsating in the same phases there is also an inertia term
attractive. in the fluid aethereal pressure, and the result is as stated above. Cf. Hicks,
*Proc. Camb. Phil. Soc.* 1880, p. 35.

with the resulting forcive are deduced from the relation of their molecules to this medium in which they have their being.

10. In the modern treatment of material dynamics, as based on the principle of energy, the notion of configuration is, as above explained, fundamental. The potential energy, from which the forces are derived, is a function of the mutual configurations of the parts of the material system. In the case of forces of elasticity the internal energy is primarily a function of the mutual configurations of the individual molecules, from which a regular or organized part (§ 49, *infra*) is separated which is expressible in terms of the change of configuration of the differential element of volume containing a great number of molecules, and from which alone is derived the stress that is mechanically transmitted. In connection with the discussion of contact action in § 6 above, the mode of this derivation and transmission becomes a subject of interest[1]. In the first place the primary notion of a *force* as acting from one point to another in a straight line, has to be generalized into a *forcive* in Lagrange's manner on the basis of the principle of virtual work: then the forcive arising from the internal strain energy of the element of volume of the material is derived by variation of this organized energy, and appears primarily as made up of definite complex bodily forcives resisting the various types of strain that occur in the element: then these forcives are rearranged, by the process of integration by parts, into a uniform translatory force acting throughout the element of volume of the material (which must compensate the extraneous applied bodily forcive) together with tractions acting over its surface. When this is done also for adjacent elements of volume, other tractions arise which must compensate the previous ones over the part of the surface that is common to the two elements; and thus the uncompensated traction is passed on from element to element until finally the boundary of the material system is reached where it remains uncompensated and must be balanced extraneously. The outstanding irregular part of the aggregate mutual potential energy of the individual molecules, which cannot be included in a function of strain of the element of volume, cannot on that account take part in the transmission of mechanical forces, and is evidenced only in local changes of the physical properties and temperature of the material. Cf. § 48, *infra*.

*Theory can extend only to those features of material configuration already recognized:*

*with the relevant organized energy,*

*leading to stress, and its mode of balance:*

*and including in the residue change of moduli.*

---

[1] It is here assumed that the direct action between the molecules is sensible only at molecular distances, which would not be the case if the material were electrically polarized. The statement also refers solely to transmitted mechanical stress of the ordinary kind: more complicated types, not expressible by surface tractions alone, are put aside, as well as molecular conceptions like the Laplacian intrinsic pressure in fluids. Cf. §§ 44–6, *infra*.

*Restriction in idea of stress.*

The other main division of the energy is the kinetic part, which is specified in terms of the rate of change of configuration of the material system with respect to an extraneous spatial framework to which its position is referred. Whatever notions may commend themselves *à priori* as to the impossibility of absolute space and absolute time, the fact remains that it has not been found possible to construct a system of dynamics which has respect only to the relative positions of moving bodies; and the reason suggests itself, that there is an underlying part of the phenomena, which does not usually explicitly appear in abstract material dynamics, namely, the aethereal medium, and that the spatial framework in absolute rest, which was introduced by Newton and was probably a main source of the great advance in abstract dynamics originated by the *Principia*, is in fact the quiescent underlying aether. In this way the purely *à priori* standpoint is pushed away a stage, and we may find justification against the reproach that a philosophical formulation of dynamics should be concerned only with relative motions.

*A frame of reference imperative:*

*such as can subsist in the aether.*

### Relation to Gas Theory: Internal Molecular Energy.

*Gas theory: the translatory part:*

11. The kinetic theory of gases is considerably affected by the view here taken of the constitution of a molecule. In those simple and satisfactory features which are concerned only with the translatory motion of the molecules, it stands intact; but it is different with problems, like that of the ratio of the specific heats, which involve the internal energy. According to the usual hypothesis of the theory of gases, all the internal kinetic energy of the molecule is taken to be thermal and in statistical equilibrium, through encounters, with the translatory energy. But on the present view, the energy of the steady orbital motions in the molecule (including therein slow free precessions) makes up both the energy of chemical constitution and the internal thermal energy; while it is only when these steady motions are disturbed that the resulting vibration gives rise to radiation by which some of the internal energy is lost. The amount of internal energy can however never fall below the minimum that corresponds to the actual conserved rotational momenta of the molecule; this minimum is the energy of chemical combination of its ultimate constituents, while the excess above it actually existing is the internal thermal energy[1]. The present view requires that the

*the atom as constituted of orbital electrons:*

*its steady internal orbits and precessions: configuration is of minimal energy when undisturbed:*

*so additions thrown off as radiation:*

[1] As a concrete illustration, we can imagine two ideal atoms, each consisting of a single gyrostat enclosed in a suitable massless case, coming into mutual encounter. We may imagine that neither of them has any internal heat; so that the internal energy of each is the minimum that corresponds to its steady gyrostatic momentum, and the axis of each gyrostat therefore keeps a fixed direction in space. The result of the encounter will be that the axis of each

energy of chemical constitution shall be very great compared with the thermal energy; but for this very reason our means of chemical decomposition are limited, so that only a part of that energy is experimentally realizable[1]. This being the case, the alteration produced by external disturbance in the state of steady internal motions of the molecule consists in the superposition on it of very slow free precessional motions, which have practically no influence on its higher free periods[2]: and this explains why change of temperature has no influence on the positions of the lines in a spectrum. As a gas at high temperature must contain molecules with all amounts of internal thermal energy from nothing upwards, we should on the other hand, on the ordinary gas theory, expect both a shift of the brightest part of a spectral line when the temperature is raised, and also a widening of its diffuse margin.

*its steady or constitutive energy very preponderant:*

*so spectra independent of temperature:*

*difficulties.*

The ordinary encounters between the molecules will influence this thermal energy or energy of slow precessional oscillation, without disturbing the state of steady constitutive motion on which it is superposed, therefore without exciting radiation, which depends on more violent disturbances involving dissociative action.

*Precessions need not emit radiation, which involves ruptures of structure.*

On this view the postulates of the Maxwell-Boltzmann theorem on the distribution of the internal energy in gases would not obtain, for the thermal energy of the molecule would not be expressible as a sum of squares. The ratio of the specific heats in a gas must still lie between $1$ and $\frac{4}{3}$; but the nature of the similarity of molecular constitution in the more permanent gases, which makes the ratio of the total thermal energy to the translatory energy either $\frac{2}{3}$ or unity for most of them, would remain to be discovered. In those gases for which the latter value obtains, the energy of precessional motion in the molecule would be negligibly small, involving small resultant angular momentum and *possibly* small paramagnetic moment.

*Effect of an orbital structure on equipartition of molecular energy.*

The necessity of a distinction such as that here drawn between the internal thermal energy and the energy of the vibratory disturbances of internal structure which maintain radiation is well illustrated by the recent recognition (foreshadowed by Dulong and Petit's researches

*Vibrational and thermal energy.*

gyrostat acquires steady wobbling or free precessional motion, so that its internal energy is increased at the expense of the energy of translation of the atoms; but in this the simplest case there will be no unsteady vibration, such as could be radiated away. If however there are also other types of momenta associated with the atom, for example if the case of the gyrostat is not massless, the encounter will leave vibrations about the new state of steady motion, which if of high enough period will lead to loss of energy by radiation.

*Encounter of gyrostatic atoms may result only in steady precessions.*

[1] Ideas somewhat similar to the above are advanced by Waterston in his classical memoir of 1845 on gas theory, recently edited by Lord Rayleigh; *Phil. Trans.* A, 1892, p. 51.

[2] Cf. Thomson and Tait, *Nat. Phil.* § 345, xxiv.

on the law of cooling) and application by Dewar of the remarkable

Cooling occurs mainly by convection:
insulating power of a vacuum jacket as regards heat. If this distinction did not exist, both conduction and convection must ultimately depend on transfer by ordinary radiation at small distances, as

as against Fourier.
Fourier imagined; and it would not appear why convection by a gas, even when highly rarefied, is so much more efficient in the transfer of heat than radiation.

12. The result obtained by Ramsay and Young, and others, that all over the gas-liquid range the characteristic equations of the substances on which they experimented proved to be very approximately

Generalized characteristic for gases:
of the form $p = aT + b$, where $a$ and $b$ are functions of the density alone, also supplies corroboration to this view. Expressing the increment of energy per unit mass $dE = Mdv + \kappa dT$, we have for the increment of heat supplied $dH = dE + pdv$; and the fact that

its thermodynamic consequences:
$dE$ and $dH/T$ are perfect differentials shows immediately that $M$ is equal to $p - b$ and $\kappa$ is independent of $v$, so that the total (non-constitutive) energy per unit mass consists of two independent parts, an energy of expansion and an energy of heating[1]. The latter part is the thermal energy of the individual molecules; it is a function of their mean states and velocities alone, and constitutes almost all the

interpreted:
energy in the gaseous state. The former part is the energy of mutual actions between the molecules; it is negative and bears a considerable ratio to the whole thermal energy in the liquid state, in the case of substances with high latent heats of evaporation; for all gases except hydrogen, inasmuch as they are cooled by transpiration through a porous plug, $b$ is negative at ordinary densities. Cf. § 62, *infra*.

constitutive energy persists down to absolute zero:
There would be no warrant for a view that the forces of chemical affinity fall off and finally vanish as the ultimate zero of temperature is approached. The translatory motions of the molecules would diminish without limit, and therefore also the opportunities for reaction between them, so that many chemical changes would cease to take place for the same reason that a fire ceases to burn when the

but translatory manifestations disappear.
supply of air is insufficient, or coal gas ceases to explode when too much diluted with air: but the energies of affinity exist all the time in probably undiminished strength, while the forces of cohesion are modified by the fall of temperature but not necessarily in the direction of extinction.

[1] Cf. G. F. FitzGerald, *Roy. Soc. Proc.* Vol. XLII, 1887: cf. also Clausius' early ideas on "disgregation." [Cf. also Appendix, p. 133, *infra*, on virial theory.]

*The Equations of the Aethereal Field, with Moving Matter: various
applications: influence of Motion through the Aether on the
Dimensions of Bodies.*

13. Let $(u, v, w)$ represent the total circuital current, and $(u', v', w')$
the conducted part of it, which will be taken to include the current
$(u_0, v_0, w_0)$ of migration of the free electric charge as this is in all
cases very small in comparison; let $(f', g', h')$ denote the electric
polarization of the material, and $(f, g, h)$ the aethereal elastic displacement, so that the total circuital displacement of Maxwell's
theory is their sum $(f'', g'', h'')$; let the space of reference be fixed
with respect to the stagnant aether, and $(p, q, r)$ be the velocity with
which the matter situated at the point $(x, y, z)$ is moving, and let
$\delta/dt$ represent $d/dt + p\,d/dx + q\,d/dy + r\,d/dz$; let $(P, Q, R)$ denote the
electric force, namely that which acts on the electrons, and $(P', Q', R')$
the aethereal force, that which produces the aethereal electric displacement $(f, g, h)$; let $\rho$ denote density of free electric charge. Then
the electromotive equations are[1]

Notation: specification of electro-molecular field.

$$P = qc - rb - \frac{dF}{dt} - \frac{d\Psi}{dx}, \quad P' = -\frac{dF}{dt} - \frac{d\Psi}{dx}, \quad f = \frac{1}{4\pi c^2}P',$$

where $\quad F = \int \dfrac{u}{r}\,d\tau + \int \left(B\dfrac{d}{dz} - C\dfrac{d}{dy}\right)\dfrac{1}{r}\,d\tau, \quad a = \dfrac{dH}{dy} - \dfrac{dG}{dz};$

The averaged field equations for polarized and magnetized material.

and[2]
$$u = u' + \frac{\delta f'}{dt} + \frac{df}{dt} + p\rho,$$

where
$$\rho = \frac{d\,(f' + f)}{dx} + \frac{d\,(g' + g)}{dy} + \frac{d\,(h' + h)}{dz}.$$

[1] This scheme forms an improved summary of that worked out in Part II,
§§ 15-19 [*supra*, Vol. I. pp. 567-575]; the expressions there assigned for $\rho$ and $\Psi$
have here been corrected, and $(u_0, v_0, w_0)$ is merged.

[2] *Added Sept.* 14. The term $\delta f'/dt$ in $u$ arises as follows. In addition to the
change of the polarization in the element of volume, $df/dt$, there is the electrodynamic effect of the motion of the positive and negative electrons of the
polar molecule. Now the [convective] movement of two connected positive
and negative electrons is equivalent to that of a single positive electron round
the circuit formed by joining together the ends of their paths [less the change
in the distribution of polar moment produced by annulling them in the initial
position and restoring in the final—but the gradient of the distribution of
polarization occurring in the equations as induced by the field has to be
increased by the same change, the two thus cancelling]: and a similar statement holds when there are more than two electrons in the molecule. Hence
the motion of a polarized medium with velocity $(p, q, r)$, which need not be
constant from point to point, produces [in addition to the change of induced
polarization] the electrodynamic effect of a magnetization

Effects of convected polarization.

$$(rg' - qh', \; ph' - rf', \; qf' - pg')$$

distributed throughout the volume: cf. Part I, § 125. And it has been shown
in Part II, § 31, that any distribution of magnetism $(A, B, C)$ may be

From the formula for $(P, Q, R)$ Faraday's law follows that the line integral of electric force round a circuit *in uniform motion with the matter* is equal to the time rate of diminution of the magnetic flux through its aperture. The line integral of the aethereal force $(P', Q', R')$ round a circuit *fixed in the aether* has the same value. Again if $(F', G', H')$ be defined as the first part of $(F, G, H)$ so that

$$F' = \int \frac{u}{r} d\tau, \text{ and therefore } \frac{dH'}{dy} - \frac{dG'}{dz} = \int \frac{1}{r}\left(-\frac{dw}{dz} + \frac{dv}{dy}\right), \text{ etc.,}$$

we derive

$$a = \frac{dH'}{dy} - \frac{dG'}{dz} + 4\pi A - \frac{d}{dx}\int\left(A\frac{d}{dx} + B\frac{d}{dy} + C\frac{d}{dz}\right)\frac{1}{r} d\tau,$$

so that

$$a + \frac{dV'}{dx} = \frac{dH'}{dy} - \frac{dG'}{dz},$$

where $(a, \beta, \gamma)$ is magnetic force and $V'$ is the potential of the magnetism: hence Ampère's law follows that the line integral of the magnetic force round *any* circuit is equal to $4\pi$ times the total current that flows through its aperture. These two circuital relations are coextensive [for the general magnetized medium] with the previous equations involving the vector potential, and can thus replace them, inherent in
present
molecular
scheme of
wider
conditions. when the difference between $(P, Q, R)$ and $(P', Q', R')$ is inessential, that is (i) when the displacement currents are negligible, or (ii) when the matter is at rest; the quantity $\Psi$ then enters as an arbitrary function in the integration of the equations.

represented as a volume distribution of electric current equal to curl $(A, B, C)$, which is necessarily circuital, together with a surface current sheet equal to $(Bn - Cm, Cl - An, Am - Bl)$. Thus, *when $(p, q, r)$ is uniform and $(f', g', h')$ is circuital*, the above magnetic distribution is equivalent to a current system $(pd/dx + qd/dy + rd/dz) (f', g', h')$ together with current sheets on interfaces of discontinuity: this system is to be added on to $d/dt (f', g', h')$ in order to give the full electrodynamic effect. Thus in these special circumstances the formulation in the text is correct in so far as it leads to the correct *differential* equations for the element of the medium: the *integral* expression there given for $F$ is however only correct either when it is reduced to the differential form $-\nabla^2 F/4\pi = u + dC/dy - dB/dz$, which is derivable on integration of its second term by parts, or else when, the velocity of the matter still being uniform, discontinuous interfaces are replaced in the analysis by gradual though rapid transitions. These conditions are satisfied in all the applications that follow: but they would not be satisfied for example in the problem of the reflexion of radiation from the surface of moving matter.

But a formulation which is preferable to the above, in that it is *absolutely general*, is simply to include implicitly the above virtual magnetization directly in $(A, B, C)$ and consequently change from $\delta f'/dt$ to $df'/dt$ in the expression for $u$: this will also involve that the relation $A = \kappa a$ which occurs lower down shall be replaced by $A = \kappa a + rg' - qh'$, but there will be no further alteration in the argument of the text. The boundary conditions of the text are unaltered.

[Cf. *Aether and Matter* (1900), § 63, which also needs the modification of argument here inserted in square brackets.]

The mechanical force acting on the matter, or ponderomotive force, is $(X, Y, Z)$ per unit volume, where (§ 38, *infra*)

$$X = \left(v - \frac{dg}{dt}\right)\gamma - \left(w - \frac{dh}{dt}\right)\beta + A\frac{d\alpha}{dx} + B\frac{d\alpha}{dy} + C\frac{d\alpha}{dz}$$

$$+ f'\frac{dP}{dx} + g'\frac{dP}{dy} + h'\frac{dP}{dz} + \rho P.$$

Distribution of ponderomotive force.

The mechanical traction on an interface will be considered later (§ 39). In a magnetic medium the magnetic force $(\alpha, \beta, \gamma)$ differs from the magnetic flux $(a, b, c)$ simply by not including the influence of the local Amperean currents; thus $\alpha = a - 4\pi A$.

Magnetism.

When there is no conductivity, the free charge must move along with the matter, so that

$$\frac{d\rho}{dt} + \frac{d\rho p}{dx} + \frac{d\rho q}{dy} + \frac{d\rho r}{dz} = 0;$$

Convection of free charge:

therefore, from the circuitality of the total current, we must have, identically [cf. footnote, p. 66, *infra*],

$$\frac{d\rho}{dt} = \frac{d}{dx}\left(\frac{\delta f'}{dt} + \frac{df}{dt}\right) + \frac{d}{dy}\left(\frac{\delta g'}{dt} + \frac{dg}{dt}\right) + \frac{d}{dz}\left(\frac{\delta h'}{dt} + \frac{dh}{dt}\right).$$

The latter is the same as the convergence of $(\delta/dt - d/dt)$ $(f', g', h')$, which asserts (for the case of uniform motion that is contemplated) that mere convection of the polarized medium does not involve separation of free electricity. The relation between $(f', g', h')$ and $(P, Q, R)$ must be such as to strictly satisfy this equation. The quantity $\Psi$ occurs in the equations of the field as an undetermined potential which is sufficient in order to conserve the condition of bodily circuitality $du/dx + dv/dy + dw/dz = 0$.

and of polarization.

In order to express the conditions that must hold at an interface of transition, we notice that by definition $F, G, H$ are continuous everywhere; but it is only when the media are non-magnetic that their rates of change along the normal (and therefore all their first differential coefficients) are also completely continuous. Across an interface the traction in the aether must be continuous, so that the tangential component of the aethereal force $(P', Q', R')$ must be continuous, which is satisfied by continuity of $\Psi$. The continuity of the total electric current secures itself without further condition by a compensating distribution of electric charge on the interface, that is, by a discontinuity of $d\Psi/dn$. The tangential continuity of the elastic aether requires that the tangential component of the magnetic force $(\alpha, \beta, \gamma)$ must be continuous; the normal continuity of the magnetic flux is assured by the continuity of $(F, G, H)$. It might be argued that if the electric force $(P, Q, R)$ were not continuous tangentially, a perpetual motion could arise by moving an electron along

Analysis of interfacial continuities:

of tangential electric force for moving media.

one side of the interface and back again along the other side. But this reasoning requires that $(p, q, r)$ shall be continuous across the interface, as otherwise the circuit returning on the other side could not be complete; and it also requires that there shall be no magnetization, as otherwise the mechanical force on the electrons in an element of volume, which is what we are really concerned with in the perpetual motion axiom, is different from the sum of the electric forces on the individual electrons, by involving $(\alpha, \beta, \gamma)$ instead of $(a, b, c)$. We can thus assert continuity of the tangential electric force only in the cases in which it is already involved in that of the tangential aethereal force; and consistency is verified. The aggregate of all these interfacial electromotive conditions is thus continuity of the vector potential $(F, G, H)$, and of $\Psi$, and of the tangential components of the magnetic force; they *formally* involve continuity of the tangential components of the aethereal force $(P', Q', R')$, and of the electric and magnetic fluxes. But further, in the equations from which Ampère's circuital relation is derived above, it is only the normal space-variation of $V'$ that is discontinuous; hence continuity of the tangential magnetic force is involved in that of $F, G, H, \Psi$ by virtue of the mode of expression of $(F, G, H)$ in terms of the currents and the magnetism. Thus there are in all cases only four independent interfacial conditions to be satisfied.

*Reduction to four interfacial conditions.*

The scheme is thus far absolute, in the sense that the relations between the variables are independent of the special molecular constitution of the matter that is present. The system of equations must now be completed for material media by joining to it the relations which connect the conduction current in the matter with the electric force, and the electric polarization of the matter with the electric force, and the magnetic polarization of the matter with the magnetic force, in the cases in which these relations are definite and can be experimentally ascertained. In the simplest case of isotropic matter, polarizable according to a linear law, they are of types

*Constitutive relations in addition to the intrinsic electric.*

$$u' = \sigma P, \quad f' = (K - 1)/4\pi c^2 . P, \quad A = \kappa a.$$

The expression for $\rho$ leads in homogeneous isotropic media to

$$KV^2\Psi = -4\pi c^2 \rho + (K - 1)\left\{\frac{d}{dx}(cq - br) + \frac{d}{dy}(ar - cp) + \frac{d}{dz}(bp - aq)\right\}$$

so that $\Psi$ is only in part an electrostatic potential. Inside a uniform isotropic conductor *at rest*, the condition of circuitality becomes $\sigma V^2\Psi = d\rho/dt$; substituting this, we have $d\rho/dt + 4\pi c^2\sigma K^{-1}\rho = 0$, so that $\rho = \rho_0 \exp(-4\pi c^2\sigma K^{-1}t)$, showing that an initial volume density of free electricity would in that case be instantly driven to the

*A conductor clears itself of free charge.*

boundary owing to the dielectric action. This proposition may be extended to aeolotropic media:

14. The nature of the foregoing electric scheme may be elucidated by aid of some simple applications.

(i) When a conducting system is in steady motion so that there is no *conduction* current flowing into it, the electric force $(P, Q, R)$ must be null throughout its substance. Thus for the case of a solid conductor rotating round an axis of symmetry in a uniform magnetic field parallel to that axis, with steady angular velocity $\omega$, the electric force in it, namely $(\omega c x - d\Psi_1/dx, \omega c y - d\Psi_1/dy, - d\Psi_1/dz)$, must be null, so that $\Psi_1 = \tfrac{1}{2}\omega c (x^2 + y^2) + A$; the polarization in it is therefore null, but there is in it an aethereal displacement

Conductor spinning in uniform magnetic field.

$$- (4\pi c^2)^{-1} (d/dx, d/dy, d/dz) \Psi_1.$$

In outside space, the electric force and aethereal force are each $- (d/dx, d/dy, d/dz) \Psi_2$, where $\Psi_2$ is that free electrostatic potential which is continuous with the surface value $\tfrac{1}{2}\omega c (x^2 + y^2) + A$ at the conductor. Inside the conductor this purely aethereal displacement involves an electrification of volume density $\rho = - \omega c/2\pi c^2$, which will be a density of free electrons or ions as all true electrifications are; while there is a compensating surface density $\sigma$ equal to the difference of the total normal electric displacements on the two sides, that is to $(4\pi c^2)^{-1} (d\Psi_2/dn_2 + d\Psi_1/dn_1)$, where $dn_2$, $dn_1$ are both measured towards the surface, the outside medium being air for which $K$ is unity. The value of the constant $A$ is determined by the circumstance that the aggregate of this volume and surface charge shall be null when the conductor is insulated and unelectrified, or equal to the given total charge when it is insulated and charged: when it is uninsulated, the constant is determined by the position of the point on it that is connected to Earth, and therefore at zero potential. The procedure of Part II, § 25 is thus justified, because there is in fact no dielectric polarization in the conductor, but only aethereal displacement.

Case of axial symmetry, no currents involved:

It remains to consider whether the parts of this volume density $\rho$ and surface density $\sigma$ of electrification are carried round with the conductor in its motion, or slip back through its volume and over its surface so as to maintain fixed positions in space. It is clear (as in Part II, § 27) that the same cause, namely, viscous diffusion of momentum among moving ions and molecules, which produces ohmic resistance to a steady current, will lead to the electrons constituting electric densities being wholly carried on by the matter whenever a steady state is attained. This necessary consequence of the theory is in keeping with Rowland's classical experiments on

but a slight charge, its convection producing a magnetic field.

convection currents. The excessively minute magnetic field due to these convection currents themselves has been neglected in the above analysis, which has enabled us to specify the slight redistribution of free charge on the rotating conductor when under the influence of a powerful extraneous magnetic field: when the magnetic field is due solely to its own motion the redistribution is of course absolutely negligible.

*Dielectric body spinning in magnetic field.*  (ii) In the case of a dielectric (as also in the above) the restriction to a steady state and permanent configuration may be dispensed with; for the magnetic field arising from induced displacement currents can always be neglected in comparison with the inducing field. Thus, $(a, b, c)$ being the extraneous inducing field, the electric forces inside and outside a rotating mass are

$$\left(\omega c x - \frac{d\Psi_1}{dx},\ \omega c y - \frac{d\Psi_1}{dy},\ -\omega a x - \omega b y - \frac{d\Psi_1}{dz}\right) \text{ and } -\left(\frac{d}{dx},\ \frac{d}{dy},\ \frac{d}{dz}\right)\Psi_2.$$

As there can be no free electrification,

$$\nabla^2\Psi_1 = (1 - K^{-1})\,\omega\left\{2c + x\left(\frac{dc}{dx} - \frac{da}{dz}\right) + y\left(\frac{dc}{dy} - \frac{db}{dz}\right)\right\}, \quad \nabla^2\Psi_2 = 0;$$

while at the surface

$$\Psi_1 = \Psi_2, \quad K\frac{d\Psi_1}{dn} - (K-1)\,\omega\,\{cxl + cym - (ax+by)\,n\} = \frac{d\Psi_2}{dn},$$

*Case of sphere.*  the outside medium being air. If the dielectric body is a sphere rotating in a uniform field $(0, 0, c)$ parallel to the axis, this gives by the usual harmonic analysis

$$\Psi_1 = \tfrac{1}{3}\,(1 - K^{-1})\,\omega c r^2 + A r^2\,(\cos^2\theta - \tfrac{1}{3}) + A',$$
$$\Psi_2 = B r^{-3}\,(\cos^2\theta - \tfrac{1}{3}) + B' r^{-1},$$

where [the constants are fixed by the surface conditions]; thus determining the electric potential $\Psi_2$ in the space surrounding the rotating sphere.

15. More generally, let us consider steady distributions of electric charges on a system of conductors and dielectric bodies in motion *General conditions for a steady field without currents.*  through the aether. That there may be a steady state, without conduction currents, it is necessary that the configuration of the matter shall be permanent, and that its motion shall be the same at all times relative to this configuration and to the aether, and also to the extraneous magnetic field if there is one: this confines it to uniform spiral motion on a definite axis fixed in the aether. Referring to axes fixed in the material system, the vector potential has in the steady motion no time variation: hence

$$(P, Q, R) = -\left(\frac{d}{dx},\ \frac{d}{dy},\ \frac{d}{dz}\right) V,$$
$$(P', Q', R') = (P - qc + rb,\ Q - ra + pc,\ R - pb + qa),$$

for the magnetic induction through any circuit moving with the matter being constant, $(P, Q, R)$ is derived (§ 12) from an electric potential function $V$. Inside a conductor the electric force must vanish, otherwise electric separation would be going on; therefore $V$ must there be constant.

When the surrounding dielectric is free space, the total current in it, referred (§ 13) to these axes moving with the matter, is

$$- (p\,d/dx + q\,d/dy + r\,d/dz)\,(f, g, h).$$

When the velocity $(p, q, r)$ of the matter is uniform, it then follows from Ampère's circuital relation that

$$(a, b, c) = 4\pi\,(qh - rg,\ rf - ph,\ pg - qf).$$

Hence $(f, g, h)$, given by $4\pi c^2 f = P - qc + rb$, is expressed in terms of $(P, Q, R)$ by equations of type

$$(c^2 - p^2 - q^2 - r^2)\,f = \frac{P}{4\pi} - \frac{p}{4\pi c^2}\,(pP + qQ + rR).$$

The circuital quality of $(f, g, h)$ thus gives the characteristic equation of the single independent variable $V$ of the problem, in the form

$$\nabla^2 V = c^{-2}\left(p\,\frac{d}{dx} + q\,\frac{d}{dy} + r\,\frac{d}{dz}\right)^2 V,$$

the boundary condition being that $V$ is constant over each conductor.

Thus in the case of a system of conductors moving steadily through space with uniform velocity $v$ in the direction of the axis of $x$, $\epsilon$ denoting $(1 - v^2/c^2)^{-1}$ we have

$$(f, g, h) = (4\pi c^2)^{-1}\,(P, \epsilon Q, \epsilon R),\ \text{so}\ \left(\frac{d^2}{dx^2} + \epsilon\,\frac{d^2}{dy^2} + \epsilon\,\frac{d^2}{dz^2}\right) V = 0.$$

its field:

The distribution of electric force is therefore precisely the same as if the system were at rest, and the isotropic dielectric constant unity of the surrounding space changed into an aeolotropic one $(1, \epsilon, \epsilon)$, cf. Part I, § 115; and so would the surface density of true charge, which is the superficial discontinuity of total displacement, be the same, were it not that there is aethereal displacement *inside* the conductors which must be subtracted. The internal displacement current thence arising is

$$- \frac{v}{4\pi c^2}\frac{d}{dx}\,(0, -vc, vb);$$

hence $\qquad (a, b, c) = \left\{\dfrac{d}{dx},\ \left(1 + \dfrac{v^2}{c^2}\right)^{-1}\dfrac{d}{dy},\ \left(1 + \dfrac{v^2}{c^2}\right)^{-1}\dfrac{d}{dz}\right\}\phi,$

by Ampère's circuital relation: the circuitality of $(a, b, c)$ then leads to a characteristic equation for $\phi$, which must be solved so as to give at the surface of the conductor a value for the normal component of $(a, b, c)$ continuous with the already known outside value, and the internal displacement is thereby determined. There is no bodily

Uniform convection of a general charged system:

electrification inside the conductors, since this displacement is circuital.

the exact
Lorentz cor-
respondence,
We can restore the above characteristic equation of $V$, the potential of the electric force, to an isotropic form by a geometrical strain of the system and the surrounding space, represented by

$$(x', y', z') = (\epsilon^{\frac{1}{2}}x, y, z):$$

the actual distribution of potential around the original system in motion corresponds then to that isotropic distribution of potential round the new system at rest which has the same values over the conductors. The aethereal displacements through related elements of area $\delta S$ and $\delta S'$, of direction cosines $(l, m, n)$ and $(l', m', n')$ in the two spaces, multiplied by $4\pi c^2$, will be

$$-\left(l\frac{d}{dx} + \epsilon m\frac{d}{dy} + \epsilon n\frac{d}{dz}\right)V\delta S, \quad -\left(l'\frac{d}{dx'} + m'\frac{d}{dy'} + n'\frac{d}{dz'}\right)V'\delta S';$$

of these the second is always $\epsilon^{-\frac{1}{2}}$ times the first; thus the elements of surface for which the total displacement is null correspond in the two systems, and therefore the lines and tubes of total displacement also correspond, the flux of displacement in these tubes being $\epsilon^{-\frac{1}{2}}$ times greater in the second system than in the first. But on account of the aethereal displacement in the interior, the outside tubes do not mark out the distribution of the charge on each conductor. If then
with its
shrinkage
relative to
the frame,
but as *infra.*
a system of charged conductors has a velocity of uniform translation $v$ through the aether: and an auxiliary system at rest is imagined consisting of the original system and its space each uniformly expanded in the ratio $\epsilon^{\frac{1}{2}}$ or $(1 - v^2/c^2)^{-\frac{1}{2}}$ in the direction of the motion, and the charges on this new system are $\epsilon^{\frac{1}{2}}$ times those on the actual system: then the fields of aethereal displacement of the two systems agree in the surrounding spaces so as to be the same across corresponding areas, but the distributions of the charges on the conductors do not thus exactly correspond. (These results are obtained on the supposition that the structure of the matter is not affected by its motion. The conductors on which these charges are situated will, however, if the results of the more fundamental analysis of § 16 are admitted, change their actual forms by the corresponding slight shrinkage when they are put in motion, and this change will restore the distribution of charges and displacements, as there given.)

16. The circumstances of propagation of radiation in a material medium moving with uniform velocity $v$ parallel to the axis of $x$ will form another example. We may here (§ 13) employ the circuital relations, of types

$$4\pi u = \frac{d\gamma}{dy} - \frac{d\beta}{dz}, \quad -\frac{\delta\alpha}{dt} = \frac{dR}{dy} - \frac{dQ}{dz},$$

where $\qquad u = \dfrac{df}{dt} + \dfrac{\delta f'}{dt}$, $\quad (f', g', h') = \dfrac{K-1}{4\pi c^2}(P, Q, R)$,

$$(f, g, h) = \frac{1}{4\pi c^2}(P, Q + vc, R - vb).$$

There readily results, on eliminating the electric force $(P, Q, R)$,

*The partial convection of radiation.*

$$4\pi(u, v, w) = \text{curl}(\alpha, \beta, \gamma), \quad \frac{D^2}{dt^2}(a, b, c) = 4\pi c^2 \, \text{curl}(u, v, w),$$

where $\qquad \dfrac{D^2}{dt^2} = \dfrac{d^2}{dt^2} + (K-1)\left(\dfrac{d}{dt} + v\dfrac{d}{dx}\right)^2$;

which agrees with the equation obtained in Part I, § 124, and Part II, § 13, leading to Fresnel's law of alteration of the velocity of propagation.

Now let us consider the free aether for which $K$ and $\mu$ are unity, containing a definite system of electrons which are grouped into the molecules of a material body moving across the aether with uniform velocity $v$ parallel to the axis of $x$; and let us remove the restriction to steadiness of § 15. We refer the equations of free aether, in which these electrons are situated, to axes moving with the body: the alteration thus produced in the fundamental aethereal equations

*The exact Lorentz transformation for the convected field:*

$$4\pi\frac{d}{dt}(f, g, h) = \text{curl}(a, b, c), \quad -\frac{d}{dt}(a, b, c) = 4\pi c^2 \, \text{curl}(f, g, h)$$

is change of $d/dt$ into $d/dt - v\,d/dx$, leading to the forms

$$4\pi\frac{d}{dt}(f, g, h) = \text{curl}(a', b', c'), \quad -\frac{d}{dt}(a, b, c) = 4\pi c^2 \, \text{curl}(f', g', h'),$$

where $\qquad (a', b', c') = (a, b + 4\pi v h, c - 4\pi v g)$,

$$(f', g', h') = (f, g - vc/4\pi c^2, h + vb/4\pi c^2);$$

from which eliminating the unaccented letters, *neglecting* $(v/c)^{3*}$,

---

* Nothing need be neglected: the transformation is *exact* if $v/c^2$ is replaced by $\epsilon v/c^2$ in the equations and also in the change following from $t$ to $t'$, as is worked out in *Aether and Matter* (1900), p. 168, and as Lorentz found it to be in 1904, thereby stimulating to modern schemes of intrinsic relational relativity. The main steps in the deduction (*loc. cit.* p. 168) are

$$g = g' + \frac{v}{4\pi c^2}(c' + 4\pi v g), \qquad b = b' - 4\pi v\left(h' - \frac{v}{4\pi c^2}b\right),$$

$$\epsilon^{-1}g = g' + \frac{v}{4\pi c^2}c', \qquad \epsilon^{-1}b = b' - 4\pi v h',$$

leading to the reversed forms of the relations in the text, as now made exact,

$$\epsilon^{-1}(a, b, c) = (\epsilon^{-1}a', b' - 4\pi v h', c' + 4\pi v g'),$$

$$\epsilon^{-1}(f, g, h) = \left(\epsilon^{-1}f', g' + \frac{v}{4\pi c^2}c', h - \frac{v}{4\pi c^2}b'\right).$$

On substitution of these values on the first sides of the modified circuital relations just preceding, the following forms with $\epsilon v$ replacing $v$, which is restored finally, are obtained.

and writing as before $\epsilon$ for $(1 - v^2/c^2)^{-1}$, we derive the system

$$4\pi \frac{df'}{dt} = \frac{dc'}{dy} - \frac{db'}{dz} \qquad\qquad -(4\pi c^2)^{-1} \frac{da'}{dt} = \frac{dh'}{dy} - \frac{dg'}{dz}$$

$$4\pi\epsilon \frac{dg'}{dt} = \frac{da'}{dz} - \left(\frac{d}{dx} + \frac{v}{c^2}\frac{d}{dt}\right)c' \qquad -(4\pi c^2)^{-1}\epsilon \frac{db'}{dt} = \frac{df'}{dz} - \left(\frac{d}{dx} + \frac{v}{c^2}\frac{d}{dt}\right)h'$$

$$4\pi\epsilon \frac{dh'}{dt} = \left(\frac{d}{dx} + \frac{v}{c^2}\frac{d}{dt}\right)b' - \frac{da'}{dy} \qquad -(4\pi c^2)^{-1}\epsilon \frac{dc'}{dt} = \left(\frac{d}{dx} + \frac{v}{c^2}\frac{d}{dt}\right)g' - \frac{df'}{dy}.$$

Now change the time variable from $t$ to $t'$, equal to $t - vx/c^2$, so that

$$\frac{d}{dx} + \frac{v}{c^2}\frac{d}{dt} \text{ becomes } \frac{d}{dx}, \text{ and } \frac{d}{dt} \text{ becomes } \frac{d}{dt'},$$

and these equations assume the form of an electric scheme for a crystalline medium at rest. Finally write

*its spatial and electric features.*

$$x_1 \text{ for } x\epsilon^{\frac{1}{2}}, a_1 \text{ for } a'\epsilon^{-\frac{1}{2}}, f_1 \text{ for } f'\epsilon^{-\frac{1}{2}}, dt_1 \text{ for } dt'\epsilon^{-\frac{1}{2}},$$

keeping the other variables unchanged, and the system comes back to its original isotropic form for free aether. Thus the final variables $(f_1, g_1, h_1)$ and $(a_1, b_1, c_1)$ will represent the aethereal field for a correlative system of electrons forming the molecules of another material system at rest in the aether, of the form of the original one pulled out uniformly in the ratio $\epsilon^{\frac{1}{2}}$ along its direction of movement; the electric displacements through corresponding areas in the two systems are not equal, but their molecules are composed of equal electrons* and are situated at corresponding points, and the individual electrons describe corresponding parts of their orbits in times shorter

---

* The electrons do not here remain unchanged, but all are diminished by the same factor $\epsilon^{-\frac{1}{2}}$: to restore the values of the electrons, the electric and magnetic fields, as expressed immediately below, must each be multiplied by $\epsilon^{\frac{1}{2}}$ as in *Aether and Matter* (1900), p. 176, thus giving the correct Lorentz transformation, as valid (*loc. cit.*) on an electric orbital theory of atoms certainly up to the order $v^2/c^2$.

There may be superposed any arbitrary change of scales of space and time, the same for both, which determines the units of field measurement and must be settled once for all. But more is involved: it has been utilized to determine the dynamics of an electron, constrained internally if that were possible, to constant volume. It is not possible because, after Poincaré and Lorentz, a convection would not be reversible. Permitted frames of convection must form a group so that the result of a succession of convections must be a single convection. This absence of any privileged frame within the group is not *The aether,* evidence against an aether: it is a consequence of the permanence of the electron, which requires it to return to the same configuration and field every time the same conditions recur.

*In relation to matter.* The implication throughout is that there is an external world, which can be identically expressed in various frames in space and time involving each its own system of physical units of measurement: and the problem is to consolidate this group of frames into a connected unity.

The alternative view, that of the new relativity, seems to be that only the forms of relations persist as expressed in terms of four variables of space and time connected so as to belong to a spatial continuum, that the subjects of

for the latter resting system in the ratio $\epsilon^{-\frac{1}{2}}$ or $(1 - \frac{1}{2}v^2/c^2)$, while those less advanced in the direction of $v$ are also relatively very slightly further on in their orbits on account of the difference of time reckoning. Thus we have here two correlative systems each governed by the circuital relations of the free aether:

(i) A system in which the electric and magnetic displacements are $(f, g, h)$ and $(a, b, c)$, moving steadily with uniform velocity $v$ parallel to the axis of $x$.

(ii) The same system expanded in the direction of $x$ in the ratio $\epsilon^{\frac{1}{2}}$ and at rest, the displacements [electric and magnetic] at the corresponding points being*

$$(\epsilon^{-\frac{1}{2}}f,\ g - vc/4\pi c^2,\ h + vb/4\pi c^2)\ \text{and}\ (\epsilon^{-\frac{1}{2}}a,\ b - 4\pi vh,\ c + 4\pi vg),$$

and the molecules being situated in the corresponding positions with due regard to the varying time origin. Inasmuch as the circuital relations form a differential scheme of the first order which determines step by step the subsequent stages of a system when its initial state is given, it follows that if these two aethereal systems are set free at any instant in corresponding states, they will be in corresponding states at each subsequent instant, their electrons or singularities being at corresponding points. If then the latter collocation represent a fixed solid body, the former will represent the same body in uniform motion; one consequence of the motion thus being that the body is shrunk in the direction of its velocity $v$ in the ratio $\epsilon^{-\frac{1}{2}}$ or $1 - \frac{1}{2}v^2/c^2$. It may be observed that there is here no question of verifying that the mechanical forces acting on the single electrons in the two cases are such as to maintain this correspondence; for in the present complete survey of the individual atoms there is no such entity as mechanical force, any more than there is on a free vortex ring in fluid; the notion of mechanical forces enters at a subsequent stage when we are treating of molecular aggregates considered as continuous bodies, and are examining the relations between the different groups into which our senses analyze their interactions (§ 48).

*[margin: Perfect correspondence of moving with resting system: material identity: provided the inertia is wholly electric.]*

If this argument is valid, it will confirm the hypothesis of Fitz-Gerald and Lorentz, to which they were led as the ultimate resource for the explanation of the negative result of Michelson's optical the relations have no permanence, and being merely formal may be altered at will to bring this invariance about. Thus it is found that mass and energy have to be changeable, even the atoms themselves have to change with circumstances so as to be adapted to the principle: the material world is a fugitive affair, only the charge of the electron and the modulus of gravitation being unalterable, in addition to the multiplier required to reduce time to spatial measure.

* These values must each be multiplied by $\epsilon^{\frac{1}{2}}$, as in the preceding footnote, to ensure that the values of the electrons remain invariant.

experiments; and conversely it will involve evidence that the constitution of a molecule is wholly electric, as here represented. The reasoning given in Part II, § 13, was insufficient, because the correlation between the two systems was not there pushed to their individual molecules.

### Consideration of a possible Limitation of the Rotational Aether Scheme.

The field equations independent of rotational aether hypothesis.

17. In the preceding sections the equations of the aethereal field have been expressed (as they were in Part II) without reference to the dynamical hypothesis of a rotational aether which suggested their present form. Reverting now to that hypothesis, let us examine whether a limitation of the kind that was unavoidable in the material model of § 3, may not also be involved in the general scheme of a rotational aether. In the first place, it is natural to take the elastic rotation in the medium as very small, so that its translatory velocity which is connected therewith is also very small, though the velocities of the strain centres which flit across it, and represent the matter, may have any values; this is in agreement with the conclusion derived from optical experiments that the aether is practically stagnant. But there is one conceivable class of cases in which the changes of position of the elements of the medium go on accumulating, that namely of a steady magnetic field kept up say by a current of electrons constrained to flow permanently round a circuit. On account of the smallness of the velocity of the aether, the corrections to the dynamical equations which arise from the velocity of convection of the elastic strain may always be left out of account, being utterly insignificant in ordinary electrodynamics, and actually beyond

A magnet could not be stationary for ever in a rotational aether.

the limits of experiment in optics: yet in a magnetic field continuing steady *for an unlimited time* the elements of volume of the aether will ultimately have wandered far from their original positions, and a difficulty presents itself[1]. To cover such a case, the definition of the

---

[1] I am indebted to Lord Rayleigh for drawing my attention to this point, as one requiring further consideration [with reference to Thomson and Tait, *Nat. Phil.*, on cumulation of rotation in an isolated region of fluid in differentially irrotational motion, as *infra*. It is to be noted that the high inertia does not arise here merely as a device *ad hoc* to save permanent magnets. The elastic rotation of the medium has to remain small even close to an electron: this involves a density exceeding the enormous value $10^{15}$.]

(A steady magnetic field involves a cyclical motion of the aether; thus in a very great time even a very small velocity will produce large changes of position. It is true that any motion of electrons whatever will produce change of strain, and therefore movement in the aether, but that movement will be very slight, and will not be cumulative except in the one case of permanent cyclical motion which represents absolutely permanent magnets. If there were no other way out of the difficulty described in the text, it might be turned

elastic rotation of the medium must be made more precise. For the motion of a perfect fluid, which is differentially irrotational at each instant, will yet result after a time in finite rotations of its elements of volume; for example it is known that if a rigid ellipsoidal shell be filled with perfect fluid, and be set rotating about a fixed axis, then after a certain interval of time the parts of the fluid will have returned to the original configuration with respect to the shell, so that the fluid will have been rotated bodily in space just like a solid, although its motion at each instant has been differentially irrotational. In fact when the change of position of the element of volume is finite we can no longer analyze it definitely into rotations and pure strains, in such wise that the order of their application shall be indifferent; thus we obtain no longer in that way an analytical specification of the aethereal elastic rotation, and a more precise formulation must be made. The rough material model of § 3 indicates the necessary modification: in that model a differential pure strain of the element of volume does not tend to rotate the sub-element which is elastically effective; thus the efficient elastic rotation is the vector sum of the series of differential rotations which the element of the aether has experienced in its previous history. This is therefore the more precise definition of the total rotation, proportional to $(f, g, h)$, from which the electrostatic forcive is derived as in Part II, § 18: it makes the rotation equal to the curl of the linear displacement when these quantities are both small so that their squares and products can be neglected, but not after the long-continued cumulative effect of a permanent magnetic field has come in. In that case, however, the small irrotational velocity, say for the moment $(p, q, r)$, which constitutes the magnetic field, will contribute to $(f, g, h)$ only by shifting by convection the element of the medium along with its rotation, while the rotation so transferred will be continually re-adjusting itself by elastic action into the new equilibrium configuration: the relation between elastic rotation and magnetic force will then be of the type

$$\frac{df}{dt} = \frac{d\gamma}{dy} - \frac{d\beta}{dz} - \left(p\frac{d}{dx} + q\frac{d}{dy} + r\frac{d}{dz}\right)f,$$

where $(p, q, r)$ is equal to $(\alpha, \beta, \gamma)$ multiplied by a very small scalar factor. Unless the velocity $(p, q, r)$ is uniform,

$$\frac{d}{dt}\left(\frac{df}{dx} + \frac{dg}{dy} + \frac{dh}{dz}\right)$$

by simply asserting that absolutely permanent magnets do not exist [stationary in the aether].

The nature of the constraints which may be necessary to prevent the nucleus of an electron from ever becoming sub-divided is a different question, and wholly outside the scope of the present theory, which simply takes these nuclei to exist as it finds them without inquiring in detail into their structure.)

*Marginal notes:*

Finite rotational displacement producible by motion differentially irrotational.

Reference back to the Kelvin molecular model (unnecessary):

consequences that might have to ensue:

The electron unknown except as a point charge.

will not be exactly null; so that the movement of the aether by the steady magnetic field will lead to a development of electric charge, extremely slow and gradual, throughout the volume concerned. On the other hand, the combination of permanent electric and magnetic fields which is the origin of such a creation of electricity must be confined to a limited region, beyond which the aether is in equilibrium; therefore the electrification thus developed consists of com- *evaded* pensating amounts of positive and negative signs. These diffuse *locally,* charges, of the second order of small quantities, will subsequently by their mutual attractions drift together again and neutralize each other, by moving as strain forms across the aether without sensibly interfering with the motion of the medium itself (§ 6). Thus a steady magnetic field of unlimited duration would not theoretically get inter-locked with a concomitant electrostatic field, but would relieve itself by very slowly developing a very minute diffuse electrification which will simultaneously gradually fade away by its own natural actions, so that no sensible effect would ever be accumulated. The rotational aether scheme therefore would not break down in this limiting case, the consequence of long-continued cumulation being obviated by a process which is at each instant so insignificant as to be far below the *without* reach of experience: electrons may, it is true, also conceivably *obliteration* obliterate each other in the same way as these diffuse electrifications, *of electricity or matter.* but that is a contingency of negligible probability with which we are familiar in all kinds of molecular theory.

### *Relations of Inductive Capacity and Optical Refraction to Density.*

18. Let the medium be free aether containing $n$ similar molecules per unit volume; and suppose each molecule to be polarized to moment $\mu$ by the field of electric force. This field is made up of the extraneous exciting field and that of the polarized molecules themselves; the latter again consists of a part arising from the polarized medium as a whole and a part involving only the immediate surroundings of the point considered. If $h$ denote this local part, and $H$ the remainder of the total electric field, we have relations of the types $\mu = k (H + h)$, $i' = n\mu$, $h = \lambda . n\mu$, $i'$ denoting the intensity of the polarization, $k$ a constant independent of the density of the material medium, and $\lambda$ a parameter which, as will appear, must be nearly independent of the density. These relations lead to

$$i' = kn (H + \lambda i'),$$

that is, since by the definition of $K$ the inductive capacity,

$$i' = (K - 1)/4\pi . H$$

with electrostatic units, they lead to

$$\frac{3}{4\pi} kn = \frac{K - 1}{K - 1 + 4\pi/\lambda};$$

so that, $\rho$ denoting density, $(K - 1)/(K - 1 + 4\pi/\lambda)\,\rho$ is constant for the same material medium. For fluid media at any rate, it will appear that $\lambda$ must be very nearly equal to $\frac{4}{3}\pi$, so that for them Lorentz's expression $(K - 1)/(K + 2)\,\rho$ should be approximately constant.

When the dielectric is a compound one consisting of $n$ molecules of one kind and $n'$ of another per unit volume, we have $i' = n\mu + n'\mu'$, $h = \lambda i'$; and $\mu = k (H + h)$, $\mu' = k' (H + h)$, so that

$$\mu/k = \mu'/k' = i'/(kn + k'n').$$

Thus $i' = (kn + k'n') (H + \lambda i')$, so that, with the above value $\frac{4}{3}\pi$ for $\lambda$,

$$\frac{3}{4\pi} (kn + k'n') = \frac{K - 1}{K + 2}.$$

This formula gives an additive character to the refraction equivalent for a mixture; and also for a chemical compound, provided in the latter case the moment $\mu$ belongs to the individual atom, and is not sensibly affected by the molecular grouping of the atoms*.

This investigation is of course not absolutely exact; but it is the first approximation in a statistical theory, and the question presents itself how far it is a sufficient approximation†. On examination, it will appear that the coefficients $k$ and $k'$ are rightly taken to be numerical quantities independent of $n$ and $n'$, provided the distance between the effective poles of an atom or molecule is not a considerable fraction of the mean distance between adjacent molecules. The constancy of the value of $\lambda$, when the component densities are altered, appears from considerations of dimensions. For the force due to a polarized molecule varies as $\mu \times (\text{distance})^{-3}$: thus, as on change of density $(\text{distance})^{-3}$ varies directly as density, the character of the arrangement of the molecules being supposed unaffected, the force due to the molecules surrounding the point is proportional to $\mu \times \text{density}$, that is, it is equal to $\lambda n\mu$, as assumed. For the case of a mixture $\lambda$ is the same for both constituents; a result which may or may not hold good for a solution or a chemical compound.

* The relevant dates for this type of relation appear to be Mossotti (1850), Lorentz, Lorenz (refractive, 1868), Maxwell (conductance, 1872), Clausius (inductance, 1878), Debye (structural and experimental, recent). Cf. *infra*, p. 47.

† Isotropy is assumed: even for cubic crystals a special kind of minute anisotropic refraction arises if the ionic separation in the molecule is comparable with distance between adjacent molecules, which Lorentz has detected in actuality. On the practical limitations of refractive invariants see a report by K. Fajans, and resulting discussion, in *Trans. Faraday Soc.* Aug. 1927.

19. The value of $\lambda$, namely $\frac{4}{3}\pi$, which has here been assumed, is not merely dictated by the form desired for the final result. That value has in fact already been specified as the first approximation in quite another connection[1]. As this is the crucial point of the theory, it may be allowable to present the argument in detail. The total electric force acting on a single molecule is derived from the aggregate potential

*The general local molecular field.*

$$V = \Sigma \left( \mu_x \frac{d}{dx} + \mu_y \frac{d}{dy} + \mu_z \frac{d}{dz} \right) r^{-1},$$

where $\mu_x$, $\mu_y$, $\mu_z$ are the components of the moment $\mu$ of a polarized molecule. This potential, when the point considered is inside the polarized medium, involves the actual distribution of the surrounding molecules; and thus the force derived from it changes rapidly at any instant of time, in the interstices between the molecules. But when the point considered is outside the polarized medium, or inside a cavity formed in it whose dimensions are considerable compared with molecular distances, the summation in the expression for $V$ may be replaced by continuous integration; so that, $(f', g', h')$ being the intensity of polarization in the molecules of the dielectric,

$$V = \int \left( f' \frac{d}{dx} + g' \frac{d}{dy} + h' \frac{d}{dz} \right) r^{-1} d\tau;$$

and the force thence derived is perfectly regular and continuous. This expression may be integrated by parts, since, the origin being outside the region of the integral, no infinities of the function to be integrated occur in that region. Thus

*Integration by parts can smooth away irregular features,*

$$V = \int (lf' + mg' + nh') \, r^{-1} dS - \int \left( \frac{df'}{dx} + \frac{dg'}{dy} + \frac{dh'}{dz} \right) r^{-1} d\tau;$$

that is, the potential at points in free aether is due to Poisson's ideal volume density $\rho$, equal to $-\left( df'/dx + dg'/dy + dh'/dz \right)$, and surface density $\sigma$, equal to $lf' + mg' + nh'$. When the point considered is in an interior cavity, this surface density is extended over the surface of the cavity as well as over the outer boundary. Now when it is borne in mind that, at any rate in a fluid, the polar molecules are in rapid movement, and not in fixed positions which would imply a sort of crystalline structure*, it follows that the electric force on a molecule in the interior of the material medium, with which we are concerned, is an average force involving the average distribution of these polar molecules, and is therefore properly due to an ideal

*introducing boundary integrals instead.*

---

[1] "On the Theory of Electrodynamics," *Proc. Roy. Soc.* Vol. LII (1892), p. 64 [*supra*, Vol. I, p. 284].

* The recent calculations for a theory of polarization (magnetic or electric) in crystals, with the polarizable molecules arranged definitely in the lattices, contrast with this separation of an averaged local interaction for amorphous bodies, especially in the view afforded of ferromagnetic excitation. Cf. §§ 45, 73, *infra*.

*Crystals.*

continuous density like Poisson's, even as regards elements of volume which are very close up to the point considered. To compute the average force which causes the polarization of a given molecule we have thus to consider that molecule as situated in the centre of a spherical cavity whose radius* is of the order of molecular distances; and we have to take account of the effect of a Poisson distribution on the surface of this cavity, or more precisely of the result of an averaged continuous local polarization, surrounding the molecule, whose intensity increases from nothing at a certain distance from the centre up to the full amount $i'$ at the limit of the molecular range, this intensity being uniform in direction and a function of the distance only. The force due to this is $\frac{4}{3}\pi i'$ along the direction of the polarization $i'$; which is therefore the local part to be added on to the electric force as ordinarily defined—namely, to that arising from the density $\rho$ throughout the medium and the density $\sigma$ on its external surface, and so everywhere derivable from a potential by the theory of gravitational mass distributions. The value of the coefficient $\lambda$ of the above analysis should thus be $\frac{4}{3}\pi$ for a fluid[1]; but it may deviate from this value somewhat in the case of a solid, especially of course if it be crystalline.

*Effectively spherical cavity.*

*Equivalent cavity is ellipsoidal for crystals.*

20. The mathematical principles, on which the above formula for the relation between inductive capacity and density is based, were first given by Poisson for the corresponding problem in magnetic polarization. The explicit application to electric polarization, on the lines of Faraday's ideas, was made by Lord Kelvin and Mossotti. The investigation of the formula which has been implicitly given by Maxwell (*Treatise*, § 313), expressed in terms of the cognate problem of conduction, is however valid only for the case in which the coefficient of polarization of the medium is small compared with unity, that is, only for gaseous media. The same formula, viewed as a relation between refractive index and density for transparent media, was obtained by Lorentz[2] and was shown by him to be experimentally valid in an approximate way over the wide range of density including

*Poisson-Kelvin-Mossotti-Maxwell:*

*Lorentz:*

---

* The circumstance that the radius is inessential is of course a confirmation of the general validity of this procedure, which ignores the polarized matter supposed to be abstracted from the cavity: if the attraction were not of the type of inverse square there could be no such complete elimination of local influence. Cf. p. 45.

*Restriction of the theory.*

[1] The fact that the values of the refractive index for liquids are slightly in excess of what Lorentz's formula would give by computation from the values for their vapours, may be an indication that this averaged field of molecular action is slightly elongated instead of spherical.

*Test of form of effective cavity.*

[2] H. A. Lorentz, "Ueber die Beziehung zwischen der Fortpflanzungsgeschwindigkeit des Lichtes und der Körperdichte," Wied. *Ann.* IX, 1879, p. 641.

the liquid and gaseous states; though for the small changes of density induced in a liquid by alterations of pressure and temperature the effect of the change in the internal energy and mutual configuration of the molecules may considerably mask the direct effect of the slight change of density[1]. For gases, however, in which the molecules are more isolated and the changes of density greater, the refraction is found to be in accordance with the formula. The investigation of Lorentz[2] was probably the first effective attempt to introduce the molecular constitution of the medium into the electric theory of light, and so arrive at laws of refraction and dispersion. The *form* of the refraction constant was really settled by statical considerations akin to those here given*; but the theory of electric propagation employed by him at that time was the one developed in von Helmholtz's early memoirs on electrodynamics, and it would appear that discrepancies come in through treating the aether as polarized like a material dielectric; at any rate his final result (*loc. cit.* p. 654) seems to give the refractive index a value greater than unity for free aether, and one only infinitesimally different for a ponderable medium. A mathematical investigation has been given by Lord Rayleigh[3], in which the range of density over which these statical computations are valid is tested by finding for certain cases the complete expressions in a statical theory, of which they form the first rough approximations. The result is rather unfavourable to Lorentz's formula, so much so as perhaps to excite surprise at its close agreement with the facts when the range of density is so great as that between the liquid and gaseous states of the same substance. There is thus room for the statistical method under which the subject has here been approached[4], in that it explains the wide range through which the formula proves

*closeness of optical verification.* (margin)

*Rayleigh's scrutiny for crystals.* (margin)

*Present statistical procedure.* (margin)

---

[1] For these small changes, the Lorentz refraction function $(m^2 - 1)/(m^2 + 2)$ is approximately proportional to that of Gladstone and Dale, their ratio $(m + 1)/(m^2 + 2)$ being nearly constant; but it does not appear why the latter function happens to be usually more nearly proportional to the density than the former. The results of Röntgen and Zehnder, Wied. *Ann.* XLIV. 1891, on the effects of pressure on various fluids, make the two formulae in default in opposite directions by about equal amounts.

[2] *Verhandl. der Akad. Amsterdam*, XVIII; abstract in Wied. *Ann.* IX, 1872, pp. 641–665.

* As also the contemporary deduction of the same formula by an analysis expressed in terms of solid elasticity by L. Lorenz of Copenhagen.

[3] Rayleigh, "On the influence of obstacles arranged in rectangular order on the properties of a medium," *Phil. Mag.* Vol. XXXIV, 1892 (2), p. 481. [Also foreshadowing the recent lattice theories with polar elements: cf. Mahajani, *Proc. Camb. Phil. Soc.* 1926.]

[4] Since this was written, I have found that the analytical method here employed is essentially the same as that of Clausius (*Mechanische Wärmetheorie*, II, 1879); the fundamental importance of the ideas involved, and the

*Gladstone-Dale form of the refraction invariant.* (margin)

*Clausius.* (margin)

to be valid as a first approximation; while at the same time it recognizes that when the change of density is itself small, but is accompanied by other kinds of physical change, the influence of the latter on the polar molecule may be sufficiently important to prevent its exact verification[1]. On the specific influence of temperature, cf. § 72, *infra*.

21. In thus basing a theory of refraction equivalents on the value of the inductive capacity, it has been tacitly assumed that the dispersion of the medium is small; hence the results apply certainly only in the cases in which there is approximate agreement between the inductive capacity and the square of the refractive index[2]. When dispersion in *absolutely* non-conducting media is taken into account, as in the previous memoir, § 11, and *infra*, § 23, the formula however still holds, the constant $\kappa$, equal to $f'/P$ of § 24, now involving the period of the light. *Influence of dispersion.*

The fact that for gases, and a large class of denser bodies as well, the inductive capacity is approximately equal to the square of the refractive index, shows that in them the polarization of the molecules can completely follow the rapid alternations of electric force which belong to the light waves. Thus we can conclude that when the polarization of a molecule is upset by an encounter with another molecule, it is instantly restored to its normal value, as soon as the violence of the encounter is over; so that, the relative times spent by the molecule in encounters being small in every case, they hardly affect the inductive capacity of the medium; or in other terms, the density by itself hardly affects the molecular refraction equivalent (except in so far as the restoration of the steady state may involve absorption, § 28, *infra*), and the constancy of the coefficient $k$ is further justified[3]. *Refraction equivalents and polarization equivalents.*

discussion here given of the value of $\lambda$, in the case of fluid media, may perhaps justify the retention of the above independent statement. [Also the virtually identical formula for conductance in Maxwell, *Elec. and Mag.* 1873, § 314.]

[1] A theory precisely similar to the above of course applies to determinations of molecular magnetization in solutions of iron or other salts; strictly it is not the coefficient of magnetization $\kappa$, but $\kappa/(1 + \frac{4}{3}\pi\kappa)$ that is proportional to the density of the magnetic molecules. The values of $\kappa$ are however usually so small that this constant is practically equal to $\kappa$ [cf. § 72, *infra*]. *Magnetization is correlative.*

[2] This accords with the conclusion drawn by Linde from an experimental examination of the subject, Wied. *Ann.* LVI, 1895, pp. 546–70 (see p. 566). (Philip, *Zeitsch. Phys. Chem.* 1897, finds that the Clausius formula is quite inapplicable to mixtures of substances with abnormally high values of $K$.) *Limitations.*

[3] The analysis of this section does not agree with a theoretical investigation of the inductive capacities of mixtures of non-conducting liquids which do not exhibit change of volume in mixing, given by Silberstein (Wied. *Ann.* 1895); his result is that $K$, or what comes to the same under these conditions,

22. The molecular theory leads to the conclusion that the electric
Double refraction might be conditioned by form of lattice alone: aeolotropy of crystals in which the dielectric constant differs much from unity, may be in part due to the distribution of the molecules in space and in part to the orientation of the individual molecules; and that therefore the same applies to double refraction. The intrinsic
piezoelectric effect otherwise: polarity which is revealed by pyroelectricity and piezoelectricity also shows that orientation is a real cause. But magnetic aeolotropy must
also magnetic: practically be wholly due to orientation of the molecules, as the smallness of the susceptibility makes the effect of arrangement inappreciable. The double refraction induced in dielectrics in a strong
distortional refraction: electric field is possibly mainly due to molecular orientation, as also that arising from mechanical strain.

The difference of absorption in different directions in a crystal like
pleochroic effects, tourmaline must be of an order of numerical magnitude not higher than the difference of the refractions: an easy computation shows
slight, that it is really of a considerably lower order. This crystalline absorption can only be due to molecular orientation: it is of course exces-
not metallic. sively smaller than the absorption in metals, which is comparable with the whole refractive index; it would not therefore sensibly affect the laws of reflexion.

### *General Theory of Optical Dispersion* [: *Simplest Formal Dynamical Scheme** ].

Case of single free period. 23. A formula for optical dispersion was obtained in § 11 of the second part of this memoir, on the simple hypothesis that the electric

$K - 1$, is an additive physical constant, whereas the formula of Clausius and Lorentz makes $(K - 1)/(K + 2)$ additive. The determinations made by Silberstein for mixtures of benzol and phenylethylacetate give results for the Lorentz constant which are always in excess of the theoretical value, by amounts ranging up to 8 per cent.; the discrepancies for his own constant $K - 1$ are rather smaller, and are in both directions.

* This analysis aimed at developing the vibrational qualities of an atom consisting of a set of structural electrons, imbedded in constraints such as are required to ensure stability, when it is excited by an incident aethereal field.
Scope still limited. But the present constraints are of restricted type, in that they are expressible in terms of continuous linear equations. The recent classifications of natural spectra, and especially of their minute constitution and perturbations, indicate a more complex type of structure. The discussions, apart from direct
The recent atomic models. computations over orbital atomic models, possibly point to cyclic concatenation of the dynamical coordinates in the atomic field, in some way not yet definitely made out, the cyclic constants being the absolute quanta of the schemes of spectral expression. A hydrodynamic illustration which enlarges the field of imagination in relevant directions might for example be a system
Ideas from linked vortex rings. consisting of several vortex rings permanently linked together: the vibrational properties of such a system would involve the cyclic constants of all the rings, which would operate through relations of cyclic momenta not remotely different. The present analysis can remain, however, as the solution of the

polarization of the molecules vibrated as a whole in unison with the electric field of the radiation. The kinetic molecule of § 11, *supra*, with its steady momenta, will however usually have various free periods, and as many absorption bands; to take account of them, and also for other reasons which will appear, it is desirable to have a more complete dynamical theory.

The problem of dispersion, in its general form, is thus that of the transmission of radiation across a medium permeated by molecules, each consisting of a system of electrons in steady orbital motion, and each capable of free oscillations about the steady state of motion with definite free periods analogous to those of the planetary inequalities of the Solar System; and its analysis will in fact resemble Laplace's general investigation of the latter problem. If $\theta_1$, $\theta_2$, ... represent small deviations from the state of steady motion of a molecule, so that the coordinates of the system are $f_1(t) + \theta_1$, $f_2(t) + \theta_2$, ..., the kinetic and potential energy of the molecule when expanded in powers of these small quantities will assume the forms

*[margin: General analogue by orbital system:]*

*[margin: its oscillations around a state of steady motion:]*

$$T = \text{const.} + [\theta_1, \theta_2, ...]_1 + [\dot\theta_1, \dot\theta_2, ...]_1 + [\dot\theta_1, \dot\theta_2, ...]_2$$
$$+ [\theta_1, \theta_2, ...]_2 + [\{\theta_1, \theta_2, ...\}\{\dot\theta_1, \dot\theta_2, ...\}]$$
$$W = \text{const.} + [\theta_1, \theta_2, ...]_1 + [\theta_1, \theta_2, ...]_2,$$

where the terms in $T$ and $W$ denote functions of the various degrees of these velocities and displacements, the last term in $T$ being a lineolinear function of them jointly. From these expressions the free vibrations are determined by the Lagrangian method. As the undisturbed motion is steady, the type of a free vibration must be the same at whatever time it is excited, therefore the coefficients in $T$ and $W$ are all independent of the time; indeed if they were not constant the system could have no free periodic vibrations at all. The equations of the steady motion show that there can be no terms in $T - W$ of the first degree in the displacements, when the coordinates are properly chosen[1]. At the present stage we may conveniently by transformation of coordinates express the Lagrangian function, on which the motion in the molecule depends, in the form

*[margin: the Action function,]*

*[margin: is quadratic:]*

$$T - W = [\dot\theta_1, \dot\theta_2, ... \dot\theta_n]_1 + \tfrac{1}{2}\{A_1\dot\theta_1{}^2 + A_2\dot\theta_2{}^2 + ... + A_n\dot\theta_n{}^2\}$$
$$- \tfrac{1}{2}\{a_1\theta_1{}^2 + a_2\theta_2{}^2 + ... + a_n\theta_n{}^2\} + b_{11}\dot\theta_1\theta_1 + ... + b_{12}\dot\theta_1\theta_2 + b_{21}\dot\theta_2\theta_1 + ...,$$

problems belonging to a distribution of vibrators of a definite kinetic structure imbedded in an elastic medium.

The idea that occurs in various places in the text may be stressed, that such a system may have configurations of minimal energy, which thus cannot radiate away any more of its energy, though its constitutive cyclic motions, possibly connoting magnetic quality, would continue steady. For a *free* orbital system, as above mentioned, this is of course not possible, until it is helped by constraints, which may be slight, linking together the electrons.

*[margin: Nonradiating states.]*

[1] This analysis, so far, is as given by Routh, *Advanced Rigid Dynamics*, § 111.

simplified by
rejecting an
exact
differential: from which, by a property which is an immediate corollary of the Action principle, we may subtract any perfect differential coefficient with respect to time, for example here

$$[\theta_1, \theta_2, \dots \theta_n]_1 + d/dt \{\tfrac{1}{2}b_{11}\theta_1{}^2 + \dots + \tfrac{1}{2}(b_{12} + b_{21})\theta_1\theta_2 + \dots\},$$

without affecting the course of the motion, leaving thus an *effective*

the ultimate
type form. *Lagrangian function*

$$L = \tfrac{1}{2}\{A_1\theta_1{}^2 + A_2\theta_2{}^2 + \dots + A_n\theta_n{}^2\} - \tfrac{1}{2}\{a_1\theta_1{}^2 + a_2\theta_2{}^2 + \dots + a_n\theta_n{}^2\}$$
$$+ \tfrac{1}{2}\{e_{12}(\theta_1\theta_2 - \theta_2\theta_1) + \dots\}.$$

Its disturb-
ance by an
alternating
field: 24. We have now to determine the vibrations forced on this molecule by the uniform alternating field of electric force, say $P$ parallel to the $x$ axis, belonging to the radiation which is traversing the medium. Bearing in mind that the wave-length covers about $10^3$ molecules, it appears that if $f'$ denote the total intensity per unit volume of polarization of the molecules, the electric force acting on a single molecule will, as in § 19 but now using electrodynamic units, be $P_1 = P + \lambda c^2 f'$; this force will maintain vibratory motion in the polar molecule, but will not cause any oscillation of its centre of mass. The interaction of the electric field with the internal co-

the potential
energy of
disturbance, ordinates of the molecule will thus introduce an extraneous potential energy function of the form

$$W' = F(t) - (c_1\theta_1 + c_2\theta_2 + \dots + c_n\theta_n)P_1,$$

higher powers of the small internal coordinates $\theta_1, \theta_2, \dots \theta_n$ being, as usual in problems of vibration, omitted; and here again the coefficients $c_1, c_2, \dots c_n$ must be independent of the time. There will also be terms in the kinetic energy involving the interaction of the magnetic intensity of the field with the component velocities of the molecular vibration: now in a train of waves of type

$$\exp q\,(t - K'^{\frac{1}{2}}c^{-1}z),$$

and kinetic. the magnetic induction $b$, being derived from the electric force $P$, both in the plane $xy$ of the wave-front, by the relation

$$-\frac{db}{dt} = \frac{dP}{dz},$$

is equal to $K'^{\frac{1}{2}}c^{-1}P$: hence this part of the total kinetic energy will be of the form

$$T' = f(t) + (c'_1\theta_1 + c'_2\theta_2 + \dots + c'_n\theta_n)K'^{\frac{1}{2}}c^{-1}P,$$

where $c'_1, c'_2, \dots c'_n$ are coefficients independent of the time.

The polariza-
tion pro-
duced. The form of $W'$ shows that $c_1\theta_1 + c_2\theta_2 + \dots + c_n\theta_n$ is equal to the electric polarization $f_1$ in the molecule on which the electric force $P_1$ acts. If unit volume of the medium contains $n_1$ molecules of one kind,

$n_2$ of another and so on, and the polarizations in each molecule are respectively $f_1, f_2$ and so on, then

$$f' = n_1 f_1 + n_2 f_2 + \dots.$$

25. To obtain the general equation of propagation in the aether, let $\mathfrak{F}$ denote the electric force, or the torque acting on the aether; and we have, as in Part II, § 11, the kinematic relation

$$(4\pi)^{-1} \operatorname{curl} \mathfrak{B} = d/dt\,(\mathfrak{D} + \mathfrak{D}') + \mathfrak{C},$$

and also the dynamical equation

$$- d\mathfrak{B}/dt = \operatorname{curl} \mathfrak{F}, \quad \text{where } 4\pi c^2 \mathfrak{D} = \mathfrak{F}.$$

It is to be observed that this dynamical equation leaves out the purely local part of the electric force. The propagation of radiation of ordinary wave-length is in fact an action involving the medium in bulk, and not one of molecular type; thus in accordance with the Young-Poisson principle (*infra*, § 47) the local part of the electric force, arising from the surrounding molecules, is compensated intermolecularly by an influence on the physical properties of the material medium which thereby become functions of the density and strain, and this part therefore does not enter into the molar electric forcive maintaining the radiation. These equations lead to

$$c^2 \nabla^2 \mathfrak{D} = d^2/dt^2\,(\mathfrak{D} + \mathfrak{D}') + d\mathfrak{C}/dt.$$

Hence, when the current of conduction $\mathfrak{C}$ is non-existent,

$$K' = 1 + \mathfrak{D}'/\mathfrak{D};$$

whilst here $\mathfrak{D}'$ is $f'$, and $4\pi c^2 \mathfrak{D}$, or $P$, is $P_1 - \lambda c^2 f'$; so that

$$\frac{K' - 1}{K' - 1 + 4\pi/\lambda} = \lambda c^2 \frac{f'}{P_1},$$

or with $\lambda$ equal to $\tfrac{4}{3}\pi$,

$$\frac{K' - 1}{K' + 2} = \tfrac{4}{3}\pi c^2 \frac{f'}{P_1}.$$

26. The value of $f'/P_1$ is to be obtained from the equations of forced vibration of the molecules. By the Lagrangian method, these equations expressed for a molecule of the first kind and for radiation of the above type $e^{qt}$, form a system, skew symmetric in so far as $e_{21} = -e_{12}$, of type

$$(A_1 q^2 + a_1)\,\theta_1 + e_{12} q \theta_2 + \dots + e_{1n} q \theta_n - c_1 P_1 + c'_1 K'^{\frac{1}{2}} c^{-1} q P = 0,$$

wherein $\qquad c_1 \theta_1 + c_2 \theta_2 + \dots + c_n \theta_n - f_1 = 0.$

They give the relation

$$f_1 = [\{A_1 q^2 + a_1,\, e_{12} q,\, \dots,\, e_{1n} q,\, -c_1\}\,P_1$$
$$+ K'^{\frac{1}{2}} c^{-1} q\,\{A_1 q^2 + a_1,\, e_{12} q,\, \dots,\, e_{1n} q,\, c'_1\}\,P] \div \{A_1 q^2 + a_1,\, e_{12} q,\, \dots,\, e_{1n} q\},$$

in which the denominator represents a skew determinant, and each

---

*Margin notes:*

The interaction with aether:

of simple type for long light waves:

constitutive effects reserved.

Vibrational generalization of refraction invariant.

Polarization of atomic vibrations:

of the two coefficients in the numerator the same determinant bordered. The denominator involves when expanded only even powers of $q$, and when equated to zero it gives the periods of the free vibrations in the molecule; as these are all real the roots in $q^2$ must be all real and negative. The second term in the numerator has $c^{-1}$ as a factor; we may therefore neglect it as has been done in the previous paper; this means that the elasticity of the aether is so high compared with its inertia that the pull exerted by it on the molecule will be important while the interaction of its kinetic energy will be negligible. The remaining determinant in the numerator, when expanded, contains only even powers of $q$ and is of order lower by two than the denominator. Hence writing $-p^2$ for $q^2$, so that $2\pi/p$ is the period of the radiation, and expanding in partial fractions, we can express the equation in the form

*formulae simplified for high elasticity.*

*General formula for dispersion:*

$$4\pi c^2 \frac{f_1}{P_1} = \frac{g_1}{p_1{}^2 - p^2} + \frac{g_2}{p_2{}^2 - p^2} + \dots + \frac{g_n}{p_n{}^2 - p^2},$$

in which $g_1, g_2, \dots g_n$ are real quantities, positive or negative.

Now the index of refraction $\mu$ or $K'^{\frac{1}{2}}$ of the compound medium is given as above by the formula

$$\frac{K' - 1}{K' + 2} = n_1 \frac{\frac{4}{3}\pi c^2 f_1}{P_1} + n_2 \frac{\frac{4}{3}\pi c^2 f_2}{P_2} + \dots.$$

The final result is thus

$$\frac{K' - 1}{K' + 2} = \Sigma nm,$$

*for compound medium.*

where

$$m = \frac{g_1}{p_1{}^2 - p^2} + \frac{g_2}{p_2{}^2 - p^2} + \dots + \frac{g_n}{p_n{}^2 - p^2};$$

*Dispersion invariant in terms of free periods and their susceptibilities:*

so that it is $m$ and not $\mu^2$ that has an infinity at each free period of the molecule. We here again arrive at Lorentz's refraction equivalent, and the theorem that it is an additive physical constant; but with the important addition that it is the law of dispersion of the molecular refraction equivalent $m$, equal to $(\mu^2 - 1)/(\mu^2 + 2)\,\rho$, of each constituent of the medium, not that of the refractive index of the

*is additive.*

aggregate, which admits of simple theoretical expression. In physical investigations concerning laws of dispersion, it is thus essential to deal with simple substances; the dispersion in the molecular refraction constant of a mixture, and no doubt also to some extent of a solution or chemical compound, is made up, according to this formula, of the aggregate of those of its constituents[1].

[1] In cases however in which a formula of the Cauchy type is sufficiently exact, so that $(\mu^2 - 1)/(\mu^2 + 2)\,\rho = A + B/\lambda^2 + C/\lambda^4 + \dots$, not only is $A$ an additive refraction equivalent, but there will also be additive dispersion equivalents $B, C, \dots$.

*Dispersion equivalents.*

27. Let us consider briefly the case of a perfectly transparent substance whose dispersion is dominated by a single free period, say $2\pi/p_1$: the equation is

$$\frac{\mu^2 - 1}{\mu^2 + 2} = \frac{ng_1}{p_1{}^2 - p^2},$$

that is,

$$\mu^2 = 1 + 3\frac{ng_1}{p_1{}^2 - p^2}\bigg/\left(1 - \frac{ng_1}{p_1{}^2 - p^2}\right).$$

The local field of influences of a free period:

It will be convenient to form a graph of the formula for $\mu^2$; when $p$ is small, $\mu^2$ has a positive value, which should be the statical dielectric constant of the material; as $p$ increases, $\mu^2$ increases until it becomes infinite when $p^2 = p_1{}^2 - ng_1$; it then becomes negative, but again attains a positive value after $p^2 = p_1{}^2 + 2ng_1$ which corresponds to value zero. Thus there is a band of absorption, which is absolutely complete for some distances on both sides of the bright spectral line corresponding to the substance in the gaseous state, but which extends about twice as far on the upper side of that line as it does on the lower when $ng_1$ is positive, as will be the case when $\mu$ exceeds unity and the dispersion is in the normal direction. When, as in all ordinary media, the dispersion of the visible light is small, being for example of the order of one per cent. for glass, $p_1$ must be great compared with $p$, and the range of this single dominant ultra-violet band of absolutely complete absorption would be measured by an interval $(\delta p/p)$ equal to $\frac{1}{2}(\mu^2 - 1)/(\mu^2 + 2)$ below the free period, and one equal to $(\mu^2 - 1)/(\mu^2 + 2)$ above it, where $\mu$ is the index for luminous rays *.

finite band of absorption.

A dominant ultra-violet absorption, its range and trend:

its breadth.

28. For a substance such as a gas, with numerous narrow bands of absorption, in the immediate neighbourhood of any one of them the value of $\mu^2$ depends on that one alone; the breadth of the complete absorption thus corresponds to a total interval $(\delta p/p$ or $- \delta\lambda/\lambda)$ equal to $3ng_1/2p_1{}^2$, which should thus be proportional to the density of the gas. The distance on each side of the band to which the anomalous dispersion extends, which may possibly be observed as has been done by Kundt for sodium vapour, ought also to be of the order of magnitude of $ng_1/p_1{}^2$. The law of Janssen, that the [intensity] of the absorption in a compressed gas is roughly proportional to the square of the density, seems to show that in dense media most of the actual specific absorption is outside these limits of complete blackness, and is conditioned by the molecular encounters deranging the states of steady directed synchronous vibration, say by rotation of the molecule, and so necessitating absorption of fresh energy from the radiation in order to re-establish them. It is to be observed that this process would be a true absorption of radiation which would go to

Region of absorption close to each line.

Absorption varies square of density:

but not scattering:

* Actually the asymptotes must be smoothed away, else the scattering of the radiation, proportional to $\int(\mu^2 - 1)^2\, d\lambda$, would assume an infinite value.

heating the gas, as contrasted with mere refusal of a perfectly trans-
significance: parent gas to transmit radiation in a region in which $\mu^2$ is negative[1].
The gradual change from an emission spectrum of definite lines to
a continuous spectrum, with increasing density, would thus be due,
possible not to any want of definiteness of the free periods, but to changes in
transition to
continuous the orientation of the vibrating molecules arising from increased
spectrum. frequency of encounters, the corresponding rather abrupt changes
in the radiation received at any point not being analysable into the
regular Fourier periods.

29. The possible characteristics of the dispersion of an ideal per-

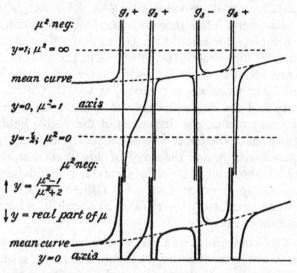

Type scheme fectly transparent medium may be very simply represented by a
for trans-
parent graph* of the general formula $(\mu^2 - 1)/(\mu^2 + 2) = \Sigma nm$. In a curve
dispersion: whose ordinate is $(\mu^2 - 1)/(\mu^2 + 2)$ and abscissa the frequency $p/2\pi$,
all parts which lie outside the two horizontal dotted lines corre-

---

[1] The validity of the general formulae is not vitiated by this circumstance
that the molecules are in various orientations which change from time to time
owing to encounters. The effect of this is that the coefficients which represent
Averaged the interaction between the aggregate of the matter and the aether, in the
atomic element of volume, are now the steady aggregates of the coefficients $c_1, c_2, \dots c_n$
moduli: which belong to the various simultaneous orientations of the molecules. Thus
the analysis remains intact provided $c_1, c_2, \dots c_n$ represent average values, and,
where necessary, a coefficient of absorption is introduced to represent the
abstraction of energy from the waves owing to the continual changes of
molecular orientation. After each such change of orientation of a molecule,
degradation the energy of its previously accumulated synchronous vibration is radiated
into heat. away or degraded into heat.

* Molecular scattering of the radiation would make the curve continuous,
smoothing away the asymptotes, for otherwise the energy scattered would be
infinite.

sponding to ordinates $+ 1$ and $- \frac{1}{2}$ belong to regions of complete opacity; the points where the curve crosses the axis represent the free periods or bright lines. A mean continuous curve of dispersion may be sketched in, by a dotted line, which coincides with the actual curve in the parts where the dispersion is normal, and may be considered as gradually rising towards a band of intense absorption in the ultra-violet, which dominates the mean dispersion; near an absorption band the dispersion is anomalous, but if the band is narrow as in the case of gases, the anomaly is confined to very narrow range. The diagram here given represents a case of four free molecular periods, for the third of which $g$ is negative while it is positive for the others. The refractive index that is determined by prismatic deviation is the real part of $\mu$, taken positively (§ 34, *infra*). A graph of this quantity is represented by the thick broken line of the lower curve: thus near a free period $2\pi/p_1$ the ordinate rises to infinity when $p^2 = p_1^2 - ng_1$, then sinks instantly to zero, and remains zero until $p^2 = p_1^2 + 2ng_1$, when it becomes positive again. Slight general absorption would ease off the corners of this graph so that it would not go up to infinity nor go down to zero: but there is nothing in its general aspect, at any rate for $g$ positive, as for example given by von Helmholtz, and verified by Pflüger for anomalously dispersive solid dye stuffs, which specially favours any one theory of dispersion. features are independent of any special theory.

30. The medium has hitherto been taken as absolutely transparent, that is, no degradation of energy occurs in it, the absorption bands above so called being really produced by total and nearly total reflexion of the radiation, which thus is not absorbed by the medium, but simply cannot get into it. Suppose that there is present a conduction current, which may be considered to include all causes which put $\mathfrak{D}'$ out of phase with $\mathfrak{D}$ and so lead to regular absorption of the energy of the waves: it may be represented, as in Part II, § 11, by the formula Opacity operating by total reflexion:conductance introduces absorption.

$$\mathfrak{C} = k \left( m' d/dt + \sigma' \right) \mathfrak{D};$$

we now have $\qquad K' = 1 + \{\mathfrak{D}' + (d/dt)^{-1}\,\mathfrak{C}\}/\mathfrak{D},$

thus simply adding to the formula for the square of the index of refraction the terms $- (km' + kp^{-1}\sigma'\iota)/(m'^2p^2 + \sigma'^2)$, which satisfactorily represent the general features of metallic propagation as was shown in Part II, § 11[1].

[1] For example, it explains why the real part of $\mu^2$ is negative for metals. In Part II the generally received contention (based on a narrower theory) that $\mu^2$ cannot be negative for purely elastic, that is dielectric, media was admitted without examination: but it is obviously inconsistent with the discussion above. As a concrete illustration, for a stretched thread weighted with equidistant particles, the square of the velocity of propagation of transverse waves No metallic optical instability involved, as the medium is compound.

It is noteworthy that as the period becomes very rapid the effective index of refraction always approximates to unity; so that very short waves will not suffer sensible reflexion, refraction, or diffraction even while their length may include many molecules of the material medium. In fact when the period of the radiation is sufficiently high, the free periods of the polar molecules are not quick enough to enable them to respond[1], while the comparatively free ions are prevented by their inertia from attaining any sensible velocity before the force is reversed; so that in neither way can the propagation of electric displacement across the medium be sensibly affected by the presence of the molecules. The formula shows, however, that the damping effect of conductivity usually persists to higher periods than the simple refracting effect of the excited vibrations.

*Shortest waves not refracted*

*though they are damped.*

The theory of dispersion would assume a simpler form if the molecules were systems vibrating about positions of equilibrium, instead of about states of steady motion. In that case the coefficients $e_{12}, e_{13}, \ldots e_{1n}$ are null: the restoring forces, proportional to the velocities, to which these belong, are in fact introduced by the steady motions, and may be named, after Lord Kelvin, motional gyrostatic

*Contrast of this kinetic dispersion with previous theory for static molecules:*

of sufficiently short periods is negative: yet no inference follows as regards instability.

[The latest experimental discussion of the optical constant of metals and the influence of surface contamination is in a Jena dissertation by G. Pfestorf, *Ann. der Phys.* Dec. 1926, pp. 906–28.]

[1] For a similar reason, the periods of luminous radiation are already too high to allow magnetic polarization to play any part in its propagation.

The statement in the text involves the currently received explanation of the Röntgen radiation. The different view has been recently advanced by Sir George Stokes that it may consist of sudden shocks transmitted through the aether from impacts by the molecules of the cathode streams. The molecules of matter lying in the track of the rays would not have time to be sensibly polarized by a sudden pulse which is over in a small fraction of their natural periods, and thus the pulse would pass across in the spaces between them, like sound through a grove of trees, without sensible refraction or diffraction: on the other hand the disruptive effect would resemble that of an explosive wave. Such pulses could hardly be other than the irregular beginnings of regular wave-trains sent out by the individual vibrating molecules; and as all radiation consists of such intermittent trains each with its irregular beginning, it would be assumed that the initial pulse is very much more intense in the electric bulb than in ordinary light, though still perhaps representing but a small portion of the total energy of the radiation. That the bombardment by the cathode streams is of a very disruptive, so to speak explosive, character, compared with ordinary molecular encounters, is in keeping with the rapid disintegration and evaporation of metallic plates under its influence. It is conceivable that the long-continued Becquerel radiation from fatigued phosphorescent substances arises in like manner from very sudden release of their molecules into new groupings, in the course of their gradual return to a natural or unfatigued configuration.

*Pulse theory of X-rays,*

*implies powerful atomic resilience:*

*so must involve also characteristic wave-trains,*

*perhaps slight if the shock is disruptive.*

*Radiation by sudden release in the molecule.*

forces; they evidence themselves by causing slow precessional oscil- precessions.
lations. The positional gyrostatic forces, or centrifugal forces proper, Positional
are merged in $T - W$ along with the forces arising from the potential kinetic forces merged with
energy[1]. When these motional forces are absent, we have potential energy.

$$\theta_1 = c_1 P_1 / (a_1 - A_1 p^2)$$

and similarly, so that

$$\frac{f_1}{P_1} = \Sigma \frac{c_1{}^2}{a_1 - A_1 p^2};$$

Law of dispersion:

and as before

$$\frac{K' - \mathrm{I}}{K' + 2} = \tfrac{4}{3}\pi c^2 \frac{\Sigma nf}{P_1}.$$

Thus the values of $g_1, g_2, \ldots g_n$ are in this case all positive; so that, if this represented the facts, the fragments of a horizontal spectrum, with red on the left, would after further refraction by a prism of anomalous material with its edge horizontal and uppermost, all a special feature,
slope upwards from left to right. On the other hand, each change of sign from positive to negative among the successive values of $g_1, g_2, \ldots g_n$ would give two fragments of the spectrum which would be curved back so as to be highest and lowest respectively near the middle, while a negative following a negative would imply slope upwards from right to left. According to Kundt's law the index is generally obtains in
abnormally great on the lower side and abnormally small on the fact.
upper side of an absorption band; and if this generalization is universally valid\*, it will follow that $g_1, g_2, \ldots g_n$ are actually all positive.

31. It is thus fundamentally desirable on various grounds to obtain information as to how the signs of these quantities depend on the gyrostatic coefficients; in particular because the present theory Wide scope of this
of gyrostatic molecules is a very wide one, and for example includes formal
as a limiting case the hydrodynamical vortex atoms of Lord Kelvin, theory.
in which the constitution is purely gyrostatic but the number of degrees of freedom is infinite. It will be convenient for this purpose to change to new coordinates of which $c_1\theta_1 + c_2\theta_2 + \ldots + c_n\theta_n$ is one, say the one with suffix unity: and to choose them semi-normal, so that the potential energy is represented by a sum of squares, which would be all necessarily positive if there were no gyrostatic influence. It may then be shown in the manner of the preceding analysis that $f_1/P_1$ is equal to a fraction of which the denominator is the period determinant of the molecule and the numerator is the minor of its leading term; the numerator is therefore the period determinant of

---

[1] The periods of small free vibrations, and the amplitudes of small forced ones, would not be affected by reversal of *all* the gyrostatic momenta in the Chiral
system: in fact this reversal would just change the system into its optical asymmetry.
image.

\* Cf. *infra*, p. 299.

the same molecule when its leading coordinate is prevented by constraint from varying. Thus we have the theorem

$$\frac{f_1}{P_1} = \frac{(p^2 - \alpha'_2{}^2)\,(p^2 - \alpha'_3{}^2) \ldots (p^2 - \alpha'_n{}^2)}{A_1\,(p^2 - \alpha_1{}^2)\,(p^2 - \alpha_2{}^2) \ldots (p^2 - \alpha_n{}^2)},$$

Law of excitation interpreted in terms of static effect of a constraint.

where $(\alpha_1, \alpha_2, \ldots \alpha_n)/2\pi$ are the natural frequencies of the molecule, while $(\alpha'_2, \alpha'_3, \ldots \alpha'_n)/2\pi$ are its frequencies when it is subjected to that particular constraint (namely on $\alpha_1$) which would prevent it from vibrating under the influence of the incident radiation; also $A_1$ is the coefficient of inertia of that particular vibration, so that its kinetic energy is $\frac{1}{2}A_1\theta_1{}^2$.

Gradual transition of system towards a constraint.
Peculiarity of gyrostatic stability.

This constraint may be represented analytically by making the elastic coefficient $a_1$ infinite; we may therefore attempt to trace, by the examination of graphs of the separate terms involved, the effects on the free periods of continuously varying this constant. The behaviour of a gyrostatic system may be very different from what experience teaches as to vibrations about configuration of rest, for the mere imposition of constraint to limit the vibrations of one coordinate may upset the stability of others: thus if $x$ represent $p^2$, the present period equation is of type $\phi(x) + a_1\psi(x) = 0$, in which all the $n$ roots of $\phi(x)$ are real and positive, while the same may not be true of the $n - 1$ roots of $\psi(x)$. It follows, easily, however, by

Kelvin theorem:

application of the principle of energy[1], that if the system be completely stable when all the gyrostatic motional forces are removed, then it will remain stable when these forces are restored; and stability will therefore also be maintained when elastic connections are strengthened or constraints are introduced[2]. In that case the roots

points toward Kundt's law for anomalous refraction.

of $\psi(x)$ will all be real and positive: and it is easy to deduce that they will separate those of $\phi(x)$, and that in consequence $g_1, g_2, \ldots g_n$ will be all positive, so that Kundt's law will hold good. The further conclusion is thus also somewhat probable that if the constitution of the gyrostatic molecule is thoroughly stable so that the imposition of mere constraint could not upset it, then Kundt's law will hold.

32.   The specific refraction $(\mu^2 - 1)/(\mu^2 + 2)$ always increases along with the index $\mu$: if the dispersion were controlled solely by powerful

Symptoms indicating ultra-violet absorption.

absorption bands in the ultra-violet, with positive $g$, the trend of the index would always be in the same direction as the frequency increases. Hence in the large class of substances with normal dispersion of visible light for which $K$ exceeds $\mu^2$, there must also be strong

---

[1] Cf. Thomson and Tait's *Nat. Phil.* 2nd ed., Part I, p. 409: the relations of the free periods of a gyrostatic system are there discussed at length in pp. 370–415.

[2] All the relations as to the march of the periods developed by Rayleigh (*Theory of Sound*, 2nd ed., § 92 a), from Routh's analysis will then hold good.

absorption in the ultra-red[1]. The specific refraction however always tends to the limit unity as $\mu^2$ or $K'$ increases; so that the large dielectric constants of water and alcohol (at ordinary temperatures) are not so abnormal in their optical as in their electrical aspect. These large values are an indication that the constituents of the molecules are distantly and loosely connected together, which may be related to the powerful action of these substances as solvents[2]; it has been

High dielectric constants:

why associated with powerful solvent property.

[1] In von Helmholtz's memoir on the electric theory of dispersion, he found satisfactory agreement between the formula with one ultra-violet absorption band and the observations for glycerine, and he suggested that agreement might also be established for carbon bisulphide by assuming slight dissipation such as would not sensibly modify the laws of reflexion: but he apparently omitted to notice that this verification is defective in not making the square of the index equal to the statical dielectric constant for the case of very long waves. To amend it, another region of absorption would have to be assumed in the ultra-red, far down so as not to sensibly affect the visible radiations, thus leading to the Ketteler type of formula, which is approximately $\mu = A + B\lambda^2 + C\lambda^{-2}$ for slightly dispersive substances: in the case of glycerine $C$ would be small. When there is only one absorption band, the dispersion formula common to these discussions is in form the same as one derived by von Lommel (Wied. *Ann.* XIII. p. 353) from a mechanical theory, and compared by him with observation for a considerable number of media. In all media whose dispersion is effectively controlled by one absorption band, that band must be far in the ultra-violet or else the dispersion of the visible radiation will be excessive, so that the formula must approximately coincide with Cauchy's: thus it is only substances for which $\mu^2$ is approximately equal to but slightly greater than $K$, which can have any chance of coming into that class.

Historical.

It appears from the above that in this formula for dispersion in a medium dominated by one main band of absorption, as given in Part II, § II, we must make the distinction, that the value of $2\pi/p$ for which $\mu^2$ is infinite is not the free period of a single molecule by itself, or that of the bright line in a gaseous spectrum, but is the period when it is vibrating in step with all the surrounding molecules under whose influence it lies. [On consequent broadening of the lines see *Astrophys. Journal*, Dec. 1916: as *infra*.]

Free period modified by neighbours.

For very slow periods there is no dispersion in a transparent medium and the refraction depends wholly on the statical character of the medium, including its density. For higher periods of the incident radiation, the free periods of the molecules introduce dispersion and also absorption bands; but the position of these bands depends not merely on the free periods, but also slightly on the density of the medium, through the influence of the latter on its statical inductive capacity. It is not impossible that the free molecular periods, as well as the absorption bands, may be affected in this way. An influence of density of the medium on the position of the lines in the spectrum has been found and investigated by the Baltimore spectroscopists. (Cf. FitzGerald, *Astrophys. Journal*, 1896.)

Dispersion falls off at low frequencies.

Pressural displacement in spectra.

[2] The reason here assigned is different from the one that has been given by various writers [Nernst, Thomson], that high inductive capacity of an intervening medium weakens the electric forces between the ions in the dissolved molecule. Here it is taken as an indication that the effective ions are far apart in the molecules of the solvent, so that a dissolved molecule can come under

With large $K$ the ions are weakly associated.

noticed that high inductive capacity is usually associated with conductivity.

Contrast with a mechanical theory of dispersion:

33. It is of interest to contrast these results with the ones that flow from a purely mechanical theory of dispersion. If the molecule consist of a dynamical system, simple or gyrostatic, of dimensions small compared with the wave-length, joined on to the aether by mechanical connections, the uniform oscillatory displacement of the aether will exert no differential statical force on the molecule, but the kinetic energy of the whole compound system will contain terms involving products of the displacement $u$ of the aether and the coordinates $u$, $\theta_1$, $\theta_2$, ... $\theta_n$ of the small disturbance of the molecule, say terms $(c_0\dot{u} + c_1\dot{\theta}_1 + c_2\dot{\theta}_2 + ... + c_n\dot{\theta}_n)\,\dot{u}$. In the equation of propagation, formed in the Lagrangian manner, $\rho d^2u/dt^2$ will now be replaced by $d^2/dt^2\,(\rho u + c_0 u + c_1\theta_1 + c_2\theta_2 + ... + c_n\theta_n)$; while in the equation of vibration of the molecule terms of type $c_1 d^2u/dt^2$ will occur. The square of the index of refraction is thus given by

$$\mu^2 = 1 + (c_0 u + c_1\theta_1 + c_2\theta_2 + ... + c_n\theta_n)/\rho u;$$

its dispersion formula:

and this leads by analysis similar to the above to a dispersion formula $\mu^2 = A + \Sigma g_1/(p_1^2 - p^2)$[1]. It is to be noticed that, on a mechanical

refraction would persist for very short waves:

theory, the index does not finally tend to unity as the frequency $p/2\pi$ rises, for when the waves have ceased to excite internal vibrations in the molecules the aether is still loaded by their inertia; an exception occurs when the attachment of the molecule to the aether is such that, when owing to the high period it is not internally vibrating, the aether does not sensibly displace its centre of mass, in which case the constant $A$ is unity and there is no effective load

molecule anchored to a large inertia,

on the aether. If we suppose that each molecule has an attachment to a very large mass, so as to be practically anchored to it in space, this will require us to take one of the natural frequencies to be

gives the Kelvin formula.

infinite in the above analysis, so that say $p_1$ is zero. When both these characteristics are present, we arrive at Lord Kelvin's formula[2]. If,

the influence of one of these ions alone, without much counteracting effect from the other ion.

[1] An equation equivalent to this, with $g_1$, $g_2$, ... all positive, appears in Sellmeier's original paper (Wied. *Ann.* 1872), based however on a much more special hypothesis.

Absorption attainable without intrinsic friction:

[2] Baltimore Lectures, 1884: cf. also present memoir, Part I, *Phil. Trans.* A, 1894, p. 820. In these lectures Lord Kelvin, with a view to explaining true absorption without introducing frictional forces into ultimate theory, contemplates the molecules as able to take up a vast amount of energy, near certain periods, before they attain to a steady state of synchronous vibration; as however that state must come after at any rate some millions of vibrations, and absorption would then cease, it is presumably part of the theory that the absorbed energy is constantly being degraded in the molecule by a process analogous to fluorescence, and so being got rid of by radiation at a lower period

on the other hand, we take the medium to be like an elastic jelly, permeated by spherical portions of different inertia and elasticity, the problem is a quite different one, which forms in fact a rude mechanical analogue of the electric theory; and it was in this way that L. Lorenz independently arrived at the specific refraction formula above discussed.

*Contrast with heterogeneous elasticity theory.*

### Prismatic Deviation by Opaque Media.

34. The fact that light preserves its period shows (Part 1, § 90) that the circumstances of its propagation across opaque media are determined simply by a complex index of refraction, of which the imaginary part represents the absorption. Measures of deviation by opaque prisms, such as those made by Kundt, yield directly the value of this complex index, by simple consideration of the geometrical continuity of the traces of the waves along the interfaces, without

*Optical equations necessarily linear.*

*Index for metals as determined by prisms,*

—or it may be simply scattered owing to change of orientation of the steady state of vibration at which the molecule has arrived, due to encounters with other molecules, as indicated by Janssen's law [p. 55] for gases. In the electric theory, metallic absorption is here taken to be chiefly due to the presence of free ions or electrons; but in weakly absorbing media it is probable that the former cause is the effective one. The only analytical way open for representing it is to introduce an absorption coefficient expressing the averaged rate at which the energy of the radiation is being exhausted.

*its relation to radiation:*

The synchronously vibrating material molecules would not in any case give rise to further absorption by sending out energy in regular secondary waves: their uniformity in distribution and phase prevent this, just as they prevent the separate elements of the continuous aether from acting in the same way. The dust particles which give rise to the blue sky are irregularly distributed, and the individual secondary waves thence originating have irregular and independent phase differences with reference to the primary exciting wave. The analogous medium for sound, filled with fixed attuned resonators, is absorbent solely on account of the secondary radiation of the resonators: consequently if they were all alike and regularly distributed and they occupied a very large number of wave-lengths, there would be no absorption and the medium would be transparent to sound, unless it has the same period as the resonators when it could not penetrate into the medium at all. The correlative absorption of light would thus be a process special to it, arising from ionization and molecular impact. Unless the absorption in iodine vapour is accomplished by ionization so that it goes mainly into heat, there must be scattered light accompanying it and representing part of it.

*can the blue sky be due to molecular scattering?*

*True absorption.*

[This is in some respects at variance with Lord Rayleigh's later ascription (1899) of the blue sky to the molecules of dustless air. The scattering effect is however proportional statistically, on his principles, to $\sqrt{n}$, where $n$ is the number of scatterers per wave-length, so that it is the number of the air molecules that counteracts the small effect of each. The effect of each is diminished by the presence of adjacent ones, and for a uniform medium when the subdivision into elements has no superior limit it would not arise at all. In any case, crystals, owing to the regularity of spacing, should scatter only very slightly, as Lorentz and Rayleigh have recently remarked. But on this subject see Vol. 1, *supra*, Appendix VII.]

*Reconciliation.*

without
ambiguity. the necessity of the intervention of any dynamical theory whatever and therefore free of all ambiguity of interpretation. The thickness of the portion of the prism that is traversed does not affect the deviation of the light; so it may be taken as null, and we have only to consider refraction into the prism, and then out of it at a plane inclined at an angle differing by $\alpha$, the angle of the prism, without changing the point of incidence. Let the axis of $x$ be in the first face of the prism, towards the edge, that of $y$ normal to it and that of $z$

The theory for an opaque prism: parallel to the edge. Then for the incident, reflected, and refracted waves, of period $2\pi/p$, the vibration vectors are proportional respectively to

$$\exp \iota\,(lx + my - pt), \quad A \exp \iota\,(lx - my - pt), \quad A' \exp \iota\,(lx + m'y - pt)$$

where
$$l^2 + m^2 = c^{-2}p^2, \quad l^2 + m'^2 = K'c^{-2}p^2,$$

c being the velocity in free space, and $K'$ the complex value of the square of the index of refraction. It will now be convenient to refer the second refraction to corresponding axes $\xi$, $\eta$, $\zeta$ related to the second face of the prism; thus

$$x = \xi \cos \alpha - \eta \sin \alpha, \quad y = \xi \sin \alpha + \eta \cos \alpha, \quad z = \zeta.$$

The vectors of the incident, reflected, and emergent beams are then proportional to

$$A' \exp \iota\,(\lambda\xi + \mu\eta - pt), \quad B \exp \iota\,(\lambda\xi - \mu\eta - pt),$$
$$B' \exp \iota\,(\lambda\xi + \mu'\eta - pt),$$

where
$$\lambda = l \cos \alpha + m' \sin \alpha, \quad \mu = -l \sin \alpha + m' \cos \alpha, \quad \text{and} \quad \lambda^2 + \mu'^2 = c^{-2}p^2.$$

for a thin prism: When the refracting angle $\alpha$ is small, this gives approximately

$$\mu'^2 = m^2 - 2lm'\alpha, \quad \text{or} \quad \mu' = m - l/m \,.\, m'\alpha,$$

where
$$m' = \{(l^2 + m^2)\, K' - l^2\}^{\frac{1}{2}} = j + \iota k \quad \text{say}.$$

Thus the emergent vibration vector is represented by

$$B' \exp \iota\,\{(l + \alpha m')\, \xi + (m - l/m \,.\, \alpha m')\, \eta - pt\},$$

that is

$$B' \exp - \alpha k\,(\xi - l/m \,.\, \eta) \exp \iota\,\{(l + \alpha j)\, \xi + (m - l/m \,.\, \alpha j)\, \eta - pt\}.$$

It therefore emerges at an angle of refraction $\psi$, away from the edge, given by

$$\cot \psi = (m - l/m \,.\, \alpha j)/(l + \alpha j) = m/l \,.\, \{1 - \alpha j l\,(l^{-2} + m^{-2})\},$$

and direct incidence: the angle of incidence being $\phi$ where $\cot \phi = m/l$. Thus $\psi = \phi + j/m \,.\, \alpha$, and the deviation is $\psi - \phi - \alpha$, that is $\alpha\,(j' \sec \phi - 1)$, where $j'$ is the

usual deviation formula holds in terms of real part of index: real part of $(K' - \sin^2 \phi)^{\frac{1}{2}}$. When the angle of incidence $\phi$ is small, the deviation is thus $(n - 1)\,\alpha$, where $n$ is the real part of the complex refractive index $K'^{\frac{1}{2}}$. Thus the experiments of Kundt on metallic prisms, and of Pflüger on anomalously refracting media[1], determine

---

[1] A. Pflüger, Wied. *Ann.* LVI, 1895, p. 412.

the march of $n$. Although $\sin^2 \phi$ is not usually very considerable compared with $K'$, and thus oblique incidence on the prism does not very greatly affect the deviation[1], yet it would seem desirable to have observations at oblique incidence, as they would give data for determining the imaginary part of the index also by this uniform method, and thus its complete value. If this were known for the neighbourhood of an absorption band, we should possess all the data requisite to guide and correct theory in the matter of optical dispersion; but a knowledge of $n$ by itself is not of much service in this respect. The value of this method of prismatic deviation lies in the fact that the complex index is determined without the intervention of any considerations as to dynamical theory or the effect of surface contamination on polarization, which must enter into the interpretation of experiments on reflexion.

*oblique incidence would determine imaginary part also,*

*without dynamical hypothesis.*

### The Mechanical Tractions on Dielectric Interfaces: and the Mechanical Bodily Forcive.

35. When the local part of the forcive on the polarized molecules of the medium, arising from their interaction with the neighbouring polarized molecules, is left out of account, the remainder, which is the mechanical force on the element of volume, is derived from the energy function $-(f'P + g'Q + h'R)$; this would be also a potential function of the forces were it not that in it only the electric force $(P, Q, R)$ is to be varied. When however the dielectric is homogeneous, the negation of perpetual motions requires that

*Expression for mechanical energy.*

$$f'dP + g'dQ + h'dR$$

shall be a complete differential; thus when the law of induced polarization is linear, the force will be derived from a potential function $-\frac{1}{2}(f'P + g'Q + h'R)$, and so will be balanced, as regards the interior of the medium and as regards the translatory part, by a hydrostatic pressure $\frac{1}{2}(f'P + g'Q + h'R) + \text{const.}$; and when the medium extends continuously to a distance from the seat of the electric action, the constant in this expression must be null. When the medium is isotropic, the translatory force is all, there being no torque on the element of volume. In a fluid dielectric this compensating hydrostatic pressure actually exists, and has been measured; in a solid it is merely a compendious expression for the material reaction per unit volume against the electric forces transmitted by the aether from other matter at a distance. If however the fluid dielectric is heterogeneous there will not be a potential function, and it can only be in equilibrium when stratified in a certain manner; if

*Transformation for linear polarization into a potential function: compensated by a hydrostatic pressural stress, which is its actual reaction for a fluid.*

*Stratification if heterogeneous.*

---

[1] Cf. the measures of Shea, Wied. *Ann.* LVI.

gravity did not operate the surfaces of stratification would be the equipotentials of the field of force.

Traction on an interface. 36. When there are in the electric field interfaces of transition between different dielectrics, there will also exist surface tractions on them which may be evaluated by considering an actual, somewhat abrupt, interface to be the limit of a rapid continuous variation of the properties of the medium which takes place across a layer of finite though insensible thickness. Let then the total displacement $(f'', g'', h'')$, with its circuital characteristic where there is no free charge, be made up of the dielectric material polarization $(f', g', h')$, and the displacement proper $(f, g, h)$ which is the aethereal elastic rotation $(P, Q, R)/4\pi$. Thus if we neglect now the minute difference between the aethereal force $(P', Q', R')$ and the electric force Equivalent electric densities, $(P, Q, R)$,

$$\frac{df'}{dx} + \frac{dg'}{dy} + \frac{dh'}{dz} = -\rho', \quad \frac{df}{dx} + \frac{dg}{dy} + \frac{dh}{dz} = \rho + \rho',$$

free and polarized. where $\rho'$ is the Poisson ideal volume density corresponding to the polarization, and $\rho$ is the volume density of free electrons, surface distributions being now by hypothesis non-existent[1]. The mechanical forcive acting in the dielectric is, per unit volume, a force $(X', Y', Z')$ and a torque $(L', M', N')$, where

Mechanical forcive: $$X' = f'\frac{dP}{dx} + g'\frac{dP}{dy} + h'\frac{dP}{dz} + \rho P, \quad L' = g'R - h'Q.$$

The component parallel to $x$ of the aggregate force acting on the whole transitional layer is the value of $\int X' d\tau$ integrated throughout it. This integral is finite, although the volume of integration is small, on account of the large values of the differential coefficients which occur in the expression for $X$. To evaluate it, we endeavour by integration by parts to reduce the magnitude of the quantity that

---

[1] The notation of Part II is here maintained; thus $(f'', g'', h'')$ represents the $(f, g, h)$ of Maxwell's *Treatise*. Electrostatic units are here employed. It may be well to recall the relations of these quantities. As the aethereal elemental rotation is from its nature circuital, the increment in its outward flux across any closed surface is equal to the amount of electrons that have crossed that surface into the enclosed region, arising partly from movement of free electrons, Free and bound electrons. and partly from orientation of polar molecules over the surface so that one pole is inside and the other outside. Thus, $(l, m, n)$ being the direction vector of the normal, and $\Delta$ representing a finite increment,

$$\Delta \int (lf + mg + nh) \, dS = \Delta \int \rho \, d\tau - \Delta \int (lf' + mg' + nh') \, dS;$$

so that $$\int (lf'' + mg'' + nh'') \, dS = \int \rho \, d\tau,$$

which gives $$df''/dx + dg''/dy + dh''/dz = \rho.$$

remains under the sign of volume integration, so that in the limit reduced for an interfacial stratum, we may be able to neglect that part; thus we obtain

$$\int X' d\tau = \int (lf' + mg' + nh')\, P dS + \int (\rho' + \rho)\, P d\tau.$$

By the definition of electric force, $(P, Q, R)$ is the force due to a volume distribution of density $\rho + \rho'$ and to extraneous causes; so that in the limit when the transitional layer is indefinitely thin, we have by Coulomb's principle

$$\int (\rho + \rho')\, P d\tau = \tfrac{1}{2} \int (\sigma' + \sigma)\, (P_2 + P_1)\, dS$$

$$= (8\pi)^{-1} \int (N_2 - N_1)\, (P_2 + P_1)\, dS,$$

$P_1$, $P_2$ being the values of the $x$ component $P$, and $N_1$, $N_2$ those of the normal component $N$, of the electric force $(P, Q, R)$ on the two sides of the layer, all measured towards the side 2, while $\sigma'$ and $\sigma$ are the surface densities constituted in the limit by the aggregates of $\rho'$ and $\rho$ respectively taken throughout the layer. Hence in the limit

$$\int X' d\tau = \left| \int (lf' + mg' + nh')\, P dS \right|_1^2 + (8\pi)^{-1} \int (N_2 - N_1)\, (P_2 + P_1)\, dS.$$

Thus the electric traction on the interface of transition may be represented by a pull towards each side, along the direction of the resultant to a traction specification. electric force $F$; this pull is on the side 2 of intensity

$$n_2' F_2 - \tfrac{1}{2} (n_2' - n_1' - \sigma)\, F_2, \text{ that is } \tfrac{1}{2} (\sigma + n_2' + n_1')\, F_2$$

in the direction of $F_2$, where $n'$ is the normal component of the polarization of the medium measured positive towards the side 2; on the face 1 the pull is $\tfrac{1}{2} (\sigma - n_2' - n_1')\, F_1$

now in the direction of $F_1$, $n'$ being measured positive as before.

As the tangential component of the electric force $F$ is under all circumstances continuous across the interface, the total traction on The resultant traction. both sides combined is $\tfrac{1}{2} (n_2' + n_1')\, (N_2 - N_1)$ along the normal towards 2, together with the tractions $\tfrac{1}{2}\sigma F_2$, $\tfrac{1}{2}\sigma F_1$ acting on the true charge $\sigma$ along $F_2$, $F_1$. If $n''$ denote the normal component of the total displacement $(f'', g'', h'')$, so that $n'' = N/4\pi + n'$, $n_2'' - n_1'' = \sigma$, the first part of this total traction is

$$- 2\pi n_2'^2 + 2\pi n_1'^2 + 2\pi (n_2' + n_1')\, \sigma.$$

When there is no charge the whole is a normal traction

$$- 2\pi n_2'^2 + 2\pi n_1'^2$$

towards the side 2.[1] When the interface is between a dielectric 1

[1] It may be recalled that in the terminology here employed, the *true* electrification $\sigma$ is the density of unpaired electrons; while the *true* electric current Maxwell's terminology. arises from the movements of all the electrons, free and paired, but does not include the change of aethereal strain which must be added in order to make up the *total* circuital current of Maxwell.

Traction on a conductor.
and a conductor 2, the traction is only towards the side 1 and is equal to $\frac{1}{2}(n_1' + \sigma) F_1$, or $\frac{1}{2}n_1''F_1$, per unit area, along the normal which is now the direction of the resultant force.

Law of polarization not involved.
All this is quite independent of the law of the connection between the polarization and the electric force in the material medium. Thus, under the most general circumstances as regards electric field, whether there is material equilibrium or not, the forcive on the material due to its electric excitation consists of the interfacial tractions thus specified, together with a force $(X', Y', Z')$ and a torque $(L', M', N')$ per unit volume, given by the formulae

$$(X', Y', Z') = \left(\frac{d}{dx}, \frac{d}{dy}, \frac{d}{dz}\right)(f'P + g'Q + h'R),$$

$$(L', M', N') = (g'R - h'Q, \ h'P - f'R, \ f'Q - g'P),$$

in the former of which $(f', g', h')$ is not to be differentiated.

Nature of transition not involved.
The assumption underlying this analysis, that the transitions are gradual, will be sufficiently satisfied even if the intermediate layer is only one or two molecules in thickness; for as these molecules are arranged slightly in and out, and not in exact rows along the interface, their polarity can still be averaged into a continuous density as above. The aggregate tractions over a thin layer of transition thus do not depend sensibly on the nature of the transition, but only on the circumstances on the two sides of the layer.

37. In the case of a fluid medium, the bodily part of the forcive produces and is compensated by a fluid pressure $\int i'dF$, where $i'$, being the polarization induced by the electric force $F$, is for a fluid in the same direction as $F$ and a function of its magnitude. This pressure will be transmitted statically in the fluid to the interfaces[1]; combining it there with the surface traction proper, it appears that

For fluids interfacial tractions alone,
the material equilibrium of fluid media is secured as regards forces of electric origin if extraneous force is provided to compensate a total normal traction towards each side of each interface*, of intensity $- 2\pi n'^2 - \int i'dF$. In the case usually treated, in which a linear law of induction is assumed so that the relation between $i'$ and $F$ is

for linear law of induction derivable from the Maxwell stress.
$i' = (K - 1) F/4\pi$, the mechanical result of the electric excitation of the fluid medium is easily shown to be the same[2] as if each interface were pulled towards each side by a Faraday-Maxwell stress,

---

[1] That is, a reacting pressure $\int i'dF$ exerted by the interface will keep the medium in internal equilibrium: no constant term is added because the pressure must vanish along with the polarization.

* This supposes that there is no surface charge $\sigma$: if there is, it must adjust itself by flow so that at the interface $F_2$ and $F_1$ are both normal.

[2] It is a normal traction equal to $- (K - 1)(KN^2 + T^2)/8\pi$ towards each medium, or in all a single traction $(K_2 - K_1)(2\pi n''^2/K_1 K_2 - T^2/8\pi)$ towards the medium 1 [, whether the uncharged media are fluid or not].

made up of a pull $KF^2/8\pi$ along the lines of force and an equal pressure in all directions at right angles to them. But this imposed geometrical self-equilibrating stress system would not be an adequate representation of the mechanical forcive in a solid medium; for then the bodily forcive, instead of being wholly transmitted, is in part balanced on the spot by reactions depending on the elasticity of the material. The forcive on uncharged isotropic material may however in every case, whether the induction follows a linear law or not, be expressed as an extraneous or imposed system, made up of bodily hydrostatic pressure $\int i'dF$ (which in the case of a fluid only relieves the ordinary fluid pressure and so diminishes the compression, § 79, *infra*) together with normal tractions on the interfaces between dielectric media, of intensity $-2\pi n'^2 - \int i'dF$ acting towards each side, and tractions $\frac{1}{2}n''F - \int i'dF$ on the surfaces of conductors acting towards the dielectric.

Force distribution for general solid.

38. A similar analysis applies to the electromagnetic forcive acting on a magnetically polarized medium. Excluding as before the part arising wholly from the interaction of neighbouring molecules, which (§ 44, *infra*) is not transmitted by material stress, but is compensated on the spot by molecular action due to change of physical state induced by it, the electromagnetic forcive proper is made up of a bodily force $(X, Y, Z)$ and torque $(L, M, N)$, where, $(u', v', w')$ representing the true current,

Electro-magnetic force distribution.

$$X = v'\gamma - w'\beta + A\frac{d\alpha}{dx} + B\frac{d\alpha}{dy} + C\frac{d\alpha}{dz}$$

$$= vc - wb + A\frac{d\alpha}{dx} + B\frac{d\beta}{dx} + C\frac{d\gamma}{dx} - \gamma\frac{dg}{dt} + \beta\frac{dh}{dt},$$

$$L = B\gamma - C\beta.$$

Under the usual circumstances, in which the aethereal displacement current can be neglected, these expressions are identical with the ones given without valid demonstration in Maxwell's *Treatise*[1]. The remarkable property is there established (*loc. cit.* § 643) that, independently

---

[1] Vol. II, § 640. It will be observed that the force acting on the moving electrons which constitute the true current is here taken to be $(v'\gamma - w'\beta, \ldots, \ldots)$. In the investigation of Part II, § 15, which determines the motional force on a single electron, the expression for $T$ represents the kinetic energy of the aether; it is transformed so as to be expressed in terms of the electric displacement of the aether and the electrons of the materials by introducing $(F, G, H)$ whose curl gives the actual velocity of the aether near the electron; and finally, after the forces acting on the electrons and on the aethereal displacement have thus been separated out, $(F, G, H)$ is eliminated by the same relation. Thus the force acting on the single moving electron comes out as $e(\dot{y}\zeta - \dot{z}\eta - \dot{F}, \ldots, \ldots)$, where $(\dot{\xi}, \dot{\eta}, \dot{\zeta})$ is the velocity of the medium; and the average force acting on the electrons in the element of volume, that

The mechanical force acts on true current only:

how deduced:

Slight dis-
crepancy
from the
Maxwell
electro-
magnetic
stress,
of the form of the relation between magnetic induction and magnetic force in the medium and whether there is permanent magnetism or not, this bodily forcive (with the last terms neglected) can be formally represented in explicit terms as equivalent to an imposed stress: viz. $\mathfrak{H}$ denoting magnetic force and $\mathfrak{B}$ magnetic induction, the bodily forcive is the same as would arise from

(i) a hydrostatic pressure $\mathfrak{H}^2/8\pi$,

(ii) a tension along the bisector of the angle $\epsilon$ between $\mathfrak{H}$ and $\mathfrak{B}$, equal to $\mathfrak{H}\mathfrak{B} \cos^2 \epsilon/4\pi$,

as here
expressed.
(iii) a pressure along the bisector of the supplementary angle, equal to $\mathfrak{H}\mathfrak{B} \sin^2 \epsilon/4\pi$, together with an outstanding bodily torque turning from $\mathfrak{B}$ towards $\mathfrak{H}$ and equal to $\mathfrak{H}\mathfrak{B} \sin 2\epsilon/4\pi$.

When $\mathfrak{B}$ and $\mathfrak{H}$ are in the same direction, the torque vanishes, and a pure stress remains in the form of a tension $(\mathfrak{H}\mathfrak{B} - \frac{1}{2}\mathfrak{H}^2)/4\pi$ along the lines of force and a pressure $\mathfrak{H}^2/8\pi$ in all directions at right

The stress
cannot be
a physical
reality
(unless
supplemented
by a
momentum).
angles to them. There is no warrant for taking this stress to be other than a mere geometrical representation of the bodily forcive. It is however a convenient one for some purposes[1]. Thus the traction acting on the layer of transition between two media, in which $(\alpha, \beta, \gamma)$ changes very rapidly, which might be directly deduced in the same manner as the electric traction above (§ 35), may also be expressed directly as the resultant of these Maxwellian tractions towards the two sides of the interface. As there cannot be free magnetic surface density or purely superficial current sheets, the traction on the

It involves
traction on
a layer of
transition:
interface is represented, under the most general circumstances, whatever extraneous magnetic field may there exist, by purely normal pull of intensity $2\pi\nu^2$ towards each side, where $\nu$ is the normal com-

is, the induced electric force causing electric separation in the element, is $e(\dot{y}c - \dot{z}b - \dot{F}, ..., ...)$, as there given. But in computing, as in Part II, § 23, the electromagnetic force on an element of volume carrying a current, it must be borne in mind that part of the above force on the single electron arises from the magnetism in this element of volume itself; and the principle of energy forbids that any part of the forcive on the mechanical element of volume of the material can arise from mutual actions inside the element, so that this part must be compensated by a reciprocal action of the moving electrons which

involves the
magnetic
force not
induction.
constitute the current on those which constitute the magnetism, in a manner which might be expressed if necessary. Hence, when this local part is omitted in accordance with the general principle, the transmitted electromagnetic force is $(v'\gamma - w'\beta, ..., ...)$ as above, not $(v'c - w'b, ..., ...)$ as previously stated in closer accordance with the Ampère-Maxwell formula. Cf. § 44, *infra*. (Observe, however, that in quoting Part II, § 15, $(\xi, \eta, \zeta)$ must now represent the velocity of the aether multiplied by the square root of $4\pi$ times its very

Stress scheme
convenient
for expressing
an aggregate
forcive.
high coefficient of inertia: the unit of time was there tacitly chosen so that this factor should be unity.)

[1] (For example, the repulsion exerted by alternating currents on pieces of copper or other conducting masses may thus most conveniently be represented.)

ponent of the magnetization at that side. When the medium is non-magnetic, there is no such superficial traction, but only the bodily electromagnetic forcive on the *true* electric currents of the material medium, which is represented by the above stress system.

39. The form of the mechanical forcive is identical whether the polarization is electric or magnetic, provided there are no electric currents; in the first case it is the material reaction to the static strain in the aether, in the other case it is the reaction to the motional aethereal forcive arising from the revolving electrons in the molecule. Omitting for simplicity the slight effect of the convection current in cases where any exists, the forcive arising from the electric polarization of the medium consists of a bodily force $(X', Y', Z')$ and torque $(L', M', N')$, where $[X' = f'\,dP/dx + g'\,dP/dy + h'\,dP/dz$, so that]

$$X' = f'\frac{dP}{dx} + g'\frac{dQ}{dx} + h'\frac{dR}{dx} + \rho P + g'\frac{dc}{dt} - h'\frac{db}{dt}, \quad L' = g'R - h'Q,$$

together with an interfacial traction between media 1 and 2 which is along the normal and equal to $- 2\pi n'_2{}^2 + 2\pi n_1{}'^2$ [multiplied by $c^2$] towards the medium 2, $n'$ representing the component of the polarity $(f', g', h')$ along the normal; the motional forcive arising from the magnetic polarity and the electric currents consists of a bodily force $(X, Y, Z)$ and torque $(L, M, N)$, where [in like manner]

$$X = A\frac{d\alpha}{dx} + B\frac{d\beta}{dx} + C\frac{d\gamma}{dx} + vc - wb - \gamma\frac{dg}{dt} + \beta\frac{dh}{dt}, \quad L = Bc - Cb,$$

together with a normal interfacial traction $- 2\pi v_2{}^2 + 2\pi v_1{}^2$ towards the medium 2, $v$ representing the normal component of the magnetization $(A, B, C)$. When the last terms in $X$ involving the aethereal displacement current are neglected, the latter forcive is the same as would arise from Maxwell's magnetic stress specification. It may be shown that

$$X' = \frac{d}{dx}\{f''P - \tfrac{1}{2}(fP + gQ + hR)\}$$

$$+ \frac{d}{dy}g''P + \frac{d}{dz}h''P + h\frac{db}{dt} - g\frac{dc}{dt},$$

so that the former forcive is what would arise from an analogous electric stress specification in which

$$(P, Q, R),\ (f', g', h'),\ 4\pi\,(f'', g'', h'')$$

correspond to $\quad(\alpha, \beta, \gamma),\ (A, B, C),\ (a, b, c)$

respectively, with the exception however in this case also of an outstanding bodily forcive $(hdb/dt - gdc/dt, ..., ...)$ which is not included in the stress. A theory which assumes that there is but one medium in which everything is transmitted by contact action, not two interacting media matter and aether as here, is compelled to get rid of any outstanding forcive like this, which is not expressible explicitly in terms of stress: for this reason supporters of that view have found it necessary to introduce into the electric field a purely hypothetical mechanical forcive arising from the electric field acting on the so-called magnetic current $d/dt$ $(a, b, c)$, in analogy with the Amperean forcive arising from the magnetic field acting on the electric current. The addition of this forcive $(hdb/dt - gdc/dt, ..., ...)$ to $(X', Y', Z')$ and the omission of $(- \gamma dg/dt + \beta dh/dt, ..., ...)$ from $(X, Y, Z)$, permit both to be expressed *explicitly* in terms of stress*.

*(margin: which would disappear if there could be magnetic currents amenable to electric force,)*

*(margin: but an electro-magnetic residue would remain.)*

### Maxwell's Theorem of a Representative Stress†.

40. The mechanical forcive acting in a polarized medium thus corresponds in the main to the system of bodily force and interfacial traction which is the result of Maxwell's magnetic stress (*Treatise*, § 640) considered as an extraneous system applied to the medium. The electric stress of Maxwell (*Treatise*, § 105) is something wholly different, leading in the case of homogeneous media to interfacial tractions only, without bodily force; it could thus have valid application only to unpolarized media, as for example to the theory of gravitation which passes through material bodies just as through a vacuum. The proposition really established[1] is that the mechanical forcive due to attraction at a distance, obeying the law of inverse squares, between material bodies, may be represented by a con-

*(margin: The actual stress tensor locally existent (e.g. local momentum):)*

*(margin: in contrast with a stress of pure electric propagation,)*

---

* The aggregate outstanding force not merged in the stress is the sum of the first of these residues and the second with changed sign. *When the medium is non-magnetic* it is the exact time gradient differential of a vector $(h\beta - g\gamma, ...,....)$ which has been interpreted as a distribution of momentum located in the aether. The effect of the forces on the matter would thus be expressible physically as arising from transfer of a distribution of stress and of momentum. Its specification is aethereal, independent of the matter present: thus it would be an intrinsic momentum excited somehow in the aether.

*(margin: Concomitant momentum introduced.)*

The further analytic generalization into the stress-momentum tensor of the relativity fourfold followed later. Cf. Vol. I, *supra*, Appendices v, vii.

The misfit in the stress-momentum formulation above noted for a magnetic medium should not present itself if the magnetism were replaced by Amperean atomic orbital electrons: therefore on the Amperean theory of magnetism there ought to be an additional (local) mechanical force equal per unit volume to $4\pi$ times the vector product of $(A, B, C)$ and $(df/dt, dg/dt, dh/dt)$, required to cancel it. The necessity for this interaction of magnetism and aethereal displacement current, too slight to admit of experimental scrutiny, has been discovered directly by Leathem: "On the force exerted on a magnetic particle by a varying electric field," *Proc. Roy. Soc.* 1913.

*(margin: A new mechanical force.)*

† See Appendix, "The essential nature of the Maxwell stress," inserted at the end of this paper.

[1] Maxwell, "On physical lines of force," Part I, *Phil. Mag.* Vol. xxi, 1861, especially Prop. iii[: or in *Scientific Papers*, Vol. i].

nection in the form of an imposed extraneous stress symmetrical with respect to the lines of force, acting across the intervening medium, *provided* that medium is not in any way polarized by the force. A stress restricted by this relation of symmetry involves only two variables, the principal tractions along and at right angles to the line of force; and the essence of Maxwell's theorem is that it is possible always to determine these two variables so as to satisfy the three equations of equilibrium of the element of volume of the medium. These principal tractions prove, as is well known, to be equal in magnitude but opposite in sign. The proposition is in itself so remarkable that it deserves to be formulated abstractly without reference to hypothetical applications. The representation of a given bodily forcive by a geometrical stress system is in general a widely indeterminate problem, as the six stress components have to satisfy only three equations: but the condition of symmetry with respect to lines of force restricts the stress so much that such a representation would only in special circumstances be possible.

*without resulting local force,*

*a remarkable possibility.*

*The regular local Molecular Forcive in an excited Dielectric: its Expression as a Stress system: Examples of the Principle of the Mutual Compensation of local Molecular Forcives\*.*

41. In the above estimate of the mechanical forces acting on an element of a polarized medium, the influence of the general mass of the medium on the molecules in the element has been alone included; it remains to consider the *rôle* of such terms as would arise from the

*The local interatomic field:*

---

\* The train of ideas here adumbrated can obtain more precise illustration in terms of recent formulations in regard to crystal structure. For the molecular arrangement is there definite, except as regards thermal vibrations, instead of accidental. The molecules can be taken, as an opening to precise calculation, to act on one another by mutual forces, either according to an empirical direct-distance law, or a law as between electric doublets, or as between magnetic doublets. A potential energy can thus be calculated (after Born): and the crystalline form would be one which makes it minimal under the suitable restrictions.

*The case of crystals:*

This energy and these forces are expressed by divergent summations when the crystal is taken as unlimited in extent: and one course of procedure is to throw away the divergent part. But the physical meaning of such rejection is brought out in all such cases by dividing the series that is involved into a continuous expression in the form of an integral, and a part which is rapidly divergent around each atomic position, so is the imperfect expression of local interactions which expend their influence in modifying the local structure. The continuous integral part is the basis of a theory of propagation of influences across the medium in bulk, with constitution thus determined locally and represented by elastic and other moduli, which is thus expressed by continuous differential equations: it is a feature in its development (*e.g.* Green's method for electric fields) to assign symbols for the unknown potentials in each extended homogeneous region, establish interfacial conditions between such regions, and

*transition to continuous stress.*

special forcives of neighbouring molecules. The intensity of the local part of the regular electric force acting at a molecule has already been assigned (§ 19) as $\frac{4}{3}\pi i'$, very approximately for the case of fluid media, possibly not so approximately for solids. The argument was that owing to the translational mobility of the surrounding molecules, their action on the one under consideration averages into that of the uncompensated distribution of poles which would exist on the surface of a small spherical cavity in a continuous uniformly polarized medium—or more precisely, into that of a spherical shell of poles, of thickness not indefinitely small but with this law of distribution around the centre. For the interior of a uniformly polarized medium the local part of the electric force is thus at each instant constant throughout this cavity and equal to $\frac{4}{3}\pi i'$; therefore the mechanical force exerted on the polar molecule (that is one involving equal numbers of positive and negative electrons) at the centre of the cavity is null, as it depends on the rate of variation of this electric force. But at a place where the polarization varies from point to point, the alteration in the law of surface density over the cavity will supply a local part.

*as obtained (for amorphous bodies) by averaging;*

*with residual terms,*

*analyzed for constant direction of an exciting field:*

When the polarization $i'$ changes only in magnitude and not in direction, this part will arise from a distribution of uncompensated poles over the surface of the cavity, of density

$$-\left(i_0' + x\,\frac{di_0'}{dx} + y\,\frac{di_0'}{dy} + z\,\frac{di_0'}{dz}\right)\cos\theta,$$

where the subscript zero implies the value at the centre. If the axis of $x$ is taken along the direction of $i_0'$, the electric potential $U$ in the interior due to this distribution is equal to

$$-\tfrac{4}{3}\pi\left(i_0' + \tfrac{1}{3}\frac{di_0'}{dx}\right)x - \tfrac{4}{5}\pi\left\{\tfrac{1}{3}\frac{di_0'}{dx}(2x^2 - y^2 - z^2) + \frac{di_0'}{dy}xy + \frac{di_0'}{dz}xz\right\}.$$

*result,* On a molecule of moment $\mu_x$, at the centre, this gives a force

$$-\mu_x\frac{d}{dx}\left(\frac{d}{dx}, \frac{d}{dy}, \frac{d}{dz}\right)U, \text{ that is } \tfrac{4}{5}\pi\mu_x\left(\tfrac{4}{3}\frac{d}{dx}, \frac{d}{dy}, \frac{d}{dz}\right)i_0'.$$

Thus there is a bodily force due to this cause, of intensity

*now without torque.*

$$\tfrac{2}{5}\pi\left(\tfrac{4}{3}\frac{d}{dx}, \frac{d}{dy}, \frac{d}{dz}\right)i_1'^2;$$

but there is not any bodily torque.

proceed by adjusting appropriate solutions of the characteristic spatial differential equations. Thus a discrimination between field phenomena such as are adjusted by transmission, and the constitutive residue of mutual actions which determines local structure involved in this transmission, becomes in the simpler types of problem practically precise and definite. For a body of finite extent this integration by parts introduces surface integral terms which are the expression of surface energies, as *infra*, p. 79.

42. Now let us proceed to the general case, in which the direction of the polarization $(f', g', h')$, as well as its magnitude, varies from point to point; in the hypothetical case in which the effective distance between the poles of a molecule is small compared with the average distance between neighbouring molecules, we can express the molecular part of the forcive on an element of volume by simple summation for $f'$, $g'$, and $h'$ separately, by aid of the expressions just found. Thus it consists of a bodily force $(X_1, Y_1, Z_1)$ and torque $(L_1, M_1, N_1)$, where

$$X_1 = f' \frac{dP_1}{dx} + g' \frac{dP_1}{dy} + h' \frac{dP_1}{dz}, \quad L_1 = g' R_1 - h' Q_1,$$

$(P_1, Q_1, R_1)$ being the *local* part of the electric force in the spherical cavity, so that

$$P_1 = \tfrac{4}{3}\pi \left\{ \left( \tfrac{4}{3}x \frac{df'}{dx} + y \frac{df'}{dy} + z \frac{df'}{dz} \right) + \left( y \frac{dg'}{dx} - \tfrac{2}{3}x \frac{dg'}{dy} \right) \right.$$

$$\left. + \left( z \frac{dh'}{dx} - \tfrac{2}{3}x \frac{dh'}{dz} \right) \right\}.$$

Hence, omitting terms involving $x$, $y$, $z$,

$$X_1 = \tfrac{4}{3}\pi \left\{ \left( f' \frac{d}{dx} + g' \frac{d}{dy} + h' \frac{d}{dz} \right) f' + \tfrac{1}{2} \frac{d}{dx} (f'^2 + g'^2 + h'^2) \right.$$

$$\left. - \tfrac{2}{3}f' \left( \frac{df'}{dx} + \frac{dg'}{dy} + \frac{dh'}{dz} \right) \right\}$$

$$= \tfrac{4}{3}\pi \left\{ \frac{d}{dx} (f'^2 + g'^2 + h'^2) - \tfrac{2}{3}f' \left( \frac{df'}{dx} + \frac{dg'}{dy} + \frac{dh'}{dz} \right) \right.$$

$$\left. - g' \left( \frac{dg'}{dx} - \frac{df'}{dy} \right) + h' \left( \frac{df'}{dz} - \frac{dh'}{dx} \right) \right\},$$

with similar expressions for $Y_1$ and $Z_1$; while the torque vanishes in the limit.

43. In these formulae the aim has been simply to represent as they are, the regular local forcives acting on the molecules, as a distribution of force throughout the volume and, if need be, of traction over the surfaces of the material, thus avoiding the use of any hypothetical stress system which might be a geometrical equivalent. It will presently be shown that an extension of the ideas underlying the Young-Poisson principle of the mutual compensation of molecular forcives, employed in the theory of capillary action, requires that this local forcive shall set up a purely local physical disturbance of the molecular configuration in the material, until it is thereby balanced; in the case of an isotropic medium in a steady state it must thus necessarily be expressible [to a first and often sufficient approximation] as an imposed stress symmetrical with respect to the direction of polarization.

Let us, therefore, with a view to the verification of this proposition, analyze the effects of an internal stress symmetrical with respect to the lines of some kind of polarization denoted generally by $i$ or $(f, g, h)$. Such a stress must be of the type of a tension $(p + q) i^2$ along these lines combined with a tension $pi^2$ in all directions at right angles to them; for the stresses we are examining clearly vary as the square of the polarization. Thus the stress must be made up of a hydrostatic pressure $- pi^2$ combined with a tension $qi^2$ along the lines of the polarization. The tractions exerted by the latter part on elements of interface parallel to the coordinate planes $yz, zx, xy$ are, per unit area,

$$(qf^2, qfg, qfh), \quad (qgf, qg^2, qgh) \quad \text{and} \quad (qhf, qhg, qh^2).$$

Hence the total force exerted by the stress on the element of volume $\delta x\, \delta y\, \delta z$ is, per unit volume, $(X, Y, Z)$ where

$$X = \frac{d}{dx}(pi^2) + \frac{d}{dx}(qf^2) + \frac{d}{dy}(qgf) + \frac{d}{dz}(qhf)$$

$$= (p + \tfrac{1}{2}q)\frac{d}{dx}i^2 + qf\left(\frac{df}{dx} + \frac{dg}{dy} + \frac{dh}{dz}\right)$$

$$+ qg\left(\frac{df}{dy} - \frac{dg}{dx}\right) + qh\left(\frac{df}{dz} - \frac{dh}{dx}\right);$$

and, the stress being self-conjugate, there is no torque. On comparison of this force with the local molecular, or cohesive, force on the element of volume, of electric origin, expressed above, it appears that they are of the same type provided $f\,dx + g\,dy + h\,dz$ is an exact differential, which is the case with the equilibrium electric polariza- tion $i'$ or $(f', g', h')$ induced in an isotropic medium, the electric force being always under conditions of equilibrium circuital. The material stress, which represents the regular electrostatic part of the molecular forcive by which the molecules hang together, is therefore a tension $\frac{2}{3} \cdot \frac{4}{3}\pi i'^2$ along the lines of the polarization $i'$ combined with an equal pressure $\frac{2}{3} \cdot \frac{4}{3}\pi i'^2$ in all directions at right angles to them[1]. If, however, the medium were crystalline, the stress would be of a more complex type than this, being related to the crystalline axes as well as the axis of polarization. When the interface is the surface of a conductor, the forcive on the charge of free electrons which pervades the layer of transition adds nothing to this effect beyond what has been already set down; for the electric force due to a volume distribution of single poles or electrons has no finite part depending solely on the element of volume at which its value is expressed, that is, it involves no molecular term.

---

[1] This is an example of Maxwell's theorem, § 40, *supra*.

44. The analysis here given is not however numerically applicable to a case in which the effective distance between the poles of a molecule is comparable to the distance between neighbouring molecules. The system formed by a bundle of iron nails suspended from the pole of a magnet and hanging on to each other against gravity, which has been used as an illustration of the molecular part of the forcive in the previous papers, does not come under these formulae. That system may however be employed with advantage as a real illustration of the general principles, especially if we imagine the magnetized iron nails to be connected by springs or imbedded in an elastic matrix. When no extraneous forces such as gravity act on this model of a molecular medium, it adjusts itself into a condition of internal equilibrium, in which attractions between the magnetic nails are locally balanced by repulsions exerted by the springs. The various local molecular forcives, typified here by these attractions between magnets and forces exerted by springs, precisely compensate each other in each portion of the medium. If an additional magnetic field is introduced, which alters the magnetic polarities of the nails, the parts of the medium will change their shapes and volumes until compensation again supervenes: there will thus occur an intrinsic deformation of the medium, and there may be also intrinsic changes of its physical properties, associated with the polarization and proportional in simple cases to its square. Suppose now that an extraneous force like gravity, or the magnetic field arising from the medium as a whole, begins to act, that is, a regular mechanical force on the medium in bulk so that it is in the aggregate proportional to the volume on which it acts; this will produce a further deformation, but one proportional to the first power of the exciting force. The local internal molecular forcive will again no longer be exactly balanced; but the unbalanced part will possess at each point the characteristics of an elastic stress system, because when the element of volume is small enough the tractions thus arising over its surface must equilibrate without any assistance from the then negligibly small [residual] extraneous bodily forcive. Even then however this elastic stress excited by an external field cannot be specified in terms of surface tractions unless the dimensions of the smallest element of volume which the circumstances require us to consider are large compared with the range of the intermolecular forces. Unless that is the case, the energy of elastic strain of the element of the medium, expressed in Green's manner, will involve higher fluxions of the displacement in addition to those of the first order, and the equilibrium between two contiguous portions will not depend on continuity of displacement and of surface traction alone: other quantities

*Marginal notes:*

But circumstances may be wider:

*e.g.* sheaf of coherent magnetized nails,

developed into a useful image:

adapts itself to a balance of all local forces,

including those of polarization,

by structural change.

An extraneous forcive then balanced by linear stress,

amenable to elastic theory,

only of simple standard type for fine-grained media:

*e.g.* energy may involve gradients of strain higher than the first,

complicated analysis,

also would have to be continuous for which there is no interpretation in the ordinary analysis of elastic reactions: the elastic stress would in fact not then be expressible in terms of tractions on interfaces.

evaded by changing physical moduli.

In such a case the only procedure that seems open, as the science of mechanics is now constituted, would be to transfer the effects of that part of the elastic energy which involves higher differential coefficients to the class of intrinsic or non-mechanical deformations.

45. In this theory of electric polarization the division of the forcive per unit volume into a molar and a molecular part has been made by means of the ideal volume and surface densities of Poisson, which are the equivalent as regards outside points of the actual polarization

The Poisson-Kelvin equivalent densities for polarization.

of the material. This method consists essentially in computing the forcive by combining opposed poles of neighbouring elements, instead of taking the single polarized element as the unit; it shows that these adjacent poles nearly compensate each other except as regards a simple volume density whose attraction has no molecular part, and a surface density at the outer surface and partly at the surface of the cavity which contains the point under consideration. The effect of the latter surface density, depending as it does wholly on the immediate surroundings, is the molecular or local part of the average forcive.

These principles may be enforced and illustrated by contrast with a procedure by separate molecules which would usually lead to a different result; it will suffice to consider the case in which the polarization is uniform in direction throughout the material. If the axis of $x$ be taken in the direction in which the intensity of the electric polarization changes most rapidly in the neighbourhood of the point considered, it is easy to see that the bodily force on an element due to the surrounding polar molecules is parallel to $x$ and equal to

Molecular survey in a locality:

$- (\mathfrak{a}f' df'/dx + \mathfrak{b}g' dg'/dx + \mathfrak{b}h' dh'/dx)$, and thus derivable from a potential function $- \frac{1}{2} (\mathfrak{a}f'^2 + \mathfrak{b}g'^2 + \mathfrak{b}h'^2)$, where $\mathfrak{a}, \mathfrak{b}$ are constants. In the case of an interface of rapid transition from one uniformly polarized medium to another, there is thus a forcive only in the transition layer, and its integral throughout that layer is equivalent to a traction parallel to the axis of $x$, of intensity

$$- \tfrac{1}{2} (\mathfrak{a}f'^2 + \mathfrak{b}g'^2 + \mathfrak{b}h'^2) \sin^2 \theta,$$

where $\theta$ is the angle between the interface and the axis of $x$, pulling at the interface into each medium. If the polarization $(f', g', h')$, or $i'$, is normal to the interface, this traction is $- \tfrac{1}{2}\mathfrak{a}i'^2 \sin^2 \theta$, if tangential it is $- \tfrac{1}{2}\mathfrak{b}i'^2 \sin^2 \theta$. To estimate the values of $\mathfrak{a}$ and $\mathfrak{b}$, we may consider separately the forcives exerted by the molecules of the polarized medium on $\mu_x, \mu_y, \mu_z$, the components of a molecular moment $\mu$ situated in the neighbourhood of the interface, in the case when the interface is normal to the axis of $x$. The outstanding terms in the aggregate forcive due to the surrounding molecules, which do not cancel each other by symmetry, are normal to the interface and make up

$$\mu_x \Sigma \mu_x' \frac{d^3 \gamma^{-1}}{dx^3} + \mu_y \Sigma \mu_y' \frac{d^3 \gamma^{-1}}{dx\,dy^2} + \mu_z \Sigma \mu_z' \frac{d^3 \gamma^{-1}}{dx\,dz^2};$$

or, per unit volume,

$-\frac{1}{2}\,(\mathfrak{a}f'^2 + \mathfrak{b}g'^2 + \mathfrak{b}h'^2)$ wherein $\mathfrak{a} = -2\mathfrak{b}$ because $\nabla^2\gamma^{-1} = 0$.
The point, however, to be noticed is that this expression for the *total* traction on an interface, due to both molar and molecular forcive, which makes the pull on an interface lying normal to the lines of polarization twice as great as the push on one tangential to those lines, holds only for the case in which the material is polarized in the *of restricted* same direction throughout the whole extent of its volume. We may *validity.* estimate by itself the action of the surrounding portion only, extending to any distance we please; but the action of the remaining outside part of the medium will still involve that of an inner surface density of uncompensated poles which will remain of undiminished order of magnitude*. This procedure by separate molecules is thus not suitable *Problem* for discrimination between the forcive due to the medium as a whole, *polarized* which is transmitted, and the molecular forcive which is compensated *lattice* locally. *theory.*

46. The justification of the theory here applied, which balances on the spot the molecular part of the forcive due to the electric polariza- *The cohesive* tion, by an intrinsic cohesive stress in the material which is indepen- *stress:* dent of the material elastic constants and strains at the place, may be further enforced by consideration of the ideally simple case of a gas. If a system of bodily forces act on it from a distance, they can *illustrated* always be balanced by a simple increase of pressure when they are *from a gas.* derived from a potential function; while if they were not so derived *Reacts only* the medium could not be in equilibrium. The dual phenomenon of *by pressure:* equilibrium of the element of volume maintained by a balance between two forcives, an extraneous and an internal one, is really a balance between a forcive on the element of volume acting from a distance by the mediation of the aether, and another forcive arising according to the explanations of the theory of gases from the impacts of the

---

* Further treated *Roy. Soc. Proc.* Vol. XCIX (1921), "On electro-crystalline properties as conditioned by atomic lattices": as *infra*. The surface tractions there introduced are the physical expression of the mathematically divergent terms that have to be got rid of, from the summations over the lattice, in an analytical development for the internal field, as in Mahajani, *Proc. Camb. Phil. Soc.* 1926. The very intense local fields that have been associated with ferromagnetism could only arise from extreme proximity of complementary poles in the lattice element.

molecules surrounding the element and, in the case of dense media, in part also from cohesive molecular actions. It is in this case a balance between a static bodily forcive and a steady kinetic molecular one; if the force transmitted through the aether from a distance increases, and equilibrium is to be maintained, the molecular configuration must be adjusted so that the impacts and the local molecular attractions shall continue to preserve the balance. When to the forces acting from a distance are added coordinated electric attractions between the molecules of the polarized medium, a further adjustment of molecular configuration must ensue. Now when the gas is electrically polarized, the attractions between neighbouring molecules give a forcive, not isotropic like a fluid pressure, but depending on the direction of polarization; its action will thus alter the originally fortuitous arrangement of the velocities of the molecules of the gas so as to impart to their distribution a slightly axial character, and when this has resulted in a new steady state the pressure due to the impacts will be different according to the manner in which the element of interface that is pressed is related to the line of polarization. This molecular addition of an intrinsic local stress, which has not the character of the ordinary fluid pressure, will just balance the action of the local electric attractions when the state of the system has again become steady; and being thus itself completely compensated *locally*, there will remain nothing of the molecular part of the electric mechanical forcive to be transmitted across the material medium. The argument applies with suitable modification to any isotropic medium, as well as to a gas; for an aeolotropic solid [*e.g.* an assigned type of crystal] the specification of the actual molecular stress of different origin which thus balances for each element the molecular electric forcive will be more complicated, involving the axes of aeolotropy as well as the axis of polarization.

*mainly kinetic, keeping in balance with local attractions by change of density.*

*Polarization of the gas:*

*gives axial character to the free paths,*

*thus an axial stress arises,*

*with local compensation.*

*Complex extension to a solid.*

*Compensation of internal forcives:*   47.  For present purposes the important consequence is that, under circumstances of equilibrium, that part of the forcive on an element of a material body which arises from the excitation of neighbouring molecules and is expressed in terms of them alone, is not transmitted by material stress, but forms a balance on the spot with the cognate internal molecular forcives of other types[1]. The only circumstance

*Historical.*
*Young:*
*Rayleigh:*
*van der*
*Waals.*

[1] This principle of compensating molecular forcives was briefly enunciated for capillary action and applied by Young in his Memoir "On the Cohesion of Fluids," *Phil. Trans.* 1805. It forms the basis of Poisson's *Nouvelle Théorie de l'Action Capillaire*, Paris, 1831, in which the attraction between the molecules of a fluid is balanced by a repulsion of much smaller range, supposed to be due to their caloric: cf. especially Ch. VII. Cf. also Lord Rayleigh, "On the Theory of Surface Forces," *Phil. Mag.* 1883, 1890, 1892, especially 1892 (I) pp. 209–220: and van der Waals' *Essay on Continuity of the Liquid and Gaseous States.*

that might apparently vitiate this conclusion would be that the transitions between different media may be too abrupt to be treated, from the point of view of individual molecules, as really gradual transitions, after the manner of the above analysis; but even if we could imagine such a case, the discrepancy must for fluids be made up, provided the interface is a permanent one, by capillary forces in the interfacial layer, the effect of an outstanding surface derangement of energy.

*with interfacial stress.*

In a dielectric body situated in an electric field there is thus the mechanical strain due to the field; and there are also intrinsic change of volume and other dimensions and of physical properties, proportional to the square of the local polarization. If the dielectric is solid, those changes of dimensions may not fit in with the continuity of the material without the intervention of secondary strains; but in fluid media the case is simple and precise, as no strain other than mere compression can exist.

*Example: an excited dielectric.*

*The Mutual Compensation of Local Molecular Agencies: Organized and Unorganized Energy: The Single Postulate of Thermodynamics, Available and Degraded Energy: Physical Basis of the Idea of Temperature.*

48. The scope of these molecular considerations (§§ 43–47) is wider than the special problem of polarization by which they are here precisely illustrated. To an intelligence that could follow the play of interaction between the individual molecules of matter, mechanical forces, in the ordinary sense, would not exist. The actual interactions between the molecules are however necessarily presented to us divided into various statistical groups, which are the subjects of perception by different senses; and it is the business of physical theory to follow out the relations of these different groupings to each other, and to trace them all back into the ultimate unity. The total energy of the molecules of a material body, corresponding to any kind of excitation or polarization, is thus for us made up of various parts. There is a part involving the interaction, with any molecule under consideration, of other molecules at finite distances, which integrates into an energy

*Classification of energies depends on the scale and nature of the sensual perceptions.*

*Mechanical energy in gross.*

In these illustrative discussions, in which the intermolecular forces are restricted to a non-polar character, the compensating stress is usually found in the assumption of an intrinsic fluid pressure of range much shorter than that of the attractions between the molecules: the principle however in its general form only asserts that this compensation must exist, and there is no necessity to specify its character.

*Surface tension of pure liquid:*

[The question whether, when surface tension is modified by an impurity, the foreign substance in any given case acts by a monomolecular layer, is a different one: for pure liquid the range of action is greater than a molecule and the transition at the interface probably gradual.]

*Monomolecular surface layers.*

function of applied mechanical forces of the system, such for example as gravitational or magnetic forces. Of the remainder of the energy, which arises from the mutual actions of neighbouring molecules, a regular or organized part can be separated out which represents the energy of elastic stress, and is a function of the deformation of the element of volume treated as a whole: this stress arising from the immediate surroundings in part compensates, for the element of mass under consideration, the applied mechanical forces aforesaid. The remaining, usually wholly irregular, parts of the local inter-molecular forces and motions compensate themselves mutually on the spot,—or at any rate can be considered as thus compensated by other such forces, of different origins, that are not at present under consideration[1]. The temperature depends in fact on this irregular *residuum* of forces, and so do the density and the other physical properties of the medium free from stress, which are thus affected when, owing to polarization or other excitation, this local part of the molecular forces and motions is altered. If we adhere to these principles, it will not be allowable, in deriving the applied bodily forces of a polarized material system from its organized energy of polarization, to vary such physical constants of the element of mass as occur in the expression for the energy; for we should thereby be trenching on that part of the energy whose variation is compensated molecularly without directly originating transmitted bodily forcive: cf. § 63.

*Locally bound constitutive energy, determines mechanical stress.*

*Thermal residue.*

*In usual mechanical theory physical moduli do not vary with the stress.*

49. It seems desirable to have names for the two parts into which the total energy of the molecules of a material medium is thus divided. If we agree to maintain the original precise meaning of the term mechanical (as above employed), viz. that a mechanical force is one which we can actually control for doing work for our purposes on matter in bulk, in contrast with a molecular force which we can reason about but not directly employ, we may call the regular part the *mechanical* energy, and the remaining wholly irregular part the *non-mechanical*; we may also use (as above) the terms *organized* energy and *unorganized* energy with the same meaning, the reference being now to the material medium as a continuous organic whole, transmitting applied forces by stress, not as a numerical aggregate of separate molecules. But it is to be observed that the distinction which is thus intended to be made is not the same as the thermo-

*Mechanical signifies controllable by man:*

---

*The dynamical postulate of d'Alembert:*

*its justification.*

[1] The principle of d'Alembert, which is the basis of the dynamics of finite material bodies, necessarily involves this order of ideas. That part of the aggregate forcive on the molecules in the element of volume which is spent in accelerating the motion of that element *as a whole*, is written off; and the regular part of the remainder must mechanically equilibrate. But the wholly irregular parts of the molecular motions and forces are left to take care of themselves; which they are known to do for the simple reason that the constitution of the material body is observed to remain permanent.

dynamic division into *free* and *bound* energy, employed by von Helmholtz, which is itself precisely equivalent to the earlier division into *available* [*at constant temperature*] and *dissipated* energy, formulated by Lord Kelvin and Rankine. The energy which in its actual condition is as regards direct mechanical effect unorganized, may become in part organized by aid of a physical transformation involving sifting processes of molecular fineness, which are necessarily non-mechanical and have no place in the dynamics of finite bodies. Thus the unorganized energy of two masses of different gases, at the same temperature and pressure, may be in part converted into organized energy and so into mechanical work by allowing them to transpire into each other across a porous partition*, the diameters of whose pores approach molecular dimensions; and the transformation in this case shows itself in a resulting fall of temperature, when the work has been done. In the same way mechanical work may be derived from the unorganized energy of liquids by utilizing osmotic pressure; and the stores of energy of chemical combination of electrolytic substances, which as it exists in them is unorganized, can be largely utilized by making use of the sifting agency of electric force on their dual constituents. All these unorganized energies are therefore in part thermodynamically available, and others not now available may become so by means of yet undiscovered processes. But the unavailable or bound energy of thermodynamics is the *residuum* which we cannot render mechanical by any sifting process in bulk, or by anything short of the application of constraint to the individual molecules. This *residuum* may not be absolutely irreducible, but as the knowledge of physical transformations increases, some parts of it may be raised into the domain of available energy: on the other hand the recognition of temperature coefficients in reversible processes will show that some energies previously considered as wholly available are really in part unavailable. Each such discovery in fact involves an amendment or improvement in the corresponding thermodynamic relations; a process which has happened, for example, with respect to Lord Kelvin's law of electromotive force of a voltaic cell †.

*[marginal notes: extended and generalized into thermodynamically available, thus including molecular sifting of energy, and the converse dissipation on diffusion in Nature.*

*Thermal residue is in part provisional:*

*reversible heat in voltaic cell.]*

50. Once the idea of temperature is acquired [postulated from experience prior to theory], the whole science of Thermodynamics is implicitly involved in the principle of dissipation, that the unavailable part of the energy of an isolated material system always tends

---

* That is, by the further disorganization involved in mixing.

† The relative character of thermodynamics seems to have been first emphasized by Maxwell, by help of his ideal "demons." Contrast the school which finds the chief aim of pure physical science to be the elimination of all anthropomorphic features.

to increase, never of its own accord to diminish. The inference follows directly from this principle, by the reasoning first employed by Sadi Carnot, that if the system pass from a state $A$ to a state $B$ such that it can retrace its path back to $A$, the unavailable part of the energy is not changed: thus there is a whole "plane" or *complexus* of states, with perfect continuity of transformation among them, so that any one state is freely convertible—whether the process has been actually discovered or not—with any other for which the available energy is the same [except for mechanical energy imparted], by transition through any intermediate series of these states; and we can pass continuously from one such *complexus* to the others in which the whole series of possible states are included, by additions of available energy to the system. The available energy is thus an analytical function of the physical condition of the system, including its temperature; and the trend of spontaneous change [with temperature maintained constant by exchange of heat with its surroundings] in an isolated system, is in the direction in which this function diminishes, the positions of stability, as regards mechanical and thermal and also constitutive disturbance, being those for which it is a minimum. The circumstances of all *steady* configurations of matter, whether static

General law of isothermal equilibrium.

or kinetic, are determined by this law[1]. It is more direct to state the proposition in the form that the unavailable energy tends to a maximum, the presumption being that sensible energy is available until it is shown to be otherwise. This principle, that energy tends to become mechanically disorganized, or that it never spontaneously tends to organize itself, cannot from its nature be other than axiomatic: and the formation of the available energy function for the different states of matter is then the main business of Thermodynamics. The reversible processes [thus conditioned to uniform temperature] which thermodynamic argument employs are ideal types of regular change, theoretically realizable by mechanical constraints which do not control the individual molecules—the limiting forms, it may be, of imperfectly reversible changes which we can

Chemical statics:

[1] In so far as our constitutive knowledge of material systems relates merely to comparison of different steady states, it can be wholly based in Willard Gibbs' manner, like ordinary statics, on relations of available energy of a simply additive character: it is where our knowledge becomes more intimate, and we attempt to trace the courses and rates of kinetic phenomena, for instance in material kinetics, electrodynamics, optics and vibratory phenomena in general, that the simple relations of energetics become insufficient as a mathematical

chemical genetics inscrutable.

basis for general physics. The principle of available energy suffices for tracing the relations of matter in bulk through the various steady phases in which *ex post facto* it is found to exist: but the genesis of these phases is expressly excluded from its domain.

[On the foundations of thermodynamics, cf. the account of the work of Gibbs, *infra*, and the Appendix to this volume.]

actually produce; the states of matter thus derivable from each other are shown, from the equality of their available energies, to have definite mutual relations which are independent of the ideal process (or construction, to use a geometrical analogy) by which the transitions between these states have been imagined. [Cf. p. 103, footnote, *infra*.]

The really abstruse abstract problem of the subject is that of the nature of temperature; and the principle most in need of elucidation is that, when a body $A$ is in thermal equilibrium with $B$, and also $B$ with another $C$, then $A$ would be in thermal equilibrium directly with $C$. The most definite thermal specification of a body is the quantity of energy it contains; two bodies are in thermal equilibrium when there is no tendency for energy to pass from one to the other without mechanical manifestations, independently of change of molar configuration or molecular constitution; they are then said to be at the same temperature. The *rationale* of this transfer of energy has been made out for the case of gases, where the exchange takes place in encounters between the molecules, so that there is no tendency to transfer by mechanical streaming from one mass of gas to another in contact with it if the mean translatory energy of the molecules is the same for each. This principle of temperature equilibrium shows that, for all states of matter, the equilibrium of energy between bodies in contact in a steady state involves that a definite molecular relation of the one body shall equilibrate a definite molecular relation of the other: and its universality for permanent states of matter requires that this relation, whatever it may prove to be, shall be a very fundamental one.

*The central problem of temperature.*

51. It would seem that we can make at any rate an advance towards a complete view by realizing that, even if our sensations of heat had not compelled us to assign a fundamental place to temperature in the physical scheme, the principle of negation of perpetual motions must have led to the formulation of that conception, just as it has in fact led to the conception of potentials. If thermal equilibrium between two homogeneous bodies $A$ and $B$ in contact were not conditioned merely by some physical property of $A$ alone being equal to some property of $B$ alone, then if we had $A$ in contact with $B$, and $B$ with $C$, each in a state of equilibrium, and, removing $B$ by mechanical means, moved $A$ into direct contact with $C$ but with such ideal constraint applied to the matter in bulk that chemical action is prevented, the physical state of each of these latter bodies would become changed, involving the performance of mechanical work; and a self-acting cycle could be designed by which we might thus obtain an unlimited quantity of work, that is, so long as there remained any diffused molecular energy to be converted. Hence in equilibrium

*Temperature a potential:*

tends to uniformity unless where molecular interaction is prevented:

but also equalized by radiation:

not directly affected by fields of force.

there must be a property*, namely the temperature, of each individual body in the field that has the same value for all of them: although of course this does not prevent us from imagining a partition or constraint, nearly adiathermanous, across which such equilibrium would be established as slowly as we please. It follows also that equilibrium of temperature must be the same whether it is brought about by conduction or by radiation†. Temperature, as thus introduced, has nothing to do directly with the field of force in which the body is situated: for the relations of bodies to fields of force, in which they are moved about, are treated independently in the consideration of [mechanical] energy relations, and must not be introduced twice over,—or, in other words, the perpetual motion principle can be directly applied.

Negation of all perpetual motions the ultimate test for permanence of a system.

The single fundamental principle, on which all thermodynamic and thermochemical theory rests, would thus be the axiom of the negation of perpetual motions: and this stands rather in the relation of a principle that could hardly be conceived to be otherwise on any feasible physical scheme, than of one of which we can expect to offer any formal demonstration. Various essays have been made to deduce Carnot's principle and a dynamical specification of temperature from special hypotheses as to molecular action: it may be held that, in so far as these are useful it is by way of illustration. It is even possible to conceive, but only in a highly abstract sense, that thermodynamics might have been developed in Carnot's manner out of the perpetual motion axiom alone, without the aid of Joule's demonstration of the nature and measure of heat‡; there would then have been merely no knowledge of what had become of energy that had ceased to be mechanically available. It is thus the principle of the limited conservation of available energy, rather than the complete conservation of total energy, that reigns in general non-molecular physics[1].

There is still however the complication that the available energy

---

* Temperature is a property of phases or constitutions of matter that are homogeneous, at any rate differentially: a set of discrete atoms has not a temperature. Cf. Maxwell's developments in gas theory, to radiometer effect with its stresses along a gradient of temperature. If it is impossible to formulate a temperature for any assigned constitution of matter, one infers that it will fade away: thus there cannot be any general molecular demonstration of the temperature laws.

† Cf. the modern attempts to elucidate this by taking radiation to be made up of localized *quanta* interacting with atoms.

‡ See in fact a paper "On the Nature of Heat as directly deducible from the Postulate of Carnot," *Roy. Soc. Proc.* Vol. xciv (1918); or *infra*.

[1] This seems to be substantially the position which Rankine took up in 1853 (*Scientific Papers*, p. 311): cf. also the weighty introduction to *Outlines of the Science of Energetics*, 1858, *loc. cit.* pp. 209–220. It is in fact the standpoint of Carnot's *Reflexions*.

of a system is not a function of its state alone, but involves comparison with some standard state into which it is possible for the system to be transformed. To find the extent of this undetermined element, let us simplify the relations in the ordinary manner, by adopting the scale of temperature that is given by the expansion of an ideal perfect gas, and find out how much energy is dissipated or lost to available mechanical effect, when a quantity of heat $H_1$ is abstracted at the temperature $T_1$ and of it $H$ is returned at the temperature $T$. If all possible mechanical effect were produced, only $H_1 . T/T_1$ would be thus returned instead of $H$: hence the dissipation is $H - H_1 . T/T_1$ or $T (H/T - H_1/T_1)$. Thus an operation of this kind which does not involve dissipation does not alter $H/T$: and by accumulation of such changes it follows that any two states of the system which are convertible without dissipation have $\Sigma H/T$ the same for both. The entropy function $\phi$ of Clausius thus necessarily enters into the analytical formulation of the principle of mechanical availability. Between a standard state at temperature $T_0$ and another state at $T$ the dissipation is $T (\phi - \phi_0)$; thus the available energy $A$ in the latter state is $E - T\phi + T\phi_0$, where $E$ is the total energy which involves an undetermined constant part, and $\phi_0$ is another undetermined constant which represents the entropy of the system in the standard state*. The temperature of the standard state to which the system is referred could not of course be the ideal, practically infinitely remote, temperature which is called absolute zero; that would imply that the energy is all mechanically available as in ordinary statics.

The presence of this undetermined multiple of $T$ does not really restrict the application of the theorem of minimum availability: it merely implies that when once mechanical and constitutional equilibrium has been determined at any assigned temperature by making $A$ a minimum with respect to the other independent variables, still further degradation will occur if opportunity is allowed for fall of

*Marginal notes:* Transition from available energy to entropy: / its change the same for all modes of transfer if reversible: / otherwise it rises irrecoverably.

---

* But the idea of available energy is of direct utility only, as in Kelvin's statement, *supra*, § 50 (cf. Thomson and Tait's *Nat. Phil.* 1867), for comparing states of the system at the same temperature $T$, that is, for isothermal change. It is the characteristic function at constant temperature of Gibbs. For changing temperatures the entropy $\phi$ of Clausius is, as Gibbs claimed, the basic mathematical concept: it tends without restriction to a maximum, and it is the *quasi*-dynamical complement of temperature for heat energy in that $\delta H = T\delta \phi$. For the mode of development in terms of available energy $A$ regarded as a function of temperature also, cf. footnote, p. 103, *infra*.

Near the absolute zero all the energy becomes isothermally available. According to Nernst's empirical theorem $\phi$ rises parabolically from a zero value, so that $E - A$ rises inflexionally: also the theorem involves that the expansion of $\phi$ in positive powers of $T$ is usually valid over a considerable range.

temperature by escape of energy from the system. All that it is necessary to ascertain in any problem is the equilibrium as regards physical state and chemical constitution *at each temperature*, and the capacity of the system for heat, which specifies the thermal change

*Temperature uniform in thermostatics.* that occurs when the temperature is altered. There is no restriction involved in taking the temperature the same throughout the system, for that is a necessary condition of equilibrium: when it is convenient to imagine partitions impervious to heat, the parts of the system thus separated can be treated as independent systems. The available energy, here arrived at directly from the perpetual motion postulate [for the system changing so that the temperature remains uniform], is the same as the free energy of von Helmholtz's exposition: he has explained (*Abhandlungen*, II. p. 870) how its form can be experimentally ascertained for the different phases of matter, except as regards an undetermined part, as above, of form $L + MT$, where $L$ and $M$

*Helmholtz's exposition for free energy.* are constants; that then the equilibrium state of a system of reacting bodies at any assigned temperature is the one that makes it minimum for that temperature, thereby formulating the general solution of the problem of physical and chemical equilibrium: while the other properties of the system, heat changes and heat capacities, as well as total energy and entropy, are obtained from it directly by processes of differentiation. The available energy is thus a single characteristic function which includes and determines completely the circumstances, mechanical, thermal, and constitutive, of the steady states of an inanimate material system.

### *Molecular Applications to Fluids: Laplace's Intrinsic Pressure: Law of Osmotic Pressure: Laws of Chemical Equilibrium.*

52. In an incompressible fluid medium in equilibrium no part of the bodily extraneous forcive is compensated by reaction arising from special strains produced around the element of volume itself; it is all transmitted by fluid pressure independently of the special physical

*Equilibrium in polarized fluid: on Pascal's principle.* constants of the medium. For equilibrium to subsist in a polarized fluid, the applied mechanical forcive must simply be derived from a potential. When the induced polarization follows a linear law, this potential must also be equal and opposite to the organized energy induced per unit volume in the medium on which this extraneous forcive operates; for the total organized energy that has been spent in the polarization of the element $\delta\tau$ is equal to $\delta\tau$ multiplied by the scalar product of the polarization and the polarizing force, and of this one-half is mutual energy of the polarizations of the elements of volume and one-half is mechanical work done in the process (cf. § 71). If therefore the organized energy of the internal excitation of the medium is expressible as a volume density of energy represented by a continuous function, the fluid medium will be in internal mechanical equilibrium: but if that function is discontinuous so that in crossing some interface the density of induced energy abruptly changes its value (as for example may be the case when the interface separates two different substances) then in order to maintain equilibrium the applied forcive must include a traction applied to this interface along its normal, of intensity equal to the difference of the densities of energy on its two sides, and acting towards the side of smaller density of energy. At an external boundary there must similarly be applied an outward traction along the normal, equal in intensity to the density of organized energy induced in the part of the substance that is just inside. *Traction on an interface.*

To illustrate and elucidate this by the electric phenomena, consider the interface between two dielectric fluids to be maintained in position by an applied traction: let an element $\delta S$ of it sustain a displacement $\delta n$ along the normal, of amount very slight compared with the linear dimensions of $\delta S$. If no other boundary within the range of the electric field is thereby affected,—for instance if each fluid is supposed to be continued in a narrow tube to a great distance beyond the field and simply advances or recedes in the end of this tube,—the change of organized electric energy is merely the substitution of a volume $\delta S \delta n$ of energy of the one intensity for the same volume of the intensity on the other side of the interface. The displacement of $\delta S$ of course affects the state of the field all over, but by hypothesis the electric field was in internal equilibrium, so that the change of the organized energy of any volume element of the mass arising from a slight derangement is of the second order of small quantities, and produces no sensible effect. The above change of energy is thus equal to the work done by the extraneous traction over $\delta S$; which confirms the result already obtained (§ 37) by detailed analysis of the polarization, that the traction is along the normal to the interface and equal in intensity to the difference of the densities of the organized electric energy on its two sides. *Electric surface traction:* *its law.*

53. It is advantageous in connection with this subject to form a definite conception of the transmission of ordinary mechanical pressure in a liquid. Let us imagine an ideal infinitely thin interface in the fluid: what concerns the equilibrium of the fluid on one side of it is not the pressure which that fluid exerts on the interface, but the forces that are exerted on that fluid itself both by the interface and by the molecular attraction of the fluid on the other side of it. As the range of molecular attraction is very small, these forces together

make up a pressural push on the fluid, equal in circumstances of
Synthesis of molecular equilibrium to the resistance of the interface against the
nature of
the fluid impacts of the molecules diminished by the attraction exerted on
pressure, these molecules across the interface; and this is the pressure that is
transmitted by the fluid. For imagine a canal or tube in the fluid,
with infinitely thin sides, and of diameter large compared with the
radius of molecular action, and consider the equilibrium of the mass
of fluid contained in it between two cross-sections $A$ and $B$. There will
be this pressure acting on the fluid just inside $A$, and a similar pressure
acting on the fluid just inside $B^*$; and unless these are equal, or balance
each other with the aid of extraneous applied forces such as gravity,
for which the mass of fluid cannot be in equilibrium. This is Pascal's principle,
Pascal's
transmission that the mechanical pressure is transmitted unchanged in amount,
principle
holds, except in so far as it is compensated by extraneous mechanical forces.
It is to be noticed that the argument does not assume that the fluid
between $A$ and $B$ is homogeneous, all that is required is that it be in
even across
interfaces: equilibrium; the cross-sections $A$ and $B$ may be in different fluids,
with an interface between, and, provided the diameter of this ideal
canal is large compared with the radius of molecular action, the interfacial forces will practically all be mutual ones between molecules
inside the tube, and so will not affect the transmission of pressure.
It is this transmitted pressure that is the subject of actual measure-
as illustrated
in Andrews'
experiments: ments: for example in Andrews' experiments on the compression of
carbonic acid, it is the pressure so transmitted through the mercury
into the companion manometer tube containing perfect gas that is
measured and is represented on his diagram of isothermal lines. The
two terms of which it is the difference, namely the reaction of the
interface against molecular impacts, and the molecular attraction
and in the
special
Van der
Waals'
formula. across the interface, are separately represented in Van der Waals'
characteristic equation $p = RT/(v - b) - a/v^2$. When the virial equa-
tion of Clausius $\overline{T} = \frac{3}{2}pv - \frac{1}{2}\Sigma (Xx + Yy + Zz)$ is applied to a mass
of liquid with a free surface abutting on a gaseous atmosphere, there
results the relation that the pressure of this atmosphere against an
outer boundary, which is the same as the transmitted pressure in
the liquid, is equal to two-thirds of the part of the mean density of
translatory energy in the liquid that is connected with the encounters
Surface
pressure
related to
the virial: and mutual forces between the molecules, diminished by two-thirds
of the mean virial per unit volume of these intermolecular forces, the
latter quantity being, if polar forces could be assumed absent, of
Clausius' form $- \frac{1}{2}\Sigma Rr$ when $R$ is attractive; and this without reference
to the character of the transition between liquid and gas at the free

---

* The internal forces acting between the molecules of this mass of fluid,
being mutual, are balanced, though they have a mean virial which affects the
internal pressure.

surface. When on the other hand, the virial equation is applied to a mass in the homogeneous interior of the liquid, bounded by an infinitely thin interface, the virial of each molecule vanishes because the attractions acting on it compensate each other on the average, and the result is that the pressure now of kinetic origin exerted by the fluid on either side of this interface is simply two-thirds of the mean density of kinetic energy of the bodily motions of the molecules, their internal constitutive energies being excluded[1]. It follows (as *infra*, § 54) that the mutual molecular attraction across the interface produces a pressure on the interface from each side equal to the internal virial per unit volume; as in fact would flow directly from the principle that two statically equivalent force systems have the same virial.

*internal pressure differently.*

54. Let us now, passing from the virial averages to direct dynamical considerations, imagine as above an ideal rigid tube, with infinitely thin walls which exert constraint on the molecules but no attraction, having one of its open ends $A$ in the liquid and the other $B$ outside it; and let us now suppose that the diameter of the tube is small compared with the radius of sensible molecular action, which implies that this radius extends over a very considerable number of molecules. The molecular forces acting on each molecule in this very narrow tube, whether it be near the end of it or not, are now almost wholly due to the molecules outside the tube, so are on the average self-balancing, leaving only the concentrated forces of collision operative; except in the case of molecules where the tube crosses the free surface which are subject to the whole *inward* molecular attraction of the liquid. The equilibrium of the contents of the tube, which are liquid in one end and gaseous in the other, therefore requires that this kinetic pressure on the molecules in the liquid end $A$ exceeds that on those in the gaseous end $B$ by a constant amount, namely the pressure due to the inward attraction exerted by all the local liquid on the surface molecules in the tube that are in the layer of transition. It follows that the pressure of molecular attraction across an internal interface is two-thirds of the virial per unit volume with changed sign, namely $\frac{1}{3}\Sigma Rr$/volume; it is Laplace's (very intense) intrinsic pressure $K$ in the liquid arising from the inward attraction of the surface molecules. This equality seems to involve the consequence that the layer of transition at the free surface is very thin

*Newton's ideal tube transmitting pressure.*

*Laplacian intrinsic pressure.*

*The surface transition of necessity rapid.*

---

[1] Some consideration is required as to the omission of the virials of mutual forces acting inside the separate molecules: these must be taken as wholly compensated by kinetic energy of internal motions not thermal, which is legitimate in so far as molecular encounters do not sensibly excite radiation but only slow free precessional motions, and so do not sensibly disturb the configuration of the internal dynamical system of the molecule.

*The forces inside the molecule excluded.*

compared with the radius of molecular attraction: an important conclusion of which the bases are here the statistical stability of the liquid state, the dynamical principle of the virial, and the hypothesis that the range of sensible molecular attraction extends over a con- siderable number of molecules in the liquid state. In the condensation of a vapour there is degradation of internal energy into sensible heat of amount equal to the latent heat of that change of state diminished by the work of condensation of the vapour and increased by the volume of liquid thereby produced multiplied by the Laplacian pressure $K$ [: as it is reversible these two components should cancel, thus giving a value for $K$, of the order of $10^7$ kg./cm.$^2$]*.

Heat of<br>condensation.

55.  Consider two fluids, one the pure solvent and the other a solu- tion, separated by a rigid porous partition, with extraneous pressure applied on the side of the solution to balance the osmotic pressure and so to produce equilibrium as regards transpiration through the partition. Now let a slight amount of transpiration of the solvent occur by very slightly reducing this extraneous pressure: thereby work is done against that pressure, equal to its intensity multiplied by the change of volume owing to transpiration of the solvent into the solution. The operation takes place steadily under conditions of equilibrium, so that it can be reversed either by a known process or, as we might assume, by some process not yet discovered—in this case merely by reversing the pressure, or it may be cyclically by evaporation: thus the work is done at the expense of an equivalent of available energy, partly thermal, and partly of a molecular type which would otherwise run down into heat of mixing of the liquids. Hence the osmotic pressure between two fluids is equal to the whole amount of free or available (not total) energy that would be degraded when unit volume of the pure solvent is mixed with an indefinitely great volume of the solution into which it transpires, supposing that there is no sensible change of volume in that process; if there is change of volume this value must be altered in the ratio of the final to the original volume of the transpired material: so long as the dissolved molecules are out of each others' range of influence, the change of volume, if any, must be independent of concentration. This proposition will be exactly true if the pores in the partition are so narrow, that the cross-sections of the filaments of fluid contained in them each involve so few molecules that the mutual energy of the molecules of fluid in the pores is negligible compared with that of an equal mass of fluid in bulk. Inasmuch as to excite the osmotic

*Reversible<br>work against<br>osmotic<br>pressure.*

*Osmotic<br>pressure in<br>terms of<br>energy of<br>mixture:*

*is influenced<br>by change<br>of volume<br>on mixture.<br>Law of this<br>change.*

*Conditions<br>for perfect<br>osmosis:*

* The wording of this section and the end of the last has been considerably modified, in an effort to mitigate obscurity: on the virial, see Note on p. 133.

pressure, pores or tubes of molecular fineness have to be employed, it follows that it is not an ordinary transmitted mechanical pressure; and the energy which is associated with it is not merely that organized energy from which the mechanical forcive is derived, but the whole amount of energy thermodynamically available. If the pores are wider, the mutual energy of the molecules in them ceases to be negligible; the effective osmotic pressure then diminishes, being accompanied by diffusion in the pores which involves dissipation of energy that would otherwise produce osmotic effect. Thus the proposition that the osmotic pressure between two fluids is equal to the free or available energy of mixture per unit volume of transpiration, gives only the limiting value which applies to partitions with pores sufficiently narrow. In the equilibrium stage of transpiration through a colloid membrane, operating by absorption into one face of the membrane and evaporation from the other, the limiting pressure may however be reached, provided the action does not involve irreversible thermal processes in the membrane. *is distinct from mechanical pressure: an optimum (reversible) value.*

The osmotic pressure between a solution and the pure solvent is, from another point of view, the mean aggregate of the forces that have to be applied to the individual molecules of the dissolved substance in order to prevent them from travelling across the interface into the pure solvent, whether that forcive be applied by the resistance of a material partition, or as in the case of ions diffusing across the interface between two salt solutions in contact, by the pull of the electric field which the diffusion has produced—the unmodified molecules of the solvent being in each case free to move either way. Viewed in this light, there is nothing occult or merely analogical —unless it be the presence of ions—in the principles by which Nernst determines the constitution of the layer of transition which gives rise to the potential difference between two salt solutions, and so determines the voltaic and thermoelectric differences of potential at such transitions, by balancing a bodily force arising from osmotic pressure by another arising from the electric field due to the reacting double layer generated by the diffusion. *Mechanical interpretation: extended to ionic interfacial effects.*

56. Suppose that the pressures on the two sides of a porous partition separating dielectric fluids are adjusted so that there is no flow across it. When an electric field is introduced this equilibrium is destroyed by the effective electric tractions on the interfaces of separation between the dielectric fluids in the individual pores. To re-establish equilibrium a difference of pressure at the two sides of the interface, equal to that of the electric tractions (§ 37), must be called into play: that is, an electric field influences the value of the *Electric influences on osmotic pressure of a dielectric solution,*

and on its vapour pressure. osmotic pressure between dielectric fluids. This effect is of course directly connected, through a cyclic process, with an influence on vapour tension (§ 81, *infra*). Its amount is to a great extent independent of the size of the pores; though when the pores are of molecular dimensions it mainly arises from a bodily forcive on the contained filaments of fluid. This electric osmotic pressure will then even hold good with respect to liquids which readily mix; for the obliteration of the sharpness of the interface in the narrow tubes or pores of the partition will take place very slowly, while the formulae of this memoir for electric tractions are precisely those which hold good for a gradual transition.

Electric osmosis is ionic: This action is different from the one discovered by Quincke and discussed by von Helmholtz[1], forming in fact a further extension of the scope of the principle of electrolytic dissociation, in which a stream of conducting fluid forced across a porous non-conducting partition produces an electric current across it, and conversely an electric current forced across the partition carries the fluid with it.

a view of its mechanism. Over the surface of each pore there is, on the present view, the intrinsic static potential difference between the partition and the fluid, due to strong orientations of the polar molecules of these two media which lie along the interface, under their mutual influence which stands in place of von Helmholtz's attraction of matter for electricity as the exciting cause of voltaic phenomena[2]; and this difference will be in time diminished by the presence of free ions which become attached among the outward-pointing poles, thus constituting a reverse potential difference with which electrocapillarity deals. The flow of fluid through the pores carries on some of these ions along with it, which thus constitute the observed electric current.

Electrification of breaking drops. Sudden diminution in the extent of the surface would act similarly by crushing them out, as in the observed electrification near waterfalls: rapid extension of the surface, as in the formation of drops in air, should conversely eliminate the effect of the ions bound to the polarized air film on the surface, by spreading them over a wider area, and so increase the potential difference towards the limiting statical value. On the other hand, when the media in contact are very dilute electrolytic solutions in the *same* solvent, the calculations

[1] Von Helmholtz, "Studien über electrische Grenzschichten," Wied. *Ann.* VII, 1879.

Intrinsic interfacial double sheets. [2] (Helmholtz had to be content in his analysis with the crude conception that different kinds of matter attract electricity differently. On a scheme like the present the obvious explanation is that the polar molecules of the two substances act on each other across the interface, producing a certain regularity of orientation which forms the intrinsic double layer to which the potential difference is due.)

of Nernst show that the potential difference is wholly an affair of ionic diffusion, as indeed it must be if the efficient polar molecules are all ionized; in that case the normal potential difference will require a sensible time to become established. When, in the case of a mercury electrode dropping into an electrolytic solution, sufficient time is not allowed, the part of the actual potential difference which arises from this cause and not from the intrinsic statical orientation of the molecules, will tend to a vanishing limit, except in so far as it is continually restored by a polarization current in the electrolyte.

*Time lag of ionic redistribution.*

*Dropping mercury electrode gives intrinsic potential difference.*

57. In the case of very dilute solutions it is possible to obtain a definite expression for the limiting, or maximum, osmotic pressure. After a certain stage of dilution, each dissolved molecule is effectively out of touch with its fellows and is completely environed by a collocation of molecules of the solvent: further dilution therefore does not involve any sensible change in the mutual free energy of the solvent and the dissolved molecules; all that occurs is a wider separation of the dissolved molecules in space, with such energy changes as may be directly concerned in it. Suppose now that the dissolved substance is a gas, and that the solution is separated from the pure solvent by a partition which the latter can traverse while the gas cannot: whether such partitions are known to exist is inessential to the theoretical argument, the function of the partition being merely passive constraint exerted on the *aggregate* of the dissolved molecules. The solvent will transpire across the partition into the solution, unless a definite osmotic pressure acts against it, when there will be equilibrium. Let us examine the change of free energy involved in the very slow transpiration of a certain volume; all essential that has happened has been an expansion of the molecules of the contained gas, each with its fluid environment, into a larger space. We may compare the two states of the gas, as it would exist free with these two different volumes, and then suppose that by an ideal process the fluid environment of the molecules is directly brought about in each case: that process will, as regards change of intimate molecular configuration, be essentially the same for both states of the gas, therefore the change of free energy due to the dilution of the solution is simply that which corresponds to the free gaseous expansion of the dissolved gas[1]. This

*Osmotic theory for perfect dilution:*

*can reason from ideal constructions.*

*Gas laws valid for all dilute systems.*

*Available energy specifiable by bulk properties, and only volume can change by further dilution.*

[1] The circumstance which makes this purely imaginary process legitimate is that the available energy is a function of the constitution of the matter *in bulk*, not depending on the accidental characteristics of state or motion of the individual molecules: now the only change that has occurred as regards the constitution of the substance in bulk, that can affect either the available or the total energy, is the change of volume of the solution by transpiration of the pure solvent across the partition, which by the above affects it in a manner

conclusion carries with it, by the thermodynamic principle of free or available energy, a theoretical proof of van 't Hoff's generalization that the osmotic pressure of a very dilute solution is equal to the gaseous pressure of the dissolved molecules when they are supposed to occupy the same volume in the gaseous state. The extension of this proof to dissolved liquids and solids, which form the practically important case, is at first sight barred (unless it is formulated as in the footnote) by the fact that we cannot then actually have the molecules existing free at the same volume as they occupy in the dilute solution. But when the Andrews isothermal for the dissolved substance is made into a continuous curve by inserting a supersaturated wavy part, there will always be a real point on it corresponding to the volume occupied by the substance thus existing in a homogeneous condition, and also a corresponding pressure which at the small density under consideration would practically be that of the gaseous state: thus there would be no difficulty in the extension to dissolved solids and liquids, were it not that this point on the isothermal might be on the thoroughly unstable reach, along which rise of density corresponds to fall of pressure, so that any slight accidental inequality of density would be spontaneously increased. The successful use made of the Andrews diagram for numerical calculation of the properties of substances by Van der Waals shows however that its physical reality is not destroyed by this instability: and when it is remembered that instability can be theoretically removed by slight constraint which does not sensibly affect the material transformations and does not affect the energy relations at all, it will appear that there is good reason for generalizing the law of osmotic pressure above demonstrated for gases. As before stated, what is most desirable to supplement on mechanical principles an explanation like the present one is not so much any accession of logical rigour on ordinary thermodynamic premises, as some precise notion of what is involved, as regards detailed molecular dynamics, in equality of temperature. In the present differential procedure no assumption has been made on that head; and no inference that osmotic pressure is, like gaseous pressure, due to simple molecular bombardment is warranted. When a theoretical basis is thus found for van 't Hoff's principle, the laws of the molecular influence of dissolved substances on the freezing point and vapour tension of very dilute solutions of course go along with it*.

*Marginal note:* Unstable part of Andrews' diagram utilized in theory:

*Marginal note:* for instability is removable by very slight constraining stress.

58. It may be objected that the application of the osmotic prin-

absolutely independent of the nature of the homogeneous solvent, and therefore of the existence of the solvent at all, because the relation of each molecule to the portion of the solvent within its sphere of influence is not changed.

* Cf. Willard Gibbs' general argument on "dilute systems": also *infra*, p. 101.

ciple to ionized solutions would compel us to admit the possible theoretical existence of a gas consisting of ions: but that is not really so, because the argument only compares one state of dilution with another. Yet on the other hand there is the hypothesis, supported by Brühl's work on optical equivalents, that under certain circumstances oxygen is a tetrad element, so that the molecule $H_2O$ can take up sufficient ions to form another saturated molecule of type $H_2 = O = X$, and that therein lies the cause of the regular ionization current produced by solution in water (the ions $X$ being free only when passing from one such combination to another), as contrasted with the irregular ionization of free gases. Changes of valency in an element remain unexplained, but their occurrence is now usually accepted as matter of fact[1]. The function of the osmotic diaphragm is merely passive, to prevent mixture by diffusion and consequent loss of mechanical availability. In a mutual solution of two substances, it is usually only the one that is present in large excess that gets through the diaphragm in purity: if it should prove to be a general law that the dialysing action is only complete when there is such large excess, it would be strong evidence for the view that the molecules dissolved in it form the nuclei of loose molecular complexes which are too large and permanent to get through, while the free solvent in which the molecules are not thus grouped is not so hindered.

*Transpiration of ions.*

*Why diaphragms are not permeable to the solute.*

When a solution is made more and more dilute, there comes a stage when it would seem impossible to imagine that its electrolysis, if it remains of normal type, is conducted through a mechanism like Grotthus' chains: the dissolved molecules are far out of each others' range of influence, and the very first stage of the working of a Grotthus' chain containing molecules of the solvent would produce that dis-

*Electrolytic chains ineffective.*

---

[1] The connection between the various phenomena may be pictured in neutral terms, as Poynting has recently done (*Phil. Mag.* Oct. 1896), starting from a hypothesis that pressure increases the "molecular mobility" of a fluid according to an *assumed* law equivalent to van 't Hoff's principle. In order to evade the hypothesis of partial dissociation in salt solutions, he restricts the sphere of action of a dissolved undissociated molecule to one or two or three definite molecules of the solvent, leading to correspondingly different amounts of osmotic pressure; thus a temporary chemical combination is dealt with instead of, or it may be in addition to, an extended sphere of influence. But the considerations given above show that it is the number of spheres of influence that is really effective, so that if there is chemical combination it must be in part with dissociated ions as in Brühl's view. On the special assumptions involved in the extension of the methods of gas theory to liquids, Boltzmann (*Zeits. für Phys. Chemie*, VI, p. 478) has offered a demonstration (approved by Lorentz) of the law of osmotic pressure, which seems to refer it to molecular bombardment, and require that the mean energy of translation of a molecule shall be the same in the liquid state as in the gaseous state at the same temperature.

*Temporary association invoked.*

*Kinetic aspect.*

sociation which it is the object of the chain theory to evade. Similar
considerations apply to the velocity of chemical reactions. When a
solution of K.HO neutralizes one of HCl, the heat generated is
mainly that of the union of H and HO to form $H_2O$: when the solutions are very dilute this should take a considerable time to develop,
even allowing for intimate mixture by stirring, if each H had to find
its HO partner directly. The immediate reaction must therefore be
due to a mobile equilibrium of dissociation being disturbed by the
mixing of the solutions, and then re-establishing itself[1]. Thus in the
progress of an ion H through the water under the electric force in
electrolysis, it would not be the same H that is driven on, but that
ion often gets fixed liberating another one in its place, so that it is
the mean translation of a condition of matters in which there is a
definite number of H ions in the element of volume, that is given by
Kohlrausch's law, not that of an individual ion. This accords with
Whetham's interpretation of his result, that in acetic acid solutions,
in which the conductivity is abnormally low, the ionic velocity is
abnormal to an equal extent[2].

*Dilute reactions would be slow,*

*if not effected by statistical ionization exchanges.*

*Ionic velocity a mean drift.*

59. A principle quite analogous to the one on which van 't Hoff's
law has here been based, has already been applied to a cognate
phenomenon in authoritative investigations. The transpiration of
two different gases into each other across a porous partition establishes a difference of pressure; there is thus present a store of available energy, which would be run down in the mixing of the gases;
its amount, as originally determined by Lord Rayleigh, from the
special properties of gases, is obtained by finding how much free
energy runs down when the gases are each separately expanded to
the volume of the mixture, and adding these amounts. This result,
either in the present form or expressed with reference to entropy, has
been sanctioned, explicitly or tacitly, as axiomatic by Maxwell[3] and
other authorities, when applied to gases whose molecules do not
exhibit sensible mutual attraction: the change of configuration arising
from the two mutually independent systems occupying the same
space, instead of different equal spaces, is rightly held to involve no

*The Dalton gas principle as estab-lished by Rayleigh:*

*dissipation by mixing gases.*

*Is its principle axiomatic?*

---

*Ternary reactions must be catalytic.*

[1] In the same way, if a gaseous reaction were of ternary type, so that three
atoms or ions had to unite to form a molecule, it must proceed far more slowly
than a binary reaction, and may not get established at all, except by the help
of the catalytic action of some other substance, such as water vapour, in
reducing it to binary stages or facilitating the simultaneous presence of the
three kinds of atoms in the same molecular sphere of action. [Cf. Wilde Lecture,
*Proc. Manchester Lit. and Phil. Soc.* 1908, *infra.*]

[2] W. C. D. Whetham, *Solution and Electrolysis*, 1895, pp. 142, 155.

[3] *Encyc. Brit.*, Art. "Diffusion"; *Collected Papers*, II, p. 644.

change in the available energy. The principle above employed is of precisely similar nature.

If we imagined two gases in which the molecular mass differed only infinitesimally, just the same amount of work could still be gained by mixing given volumes of them in a reversible manner as if they were gases wholly unlike; but the transpiration pressure would then be infinitesimally small and the time of transpiration infinitely great[1]. It is thus impracticable to proceed to a limit, and no paradox is here involved such as the assertion that a finite amount of work could be gained by mixing two gases which are practically identical in properties. A similar apparently paradoxical limiting case might be formulated as regards osmotic pressure of a dissolved substance very nearly identical with the solvent.

*The "Gibbs paradox" neglects the time element.*

60. The law of Henry that the density of dissolved gas is in a constant ratio $s$ to its density as it exists free in the surrounding atmosphere, is involved in the osmotic law, and conversely may be employed to verify it. In circumstances of equilibrium the potential of free energy of the dissolved gas (in Gibbs' sense) must be the same in the liquid and the atmosphere; that is, the removal of an elementary portion of the gas from the liquid to the atmosphere must not alter the free energy of the whole. Thus the difference of the free energies of the dissolved gas per unit mass, when its partial pressure in the liquid is changed from $p_1$ to $p_2$, is equal to the difference of the free energies of the gas per unit mass in the surrounding atmosphere when its partial pressure is changed from $p_1/s$ to $p_2/s$. The latter difference is by Lord Rayleigh's principle independent of what other gases may also be present in the atmosphere: it is thus $\int p\,dv$, where $pv = R'\theta$ for the unit mass of gas, and is therefore at constant temperature $R'\theta \log p_1/p_2$. This does not involve $s$, and therefore the *difference* of the free energies of the dissolved gas at two different densities is the same as if it existed in the free gaseous state at those densities: and this carries with it identity for the two states as regards all relations of available energy and work. Conversely, the law of Henry follows as above, by the principle of available energy, from the circumstance that the molecules of the dissolved gas are outside each

*Transition from liquid to vapour:*

*free energy unaltered:*

*establishes laws of dilute state of solute as regards vaporizable substance.*

[1] The assumption is involved that the gases are really different and that means exist for separating them.

The fact that the amount of available energy *at our command* depends on the control we have learned to exercise over physical processes does not detract from the objective validity of that conception as a deduction from general principles of molecular theory, as has often been suggested (cf. § 49, *supra*): any more than our possible complete ignorance of some forms of total energy would give to the idea of energy itself a subjective aspect [or now throw doubt on its conservation].

*Physical evidence always essentially inductive.*

others' spheres of molecular action, independently of any picture that we may form of the process of exchanges in evaporation and absorption*.

Historical.        It is a confirmation of the soundness of this thermodynamic theory, that the law of osmotic pressure for dissolved gases is immediately involved in, and might have been predicted from, the equations given by von Helmholtz in 1883[1], in a discussion of their energy relations in connection with the theory of galvanic polarization.

Free energy of a surface film:        * Surface osmotic force can also exist. Contamination by a skin of foreign molecules reduces surface tension: else indeed it would not occur, for there would not be a fall of available energy: and for the same reason it spreads. It may thus become a single close layer of foreign molecules, and then spread as the surface is further expanded into an open layer, until finally the adjacent molecules and their partially attached environments may get outside the range of each others' repulsive influence. In that stage one might expect the tension to remain at a constant value on further expansion, this value being the tension of the pure solvent. But although there is then no static overlap of the environments of the foreign molecules, there is still their kinetic movement over the its tension surface, exerting a tangential surface pressure, that is, a reduction of the follows the surface tension, expressed now in terms of surface density of foreign molecules gas law: by the gas law, and even with the same universal gas constant. This implies, on the usual principles of statistical molecular dynamics, that the layer of extraneous molecules, in addition to extensive excursions along the surface, also vibrates with smaller amplitude transverse to it in a way which entangles them with the other molecules.

With increasing closeness of the foreign molecules by contraction of the surface the pressure due to this momentum effect is diminished by their mutual attraction: when the latter preponderates there would be instability just as in the Andrews diagram for gases, with the cognate result that the surface abrupt layer, if not too dense, would divide into a quasi-gaseous region of uniform change density, in free contact with a dense region quasi-liquid where the molecules possible. are in mutual support: and ultimately a solid film may form.

The experimental investigations of Langmuir, N. K. Adam, and others show that this course of events can actually be followed out with considerable precision.

The presence of strata in soap bubbles (Perrin), and even the obvious sharp breaks in visible colour, invite similar mode of explanation of the free contacts of superposed surface phases.

Willard Gibbs: Helmholtz: Rayleigh.        [1] "Zur Thermodynamik Chemischer Vorgänge," III, *Monats. Berl. Akad.* May 1883, especially equations (4) and (5); *Abhandlungen*, III, pp. 101–114. The law had however been arrived at quite explicitly by Willard Gibbs as early as 1876 in his discussion of the general theory (p. 226). Recently the argument has been carefully formulated by Lord Rayleigh, *Nature*, Jan. 14, 1897: cf. also a letter by Gibbs, March 18. The pressure difference is necessitated by the circumstance that the steady state would be brought about by inter-Perfect change of individual molecules. But its amount is calculable *à priori* only dilution. when the dissolved molecules are practically out of each others' range: and then the argument in the text shows that it depends solely on the number of molecular aggregates with foreign nuclei that are present, irrespective of whether these nuclei are complete molecules or parts of dissociated molecules.

61. It is the circumstance that the available energy $A$ of § 51 is a function of the bodily configuration and constitution of the system, whose alteration by dilution is independent of the nature of the solvent provided the solution is sufficiently dilute, that makes osmotic pressure independent of the solvent and therefore the same as the corresponding gas pressure. This is of course different from asserting that the whole available energy of a dissolved substance is the same as its available energy at the same density in the free gaseous state. In fact the difference between these energies may be estimated from a knowledge of the solubility: thus the available energy per unit mass of the gas in the solution at its actual density $\rho'$ is equal to that of the same gas in the free space at the corresponding density $\rho$, as there is freedom of exchange; so that the available energy per molecule of the dissolved gas is equal to that of free gas of its own density and temperature together with $R_1 T \log \rho'/\rho$ and also $R_1 T$ for the volume occupied by the free gas. This makes in all for the excess of available energy, per molecule, of the dissolved gas $R_1 T \log e\rho'/\rho$, or $R_1 T \log es$, where $s$ is the solubility and $R_1$ is a gas constant the same for all kinds of molecules. Like information is derivable from the ratio of partition of any dissolved substance between any two solvents which do not intermix: its available energy per unit mass must in the state of equilibrium be the same in both solutions.

62. The increase of available energy involved in molecules or atoms of given species appearing in the dilute solution during chemical change is, per molecule, $a + R_1 T \log bN$ where $N$ is the number of such molecules already there per unit volume and $a$ is a function of the temperature, $b$ being a constant which depends on the standard temperature of reference (§ 51). A reaction going on in the solution involves the disappearing by breaking up of molecules of some of the types present, and the appearing of molecules of other types to an equivalent extent: when chemical equilibrium is attained, the change of available energy arising from a slight further transformation of this kind must vanish: that is,

$$n_1 \{a_1 + R_1 T \log b_1 N_1\} + n_2 (a_2 + R_1 T \log b_2 N_2) + \ldots$$

vanishes, leading to

$$R_1 T \log b_1{}^{n_1} b_2{}^{n_2} \ldots N_1{}^{n_1} N_2{}^{n_2} \ldots = -(n_1 a_1 + n_2 a_2 + \ldots),$$

where $n_1$, $n_2$, ... are the numbers of the molecules of the different types that take part in the reaction, reckoned positive when they appear, negative when they disappear; so that $N_1{}^{n_1} N_2{}^{n_2} \ldots$ is equal to $K$, a function of the temperature, which is the law of chemical equilibrium originally derived by Guldberg and Waage from statistical considerations*. Again, if $A$ is the available energy of the whole solution, and $\delta A$, equal to $\delta A_0 + R_1 T \log K'$, where $K' = K b_1{}^{n_1} b_2{}^{n_2}...$, denotes its variation per molecule of reaction without change of temperature, $\delta A$ is null as above in the equilibrium state at each temperature, so that $\delta A_0 = - R_1 T \log K'$. And, with partial differentiation,

$$\frac{d}{dT} \frac{\delta A}{T} = \frac{d}{dT} \frac{\delta A_0}{T} = - R_1 \frac{d}{dT} \log K' = - R_1 \frac{d}{dT} \log K,$$

and is independent of the unknown term $A_0$. Now reverting to the general theory, if $E$ is the energy in a system, $dH$ the heat imparted to it and $dW$ the work done to it,

$$dE = dW + dH = dW + T d\phi, \quad A = E - T\phi + T\phi_0,$$

where $A, E, T, \phi$ are all analytical functions of the state of the system. Thus, employing total differentials,

$$d \frac{A}{T} = - E \frac{dT}{T^2} + \frac{dE}{T} - d\phi = - E \frac{dT}{T^2} + \frac{dW}{T}:$$

so that in the present case

$$d \frac{\delta A}{T} = - \delta E \frac{dT}{T^2} + \frac{d}{dT} \delta W.$$

If the small amount of reaction represented by $\delta$ occurs so that no mechanical work is done on the system, $\delta W$ is null; hence $\delta E$ is equal to $\delta H$ the amount of heat taken into the system from its surroundings per molecule of the reaction when it proceeds [isothermally] without work. Thus

*van 't Hoff's heat of reaction formula.*

$$\delta H = - T^2 \frac{d}{dT} \frac{\delta A}{T} = R_1 T^2 \frac{d}{dT} \log K,$$

the thermal relation [for isothermal reactions] of van 't Hoff[1] [adumbrated by Kirchhoff, 1857].

* It seems proper to emphasize that the special developments in these sections all refer to dilute systems. They are narrower than a general statistical theory, in that the molecular encounters by which the constitution of the phases of matter is adjusted are practically all binary. The more general statistical possibilities are considered in a Wilde Lecture on "The Physical Aspect of the Atomic Theory" (1908), especially Appendix II, as *infra*.

*Restriction to case of binary encounters:*

*is necessary generally.* The formulation of statistical mechanics, as developed on the lines of the gas theory by Maxwell and Boltzmann, and systematized by Willard Gibbs in his treatise on the subject, appears to be similarly restricted. It rests on Liouville's multiple differential invariant of dynamical systems, and the application of that theorem implies definite binary interaction. The principle of equipartition of energy founded on it, which, unduly extended, has been a stumbling-block to all theories of molecular interaction, appears to be subject to the same limitation. So is, ultimately, the Boltzmann formulation of entropy in terms of statistical probability. Cf. *Roy. Soc. Proc.* 1909, as *infra*.

The formal theory of thermodynamics, on the basis of the Carnot axiom (cf. *Roy. Soc. Proc.* 1919, as *infra*) or its equivalent that of available energy, as here treated, is thus far from being replaced as yet by any complete explanation on a molecular foundation.

[1] Cf. Willard Gibbs, *loc. cit.* p. 231, where the case of gaseous reactions was treated. More directly, we can form a reversible Carnot cycle in which the

*On the Electromotive Forces established by Finite Diffusion.*

63. The function of an osmotic partition in preventing by pure constraint the diffusive degradation of energy (§ 54) is illustrated by the theory of electromotive forces of diffusion. In the concentration-cells, of which the theory was established by von Helmholtz, the solution in each cell was homogeneous, and the influence of concentration was determined by balancing different cells against each other: there being no diffusion, the process was reversible, and thermodynamic formulae were applicable. By forming an electrode of a metal surrounded by one of its insoluble salts, such as mercury surrounded by calomel, employing for the other one zinc immersed in zinc chloride solution, the net constitutive change at the mercury electrode when electricity passes through the cell is independent of the concentration of the solution, being simply the deposition of the equivalent quantity of mercury from undissolved calomel: hence that electrode accounts for a constant part of the electromotive force. On the other hand the change of free energy by dissolution of the equivalent of zinc is made up of a part arising from change of chemical constitution and another part depending on the concentration of the solution which receives the resulting chloride. The part of the electromotive force depending on the processes at the zinc electrode is thus in the case of a reversible electrode equal to const.$-RT \log p$, where $p$ is the osmotic pressure of the zinc chloride solution and $R$ is now the gas-constant belonging to an electrical equivalent per unit volume, which is 8580 in c.g.s. units: this may be expressed in the form $RT \log P/p$, where $P$ depends on the metal of the electrode and on the solvent employed in the cell and on the temperature, but not on the concentration of the solution. This quantity $P$ has been called by Nernst, on grounds of analogy, the solution pressure of the metal electrode in the solvent[1]. When the electrode is polarizable so that

*Marginal notes:*
Reversible voltaic systems.

Influence of change of concentration:

Helmholtz's experimental procedure.

Electrode effect: its solution *quasi-pressure.*

---

constitutive change $\delta A$ is made at temperature $T$ and unmade at $T - dT$. The work of the cycle must be $\delta h \cdot dT/T$, where $\delta h$ is the heat absorbed in the change when the maximum amount of mechanical work is done in it by osmotic or other appliances: thus $\delta h \cdot dT/T = dT \cdot d\delta A/dT$, so that $\delta h = T d\delta A/dT$. When no work is done in the [isothermal] change, the heat absorbed is $\delta H$, equal to $\delta h - \delta A$, which is $-T^2 d/dT (\delta A/T)$ as above. [Generally, the change of energy along the isothermal arm of the cycle is $\delta E$, equal to $\delta h + \delta A$: and this equation is then identical with the more familiar form $\delta A = \delta E + T d\delta A/dT$, in which the symbol $\delta$ may now be omitted, giving $A = E + T dA/dT$. The more compact principle of Clausius is that the value of $dA/dT$ for unchanged state of the system, that being the entropy with sign reversed, cannot increase for transformations that are spontaneous and therefore of constant energy.

The signs of terms involving $\delta A$ in the early part of this footnote, except the last one, have to be changed: see Appendix to this volume.]

[1] There appears a difficulty in imagining, in accordance with the view here taken, that the value of $P$ can be dependent on a layer of the metallic ions

*Marginal note:*
Heat of reaction obtained by a Carnot cycle.

the processes are not reversible, the difference of potential must be less than this formula would give. If now we are dealing with a two-fluid cell, in which the fluids are separated by an osmotic partition and passage of the solvent is prevented by balancing the osmotic

*Electromotive force of a cell:*

tendency by hydrostatic pressure, the processes are still reversible and the electromotive force of the cell will be $RT (\log P_1/p_1 - \log P_2/p_2)$,

*osmotic partitions become polarized:*

while passage of a current will gradually polarize the faces of the partition. If however the ions could pass through the partition, into the solution of different concentration, without diffusion of the fluids

*intrinsic electrode part:*

in bulk, the part of this electromotive force depending on concentration would be cancelled, and there would remain $RT \log P_1/P_2$ due solely to the affinity of the solvent for the materials of the electrodes. But if we are dealing with a cell, in which the fluids are in direct contact along an interface of finite dimensions so that steady diffusion

*concentration part may be diminished*

at a finite rate is going on, or in which they are even allowed to diffuse steadily across an osmotic partition, there will be loss of

*by the natural diffusion,*

availability owing to that diffusion, so that the back electromotive force arising at the junction of the fluids is less than the maximum value $- RT \log p_1/p_2$. In the absence of knowledge of the rate at which the diffusive degradation of energy is proceeding and is affected by electric transfer, the principle of availability cannot supply a

*to amount determined in the Nernst-Planck theory of diffusion of electrolytes, as infra.*

formula for this diminution of the back electromotive force, which will depend on the nature of the layer of transition: but a theory of the process of steady interdiffusion of two ionized fluids has been formulated by Nernst and Planck which involves an expression for its magnitude[1]. Thus, considering diffusion of a simple solution across a layer in which the concentration varies, when the steady state is attained both ions must diffuse together at the same rate notwith-

extending into the solution, especially as the potential difference between dielectrics could not be so explained. Cf. § 56, *supra*.

[1] I find that applications similar to the above, but on a more extensive scale and with considerable differences in the argument, especially a more prominent use of entropy, are made in Planck's later important exposition "Ueber das Princip der Vermehrung der Entropie," Wied. *Ann.* xliv, 1891, pp. 385–428. The general formula for the potential difference between two diffusing solutions is there obtained from the variation of an analytical function, which is really the available energy, on the hypothesis that the solutions are

*Non-electric diffusion neglected.*

in a permanent state of diffusion, determined by Nernst's principles, in which the concentration varies from point to point so slowly that the diffusive dissipation other than electric may be neglected. Cf. also on the history of the subject Negbaur, Wied. *Ann.* xliv, p. 737. In the text above the statements are confined to the case of binary electrolytes.

The development of the laws of chemical equilibrium given in § 60 has also been largely anticipated as to form by Planck, Wied. *Ann.* xxxii, 1887: his postulates are however different from those that enter here, where the analysis occurs as an outcome of a general view of the relations of molecules in bulk to the aether and to each other, §§ 11–12. (Cf. Planck, *Vorlesungen über Thermodynamik*, 1897.)

standing their different mobilities $u$ and $v$, measured by Kohlrausch as the values of their mean velocities of drift due to unit electric force. Now the mean steady velocity of migration of a single ion is equal to this mobility divided by its electric charge $e$ and multiplied by the force which causes its motion: this force consists of an electric part $- edV/dx$, where $V$ is the electric potential set up during the transition to the steady state of diffusion, and an osmotic part. To determine the latter, observe that when a solution is separated from the pure solvent by a permeable osmotic partition, the solvent is restrained from passing across only by an osmotic pressure acting against it: this means that to maintain the steady state without diffusion the osmotic partition must exert more pressure by the amount $p$ on the solution than on the pure solvent. If we consider a layer of the actual solution, of cross-section unity and thickness $\delta x$, there would thus have to be a bodily force $dp/dx \cdot \delta x$ exerted on it if the diffusion of its ions were prevented: therefore $- dp/dx \cdot \delta x$ is the aggregate of the forces acting on the contained ions and producing diffusion, that arise from the gradient of concentration. If $n$ be the number of ions of either kind per unit volume, the mean force per ion is thus $- n^{-1}dp/dx$: this is not a mere hypothesis founded on a vague analogy of osmotic pressure with ordinary hydrostatic pressure, but gives a precise measure of an actual force on a constituent of the medium. The number $dN/dt$ of single ions of either kind that is driven across unit area of a geometrical interface in a solution of varying concentration by these forces is thus given, after Nernst, by

$$\frac{dN}{dt} = - nu\frac{dV}{dx} - \frac{n}{e}\frac{u}{n}\frac{dp}{dx}, \quad \text{also} \ = nv\frac{dV}{dx} - \frac{n}{e}\frac{v}{n}\frac{dp}{dx}.$$

If $D$ denote the coefficient of diffusion of the solution,

$$dN/dt = - D\,dn/dx;$$

and by the gaseous law which applies to osmotic pressure of very dilute solutions $p = neRT$. Hence immediately

$$D = \frac{2uv}{u + v}RT, \quad \frac{dV}{dx} = \frac{v - u}{v + u}\frac{1}{ne}\frac{dp}{dx},$$

so that the integrated potential difference across the diffusion layer is given by

$$V_2 - V_1 = RT\frac{v - u}{v + u}\log\frac{p_2}{p_1}.$$

It follows that when steady diffusion is allowed to go on, the back electromotive force at the junction of the fluids is thereby reduced in the ratio of the difference to the sum of the ionic mobilities. The agreement with experiment of these expressions for $V_2 - V_1$, and for the ordinary diffusion coefficient $D$ of a solution as thus determined electrically, constitutes two distinct tests of the general validity of this diffusion scheme, and of the hypothesis of independent mobility of the ions [in completely ionized solutions] of which it is a corollary*.

**Extension to solutions only partially ionized,**

In the case of a solution only partially dissociated, like that of acetic acid referred to in § 58, provided the time of association of two paired ions is on the average large compared with the time of relaxa-

**with a modification.**

tion of the system, these expressions for $D$ and $V_2 - V_1$ will still hold for the dissociated portion, if $u$ and $v$ denote the actual velocities of drift of the ions when free, not the abnormally small effective velocities as determined by Whetham. Thus the total diffusion would now consist of this part belonging to the ionized portion, with coefficient independent of the degree of ionization, together with the actual diffusion of the non-ionized portion. On the same hypothesis the potential difference between the fluids would depend, as might have been foreseen, only on the actual concentrations of the ions in the two solutions, the amount of non-ionized substance being immaterial except in so far as it gives rise to an ordinary contact difference

**The limitation.**

(§ 56): but it may not be computed from the abnormal ionic velocities by the ordinary formula unless the degree of ionization is independent of the concentration.

### *Critique of von Helmholtz's Theory of Electric Stresses:*
### *Electrostriction not due to Mechanical Forcive.*

**Expression for energy in polarized dielectric:**

64. A theory of electrostatic stress in dielectric media, based on the method of energy, and avoiding molecular theory, has been originated by Korteweg[1], formulated in general terms by von Helmholtz[2], and further developed by Lorberg, Kirchhoff[3], Hertz[4] and others: it is desirable to examine the relation in which it stands to the views here set forth. The investigation of von Helmholtz postulates a dielectric medium which is effectively continuous, not molecular; also a potential function, that namely of the distribution of uncom-

---

* All this implies that the ions are in the main beyond each others direct electric influence, thus limiting to very small concentration: for greater, a tendency for the activity and the conductance to vary as the square root of the concentration (Kohlrausch) soon becomes predominant. This has been ascribed recently by Debye and Hückel, modified by Onsager and others, to a kind of ideal atmosphere of charge round each ion, formed of other ions, of the same size and so to be regarded as a time mean, which lags behind in the motion of the ion and so pulls it back somewhat as in the frictional electrophoresis of separated ionic layers of Quincke and Helmholtz. There is already an extensive and intricate literature: see reports by the main workers in *Trans. Faraday Soc.* Aug. 1927, also a critique by J. A. Gaunt, *M. N. R. Astron. Soc.* 1928.

[1] D. J. Korteweg, Wied. *Ann.* IX, 1880.
[2] H. von Helmholtz, Wied. *Ann.* XIII, 1882: *Abhandlungen*, I, p. 798.
[3] G. Kirchhoff, Wied. *Ann.* XXIV, XXV, 1885: *Abhandlungen*, Nachtrag, p. 91.
[4] H. Hertz, Wied. *Ann.* XLI, 1890: *Electric Waves*, English edition, pp. 259–268.

pensated polarity which represents the electric state of the medium, satisfying a characteristic equation, that of the Faraday-Maxwell theory. The energy per unit volume is expressed in terms of this potential, in such form that the variation of the integral which represents the energy for the whole volume leads, *on integration by parts*, to this characteristic equation as one of the conditions of internal equilibrium; the integral is then asserted to be in the normal form, which would mean, in our order of ideas, that it represents the actual distribution of the energy in the medium* as well as its total amount. Its variation with sign changed, arising from change of material configuration, should then give the extraneous forcive that must be applied in order to maintain mechanical equilibrium; the variation with respect to the electrical configuration being null, so that electric internal equilibrium is provided for, by the characteristic equation already satisfied. The variation without change of sign should thus give the mechanical forcive of electric origin that acts on the medium. But the *data* do not even on these assumptions suffice to lead to a definite stress system for the material; a certain geometrical stress system is merely assumed which yields on the element of volume a mechanical forcive the same as the one thus deduced from the energy function. An infinite number of such stress systems might in fact be specified, for there are six components of stress which need satisfy only three conditions. If however the stress system is required to be symmetrical with respect to the lines of polarization, there is in this respect no indefiniteness (§ 40); and the one given by von Helmholtz is of this kind. Thus the definite result really deduced by von Helmholtz from his energy hypothesis is an expression for the bodily mechanical forcive in the polarized medium, the $(X, Y, Z)$ of equation (4) of his memoir; while correlative formulae are applied by him and by Kirchhoff for the bodily forcive in a magnetized medium. These expressions, however, definitely contradict the formulae of Maxwell and of the other previous writers for the bodily mechanical forcive in a magnetized medium, which are in general agreement with those developed in this paper: in fact von Helmholtz translates his formulae into Maxwell's electric stress system, while Maxwell himself had to invent for the case of magnetic polarization, which was the one he considered, a different stress system, namely his magnetic stress. As recent writers have in the main tacitly accepted von Helmholtz's procedure, it is incumbent on us to assign the origin of this discrepancy; and for this purpose a summary of his method is given, the effect of alteration of the coefficient of polarization arising from strain in the material being for the present left out of account.

*its form determined by internal equilibrium: distribution of mechanical forces thence derived,*

*but not a stress system,*

*which is indeterminate,*

*unless symmetrical.*

*Discrepancies from Maxwell.*

*Constitutive change deferred.*

---

* Other than the part compensated locally by adaptation of structure.

65. The organized electric energy in the polarized medium being assumed, from other considerations, to be

$$W = \int \frac{K}{8\pi} \left( \frac{dV^2}{dx^2} + \frac{dV^2}{dy^2} + \frac{dV^2}{dz^2} \right) d\tau,$$

where in regions of electric density

$$\frac{d}{dx} \left( K \frac{dV}{dx} \right) + \frac{d}{dy} \left( K \frac{dV}{dy} \right) + \frac{d}{dz} \left( K \frac{dV}{dz} \right) + 4\pi\rho = 0,$$

in which $\rho$ is a constant associated with the element of dielectric matter, called the density of its free electric charge, the forces acting will be derived from the variation of $W$; variation with respect to $V$ leads to the electric forces, and that with respect to the material configuration leads to the mechanical ones. The problem is to determine the mechanical forces when there is electric equilibrium, that is when variation with respect to $V$ yields a null result. The form of $W$ above expressed does not lead to this null result; we can however by integration by parts derive the form $W = \frac{1}{2}\int V\rho d\tau$, the essence of this transformation being that in the new integral the distribution of the energy among the elements of volume $d\tau$ of the medium has been altered. This form does not satisfy the above requirement either, but by combining the two forms we obtain

$$W = \int \left\{ \rho V - \frac{K}{8\pi} \left( \frac{dV^2}{dx^2} + \frac{dV^2}{dy^2} + \frac{dV^2}{dz^2} \right) \right\} d\tau,$$

whose variation with respect to $V$ is null as required: although as integration by parts is employed, the variation is *not* null for each

Energy<br>adjusted<br>for electric<br>equilibrium, single element of mass. This integral is then taken to represent the actual distribution of the organized energy in the medium when in electric equilibrium, and not merely its total amount: and variation of it with respect to the material configuration should on that hypothesis give the actual bodily distribution of mechanical forcive, not merely its statical resultant on the hypothesis that the system is absolutely rigid. Now in finding the variation of $W$ arising from a virtual displacement $(\delta x, \delta y, \delta z)$ of the polarized material, we have to respect the conditions that the free charge $\rho\delta\tau$ is merely displaced, so that by the equation of continuity

$$\delta\rho + \frac{d}{dx} \rho\delta x + \frac{d}{dy} \rho\delta y + \frac{d}{dz} \rho\delta z = 0,$$

and also that each element of the material is moved on with its own $K$, so that $\delta K + dK/dx . \delta x + dK/dy . \delta y + dK/dz . \delta z = 0$; while things have been arranged so that a variation of $V$ produces no result—but only however no aggregate result on integration by parts. Unless the transitions at interfaces are supposed to be gradual as regards $K$, and the integration then to extend throughout all space, there will also be direct surface terms in the variation, because the virtual shift of the material leaves a space unoccupied on one side and occupies a new space on the other; thus finally by the ordinary process of integration by parts we obtain for any region

$$\delta W = \int \left[ \left\{ \rho \frac{dV}{dx} + \frac{1}{8\pi} \frac{dK}{dx} \left( \frac{dV^2}{dx^2} + \frac{dV^2}{dy^2} + \frac{dV^2}{dz^2} \right) \right\} \delta x + \ldots + \ldots \right] \delta \tau$$
$$- \int \frac{K}{8\pi} \left( \frac{dV^2}{dx^2} + \frac{dV^2}{dy^2} + \frac{dV^2}{dz^2} \right) \delta n \, dS.$$

The coefficient of $\delta x$ with sign changed has been taken to be the component of the bodily mechanical forcive. But to obtain the total mechanical forcive acting on an element we must retain all the terms in the variation that belong to it, so that it is illegitimate in this connection to transmit a traction from it to the boundary of the medium by the process of integration by parts. If then we consider the single element of volume by itself, so that in the formula $\delta S$ is an element of its surface, the forcive on it will be von Helmholtz's one together with a hydrostatic pressure $- K \, (\text{grad } V)^2/8\pi$ acting over its surface; and this complete specification would agree with our previous results, *except* that we have $K - 1$ in place of $K$ for reasons already assigned. But it would seem that the method thus described must be radically unsound; it would be valid if there were only one medium under consideration, of which $W$ is the energy function: but there is here, in the same space, the aether with its stress and the polarized matter with its reacting mechanical forces, and (§ 6) there is no means of disentangling from a single energy function in this way the portions of energy which are associated with these different effects*.

*[margin note: but not correctly distributed for mechanical stress.]*

*[margin note: Mode of its correction:]*

*[margin note: but still not applicable to a compound medium.]*

66. There are also subsidiary terms in von Helmholtz's formulae, involving the rate of alteration of the inductive capacity of the fluid dielectric by compression, terms which are extended in the work of Korteweg, Lorberg, and Kirchhoff to include the alterations of the inductive capacity of a solid dielectric produced by the various types of strain that it can sustain. Their *rationale* is best seen by the more elementary procedure of Korteweg, who first introduced them. He considered the following cycle; (i) move up a piece of the dielectric material from an infinite distance into an electric field, (ii) strain it and so alter its inductive capacity and therefore the electric energy, (iii) move it back to an infinite distance in the strained state, (iv) restore it to its original condition by removing the strain. In order to evade perpetual motions, the mechanical work done by electric attractions as it approaches must exceed the work absorbed

*[margin note: Constitutive change of the energy.]*

*[margin note: Cyclic deduction of the forces:]*

* The standard procedure by Action, *Aether and Matter* (1900), Ch. VI, which in free aether agrees with the Helmholtz process, effects this purpose.

as it recedes, by the loss of available electrical energy due to strain; and this leads Korteweg to terms in the mechanical forcive which depend on the rate of variation of inductive capacity [if any] with strain. The process is analytically developed for fluids by von Helmholtz, by adding on to $\delta K$, the variation of $K$, a part arising from the compression of the material which the virtual displacement involves, namely by adding

$$- \frac{dK}{d \log s} \left( \frac{d\delta x}{dx} + \frac{d\delta y}{dy} + \frac{d\delta z}{dz} \right),$$

where $s$ denotes the material density: and Kirchhoff formulates it for isotropic solid media, replacing $dK/d \log s$ by Korteweg's two coefficients which express the actual rates of change of $K$ due to elongations along the line of polarization and at right angles to it. But here again a process of integration by parts comes in, which removes part of the bodily forcive from the element of volume at which it is directly applied to the boundary, and so vitiates the result regarded as a specification of the forcive which produces the actual distribution of mechanical strain in the material.

67. Moreover, phenomena of this latter kind are more appropriately investigated as intrinsic changes of the equilibrium configuration of the material arising from molecular actions produced by the polarization, the forcive of the above argument being simply what would be originated if these changes were prevented by constraint. Such deformations of the elements of volume of the material, the result of electrostriction or magnetostriction, may not fit in with each other, and the strain thence arising will originate secondary mechanical stresses: but it appears preferable to keep these distinct from the regular stress which is the effect of the *direct* electric or magnetic action of different finite portions of the material on each other.

This separate procedure may be illustrated by an investigation of the change of intrinsic length of a bar of magnetic material, caused by its introduction into a magnetic field. Clamp the bar to its natural length when at a great distance; then introduce it into the magnetic field so as to lie along the lines of force; then unclamp in such way that it may do as much work as possible in pushing away resistances to its magnetic elongation; finally remove the unclamped bar again to a great distance. If this cycle is performed at a uniform temperature, it follows from Carnot's principle that there can be no resultant work done in it. Now the work done by the magnetic forces in introducing the bar is

$$\int I dH, \text{ that is } \int \left( \kappa + Q \frac{d\kappa}{dQ} + I \frac{d\kappa}{dI} \right) H dH,$$

per unit volume, where $\kappa$ is the magnetic susceptibility which is presumably a function of the internal longitudinal pressure $Q$ in the bar and of its intensity of magnetization $I$. The work done in un-clamping is $\frac{1}{2}Q_1 l_1$ per unit volume, where $l_1$ is the intrinsic magnetic elongation and $Q_1$ the pressure corresponding to the strength $H_1$ of the part of the field in which it is unclamped. This is on the suppo- Cycle of<br>operations. sition that the bar is long, so that there are no free magnetic poles near together which would diminish $Q$ by their mutual attraction. The work done per unit volume by the magnetic forces during the removal of the bar is $-\int(\kappa + I d\kappa/dI) H dH$. The resultant work in the cycle being null, we have

$$\frac{d\kappa}{dQ}\int QH\,dH = -\tfrac{1}{2}Q_1 l_1 = -\tfrac{1}{2}\frac{Q_1{}^2}{M},$$

where $M$ is Young's elastic modulus. This can only be satisfied if $Q$ is of the form $\lambda H^2$, where $\lambda$ is a constant, and it then gives Law of<br>elongation<br>of bar. $d\kappa/dQ = -2\lambda/M$, and the elongation $l$ is

$$-\tfrac{1}{2}\frac{d\kappa}{dQ}H^2, = \tfrac{1}{2}\frac{d\kappa^{-1}}{dQ}I^2 = -\tfrac{1}{2}\frac{d\log\kappa}{dQ}HI;$$

while the corresponding stress $Q$ is $-\tfrac{1}{2}d\kappa/dl \,.\, H^2$.[1] This result is of course valid only in the absence of hysteresis. A similar process applies where the field is transverse to the bar; and thus Kirchhoff's complete results may be obtained. A more complete enumeration Change in<br>elastic<br>moduli of possible physical changes would also take cognizance of alteration inoperative,<br>unless under of the elastic constants of the material due to the magnetic excitation; stress. but this cause (cf. § 83) will not add terms of the first order to the energy changes unless the bar is under extraneous stress, not merely constraint, while the cycle is being performed.

For dielectrics, direct experiments have not found any sensible de-pendence of inductive capacity on the pressure in the case of liquids\*; while the experimental discrepancies[2], which these terms were intro-duced by von Helmholtz to reconcile, have since been cleared up.

68. In the paper above referred to, Kirchhoff remarks (§ 3) that An inter-<br>posed film<br>alters local an expression for the traction across an ideal interface in a uniform constitutive polarized medium might be arrived at by supposing a very thin film stress. of air introduced along the interface, and computing the attraction between the two layers of opposed poles thus separated, a process which had been employed by Boltzmann. He concludes that this process must be at fault, on the ground that the specification of stress

---

[1] (In these differentiations $I$ is constant; see § 83.)

\* The slight effect has been measured by an optical method, and its cognate theory worked out, by Pauthenier, *Ann. de Phys.* (1920), *J. de Phys.* (1924): cf. Bocard, *J. de Phys.* (1927).

[2] Namely, the differences in the values of $K$ at first found by Quincke, by use of three different experimental methods (§ 78), which it is easy to see would on the usual theory involve perpetual motions.

thus obtained does not satisfy a necessary property of *mechanical* stress systems, namely that the tractions exerted over the surface of an infinitesimal element of volume of any form must balance each other; and he gives this as the reason for having to fall back on an energy method in order to obtain a specification free from objection. The preceding considerations (§§ 44–48) indicate the direct reason of the illegitimacy of that process, while they also exhibit the logical basis of the application of the method of mechanical energy in problems of molecular physics.

### Conservation of Energy in the Electric Field: Limited Validity* of Poynting's Principle.

69.  It has been explained (§ 6) that the agencies in an electric field may be in part traced by transmission through the aether after the manner of ordinary mechanical stress, and in part, namely as regards

*Flow of energy.* forces on the electrons, not so traced. As regards the former part, therefore, the increase of energy in any region must be expressible explicitly as a surface integral, representing work done by tractions exerted over its boundary. This theorem will thus have application

*Electrons do not interact by stress (footnote).* in all cases in which the configuration of the electrons is not changing; for its strict application, the bodies inside the region which carry currents or electric charges or are polarized, must thus be at rest, and there must be no change in the state of electrification of any conductor in the region. Recurring for an illustration to the simpler circumstances of a perfect fluid containing vortex rings, it is easy to show analytically that the rate of increase of energy in any region is there expressible as a surface integral, involving the velocity and the pressure, only when there are no vortex rings in the region or when

*The integral expression for the flux not localized.* the rings in it are all supposed to be held fixed by constraint[1]. This illustration also emphasizes the point that the surface integral must be taken as a whole, that an element of it does not necessarily represent the activity across the corresponding portion of the surface.

*Material system undisturbed: analysis of the play of energy,* Thus taking the material system, concerning which we need make no hypothesis as regards inductive quality or aeolotropy, to be *at rest* in the electric field so that there are no changes of energy due

---

* Namely, the flow of energy is to be across a steady material system.

[1] The reason is simply that the form of the contained vortex rings is not a function merely of the state of the fluid inside the surface, but also in part determines the simultaneous velocity of the fluid throughout all space. So also the energy associated with each atom of matter is really distributed throughout the whole aether [though only slightly if its polarity is compensated], and therefore the energy changes associated with changes in the configuration of matter cannot be represented as propagated step by step across the aether.

to the mechanical forcives, and neglecting those due to convection currents which rearrange electrifications, if $W$ and $T$ denote the organized potential and kinetic electric energies in the region, and $D$ the rate of dissipation of organized energy due to currents of conduction,

$$\frac{dW}{dt} + \frac{dT}{dt} + D$$

must be expressible as a surface integral. Now $W$ is made up, in electrostatic units, of the energy of aethereal strain

$$(8\pi)^{-1} \int (P^2 + Q^2 + R^2)\, d\tau,$$

and that of material polarization

$$\int \phi\, d\tau, \text{ where } \phi = \int (Pdf' + Qdg' + Rdh'),$$

which must be an exact differential when there is no dielectric hysteresis; thus in all

$$\frac{dW}{dt} = \int \left( P\frac{df''}{dt} + Q\frac{dg''}{dt} + R\frac{dh''}{dt} \right) d\tau.$$

Again the rate of dissipation arising from ionic migration in the conducting circuits is

$$D = \int \left\{ P\left( u - \frac{df''}{dt} \right) + Q\left( v - \frac{dg''}{dt} \right) + R\left( w - \frac{dh''}{dt} \right) \right\} d\tau.$$

Hence we must have

$$\frac{dT}{dt} = \frac{dE}{dt} - \int (Pu + Qv + Rw)\, d\tau,$$

in which $dE/dt$ is a surface integral; and this equation will give an *à priori* indication, independent of special hypothesis, of the distribution of organized kinetic energy in the medium, that of the potential energy and the dissipation being supposed known. Substituting from the kinematic relation $4\pi u = d\gamma/dy - d\beta/dz$, and integrating by parts, *à will indicate the location of the kinetic part. The Faraday circuital relation*

$$\frac{dT}{dt} = \frac{dE}{dt} - (4\pi)^{-1} \int \{l\,(\beta R - \gamma Q) + m\,(\gamma P - \alpha R) + n\,(\alpha Q - \beta P)\}\, dS$$

$$+ (4\pi)^{-1} \int \left( \alpha\frac{da}{dt} + \beta\frac{db}{dt} + \gamma\frac{dc}{dt} \right) d\tau.$$

This equation of energy can however only apply to the case in which the energy of magnetic, as well as of electric, polarization is completely organized, and not mixed up with other molecular energy of the material, as it would be if there were hysteresis or permanent magnetism*. When this condition is satisfied, the negation of perpetual motion requires that $\alpha da + \beta db + \gamma dc$ shall be an exact differential, say $d\psi$: thus we may tentatively assume $T = (4\pi)^{-1}\int\psi\, d\tau$, when the surface integral will remain as the value of $dE/dt$. In the

* The energy of the magnets is compensated locally, not transmitted: see *Roy. Soc. Proc.* 1903, as *infra*, p. 232.

points to location of the electro- dynamic energy and to Poynting's energy flow: case usually considered, in which the law of induced magnetization is linear, this gives Maxwell's formula for the distribution of the energy,

$$T = (8\pi)^{-1} \int (a\alpha + b\beta + c\gamma) \, d\tau;$$

while the value of $dE/dt$, now the surface integral, expresses Poynting's law of flux of electric energy corresponding to that hypothesis*.

70. That this law of distribution of electrokinetic aethereal energy, for a magnetic medium of constant permeability, falls in with the present scheme may be verified as follows. Let $(\alpha', \beta', \gamma')$ be proportional to the velocity of the irrotational flow of the aether, due in dissected for a magnetic medium. part when there is magnetism to the Amperean aethereal vortices, in such wise that the total kinetic energy is $(8\pi)^{-1} \int (\alpha'^2 + \beta'^2 + \gamma'^2) \, d\tau$; this is equal to $(8\pi)^{-1} \{ \int V dV/dn \, dS - \Sigma \mathfrak{k} \int dV/dn \, d\sigma \}$, where $\delta\sigma$ is an element of a barrier surface closing a magnetic vortex of strength $\mathfrak{k}$, and $\delta S$ is an element of the outer boundary of the region under consideration. As $T$ is to include only the organized energy, it is given by this expression when in it $V$ is restricted to be the potential of the magnetic force as ordinarily defined. In that case for an element of volume $\delta\tau$,

$$\Sigma \int \mathfrak{k} \frac{dV}{dn} \, d\sigma = \Sigma \left( \mathfrak{k}l \frac{dV}{dx} + \mathfrak{k}m \frac{dV}{dy} + \mathfrak{k}n \frac{dV}{dz} \right) \sigma$$

$$= -4\pi \int (A\alpha + B\beta + C\gamma) \, d\tau;$$

and therefore

$$T = (8\pi)^{-1} \int (\alpha^2 + \beta^2 + \gamma^2) \, d\tau + \tfrac{1}{2} \int (A\alpha + B\beta + C\gamma) \, d\tau$$

$$= (8\pi)^{-1} \int (a\alpha + b\beta + c\gamma) \, d\tau.$$

But although this expression locates the energy correctly as regards distribution throughout space, it still ignores the essential distinction between the energy of the translatory motions of electrons which constitute the current and that of their orbital motions which involve the magnetism; in a complete and consistent [dynamical] theory these two parts must be kept separate; cf. footnote, § 38, *supra*.

*On the Nature of Paramagnetism and Diamagnetism, as indicated by their Temperature Relations.*

71. As a result of an extensive investigation of the magnetic pro- Curie's law of magnetiza- tion: perties of matter, the law has recently been formulated by Curie[1] that in all feebly paramagnetic substances, including gases, the

---

* The form of $dE/dt$ may be varied, as has been done, by addition of any vector whose surface integral vanishes over a region not containing sources, provided such vector is isotropic and physically invariant. Its effect would be an alteration of the formula for radiation from each source, usually however introducing infinities.    [1] P. Curie, *Annales de Chimie*, 1895.

coefficient of magnetization varies inversely as the absolute temperature, with a degree of accuracy which tends to perfection at high temperatures: that in strongly magnetic substances such as iron, nickel, and magnetite, the same law is ultimately reached when the temperature is sufficiently high: while in diamagnetic substances the coefficient is usually nearly independent of temperature and also of changes in the chemical state of the material. The inference is made by Curie that this points to diamagnetism being an affair of the internal constitution of the molecule, having only slight relation to the bodily motions of the molecules on which temperature depends; which is in accordance with the modified Weberian view necessitated by the present theory.  On the other hand, paramagnetization is an affair of orientation of the molecules in space without change of internal conformation, so that alteration of the mean state of translational motion is involved in it, and we should expect a temperature effect.  A striking and probably just analogy is drawn by Curie between (i) the simple law of expansion of a gaseous substance at high temperature, and the sudden change which it undergoes on lowering the temperature beyond a critical point so that the mutual attractions of the molecules come into play and produce the liquid state, and (ii) the simple law of magnetization of a substance like iron or nickel at high temperatures, and the sudden change which it undergoes when the temperature is lowered beyond the point at which the material passes into its strongly magnetic or ferromagnetic condition. The relation between paramagnetization and temperature in the former state proves to be so simple and universal that it must be the expression of a theoretical principle. The following considerations in fact derive it from Carnot's principle: the argument is precise so long as the induced magnetization is so slight that the exciting magnetic force on the separate molecules is practically that of the inducing field, but it loses exactness as soon as, owing to diminution of energy of agitation with falling temperature, the molecules begin to exercise sensible magnetic control over each other, and thus introduce the phenomena of grouping and consequent hysteresis that are associated with the ferromagnetic state.

*his interpretation:*

*the law directly derivable from Carnot's principle.*

72. Consider a mass of paramagnetic material, moved up from a place where the intensity of the magnetic field vanishes to a place where it is $H$. The aggregate per unit volume of the total magnetic energies of its molecules is thereby altered from null to $-IH$ or $-\kappa H^2$. The mechanical work done by the mass in virtue of its attraction by the field is $\frac{1}{2}IH$, for the magnetization is at each stage of its progress proportional to the inducing force. Thus there remains a loss in the total magnetic energy of the molecules, equal to $\frac{1}{2}IH$; this

*Provided paramagnetic orientation interacts with heat:*

can only have passed into heat in the material*; for we can work on the hypothesis that the field of force $H$ is due to an absolutely permanent magnetic system, so that no energy is used up in producing magnetic displacements in the inducing magnets. Now let us apply Carnot's principle to a reversible cycle in which the material is moved up into the field at temperature $T + \delta T$ and removed at temperature $T$, with adiabatic transition between these temperatures. Let $h + \delta h$ be the thermal energy per unit volume which it must receive from without at the higher temperature, and $h$ that which it must return at the lower, in order to perform the amount of work $\delta W$, equal to $\frac{1}{2}H^2 d\kappa/dT \,.\, \delta T$, in the cycle; then, by Carnot's principle, $\delta W/\delta T = h/T$; and $h = -\frac{1}{2}\kappa H^2$ as above; so that $d\kappa/dT = -\kappa/T$, leading to

involving<br>that para-<br>magnetism<br>is molecular<br>orientation. $\kappa = A/T$ which is Curie's law. Conversely, assuming Curie's law we can deduce that in paramagnetic bodies magnetization consists in orientation of the molecules without sensible change in their internal energies. In an analytical form the argument will then run as follows:

$$dh = M dI + N dT, \quad dE = dh - \kappa^{-1} I dI\dagger;$$

whence by the thermodynamic formula

$$\frac{M}{T} = \frac{d}{dT}\,\kappa^{-1}I, \text{ so that } \frac{M}{I} = T\frac{d\kappa^{-1}}{dT}, = \kappa^{-1}$$

by Curie's law; hence $dh = H dI + N dT$, so that at constant temperature $h = \frac{1}{2}HI$, that is, the heat that the material develops during magnetization is the equivalent of the magnetic energy that is not used up in mechanical work. This is precisely what we should expect if the material is a gas; for there is then no internal work by which this energy could be used up, and the magnetization arises from the effort of the magnetic field to orientate the molecules which are spinning about as the result of the gaseous encounters‡. The law of

---

* When the temperature is kept constant as *infra*, this heat must leave the system. These equations imply, as stated, that at constant temperature $I$ varies as $H$: thus if the law of Curie here deduced fails, this could not be true.

† This equation should be $dE = dh - I dH$: for the loss $d\,(IH)$ of magnetic energy of the system is made up of $I dH$ abstracted as mechanical work and $H dI$ lost as heat. But here we assume that this is the only loss of heat. Thus $d\,(E - IH) = dh + \kappa^{-1} I dI$, which is an exact differential instead of the $dh - \kappa^{-1} I dI$ of the text: also so is the entropy $dh/T$. This requires change of sign of every term involving $\kappa$ or $H$ in what follows, and finally,

$$dH = -I T dI \,.\, d\kappa^{-1}/dT$$

at constant temperature, which agrees with the value $-H dI$ just deduced directly from a cycle only when the law of Curie $\kappa T$ constant holds. If that law fails in reversible circumstances, the total loss of energy is not solely the direct magnetic loss $d\,(IH)$. In any case there seems to be no justification other than statistical in gases for the formula that $I$ must be always a function of $H/T$. See Appendix at end of this volume.

‡ Hence the theory of Langevin (*Annales de Chimie*, 1905), which develops this thesis on the lines of the law of distribution in gas theory. For a closer scrutiny of that procedure, see Appendix at end of this volume.

Curie thus indicates that the same is sensibly true for all paramagnetic media at high temperatures: at lower temperatures they gradually pass into the ferromagnetic condition. It is the magnetization, so to speak, of an ideal perfect ferromagnetic, in which the controlling force that resists the orientating action of the field is practically wholly derived from the magnetic interaction of the neighbouring molecules*, which for this purpose form elastic systems, that is illustrated by Ewing's well-known model, which so clearly represents the hysteresis accompanying ferromagnetic excitation. In ordinary paramagnetic substances this mutual magnetic control of the molecules is insensible compared with the control due to other molecular causes; and our conclusion is that these causes are such that the magnetic energy expended in working against them is transformed into heat energy, not into internal energy of any regular elastic type. *Ewing's model of ideal ferromagnetic.*

But we have not taken account of the fact that the molecules of every substance are subject to both paramagnetic and diamagnetic influence, of which one or the other preponderates. The theoretical law† should thus be $\kappa = - B + A T^{-1}$ or $\kappa T = A - BT$; so that in a diagram of the relation between $\kappa T$ and $T$ each substance would be represented by a straight line. In paramagnetics the line should slope slightly down towards the axis: for diamagnetics it should pass not through the zero of temperature but on the positive side of it. According to Curie, his experimental results are equally well represented by this formula, on account of the preponderant influence of the paramagnetism. *Superposed diamagnetism internal to the atom.* *Curie.*

Similarly, should it turn out that for weakly electric media such as gases, $(K - 1)/\rho$ is independent of the temperature, it would follow that the electric polarization is mainly an affair of change of internal constitution of the molecules: while were the polarization mainly an affair of molecular orientation, $(K - 1)/\rho$ would vary inversely as the absolute temperature: in intermediate cases it would not vary so rapidly as this. The circumstance that in gaseous media and some others, the dielectric constant is exactly equal to the square of the refractive index, favours the former alternative (§ 21)‡. *Nature of electric polarization in gases.*

* This control by adjacent molecules, expressing the local field of Weiss, involves a mathematically definite theory when the magnetic molecules, polarizable whether by rotation or by internal change, are constrained to form a crystalline lattice: cf. Mahajani, *Proc. Camb. Phil. Soc.* 1926.

† The form now regarded as best adapted to express the experimental results is of a different type, $\kappa = A/(T - T_0)$, involving a critical temperature $T_0$, that, of ferromagnetic change, usually far down on the scale.

‡ This trend of ideas has been much developed in recent years from a formula of Debye for gases (*Physikal. Z.* 1912: also J. J. Thomson, *Phil. Mag.* 1914) which employs the very slightly more exact Maxwell-Lorentz invariant $(K - 1)/(K + 2)\rho$, and makes it equal to $b + a/T$, where $a$ represents the effect of permanent polar doublets in the atom and $b$ the purely induced (quasi-

*Mechanical Relations of Radiation reconsidered.*

73. The results given in Part II, §§ 28–29, as to the mechanical forcive exerted on a material medium by a stream of radiation passing across it, require amendment in the light of these principles, of which they also form an apt illustration. Consider, as there, two media separated by the plane of $yz$, and a system of plane-polarized waves advancing across them, with their fronts parallel to that plane, the electric vibration parallel to the axis of $y$, and the magnetic one parallel to the axis of $z$; we may generalize by taking $K$ and $\mu$ to be in each medium functions of $x$. The electrical equations are

$$4\pi v = -\frac{dy}{dx}, \quad \frac{dQ}{dx} = -\frac{dc}{dt}, \quad v = \sigma Q + \frac{K}{4\pi} \mathrm{c}^{-2} \frac{dQ}{dt}.$$

Applying the formulae found above (§ 38), the force acting on the electric polarity comes out to be null, while the electromagnetic bodily force is, per unit volume, $X = v'y$, being wholly parallel to the axis of $x$: its periodic part has double the frequency of the radiation.

Limitation on mechanical transmission. Now there is no mechanical elasticity associated with matter which is powerful enough to transmit in any degree the alternating phases of forcives connected with a phenomenon which travels so fast and with such short wave-length as radiation, long Hertzian waves being excluded. Consequently when $X$ is wholly alternating it is not transmitted by material stress at all; and it is only when its value for each element of the medium contains a non-alternating part that we can have a material forcive. When the media are perfectly transparent, and are traversed by a steady train of waves, there is therefore no transmissible *material* forcive either on surfaces of transition or anywhere else, and the $\int X dx$ previously calculated has no relation to material stress. But if we consider a stream of radiation passing across a transparent medium into an opaque one, and for simplicity take the latter to be homogeneous so that for the transmitted waves

$c = c_0 e^{-px} \cos{(nt - qx)}$, the expression for $X$, viz. $\left(v - \frac{1}{4\pi} \mathrm{c}^{-2} \frac{dQ}{dt}\right) y$,

contains a non-periodic term $\cdot \left(1 - \frac{\mu n^2 \mathrm{c}^{-2}}{p^2 + q^2}\right) \frac{p c_0^2}{8\pi \mu^2} e^{-2px}$. This when integrated over the medium gives a pressure [originating in] the opaque medium, of intensity $\left(1 - \frac{\mu n^2 \mathrm{c}^{-2}}{p^2 + q^2}\right) \frac{cqE}{2\mu n}$, where $E$ is the energy per unit volume of the incident radiation absorbed. Unless the opacity is so great that the intensity of the light is diminished in the

diamagnetic) part. "The permanent moments of many molecules have now been calculated. In most cases it is found that the moments of substances usually classed as non-polar are vanishingly small, while those of polar substances are comparatively large": quoted from a valuable report by L. F. Gilbert, *Science Progress*, 1926, p. 65.

ratio $e^{-1}$ in penetrating a few wave-lengths, that is when $p$ is negligible compared with $q$,[1] this pressure will be practically $(1 - \mu m^{-2})\, mE/2\mu$, where $m$ is the real part of the index of refraction of the medium, as measured by the ratio of the velocities in deviation experiments with prisms (p. 64), and $\mu$ is practically unity. And in general it appears that it is only absorption, not reflexion*, of radiation that is accompanied by a mechanical forcive, the force [originating in] any absorbing mass being $(1 - \mu m^{-2})\, mE'/2\mu c$ in the direction of the radiation traversing it, where $E'$ is the total radiant energy absorbed by the mass per unit time.

In the case of a transparent medium traversed by two systems of waves, direct and reflected, forming stationary undulations, the mechanical force is proportional to $\sin 2qx$, and vanishes at both the nodes and antinodes of the electric vibration.

74. The *rationale* of [this] mechanical forcive is that, owing to the absorption of energy, which can only occur when phase differences exist, a difference of phase becomes established between the two factors of $X$, the electric current and the magnetic field, so that their product contains a non-alternating part. It is known that vapours **Repulsion of comets' tails.** of complex chemical constitution are very powerful absorbers of radiation; and in this case (if not in all cases) the absorption must be a property of the single molecule. By the argument just stated, there must then be difference of phase between the electric flux (displacement of electrons) in the molecule and the magnetic field†, and this will give a mechanical forcive driving the molecule along the path of the radiation. As the tails of comets and the Solar Corona consist of very rare distributions of vaporous or other matter, in free space which exerts no retarding influence, a comparatively small absolute amount of absorption‡ by them of the Solar radiation might account for their observed repulsion from the Sun; in this way a definite and actually existing physical agency may be made to take the place of vague electrical repulsion in Bredichin's important analysis of cometary appendages.

[1] This will not usually be the case for metallic media.

The sign of this mechanical force may be negative in a region of intense absorption.

\* This is misleading. The force originating in the matter of the reflecting **Repulsion is due to momentum of waves.** layer integrates to nothing on account of its vanishing breadth: but that is not all that happens, for the configuration of the radiation becomes changed. The beam must be regarded as carrying momentum which it partly yields up to the reflector. See footnote, *loc. cit.* Part II, Vol. I, *supra*, p. 586, also Appendix V: on the nature of this momentum see *Math. Congress Lecture*, 1912, as *infra*.

† Such as could only arise from the secondary radiation of the molecule.

‡ Or reflexion: as in previous footnote. See FitzGerald, *Scientific Papers* (1882), p. 108.

This mechanical action of waves on absorbing systems placed in their path may be roughly illustrated by an arrangement in which a system of sound waves traverses a space filled with resonators approxi- mately in unison with them. The open mouth of each resonator is repelled[1], so that in case there is any regularity in their orientation, the system as a whole will be subject to mechanical force. The resonators might be suspended so that the mechanical forces may themselves produce this orientation, and thus form a sort of medium polarizable by waves. A corresponding electric illustration is the action of long Hertzian waves in orientating and repelling mobile conducting circuits which lie in their path. The very considerable repulsion of the vanes in the radiometer arises of course from a mutual stress between the vanes and walls and the rarefied gas, and so has a null resultant as regards the system as a whole.

*Illustration from sound waves:*

*from long electric waves.*

## Stresses and Deformations in Electric Condensers.

75. The elastic deformation produced in the dielectric of a spherical condenser by the mechanical forcive may be readily calculated. If $u$ denote the radial displacement, the normal and transverse principal tractions at any point in the spherical dielectric shell are

*Mechanical stress in an excited spherical condenser.*

$$P = \lambda \left(\frac{du}{dr} + 2\frac{u}{r}\right) + 2\mu \frac{du}{dr}, \quad Q = \lambda \left(\frac{du}{dr} + 2\frac{u}{r}\right) + 2\mu \frac{u}{r},$$

where $\mu$, $\lambda + \frac{2}{3}\mu$ are the moduli of rigidity and compressibility of the material. The electric force at any point is $kr^{-2}$, where $k$ is the charge on a coating: hence (§ 36) the mechanical bodily force is derived from the potential $- (K - 1)/8\pi \cdot k^2 r^{-4}$; and there is also an outward normal traction over each coating equal to $- K/8\pi \cdot k^2 r^{-4}$. The equation of equilibrium of a conical element of volume is

$$\frac{d}{dr}(Pr^2) - 2Qr = \frac{K - 1}{2\pi} \frac{k^2}{r^3},$$

so that

$$\frac{d}{dr}\left(r^2 \frac{du}{dr}\right) - 2u = \frac{K - 1}{2\pi(\lambda + 2\mu)} \frac{k^2}{r^3},$$

giving

$$u = Ar + \frac{B}{r^2} + \frac{K - 1}{\lambda + 2\mu} \frac{k^2}{8\pi r^3},$$

and therefore

$$P = (3\lambda + 2\mu)A - \frac{4\mu}{r^3}B - \frac{\lambda + 6\mu}{\lambda + 2\mu}\frac{(K - 1)k^2}{8\pi r^4},$$

$$Q = (3\lambda + 2\mu)A + \frac{2\mu}{r^3}B - \frac{\lambda - 2\mu}{\lambda + 2\mu}\frac{(K - 1)k^2}{8\pi r^4}.$$

[1] Cf. Lord Rayleigh, *Theory of Sound*, Vol. II, §§ 255 a, 319. [The stress maintaining the waves is $\delta p$, the change of pressure, but this repulsion varies as $\delta p^2$, in analogy with the electric case.]

The values of $A$ and $B$ are determined by the normal tractions at the coatings $r = r_0$ and $r = r_1$, and are, when the coatings are wholly supported by the dielectric,

$$A = \frac{1}{r_1 r_2 \,(r_1{}^2 + r_1 r_2 + r_2{}^2)} \frac{\vartheta k^2}{8\pi \,(3\lambda + 2\mu)},$$

$$B = \frac{r_1{}^3 + r_1{}^2 r_2 + r_1 r_2{}^2 + r_2{}^3}{r_1 r_2 \,(r_1{}^2 + r_1 r_2 + r_2{}^2)} \frac{\vartheta k^2}{32\pi\mu},$$

$$\vartheta = 1 - \frac{4\mu}{\lambda + 2\mu}\,(K - 1).$$

It will suffice to state the results for the case of a thin shell of radius $a$ with adhering coatings: then *(Case of thin dielectric.)*

$$A = \frac{\vartheta k^2}{24\pi \,(3\lambda + 2\mu)\, a^4}, \qquad B = \frac{\vartheta k^2}{24\pi\mu a}.$$

The coefficient of expansion of the radius of the sphere, due to the electric stress, is

$$\frac{u}{a} = \frac{\lambda + 2\mu - \mu K}{8\pi\mu \,(3\lambda + 2\mu)}\, F^2,$$

and the coefficient of expansion of the volume of the sphere is three times this. It is easily verified that when the shell is thin the stress in the material of the dielectric is made up of a pressure $KF^2/8\pi$ normal to the shell combined with a pressure $(K - 2)\, F^2/8\pi$ in all directions tangential to it[1].

(76[2]. The circumstance that these results are independent of the radius of the sphere suggests an extension of their scope. Whatever *(Thin dielectric shell of any form.)* be the form of the dielectric shell provided it is of uniform thickness, $F$ will be the same all over it; and the mechanical force acting on its substance, being derived from a potential $- (K - 1)\, F^2/8\pi$, will be directed at each point along the normal $\delta n$ to the shell. Consider the internal equilibrium of an element of volume $\delta S\, \delta n$, of which the opposite faces $\delta S$ are elements of level surfaces bounded by lines of curvature for which $R_1$, $R_2$ are the principal radii: it will be maintained if a pressure of intensity $P$, equal to $KF^2/8\pi$, act on the element across the faces $\delta S$, and another pressure $- Q$ act on it, which is the same across all perpendicular faces. For, resolving the forces along $\delta n$, we must have for equilibrium

$$- \delta n \frac{d}{dn}\,(P\delta S) - Q\, \delta n\, \delta S \left(\frac{1}{R_1} + \frac{1}{R_2}\right) + \frac{K - 1}{4\pi}\, F \frac{dF}{dn}\, \delta S\, \delta n = 0.$$

---

[1] This naturally differs from Kirchhoff's result, Wied. *Ann.* xxiv, p. 52, § 4. [Experimental data are now plentiful: see also § 79, *infra*.]

[2] Rewritten December 2, 1897.

Now, by the constancy of the induction, we have $d/dn\,(F\delta S) = 0$, leading to

$$\frac{dF}{dn} = -F\left(\frac{1}{R_1} + \frac{1}{R_2}\right) \text{ and also } \frac{d}{dn}\,(P\delta S) = \frac{K}{8\pi}\,\delta S F \frac{dF}{dn}:$$

Same result if thickness uniform: thus, on substitution, we obtain $-Q = (K-2)\,F^2/8\pi$. The constancy of $Q$ all round the edge of a flat element of volume $\delta S\,\delta n$ secures the balancing of the tangential components of the forces. Hence the mechanical stress in any condenser sheet of uniform thickness is the same as has been found above for the spherical case. If $e$ and $f$ denote the elongations of the material in the normal and tangential directions,

the strain produced.
$$\lambda\,(e+2f) + 2\mu e = -\frac{KF^2}{8\pi}, \quad \lambda\,(e+2f) + 2\mu f = -(K-2)\,\frac{F^2}{8\pi};$$

hence
$$f = \frac{2 + \lambda/\mu - K}{3\lambda + 2\mu}\,\frac{F^2}{8\pi}, \quad e = \frac{2\lambda/\mu - K}{3\lambda + 2\mu}\,\frac{F^2}{8\pi}:$$

so that the extension of the volume of the shell and the change in its thickness are the same as were found above for the spherical case. Influence of edge is local. If the shell is an open one, the presence of its free edge will disturb these relations: but that influence will be mainly local, as the forces introduced by the edge will be almost wholly of the nature of local action and reaction.)

## *Various Practical Illustrations and Applications of the Stress Theory.*

### *Refraction of a Uniform Field of Electric Force.*

Experi-mental field of straight refracted lines of force: 77. An arrangement by which these principles may be precisely verified is that of the refraction, at a plane interface $AB$, of a sheaf of parallel lines of electric force $F$, according to the Faraday-Maxwell law of tangents, $\tan \iota_1/\tan \iota_2 = K_1/K_2$, $F_1/F_2 = \operatorname{cosec} \iota_1/\operatorname{cosec} \iota_2$. This configuration of lines of force may be obtained and fixed by means of a condensing system having its plates $P_1Q_1$ and $P_2Q_2$ normal to the precaution: incident and refracted lines. Each plate may be protected from convective discharge into the fluid dielectric by a covering plate of glass or mica, which will itself produce no refraction. In such a case, when both the dielectric media are fluid, the total mechanical result of the electric excitation will be the same as that of normal tractions on the mechanical stress: the interface between them, of intensity $-2\pi n'^2 - \frac{1}{2}i'F$ towards each side, that is, in all $(K_2 - K_1)\,(2\pi n''^2/K_1K_2 - T^2/8\pi)$ towards the side 1. As the field of force in this condensing system is uniform except near its edge, the interface will simply be lifted up between the plates by the amount which corresponds to this traction, without ceasing to be horizontal. Thus the common surface will be elevated when $\tan \iota_1$ is less than $(K_1/K_2)^{\frac{1}{2}}$, while at greater incidences it will be depressed.

This principle supplies in fact a method of obtaining the inductive capacities of fluid media by angular measurement only, without the aid of an electrometer. When the condenser $P_1Q_1$, $P_2Q_2$ is charged, the interface between the fluids will usually cease to be horizontal; the upper plate $P_1Q_1$ is then to be rotated until horizontality is again obtained, as may be tested very exactly by reflexion of a beam of light; then the ratio of the tangents of the inclinations of the plates will be that of the inductive capacities of the media. The method would also apply to solids, as we might employ a prism of the material, over a horizontal face of which a sheet of a fluid dielectric is spread, and observe the deviations from horizontality of the upper surface of this sheet. An equivalent arrangement has been actually employed for solids by Pérot[1], who however adjusted the plates to uniformity of the electric field by the electric test that translation of a small piece of solid dielectric in the field between them should not affect the capacity of the condenser.

gives a method for determining dielectric constants.

### *Experiments on Electric Traction and Change of Pressure in Fluids.*

78. The direct experimental examination of the material forcives of polarized media is necessarily confined to fluids, for in the case of solids the strains produced by them could hardly be disentangled from the intrinsic changes of configuration due directly to the polarization. The field for fluids has been very fully explored by Quincke[2]. The inductive capacity of the fluid dielectric of a horizontal condenser was first determined by direct electrical measurement. The attraction between the plates was then weighed. Then, using a wide cylindrical air bubble extending across the space between the plates, and connected through an aperture in the upper one with a manometer, the increase of air pressure in the bubble produced by charging the condenser was measured. As half-way between the plates the capillary interface between air and liquid lies along the lines of force, there is by the previous formulae (§ 37) no true surface traction on that part of the interface; so that the indication of the manometer would give exactly the change of pressure in the liquid due to the electric

Stress in fluid dielectrics as explored: Quincke's methods:

[1] Pérot, *Comptes Rendus*, 1891; quoted by Drude, *Physik des Aethers*, p. 299. The conjugate condensing system, in which namely the lines of force and the lines of equal potential are interchanged, so that the plates are now bent according to the law of tangents where they cross the surface of the liquid, has recently been brought into requisition by Pellat (*Annales de Chimie*, 1895), in order to derive the law of the traction on a dielectric interface from the expression for the energy of the dielectric system. As in these cases the field of force is uniform in both media, the bodily part of the mechanical force vanishes, and the interfacial traction proper thus constitutes the whole forcive: but that would not generally be so.

The conjugate field with bent condenser plates.

[2] G. Quincke, Wied. *Ann.* XIX, 1883; or an abstract in the paper below cited, *Roy. Soc. Proc.* Vol. LII, 1892, pp. 59–62 [Vol. I, *supra*, p. 281].

excitation, were it not that the different electric conditions over other parts of the interface change the value of its curvature and so introduce a capillary change of pressure. Finally, employing a flat bubble of air resting against the upper plate alone, and maintaining the pressure in it constant, the change of curvature of its lowest part produced by the electrical excitation was measured by the optical method; the surface tension operating through this change of curvature must balance exactly the direct traction on the surface and the change of pressure in the liquid below it. To compute these, we notice that the line of force through the middle of the bubble is straight, so that if $F$ denote the electric force in air and $F/K$ that in the liquid, the traction on the surface due to these two causes is $(K-1)^2 F^2/8\pi K^2 + (K-1) F^2/8\pi K^2$, that is $(K-1) F^2/8\pi K$

agreement with theory, upwards, which is the formula employed by Quincke; while $F$ is determined by the relation $aF + bF/K = V$, provided the bubble is so broad that the middle tube of induction is practically cylindrical, $a$ and $b$ being the lengths of this tube that are in air and the liquid, and $V$ the difference of potential between the plates. After an error in the direct determination of $K$, due to an experimental oversight

after correction. whose existence was suggested by Hopkinson, had been corrected, all the results showed substantial agreement, for thirteen liquids that were examined, with these theoretical formulae. But the agreement was not quite complete; a subsequent examination[1] still showed that the attraction between the plates always came out less and the change of pressure in the liquids greater than the formulae would give, though these discrepancies were within the limits of experimental error, except for the case of rape oil in which they amounted

Further corrections. to as much as ten per cent. The neglect of the capillary correction above mentioned would account for a discrepancy in the same direction as the first; and the irregularity of the electric distribution near the edge of the plates would account for one in the same direction as the second.

In a paper on the bearing of the phenomena of electric stress on
Inter-molecular stress excluded. electrodynamic theory[2], I had previously been led to inferences militating against the possibility of dielectric polarization being of molecular type, from a comparison of these experimental results with an electric traction formula including both the molar and the molecular forcives. According to the present argument (§ 44), the latter forcive being separately compensated, the difficulty there
Mechanical energy theory. encountered does not exist. The remaining considerations in that paper retain their validity; they show for instance that the formulae for the experimental reductions can be derived from a knowledge of

[1] G. Quincke, Wied. *Ann.* xxxii, 1887, p. 537.
[2] *Roy. Soc. Proc.* Vol. lii, 1892, pp. 65–66 [Vol. i, *supra*, p. 279].

the distribution of the organized energy alone. But in the light of the present views we are no longer restricted or even allowed to consider the induction in a dielectric as all of one kind; the total circuital induction is in fact made up of a material polarity combined with an aethereal elastic displacement, giving an apparent but natural complexity which it had previously been an aim to evade.

### *Experiments on Electric Expansion in Solids and Fluids.*

79.  The results obtained in § 76 may be applied to the discussion of a very thorough series of experiments on electric expansion, made by Quincke[1], which appear hitherto not to have been correctly interpreted.  Following the early experiments of Fontana, and more recent ones by Govi and Duter, a condenser of the form of a glass thermometer bulb was used, and the expansion of volume arising from electric excitation was read directly on its tube.  By employing a long cylindrical bulb, the longitudinal expansion of the glass could also be measured microscopically at the same time.  It was found by Quincke that the coefficient of volume expansion was always about three times that of this longitudinal expansion of the glass dielectric, just as the above theory [p. 121] indicates for the elastic strain in a bulb of uniform thickness.  The erroneous deduction was however made from an imperfect theory, that electric expansion of solids is uniform in *all* directions, like expansion by heat, and so in no part due to mechanical forces of attraction. *Strains in spherical and cylindric dielectrics of glass:*

By using the formula of § 76 along with various known physical constants, a test of the order of magnitude of Quincke's determinations may be obtained.  Thus with a striking distance of ·4 centim. between brass balls 2 centims. in diameter, the expansion in volume of a flint glass condenser of thickness ·06 centim. was found[2] to be $\frac{3}{4} \times 10^{-6}$.  According to Baille's experiments[3] this striking distance corresponds to a difference of potential of 47 c.g.s. The expansion of volume of the bulb due to the mechanical force is

$$\frac{2 - K + \lambda/\mu}{\lambda + \tfrac{2}{3}\mu}\ \frac{F^2}{8\pi},$$

where, according to Everett's experiments[4] for flint glass, *require an intrinsic expansion,*

$$\mu = 25 \times 10^{10}\ \text{c.g.s. and}\ \lambda = \mu,$$

and according to Hopkinson $K$ is about 7. This gives for the thickness under consideration a coefficient of expansion equal to

[1] Abstracted in *Sitz. Akad. Berlin*, February 1880, and *Phil. Mag.* July 1880, pp. 30–39: in full in Wied. *Ann.* x, 1880, pp. 161–202, and 513–553.

[2] *Loc. cit.* Wied. *Ann.* x, p. 190.

[3] *Annales de Chimie*, 1882: quoted in J. J. Thomson's *Recent Researches*, p. 87.

[4] *Phil. Trans.* 1868, p. 369.

— $0.24 \times 10^{-6}$, while the observed value[1] was $0.75 \times 10^{-6}$. The difference between them, in so far as it does not arise from experimental uncertainties, is an intrinsic superficial expansion of the glass, which arises directly from the transverse polarization itself,

due to constitutive change. and is not due to the mechanical forces caused by it. That there is such intrinsic alteration due to electric excitation, is independently suggested by Quincke's observation that the values of the elastic constants of the material are slightly altered by that cause.

In the case of a fluid the effect of electric excitation is to [augment*]

Results for fluids. the hydrostatic pressure; consequently [contraction] should result when the plates of the condenser are fixed. [Expansion is what occurs] for most fluids: but the fatty oils form an exception. Thus for most fluids there is an intrinsic electric expansion superposed on the mechanical contraction in the electric field. [See section on Electrostriction, *Roy. Soc. Proc.* May 1898, *infra*, p. 156.]

In both these cases the intrinsic change of volume is of order of magnitude not higher than the change due to the mechanical stress.

Glass-tube condenser becomes bent: The observation of Quincke[2] that a thin glass-tube condenser, with walls thicker on one side than the other, becomes curved (in accordance with the theory above) when it is electrically charged, virtually affords a convenient method for studying the gradual rise of the charge of the condenser and its residual discharge. With such a "glass thread electrometer" it appears that the curvature takes place gradually on excitation, occupying for small charges sometimes as long as 30 seconds: and on discharge it is annulled with corre-

by creeping motion. sponding slowness. As part of this deformation is intrinsic, that is, due to molecular forces and not to the mechanical stress, it is a direct indication of gradual shaking down of the material into modified molecular groupings under the influence of the electric field.

### *Influence of Polarization of a Fluid on Surface Ripples.*

80. The last illustrations belong to cases in which the field of force

Experimental ripples on fluid: is uniform, so that the bodily mechanical forcive vanishes. A problem amenable to experimental examination, in which this is not the case, is the influence of electric polarization on ripple motion in fluids. The fluid, in a glass dish, might for example form part of the dielectric of a horizontal condenser, of which the upper coating is a wire

---

[1] According to Quincke's own determinations (*loc. cit.* p. 187) the value of $K$ would be about 12, which is a great deal higher than Hopkinson's results, and would give an expansion — $0.30 \times 10^{-6}$. It seems possible that these determinations, which involve considerable unitary complexity, may be wrongly recorded, as Quincke's reduction of them sometimes appears to give values for $K$ that are less than unity.                                    [2] *Loc. cit.* p. 394.

* The intensity of force on the medium being $- F \operatorname{grad} D'$, which is the gradient of a potential $- (K - 1) F^2/8\pi$ in electrostatic units.

grating separated from the fluid by a plate of glass or mica so as
to prevent communication of electrification to its surface.

Taking the axis of $y$ downwards into the fluid whose dielectric
constant is $K_2$, and the axis of $x$ along the interface, the electric
potentials in the two fluids, of which the upper will usually be air, are

$$V_1 = A_1 y + B_1 e^{mv} \cos mx, \quad V_2 = A_2 y + B_2 e^{-mv} \cos mx,$$

subject to the condition that at the interface

$$y = C \cos mx$$

we must have

$$V_1 = V_2, \quad K_1 \frac{dV_1}{dn} = K_2 \frac{dV_2}{dn}.$$

Thus $A_1 C + B_1 = A_2 C + B_2, \quad K_1 (A_1 + mB_1 \cos mx)$
$$= K_2 (A_2 - mB_2 \cos mx),$$

the latter involving both

$$K_1 A_1 = K_2 A_2, \quad K_1 B_1 = - K_2 B_2.$$

The velocity potentials of the wave-motions in the two fluids are

$$\phi_1 = m^{-1} \frac{dC}{dt} e^{mv} \cos mx, \quad \phi_2 = - m^{-1} \frac{dC}{dt} e^{-mv} \cos mx.$$

In addition to the hydrodynamical pressure difference acting from
the upper to the lower side of the interface, equal to

$$- g (\rho_2 - \rho_1) y + \left( \rho_2 \frac{d\phi_2}{dt} - \rho_1 \frac{d\phi_1}{dt} \right),$$

that is $\quad - \left\{ g (\rho_2 - \rho_1) C + m^{-1} (\rho_2 + \rho_1) \frac{d^2 C}{dt^2} \right\} \cos mx,$

there will act on it a downward capillary traction $T d^2 y / dx^2$, or
$- m^2 TC \cos mx$, and a downward electric normal traction[1]

$$(8\pi)^{-1} (K_1 - K_2) (K_1 K_2^{-1} N_1^2 - T_1^2),$$

in which $N_1 = A_1 + mB_1 \cos mx$ while $T_1$ is of the second order.
This electric traction is therefore equal to

$$\frac{K_1 (K_1 - K_2)}{8\pi K_2} (A_1^2 + 2mA_1 B_1 \cos mx),$$

where $\qquad\qquad B_1 = \frac{K_1 - K_2}{K_1 + K_2} C A_1;$

while the intensity of the total displacement, material and aethereal,
in the electric field is $i'' = - K_1 A_1 / 4\pi$. The balancing of these tractions
at the interface $y = C \cos mx$ requires that, in addition to the mean
statical elevation, we should have

$$(\rho_2 + \rho_1) \frac{n^2}{m^2} = (\rho_2 - \rho_1) \frac{g}{m} + mT - \frac{2 (K_2 - K_1)^2}{K_1 K_2 (K_2 + K_1)} 2\pi i''^2,$$

in which $n/m$ is the velocity of wave-propagation. The effect of the

---

[1] This is the statical equivalent as above (§ 37) of both the actual electric
traction on the surface, and the electric pressure transmitted from the interior
of the fluid to the surface.

influence of electric polarization, electric polarization is thus for ripples of length $\lambda\ (= 2\pi/m)$ the same as would be that of a diminution of the surface tension* by

$$\frac{2\,(K_2 - K_1)^2}{K_1 K_2\,(K_2 + K_1)}\,\lambda i''^2.$$

If the lower medium were conducting, we should have had $A_1 C + B_1 = 0$, and the electric downward traction would be

$$-\frac{K_1 N_1{}^2}{8\pi},\ \text{that is}\ -\frac{K_1 A_1{}^2}{8\pi} + \frac{m}{4\pi}\,A_1{}^2 C \cos mx.$$

of electric free charge. Thus in the equation giving as above the velocity of propagation, the electric term would be $-A_1{}^2/4\pi$, or $-4\pi\sigma^2/K_1{}^2$, where $\sigma$ is the density of the electrification on the interface. The effect of this electrification is thus the same as that of a diminution of the surface tension by $2\sigma^2\lambda/K_1{}^2$, where $\lambda$ is the wave-length[1].

### Relations of Electrification to Vapour Tension and Fluid Equilibrium.

81. It has already been shown (§ 52) that the possibility of mechanical equilibrium between fluid dielectrics which do not mix requires that the electric tractions on the interface shall be in the direction of the normal. There are also other dynamical relations Fluid phases in contact: deducible from the fact that such forcives, when integrated round a closed circuit in fluid media, must give a null result, in order to avoid argument from cycles. the establishment of cyclic perpetual motions. The earliest example which led the way to relations of this kind was Lord Kelvin's establishment of a connection between the vapour tension of a liquid and the curvature of its free surface: and similar balances must independently hold good between vapour tension and other causes of surface traction.

Consider in the first place a volume of conducting fluid with a large horizontal free surface. Let an electric field be established over a portion of this surface; there will be a surface density $\sigma$ of electrification induced over that portion, which will vary from point to point; while the electric forces will elevate the surface by an amount $h, = 2\pi\sigma^2/g\rho$, where $\rho$ is the density of the fluid, above the level at a distance where there is no electrification. The vapour tension over the electrified part must thus be smaller by $g\rho_0 h$ than over the unelectrified part, where $\rho_0$ is the density of the vapour. This difference of tension must be the natural steady difference produced by the electrification of the surface; for otherwise a process of distillation will set in and there could not be equilibrium, though there could

---

* The factor 2 is now inserted. See *Roy. Soc. Proc.* May 1898, as *infra*, p. 157, for the consequence as regards instability of the surface.

[1] This result was given twice too large in *Proc. Camb. Phil. Soc.* April 1890 [corrected *supra*, Vol. I, p. 202].

theoretically be perpetual generation of work while the temperature remains uniform, as the electric charge does not evaporate with the fluid. It follows that an electrification of surface density $\sigma$ must depress the equilibrium vapour tension by an amount[1] $2\pi\sigma^2\rho_0/\rho$.

Vapour pressure depressed by electric charge:

Suppose again that the fluid is a dielectric of inductive capacity $K$, and has no free charge. A similar train of reasoning shows, by the formula of § 37, that when the polarization of the material dielectric, at the surface, is made up of a normal component $n'$ and a tangential component $t'$, its vapour tension is thereby diminished by an amount $2\pi (Kn'^2 + t'^2)/(K - 1) . \rho_0/\rho$. Conversely, we can argue that, as the change of vapour tension can depend only on the state of polarization or electrification at the part of the surface which is under considera-tion, the effect of the electric excitation must be completely expres-sible by a mechanical traction over the surface which must be wholly normal and depend only on the intensity of the field of force at the place. That this is the case for fluids, but not for solids, has already been shown. And this law of dependence of vapour tension on electric state only applies to fluids, not to solids like ice; for a flow of the medium is required to complete the cycle on which the argument is based. In the case of a solid with finite vapour tension, electric excitation—as also gravity, strains, and other physical agencies— will promote evaporation, excessively slow of course, from some parts of its surface, and condensation on others, until a form suitable to equilibrium of vapour tension is attained.

by polarization.

Fluid surface adjusts so that surface traction is along normal:

different result for solids.

In expressing conditions of equilibrium for fluid media, the above total electric normal traction over each interface is simply to be added to such other forces as would exist in the material system if there were no electric field. Thus if we take for example the case of a number of dielectric fluids superposed on each other in a tall jar under the action of gravity, the form of the upper surface is obtained by equating the electric traction to the pressure difference produced by difference of level alone; and for any interior interface the same statement holds good, the form of each interface depending only on the electric field at the place and the inductive capacities of the two fluids which it separates. And this procedure is quite general what-ever extraneous forcives there may be; the form of each interface is always determined by equating the difference of electric tractions on its two sides to the difference of pressures due to other than electric causes.

Superposed polarized liquids.

[1] This agrees with a result given by Professor J. J. Thomson, *Applications of Dynamics to Physics and Chemistry*, 1888, § 86.

## Tractions on the Interfaces of a divided Magnetic Circuit.

Traction
across
magnetic
air gap:

82. An important practical deduction is that when a bar or ring, longitudinally magnetized temporarily or permanently, is divided by an air gap, the force drawing together the two halves of it consists of the attractions of the uncompensated polarities of volume which would remain if there were no air gap, together with a traction on each face of the gap, at right angles to its plane, and of intensity $2\pi\nu^2$, where $\nu$ is the normal component of the magnetization. This traction is in other respects quite independent of the character of the magnetic field that may exist at the gap; when the gap is narrow it is simply the attraction between the terminal poles on its two faces. When the gap is transverse to the magnetization, the total amount of the traction is thus $2\pi\int I^2 dS$, that is $(8\pi)^{-1}\int(B-H)^2 dS$, where $B$ and $H$ are the longitudinal components of the magnetic induction and force; when it is oblique the longitudinal pull between the halves of the bar varies as the square of the cosine of the obliquity. When the substance is magnetized by an electric coil, there may in addition be the attraction between the two halves of the coil. For the case of iron $H$ is very small compared with $B$, unless the field is far greater than is required to saturate the iron; so that the part of the mechanical

simple result
for iron.

traction across a transverse gap which is due to the polarities on its faces is practically $(8\pi)^{-1}\int B^2 dS$.

## Interaction of Mechanical Stress and Magnetization.

83. Consider a wire, magnetized to intensity $I$ by a longitudinal magnetic field $H$, and subject to an extraneous tensile force of intensity $Q$ per unit cross-section: let $M$ denote the modulus of elastic extension of its material, which will be an even function of $I$. The mechanical work expended on the wire in a slight alteration of its circumstances is per unit volume

Magnetized
wire under
changing
tension:

$$\delta W = (M^{-1}\delta Q + \delta\eta)\, Q + H\delta I$$

$$= \left(M^{-1}Q + Q\frac{d\eta}{dQ}\right)\delta Q + \left(H + Q\frac{d\eta}{dI}\right)\delta I,$$

where $\eta$ is its intrinsic magnetic elongation when magnetized to intensity $I$ under tension $Q$, this magnetization practically not altering the extraneous field in the case of a wire. To avoid perpetual motions,

energy cycle.

$\delta W$ must in the absence of hysteresis be a perfect differential of the independent variables $I$ and $Q$: hence

$$Q\frac{dM^{-1}}{dI} = \left(\frac{dH}{dQ}\right)_I + \frac{d\eta}{dI}.$$

Here $I$ is a function of $H$ and $Q$, so that to determine $dH/dQ$ when $I$ is constant we have

$$\left(\frac{dI}{dH}\right)_Q + \left(\frac{dI}{dQ}\right)_H \left(\frac{dQ}{dH}\right)_I = 0,$$

the subscript denoting the variable that is constant in the differentiation. Thus on substitution

$$\frac{d\eta}{dI} = \left(\frac{dI}{dQ}\right)_H \bigg/ \left(\frac{dI}{dH}\right)_Q + Q\frac{dM^{-1}}{dI};$$

and the total expansion is $\eta' = \eta + Q/M$: so that on writing as usual $\kappa$ for $(dI/dH)_Q$, we have

$$\frac{d\eta'}{dI} = -I\left(\frac{d\kappa^{-1}}{dQ}\right)_H + 2Q\frac{dM^{-1}}{dI}.$$

This is the exact equation which should be directly satisfied by series of observations of $\eta'$, $I$, and $M$, formed with different constant values of $H$ and $Q$, provided hysteresis is negligible. As $\eta'$ must be an even function of $I$, it follows that when $I$ and $Q$ are small, $\eta'$ is equal to $-\frac{1}{2}I^2 (d\kappa^{-1}/dQ)_H$, or $\frac{1}{2}H^2 (d\kappa/dQ)_I$, as in § 67.

For the case of a ring magnetized by a coil, there can be no free polarity except at an air gap; thus there is no stress of magnetic origin in the material. The alterations of longitudinal and transverse dimensions of rings of iron and nickel[1] are thus wholly intrinsic changes due to the magnetic polarity and in no part due to mechanical stress such as $Q$. In the neighbourhood of the origin, where $\eta'$ is proportional to $I^2$, the curves given by Bidwell expressing the relation between $\eta'$ and $I$ should be parabolic, as in fact they are. At the magnetization corresponding to a maximum or minimum ordinate $\eta'$ of the curve, the effect of a very small imposed tension on the magnetization should change sign, being null for that particular magnetization; the summit of the curve is therefore the Villari critical point. But if there is a tension $Q$ so considerable that intrinsic change of elastic modulus by magnetization contributes appreciably to the elongation, the Villari point will be displaced from the summit of the curve, backwards when magnetization increases the modulus. It appears from the experiments of Bidwell[2] that for iron tension increases the intrinsic elongation, for nickel it at first increases then diminishes and finally for stronger fields increases it, while for cobalt there is no sensible effect.

*No stress in ring magnet: its elongation intrinsic,*

*agreeing with Bidwell.*

*Displacement of Villari neutral point:*

*as determined by rings.*

## *Mechanical Stress in a Polarized Solid Sphere.*

84. The mechanical stress sustained by a sphere of soft iron situated in a uniform magnetic field $H$ can be simply expressed. The

[1] Shelford Bidwell, *Phil. Trans.* A, 1888, p. 228; *Roy. Soc. Proc.* 1894.
[2] *Roy. Soc. Proc.* Vol. XLVII, 1890, p. 480.

well-known analysis of Poisson gives a constant field $H'=3H/(\mu + 2)$ and uniform magnetization $I' = (\mu - 1) H'/4\pi$ inside the iron,

*Stress in magnetized sphere:* whether the law of induction is linear or not. Thus for this case of a sphere the mechanical forces exerted on the iron involve no distribution of forcive throughout its volume, but simply an outward normal traction of intensity $\{(\mu^2 - 1) \sin^2\theta - \mu + 1\} H'^2/8\pi$ over its *agreeing with Kirchhoff.* surface: that being so, the stresses agree with Kirchhoff's values, and the elastic strain produced in the sphere is given by his formulae[1], the result of course involving only very slight deformation. In fact, taking the axis of $x$ along the direction of $I'$, it is clear that an elastic displacement $(u, v, w)$ of the type

$$u = ax^3 + bx (y^2 + z^2) + cx, \quad v = w = a'x^3 + b'x (y^2 + z^2) + c'x$$

satisfies the conditions of the problem for the case of a sphere, the constants being determined by satisfying the equations of internal equilibrium and adjusting the surface tractions. In addition to this *Strain and intrinsic deformation.* mechanical deformation there will be the intrinsic deformation above determined (§ 83) arising from the molecular changes produced by the magnetic polarization.

Precisely similar formulae express the mechanical stress in a sphere of solid dielectric matter situated in a uniform electric field.

I desire to express, as in previous Memoirs, my obligation to the friendly criticism of Professor G. F. FitzGerald, which has enabled me to remove obscurities and in various places to make my meaning clearer.

## *Appendix* (1927).

### APPLICATIONS OF THE VIRIAL (§§ 53-4).

THE main trend of these two sections is to obtain results from comparison of the van der Waals equation of state, derived as it was by him from the virial, and the end aspects of direct equilibration of the *Why the virial gives results.* liquid contained in Newtonian ideal narrow tubes. The virial method is fruitful because it begins with knowledge that the configuration is a possible one: but it must avoid interfaces where the transition is gradual and unknown. It is proposed to pass its capabilities under review*.

The simplest subject matter for it is a crowd of particles in motion like a gas, repelling into orbits swinging round each other under

[1] Kirchhoff, *Gesamm. Abhandl.*, Nachtrag, p. 124; cf. also Love, *Treatise on Elasticity*, Vol. 1, § 168.

* There is a very complete discussion in Boltzmann's Lectures: cf. the French edition *Théorie des Gaz.*

mutual central forces according to a law $r^{-n}$: when $n$ is a considerable number the repulsion is relatively intense only at very small distances. The masses may be all different. Considering the component motions along $x$, the dynamical equation of one of the masses may be expressed as

General virial theorem.

$$m\ddot{x} = X,$$

leading to

$$\int X x \, dt = \int m\ddot{x}x \, dt = \left| m\dot{x}x \right| - \int m\dot{x}^2 dt = \left| \frac{d}{dt} \tfrac{1}{2}mx^2 \right| - \int m\dot{x}^2 dt.$$

If then the system is in a stationary state to the extent that the value of $\tfrac{1}{2}mx^2$ merely fluctuates round a mean, and bars denote mean values with regard to time, we obtain by summation for all the bodies,

$$\Sigma \overline{\tfrac{1}{2}m\dot{x}^2} = -\Sigma \tfrac{1}{2}\overline{X}x.$$

This equation is not invariant: but by summation for the three coordinate directions it gives

$$\Sigma \overline{\tfrac{1}{2}mv^2} = -\Sigma \overline{\tfrac{1}{2}(Xx + Yy + Zz)}.$$

The second side is the virial of Clausius, thus equal for the stationary system to the mean kinetic energy, and so invariant. This seems a small foundation from which to draw weighty conclusions.

Suppose our dense system of moving masses is enclosed in an outer boundary which itself exerts no attraction: part of the forces sustained by the system is the containing pressure $p$ of the shell, of which the virial is $+\tfrac{3}{2}pv$. This pressure counteracts the impacts of the masses: but if there is mutual repulsion of the masses within, the outer ones will be concentrated near the surface to an extent not yet ascertained. The cognate complication of the Laplacian theory of capillarity in the hands of Poisson may be recalled. It is only when this transition layer is negligible compared with the volume of the system that the equation of the virial reduces to practical results*.

Conditions near surface are exceptional.

Maxwell's illustration, not however an average, may be recalled, that in a strained jointed frame in which $R$ is the tension in a member of length $l$, $\Sigma R l = 0$.

In certain cases the internal virial admits of significant expression. If the forces $R$ between the particles are along the distance $r$, the contribution from each pair is $-\tfrac{1}{2}Rr$. However various the particles and their mutual forces be, if $R$ varies for every pair as the same power $-n$ of $r$, this contribution is equal to the mutual potential energy of the pair divided by $-2n+2$: the same applies to the aggregate, so

For attracting molecules,

---

* The circumstances of the reflexion and the "absorption and evaporation" of molecules of a rarefied gas by the boundary layer were closely analysed by Maxwell in an appendix to his last memoir, *Phil. Trans.* 1879, *Scientific Papers*, Vol. II, pp. 703–712.

that the total internal virial is $-W/(2n-2)$, where the potential energy $W$ is relative to the state of infinite dispersion.

The theorem that the mean kinetic energy is equal to the mean virial of the forces can be applied to an interior region in the domain of particles, separated off by an interface, if a pressure $p$ is included as a force exerted across that boundary from outside this part of the system. But this is definite (cf. p. 90) only when the effect of the forces near the interface is wholly local: and then $p$ is the hydrostatic pressure, that which has the property of transmission in accordance with Pascal's principle. For example, this condition would not be satisfied if the particles were electric ions of the same sign, for repulsion according to inverse square is not a force of molecular type: the question of a mixture of ions in compensating amounts requires deeper scrutiny.

*under restriction,*

With this understanding the virial value for $T$ gives for the constant of a volume $v$, in the present case

*as a power of distance.*

$$T = -\overline{W}/(2n-2) + \tfrac{3}{2}pv, \quad T + \overline{W} = \overline{E},$$

where $T$ and $\overline{W}$ are the time-means of kinetic and potential energies, $\overline{E}$ of the energy.

If the idea of a temperature $\theta$, as proportional to mean energy of translation*, is introduced, in a form $\overline{T} = NR'\theta$, where $N$ is the number of particles, now all similar, this can give the value of the total energy by an equation

*Equation for the energy:*

$$pv = \frac{2n-3}{3n-3} NR'\theta + \frac{1}{3n-3} E.$$

For a self-contained system, constituting a molecule, or a planetary system in astronomy, there is no $p$: thus, $E$ being now constant,

*its partition in a self-contained system:*

$$\overline{W}/(2n-2) = -\overline{T} = E/(2n-3).$$

For mutual forces varying as inverse square, however heterogeneous otherwise, this partition of energy is expressed by $\tfrac{1}{2}\overline{W} = -\overline{T} = E$, as Jacobi found for the case of a planetary system: which may be verified for example on a simple elliptic orbit. For an application to atoms cf. Vol. I, *supra*, p. 670.

*astronomical.*

* The translatory virial is alone involved, its internal part being compensated by the distribution of energy within each self-contained molecule, so long as the molecule is outside the range of others.

The general virial theorem is applicable only to the internal energy of a system relative to its centre of mass: for if the system as a whole has translatory motion $\Sigma m (\dot{x}^2 + \dot{y}^2 + \dot{z}^2)$ is not merely fluctuating.

*Virial of a system is wholly internal.*

The virial theorem however holds good for reference to a system of axes rotating with uniform angular motion relative to itself, provided the centrifugal forces are included with the applied forces.

## *Appendix* (1927).

### THE ESSENTIAL NATURE OF THE MAXWELL STRESS (§§ 6–10, 40).

THE inner significance of the Maxwellian quadratic transmitting stress in the electrodynamic field, of which his pressure of radiation is a special case, has been a puzzle for half a century. As above emphasized it cannot be the stress by which electric disturbance is transmitted in undulatory fashion, for that involves linear equations: the tractions there concerned are linear in the electric variables, not quadratic as here. *Historical.*

The misfit in the stress as regards magnetism, accounted for in the footnote p. 72, was early detected: cf. Maxwell's *Treatise*, ed. 2 (1881), Vol. II, p. 262, Appendix 2 to the Chapter XI on "Energy and Stress."

That there is an unbalanced local resultant of this transmitting stress, which is the time gradient of a quantity interpreted later as a distribution of momentum, was recognized first by Lorentz on the basis of interaction of the field with ions, in his memoir of 1895* (*Spannungen im Aether*, pp. 24–28). After an interesting discussion connecting it up with possible motion in the aether (which would be very small on the present illustrative hypothesis of a rotational aether of very great inertia) he finally leaves the stress as (perhaps with Maxwell himself) a useful analytic synthesis of the forces in the field.

The next progress was with Minkowski, who detected in the distribution of stress, momentum, and energy in the field the components of a fourfold invariant tensor in optical space time, which was later generalized to be the fundamental feature of physical reality, regarded as an extensional phenomenon, by Einstein.

From the dynamical point of view, the question however persists, whence does this apparent play of mechanical force come into the aethereal scheme? And an adequate and very significant answer can, as may be held, now flow direct from the Principle of Action as the one ultimate consolidation. When the Action density in the field is formulated in terms of the fourfold electric vector potential $(F, G, H, V)$, the variation of the Action with regard to that potential gives the equilibrated internal structure of the electrodynamic field, in the form of the Maxwellian circuital equations which regulate the pro- *Stress momentum tensor inherent in fourfold Action.*

* The field is there restricted to the empty space between the electric systems. The results *supra*, p. 72, basing on the formulae of Maxwell's *Treatise*, apply for a material medium specified by the most general types of electric and magnetic polarizations: there is then the modified Maxwell stress, but the same aethereal momentum.

pagation of electric effects through the free aether. As the variation of the Action thus vanishes for all possible variations of the potential, it vanishes for the special case when they are such as would be produced by a virtual displacement of the space time of the frame of reference itself relative to an unvaried material world to which the potential belongs, for which therefore the variation $\delta F$ at a position that is displaced with the frame is $- F_x \delta x - F_y \delta y - F_z \delta z - F_t \delta t$, including also at a boundary additional terms there accruing. Carrying through this special process of variation leads to the fourfold stress tensor, as a balanced stress system over the boundary of every region of free aether. But it is not balanced by itself if the region surrounds ions or abuts on a boundary: its resultant around them represents, from its original mode of construction by Maxwell, the forcive acting on them by transmission.

Making now a different approach, the Action of the complete
Equivalent<br>formulation<br>by singu-<br>larities. system, as thus regulated by the Maxwellian field equations, integrates to terms associated with the ions and other sources\*, the remainder, if any, being free radiation with its momentum and energy, and so fugitive. This equivalent expression for the Action as associated with the material sources can now be further varied for virtual displacement of the sources in the space time, thus determining the forcive (fourfold) on each source as arising from the influence transmitted from the others.

The two variations must give equivalent results. Therefore the forcive (fourfold) on any one of the material system of sources is the resultant integrated over any boundary around it of the aethereal stress tensor. Or, on joining the former on to the latter reversed,
Equilibrated<br>compound<br>tensor. there is a balanced combined tensor.

When the Einstein gravitational term is introduced into the Action density, annulling of its variation with respect to its potentials $(g_{rs})$ determines the internal laws of the gravitational field. The general variation thus annulled includes, as before, variation arising from every displacement of these potentials in any convenient fourfold coordinate frame, leading, in complete analogy to the case of an electrodynamic
Gravitational<br>tensor, field, to a gravitational tensor which is balanced except around the sources. It is, in like manner, the expression, as a formal field stress, of the gravitational interaction of the sources. There is, however, a com-
avoiding<br>second<br>gradients, plication here in that the assumed gravitational field Action involves second gradients, which transcend representation by ordinary stress: when they are removed by integration by parts, as is possible because
is not<br>invariant. they enter linearly, the resulting Action form is no longer invariant.

\* This holds good for all systems in which the Action density is a quadratic function of the gradients of the fourfold potential.

Thus the laws of energy, momentum, and inertia in the fourfold material field are consolidated into a formal transmitting tensor, associated with the medium and defined by its Action density, this being all inherent in the Principle of Action in what is surely a very remarkable way. It exhibits, for example, how a field of force-momentum-energy for the matter that is present can emerge from the variation, apparently foreign to it, of a spatial Action purely electrodynamic and regulated by its own variational equation. It is a tensor of displacement in the selected frame, which implies something that is displaced in the frame, in addition to the matter, even if only virtually or formally. It affords the ideal general solution of the problem of how the electromotive or aethereal relations of the field give rise to mechanical forces between the material sources. The formal mathematical verification of this train of ideas was effected for the general relativity formulation in a series of early papers by Hilbert, F. Klein and Emmy Noether in *Göttinger Nachrichten*, 1916–18: previously for general elastic propagation in a uniform fourfold by Herglotz.

*Material forces as latent in the medium:*

*the revealing procedure.*

All this is restricted to free space pervaded by ions and other local singularities. The original Maxwell stress was much wider: as developed in the text, it applies to all electrically and magnetically polarized material media as averaged out into a continuous analysis.

*The general form of stress for polarized media.*

# 50

# THE INFLUENCE OF A MAGNETIC FIELD
# ON RADIATION FREQUENCY

[*Proceedings of the Royal Society*, Vol. LX, p. 514.
Received and read February 11th, 1897.]

IN the course of the development of a dynamical hypothesis[1] I have been led to express the interaction between matter and aether as wholly arising from the permanent electrons associated with the matter; and reference was made to von Helmholtz (1893) and Lorentz (1895) as Historical. having followed up similar views. A footnote in Dr Zeeman's paper has drawn my attention to an earlier memoir of Lorentz (1892), in which it was definitely laid down that the electric and optical influences of matter must be formulated by a modified Weberian theory, in which the moving electrons affect each other, not directly by action at a distance but mediately by transmission across the aether in accordance with the Faraday-Maxwell scheme of electric relations. The development of a physical scheme in which such action can be pictured as possible and real, not merely taken as an unavoidable assumption which must be accepted in spite of the paralogisms which it apparently involves[2], was a main topic in the papers above mentioned.

The experiments of Dr Zeeman verify deductions drawn by Lorentz from this view. It might, however, be argued that inasmuch as a magnetic field alters the index of refraction of circularly polarized light, which depends on the free periods of the material molecules, it must therefore, quite independently of special theory, alter the free periods of the spectral lines of the substance. But the actual phenomena do not seem to be thus reciprocal. On the electric theory of light it is only the dispersion in material media that arises from direct influence of the free molecular periods: the main refraction arises from the static dielectric coefficient of the material, which is not connected with the periods of molecules[3]. From the phenomena of magneto-optic reflexion it may be shown that, on the hypothesis that the Faraday effect is due to regular accumulated influences of the individual molecules, it must be involved in the relation between the

---

[1] *Phil. Trans.* A, 1894, pp. 719–822; A, 1895, pp. 695–743 [as in Vol. 1, *supra*].

[2] H. A. Lorentz, "La Théorie Electromagnétique de Maxwell, et ses Applications aux Corps Mouvants," *Archives Néerlandaises*, 1892. Cf. especially § 91.

[3] *Loc. cit. Phil. Trans.* A, 1894, p. 820; and A, 1895, p. 713 [: Vol. 1, *supra*, p. 563].

electric force $(P, Q, R)$ and the electric polarization of the material $(f', g', h')$, of type

$$f' = \frac{K - 1}{4\pi} P - c_3 \frac{dQ}{dt} + c_2 \frac{dR}{dt},$$

where $(c_1, c_2, c_3)$ is proportional to the impressed magnetic field. This relation, interpreted in the view that the electric character of a molecule is determined by the orbits of its electrons, simply means that the capacity of electric polarization of the molecule depends on its orientation with regard to the imposed magnetic field, that, in fact, the static value of $K$, depending on the molecular configurations just as much as do the free periods, is altered by the magnetic field. This relation agrees with the main feature of rotatory dispersion, namely, that it roughly follows the law of the inverse square of the wave-length. The specific influence of the molecular free periods, that is, of the ordinary dispersion of the material, on the Faraday effect, is presumably a secondary one\*; though it, too, follows the same law for different wave-lengths, in the case of substances for which Cauchy's dispersion formula holds good. It is this latter part of the Faraday effect that is reciprocal to Dr Zeeman's phenomenon.

*Faraday effect mainly constitutive, not dispersional.*

*Is the Zeeman effect its converse?*

The question is fundamental how far we can proceed in physical theory on the basis that the material molecule is made up of revolving electrons and of nothing else. Certain negative optical experiments of Michelson almost require this view; at any rate, they have not been otherwise explained. It may be shown after the manner of *Phil. Trans.* A, 1894 [Vol. I. p. 524] (and Dr Zeeman's calculation, in fact, forms a sufficient indication of the order of magnitude of the result), that in an ideal simple molecule consisting of one positive and one negative electron revolving round each other, the inertia of the molecule would have to be considerably less than the chemical masses of ordinary molecules, in order to lead to an influence on the period, of the order observed by Dr Zeeman. But then a line in the spectrum may be expected to arise rather from one of the numerous epicycles superposed on the main orbits of the various electrons in the molecule than from a main orbit itself.

*Zeeman effect requires small inertia of the electrons.*

---

\* The paper next following shows how, in the standard case, it is really constitutive, as it affects all the spectral periods alike.

# 51

# ON THE THEORY OF THE MAGNETIC INFLUENCE ON SPECTRA; AND ON THE RADIATION FROM MOVING IONS

*[Philosophical Magazine* for December, 1897.]

A THEORETICAL analysis of somewhat general character can be developed in connection with Zeeman's phenomenon[1], which may help to throw light on the nature of the electric vibrations in the molecule. It will be convenient to begin with a simple case.

The simple Zeeman effect: 1. Consider a single ion $e$, of effective mass $M$, describing an elliptic orbit under an attraction to a fixed centre proportional to the distance therefrom. The equations of motion will be $(\ddot{x}, \ddot{y}, \ddot{z}) = - a^2 (x, y, z)$; and the frequency of oscillation in any direction will be $a/2\pi$. Now suppose that a uniform magnetic field $H$, in a direction $(l, m, n)$, is introduced: the equations of motion will become

$$\ddot{x} = - a^2 x + \kappa (n\dot{y} - m\dot{z}),$$
$$\ddot{y} = - a^2 y + \kappa (l\dot{z} - n\dot{x}),$$
$$\ddot{z} = - a^2 z + \kappa (m\dot{x} - l\dot{y}),$$

where $\kappa = eH/Mc^2$, in which c is the velocity of radiation[2]. To obtain the frequencies $(p/2\pi)$ of the oscillations thus modified, we make as usual $(x, y, z)$ proportional to $e^{\iota p t}$. This gives, after easy reduction, an equation for $p$,

$$(p^2 - a^2)^3 - \kappa^2 p^2 (p^2 - a^2) = 0.$$

[1] Zeeman, *Phil. Mag.* March and July, 1897; Michelson, *Phil. Mag.* May, 1897; Lodge, *Proc. Roy. Soc.* Feb. and June, 1897 [: cf. also *supra*, p. 138.]

Historical. [The earliest information on Zeeman's discovery that reached this country was a sentence in *Nature*, Dec. 24, 1926, in its Amsterdam abstracts for October 31. A tube emitting sodium light "is placed between the poles of an electromagnet. When acted on by the magnet a slight broadening of the two sodium lines is seen...." The writer, then away in the country in Ireland, had been cognizant of the results of applying a magnetic field to the orbital ionic pair discussed in *Phil. Trans.* 1894, § 118 [*supra*, Vol. 1, p. 524]. Taking the masses of the ions to be comparable with that of a hydrogen atom, the spectral effect would be inappreciable. He pointed out the circumstance to Professor Lodge, and suggested the importance of confirming the experiment, which Lodge soon succeeded in doing.

When Zeeman's papers became available later they were found to cover a wide range, including a suggestion of probable triplication and not mere broadening of the lines, a verification of the polarizations predicted by Lorentz, and an anticipation of applications to the solar atmosphere.]

[2] Cf. *Phil. Trans.* A, 1895, p. 718 [*supra*, Vol. 1, p. 569]. These equations only apply strictly (*infra*, § 10) when the velocity of the ion is small compared with c.

[In the formula for $\kappa$, $e$ and $H$ are tacitly in electrostatic units: else $c^2$ should be omitted.]

Thus, corresponding to each original period represented by $p = a$, there are three modified ones represented by $p = a$ and $p^2 \pm \kappa p - a^2 = 0$; when, as in practice, $\kappa$ is very small the two latter will be approximately $p = a + \kappa^2/8a \pm \tfrac{1}{2}\kappa$, or with sufficient accuracy $p = a \pm \tfrac{1}{2}\kappa$. Each vibration period will therefore be tripled: and the striking feature is that the modification thus produced is the same whatever be the orientation of the orbit with respect to the magnetic field. one of sharp triplication:

An inquiry into the cause of this feature enables us to generalize the result. Suppose that the original orbit is referred to a system of axes $(x, y, z)$ that are themselves revolving with angular velocity $\omega$ round an axis of which the direction is $(l, m, n)$. The component velocities $(u, v, w)$ referred to this moving space are

$$\dot{x} - y\omega n + z\omega m, \ldots, \ldots,$$

and the component accelerations are

$$\dot{u} - v\omega n + w\omega m, \ldots, \ldots.$$

Thus the component acceleration parallel to $x$ is

$$\ddot{x} - 2\omega (n\dot{y} - m\dot{z}) - \omega^2 x + \omega^2 l (lx + my + nz).$$

If, then, we take $\omega$ equal to $\tfrac{1}{2}\kappa$, and so can neglect $\omega^2$, the equations of the original orbit referred to this revolving space are identical with those of that orbit as modified by the magnetic field. In other words, the oscillation thus modified will be brought back to its original aspect if the observer is attached to a frame which revolves with angular velocity $\tfrac{1}{2}\kappa$ or $eH/2Mc^2$ round the axis of the magnetic field. In a circular orbit described one way round this axis the apparent rotation will in fact be retarded, in one described the other way round it will be accelerated, in a linear oscillation along the axis there will be no alteration: hence the three periods found above. interpreted as an orbital precession.

2. Now the argument above given still applies, whatever be the number of revolving ions in the molecule, and however they attract each other or are attracted to fixed centres on the axis, provided $\kappa$ has the same value for them all. In any such case the actual oscillation in the magnetic field is identical with the unmodified oscillation as seen from a revolving frame; or, more simply, the modification may be represented by imparting an opposite angular velocity $\tfrac{1}{2}\kappa$ to the vibrating system*. Thus the period of a principal oscillation of the system will be affected by the magnetic field in the opposite way to that of its optical image in a plane parallel to the field; and these two oscillations, previously identical as regards period, will be sepa- Generalized to orbital atomic systems, if axial:

* This precessional result has recently become the starting point for Bohr's correlation principle in this domain, whose orderly relations have proved to be more complex than the present standard case.

When the magnetic field is imposed, the system gradually changes into one of the same type but different phases, with this precession added. As $e$ is negative, the precession is positive.

rated on account of their right-handed and left-handed qualities. An oscillation which does not involve rotation round an axis parallel to the field will, however, present the same aspect to the field as its image, and will not be affected at all. This latter type of oscillation, in a compound system, will be a very special one; and when a crowd of vibrators indifferently orientated are considered, radiation of this kind will usually be practically non-existent. Thus *each* spectral line of the vibrator will be split up into two with right-handed and left-handed circular polarizations when seen along the axis, and plane-polarized with phase difference of half a wave-length when seen at right angles to it, and with differences of frequency the same for all lines in the spectrum, as in the special case above. This simple statement applies to all systems in which the electric charge of each mobile ion in the vibrator is proportional to its effective mass, which implies that the charges of the mobile ions are all of the same sign.

polarization.

But requires absence of mobile positive electrons.

3. The characters of the three principal oscillations in § 1 may be determined in the usual manner by substituting in the equations of motion $(x, y, z) = (x_0, y_0, z_0)\, e^{\iota p t}$ and determining $(x_0, y_0, z_0)$ from the resulting system of linear equations. But algebraic reductions will be avoided by taking the magnetic field to be along the axis of $z$, so that $(l, m, n) = (0, 0, 1)$, as might in fact have been done from the beginning. The equations of motion are then

$$\ddot{x} = -a^2 x + \kappa \dot{y}, \quad \ddot{y} = -a^2 y - \kappa \dot{x}, \quad \ddot{z} = -a^2 z.$$

They show at once that the unmodified principal vibration is a linear oscillation parallel to the $z$-axis. As regards the others, writing $(x, y) = (x_0, y_0)\, e^{\iota p t}$ we have

$$(a^2 - p^2)\, x = \iota \kappa p y, \quad (a^2 - p^2)\, y = \iota \kappa p x;$$

thus $p^2 \pm \kappa p - a^2 = 0$, or very approximately $p = a \pm \tfrac{1}{2}\kappa$ as before; and separating the real parts of this solution

$$x = \frac{A}{a^2 - p^2} \cos pt, \quad y = \frac{A}{\kappa p} \sin pt,$$

which to our order of approximation represents motion round a circle in the plane of $(x, y)$, right-handed or left-handed according to the value of $p$ that is taken[1]. The character of the radiation from such a vibrator is thus precisely independent of the orientation of its orbit with respect to the magnetic field. With a large number of such vibrators, orientated indifferently, every spectral line seen in a direction at right angles to the magnetic field would be split up into three lines, each of the same breadth as the original, the middle one plane-polarized at right angles to the magnetic field, the outer ones in the

The resolved spectral lines are sharp:

---

[1] This is no doubt the analysis recently indicated by Professor FitzGerald in *Nature*, Sept. 1897.

direction of the field: as the aggregate light must be unpolarized, the intensity of the middle line would be twice that of either of the outer ones. But viewed along the field the middle line would disappear, as the exciting vibration would be end on to the observer and could not therefore send out transverse radiation: the other lines (equally sharp as before) would be circularly polarized, and their directions of polarization would, as Zeeman remarks, determine whether the vibrator involves a positive or a negative electron.

*intensities.*

*Negative electrons the effective ones.*

4. A view has been enunciated that it is only one kind of ions, namely the negative ones, that are mobile and free to vibrate in the atom or molecule, the other kind being fixed to the matter and immobile. On such an hypothesis, if the charges of these negative ions are proportional to their effective masses, for example if they are simple electrons without inertia other than that of the electric charge, the intervals (measured in difference of frequency) between magnetic doublets and the outside lines of magnetic triplets in the spectrum should be the same for all lines. Moreover, they should be the same in different spectra. Thus an hypothesis of that kind can be definitely put to the test.

*A spectral test of the identity of all electrons in the atoms.*

5. When there are ions of different kinds describing orbits in the molecule, these exact results no longer hold: but even then we can assert that the difference of frequency between the lines of a magnetic doublet is of the order $eH/2\pi Mc^2$, and the order of magnitude of $e/M$ can be thence derived. Thus Zeeman concludes from his experiments that the effective mass of a revolving ion, supposed to have the full unitary charge or electron, is about $10^{-3}$ of the mass of the atom. This is about the same as Professor J. J. Thomson's estimates of the masses of the electric carriers in the cathode rays. If we took these carriers to be simply electrons, as their constancy under various environments tends to indicate, there would thus be about $10^3$ electrons in the molecule*.

*Mass of electron.*

6. In view of the above considerations, the circumstance that in a magnetic field certain lines, viewed transversely, are divided into sharp triplets with perfect plane polarizations, which has been described by Zeeman and assented to by Lodge and by Michelson[1], is an important clue to the character of the principal oscillations which emit those lines. In an oscillating molecule undisturbed by a magnetic field there must be three types of vibration which all have

*General atomic type admitting of standard resolution.*

* This is on the assumption, simplest and so natural at the time, that positive and negative ultimate ions are exact mirror images of each other. This positive electron has not yet been found.

[1] In Cornu's experiments, *Comptes Rendus*, Oct. 18, the application of the polarizing apparatus seems to have been required to divide the lines.

the period belonging to that line, namely two types which differ only by involving rotations in opposite directions round the axis of that magnetic field and would naturally have the same period, and a third type which does not involve any rotation with respect to that axis. Now that extraneous axis may have any direction with reference to the molecule. Hence a principal oscillation which is thus magnetically tripled must be capable of being excited with reference to any axis in the molecule: otherwise there would be merely hazy broadening or duplication instead of definite triplication.

*Electric surges on an ideal conductor,*

7. A system of electrons or ions of the same sign, confined to a surface over which they are free to move and constituting an electric charge on it, is an artificial vibrator whose periods illustrate these results. The free periods of such a vibrating system in which the forces acting on the electrons work against the inertia of the moving electrons, would [not] be only theoretically different from the free periods of an actual electric charge on a metallic conductor; although in the latter

*itself non-ionic,*

case the forces acting on the ions work mainly against the ohmic diffusive resistance to their transfer either actual or electrolytic through the crowd of neighbouring molecules, which is far greater than the reaction arising from inertia alone unless it is rapid optical vibrations that are dealt with. In either case the forces acting on the ions are so great compared with the possible kinetic reactions to their

*and without surface friction.*

motion that their distribution is at each instant practically in equilibrium on the surface, so that there is no electric force along it: and either set of conditions simply reduces to the condition that there shall be no electric or magnetic field in the interior. It follows that the oscillations of an electric charge on any conductor of the form of a surface of revolution are modified by the introduction of a magnetic field along the direction of the axis of the surface, just as if the angular velocity $\frac{1}{2}\kappa$ above given were imparted to the vibrating system. All the free periods except those of zonal oscillations would be duplicated in the manner above explained, the interval measured in difference of frequencies between the components of the doubled vibration being the same for all. In the special case of zonal oscillations on a sphere they would be triplicated, because the period of a zonal vibration along the axis would not be modified: but the middle line would be very weak compared with the flanking ones[1].

*Contrast with actual metal.*

[1] The radiation from a continuously distributed electric charge is, however, known to be so great as to make its oscillations dead-beat; hence these conclusions could only be applied if (i) there are only a limited number of discrete ions moving [so as to slide freely, cf. Vol. v, *supra*, Appendix v] on the surface, or (ii) there are material forces other than electric imagined to act between the ions, whose energy could maintain the vibration for a large number of periods. See § 10 [: also Vol. i, *supra*, Appendix v].

8. This analysis gives a hint as to one way in which a series of double lines in a spectrum, with equidistant frequencies, might be originated. Suppose, as a very rough illustration, that a polar molecule is constituted of a system of positive electrons around one pole and a system of negative ones around the other, the two systems being so far apart as to have practically separate sets of periods for their orbital motions, each of course disturbed by the presence of the other. Each of these systems moves in the magnetic field, more or less constant, arising from the other; and the effect of this disturbing field will, as above, be to duplicate all the periods of that set in the above regular way.

*Paired conjugate systems giving doubled spectral lines.*

9. It is desirable to formulate precisely the relation between the motions of the electrons and the radiation emitted by them, which has been tacitly employed in the foregoing discussion. The specification of that radiation may be readily assigned by a summation over the different elements of the paths of the oscillating ions. Suppose that an ion $e$ is at $A$ and after a time $\delta t$ is at $B$, where $AB = v\delta t$, $v$ being its velocity; the effect of its displacement is the same as that of the creation of an electric doublet $AB$ of moment $ev\delta t$; thus we have only to find the influence propagated from such a doublet, and then integrate along the paths of the ions of the molecule.

*Radiation from a single moving ion,*

*determinable by summation over path.*

Consider now such a doublet at the origin, lying along the axis of $z$; for it, or indeed for any distribution symmetrical with respect to that axis, the lines of magnetic force will be circles round the axis, and the force will be specified by a single variable, its intensity $H$. The current, whether in dielectric or in conducting media, will circulate in wedge-shaped sheets with their edges on the axis, and may be specified by a stream function, as in fact will appear below. If we employ cylindrical coordinates $\rho$, $\theta$, $z$, and apply the Amperean circuital relation (viz. circulation of magnetic force equals $4\pi$ times current) to the faces of the element of volume $\delta\rho \, . \, \rho\delta\theta \, . \, \delta z$, we obtain for the components P, o, $R$ of the electric force

*Its momentary field:*

*of axial type.*

$$\frac{d\mathrm{P}}{dt} = -\, \mathrm{c}^2 \frac{dH\rho}{\rho \, dz}, \quad \frac{dR}{dt} = \mathrm{c}^2 \frac{dH\rho}{\rho \, d\rho},$$

so that $H\rho$ plays the part of a stream function; while by the circuital relation of Faraday

$$\frac{d\mathrm{P}}{dz} - \frac{dR}{d\rho} = -\, \frac{dH}{dt}.$$

Thus the characteristic equation for $H$ is

$$\frac{d}{d\rho} \frac{1}{\rho} \frac{d}{d\rho} \rho H + \frac{d^2H}{dz^2} = \mathrm{c}^2 \frac{d^2H}{dt^2},$$

which is

$$(\nabla^2 - \rho^{-2}) \, H = \mathrm{c}^2 d^2H/dt^2,$$

The characteristic equation,

where $\nabla^2$ is Laplace's operator. But a more convenient reduction comes on substituting $H = dY/d\rho$, and then neglecting an irrelevant operator $d/d\rho$ along the equation: this gives

$$\nabla^2 Y = c^2 d^2 Y/dt^2.$$

solved in terms of its source,

We can now express the disturbance emitted by an electric doublet situated along the axis of $z$ at the origin, and vibrating so that its moment $M$ is an arbitrary function of the time. As regards places at a finite distance, the doublet may be treated as a linear current element of strength $dM/dt$. Close up to such an element in its equatorial plane, the magnetic force $H$ due to it is $- r^{-2}dM/dt$. The appropriate solution for $Y$ for this simplest case is $Y = r^{-1}f(t - r/c)$, so that

$$H = - \sin \theta \left\{ \frac{f(t - r/c)}{r^2} + \frac{f'(t - r/c)}{cr} \right\},$$

giving when $\theta$ is $\frac{1}{2}\pi$ and $r$ is very small, $H = - r^{-2}f(t)$: thus

regarded as a vibrating doublet.

$dM/dt = f(t)$. That is, if the moment of the oscillating doublet is given in the form $dM/dt = f(t)$, the magnetic force thus originated at the point $(r, \theta)$ is

$$H = - \sin \theta \left\{ \frac{f(t - r/c)}{r^2} + \frac{f'(t - r/c)}{cr} \right\},$$

or

$$\sin \theta \frac{d}{dr} r^{-1}f(t - r/c).$$

The intrinsic field and the radiation field.

The second term is negligible for movements of slow period, as it involves the velocity $c$ of radiation in the denominator. The components of the magnetic field due to a vibrating doublet $M$ at the origin whose direction cosines are $(l, m, n)$ are then

$$(mz - ny,\ nx - lz,\ ly - mx)\ r^{-1}d/dr\ r^{-1}f(t - r/c),$$

where

$$dM/dt = f(t);$$

and the components of the magnetic field, and therefore of the radiation emanating from any system of electric oscillators vibrating in any given manner can thence be expressed in a general form by integration.

Doublet suddenly established emits a pulse,

At present we only want the effect of suddenly establishing the doublet $M = ev\delta t$ at the origin. This comes by integration over the very small time of establishment; there is a thin spherical shell of magnetic force propagated out with velocity $c$, the total force integrated across the shell being exactly $- Mr^{-2} \sin \theta$ whatever be its

travelling as a concentrated shell,

radius, for the integral of the second term in $H$ vanishes because $dM/dt$ is null at the beginning and end of the operation. The aggregate amount of magnetic force thus propagated in the spherical sheet is the same as the steady magnetic force due to a permanent steady current element of intensity equal to $M/\delta t$, or $ev$: it is clear, in fact, that this must be so, if we consider a sudden creation of this current element and remember that its magnetic field establishes itself by spreading out ready formed with the velocity of radiation.

The magnetic force at a point at distance $r$ due to a moving ion thus depends on the state of the ion at a time $r/c$ previously; for near points it is in the plane perpendicular to $r$, at right angles to the projection $v$ of the velocity of the ion on that plane, and equal to $evr^{-2}$. For vibrations whose wave-length in free aether is very great compared with the dimensions of the molecular orbit, if we interpret magnetic force as velocity of the aether, the vibration path of a point attached to the aether, and close to the vibrator, will be in the plane transverse to $r$, and similar to the projection of the orbit of the electron on that plane when turned round through a right angle[1]. If the condition of wave-length very large compared with molecular magnitude were not satisfied, phase differences would sensibly disturb this result, and in effect each spectral line would be accompanied, more or less, [if things went thus,] by its system of harmonics.

*the same as that for a moving ion.*

*Graphical representation of its radiation:*

As the vibration of a near point in the aether is thus similar to the projection on the wave-front of the vibration of the electrons in the molecule, it is verified that the free periods of the radiation are those of the system of ions*.

*its mode of polarization.*

10. It might appear also at first sight that every steady orbital motion must rapidly lose its energy by radiation just as vibrations on the conductor in § 7 would do if the ions on it formed a continuous charge. On the other hand, it might be argued that what we have really been calculating is the amount added on to the previous motion in the medium by the successive displacements of the electrons; and, in the cases of steady motion, it is just this amount that is needed to maintain the permanency of the motion in the aether, which of itself has a tendency to be carried away. Thus in the parallel case of the movement of a very long stretched cord when an end of it has a steady circular motion imparted to it, an analysis in the ordinary way leads to a train of circular waves running along the cord; but there exists a steady motion in which the cord whirls round bodily, and which will be generated when the velocity of the motion imposed on the end is gradually increased from a very small initial value to its final amount.

*Non-radiating systems may exist.*

*A possible mechanical analogy:*

The difficulty is, however, not thus surmounted; for this steady motion which does not involve radiation is really a state of stationary undulation arising from the superposition of a wave-train travelling

*implies inward waves.*

---

[1] For a different treatment of similar topics cf. H. A. Lorentz, "La Théorie Electromagnétique de Maxwell," §§ 112–119, *Archives Néerlandaises*, 1892; "Versuch...," 1895; quoted by Zeeman, *Phil. Mag.* March, 1897.

* Developed more fully in *Aether and Matter* (1900), Chap. XIV.

outwards on another travelling inwards, and the genesis of the latter one would have to be accounted for. We might assume that these non-radiating vibrations consisted of stationary waves reflected backwards and forwards between two vibrating molecules, or between two ions in the same molecule; but even that would not be satisfactory. As a matter of fact, however, no explanation of this kind is needed.

*The radiation is largely suppressed,* The effective electric inertia of an ion $e$ by itself is $\frac{2}{3}e^2a^{-1}$,[1] where $a$ is the radius of its nucleus supposed spherical: the rate at which it loses energy by radiation is proportional to $e^2$, and involves its motion, but does not depend on $a$ at all. The kinetic reaction to change of its velocity which is connected with loss of energy by radiation can thus be made negligible in comparison with the kinetic reaction arising from its inertia\*. In fact, the energy of the aethereal motion carried along by the moving ion depends on the first term in $H$, involving $r^{-2}$, and the radiated energy depends on the second term, involving *but not abolished.* $r^{-1}$. But in types of oscillation in which there are crowds of ions moving close together in step, loss of energy by radiation is an important feature in the dynamics of free vibrations. [Cf. § 11: also for verification from the facts, Vol. I, *supra*, Appendix, V, VII.]

These considerations can be developed by aid of the analysis of § 9 *The field of the radiation:* above. In consequence of the stream-function property of $H\rho$, the components of $d/dt$ of the electric force, along $\delta r$ and along $r\delta\theta$, are respectively

$$\frac{c^2}{\rho}\frac{dH\rho}{rd\theta} \quad \text{and} \quad -\frac{c^2}{\rho}\frac{dH\rho}{dr},$$

$\rho$ being $r\sin\theta$; thus the time gradients of the electric force are

$$-2c^2\cos\theta\left\{\frac{f(t-r/c)}{r^3}+\frac{f'(t-r/c)}{cr^2}\right\},$$

and $$-c^2\sin\theta\left\{\frac{f(t-r/c)}{r^3}+\frac{f'(t-r/c)}{cr^2}+\frac{f''(t-r/c)}{c^2r}\right\};$$

and the electric force is obtained by integrating with respect to $t$.

*interpreted.* At a very great distance the electric force (as well as the magnetic force) is thus perpendicular to $r$, and is equal to $-r^{-1}\sin\theta f'(t-r/c)$; and the flow of energy is thus by Poynting's principle radial. For the case of an ion $e$ moving with velocity $v$, $f(t)$ is equal to $ev$; and in $f(t-r/c)$ the value of the function $f$ belongs to the position of the

---

*Mass changes at high velocities.* [1] *Phil. Trans.* A, 1894, p. 812. This inertia [which should be $\frac{1}{2}e^2a^{-1}$, *supra*, Vol. I, p. 670] is no longer quite constant when the velocity of the ion is considerable compared with that of radiation. In that case also the simple computation of the radiation here given would not be exactly applicable; and the problem would have to be treated by continuous differential analysis after the manner of *Phil. Trans.* A, 1894, p. 808 [: *supra*, Vol. I, p. 517].

\* But experiment assigned a mass of the electron too small for this.

molecule at a time $r/c$ previous, where $r$ is its distance at that time. The rate of loss of energy by radiation may be computed by Poynting's formula as $(4\pi)^{-1}$ times the product of the above electric and magnetic forces integrated over an infinite sphere: it is thus

$$(4\pi r^2 c)^{-1} \{f'(t - r/c)\}^2 \int \sin^2 \theta \, dS, \quad \text{or} \quad \tfrac{2}{3} e^2 c^{-1} \dot{v}^2.$$

Formula for radiation from a travelling ion:

In the process of getting up a velocity $v$ of the ion from rest, there is a loss of energy equal to $\tfrac{2}{3} e^2 c^{-1} \int \dot{v}^2 dt$. In motion with uniform velocity there is no loss; during uniformly accelerated motion the rate of loss is constant.

As the electric and magnetic forces at a great distance are each proportional to the acceleration of the ion and do not involve its velocity, and as we can combine the components of its motion in fixed directions, it follows generally that the rate of loss of energy by radiation is $\tfrac{2}{3} e^2 c^{-1} \times$ (acceleration)$^2$.

generalized:

The store of kinetic energy belonging to the ion is $\tfrac{2}{3} e^2 a^{-1} v^2$. Thus the loss of energy by radiation from an undisturbed vibrating molecule would not be sensible compared with its whole intrinsic kinetic energy, when the velocities of the ions are not of the order of magnitude of that of radiation: while for higher velocities the importance of the radiation is, in part at any rate, counteracted by the increase of the inertia coefficient.

may be small compared with its energy of motion.

11. Finally, it is to be observed that the law of the magnetic vibration excited by a moving ion was analysed in § 9 only when $r$ is small compared with the wave-length. Further away from the ion the law of variation of the magnetic force with distance is $ev/r^2 + ev/cr$ instead of $ev/r^2$. Thus at a distance of a large number of wave-lengths, the vibration curve, then of the radiation proper, is similar to the projection of the hodograph of the orbit of the ion on the wave-front, instead of the projection of the orbit itself.

Form of the radiation vector,

as a hodograph.

It would thus appear that when the steady orbital motions in a molecule are so constituted that the vector sum of the accelerations of all its ions or electrons is constantly null, there will be no radiation, or very little, from it [for waves long compared with its dimensions], and therefore this steady motion will be permanent. But this is just the condition which holds good so long as the molecule is free from extraneous disturbance*.

The condition for no radiation:

is a natural law of the system.

* An important element in the residual radiation would arise from Doppler fluctuations in component wave-lengths. It is cognate that for atoms in a stream of $x$-ray radiation the electrons are found (Barkla and Thomson) to scatter the radiation individually: while the close verification of the Rayleigh formula for the blue sky shows that for the long wave-lengths of light each atom scatters as a whole. See however, *supra*, Vol. I, Appendix VII.

# 52

# NOTE ON THE COMPLETE SCHEME OF ELECTRO-DYNAMIC EQUATIONS OF A MOVING MATERIAL MEDIUM: AND ON ELECTROSTRICTION

[*Proceedings of the Royal Society*, Vol. LXIII, May 26th, 1898, pp. 365–372.]

THIS note forms a supplement to my third memoir on the "Dynamical Theory of the Aether[1]," to the sections of which the references are made.

1. It is intended in the first place to express with full generality the electrodynamic equations of a material medium moving in any manner, thus completing the scheme which has been already developed subject to simplifying restrictions in the memoirs referred to. To obtain a definite and consistent theoretical basis it was necessary to contemplate the material system as made up of discrete molecules, involving in their constitutions orbital systems of electrons, and moving through the practically stagnant aether. It is unnecessary, for the mere development of the equations, to form any notion of how such translation across the aether can be intelligibly conceived: but, inasmuch as its strangeness, when viewed in the light of motion of bodies through a material medium and the disturbance of the medium thereby produced, has often led to a feeling of its impossibility, and to an attitude of agnosticism with reference to aethereal constitution, it seems desirable that a kinematic scheme such as was there explained, depending on the conception of a rotationally elastic aether, should have a place in the foundations of aether theory. Any hesitation, resting on *à priori* scruples, in accepting as a working basis such a rotational scheme, seems to be no more warranted than would be a diffidence in assuming the atmosphere to be a continuous elastic medium in treating of the theory of sound. It is known that the origin of the elasticity of the atmosphere is something wholly different from the primitive notion of statical spring, being in fact the abrupt collisions of molecules: in the same way the rotational quality of the incompressible aether, which forms a sufficient picture of its effective constitution, may have its origin in something more fundamental that has not yet even been conceived. But in each case what is important for immediate practical purposes is a condensed and definite basis from which to develop the interlacing ramifications of a physical scheme: and in each case this is obtained by the use of

*The final general theory:*

*with mobile electric atoms.*

*The function of a model:*

*illustration from sound waves in air:*

*the model may separate off the relevant features.*

[1] *Phil. Trans.* A, 1897: *supra*, p. 11.

a representation which a deeper knowledge may afterwards expand, transform, and even modify in detail. Although, however, it is possible that we may thus be able ultimately to probe deeper into the problem of aethereal constitution, just as the kinetic theory has done in the case of atmospheric constitution, yet there does not seem to be at present any indication whatever of any faculty which can bring that medium so near to us in detail as our senses bring the phenomena of matter: so that from this standpoint there is much to be said in favour of definitely regarding the scheme of a continuous rotationally elastic aether as an ultimate one*.

*Possibilities for further progress.*

### Transition from Molecular to Continuous Scheme.

A formal scheme of the dynamical relations of free aether being postulated after the manner of Maxwell and MacCullagh, and a notion as clear as possible obtained of the aethereal constitution of a molecule and its associated revolving electrons, by aid of the rotational hypothesis, it remains to effect with complete generality the transition between a molecular theory of the aethereal or electric field which considers the molecules separately, and a continuous theory expressed by differential equations which take cognizance only of the properties of the element of volume, the latter alone being the proper domain of mechanical as distinct from molecular theory. This transformation is, as usual, accomplished by replacing summations over the distribution of molecules by continuous integrations over the space occupied by them. In cases where the integrals concerned all remain finite when the origin to which they refer is inside the matter so that the lower limit of the radius vector is null, there is no difficulty in the transition: this is for example the case with the ordinary theory of gravitational forces. But in important branches of the electric theory of polarized media, some of the integral expressions become infinite under these circumstances; and this is an indication that it is not legitimate to replace the effect of the part of the discrete distribution of molecules which is adjacent to the point considered by that of a continuous material distribution. The result of the integration still, however, gives a valid estimate of the effect of the material system *as a whole*, if we bear in mind that the infinite term coming in at the inner limit really represents a finite part of the result depending *solely* on the local molecular configuration, a part whose actual magnitude could be determined only when that configuration is exactly assigned or known. The consideration of this indeterminate part is altogether evaded by means of a general mechanical principle

*Mechanical v. molecular theory:*

*difficulty of the transition.*

*Merger of intractable part in local structure,*

* *I.e.* regarded as a visualized kinematic description of the differential equations of structure.

which I have called the principle of mutual compensation of molecular forcives. This asserts that in such cases, when a finite portion of the effect on a molecule arises from the action of the neighbouring molecules, this part must be omitted from the account in estimating the *mechanical* effect on an element of volume of the medium*; indeed otherwise mechanical theory would be impossible. The mutual, statically equilibrating, actions of *adjacent* molecules determine the structure of the medium, and any change therein involves change in its *local* physical constants and properties, which may or may not be important according to circumstances: but such local action contributes nothing towards polarizing or straining the element of mass whose structure is thus constituted, and therefore nothing to mechanical excitation, unless at a place where there is abrupt change of density[1]. In the memoir above mentioned this molecular principle was applied mainly to determine the mechanical stress in a polarized material medium. It necessarily also enters into the determination of the electrodynamic equations of a moving medium treated as a continuous system, and even of a magnetized medium at rest, from consideration of its molecular constitution.

*so affecting the material moduli.*

*Applications: interface:*

*polarization: moving electric medium.*

It is here intended only to record in precise form the general scheme that results from it, details of demonstration being for the present reserved. Everything being expressed in a continuous scheme per unit volume, let $(u', v', w')$ denote the current of conduction, $(u, v, w)$ the total current of Maxwell, $(f, g, h)$ the electric displacement in the aether and $(f', g', h')$ the electric polarization of the molecules so that the total co-called displacement flux of Maxwell is $(f + f', g + g', h + h')$; let $\rho$ be the volume density of uncompensated electrons or the density of free charge, let $(A, B, C)$ be the magnetization, and $(p, q, r)$ the velocity of the matter with respect to the stagnant aether. As before explained [p. 31, footnote], the convection of the material polarization $(f', g', h')$ produces a *quasi*-magnetization $(rg' - qh', ph' - rf', qf' - pg')$ which adds on to $(A, B, C)$. Also, as before shown, the vector potential of the aethereal field, so far as it comes from the molecular electric whirls which constitute magnetization, is given, for a point outside the magnetism, by

*Formulation of general scheme:*

*dielectric convection:*

*vector potentials of magnetism,*

$$F = \int \left( B \frac{d}{dz} - C \frac{d}{dy} \right) \frac{1}{r} d\tau$$

$$= \int (Bn - Cm) \, r^{-1} dS + \int \left( \frac{dC}{dy} - \frac{dB}{dz} \right) \frac{1}{r} d\tau,$$

---

* Cf. d'Alembert's principle in dynamics.

[1] This exception explains why the mechanical tractions on an interface, determined in [p. 66] as the limit of a gradual transition, are different from the forces on the Poisson equivalent interfacial distribution.

$(l, m, n)$ being the direction vector of $\delta S$, and therefore is that due to a bodily current system curl $(A, B, C)$ together with current sheets on the interfaces. When the point is inside the magnetism, there are still no infinities in the integral expressing $F$, and this transformation of it by partial integration is still legitimate. But the spatial differential coefficients of $(F, G, H)$ are also involved in the forcives of the aethereal field, and with them the case is different: the transformation by parts is then analytically wrong, owing to neglect of the infinite elements at the origin, while in actuality a finite portion of the whole effect arises from the influence of the neighbouring molecules. We have, therefore, by the molecular principle, to separate the infinite elements from the integrals and leave them out of account; and this is effected by employing the second form above for $F$, which differs from the first form only in having got rid of the local terms at the origin in its differential coefficients. Thus it is not merely convenient, but even necessary for a mechanical theory, which considers distributions instead of individual molecules, to replace magnetism by its equivalent continuous current system as here. The *quasi*-magnetism arising from electric convection adds to this equivalent current system the additional bodily terms

*(marginal notes:)* local part transferred,

thus also avoiding analytical fallacies:

*e.g.* magnetism replaced by its originating current, electric flux:

$$\left\{ \frac{d}{dy} (qf' - pg') - \frac{d}{dz} (ph' - rf'), \ldots, \ldots \right\}$$

*(marginal note:)* dielectric convection.

[to the polarization current,] together with surface sheets: thus the volume current so added [to $df'/dt$] has for $x$ component

$$\frac{\delta f'}{dt} - \frac{df'}{dt} - p \left( \frac{df'}{dx} + \frac{dg'}{dy} + \frac{dh'}{dz} \right) - f' \frac{dp}{dx} - g' \frac{dp}{dy} - h' \frac{dp}{dz},$$

where $\dfrac{\delta f'}{dt}$ represents $\dfrac{df'}{dt} + \dfrac{dpf'}{dx} + \dfrac{dqf'}{dy} + \dfrac{drf'}{dz}$, or the rate of change of

$f'$ supposed associated with the moving matter. Combining all these parts, the current and magnetism together are completely represented as regards determination of electric effect by what we may call the *total effective current* $(u_1, v_1, w_1)$, where

*(marginal note:)* Effective total current,

$$u_1 = u' + \frac{dC}{dy} - \frac{dB}{dz} + \frac{df}{dt} + \frac{\delta f'}{dt} - \left( \frac{dpf'}{dx} + \frac{dpg'}{dy} + \frac{dph'}{dz} \right) + p\rho,$$

together with superficial current sheets arising from the true magnetism $(A, B, C)$ and the electric convection. Since $\rho$ is equal to

*(marginal note:)* and surface sheets:

$$\frac{d (f + f')}{dx} + \frac{d (g + g')}{dy} + \frac{d (h + h')}{dz},$$

we may write

$$u_1 = u' + \frac{dC}{dy} - \frac{dB}{dz} + \frac{df}{dt} + \frac{\delta f'}{dt} - \left( f' \frac{dp}{dx} + g' \frac{dp}{dy} + h' \frac{dp}{dz} \right)$$

$$+ p \left( \frac{df}{dx} + \frac{dg}{dy} + \frac{dh}{dz} \right)$$

in which the last term[*] may be expressed as $- \nabla^2 \Psi / 4\pi c^2$.

*is circuital.*   It is to be observed that this effective current satisfies the condition of incompressible flow[1], which by definition (or rather by the aethereal constitution) is necessarily satisfied by the *total current* $(u, v, w)$ of the previous memoirs; for the additional terms which represent the magnetism clearly satisfy the stream relation. The remainder of the scheme of electrodynamic relations is established as in the previous memoirs.

*The un-retarded vector potential:*   Thus, $(F, G, H)$ now representing simply $\int(u_1, v_1, w_1) \, r^{-1} d\tau$ which satisfies the stream relation $dF/dx + dG/dy + dH/dz = 0$ because $(u_1, v_1, w_1)$ is a stream vector, we deduce an electric force $(P, Q, R)$ acting on the electrons, where

$$P = cq - br - \frac{dF}{dt} - \frac{d\Psi}{dx},$$

also an aethereal force $(P', Q', R')$ straining the aether, where

$$P' = (4\pi c^2)^{-1} f = - \frac{dF}{dt} - \frac{d\Psi}{dx},$$

*and electro-static potential.*   the function $\Psi$ being determined in each problem so as to avoid aethereal compression.

*Interfacial conditions.*   Across an abrupt transition, $F, G, H$ and the normal component of $(u_1, v_1, w_1)$ must be continuous, thus making up the *four* necessary and sufficient interfacial conditions. The gradients of $F, G, H$ are, however, not continuous when there is magnetization or dielectric convection, on account of the effective interfacial current sheets before mentioned.

---

[*] This last term represents the convection of the free charge, say $p\rho_0$. For the special case of uniform convection, $(p, q, r)$ constant, the results agree with footnote, *supra*, p. 31.

*Names for types of vectors.*   [1] It is proposed to call a flow vector which obeys this condition a *stream*, the more general term *flow* or *flux* including cases like the variable stage of the flow of heat in which the condition of absence of convergence is not satisfied. The two main classes of physical vectors may be called *fluxes* and *gradients*, the later name including such entities as forces and being especially appropriate when the force is the gradient of a potential. Lord Kelvin's term *circuital* flux has previously been used to denote a *stream* vector; but it is perhaps better to extend it to a general vector which is directed along a system of complete circuits.

The exact value of the mechanical force $(X, Y, Z)$ per unit volume, comes out as

$$X = \left(v - \frac{dg}{dt}\right)\gamma - \left(w - \frac{dh}{dt}\right)\beta + A\frac{da}{dx} + B\frac{da}{dy} + C\frac{da}{dz}$$
$$+ f'\frac{dP'}{dx} + g'\frac{dP'}{dy} + h'\frac{dP'}{dz} + \rho P',$$

where $a = dH/dy - dG/dz - 4\pi A$.

In these formulae, with the exception of the one for $(u_1, v_1, w_1)$ above, $(A, B, C)$ includes the *quasi*-magnetism arising from electric convection, while $(u, v, w)$ is the total electric current that remains after all magnetic effect of whatever type has been omitted. It is to be noted that the final terms in $X$ involve in strictness the aethereal force, instead of the electric force as in [p. 71].

It follows from the formula for $(P, Q, R)$ that

$$\frac{dR}{dy} - \frac{dQ}{dz} = -\frac{\delta a}{dt} + \left(a\frac{d}{dx} + b\frac{d}{dy} + c\frac{d}{dz}\right)p;$$

hence *Faraday's circuital relation holds good provided the velocity $(p, q, r)$ of the matter is uniform in direction and magnitude.*

Again, since $(F, G, H)$ is a stream vector,

$$\frac{dc}{dy} - \frac{db}{dz} = -\nabla^2 F = 4\pi\left(u + \frac{dC}{dy} - \frac{dB}{dz}\right),$$

where $(u, v, w)$ represents the total current of Maxwell, and $(A, B, C)$ the whole of the magnetism and the *quasi*-magnetism of convection: hence

$$\frac{dy}{dy} - \frac{d\beta}{dz} = 4\pi u,$$

so that *Ampère's circuital relation holds, with the above definition of $(a, \beta, \gamma)$, under all circumstances.*

But in circumstances of electric convection these two circuital relations would not usually by themselves form the basis of a complete scheme of equations, as they do when the material medium is at rest.

To complete the scheme, the above dynamical equations must be supplemented by the observational relations connecting the conduction current with the electric force, the electric polarization with the electric force, and the magnetism with the magnetic force. In the simplest case of isotropy these relations are of types

$$u' = oP, \quad f' = (K - 1)/4\pi c^2 P, \quad A = \kappa a + (rg' - qh').$$

It is to be observed that the physical constants which enter into the expression of these relations will presumably be altered by motion through the aether of the material system to which they belong:

their moduli not influenced up to first order by convection.

but because there is nothing unilateral in the system, a reversal of this motion should not change the constants, therefore their alteration must depend on the square of the ratio of the velocity of the system to that of radiation, and would only enter in a second approximation.

The various problems relating to electric convection and optical aberration worked out in §§ 14–16, [pp. 35–42,] will be found to fit into this scheme. I take the opportunity of correcting an erratum in p. 36, lines 22, 23, which should read [as *supra*].

$$\Psi_1 = \tfrac{1}{3}\left(1 + K^{-1}\right) \omega c r^2 + A r^2 \left(\cos^2\theta - \tfrac{1}{3}\right) + A',$$
$$\Psi_2 = B r^{-3} \left(\cos^2\theta - \tfrac{1}{3}\right) + B' r^{-1},$$

with of course different values of the constants.

### *Electrostriction.*

2. In a material dielectric the bodily mechanical forcive is derived from a potential $- (K - 1) F^2/8\pi$, and there is also a normal inward

Formal electric stress in condenser sheet:

traction $KF^2/8\pi$ where it abuts on conductors. For the thin dielectric shell of a condenser this forcive could be balanced by a hydrostatic pressure $(K - 1) F^2/8\pi$ together with a Maxwell stress consisting of a pressure $F^2/8\pi$ along the lines of force and an equal tension at right angles to them: in fact this reacting system gives the correct traction over the faces of the sheet and the correct forcive throughout its substance. If the sheet has an open edge the tractions on that

edge effects local,

edge are however not here attended to; when the sheet is thin these are of small amount, and their effect is usually local, as otherwise the nature of the edge would be an important element. Moreover, in

and balanced.

the most important applications of the formula the edge is of small extent, so that they form a local statically balanced system. The stress

Possible intrinsic stress.

above specified will thus represent the material elastic reaction, provided the strains in the different elements of volume, which correspond to it, can fit together without breach of continuity of the solid material. This condition will be secured if the shell is of uniform

The mechanical elastic reaction: for uniform sheet:

thickness so that $F$ is constant all over it: in that case, therefore, the elastic reaction in the material will make up a pressure $KF^2/8\pi$ along the lines of force and a pressure $(K - 2) F^2/8\pi$ in all directions at right angles to them, which is the result obtained for solids in [p. 121].

depends on nature of its support.

If, however, the coatings of the condenser are not supported by the dielectric shell, the elastic reaction in the shell will be simply a pressure $(K - 1) F^2/8\pi$ uniform in all directions. This is what actually occurs in the case of a fluid dielectric, where such support is not mechanically possible.

It appeared from § 79 that in glass there is actually an increase of Intrinsic expansion in electric field. volume under electric excitation, while the mechanical forces would produce a diminution: and the same is true for most dielectric liquids, the fatty oils being exceptions[1], though by a confusion between action and reaction the result was there stated as the opposite [here amended]. It thus appears that in general an intrinsic expansion, in addition to the effects of the mechanical force, accompanies electric excitation of material dielectrics. This circumstance will perhaps recall to mind Osborne Reynolds' theory of the dilatancy of granular media, which O. Reynolds' principle. explains that the discrete elements of such media tend to settle down *under the mutual influences of their neighbours* so as to occupy the smallest volume, and therefore any disturbing cause has a tendency to increase the volume.

In [p. 127], on the influence of electric polarization on ripple velocity, the result stated for dielectrics should be doubled [as has here been done]. It is to be remarked that a horizontal dielectric liquid surface Instability of surface of a liquid dielectric: becomes unstable in a uniform vertical electric field when the square of the total continuous vertical electric displacement exceeds the moderate value $\dfrac{K_1 K_2 (K_2 + K_1)}{4\pi (K_2 - K_1)^2} \{(\rho_2 - \rho_1)\, gT\}^{\frac{1}{2}}$ electrostatic units, $T$ may be prominent: being the capillary tension. For a conducting liquid instability ensues when the square of the surface electric density exceeds $\dfrac{K_1^2}{4\pi} \{\rho_2 - \rho_1)\, gT\}^{\frac{1}{2}}$ of surface of a conductor. electrostatic units[*].

In exciting a dielectric liquid by the approach of an electrified rod it must often have been noticed that when the rod is brought too near, the liquid spurts out vigorously in extremely fine filaments or Spurting of fine jets from liquid dielectrics. jets[†]: the fineness of the filaments may be explained, in part at any rate, after Lord Rayleigh[‡], without assuming an escape of electricity into the liquid, as arising from the circumstance that it is only narrow crispations of the surface, and not extensive deformations, that become unstable[§].

[1] In the cognate case of magnetization of ferrous sulphate solution, Hurmuzescu finds a contraction of volume.

[*] When $T$ is negligible all wave-lengths below $F^2/2g\,(\rho_1 - \rho_2)$ are unstable.

[†] This remarkable phenomenon in fact defeated an attempt by the writer to verify these formulae.

[‡] "Theory of Sound," § 364, *Phil. Mag.* 1882.

[§] The more static phenomena concerned with very viscous dielectrics, and the effect of moisture, which are important for industrial applications such as insulation of cables, are treated by G. L. Addenbrooke, *Phil. Mag.* May 1927, pp. 1166–1184, referring back to J. W. Swan, *Roy. Soc. Proc.* 1897, with beautiful illustrations, and ultimately to the discharge figures of J. Brown on photographic films and to Lichtenberg's figures.

# 53

# ON THE DYNAMICS OF A SYSTEM OF ELECTRONS OR IONS: AND ON THE INFLUENCE OF A MAGNETIC FIELD ON OPTICAL PHENOMENA.

[From *Transactions of the Cambridge Philosophical Society*, Vol. XVIII (1900), pp. 380–407: being the volume in commemoration of the University Jubilee of Sir George Gabriel Stokes, Lucasian Professor. Received 24th January, 1900.]

*The Dynamics of a System of interacting Electrons or Ions.*

1. In the usual electrodynamic units the kinetic and potential energies of a region of aether are given by

$$T = (8\pi)^{-1} \int (\alpha^2 + \beta^2 + \gamma^2)\, d\tau,$$

$$W = 2\pi c^2 \int (f^2 + g^2 + h^2)\, d\tau,$$

*The field energy in electrodynamics:* wherein $\delta\tau$ represents an element of volume, $(\alpha, \beta, \gamma)$ is the magnetic force which specifies the kinetic disturbance, and $(f, g, h)$ is the aethereal "displacement" which is of the nature of elastic strain. These two vector quantities cannot of course be independent of each other: the constitutive relation between them is, with the present units,

$$\left( \frac{d\gamma}{dy} - \frac{d\beta}{dz}, \frac{d\alpha}{dz} - \frac{d\gamma}{dx}, \frac{d\beta}{dx} - \frac{d\alpha}{dy} \right) = 4\pi \frac{d}{dt}(f, g, h),$$

or say

$$\operatorname{curl}(\alpha, \beta, \gamma) = 4\pi \frac{d}{dt}(f, g, h),$$

*aethereal electric displacement is circuital:* which restricts $(f, g, h)$ to be a stream vector satisfying the equation of continuity: it also confirms the view that $(\alpha, \beta, \gamma)$ is of the nature of a time fluxion or velocity. It is assumed that $(\alpha, \beta, \gamma)$ is itself a *magnetic is assumed to be so.* stream vector, which must be the case if electric waves are of wholly transverse type. On substituting in these expressions $(\xi, \eta, \zeta)$, the independent variable or coordinate of position, of which $(\alpha, \beta, \gamma)$ is the velocity, so that $(\alpha, \beta, \gamma) = d/dt\,(\xi, \eta, \zeta)$, the dynamical equations of the free aether can be directly deduced from the Action formula

$$\delta \int (T - W)\, dt = 0.$$

*The relevant Action in terms of magnetic field.* It is well known that they are identical with MacCullagh's equations for the optical aether, and represent vibratory disturbance propagated by transverse waves.

It will now be postulated that the origin of all such aethereal dis- turbances consists in the motion of electrons, an electron being de- fined as a singular point or nucleus of converging intrinsic strain in the aether, such for example as the regions of intrinsic strain in unannealed glass whose existence is revealed by polarized light, but differing in that the electron will be taken to be freely mobile through- out the medium. For all existing problems it suffices to consider the nucleus of the electron as occupying so small a space that it may be taken to be a point, having an electric charge $e$ associated with it whose value is the divergence of $(f, g, h)$, that is, the aggregate normal displacement $\int (lf + mg + nh)\, dS$ through any surface $S$ en- closing the electron: over any surface not enclosing electrons this integral of course vanishes, by the stream character of the vector involved in it. Faraday's laws of electrolysis give a substantial basis for the view that the value of $e$ is numerically the same for all electrons, but may be positive or negative.

As our main dynamical problem is not the propagation of disturb- ances in the aether, but is the interactions of the electrons which originate these disturbances, it will be necessary to express the kinetic and potential energies of the aether as far as possible in terms of the motions and positions of the electrons. The reduction of $T$ may be effected by introducing the auxiliary variable $(F, G, H)$ defined by

$$\mathrm{curl}\,(F, G, H) = (\alpha, \beta, \gamma).$$

Thus

$$T = (8\pi)^{-1} \int \left\{ \left(\frac{dH}{dy} - \frac{dG}{dz}\right)\alpha + \left(\frac{dF}{dz} - \frac{dH}{dx}\right)\beta + \left(\frac{dG}{dx} - \frac{dF}{dy}\right)\gamma \right\} d\tau$$

$$= (8\pi)^{-1} \int \{(\gamma G - \beta H)\, l + (\alpha H - \gamma F)\, m + (\beta F - \alpha G)\, n\}\, dS$$

$$+ (8\pi)^{-1} \int \left\{ F\left(\frac{d\gamma}{dy} - \frac{d\beta}{dz}\right) + G\left(\frac{d\alpha}{dz} - \frac{d\gamma}{dx}\right) + H\left(\frac{d\beta}{dx} - \frac{d\alpha}{dy}\right) \right\} d\tau$$

$$= (8\pi)^{-1} \int \begin{vmatrix} l, & m, & n \\ F, & G, & H \\ \alpha, & \beta, & \gamma \end{vmatrix} dS + \tfrac{1}{2} \int \left( F\frac{df}{dt} + G\frac{dg}{dt} + H\frac{dh}{dt} \right) d\tau.$$

Now it follows from the definition of $(F, G, H)$ that

$$\nabla^2 F - \frac{d}{dx}\left(\frac{dF}{dx} + \frac{dG}{dy} + \frac{dH}{dz}\right) = -\left(\frac{d\gamma}{dy} - \frac{d\beta}{dz}\right)$$

$$= -4\pi\frac{df}{dt},$$

with two similar equations. Solutions of these equations can be at once obtained by taking $dF/dx + dG/dy + dH/dz$ to be null: this makes $F, G, H$ the potentials of volume distributions throughout the medium of densities $\dot{f}$, $\dot{g}$, $\dot{h}$, together with contributions as yet un-determined from the singular points or electrons. The most general possible solution adds to this one a part $(F_0, G_0, H_0)$ which is the gradient of an arbitrary function of position $\chi$: but this part does not affect the value of $(\alpha, \beta, \gamma)$ through which $(F, G, H)$ has been intro-duced into the problem, so that the definite particular solution is all that is required.

Now the motions of the electrons involve discontinuities, or rather singularities, in this scheme of functions. One mode of dealing with them would involve cutting each electron out of the region of our analysis by a surface closely surrounding it. But a more practicable *Field effect of motion of an electron:* method can be adopted. The movement of an electron $e$ from $A$ to an adjacent point $B$ is equivalent to the removal of a nucleus of outward radial displacement from $A$ and the establishment of an equal one at $B$: in other words it involves a transfer of displacement in the medium by flow out of the point $B$ into the point $A$: now this transfer can be equally produced*, on account of the stream character of the displacement, by a constrained transfer of an equal amount $e$ of dis-placement directly from $A$ to $B$. Hence as regards the dynamics [kinematics] of the surrounding aether, the motion of such a singular point or electron is equivalent to a constrained flow of aethereal dis-placement along its path. The advantage of thus replacing it will be great on other grounds: instead of an uncompleted flow starting from $B$ and ending at $A$, there will now be a continuous stream from $B$ through the surrounding aether to $A$ and back again along the direct line from $A$ to $B$: in other words the displacement will be strictly a stream vector, and in passing on later to the theory of a distribution *a total cir-cuital flux introduced.* of electrons considered as a volume density of electricity, the strictly circuital character of the electric displacement, when thus supple-mented by the flow of the electrons, will be a feature of the analysis.

For greater precision, let us avoid for the moment the limiting idea of a point singularity at which the functions become infinite. An electron will now appear as an extremely small volume in the aether possessing a proportionately great density $\rho$ of electric charge. Its motion will at each instant be represented by an electric flux of intensity $\rho\,(\dot{x}, \dot{y}, \dot{z})$ distributed throughout this volume, which when added to the aethereal displacement now produces a *continuous* cir-cuital aggregate. For present purposes for which the electron is treated as a point and the translatory velocities of its parts are very *Electrons effectively points:* great compared with their rotational velocities, this continuous flow may be condensed into an aggregate flux of intensity $e\,(\dot{x}, \dot{y}, \dot{z})$, concentrated at the point $(x, y, z)$.

* This should read "can be compensated."

At each point in the free aether, outside such nuclei of electrons, the original specification of magnetic force, namely that its curl is equal to $4\pi d/dt$ of the aethereal displacement, remains strictly valid. It has been seen that the effect of the motion of any specified electron, as regards the surrounding aether, is identical with the effect of an impressed change in the stream of aethereal displacement at the place where it is situated: thus the interactions between this electron and the aether will be correctly determined by treating its motion as such an impressed change of displacement. This transformation, however, considers the nucleus as an aggregate: it will not be available as regards the interactions between different parts of the nucleus: thus in the energy function constructed by means of it, all terms involving interaction between the electron as a whole and the aether which transmits the influence of other electrons will be involved; but the intrinsic or constitutive energy of the electron itself, that is the total mutual energy of the constituent parts of the electron exclusive of the energy involved in its motion as a whole through the aether, will not be included: this latter part is in fact supposed (on ample grounds) to be unchangeable as regards all the phenomena now under discussion, the nuclei of the electrons being taken to occupy a volume extremely small in comparison with that of the surrounding aether[1]. *their motion involving an impressed aethereal displacement:*

*internal structure not involved:*

This principle leads to an expression for the force acting on each individual moving electron, which is what is wanted for our present purpose. But the equations of ordinary electrodynamic theory belong to a dense distribution of ions treated by continuous analysis, and we have there to employ the average equations that will obtain for an effective element of volume of the aether containing a number of electrons that practically is indefinitely great. *the resulting dynamics.*

We derive then the equations of the aether considered as containing electrons from those of the uniform aether itself by adding to the changing aethereal displacement $(\dot{f}, \dot{g}, \dot{h})$ the flux of the electrons of type $e\ (\dot{x}, \dot{y}, \dot{z})$ wherever electrons occur. In the transformed expression for $T$ we can, as already explained, treat the part of the surface integral belonging to the surface cutting an electron out of the region of integration (as well as any energy inside that surface) as intrinsic energy of the electron, of unchanging amount[2], which is not concerned in the phenomena because it does not involve the state of any other electron. The contribution from the surface integral over the infinite sphere we can take to be zero if we assume that all the disturbances *The kinetic energy in their aethereal field,*

---

[1] For a treatment on somewhat different lines cf. *Phil. Trans.* 1897 A [*supra*, p. 39], or *Aether and Matter*, Ch. vi, Camb. Univ. Press, 1900.

[2] It may be formally verified, after the manner of the formula for $T$ in § 2, that this amount tends to a definite limit as the surface surrounds the electron more and more closely.

of electrons are in a finite region: the truth of this physical axiom can of course be directly verified.

We have therefore generally

$$T = \tfrac{1}{2} \int (Fu + Gv + Hw)\, d\tau,$$

wherein

$$(F, G, H) = \int (u, v, w)\, r^{-1} d\tau:$$

and in these expressions the total electric current $(u, v, w)$ will consist of a continuous part $(\dot{f}, \dot{g}, \dot{h})$ which is not electric flow at all, and a discrete electric flux or *true* current of amount $e\,(\dot{x}, \dot{y}, \dot{z})$ for any as expressed in terms of the various types of electric flux, including a factitious past. electron $e^*$. When the electrons are considered as forming a volume density of electrification, this latter will be considered as continuous true electric flow constituted as an aggregate of all the different types of conduction current, convection current, polarization current, etc., that can be recognized in the phenomena, each being connected by an experimental constitutive relation with the electric force which originates it. The orbital motions of the electrons in the molecule cannot, however, be thus included in an electric flux, but must be averaged separately as magnetization. Neither the true current nor the aethereal displacement current taken separately need satisfy the condition of being a stream, but their sum, the total current of Maxwell, always satisfies this condition.

2.   The present problem being that of the interactions of individual electrons transmitted through the aether, it will be necessary to retain

---

* When the subject is approached more closely, and also true current of electrons is introduced in place of the total current of the text, the element of time has to come into the potentials. The dynamics of the system is set in Potentials of convected true sources: a frame of space and time with regard to which the potentials of influencing bodies are reckoned: these potentials, now retarded, will depend on the velocities of their sources relative to the frame of reference. If an ion is moving away with velocity $v_r$ relative to the frame of the analysis, its vector potential at the origin is $ev/r'$, where $r'$ is its effective distance at the time $t$ of the frame, so $r' = r\,(1 - v_r/c)$. Moreover, as it is receding it remains longer in touch: instead of being brought to bear for a time $dt$ as if it were at rest it is operative in this frame for a time $dt/(1 - v_r/c)$. Or if it is regarded as an extended source, instead of the same density of volume elements coming to bear in the frame for a longer time, we might assert that increased density of the source is brought to bear on the influenced region at the origin for unchanged time. the Liénard formula: includes a factor transferred from the Action. The latter is the usual mode of procedure in obtaining the Liénard potential formula. But however the argument be expressed, it is the element of time of the Action formulation relative to the frame that is operative in it at bottom. Cf. Vol. I, *supra*, Appendix IV. Electrodynamic relativity requires also that the phenomena are strictly independent of the frame: but that is an affair only of the second order, of the sqaure of $v/c$, as regards these potentials, and is far more intricate.

these electrons as distinct entities. The value of $(F, G, H)$ at any point is therefore of type

$$F = \int \frac{1}{r} \frac{df}{dt} d\tau + \Sigma \frac{e\dot{x}}{r},$$

in which $r$ represents the distance of the point from the element of volume in the integral and from the electron respectively. Thus

$$T = \tfrac{1}{2} \int\!\!\int (\dot{f}_1\dot{f}_2 + \dot{g}_1\dot{g}_2 + \dot{h}_1\dot{h}_2)\, r_{12}^{-1} d\tau_1 d\tau_2$$

$$+ \Sigma e\dot{x} \int \dot{f}_2 r_{12}^{-1} d\tau_2 + \Sigma e\dot{y} \int \dot{g}_2 r_{12}^{-1} d\tau_2 + \Sigma e\dot{z} \int \dot{h}_2 r_{12}^{-1} d\tau_2$$

$$+ \Sigma\Sigma e_1 e_2\, (\dot{x}_1\dot{x}_2 + \dot{y}_1\dot{y}_2 + \dot{z}_1\dot{z}_2)\, r_{12}^{-1},$$

in which each pair of electrons occurs only once in the double summation.

Also $$W = 2\pi c^2 \int (f^2 + g^2 + h^2)\, d\tau.$$

In omitting the intrinsic energy of an electron and only taking into account the energy terms arising from the interaction of its electric flux with the other electric fluxes in the field, we have, however, neglected a definite amount of kinetic energy arising from the motion of the strain configuration constituting the electron and proportional to the square of its velocity: this will be the translational kinetic energy

$$T_0 = \tfrac{1}{2} Le^2 (\dot{x}^2 + \dot{y}^2 + \dot{z}^2):$$

or we may write

$$T_0 = \tfrac{1}{2} m (\dot{x}^2 + \dot{y}^2 + \dot{z}^2),$$

where $m$ is thus the coefficient of inertia or "mass" of the electron[*], which may either be wholly of electric origin or may contain elements arising from other sources.

This transformation has introduced the positions of the electrons and the aether strain $(f, g, h)$ as independent variables. It is *necessary*, for the dynamical analysis, thus to take the aether strain as the independent variable, instead of the coordinate of which $(\alpha, \beta, \gamma)$ is the velocity, which at first sight appears simpler. For part of this strain is the intrinsic strain around the electrons; and the deformations of the medium by which it may be considered to have been primordially produced must have involved the discontinuous processes required to fix the strain in the medium, as otherwise it could not be permanent or intrinsic. If the latter coordinates were adopted the complete specification of the deformation of the medium must include these processes of primary creation of the electrons, and the medium would

---

[*] Assumed to be of unchanging configuration relative to its direction of motion, or else adaptable practically instantly thereto.

have to be dissected in order to reveal the discontinuities, after the manner of a Riemann surface in function theory[1].

3. We have now to apply dynamical principles to the specification of the energies of the medium thus obtained. The question arises as to what *are* dynamical principles. It may reasonably be said that an answer for the dynamics of known systems constituted of ordinary matter is superfluous, as the Laws of Motion formulated by Newton practically cover the case. Waiving for the present the question *Generalized dynamics:* whether the foundations of that subject are so simple as may appear, the present case is one not of ordinary matter but of a medium unknown to direct observation: and its disturbance is expressed in terms of vectors as to the kinematic nature of which we have here abstained from making any hypothesis.

Now the dynamics of material systems was systematized by Lagrange in 1760 into equations which amount to the single variational formula

$$\delta \int (T - W)\, dt = 0,$$

in which the variation is to be taken subject to constant time of *as subsumed under the* passage from the initial to the final configuration, and subject to *Action* whatever relations, involved in the constitution of the system, there *principle,* may be connecting the variables when these are not mutually independent—the only restriction being that these latter relations are really constitutive, and so do not involve the actual velocities of the motion although they may involve the time. This equation is known to include the whole of the dynamics of material systems in the most general and condensed manner that is possible. It will now be introduced as a *hypothesis* that the cognate equation is the complete expression of the dynamics of the *ultra*-material systems here under consideration. Even in the case of ordinary dynamics it can be held that there is no final resting-place in the effort towards exact formulation of dynamical phenomena, short of this Action principle: in our present *by a formu-* more general sphere of operations the very meaning of a dynamical *lation purely* principle must be that it is a deduction from the Action principle. *scalar.* This attitude will not be uncongenial to the school of physicists which recognizes in dynamical science only the shortest and most compact specification of the actual course of events.

[1] More concretely, the relation curl $(a,\ \beta,\ \gamma) = 4\pi\ (\dot{f},\ \dot{g},\ \dot{h})$ involves

$$\int (l\dot{f} + m\dot{g} + n\dot{h})\, dS = 0:$$

now $\int (lf + mg + nh)\, dS$ is not zero but is equal to $\Sigma e$: hence the displacements, of the kind whose velocity is $(a,\ \beta,\ \gamma)$, that are required to introduce the existing intrinsic strain must involve discontinuous processes. Cf. *Aether and Matter,* Appendix E.

We have then to apply the Principle of Action to the present case. In the first place the coordinates in terms of which $T$ and $W$ are expressed are not all independent, for when the distribution of $(f, g, h)$ is given that of the electrons is involved. The connection between them is completely specified by the relation

$$\int \left( \frac{df}{dx} + \frac{dg}{dy} + \frac{dh}{dz} \right) d\tau = \Sigma e :$$

provided this is supposed to hold for every domain of integration, great or small, it will follow that the electrons are the poles of a circuital or stream vector $(f, g, h)$. If then we write

$$\Omega = \int \Psi \left( \frac{df}{dx} + \frac{dg}{dy} + \frac{dh}{dz} \right) d\tau - \Sigma e \Psi,$$

the variational equation will by Lagrange's method assume the form

$$\delta \int (T + T_0 - W + \Omega) \, dt = 0,$$

in which $\Psi$ is a function of position, initially undetermined but finally to be determined so as to satisfy the above condition restricting the independence of the coordinates.

We have to vary this equation with respect to the displacement $(f, g, h)$ belonging to each element of the aether, supposed on our theory to be effectively at rest, and with respect to the position $(x, y, z)$ of each electron. All these variations being now treated as independent, the coefficient of each of them must vanish, at all points of the aether and for all electrons involved in it.

We now proceed to the variation. Bearing in mind that so far as regards aethereal displacement

$$\tfrac{1}{2} \int F f \, d\tau \text{ involves } \tfrac{1}{2} \int \int f_1 f_2 r_{12}^{-1} d\tau_1 d\tau_2, \text{ that is } \Sigma \Sigma f_1 f_2 r_{12}^{-1} \delta\tau_1 \delta\tau_2,$$

because each pair of elements appear together twice in the double integral of a product, but only once in a double summation, we obtain as the terms involving $f$ in the complete variation

$$\delta \int dt \int F f \, d\tau - 4\pi c^2 \int dt \int f \delta f \, d\tau + \int dt \int \Psi \frac{d\delta f}{dx} \, d\tau,$$

leading, through the usual integration by parts, to

$$\left| \int F \delta f \, d\tau \right|_t - \int dt \int \dot{F} \delta f \, d\tau - 4\pi c^2 \int dt \int f \delta f \, d\tau$$

$$+ \left| \int dt \int \int \Psi \delta f \, dy \, dz \right|_x - \int dt \int \frac{d\Psi}{dx} \delta f \, d\tau.$$

The coefficient of $\delta f$ must vanish in the volume integral, giving

$$4\pi c^2 f = -\frac{dF}{dt} - \frac{d\Psi}{dx}. \qquad (i)$$

The marginal notes read:

The electro-dynamic Action:

including the electronic restricting condition:

process of its variation:

Similar expressions hold for $g$ and $h$. Again, the terms in the variation involving the electron $e$ at $(x, y, z)$ are

$$\delta \int dt\, e\, (\dot{x}F + \dot{y}G + \dot{z}H) + \tfrac{1}{2}m\,\delta \int dt\, (\dot{x}^2 + \dot{y}^2 + \dot{z}^2) - \delta \int dt\, e\Psi,$$

yielding as regards variation of the position of this electron

$$\int dt\, e\, (F\delta\dot{x} + G\delta\dot{y} + H\delta\dot{z} + \dot{x}\delta F + \dot{y}\delta G + \dot{z}\delta H)$$

$$+ m \int dt\, (\dot{x}\delta\dot{x} + \dot{y}\delta\dot{y} + \dot{z}\delta\dot{z}) - \int dt\, e\delta\Psi,$$

in which $\delta\dot{x}$ means the change of the velocity of the electron, so that we have on integration by parts

$$e\left| F\delta x + G\delta y + H\delta z \right|_t - \int dt\, e\left\{ \left( \frac{DF}{dt}\delta x + \frac{DG}{dt}\delta y + \frac{DH}{dt}\delta z \right) \right.$$

$$\left. - \dot{x}\left( \frac{dF}{dx}\delta x + \frac{dF}{dy}\delta y + \frac{dF}{dz}\delta z \right) + \dots \right\} + m\left| \dot{x}\delta x + \dot{y}\delta y + \dot{z}\delta z \right|_t$$

$$- m \int dt\, (\ddot{x}\delta x + \ddot{y}\delta y + \ddot{z}\delta z) - \int dt\, e\left( \frac{d\Psi}{dx}\delta x + \frac{d\Psi}{dy}\delta y + \frac{d\Psi}{dz}\delta z \right),$$

where $DF/dt$ must represent the rate of change of $F$ at the electron as it moves, namely

$$\frac{DF}{dt} = \frac{dF}{dt} + \dot{x}\frac{dF}{dx} + \dot{y}\frac{dF}{dy} + \dot{z}\frac{dF}{dz}.$$

leading to the force on a moving ion,
The vanishing of the coefficient of $\delta x$ for each element of volume gives

$$m\ddot{x} = e\left( -\frac{DF}{dt} + \dot{x}\frac{dF}{dx} + \dot{y}\frac{dG}{dx} + \dot{z}\frac{dH}{dx} - \frac{d\Psi}{dx} \right)$$

$$= e\left( \gamma\dot{y} - \beta\dot{z} - \frac{dF}{dt} - \frac{d\Psi}{dx} \right). \qquad (ii)$$

Similar expressions hold good for $m\ddot{y}$ and $m\ddot{z}$.

The form of $W$ shows that $4\pi c^2$ is the coefficient of aethereal elasticity corresponding to the type of displacement $(f, g, h)$: the right-hand sides of equations (i) are therefore the expressions for the components of the forcive $(P', Q', R')$ inducing aethereal displace-

and to aethereal force in the field:
ment: thus this force, which will be called the aethereal force, is given by equations of type

$$P' = -\frac{dF}{dt} - \frac{d\Psi}{dx}.$$

The form of equation (ii) shows that the right-hand side is the component of the force $e\,(P, Q, R)$ inducing movement of an electron $e$: this force reckoned per unit electric charge is called the electric force $(P, Q, R)$ and is given by

$$P = \gamma\dot{y} - \beta\dot{z} - \frac{dF}{dt} - \frac{d\Psi}{dx},$$

the relation between them.
or, in terms of physical quantities only, by

$$P = \gamma\dot{y} - \beta\dot{z} + 4\pi c^2 f.$$

We do not now go into the case of a magnetically polarized material system, for which *in certain connections*[1] $(a, b, c)$ replaces $(\alpha, \beta, \gamma)$ in this formula.

These expressions for the aethereal force and the electric force, together with a complete specification of the electric current and the experimentally determined constitutive relations of the medium, form the foundation of the whole of electrical theory.

### Motion in an Impressed Magnetic Field.

When the electrons or ions constituting a molecule describe their orbital motions in a uniform field $(\alpha_0, \beta_0, \gamma_0)$, its influence is represented by an addition to the vector potential $(F, G, H)$ of the term

$$- (\gamma_0 y - \beta_0 z, \ \alpha_0 z - \gamma_0 x, \ \beta_0 x - \alpha_0 y).$$

Thus

$$T_0 + T = \tfrac{1}{2} \Sigma m \, (\dot{x}^2 + \dot{y}^2 + \dot{z}^2) + \tfrac{1}{2} \Sigma e \, (F\dot{x} + G\dot{y} + H\dot{z})$$

$$+ \tfrac{1}{2} \int (F\dot{f} + G\dot{g} + H\dot{h}) \, d\tau.$$

$$- \tfrac{1}{2} \Sigma e \begin{vmatrix} \dot{x} & \dot{y} & \dot{z} \\ x & y & z \\ \alpha_0 & \beta_0 & \gamma_0 \end{vmatrix} - \tfrac{1}{2} \int \begin{vmatrix} \dot{f} & \dot{g} & \dot{h} \\ x & y & z \\ \alpha_0 & \beta_0 & \gamma_0 \end{vmatrix} d\tau.$$

As the aether is stagnant, so that the position of the element of volume $\delta\tau$ is fixed, these new terms will not modify the formula for the aethereal force $(P', Q', R')$ unless the impressed magnetic field varies with the time: but they will modify the electric forces acting on the ions by the addition of the term

*Modification by an extraneous magnetic field:*

$$- (\gamma_0 \dot{y} - \beta_0 \dot{z}, \ \alpha_0 \dot{z} - \gamma_0 \dot{x}, \ \beta_0 \dot{x} - \alpha_0 \dot{y}).$$

### The System referred to a Rotating Frame.

It is part of the Action principle, of which the validity is at the foundation of this analysis, that its formal expression is not affected by constitutive relations involving the time explicitly, provided they do not involve the velocities of the actual motion. Let then the system be referred to axes of coordinates rotating with angular velocity $(\omega_x, \omega_y, \omega_z)$ measured with reference to their instantaneous positions, these quantities being either constant or assigned functions of the time. For the velocity, instead of $(\dot{x}, \dot{y}, \dot{z})$ there must now be substituted, in the formula for $T - W$,

*the Action principle still available:*

$$(\dot{x} - y\omega_z + z\omega_y, \ \dot{y} - z\omega_x + x\omega_z, \ \dot{z} - x\omega_y + y\omega_x),$$

and for $(\dot{f}, \dot{g}, \dot{h})$ there must be substituted

$$(\dot{f} - g\omega_z + h\omega_y, \ \dot{g} - h\omega_x + f\omega_z, \ \dot{h} - f\omega_y + g\omega_x),$$

[1] Cf. *loc. cit. ante*, p. 161.

while $(x, y, z)$ remain unchanged. Referred to these moving axes the kinetic energy, which was, so far as it involves the ion $e_1 (\dot{x}_1, \dot{y}_1, \dot{z}_1)$, given by

$$T_0 + T = \tfrac{1}{2} m_1 (\dot{x}_1{}^2 + \dot{y}_1{}^2 + \dot{z}_1{}^2) + e_1 (F_1 \dot{x}_1 + G_1 \dot{y}_1 + H_1 \dot{z}_1) + \dots,$$

where $(F_1, G_1, H_1)$ is the value of the vector potential at the point $(x_1, y_1, z_1)$, has now additional terms which on neglecting the square of the angular velocity are

$$- m_1 \begin{vmatrix} \dot{x}_1 & \dot{y}_1 & \dot{z}_1 \\ x_1 & y_1 & z_1 \\ \omega_x & \omega_y & \omega_z \end{vmatrix} + e_1 \begin{vmatrix} x_1 & y_1 & z_1 \\ F_1 & G_1 & H_1 \\ \omega_x & \omega_y & \omega_z \end{vmatrix} + e_1 (\dot{x}_1 \delta' F_1 + \dot{y}_1 \delta' G_1 + \dot{z}_1 \delta' H_1),$$

wherein
$$\delta' F_1 = \delta' \left( \Sigma \frac{e_2 \dot{x}_2}{r_{12}} + \int \frac{f_2 d\tau_2}{r_{12}} \right)$$

$$= \Sigma \frac{e_2 (\omega_y z_2 - \omega_z y_2)}{r_{12}} + \int \frac{\omega_y h_2 - \omega_z g_2}{r_{12}} d\tau.$$

The exact dynamical equations referred to moving axes may now be directly obtained by application of the Action principle.

As regards the electron $e_1$, the first of these terms is the same as
result.
that due to an impressed magnetic field given by

$$(\alpha_0, \beta_0, \gamma_0) = \frac{2m_1}{e_1} (\omega_x, \omega_y, \omega_z).$$

The others give rise to terms in the electric forces which are small compared with the internal electrodynamic forces of the system itself when the angular velocity is small: and in our applications these latter will be themselves negligible compared with the electrostatic forces.

### Mutual Forces of Electrons.

When a system of electrons or ions is moving in any manner, with velocities of an order lower than that of radiation, the surrounding aether strain may be taken as at each instant in an equilibrium con-
*Mutual ionic* formation: thus the positional forces between the electrons are simply
*forces mainly*
*electrostatic:* their mutual electrostatic attractions. As regards kinetic effects, the disturbance in the aether can be considered as determined by the motion of the electrons at the time considered, so that the kinetic energy can be expressed entirely in terms of the motions of the elec-
*the potential* trons; and the motional forces between two of them are derived in
*of the kinetic* the Lagrangian manner from the term in this total kinetic energy
*part,*

$$\frac{e_1 e_2}{r_{12}} (\dot{x}_1 \dot{x}_2 + \dot{y}_1 \dot{y}_2 + \dot{z}_1 \dot{z}_2) + \tfrac{1}{2} e_1 v_1 e_2 v_2 \frac{d^2 r_{12}}{ds_1 ds_2},$$

where $ds_1, ds_2$ are elements of their paths described with velocities $v_1, v_2$. The Weberian theory of moving electric particles involves on the other hand a kinetic energy term $\frac{1}{2}e_1 e_2 r_{12}^{-1}(dr_{12}/dt)^2$: in the field simplification for a cyclic flux. of the electrodynamics of ordinary currents it however yields equivalent results as regards mechanical force, and the electromotive force induced round a circuit, though not as regards the electric force at a point.

### The Zeeman Effect.

4. On the hypothesis that a molecule is constituted of a system of revolving ions, a magnetic field $H$ impressed in a direction $(l, m, n)$ adds to the force acting on an ion of effective mass $m$ and charge $e$, situated at the point $(x, y, z)$, the term

$$eH\,(n\dot{y} - m\dot{z},\ l\dot{z} - n\dot{x},\ m\dot{x} - l\dot{y}),$$

so that its dynamical equations are modified by change of $\ddot{x}, \ddot{y}, \ddot{z}$ into

$$\ddot{x} - \kappa\,(n\dot{y} - m\dot{z}),\ \ddot{y} - \kappa\,(l\dot{z} - n\dot{x}),\ \ddot{z} - \kappa\,(m\dot{x} - l\dot{y}),$$

where $\kappa = eH/m$, $e$ being in electromagnetic units.

If the ratio $e/m$ is the same for all the ions concerned in the motion, so is $\kappa$, and this alteration of the dynamical equations of the molecule will be, to the first order of $\kappa$, the same as would arise from a rotation of the axes of coordinates to which the system is referred, with angular velocity $\frac{1}{2}\kappa$ around the axis of the impressed magnetic field. Hence the alteration produced in the orbital motions is simply equivalent to a rotation, equal and opposite to this, imposed on the whole system*. Each line in the spectrum would thus split up into two lines consisting of radiations circularly polarized around the direction of the magnetic field, and with difference of frequencies constant all along the spectrum, namely $-\frac{1}{2}\kappa/2\pi$, together with a third line polarized so that its electric vibration is along the same axis while the frequency is unaltered. In fact each Fourier vibration of an ion, which previously consisted of a component disturbance of the type of an elliptic harmonic motion, is no longer of harmonic type when the precessional rotation $-\frac{1}{2}\kappa$ is imposed on it—this precession being imposed additively on the different constituents of the total motion: but it can be resolved into a rectilinear vibration parallel to the axis, and two circular ones around it, each of which maintains its harmonic type after the rotation is impressed and thus corresponds to a spectral line, and which are differently modified as stated. These three spectral lines would be expected to be of about equal intensities[1].

The Zeeman magnetic effect admits of analysis for a wide class of atomic types: there equivalent to a simple precession of the atom.

It is, however, essential to this simple state of affairs that the charges belonging to all the ions that are in orbital motion under their mutual influences should be of the same sign, as otherwise $e/m$ Restrictions on the relevant atomic model:

---

* Not on the actual motional system, but on a free system of identical type in different phases. As $e$ is negative, so is $\kappa$. See § 6, *infra*.

[1] For more detailed statement, cf. *Phil. Mag.* Dec. 1897: *supra*, p. 140.

could not be the same for all. It is also essential that the ions of opposite sign, or the other centres of attraction under which the orbits are described, should be carried round as well as the orbits with this small angular velocity $-\frac{1}{2}\kappa$ in so far as they are not symmetrical with regard to its axis.

*the effect requires massive positive nuclei:* If we admit the hypothesis that the effective masses of these positive ions, or other bodies to which the negative ions are attracted, are large compared with those of the negative ions themselves, this state of superposed uniform rotation of the whole system may still be expected practically to ensue from the imposition of the magnetic field. For under the action of the mutual constitutive forces in the molecule, the orbital motions of the larger masses will take place with smaller velocities. As the additional forces introduced by the magnetic field are proportional to the velocities, they will thus also be smaller for the positive ions. Let us then suppose these larger masses to be constrained to the above exact uniform rotation, with angular velocity $\omega'$, along with the negative ions, and find the order of magnitude of the forces that must be impressed on them in order to maintain this constraint. The motion of the negative ions will, as has been seen, be entirely free, the forces due to the magnetic field exactly sufficing to induce the additional rotational motion. As *deviations arising when there are several such nuclei,* regards a positive ion of effective mass $m$, the radial and transversal forces, in the plane perpendicular to the axis of the magnetic field, that are required to maintain the motion will be altered from

$$m\left(\ddot{r} - r\omega^2\right) \text{ and } \frac{m}{r}\frac{d}{dt}\left(r^2\omega\right)$$

to

$$m\left\{\ddot{r} - r\left(\omega + \omega'\right)^2\right\} \text{ and } \frac{m}{r}\frac{d}{dt}\left\{r^2\left(\omega + \omega'\right)\right\}.$$

Thus, $\omega'$ being small compared with $\omega$, the new forces required will be

$$-2mr\omega\omega' \text{ and } \frac{m}{r}\frac{d}{dt}\left(r^2\omega'\right);$$

whereas the force arising from the magnetic field acting on an ion moving with velocity $v$ is $2mv\omega'$ at right angles to its path. These two systems of forces are for each ion of the same order of magnitude: thus the forces required to maintain the imposed uniform rotation *may be slight.* in the case of the massive positive ions are small* compared with the

*Atomic model with both planets and satellites.*    * This regards them as slow and concentrated. The dynamical virial (*supra*, p. 134) appears to have application here. If the velocities of the massive positive nuclei were not thus small, the relation that the mean potential energy is equal by the virial theorem to *minus* twice mean kinetic energy, would involve large negative potential energy, so that the negative electrons would have to be concentrated on the positive nuclei in close satellite systems. Thus the

magnetic part of the forces acting on the negative ions. If these maintaining forces are absent, the system can still be regarded as a molecule in its undisturbed motional configuration rotating with uniform angular velocity, but subject to disturbing forces equal and opposite to those required to thus maintain it. Now this undisturbed motional configuration is a stable one: thus the effect of these slight disturbing forces is to modify it, but to an extent much smaller than the uniform rotation induced by the magnetic field. *(margin: Protection of the stability.)*

Our proposition is thus extended to a molecule consisting of an interacting system, constituted of equal negative ions together with much more massive positive ions, and also if so demanded of other massive sources of attraction. It would, however, be wrong to con- sider each negative electron as describing an independent elliptic orbit of its own, unaffected by the mutual attractions exerted between it and the other moving negative electrons: for the attractions between ions constitute the main part, if not the whole, of the forces of chemical affinity. But without requiring any knowledge of the con- stitution of the molecular orbital system, the Zeeman triplication of the lines, with equal intervals of frequencies for each line, will hold good wherever the conditions here stated obtain. *(margin: General planetary atomic model is fundamental.)*

It appears from the observations that the difference of frequencies of the components magnetically separated is not constant for all lines of the spectrum: so that this simple state of affairs does not hold in the molecule. The difference of frequencies seems, however, to be sensibly constant for those lines of any element which belong to the same series, as well as for those lines of homologous elements which belong to corresponding series[1]; a result which cannot fail to be fundamental as regards the dynamical structure of molecules, and which supports the suggestion that in a general way the lines of the same series arise from the motions of the same ion or ionic group in the molecule, executed under similar conditions. The directions of the circular polarizations of the constituent lines were shown by Zeeman to be in general such as would correspond in this kind of way to the motions of a system of negative ions in a steady field of force. *(margin: Preston's rule of chemical types: its sig- nificance.)*

It remains to be considered whether we are right in thus taking the stresses transmitted between the electrons, through the aether, as those arising from the configuration of the electrons alone, and in neglecting altogether the motional forces between them. The former *(margin: The internal forces pre- dominantly electrostatic:)*

system would be an orbital aggregation of standard systems in slow motion each with one positive nucleus.

The Zeeman ordered multiples are, however, more deep-seated, and are now interpreted into terms of atomic *quanta* of Action.

[1] Preston, *Phil. Mag.* Feb. 1899.

assumption is equivalent to taking the strain in the surrounding aether to be at each instant in an equilibrium state: this will be legitimate, because an aethereal disturbance will travel over about $10^3$ diameters of the molecule in one of the periods concerned—the error is in fact of order $10^{-6}$. The motional forces between two electrons are of type [Vol. I, *supra*, p. 524], as regards one of them,

$$\left(\frac{\delta}{dt}\frac{d}{d\dot{x}_1} - \frac{d}{dx_1}\right) e_1 e_2 \left(\frac{\dot{x}_1\dot{x}_2 + \dot{y}_1\dot{y}_2 + \dot{z}_1\dot{z}_2}{r_{12}} + \tfrac{1}{2}v_1 v_2 \frac{d^2 r_{12}}{ds_1 ds_2}\right).$$

To obtain a notion of orders of magnitude, let us consider the special case of two electrons $+ e, - e$ describing circular orbits round each other with radius $r$. Then $mv^2/\tfrac{1}{2}r = c^2 e^2/r^2$, while Zeeman's measurements give $e/m = 10^7$ : thus $v^2 = \tfrac{1}{2}c^2 e^2/mr$, so that, taking $r$ to be $10^{-8}$, $e = 10^{-21}$, we obtain $v = 10^{-3}c$; thus the orbital period comes out just

giving periods of optical order: of the order of the periods of ordinary light, which is an independent indication that the general trend of this way of representing the phenomena is legitimate. With these orders of magnitude, the terms in the motional forces between two electrons are of orders $e_1 e_2 \ddot{x}/r$, $e_1 e_2 \dot{x}^2/r^2$ as compared with their statical attraction of order $c^2 e_1 e_2/r^2$ and the forces arising from the impressed magnetic field $H$ of order

but the electro-kinetic are comparable with the Zeeman forces. $e\dot{x}H$; the ratios are thus of the order of $10^{-6}$ to $1$ to $3.10^{-9}\, H$. Thus when $H$ exceeds $10^8$, the forces of the impressed magnetic field are more important than the motional forces between the ions; and in all cases the effects arising from these two causes are so small that they can be taken as independent and simply additive.

### The Zeeman Effect of Gyrostatic Type.

5. Sensible damping of the vibrations of the molecule owing to radiation cannot actually come into account, because the sharpness and fixity of position of the spectral lines show that the vibrations subsist for a large number of periods without sensible change of type. In fact it has been seen above that the motion of the system of electrons, on the most general hypothesis, is determined by the principle of Action in the form

$$\delta \int (T' - W)\, dt = 0,$$

where $\qquad T' = \tfrac{1}{2}\Sigma m\, (\dot{x}^2 + \dot{y}^2 + \dot{z}^2) + \tfrac{1}{2}\Sigma\kappa \begin{vmatrix} \dot{x} & \dot{y} & \dot{z} \\ x & y & z \\ l & m & n \end{vmatrix}$ :

thus it comes under the same class as the motion of a dynamical system involving latent constant cyclic momenta, the Lagrangian function for such a system, as modified through the elimination of the velocities corresponding to these momenta by Routh, Kelvin, and von Helmholtz, being of this type. The influence of the impressed magnetic field is thus of the same character as that of gyrostatic quality imposed on a free system: and the problem comes under the general dynamical theory of the vibrations of cyclic systems[1]. In the special case above considered of massive positive ions, we can thus assert that the motion relative to the moving axes is the same as the actual motion of the system with its period altered through slight gyrostatic attachments to these positive ions. It is, moreover, known from the general theory of cyclic systems that each free period is either wholly real or else a pure imaginary, whenever the unmodified system is stable so that its potential energy is essentially positive: thus on no view can a magnetic field do anything towards extinguishing or shortening the duration of the free vibrations of the molecule, it only modifies their periods and introduces differences of phase between the various coordinates into the principal modes of vibration of the system. *[margin: The Zeeman problem one of cyclic systems: inferences involved.]*

In the general case when $\kappa$ is not the same for each ion in an independently vibrating group in the molecule, the simple solution in terms of a bodily rotation fails, and it might be anticipated that the equation of the free periods would involve the orientation of the molecule with regard to the magnetic field. But if that were so, these periods would not be definite, and instead of a sharp magnetic resolution of each optical line there would be only broadening with the same general features of polarization. To that extent the phenomenon was in fact anticipated from theory, except as regards its magnitude. The definite resolution of the lines is, however, an addition to what would have been predicted on an adequate theory, and thus furnishes a clue towards molecular structure. *[margin: The sharp resolution.]*

### A Possible Origin of Series of Double Lines.

The definiteness and constancy in the mode of decomposition of a molecule into atoms show that these atoms remain separate structures when combined under their mutual influence in the molecule, instead of being fused together. Each of them will therefore preserve its free periods of vibrations, slightly modified, however, by the proximity of the other one. For the case of a molecule containing two identical atoms revolving at a distance large compared with their own dimensions, each of these identical periods would be doubled[2]: *[margin: A possible type of molecular spectra.]*

---

[1] Cf. Thomson and Tait, *Nat. Phil.* ed. 2, Part I, pp. 370–416.

[2] In illustration of the way this can come about, consider two parallel cylindrical vortex columns of finite section in steady rotation round each other. Each by itself has a system of free periods for crispations running round its section: when one of them is rotating round the other, the velocity of the crispations which travel in the direction of rotation is different from the

thus the series of lines belonging to the atom would become double lines in the spectrum of the molecule. It has been remarked that the series in the spectra of inactive elements like argon and helium consist of single lines, those of univalent elements such as the sodium group where the molecule consists of two atoms, of double lines, while those of elements of higher valency appear usually as triple lines.

In other words, a diad molecule consists of the two atoms rotating round each other with but slight disturbance of the internal constitution of each of them. Their vibrations relative to a system of axes of reference rotating along with them will thus be but slightly modified: relative to axes fixed in space there must be compounded with each vibration the effect of the rotation, which may be either right-handed or left-handed with respect to the atom: thus on the same principles as above each line will be doubled. If the lines of a spectral series are assumed to belong to a definite atom in the molecule, those of a molecule consisting of two such atoms would thus [might then conceivably] be a system of double lines with intervals equidistant all along the series, but in this case without definite polarizations*.

*The Zeeman effect as an indicator of atomic structure.*

But if the constituents of the double lines of a series were thus two modifications of the same modes of the simpler atomic system, it would follow that they should be similarly affected by a magnetic field. This is not always the case, so that this kind of explanation cannot be of universal application: it would be interesting to ascertain whether the Zeeman effect is the same for the two sets of constituents of a double series such that the difference of frequencies is

*Vortex motion analogy.*

velocity of those that travel in the opposite direction: thus the period of revolution is different, and each single undisturbed period becomes two adjacent disturbed periods. Analogous considerations apply to the interaction of the two atoms of the molecule, rotating round each other.

According however to Smithells, Dawson, and Wilson, *Phil. Trans.* 1899, A, it is the molecule of sodium that gives out the yellow light, that of sodium chloride not being effective.

* It is the banded spectra with numerous close lines that are now referred to molecules.

The structure of such band spectra may perhaps be illustrated roughly from a row of equidistant mass points exerting slight electric mutual influences, threaded for example on a cord, which may join together at the ends into a cyclic band of length $l$. The question of vibration would be in a general way of type $\frac{d^2\xi}{dt^2} + k^2\xi = -g\frac{d^2\xi}{dx^2}$: and the appropriate type of solution $\xi \propto \sin \mu x \sin pt$, so that $\mu l = m\pi$: also $-p^2 + k^2 = g\mu^2$. Thus the frequencies $p/2\pi$ are given in terms of integer numbers $m$ by a relation $p^2 = k^2 - g\frac{\pi^2}{l^2}m^2$, which is of the type $A - Bm^2$ that has been assigned as the law of the constituent lines of bands, as distinct from the very different structure of the series of lines. Cf. *Encyclopaedia Britannica*, Supplementary Volume, 32 (1902), article "Radiation," p. 128.

the same all along it. At any rate, uniformity in the Zeeman effect along a series of lines is evidence that they are all connected with the same vibrating group: identity of the effect on the two constituents of a doublet is evidence, as Preston pointed out, that these belong to modifications of the same type of vibration.

### Nature of Magnetization.

6. The proposition above given determines the changes in the periods of the vibrations of the molecule in the circumstances there defined. But it is not to be inferred from it that the imposition of the magnetic field merely superposes a slight uniform precessional motion on the previously existing orbital system. That orbital system will be itself slightly modified in the transition. For instance, in the ideal case of the magnetic field being imposed instantaneously, the velocities of all the electrons in the system will be continuous through that instant: hence the new orbital system on which the precession is imposed will be the one corresponding to velocities in that configuration which are equal to the actual velocities diminished by those connected with the precessional motion. *[Nature of the Zeeman orbital transition:]*

On the usual explanation of paramagnetic induction, the steady orbital motion of each electron is replaced by the uniform electric current circulating round the orbit which represents the averaged effect: the circuit of this current is supposed to be rigid so that the averaged forcive acting on it is a steady torque tending to turn it across the imposed magnetic field. This mode of representation must however *à priori* be incomplete: for example it would make the coefficient of magnetization per molecule in a gas increase markedly with length of free molecular path and therefore with fall of density, because this torque would have the longer time to orientate the molecule before the next encounter took place*. It appears from the above that the true effect of the imposed magnetic field is not a continued orientation of the orbits but only a slight change in the orbital system, which is proportional to the field, and in the simple circumstances above discussed is made up of a precessional effect of paramagnetic type, accompanied by a modification of the orbital system which is generally of diamagnetic type, both presumably of the same order of magnitude and thus very small. *[relevance to law of magnetization in gases: partial:]*

The recognition of this mode of action of the magnetic field also avoids another discrepancy. If the field acted by orientating the molecules it must induce dielectric polarization as well as magnetic: for each molecule has its own averaged electric moment, as revealed *[not involving dielectric polarization.]*

---

* Cf. Langevin's formula (1906) developed on the basis of gas theory. For the general thermodynamics of magnetism see *supra*, pp. 117, 130: also Appendix to this volume.

by piezoelectric phenomena, and regular orientation would accumulate the effects of these moments which would otherwise be mutually destructive. But there is nothing either in the disturbance of the free orbital system into a slightly different *free* system, or in the precession imposed on that new system—nor in a more general kind of action of the same type—which can introduce electric polarization.

The polarization of a dielectric medium by an imposed electric field is effected in a cognate manner. The electric force slightly modifies the orbital system by exerting opposite forces on the positive and negative ions. In this case these forces are independent of the velocities or masses of the ions. The fact that the polarization is proportional to the inducing field shows that the influence produced by the field on the orbital system is always a slight one. Yet the numerical value of the coefficient of electric polarization is always considerable, in contrast with the very small value of the magnetic coefficient; which arises from the very great intrinsic electric polarity of the molecule, due to the magnitude of the electric charge $e$ of an ion. Taking the effective molecular diameter as of the order $10^{-8}$ cm., there will be $10^{24}$ molecules per unit volume in a solid or liquid, and the aggregate of their intrinsic electric polarities may be as high as $10^{24} \cdot 10^{-8} \, ec$ electrostatic units, where $ec$ is $3 \cdot 10^{-11}$. Now the moment of polarization per unit volume for an inducing field $F$ is $(K-1)\, F/8\pi$; thus even for very strong fields this involves very slight change in the orbital configuration. A similar remark applies to the polarization induced by mechanical pressure in dielectric crystals. It would be unreasonable to expect any aggregate rotational effect around an axis, such as constitutes magnetization, from the polarizing action of an electric field; in fact if it were present, reversal of the direction of the field could not affect its total amount considered as arising from molecules orientated in all directions.

The possibilities as regards the aggregate intrinsic magnetic polarities of all the molecules are of the same high order, viz. $eAn/\tau$, where $A$ is the area and $\tau$ the period of a molecular orbit, which is $elnv$ or $10^{-5} \, v$ per cubic centimetre, where $v$ is the velocity in a molecular orbit whose linear dimension $l$ is $10^{-8}$. Thus the superior limit of the magnetization if the molecules were all completely orientated would be of the order $10^{-5} v$, which is large enough to include even the case of iron if $v$ were as much as one per cent. of the velocity of radiation.

In the case of iron a marked discrepancy exists between the enormous Faraday optical effect of a very thin sheet in a magnetic field on the one hand, and the slight Zeeman effect of the radiating molecule, as also the absence of peculiarity in optical reflexion from iron, and the absence of special influence on Hertzian waves, on the other:

*Electric polarization,*

*large yet involving slight configurational change:*

*as also magnetic.*

*The ferromagnetic anomalies,*

which must be in relation with the circumstance that at a moderately high temperature the iron loses its intense magnetic quality and comes into line with other kinds of matter. This suggests the explanation *as suggesting internal hysteretic cycles.* that the magnetization of iron at ordinary temperatures depends essentially on retentiveness, owing to facility possessed by groups of molecules for hanging together when once they are put into a new configuration. This is the well-known explanation of the phenomena of hysteresis, which can be effectively diminished by mechanical disturbance of the mass. In soft iron the magnetic cohesion would be less strong and more plastic, and thus readily shaken down by slight disturbance in the presence of a demagnetizing field, so that retentiveness would not be prominent. It is conceivable that the primary effect of an inducing field is to slightly magnetize the different molecules: that then the molecules thus altered change their condition of aggregation, and so are retained mutually in new positions independently of the field, the effect persisting if the field is gently removed: that the field can then act afresh on the molecules thus newly aggregated: and so on by a sort of regenerative process, the inducing field and the retentiveness mutually reinforcing each other, until large polarizations are reached before it comes to a limit. For hard iron these accommodations take place more rapidly than for soft iron, when the field is weak, and thus are of sensibly elastic character over a wider range: cf. Ewing, *Magnetic Induction*, 1892, Chap. vi.

### On the Origin of Magneto-optic Rotation.

7. The Faraday magneto-optic rotation is obviously connected, through the theory of dispersion, with the different alterations of the free periods of right-handed and left-handed vibrational modes of the molecules, that are produced by the impressed magnetic field. The ascertained law (*infra*) that the mean of the velocities of the two *Magnetic field does not alter the atom.* kinds of wave-trains is equal to that of the unaltered radiation, shows that the phenomenon in fact arises wholly from this difference, and is not accompanied by temporary structural change in the molecule such as would involve alteration of the physical constants of the medium.

The general relation connecting the refractive index $\mu$ of a transparent medium with the frequencies $(p_1, p_2, \ldots p_n)/2\pi$ of the principal free vibrations of its molecules, which are so great that radiation travels over $10^3$ molecular diameters in one period, is of type [p. 54]

$$\frac{\mu^2 - 1}{\mu^2 + 2} = \Sigma \frac{A_r}{p^2 - p_r^2},$$

in which $A_r$ is a constant which is a measure of the importance, as regards dispersion, of the free principal period $2\pi/p_r$. The quantity on the right-hand side of this equation, of form $f(p^2)$, is a function of the averaged configuration of the molecule relative to the aethereal wave-train that is passing over it. Now consider a circular wave-train, say a right-handed one, passing along the direction of the magnetic field: on the hypothesis that the spectrum consists of a single series of lines for all of which $\kappa$ is the same, the influence of this train on the corresponding right-handed vibrations that it excites in the molecule will be to superadd a rotation of the molecule as a whole with angular velocity $\tfrac{1}{2}\kappa$*. This will modify the configuration of the vibrating system relative to the circular wave-train passing over it in the same way as if an equal and opposite angular velocity were instead imparted to the wave-train. Thus the actual effect of the magnetic field on the light will be the same as would be that of a change in the frequency of the light from $p/2\pi$ to $p/2\pi + \kappa/4\pi$, the latter term arising from this imposed angular velocity: the value of the magneto-optic effect may therefore in such a case be derived from inspection of a table of the ordinary dispersion of the medium.

General formula for the Faraday rotations along the spectrum.
The velocity of propagation of the train of circular waves will, on this hypothesis, be derived by writing $p - \tfrac{1}{2}\kappa$ or $p + \tfrac{1}{2}\kappa$ for $p$ according as the train is right-handed or left-handed, thus giving when $\kappa^2$ is neglected,

$$\frac{\mu^2 - 1}{\mu^2 + 2} = \Sigma\, \frac{A_r}{p^2 \mp \kappa p - p_r^2}.$$

Inverse Zeeman effect in absorption.
For the case when there is only a single free period this result coincides with FitzGerald's formula (*Roy. Soc. Proc.* 1898), which has been shown by him to give the actual order of magnitude for a Faraday effect as thus deduced from the Zeeman effect.

If we were to consider that each system of lines in the spectrum arises from an independently vibrating group of ions in the molecule, as (*supra*) there may be some temptation to do, then the value of $(\mu^2 - 1)/(\mu^2 + 2)$ in this formula would be obtained by addition of the effects of these independent groups: thus if the value of the Zeeman effect were known for each line of the spectrum of any substance, and the law of dispersion of the substance were known, the Faraday effect could be deduced by calculation. To our order of approximation we should have

Independent atomic electron groups.

$$\delta\left(\frac{\mu^2 - 1}{\mu^2 + 2}\right) = \Sigma\, \frac{\pm\,\kappa_r A_r p}{(p^2 - p_r^2)^2}.$$

* This precessional rotation $\tfrac{1}{2}\kappa$ is (§ 4) equal to $eH/2m$. More strictly, the orbital motion in this model of an atom is not unique as regards phases: the impressed magnetic field changes it into another of the possible forms with this precession superposed. As the rotating electric field of the transmitted radiation waxes and wanes, there is room here for a continuous change of form such as could account for the adjusting factor in the Becquerel formula, *infra*.

the circumstance that the mean of the velocities of propagation is unaltered by the impressed field points to the $A$ coefficients being unaffected by the magnetism, thus suggesting absence of change in the mean conformation, as already remarked.

For the case in which the free periods that effectively control the dispersion all belong to the same series of spectral lines, so that $\kappa$ is the same for all of them, the formula for the dispersion need not come into the argument. The influence of the impressed magnetic field on the index of refraction of circularly polarized light is then the same as the change of $p$ to $p \pm \frac{1}{2}\kappa$ according as the polarization is left-handed or right-handed. Because that influence is equivalent to rotation of the optically vibrating molecule with angular velocity $\frac{1}{2}\kappa$, the molecule will now be related in the same way to a wave-train with angular velocity $p \pm \frac{1}{2}\kappa$ as it was previously to one with angular velocity $p$. Thus light corresponding to angular velocity $p$ is now propagated with velocity $V \pm \frac{1}{2}\kappa \dfrac{dV}{dp}$ instead of $V$. Now if $\lambda$ be the wave-length in a vacuum and $\mu$ the refractive index, we have $V = c/\mu$, $p = 2\pi c/\lambda$: and the rotation of a plane of polarization for a length $l$ of the medium, being $\frac{1}{2}p$ multiplied by the difference of times of transit, is

Dispersional theory not directly involved.

$$\frac{1}{2}\left(\frac{l}{V_1} - \frac{l}{V_2}\right)\frac{2\pi c}{\lambda}, \quad \text{which is} \quad \pi l c\, \frac{\delta V}{V^2 \lambda},$$

where

$$\delta V = \kappa\, \frac{dV}{dp} = \frac{\kappa}{2\pi}\, \frac{d\mu^{-1}}{d\lambda^{-1}},$$

so that the result is

$$\frac{\kappa l}{2c}\, \lambda\, \frac{d\mu}{d\lambda}.$$

This expression, $\frac{1}{2}\dfrac{\kappa}{c}\lambda \dfrac{d\mu}{d\lambda}$, for the coefficient of magnetic rotation as a function of the wave-length, has been given by H. Becquerel[1] and shown by him to be in good agreement with actual values as regards order of magnitude, and also with Verdet's detailed observations

Becquerel's formula for Faraday effect.

[1] *Comptes Rendus*, Nov. 1897: it was based on the assumption that the magnetic field involves rotation of the aether with velocity $\frac{1}{2}\kappa$.

[It has been found recently by C. G. Darwin (*Roy. Soc. Proc.* 1927), by examination of extensive experimental records, that the formula is of very wide application, with, however, an adjusting factor ranging around the value $\frac{3}{4}$ for different classes of substances. The formula is perhaps the most direct evidence that even in solid bodies the transmission of light is controlled by the electrons in their atoms.

When the formula for $\mu$, only very slightly affected by the impressed magnetic field, given above, is substituted in the Becquerel formula, it appears that Drude's type of expression for the magnetic rotation, $\Sigma B_r/(\lambda^2 - \lambda_r^2)$, would on that basis be appropriate only when the dispersion is due to bands of absorption far away in the ultra-violet.]

along the spectrum in the cases of carbon disulphide and creosote. The restriction on which it is here based, namely that the dispersion is controlled by free periods for all of which the Zeeman constant is the same, can be neglected for the case of the anomalous dispersion close to an absorption band, because there the dispersion is controlled by that band alone[1]: thus the Faraday effect is there very large and of anomalous character*, in correspondence with the experimental discovery of Macaluso and Corbino. From another aspect of the same effect, we can conclude that light of any given period, very near a natural free period of the medium, will travel in it with sensibly different velocities according as its mode of vibration corresponds to one or other of two principal types, elliptically (or in a special case circularly) polarized in opposite directions, and thus will exhibit phenomena of double refraction.

*Anomalous Faraday rotation near absorption lines:*

*a slight double refraction involved.*

### The Influence of Rotational Terms on Optical Propagation.

8. The purely formal, *i.e.* non-molecular, theory of the magnetic influence on optical propagation may be developed in a simple and direct manner, by use of the device of a revolving coordinate system as above employed. In a non-magnetizable medium the exact relations connecting the magnetic force $(\alpha, \beta, \gamma)$, the electric force $(P, Q, R)$, and the electric current $(u, v, w)$, are of types

$$\frac{d\gamma}{dy} - \frac{d\beta}{dz} = 4\pi u, \quad \frac{dR}{dy} - \frac{dQ}{dz} = -\frac{d\alpha}{dt}.$$

Thus
$$\nabla^2 P - \frac{d}{dx}\left(\frac{dP}{dx} + \frac{dQ}{dy} + \frac{dR}{dz}\right) = 4\pi\frac{du}{dt},$$

which will lead to the differential equations of the propagation when in it $(u, v, w)$ is expressed in terms of $(P, Q, R)$ by means of the constitutive relation connecting them.

Now for the aethereal elastic displacement we have

$$(f, g, h) = (4\pi c^2)^{-1} (P, Q, R).$$

*Energy restriction on chiral quality,*

To determine the nature of the most general formal connection between the material polarization $(f', g', h')$ and the electric force, that we are at liberty to assume without implying perpetual motions,

---

[1] Cf. *Proc. Camb. Phil. Soc.* March 1899, pp. 181–2: for similar explanations but restricted to anomalous dispersion, cf. Macaluso and Corbino, *Rend. Lincei*, Feb. 1899.

Reference should also be made to the converse procedure of Voigt (cf. *Annalen der Physik*, Vol. 1, 1900, p. 390), who, by introducing dispersional terms of a certain simple type including a frictional part into the equations of optical propagation in a rotational medium, finds that each absorption line is tripled, but with an asymmetry introduced by the frictional term.

* Cf. a detailed discussion in *Aether and Matter* (1900), p. 353: at the end of it "opposite" was inadvertently written instead of "same."

we must make use of the method of energy. The energy of this electric polarization in any region is

$$W = \tfrac{1}{2}\int (Pf' + Qg' + Rh')\,d\tau,$$

where $\delta\tau$ is an element of volume: thus its intensity per unit volume is a quadratic function of $(P, Q, R)$, and possibly also of $d/dt\,(P, Q, R)$ and of the spatial gradient of $(P, Q, R)$, and it may be of gradients of higher orders as well: if the first time gradients alone are included we thus have the expression

$$F_2\,(P, Q, R) + a_{11}P\,dP/dt + \ldots + a_{12}P\,dQ/dt + a_{21}Q\,dP/dt + \ldots,$$

$F_2$ denoting a quadratic function. The variation of this energy must, from the definition of $(P, Q, R)$ as the force moving the electrons, be

$$\delta W = \int (P\delta f' + Q\delta g' + R\delta h')\,d\tau$$

$$= \delta\int (Pf' + Qg' + Rh')\,d\tau - \int (f'\delta P + g'\delta Q + h'\delta R)\,d\tau,$$

so that, transposing,

$$\delta W = \int (f'\delta P + g'\delta Q + h'\delta R)\,d\tau,$$

in which the independent variable is now $(P, Q, R)$.

On conducting the variation in the usual manner, and reducing from $d\delta P/dt$ to $\delta P$ by partial integration with respect to time (such as necessarily enters in the reduction of the fundamental dynamical equation of Action), this leads to a relation of type leading to widest type of relation.

$$f' = \frac{dF_2}{dP} + \frac{a_3}{4\pi c^2}\frac{dQ}{dt} - \frac{a_2}{4\pi c^2}\frac{dR}{dt},$$

where $\qquad (a_1,\, a_2,\, a_3)/4\pi c^2 = (a_{23} - a_{32},\, a_{31} - a_{13},\, a_{12} - a_{21}).$

When the system is referred to its principal dielectric axes,

$$F_2 = \frac{K_1 - 1}{8\pi c^2}P^2 + \frac{K_2 - 1}{8\pi c^2}Q^2 + \frac{K_3 - 1}{8\pi c^2}R^2.$$

This analysis shows that rotational quality in the relation connecting $(f', g', h')$ and $(P, Q, R)$ can come in through terms in the energy function that involve the time gradients: or, as may be shown in a similar manner, it may enter through terms involving the space gradients: but not otherwise. The latter terms introduce rotational quality of the structural type, with which we are not now concerned. The former terms lead to the magnetic type of rotation, here related to the vector $(a_1, a_2, a_3)$, which must be determined by the impressed magnetic field or other exciting cause of vector character: the existence of such mixed terms, involving $(P, Q, R)$ and $d/dt\,(P, Q, R)$, in fact adds to the polarization a part at right angles to $d/dt\,(P, Q, R)$ Energy foundation of the rotational qualities: the magnetic type.

and to this vector $(a_1, a_2, a_3)$, and equal to their vector product divided by $4\pi c^2$, which is in all cases entirely of rotational character. Terms of the form of a quadratic function of the gradients of $(P, Q, R)$ by themselves would merely modify the form of the function $F_2$ so that its coefficients depend in part on the period of the vibration, that is, they would be merged in optical dispersion of the ordinary type. The question also arises whether the ordinary dielectric constants, namely the coefficients of the function $F_2(P, Q, R)$, are sensibly

Dielectric
quality
cannot be
affected: altered by an impressed magnetic field. This point can be settled by aid of the principle of reversal. When the electric force and the impressed magnetic field and the time are all reversed, the effect on the induced electric polarity must be simple reversal: hence a reversal of the magnetic field cannot affect the coefficients in $F_2(P, Q, R)$: hence these coefficients must depend on the square or other even power of the impressed magnetic field: but the rotational terms depending on its first power are actually very small, therefore any

as confirmed. terms depending on its second power are wholly negligible. This is in accord with Mascart's experimental result.

The right-hand sides of the equations of propagation in the material medium, as above indicated, can thus, for light of period $2\pi/p$, be expressed in the form

$$p^2 c^{-2} \left( K_1 p^{-2} \frac{d^2 P}{dt^2} - a_3 \frac{dQ}{dt} + a_2 \frac{dR}{dt}, \right.$$

$$\left. K_2 p^{-2} \frac{d^2 Q}{dt^2} - a_1 \frac{dR}{dt} + a_3 \frac{dP}{dt}, \quad K_3 p^{-2} \frac{d^2 R}{dt^2} - a_2 \frac{dP}{dt} + a_1 \frac{dQ}{dt} \right).$$

In the case of an isotropic medium for which $K_1$, $K_2$, $K_3$ are each equal to $K$, these equations of vibration can be restored to their normal form, when the square of the magnetic effect is neglected, by employing a coordinate system rotating with angular velocity

$$\tfrac{1}{2} K^{-1} p^2 (a_1, a_2, a_3).$$

Direct
general
synopsis as
imposed
rotation: Thus the effect of the impressed magnetic field is that the vibrations of the electric force, propagated as if that field were absent, are at the same time carried on by a motion of uniform rotation around its axis: so also, in virtue of the second of the above circuital relations, are the vibrations of the magnetic force. The electric force is not exactly on the wave-front because under the magnetic conditions it is not exactly circuital: the magnetic force is exactly on the wave-front. Thus we have the direct result that a plane-polarized train of electric vibrations, of wave-length $\lambda$, travelling along the direction of the impressed magnetic field $H$, is rotated around its direction of propagation through an angle proportional to $\epsilon H/K\lambda^2$ per unit time, so that the rotational coefficient per unit distance is proportional to $\epsilon H/K^{\frac{1}{2}}\lambda^2$, where $\epsilon$ is itself affected by dispersion and is thus to a slight

extent a function of the wave-length. When the wave-train is not travelling in the direction of the magnetic field, it is the component of $H$ along the normal to the wave-front that is effective: the other component of the rotation, around an axis in the plane of the wave-front, then gradually deflects the front so as to produce curvature of the rays, but so excessively slight as to be of no account. The magnetic effect is thus a purely rotational one whatever be the direction of the wave-train with respect to the field: and the phenomena in an isotropic medium may be completely described kinematically on that basis.

When the medium is crystalline, its rotational quality is mixed up with its double refraction: yet in ordinary crystals the differences between $K_1$, $K_2$, $K_3$ are slight, so that the phenomena are still approximately represented by each permanent wave-train, polarized in the manner corresponding to its direction of propagation, rotated around that direction with velocity proportional to the cosine of the angle it makes with an axis which need not now be the axis of the impressed magnetic field. extended to crystalline media.

This direct method of exhibiting the nature of the effects may also be applied to the case of structural rotation, in which by an argument similar to the above, but dealing with energy terms involving space gradients of the electric force, we obtain for the material medium a constitutive relation of type Extension to structural optical rotation:

$$4\pi c^2 \left(f', g', h'\right) = \left(K_1 P + a_3 \frac{dQ}{dz} - a_2 \frac{dR}{dy},\right.$$

$$\left.K_2 Q + a_1 \frac{dR}{dx} - a_3 \frac{dP}{dz}, \quad K_3 R + a_2 \frac{dP}{dy} - a_1 \frac{dQ}{dx}\right),$$

when the principal axes of the rotational quality coincide with those of the ordinary dielectric quality. For a plane wave-train travelling in the direction $(l, m, n)$, for which

$$(P, Q, R) \propto \exp \iota \frac{2\pi}{\lambda'} (lx + my + nz - Vt),$$

$$p = 2\pi V/\lambda', \quad V = cK^{-\frac{1}{2}},$$

this may be expressed in the form

$$4\pi c^2 \left(f', g', h'\right) = -\left(K_1 p^{-2} \frac{d^2 P}{dt^2} + \frac{na_3}{V} \frac{dQ}{dt} - \frac{ma_2}{V} \frac{dR}{dt}, \dots, \dots\right),$$

so that, when $K_1$, $K_2$, $K_3$ are each equal to $K$, the equations of propagation are reducible to the normal form for a non-rotational medium by imparting to the coordinate axes a velocity of rotation $2\pi^2 cK^{-\frac{3}{2}}\lambda^{-2} (la_1, ma_2, na_3)$, which implies a coefficient of rotation of a plane-polarized wave equal per unit distance to the general result:

$$2\pi^2 K^{-1}\lambda^{-2} (la_1, ma_2, na_3),$$

where $\lambda$ is the wave-length in vacuum. This is the law of rotation for wave-trains travelling in various directions in a simply refracting medium with aeolotropic rotational quality. This law also applies approximately to crystals such as quartz, inasmuch as the difference between the principal refractive indices is not considerable: in quartz the vector $(a_1, a_2, a_3)$ must by symmetry coincide with the axis of symmetry of the crystal: thus the coefficient of the effective component, that normal to the wave-front, of the imposed rotation for *including* a wave-train that travels in a direction making an angle $\theta$ with that *Airy's law* *for quartz:* axis is proportional to $\cos^2 \theta$, not to $\cos \theta$ as in the magnetic case. *concomitant* In this case the rotational effect is superposed on the double refraction, *double* *refraction.* so that a plane-polarized wave instead of being simply rotated will acquire varying elliptic polarization: it is, however, a simple problem in kinematics[1] to determine the types and the velocities of the two elliptically polarized wave-trains that will be propagated without change of form under the two influences, each supposed slight.

It appears from this discussion that magneto-optic rotation is a phenomenon of kinetic origin, related to the free periods of the molecules and not at all to their mean polarization under the action of steady electric force: it is therefore entirely of dispersional character.

Again the intrinsic optical rotation of isotropic chiral media is represented by a constitutive relation of type

$$f = \frac{K}{4\pi c^2} P + C \left( \frac{dQ}{dz} - \frac{dR}{dy} \right),$$

*Restriction* showing that the rotational term is proportional to the time gradient *on structural* *rotation.* of the magnetic field: this effect would therefore be entirely absent in statical circumstances, and only appears sensibly in vibratory motion of very high frequency. In this case no physical account of the origin of the term has been forthcoming: we have to be content with the knowledge that the form here stated is the only one that is admissible in accordance with the principles of dynamics.

*Rotations not* As the rotatory power, of both types, is thus connected with the *definitely* *constitutive.* dispersion as well as the density of the material, it is not strange that attempts, experimental and theoretical, to obtain a simple connection with the density alone, have not led to satisfactory results. The existence of a definite rotational constant for each active substance has formed the main experimental resource in the advance of stereochemical theory: but the present considerations prepare us for the fact that no definite relations connecting rotational power

[1] Cf. Gouy, *Journ. de Phys.* 1885; Lefebvre, *loc. cit.* 1892; O. Wiener, *Wied. Ann.* 1888.

with constitution have been found to exist—that the quality, though definite, is so to speak a slight and accidental one, or rather one not directly expressible in terms either of crystalline structure or of the main constitutive relations with which chemistry can deal.

### General Vibrating System in which the Principal Modes are Circular*.

9.  We are entitled to assert, on the basis of Fourier's theorem, that any orbital motion which exactly repeats itself with a definite period can be resolved into constituent simple elliptic oscillations whose periods are equal to its own and submultiples thereof. Such a motion would therefore correspond to a fundamental spectral line and its system of harmonics. The ascertained absence of harmonics in actual spectra shows either that the period corresponding to the steady orbit is outside the optical range, or else that the steady motion emits very little radiation as in fact its steadiness demands. The radiation would then arise from the various independent modes of disturbance, each of elliptic type on account of the absence of harmonics, that are superposed on the steady orbital motion.

*Absence of optical harmonics in spectra.*

To ascertain the nature of the polarization of the vibrations when in a magnetic field, we have first to decompose each orbital motion into its harmonic constituents, which are elliptic oscillations: each of the latter can be resolved into a linear oscillation parallel to the axis of the magnetic field, another at right angles to it, and a circular oscillation around it; and of these the second linear oscillation can be resolved into two equal circular oscillations in different senses around it. Now when the uniform rotation around the axis is superposed on the components they all continue to be of the requisite simple harmonic type, but the periods of the two circular species— which as has been seen are of amplitudes different as regards the various molecules but equal in the aggregate—become different: they are the three Zeeman components.

Nothing short of complete circular polarization of the constituent vibrations of permanent type in each molecule will account for the complete circular polarization of each of the flanking Zeeman lines. If these vibrations were only elliptical, but propagated with different velocities according to the sense in which the orbit is described, each would be equivalent to a circular vibration together with a linear one: and as the total illumination is the sum of the contributions from the independent molecules, the circularly polarized light would then be accompanied by unpolarized light of the same order of intensity.

*Vibrations in magnetic field must be of circular type:*

* This section can stand as a chapter in general dynamics of cyclic systems in its own right, though remote from the present semi-empirical trend of magneto-optics.

the suitable variables: This restriction of type of vibration suggests the employment in the analysis of variables each of which corresponds to a circular vibration, as do the $\xi$, $\eta$ variables in what follows.

For simplicity let us take the axis of $z$ parallel to the impressed magnetic field, and let $(X, Y, Z)$ represent the statical forces transmitted by aether strain from the other ions in the molecule to a specified one. The equations of motion of that ion are

$$m(\ddot{x} - \kappa\dot{y}) = X, \quad m(\ddot{y} + \kappa\dot{x}) = Y, \quad m\ddot{z} = Z.$$

We now make no assumption with regard to the magnitude of the electric charges and effective masses of the various ions, which may differ in any manner. In this ion let us change the variables to

$$\xi = x + \iota y, \quad \eta = x - \iota y,$$

so that

$$2x = \xi + \eta, \quad 2\iota y = \xi - \eta,$$

and therefore

$$2\frac{d}{d\xi} = \frac{d}{dx} - \iota\frac{d}{dy}, \quad 2\frac{d}{d\eta} = \frac{d}{dx} + \iota\frac{d}{dy};$$

the equations become

$$m(\ddot{\xi} + \iota\kappa\dot{\xi}) = X + \iota Y,$$
$$m(\ddot{\eta} - \iota\kappa\dot{\eta}) = X + \iota Y,$$
$$m\ddot{z} \qquad = Z.$$

the reduced equations: If therefore $X + \iota Y$ is a function only of the $\xi$ coordinates of the electrons, and $X - \iota Y$ a function only of the $\eta$ coordinates, and $Z$ only of the $z$ coordinates, these groups of coordinates will be determined from three independent systems of equations.

On our hypothesis of ions moving with velocities of an order below that of radiation, the mutual forces acting on them are derived from a potential energy function: thus

$$(X, Y, Z) = -k\left(\frac{d}{dx}, \frac{d}{dy}, \frac{d}{dz}\right)W,$$

where $k$ may be supposed to vary from one ion to another, being equal to the electric charge when the mutual forces are considered to be wholly of electric origin. Then

$$\frac{1}{m}(X + \iota Y) = -\frac{2k}{m}\frac{dW}{d\eta}, \quad \frac{1}{m}(X - \iota Y) = -\frac{2k}{m}\frac{dW}{d\xi}.$$

The solution of the complete system of equations, three for each ion, will in any case involve the expression of $\xi$, $\eta$, $z$ for each ion as a sum of harmonic terms of the form $e^{\iota pt}$ each with a complex numerical coefficient; but when the coefficients of one of them are assigned those of the others are determined. The vibration for each ion is thus com-now descriptively solved.pounded of a system of elliptic harmonic motions of definite forms and phases. Their components in the plane $\xi$, $\eta$ will be circular vibrations only when the $\xi$ and $\eta$ coordinates vary independently of each other, that is when $dW/d\eta$ is a function of the $\xi$ coordinates of the ions alone and $dW/d\xi$ a function of the $\eta$ coordinates alone. This condition can only be satisfied, $W$ being real, when it is a linear function of $z^2$ and of products of the form $\xi_r\eta_r$ or $\xi_r\eta_s$: it may thus be any quadratic function of the coordinates which is invariant in form as regards rotation of the axes of $x$, $y$ around the axis of $z$. Under these circumstances the free periods for $\xi$ coordinates, $\eta$ coordinates, and $z$ coordinates will all be independent, and either real or pure imaginary[1]: in an actual molecule they will be real. For example a permanent vibration of $\xi$ type will be represented by

Conditions<br>for such<br>circular<br>separation<br>of periods:

$$\xi_r = \Sigma A_r e^{\iota p_r t + \iota a_r},$$

$a_r$ being chosen so that $A_r$ is real: thus

$$x_r = \Sigma A_r \cos (p_r t + a_r), \; y_r = \Sigma A_r \sin (p_r t + a_r),$$

representing a series of right-handed circular vibrations, each series having definite phases and also amplitudes in definite ratios for the various ions. Again for the $\eta$ type we have

$$\eta_r = \Sigma B_r e^{\iota q_r t + \iota \beta_r},$$

so that $\quad x_r = \Sigma B_r \cos (q_r t + \beta_r), \; y_r = - \Sigma B_r \sin (q_r t + \beta_r),$

which represents similarly a series of left-handed circular vibrations. The vibrations of $z$ type will of course be linear in form.

Thus supposing the effective masses and charges of the various ions to be entirely arbitrary, the effect of an impressed magnetic field will be to triple the periods and polarize the constituents in the Zeeman manner, provided the potential energy of the mutual forces of the ions is any *quadratic function* of the coordinates of the vibrations which satisfies the condition of being invariant in form with respect to rotation of the axes of coordinates around the axis of the magnetic field.

interpreted<br>generally.

The essential difference between the type of this system and that of the one previously considered will appear when the latter is derived on the lines of the present procedure. The equations are

Comparison<br>with previous<br>special case<br>of exact<br>precession.

$$\ddot{\xi} + \iota\kappa\dot{\xi} = - \frac{2k}{m} \frac{dW}{d\eta},$$

$$\ddot{\eta} - \iota\kappa\dot{\eta} = - \frac{2k}{m} \frac{dW}{d\xi},$$

$$\ddot{z} \quad\quad = - \frac{k}{m} \frac{dW}{dz}.$$

[1] Routh, *Essay on Stability*, 1887, p. 78; *Dynamics*, Vol. II, § 319.

On writing $\qquad \xi' = e^{\frac{1}{2}i\kappa t}\,\xi,\ \eta' = e^{-\frac{1}{2}i\kappa t}\,\eta,$

they become $\qquad \ddot{\xi}' + \tfrac{1}{4}\kappa^2\xi' = -\dfrac{2k}{m}\dfrac{dW}{d\eta'},$

$$\ddot{\eta}' + \tfrac{1}{4}\kappa^2\eta' = -\dfrac{2k}{m}\dfrac{dW}{d\xi'},$$

$$\ddot{z} \qquad\qquad = -\dfrac{k}{m}\dfrac{dW}{dz}.$$

The form $W$ will be unaltered when it is expressed in terms of $\xi'$, $\eta'$, provided it depends only on the mutual configuration of the ions, *and $\kappa$ is the same for all of them*; hence when $\kappa^2$ is negligible compared with unity, $(\xi', \eta', z)$ are determined by the same equations as would give $(\xi, \eta, z)$ on the absence of a magnetic field: and from this the previous results follow.

The general dynamical procedure for orbital atoms: 10. We have thus reached the following position. Let the co-ordinates $(x, y, z)$ of an ion be resolved into two parts, namely $(x_1, y_1, z_1)$ which are known functions of the time and represent its mean or steady motion, and $(x', y', z')$ which are the small disturbance of the steady motion constituting the optical vibrations. When this substitution is made in the dynamical equations the quantities relating to the steady motion should cancel each other, as usual; and there will remain equations, of the original form, involving $(x', y', z')$ from which the accents may now be removed. The forces relating to these new coordinates will still be derivable from a potential energy function: and as by hypothesis the vibrations are all "cycloidal" or simple harmonic, this function must be homogeneous and quadratic implies the requisite condition, in these coordinates. The total potential energy must be determined by the instantaneous configuration of the system, and will therefore remain of the same form when referred to new axes of coordinates. This confines the quadratic part representing the energy of the disturbance to the form given above: the vibration of each ion will then in general consist of a system of elliptic oscillations of all the various free periods, equal in number to the ions: and the effect of an impressed magnetic field will be to triple each vibration period and to giving Zeeman separation and polarizations. polarize the constituents in the Zeeman manner. The steady or constitutive motion of the system must be [regarded as] so adjusted that it does not sensibly radiate: otherwise it would gradually alter by loss of its energy.

As the axis of the magnetic field may be any axis in the molecule, the function which represents the potential energy must thus be such that the vibrations resolved parallel to any axis form an independent

system: hence it is confined to the form

$$W = - \tfrac{1}{2}\Sigma A_{rs}\{(x_r' - x_s')^2 + (y_r' - y_s')^2 + (z_r' - z_s')^2\}$$
$$+ \Sigma B_{rs}(x_r'x_s' + y_r'y_s' + z_r'z_s'),$$
$$= - \tfrac{1}{2}\Sigma A_{rs}\{(\xi_r' - \xi_s')(\eta_r' - \eta_s') + (z_r' - z_s')^2\}$$
$$+ \tfrac{1}{2}\Sigma B_{rs}(\xi_r'\eta_s' + \xi_s'\eta_r' + 2z_r'z_s').$$

Thus in the absence of a magnetic field the vibrations of the $x$ coordinates, of the $y$ coordinates, and of the $z$ coordinates of the ions

will form independent systems of precisely similar character. It is

in fact only under this condition that it is possible for the components, parallel to any plane, of the elliptic harmonic vibrational types of the various ions, to form a system of circular vibrations with common sense of rotation.

If $m/k = \lambda$ and $m\kappa/k = \lambda'$, the equations of motion are of type

$$\lambda\ddot{\xi} + \iota\lambda'\dot{\xi} + 2\frac{dW}{d\eta} = 0, \quad \lambda\ddot{\eta} - \iota\lambda'\dot{\eta} + 2\frac{dW}{d\xi} = 0, \quad \lambda\ddot{z} + \frac{dW}{dz} = 0.$$

The periods of the right-handed circular vibrations, of type $\xi \propto e^{\iota pt}$, period $2\pi/p$, will be given by the equation

$$\begin{vmatrix} -\lambda_1 p^2 - \lambda_1' p - \Sigma A_{1r}, & C_{12}, & C_{13}, & C_{14}, \ldots C_{1n} \\ C_{21}, & -\lambda_2 p^2 - \lambda_2' p - \Sigma A_{2r}, & C_{23}, & C_{24}, \ldots C_{2n} \\ C_{31}, & C_{32}, & -\lambda_3 p^2 - \lambda_3' p - \Sigma A_{3r}, & C_{34}, \ldots C_{3n} \\ \ldots & \ldots & \ldots & \ldots \quad \ldots \end{vmatrix} = 0,$$

in which $C_{rs} = A_{rs} + B_{rs}$: those of the left-handed circular vibrations

by changing the sign of each $\lambda'$ in this equation: those of the plane-polarized vibrations, which are the natural periods of the molecule, by making $\lambda'$ null. On account of the great number of the constants, compared with the number of free periods, simple relations among

the periods can only arise from limitations of the generality of the system.

The duplication or triplication observed in the constituent Zeeman

lines would on this theory arise from the presence of two or three equal roots in the period equation for natural vibrations of the system, which would be differently affected and therefore separated by the impressed magnetic field.

This analysis is wide enough to apply to a system consisting of a

continuous electrical distribution, whose parts are held together in their relative positions either by statical constraint or by kinetic stability: for then the potential energy still depends on the relative configurations of the elements of mass of the system.

We have, however, not arrived at any definite representation of the dynamical system constituting a molecule, except that it consists of moving electric points either limited in number or so numerous as to form a practically continuous distribution: but reasoning from the definiteness and sharpness of the periods in the spectrum, and the the essential facts of polarization of light, it has been inferred that the vibrations features, of the molecule form a "cycloidal" system and therefore arise from a quadratic potential energy function: the total potential energy function must therefore consist of two independent parts, that belonging to the steady motion, in which the coordinates of the vibrations do not occur, and this part belonging to the disturbance which is quadratic in its coordinates: as a whole it must depend on the configuration of the system and not on the axes of coordinates, hence this quadratic part is invariant with regard to change of axes: this confines it to the form given above—which had been found to be demanded by the existence of the Zeeman phenomena.

It has thus been seen that the fact that the vibrations belonging to the Zeeman constituent lines are exactly circular, and not merely elliptic with a definite sense of rotation, requires that the right-handed and left-handed groups of vibrations shall form two independent systems: as the magnetic field may be in any direction as regards the molecule, this requires that its vibrations, when the magnetic field is absent, can be resolved into three independent systems of parallel linear vibrations directed along any three mutually rectangular axes. This again involves that an electric force acting on the molecule will *induce* a polarization exactly in the direction of the including force, and proportional to it[1]: that in fact notwithstanding its numeisotropy. rous degrees of freedom the molecule is isotropic. Thus the source of The essentials double refraction in crystals or strained isotropic substances would for double reside in the aeolotropic arrangement of the molecules and not in refraction. their orientation: but there can also be an independent intrinsic electric polarity in the molecule depending on its orientation and not on the electric field, such as is indicated by piezoelectric effects in crystals.

If the molecules were not thus isotropic as regards induced electric polarity, the electric vibration induced in the molecules, when a train of radiation passes across a medium such as air, would not be wholly Bearing on in the wave-front. In the theory of optical dispersion the coefficients[2] optical would then be averages taken for a large number of molecules scattering. orientated in all directions, such as may be considered to exist in an

Evidence    [1] Cf. Kerr's striking result, *Phil. Mag.* 1895, that in the double refraction from arti- produced in a liquid dielectric by an electric field, it is only the vibration ficial double polarized so that its electric vector is parallel to the electric field that has its refraction. velocity of propagation affected.

[2] E.g. $K$, $c_1$, $c_2$, ... $c_1'$, $c_2'$, ... in *Phil. Trans.* 1897, A, p. 238[: *supra*, p. 53].

effective element of volume of the medium: and this averaging would constitute the source of its isotropy. But there would remain a question as to whether, when a plane-polarized wave-train is passing, those fortuitous components of the polarization of the molecules that are not in the direction óf the electric vibration of the wave-train would not send out radiation as independent sources and thus lead to [confusion and undue] extinction of the light. The definite features of polarization of the light scattered from a plane-polarized train by very minute particles or molecular aggregations seem also to suggest in a similar manner that the individual molecule is isotropic*.

* The effects of want of isotropy have been determined in experiment by the present Lord Rayleigh, and analysed by his father on his principle of independent scattering, *Phil. Mag.* 35 (1918), *Sci. Papers*, Vol. VI pp. 540–6: also pushed further by Cabannes and by Raman. Cf. Appendix at end of this volume: also Vol. I, Appendix VII.

# 54

# THE METHODS OF MATHEMATICAL PHYSICS.

[Address to the Mathematical and Physical Section of the
British Association: Bradford, 1900.]

It is fitting that before entering upon the business of the Section we
should pause to take note of the losses which our department of
science has recently sustained. The fame of Bertrand, apart from his
official position as Secretary of the French Academy of Sciences, was
long ago universally established by his classical treatise on the
Infinitesimal Calculus: it has been of late years sustained by the
luminous exposition and searching criticism of his books on the
Theory of Probability and Thermodynamics and Electricity. The debt
which we owe to that other veteran, G. Wiedemann, both on account
of his own researches, which take us back to the modern revival of
experimental physics, and for his great and indispensable thesaurus
of the science of electricity, cannot easily be overstated. By the
death of Sophus Lie, following soon after his return to a chair in his
native country Norway, we have lost one of the great constructive
mathematicians of the century, who has in various directions fundamentally expanded the methods and conceptions of analysis by
reverting to the fountain of direct geometrical intuition. In Italy the
death of Beltrami has removed an investigator whose influence has
been equally marked on the theories of transcendental geometry and
on the progress of mathematical physics. In our own country we
have lost in D. E. Hughes one of the great scientific inventors of the
age; while we specially deplore the removal, in his early prime, of one
who has recently been well known at these meetings, Thomas Preston,
whose experimental investigations on the relations between magnetism
and light, combined with his great powers of lucid exposition, marked
out for him a brilliant future.

*Obituary.*

Perhaps the most important event of general scientific interest
during the past year has been the definite undertaking of the great
task of the international coordination of scientific literature; and it
may be in some measure in the prolonged conferences that were
necessitated by that object that the recently announced international
federation of scientific academies has had its origin. In the important
task of rendering accessible the stores of scientific knowledge, the
British Association, and in particular this Section of it, has played

*International
collaboration.*

the part of pioneer. Our annual volumes have long been classical, through the splendid reports of the progress of the different branches of knowledge that have been from time to time contributed to them by the foremost British men of science; and our work in this direction has received the compliment of successful imitation by the sister Associations on the Continent. Reports on progress of science.

The usual conferences connected with our department of scientific activity have been this year notably augmented by the very successful international congresses of mathematicians and of physicists which met a few weeks ago in Paris. The three volumes of reports on the progress of physical science during the last ten years, for which we are indebted to the initiative of the French Physical Society, will provide an admirable conspectus of the present trend of activity, and form a permanent record for the history of our subject. Congresses.

Another very powerful auxiliary to progress is now being rapidly provided by the republication, in suitable form and within reasonable time, of the collected works of the masters of our science. We have quite recently received, in a large quarto volume, the mass of most important unpublished work that was left behind him by the late Professor J. C. Adams; the zealous care of Professor Sampson has worked up into order the more purely astronomical part of the volume; while the great undertaking, spread over many years, of the complete determination of the secular change of the magnetic condition of the Earth, for which the practical preparations had been set on foot by Gauss himself, has been prepared for the press by Professor W. G. Adams. By the publication of the first volume of Lord Rayleigh's papers a series of memoirs which have formed a main stimulus to the progress of mathematical physics in this country during the past twenty years has become generally accessible. The completed series will form a landmark for the end of the century that may be compared with Young's *Lectures on Natural Philosophy* for its beginning. Collected memoirs: J. C. Adams, Rayleigh.

The recent reconstruction of the University of London, and the foundation of the University of Birmingham, will, it is to be hoped, give greater freedom to the work of our University Colleges. The system of examinations has formed an admirable stimulus to the effective acquisition of that general knowledge which is a necessary part of all education. So long as the examiner recognizes that his function is a responsible and influential one, which is to be taken seriously from the point of view of moulding the teaching in places where external guidance is helpful, test by examination will remain a most valuable means of extending the area of higher education. Except for workers in rapidly progressive branches of technical science, a broad education seems better adapted to the purposes of Value of examinations: early education should be broad:

life than special training over a narrow range; and it is difficult to see how a reasonably elastic examination test can be considered as a hardship. But the case is changed when preparation for a specialized scientific profession, or mastery of the lines of attack in an unsolved problem, is the object. The general education has then been presumably finished; in expanding departments of knowledge, variety rather than uniformity of training should be the aim, and the genius of a great teacher should be allowed free play without external trammels. It would appear that in this country we have recently been liable to unduly mix up two methods. We have been starting students on the special and lengthy, though very instructive, processes which are known as original research at an age when their time would be more profitably employed in rapidly acquiring a broad basis of knowledge.

specialized research a life-work.

As a result, we have been extending the examination test from the general knowledge to which it is admirably suited into the specialized activity which is best left to the stimulus of personal interest. Informal contact with competent advisers, themselves imbued with the scientific spirit, who can point the way towards direct appreciation of the works of the masters of the science, is far more effective than detailed instruction at second hand, as regards growing subjects that have not yet taken on an authoritative form of exposition. Fortunately there seems to be now no lack of such teachers to meet the requirements of the technical colleges that are being established throughout the country.

Gilbert, *De Magnete.*

The famous treatise which opened the modern era by treating magnetism and electricity on a scientific basis appeared just 300 years ago. The author, William Gilbert, M.D., of Colchester, passed from the Grammar School of his native town to St John's College, Cambridge: soon after taking his first degree, in 1560, he became a Fellow of the College, and seems to have remained in residence, and taken part in its affairs, for about ten years. All through his subsequent career, both at Colchester and afterwards at London, where he attained the highest position in his profession, he was an exact and diligent explorer, first of chemical and then of magnetic and electric phenomena. In the words of the historian Hallam, writing in 1839, "in his Latin treatise on the Magnet he not only collected all the knowledge which others had possessed, but he became at once the father of experimental philosophy in this island"; and no demur would be raised if Hallam's restriction to this country were removed. Working nearly a century before the time when the astronomical discoveries of Newton had originated the idea of attraction at a distance, he established a complete formulation of the interaction of magnets by what we now call the exploration of their fields of force. His analysis of the facts of magnetic influence, and incidentally of the points in which it differs from electric influence, is virtually the one which Faraday re-introduced. A cardinal advance was achieved, at a time when the Copernican Astronomy had still largely to make its way, by assigning the behaviour of the compass and the dip needle to the fact that the Earth itself is a great magnet, by whose field of influence they are controlled. His book passed through many editions on the Continent within forty years: it won the high praise of Galileo. Gilbert has been called "the father of modern electricity" by Priestley, and "the Galileo of magnetism" by Poggendorff.

When the British Association last met at Bradford in 1873 the modern theory which largely reverts to Gilbert's way of formulation, and refers electric and magnetic phenomena to the activity of the aether instead of attractions at a distance, was of recent growth: it had received its classical exposition only two years before by the publication of Clerk Maxwell's treatise. The new doctrine was already widely received in England on its own independent merits. On the Continent it was engaging the strenuous attention of Helmholtz, whose series of memoirs, deeply probing the new ideas in their relation to the prevalent and fairly successful theories of direct action across space, had begun to appear in 1870. During many years the search for crucial experiments that would go beyond the results equally explained by both views met with small success; it was not until 1887 that Hertz, by the discovery of the aethereal radiation of long wave-length emitted from electric oscillators, verified the hypothesis of Faraday and Maxwell and initiated a new era in the practical development of physical science. The experimental field thus opened up was soon fully occupied both in this country and abroad; and the border-land between the sciences of optics and electricity is now being rapidly explored. The extension of experimental knowledge was simultaneous with increased attention to directness of explanation; the expositions of Heaviside and Hertz and other writers fixed attention, in a manner already briefly exemplified by Maxwell himself, on the inherent sim-plicity of the completed aethereal scheme, when once the theoretical scaffolding employed in its construction and dynamical consolidation is removed; while Poynting's beautiful corollary specifying the path of the transmission of energy through the aether has brought the theory into simple relations with the applications of electrodynamics.

Equally striking has been the great mastery obtained during the last twenty years over the practical manipulation of electric power. The installation of electric wires as the nerves connecting different regions of the Earth had attained the rank of accomplished fact so long ago as 1857, when the first Atlantic cable was laid. It was largely

*Progress of electric theory:*

*of electro-technics.*

the theoretical and practical difficulties, many of them unforeseen, encountered in carrying that great undertaking to a successful issue, that necessitated the elaboration by Lord Kelvin and his coadjutors of convenient methods and instruments for the exact measurement of electric quantities, and thus prepared the foundation for the more recent practical developments in other directions. On the other hand, the methods of theoretical explanation have been in turn improved and simplified through the new ways of considering the phenomena which have been evolved in the course of practical advances on a large scale, such as the improvement of dynamo armatures, the conception and utilization of magnetic circuits, and the transmission of power by alternating currents. In our time the relations of civilized life have been already perhaps more profoundly altered than ever before, owing to the establishment of practically instantaneous electric communication between all parts of the world. The employment of the same subtle agency is now rapidly superseding the artificial reciprocating engines and other contrivances for the manipulation of mechanical power that were introduced with the employment of

Electric transmission of power.
steam. The possibilities of transmitting power to great distances at enormous tension, and therefore with very slight waste, along lines merely suspended in the air, are being practically realized; and the advantages thence derived are increased manifold by the almost automatic manner in which the electric power can be transformed into mechanical rotation at the very point where it is desired to apply it. The energy is transmitted at such lightning speed that at a given instant only an exceedingly minute portion of it is in actual transit. When the tension of the alternations is high, the amount of electricity that has to oscillate backwards and forwards on the guiding wires is proportionately diminished, and the frictional waste reduced. At the terminals the direct transmission from one armature of the motor to the other, across the intervening empty space, at once takes us beyond the province of the pushing and rubbing contacts that are unavoidable in mechanical transmission; while the perfect symmetry and reversibility of the arrangement by which power is delivered from a rotatory alternator at one end, guided by the wires to another place many miles away, where it is absorbed by another alternator with precise reversal of the initial stages, makes this process of distribution of energy resemble the automatic operations of nature rather than the imperfect material connections previously in use. We are here dealing primarily with the flawless continuous medium which is the transmitter of radiant energy across the celestial spaces; the part played by the coarsely constituted material conductor is only that of a more or less imperfect guide

which directs the current of aethereal energy. The wonderful nature of this theoretically perfect, though of course practically only approximate, method of abolishing limitations of locality with regard to mechanical power is not diminished by the circumstance that its principle must have been in some manner present to the mind of the first person who fully realized the character of the reversibility of a Gramme armature.

In theoretical knowledge a new domain, to which the theory as expounded twenty years ago had little to say, has recently been acquired through the experimental scrutiny of the electric discharge in rarefied gaseous media. The very varied electric phenomena of vacuum tubes, whose electrolytic character was first practically established by Schuster, have been largely reduced to order through the employment of the high exhaustions introduced and first utilized by Crookes. Their study under these circumstances, in which the material molecules are so sparsely distributed as but rarely to interfere with each other, has conduced to enlarged knowledge and verification of the fundamental relations in which the individual molecules stand to all electric phenomena, culminating recently in the actual determination, by J. J. Thomson and others following in his track, of the masses and velocities of the particles that carry the electric discharge across the exhausted space. The recent investigations of the circumstances of the electric dissociation produced in the atmosphere and in other gases by ultra-violet light, the Röntgen radiation, and other agencies, constitute one of the most striking developments in experimental molecular physics since Graham determined the molecular relations of gaseous diffusion and transpiration more than half a century ago. This advance in experimental knowledge of molecular phenomena, assisted by the discovery of the precise and rational effect of magnetism on the spectrum, has brought into prominence a modification or rather development of Maxwell's exposition of electric theory, which was dictated primarily by the requirements of the abstract theory itself; the atoms or ions are now definitely introduced as the carriers of those electric charges which interact across the aether, and so produce the electric fields whose transformations were the main subject of the original theory.

*Discovery of the free electron.*

We are thus inevitably led, in electric and aethereal theory, as in the chemistry and dynamics of the gaseous state which is the department of abstract physics next in order of simplicity, to the consideration of the individual molecules of matter. The theoretical problems which had come clearly into view a quarter of a century ago, under Maxwell's lead, whether in the exact dynamical relations of aethereal transmission or in the more fortuitous domain of the statistics of

*Interaction with Atomic theory:*

interacting molecules, are those around which attention is still mainly concentrated; but as the result of the progress in each, they are now tending towards consolidation into one subject. I propose—leaving further review of the scientific aspect of the recent enormous development of the applications of physical science for hands more competent to deal with the practical side of that subject—to offer some remarks on the scope and validity of this molecular order of ideas, to which the trend of physical explanation and development is now setting in so pronounced a manner.

If it is necessary to offer an apology for detaining the attention of the Section on so abstract a topic, I can plead its intrinsic philosophical importance. The hesitation so long felt on the Continent in regard to discarding the highly developed theories which analyzed all physical actions into direct attractions between the separate elements of the bodies concerned, in favour of a new method in which our ideas are carried into regions deeper than the phenomena, has now given place to eager discussion of the potentialities of the new standpoint. There has even appeared a disposition to consider that the Newtonian dynamical principles, which have formed the basis of physical explanation for nearly two centuries, must be replaced in these deeper subjects by a method of direct description of the mere course of phenomena, apart from any attempt to establish causal relations; the initiation of this method being traced, like that of the Newtonian dynamics itself, to this country. The question has arisen the aether, as to how far the new methods of aethereal physics are to be considered as an independent departure, how far they form the natural development of existing dynamical science. In England, whence the innovation came, it is the more conservative position that has all along been occupied. Maxwell was himself trained in the school of physics established in this country by Sir George Stokes and Lord Kelvin, in which the dominating idea has been that of the strictly its dynamical dynamical foundation of all physical action. Although the pupil's implications. imagination bridged over dynamical chasms, across which the master was not always able to follow, yet the most striking feature of Maxwell's scheme was still the dynamical framework into which it was built. The more advanced reformers have now thrown overboard the apparatus of potential functions which Maxwell found necessary for the dynamical consolidation of his theory, retaining only the final result as a verified descriptive basis for the phenomena. In this way all difficulties relating to dynamical development and indeed consistency are avoided, but the question remains as to how much is thereby lost. In practical electromagnetics the transmission of power is now the most prominent phenomenon; if formal dynamics is put aside in the general theory, its guidance must here be replaced by some more empirical and tentative method of describing the course of the transmission and transformation of mechanical energy in the system.

The direct recognition in some form, either explicitly or tacitly, of the part played by the aether has become indispensable to the development and exposition of general physics ever since the discoveries of Hertz left no further room for doubt that this physical scheme of Maxwell was not merely a brilliant speculation, but constituted, in spite of outstanding gaps and difficulties, a real formulation of the underlying unity in physical dynamics. The domain of abstract physics is in fact roughly divisible into two regions. In one of them we are mainly concerned with interactions between one portion of matter and another portion occupying a different position in space; such interactions have very uniform and comparatively simple relations; and the reason is traceable to the simple and uniform constitution of the intervening medium in which they have their seat. The other province is that in which the distribution of the material molecules comes into account. Setting aside the ordinary dynamics of matter in bulk, which is founded on the uniformity of the properties of the bodies concerned and their experimental determination, we must assign to this region all phenomena which are concerned with the unco-ordinated motions of the molecules, including the range of thermal and in part of radiant actions; the only possible basis for detailed theory is the statistical dynamics of the distribution of the molecules. The far more deep-seated and mysterious processes which are involved in changes in the constitution of the individual molecules themselves are mainly outside the province of physics, which is competent to reason only about permanent material systems; they must be left to the sciences of chemistry and physiology. Yet the chemist proclaims that he can determine only the results of his reactions and the physical conditions under which they occur; the character of the bonds which hold atoms in their chemical combinations is at present unknown, although a large domain of very precise knowledge relating, in some diagrammatic manner, to the topography of the more complex molecules has been attained. The vast structure which chemical science has in this way raised on the narrow foundation of the atomic theory is perhaps the most wonderful existing illustration both of the rationality of natural processes and of the analytical powers of the human mind. In a word, the complication of the material world is referable to the vast range of structure and of states of aggregation in the material atoms; while the possibility of a science of physics is largely due to the simplicity of constitution of the universal medium through which the individual atoms interact on each other.

*[margin notes]* Mechanical dynamics,

and atomic dynamics.

Atomic structure.

The reference of the uniformity in the interactions at a distance between material bodies to the part played by the aether is a step towards the elimination of extraneous and random hypotheses about laws of attraction between atoms. It also places that medium on a different basis from matter, in that its mode of activity is simple and regular, whereas intimate material interactions must be of illimitable complexity. This gives strong ground for the view that we should not be tempted towards explaining the simple group of relations which have been found to define the activity of the aether, by treating them as mechanical consequences of concealed structure in that medium; we should rather rest satisfied with having attained to their exact dynamical correlation, just as geometry explores or correlates, without explaining, the descriptive and metric properties of space. On the other hand, a view is upheld which considers the pressures and thrusts of the engineer, and the strains and stresses in the material structures by which he transmits them from one place to another, to be the archetype of the processes by which all mechanical effect is transmitted in Nature. This doctrine implies an expectation that we may ultimately discover something analogous to structure in the celestial spaces, by means of which the transmission of physical effect will be brought into line with the transmission of mechanical effect by material framework.

*Knowledge of the aether, descriptive?*

*or dynamical?*

*Optical aether may have very small density:*

At a time when the only definitely ascertained function of the aether was the undulatory propagation of radiant energy across space, Lord Kelvin pointed out that, by reason of the very great velocity of propagation, the density of the radiant energy in the medium at any place must be extremely small in comparison with the amount of energy that is transmitted in a second of time: this easily led him to the very striking conclusion that, on the hypothesis that the aether is like material elastic media, it is not necessary to assume its density to be more than $10^{-18}$ of that of water, or its optical rigidity to be more than ten $10^{-8}$ of that of steel or glass. Thus far the aether would be merely an impalpable material atmosphere for the transference of energy by radiation, at extremely small densities but with very great speed, while ordinary matter would be the seat of practically all this energy. But this way of explaining the absence of sensible influence of the aether on the phenomena of material dynamics lost much of its basis as soon as it was recognized that the same medium must be the receptacle of very high densities of energy in the electric fields around currents and magnets. The other mode of explanation is to consider the aether to be of the very essence of all physical actions, and to correlate the absence of obvious mechanical evidence of its intervention with its regularity and universality.

*not so electric.*

On this plan of making the aether the essential factor in the transformation of energy as well as its transmission across space, the material atom must be some kind of permanent nucleus that retains around itself an aethereal field of physical influence, such as, for example, a field of strain. We can recognize the atom only through its interactions with other atoms that are so far away from it as to be practically independent systems; thus our direct knowledge of the atom will be confined to this field of force which belongs to it. Just as the exploration of the distant field of magnetic influence of a steel magnet, itself concealed from view, cannot tell us anything about the magnet except the amount and direction of its moment, so a practically complete knowledge of the field of physical influence of an atom might be expressible in terms of the numerical values of a limited number of physical moments associated with it, without any revelation as to its essential structure or constitution being involved. This will at any rate be the case for ultimate atoms if, as is most likely, the distances at which they are kept apart are large compared with the diameters of the atomic nuclei; it in fact forms our only chance for penetrating to definite dynamical views of molecular structure. So long as we cannot isolate a single molecule, but must deal observationally with an innumerable distribution of them, even this kind of knowledge will be largely confined to average values. But the last half-century has witnessed the successful application of a new instrument of research, which has removed in various directions the limitations that had previously been placed on the knowledge to which it was possible for human effort to look forward. The spectroscope has created a new astronomy by revealing the constitutions and the unseen internal motions of the stars. Its power lies in the fact that it does take hold of the internal relations of the individual molecule of matter, and provides a very definite and detailed, though far from complete, analysis of the vibratory motions that are going on in it; these vibrations being in their normal state characteristic of its dynamical constitution, and in their deviations from the normal giving indications of the velocity of its movement and the physical state of its environment. Maxwell long ago laid emphasis on the fact that a physical atomic theory is not competent even to contemplate the vast mass of potentialities and correlations of the past and the future, that biological theory has to consider as latent in a single organic germ containing at most only a few million molecules. On our present view we can accept his position that the properties of such a body cannot be those of a "purely material system," provided, however, we restrict this phrase to apply to physical properties as here defined. But an exhaustive discovery of the intimate nature of the atom is beyond the scope of

*Aether essential to atomic matter:*

*provides its field of activity.*

*The spectroscopy of atoms.*

*Limitations of atomic theory.*

physics; questions as to whether it must not necessarily involve in itself some image of the complexity of the organic structures of which it can form a correlated part must remain a subject of speculation

Mental
structure of
knowledge.
outside the domain of that science. It might be held that this conception of discrete atoms and continuous aether really stands, like those of space and time, in intimate relation with our modes of mental apprehension, into which any consistent picture of the external world must of necessity be fitted. In any case it would involve abandonment of all the successful traditions of our subject if we ceased to hold that our analysis can be formulated in a consistent and complete manner, so far as it goes, without being necessarily an exhaustive account of phenomena that are beyond our range of experiment. Such phenomena may be more closely defined as those connected with the processes of intimate combination of the molecules: they include the activities of organic beings which all seem to depend on change of molecular structure.

Mode of
advance of
knowledge of
the aether.
If, then, we have so small a hold on the intimate nature of matter, it will appear all the more striking that physicists have been able precisely to divine the mode of operation of the intangible aether, and to some extent explore in it the fields of physical influence of the molecules. On consideration we recognize that this knowledge of fundamental physical interaction has been reached by a comparative process. The mechanism of the propagation of light could never have been studied in the free aether of space alone. It was possible, however, to determine the way in which the characteristics of optical propagation are modified, but not wholly transformed, when it takes place in a transparent material body instead of empty space. The change in fact arises on account of the aether being entangled with the network of material molecules; but inasmuch as the length of a single wave of radiation covers thousands of these molecules the wave-motion still remains uniform and does not lose its general type. A wider variation of the experimental conditions has been provided for our examination in the case of those substances in which the phenomenon of double refraction pointed to a change of the aethereal properties which varied in different directions; and minute study of this modification has proved sufficient to guide to a consistent appreciation of the nature of this change, and therefore of the mode of aethereal propagation that is thus altered. In the same way, it was the study and development of the manner in which the laws of electric phenomena in material bodies had been unravelled by Ampère and Faraday, that guided Faraday himself and Maxwell—who were impressed with the view that the aether was at the bottom of it all— in their progress towards an application of similar laws to aether

devoid of matter, such as would complete a scheme of continuous action by consistently interconnecting the material bodies and banishing all untraced interaction across empty space. Maxwell in fact chose to finally expound the theory by ascribing to the aether of free space a dielectric constant and a magnetic constant of the same types as had been found to express the properties of material media, thus extending the seat of the phenomena to all space on the plan of describing the activity of the aether in terms of the ordinary electric ideas. The converse mode of development, starting with the free aether under the directly dynamical form which has been usual in physical optics, and introducing the influence of the material atoms through the electric charges which are involved in their constitution[1], was hardly employed by him; in part, perhaps, because, owing to the necessity of correlating his theory with existing electric know-ledge and the mode of its expression, he seems never to have reached the stage of moulding it into a completely deductive form.

The dynamics of the aether, in fact the recognition of the existence of an aether, has thus, as a matter of history, been reached through study of the dynamical phenomena of matter. When the dynamics of a material system is worked up to its purest and most general form, it becomes a formulation of the relations between the succession of the configurations and states of motion of the system, the assistance of an independent idea of force not being usually required. We can, however, only attain to such a compact statement when the system is self-contained, when its motion is not being dissipated by agencies of frictional type, and when its connections can be directly specified by purely geometrical relations between the coordinates, thus ex-cluding such mechanisms as rolling contacts. The course of the system is then in all cases determined by some form or other of a single fundamental property, that any alteration in any small portion of its actual course must produce an increase in the total "Action" of the motion. It is to be observed that in employing this law of mini-mum as regards the Action expressed as an integral over the whole time of the motion, we no more introduce the future course as a determining influence on the present state of motion than we do in drawing a straight line from any point in any direction, although the length of the line is the minimum distance between its ends. In drawing the line piece by piece we have to make tentative excursions

*Pure dynamics of a medium:*

*restriction,*

*to permit an Action theory:*

*its essence,*

---

[1] In 1870 Maxwell, while admiring the breadth of the theory of Weber, which is virtually based on atomic charges combined with action at a distance, still regarded it as irreconcilable with his own theory, and left to the future the question as to why "theories apparently so fundamentally opposed should have so large a field of truth common to both," *Scientific Papers,* Vol. II, p. 228.

into the immediate future in order to adjust each element into straightness with the previous element; so in tracing the next stage of the motion of a material system we have similarly to secure that it is not given any such directions as would unduly increase the Action. But whatever views may be held as to the ultimate significance of this

not meta-
physical:

principle of Action, its importance, not only for mathematical analysis, but as a guide to physical exploration, remains fundamental. When the principles of the dynamics of material systems are refined down to their ultimate common basis, this principle of minimum is what remains. Hertz preferred to express its contents in the form of a principle of straightness of course or path. It will be recognized, on the lines already indicated, that this is another mode of statement of the same fundamental idea; and the general equivalence is worked out by Hertz on the basis of Hamilton's development of the principles of dynamics. The latter mode of statement may be adaptable so as to avoid the limitations which restrict the connections of the system, at the expense, however, of introducing new variables; if, indeed, it does not introduce gratuitous complexity for purposes of physics to attempt to do this. However these questions may stand, this principle of straightness or directness of path forms, wherever it applies, the most general and comprehensive formulation of purely dynamical action: it involves in itself the complete course of events. In so far as we are given the algebraic formula for the time integral which constitutes the Action, expressed in terms of any suitable coordinates, we know implicitly the whole dynamical constitution and history of the system to which it applies. Two systems in which the Action is expressed by the same formula are mathematically identical, are

appropriate physically precisely correlated, so that they have all dynamical pro-
for latent perties in common. When the structure of a dynamical system is
structure. largely concealed from view, the safest and most direct way towards an exploration of its essential relations and connections, and in fact towards answering the prior question as to whether it is a purely dynamical system at all, is through this order of ideas. The ultimate test that a system is a dynamical one is not that we shall be able to trace mechanical stresses throughout it, but that its relations can be in some way or other consolidated into accordance with this principle of minimum Action. This definition of a dynamical system in terms of the simple principle of directness of path may conceivably be subject to objection as too wide; it is certainly not too narrow; and it is the conception which has naturally been evolved from two centuries of study of the dynamics of material bodies. Its very great generality may lead to the objection that we might completely formulate the future course of a system in its terms, without having obtained

a working familiarity with its details, of the kind to which we have become accustomed in the analysis of simple material systems; but our choice is at present between this kind of formulation, which is a real and essential one, and an empirical description of the course of phenomena combined with explanations relating to more or less isolated groups. The list of great names, including Kelvin, Maxwell, Helmholtz, that have been associated with the employment of the principle for the elucidation of the relations of deep-seated dynamical phenomena, is a strong guarantee that we shall do well by making the most of this clue.

Are we then justified in treating the material molecule, so far as revealed by the spectroscope, as a dynamical system coming under this specification? Its intrinsic energy is certainly permanent and not subject to dissipation; otherwise the molecule would gradually fade out of existence. The extreme precision and regularity of detail in the spectrum shows that the vibrations which produce it are exactly synchronous, whatever be their amplitude, and in so far resemble the vibrations of small amplitude in material systems. As all indications *The kinetic* point to the molecule being a system in a state of intrinsic motion, *atom,* like a vortex ring, or a stellar system in astronomy, we must consider these radiating vibrations to take place around a steady state of motion which does not itself radiate, not around a state of rest. Now not the least of the advantages possessed by the Action principle, as *with cyclic* a foundation for theoretical physics, is the fact that its statement *motions.* can be adapted to systems involving in their constitution permanent steady motions of this kind, in such a way that only the variable motions superposed on them come into consideration. The possibilities as regards physical correlation of thus introducing permanent motional states as well as permanent structure into the constitution of our dynamical systems have long been emphasized by Lord Kelvin[1]; the effective adaptation of abstract dynamics to such systems was made independently by Kelvin and Routh about 1877; the more recent exposition of the theory by Helmholtz has directed general attention to what is undoubtedly the most significant extension of dynamical analysis which has taken place since the time of Lagrange.

Returning to the molecules, it is now verified that the Action prin- ciple forms a valid foundation throughout electrodynamics and optics; *Scope of* the introduction of the aether into the system has not affected its *molecular* *Action* application. It is therefore a reasonable hypothesis that the principle *theory:* forms an allowable foundation for the dynamical analysis of the radiant

[1] For a classical exposition see his British Association Address of 1884 on "Steps towards a Kinetic Theory of Matter," reprinted in *Popular Lectures and Addresses*, Vol. I.

vibrations in the system formed by a single molecule and surrounding aether; and the knowledge which is now accumulating, both of the orderly grouping of the lines of the spectrum and of the modifications impressed on these lines by a magnetic field or by the density of the matter immediately surrounding the vibrating molecule, can hardly fail to be fruitful for the dynamical analysis of its constitution. But let it be repeated that this analysis would be complete when a formula for the dynamical energy of the molecule is obtained, and would go no deeper. Starting from our definitely limited definition of the nature of a dynamical system, the problem is merely to correlate the observed relations of the periods of vibration in a molecule, when it has come into a steady state as regards constitution and is not under the influence of intimate encounter with other molecules.

constitutive molecular energy. It may be recalled incidentally that the generalized Maxwell-Boltzmann principle of the equable distribution of the acquired store of kinetic energy of the molecule, among its various possible independent types of motion, is based directly on the validity of the Action principle for its dynamics. In the demonstrations usually offered the molecule is considered to have no permanent or constitutive energy of internal motion. It can, however, be shown, by use of the generalization aforesaid of the Action principle, that no discrepancy will arise on that account. Such intrinsic kinetic energy virtually adds on to the potential energy of the system; and the remaining or acquired part of the kinetic energy of the molecule may be made the subject of the same train of reasoning as before.

Aether stress: Let us now return to the general question whether our definition of a dynamical system may not be too wide. As a case in point, the single principle of Action has been shown to provide a definite and sufficient basis for electrodynamics; yet when, for example, one armature of an electric motor pulls the other after it without material contact, and so transmits mechanical power, no connection between them is indicated by the principle such as could by virtue of internal stress transmit the pull. The essential feature of the transmission of a pull by stress across a medium is that each element of volume of the medium acts by itself, independently of the other elements. The stress excited in any element depends on the strain or other displacement occurring in that element alone; and the mechanical effect that is transmitted is considered as an extraneous force applied at one place in the medium, and passed on from element to element through these internal pressures and tractions until it reaches another place. We have, however, to consider two atomic electric charges as being themselves some kind of strain configurations in the aether; each of them already involves an atmosphere of strain in the surrounding aether which is part of its essence, and cannot be considered apart from it; each of them essentially pervades the entire space, though on account of its invariable character we consider it as a unit. Thus we appear to be debarred from imagining the aether to act as an elastic connection which is merely the agent of transmission of a pull from the one nucleus to the other, because there are already stresses belonging to and constituting an intrinsic part of the terminal electrons, which are distributed all along the medium. Our Action criterion of a dynamical system, in fact, allows us to reason about an electron as a single thing, notwithstanding that its field of energy is spread over the whole medium; it is only in material solid bodies, and in problems in which the actual sphere of physical action of the molecule is small compared with the smallest element of volume that our analysis considers, that the familiar idea of transmission of force by simple stress can apply. Whatever view may ultimately commend itself, this question is one that urgently demands decision. A very large amount of effort has been expended by Maxwell, Helmholtz, Heaviside, Hertz, and other authorities in the attempt to express the mechanical phenomena of electrical action in terms of a transmitting stress. The analytical results up to a certain point have been promising, most strikingly so at the beginning, when Maxwell established the mathematical validity of the way in which Faraday was accustomed to represent to himself the mechanical interactions across space, in terms of a tension along the lines of force equilibrated by an equal pressure preventing their expansion sideways. According to the views here developed, that ideal is an impossible one*; if this could be established to general satisfaction the field of theoretical discussion would be much simplified.

*margin note: must be merely formal.*

This view that the atom of matter is, so far as regards physical actions, of the nature of a structure in the aether involving an atmosphere of aethereal strain all around it, not a small body which exerts direct actions at a distance on other atoms according to extraneous laws of force, was practically foreign to the eighteenth century, when mathematical physics was modelled on the Newtonian astronomy and dominated by its splendid success. The scheme of material dynamics, as finally compactly systematized by Lagrange, had therefore no direct relation to such a view, although it has proved wide enough to include it. The remark has often been made that it is probably owing to Faraday's mathematical instinct, combined with his want of acquaintance with the existing analysis, that the modern theory of the aether obtained a start from the electric side. Through his teaching and the weight of his authority, the notion of two electric currents exerting

*margin note: Historical evolution of atoms.*

* See an elucidation from Action in Vol. I, Appendix IV.

their mutual forces by means of an intervening medium, instead of by direct attraction across space, was at an early period firmly grasped in this country. In 1845 Lord Kelvin was already mathematically formulating, with most suggestive success, continuous elastic connections, by whose strain the fields of activity of electric currents or of electric distributions could be illustrated; while the exposition of Maxwell's interconnected scheme, in the earlier form in which it relied on concrete models of the electric action, goes back almost to 1860. Corresponding to the two physical ideals of isolated atoms exerting attraction at a distance, and atoms operating by atmospheres of aethereal strain, there are, as already indicated, two different developments of dynamical theory. The original Newtonian equations of motion determined the course of a system by expressing the rates at which the velocity of each of its small parts of elements is changing. This method is still fully applicable to those problems of gravitational astronomy in which dynamical explanation was first successful on a grand scale, the planets being treated as point masses, each subject to the gravitational attraction of the other bodies. But the more

**Modern dynamics:** recent development of the dynamics of complex systems depends on the fact that analysis has been able to reduce within manageable limits the number of varying quantities whose course is to be explicitly traced, through taking advantage of those internal relations of the parts of the system that are invariable, either geometrically or dynamically. Thus, to take the simplest case, the dynamics of a solid body can be confined to a discussion of its three components of translation and its three components of rotation, instead of the motion of each element of its mass. With the number of independent coordinates thus diminished, when the initial state of the motion is specified, the subsequent course of the complete system can be traced; but the course of the changes in any part of it can only be treated in relation to the motion of the system as a whole. It is just this mode of treatment of a system as a whole that is the main characteristic of modern physical analysis. The way in which Maxwell analyzed the interactions of a system of linear electric currents, previously treated as if each were made up of small independent pieces or elements, and accumulated the evidence that they formed a single dynamical system, is a trenchant example. The interactions of vortices in fluid form a very similar problem, which is of special note in that the constitution of the system is there completely known in advance, so that the two modes of dynamical exposition can be compared. In this case the older method forms independent equations for the motion of each material element of the fluid, and so requires the introduction of the stress—here the fluid pressure—by which dynamical effect is passed on to it from the surrounding elements: it corresponds to a method of contact action. But Helmholtz opened up new ground in the abstract dynamics of continuous media when he recognized (after Stokes) that, if the distribution of the velocity of spin at those places in the fluid where the motion is vortical be assigned, the motion in every part of the fluid is therein kinematically involved. This, combined with the theorem of Lagrange and Cauchy, that the spin is always confined to the same portions of the fluid, formed a starting point for his theory of vortices, which showed how the subsequent course of the motion can be ascertained without consideration of pressure or other stress.

The recognition of the permanent state of motion constituting a vortex ring as a determining agent as regards the future course of the system was in fact justly considered by Helmholtz as one of his greatest achievements. The principle had entirely eluded the attention of Lagrange and Cauchy and Stokes, who were the pioneers in this fundamental branch of dynamics, and had virtually prepared all the necessary analytical material for Helmholtz's use. The main import influenced by vortex atoms: of this advance lay, not in the assistance which it afforded to the development of the complete solution of special problems in fluid motion, but in the fact that it constituted the discovery of the types of permanent motion of the system, which could combine and interact with each other without losing their individuality[1], though each of them pervaded the whole field. This rendered possible an entirely new mode of treatment; and mathematicians who were accustomed, as in astronomy, to aim directly at the determination of all the details of the special case of motion, were occasionally slow to apprehend the advantages of a procedure which stopped at formulating a description of the nature of the interaction between various typical groups of motions into which the whole disturbance could be resolved.

The new train of ideas introduced into physics by Faraday was thus consolidated and emphasized by Helmholtz's investigations of 1858 in the special domain of hydrodynamics. In illustration let us consider the fluid medium to be pervaded by permanent vortices circulating round solid rings as cores: the older method of analysis would form equations of motion for each element of the fluid, involving the fluid pressure, and by their integration would determine the distribution of pressure on each solid ring, and thence the way it moves. This method is hardly feasible even in the simplest cases. The natural plan is to make use of existing simplifications by regarding each vortex as a permanent reality, and directly attacking the problem of its

---

[1] We may compare G. W. Hill's more recent introduction of the idea of permanent orbits into physical astronomy.

interactions with the other vortices. The energy of the fluid arising from the vortex motion can be expressed in terms of the positions and strengths of the vortices alone; and then the principle of Action, in the generalized form which includes steady motional configurations as well as constant material configurations, affords a method of deducing the motions of the cores and the interactions between them. If the cores are thin they in fact interact mechanically, as Lord Kelvin and Kirchhoff proved, in the same manner as linear electric currents would do; though the impulse thence derived towards a direct hydro-kinetic explanation of electromagnetics was damped by the fact that repulsion and attraction have to be interchanged in the analogy. The conception of vortices, once it has been arrived at, forms the natural physical basis of investigation, although the older method of determining a distribution of pressure stress throughout the fluid and examining how it affects the cores is still possible; that stress, however, is not simply transmitted, as it has to maintain the changes of velocity

some features. of the various portions of the fluid. But if the vortices have no solid cores we are at a loss to know where even this pressure can be considered as applied to them; if we follow up the stress, we lose the vortex; yet a fluid vortex can nevertheless illustrate an atom of matter, and we can consider such atoms as exerting mutual forces, only these forces cannot be considered as transmitted through the agency of fluid pressure. The reason is that the vortex cannot now be identified with a mere core bounded by a definite surface, but is essentially a configuration of motion extending throughout the medium.

The idea of transmission is incomplete. Thus we are again in face of the fundamental question whether all attempts to represent the mechanical interactions of electrodynamic systems, as transmitted from point to point by means of simple stress, are not doomed to failure; whether they do not, in fact, introduce unnecessary and insurmountable difficulty into the theory. The idea of identifying an atom with a state of strain or motion, pervading the region of the aether around its nucleus, appears to demand wider views as to what constitutes dynamical transmission. The idea that any small portion of the primordial medium can be isolated, by merely introducing tractions acting over its surface and transmitted from the surrounding parts, is no longer appropriate or consistent: a part of the dynamical disturbance in that element of the medium is on this hypothesis already classified as belonging to, and carried along with, atoms that are outside it but in its neighbourhood—and this part must not be counted twice over. The law of Poynting relating to the paths of the transmission of energy is known to hold in its simple form only when the electric charges or currents are in a steady state; when they are changing their positions or configurations their own fields of intrinsic energy are carried along with them.

It is not surprising, considering the previous British familiarity with this order of ideas, that the significance for general physics of Helmholtz's doctrine of vortices was eagerly developed in this country, in the form in which it became embodied through Lord Kelvin's famous illustration of the constitution of the matter, as consisting of atoms with separate existence and mutual interactions. This vortex-atom theory has been a main source of physical suggestion because it presents, on a simple basis, a dynamical picture of an ideal material system, atomically constituted, which could go on automatically without extraneous support. The value of such a picture may be held to lie, not in any supposition that this is the mechanism of the actual world laid bare, but in the vivid illustration it affords of the fundamental postulate of physical science, that mechanical phenomena are not parts of a scheme too involved for us to explore, but rather present themselves in definite and consistent correlations, which we are able to disentangle and apprehend with continually increasing precision. *Expanded view.*

It would be an interesting question to trace the origin of our preference for a theory of transmission of physical action over one of direct action at a distance. It may be held that it rests on the same order of ideas as supplies our conception of force; that the notion of effort which we associate with change of the motion of a body involves the idea of a mechanical connection through which that effort is applied. The mere idea of a transmitting medium would then be no more an ultimate foundation for physical explanation than that of force itself. Our choice between direct distance action and mediate transmission would thus be dictated by the relative simplicity and coherence of the accounts they give of the phenomena: this is, in fact, the basis on which Maxwell's theory had to be judged until Hertz detected the actual working of the medium. Instantaneous transmission is to all intents action at a distance, except in so far as the law of action may be more easily formulated in terms of the medium than in a direct geometrical statement.

In connection with these questions it may be permitted to refer to the eloquent and weighty address recently delivered by M. Poincaré to the International Congress of Physics. M. Poincaré accepts the principle of Least Action as a reliable basis for the formulation of physical theory, but he imposes the condition that the results must satisfy the Newtonian law of equality of action and reaction between each pair of bodies concerned, considered by themselves; this, however, he would allow to be satisfied indirectly, if the effects could be *The seat of action and reaction.*

traced across the intervening aether by stress, so that the tractions
on the two sides of each ideal interface are equal and opposite[1]. As
above argued, this view appears to exclude *ab initio* all atomic
theories of the general type of vortex atoms, in which the energy of
the atom is distributed throughout the medium instead of being
concentrated in a nucleus; and this remark seems to go to the root
of the question. On the other hand, the position here asserted is that
recent dynamical developments have permitted the extension of the
principle of Action to systems involving permanent motions, whether
obvious or latent, as part of their constitution; that on this wider
basis the atom may itself involve a state of steady disturbance
extending through the medium, instead of being only a local structure
acting by push and pull. The possibilities of dynamical explanation
are thus enlarged. The most definite type of model yet imagined of
the physical interaction of atoms through the aether is, perhaps, that
which takes the aether to be a rotationally elastic medium after the
manner of MacCullagh and Rankine, and makes the ultimate atom
include the nucleus of a permanent rotational strain configuration,
which as a whole may be called an electron. The question how far
this is a legitimate and effective model stands by itself, apart from
the dynamics which it illustrates; like all representations it can only
The electron. cover a limited ground. For instance, it cannot claim to include the
internal structure of the nucleus of an atom or even of an electron;
for purposes of physical theory that problem can be put aside, it may
even be treated as inscrutable. All that is needed is a postulate of
free mobility of this nucleus through the aether. This is definitely
hypothetical, but it is not an unreasonable postulate because a
rotational aether has the properties of a perfect fluid medium except
where differentially rotational motions are concerned, and so would
not react on the motion of any structure moving through it except
after the manner of an apparent change of inertia. It thus seems possible to hold that such a model forms an allowable representation of
the dynamical activity of the aether, as distinguished from the complete constitution of the material nuclei between which that medium
establishes connection.

At any rate, models of this nature have certainly been most helpful
in Maxwell's hands towards the effective intuitive grasp of a scheme
of relations as a whole, which might have proved too complex for
abstract unravelment in detail. When a physical model of concealed

[1] Cf. also Hertz on the electromagnetic equations, § 12, Wied. *Ann.* 1890.
The problem of merely replacing a system of forces by a statical stress is widely
indeterminate, and therefore by itself unreal; the actual question is whether
any such representation can be coordinated with existing dynamics.

dynamical processes has served this kind of purpose, when its content has been explored and estimated, and has become familiar through the introduction of new terms and ideas, then the ladder by which we have ascended may be kicked away, and the scheme of relations which the model embodied can stand forth in severely abstract form. Indeed many of the most fruitful branches of abstract mathematical analysis itself have owed their start in this way to concrete physical conceptions. This gradual transition into abstract statement of physical relations, in fact, amounts to retaining the essentials of our working models while eliminating the accidental elements involved in them; elements of the latter kind must always be present because otherwise the model would be identical with the thing which it represents, whereas we cannot expect to mentally grasp all aspects of the content of even the simplest phenomena. Yet the abstract standpoint is always attained through the concrete; and for purposes of instruction; such models, properly guarded, do not perhaps ever lose their value: they are just as legitimate aids as geometrical diagrams, and they have the same kind of limitations. In Maxwell's words, "for the sake of persons of these different types scientific truth should be presented in different forms, and should be regarded as equally scientific whether it appear in the robust form and the vivid colouring of a physical illustration, or in the tenuity and paleness of a symbolical expression." The other side of the picture, the necessary incompleteness of even our legitimate images and modes of representation, comes out in the despairing opinion of Young (*Chromatics*, 1817), at a time when his faith in the undulatory theory of light had been eclipsed by Malus' discovery of the phenomena of polarization by reflexion, that this difficulty "will probably long remain, to mortify the vanity of an ambitious philosophy, completely unresolved by any theory": not many years afterwards the mystery was solved by Fresnel.

*Physical models provisional:*

*necessarily incomplete.*

This process of removing the intellectual scaffolding by which our knowledge is reached, and preserving only the final formulae which express the correlations of the directly observable things, may moreover readily be pushed too far. It asserts the conception that the universe is like an enclosed clock that is wound up to go, and that accordingly we can observe that it is going, and can see some of its more superficial movements, but not much of them; that thus, by patient observation and use of analogy, we can compile, in merely tabular form, information as to the manner in which it works and is likely to go on working, at any rate for some time to come; but that any attempt to probe the underlying connection is illusory or illegitimate. As a theoretical precept this is admirable. It minimizes the danger of our ignoring or forgetting the limitations of human

*The problem of dynamical physics.*

faculty, which can only utilize the imperfect representations that the external world impresses on our senses. On the other hand, such a reminder has rarely been required by the master minds of modern science, from Descartes and Newton onwards, whatever their theories may have been. Its danger as a dogma lies in its application. Who is to decide, without risk of error, what is essential fact and what is intellectual scaffolding? To which class does the atomic theory of matter belong? That is, indeed, one of the intangible things which it is suggested may be thrown overboard, in sorting out and classifying our scientific possessions. Is the mental idea or image, which suggests, and alone can suggest, the experiment that adds to our concrete knowledge, less real than the bare phenomenal uniformity which it has revealed? Is it not, perhaps, more real in that the uniformities might not have been there in the absence of the mind to perceive them?

Molecular dynamics:    No time is now left for review of the methods of molecular dynamics. Here our knowledge is entirely confined to steady states of the molecular system: it is purely statical. In ordinary statics and the dynamics of undisturbed steady notions, the form of the energy formal theory: function is the sufficient basis of the whole subject. This method is extended to thermodynamics by making use of the mechanically available energy of Rankine and Kelvin, which is a function of the bodily configuration and chemical constitution and temperature of the system, whose value cannot under any [isothermal] circumstances spontaneously increase, while it will diminish in any operation which is not reversible. In the statics of systems in equilibrium or in steady motion, this method of energy is a particular case of the method of Action; but in its extension to thermal statics it is made to include thermal statics, chemical as well as configurational changes, and a new point appears to arise. Whether we do or do not take it to be possible to trace the application of the principle of Action throughout the process of chemical combination of two molecules, we certainly here postulate that the static case of that principle, which applies to steady systems, its limits. can be extended across chemical combinations. The question is suggested whether extension would also be valid to transformations which involve vital processes. This seems to be still considered an open question by the best authorities. If it be decided in the negative a distinction is involved between vital and merely chemical processes.

It is now taken as established that vital activity cannot create energy, at any rate in the long run, which is all that can from the nature of the case be tested. It seems not unreasonable to follow the analogy of chemical actions, and assert that it cannot in the long run increase the mechanical availability of energy—that is, considering the organism as an apparatus for transforming energy without being itself *The living organism as an engine.* essentially changed. But we cannot establish a Carnot cycle for a portion of an organism, nor can we do so for a limited period of time; there might be creation of availability accompanied by changes in the organism itself, but compensated by destruction and the inverse changes a long time afterwards. This amounts to asserting that where, as in a vital system or even in a simple molecular combination, we are unable to trace or even assert complete dynamical sequence, exact thermodynamic statements should be mainly confined to the activity of the existing organism as a whole; it may transform inorganic material without change of energy and without gain of availability, although any such statements would be inappropriate and unmeaning as regards the details of the processes that take place inside the organism itself.

In any case it would appear that there is small chance of reducing these questions to direct dynamics; we should rather regard Carnot's *Carnot's principle not dynamical,* principle, which includes the law of uniformity of temperature and is the basis of the whole theory, as a property of statistical type confined to stable or permanent aggregations of matter. Thus no dynamical proof from molecular considerations could be regarded as valid unless it explicitly restricted the argument to permanent systems; yet the conditions of permanency are unknown except in the simpler cases. The only mode of discussion that is yet possible is the method of dynamical statistics of molecules introduced by Maxwell. *but statistical for permanent systems.* Now statistics is a method of arrangement rather than of demonstration. Every statistical argument requires to be verified by comparison with the facts, because it is of the essence of this method to take things as fortuitously distributed except in so far as we know the contrary; and we simply may not know essential facts to the contrary. For example, if the interaction of the aether or other cause produces no influence to the contrary, the presumption would be that the kinetic energy acquired by a molecule is, on the average, equally distributed among its various independent modes of motion, whether vibrational or translational. Assuming this type of distribution to be once established in a gaseous system, the dynamics of Boltzmann and Maxwell show that it must be permanent. But its assumption in the *Statistics necessarily provisional.* first instance is a result rather of the absence than of the presence of knowledge of the circumstances, and can be accepted only so far as it agrees with the facts; our knowledge of the facts of specific heat shows that it must be restricted to modes of motion that are homologous. In the words of Maxwell, when he first discovered in 1860, to his great surprise, that in a system of colliding rigid atoms the energy would always be equally divided between translatory and rotatory motions, it is only necessary to assume, in order to evade this unwelcome conclusion, that "something essential to the complete statement of the physical theory of molecular encounters must have hitherto escaped us."

Our survey thus tends to the result, that as regards the simple and uniform phenomena which involve activity of finite regions of the universal aether, theoretical physics can lay claim to constructive

Physics accepts the world as existing.

functions, and can build up a definite scheme; but in the domain of matter the most that it can do is to accept the existence of such permanent molecular systems as present themselves to our notice, and fit together an outline plan of the more general and universal features in their activity. Our well-founded belief in the rationality of natural processes asserts the possibility of this, while admitting that the intimate details of atomic constitution are beyond our scrutiny and provide plenty of room for processes that transcend finite dynamical correlation.

# 55

## ON THE RELATIONS OF RADIATION
## TO TEMPERATURE.

[Report of the British Association: Bradford, 1900.]

THE key to this subject is the principle, arrived at independently by Balfour Stewart and Kirchhoff about the year 1857, that the constitution and intensity of the steady radiation in an enclosure is determined by the temperature of the surrounding bodies, and involves no other element. It was pointed out by Stewart[1] that if the enclosure contains a radiating and absorbing body which is put in motion, the temperature being uniform throughout, then the constitutions of the radiation in front of it and behind it will differ on account of the Doppler effect, so that there will be a chance of gaining mechanical work in the restoration of a uniform state. There must thus be some kind of thermodynamic compensation, which might arise from aethereal friction, or from work required to produce the motion of the body against pressure excited by the surrounding radiation. The hypothesis of friction is now out of court in ultimate molecular physics; while the thermodynamic bearing of a pressure produced by radiation has been developed by Bartoli and Boltzmann (1884), and that of the Doppler effect by Wien (1893).

*Application of the Doppler Principle.* The procedure of Wien amounts to isolating a region of radiation within a perfectly reflecting enclosure, and estimating the average shortening of the constituent wave-lengths produced by a very slow shrinkage of its volume. The argument is, however, much simplified if the enclosure is take to be spherical and to remain so; for it may then be easily shown that each individual undulation is shortened in the same ratio as is the radius of the enclosure, so that the undulatory content remains similar to itself, with uniformly shortened wave-lengths, whether it is uniformly distributed as regards direction or not, and whatever its constitution may be. But if there is a very small piece of a material radiator in the enclosure, the radiation initially inside will have been reduced by its radiating and absorbing action to that corresponding to its temperature. In that case the shrinkage must retain it always, at each stage of its transformation, in the constitution corresponding

The principle of B. Stewart and Kirchhoff:

extended by Stewart to convection:

implying compensation (Bartoli),

found in the pressure of radiation.

Wien's shrinking enclosure:

much simplified if taken to be spherical.

Internal temperature changes,

---

[1] *Brit. Assoc. Report*, 1871; cf. also *Encyc. Brit.* art. "Radiation" (1886), by Tait.

to some temperature. Otherwise differences of temperature would be effectively established between the various constituents of the radiation in the enclosure; these could be permanent in the absence of

*must be definite.* material bodies; but if the latter are present this would involve degradation of their energy, for which there is here no room, because, on the principles of Stewart and Kirchhoff, the state corresponding

*Hence Wien's displacement law.* to given energy and volume and temperature is determinate. Thus we infer that if the wave-lengths of the steady radiation corresponding to any one temperature are all altered in the same ratio, we obtain a distribution which corresponds to some other temperature in every respect except absolute intensities.

*Radiation pressure demanded:* *Direct Transformation of Mechanical Energy into Radiation*[1]. There is one point, however, that rewards examination. When undulations of any kind are reflected from an advancing wall, there is slightly more energy in the reflected beam than there was in the incident beam, although its length is shorter on account of the Doppler effect.

*can change mechanical work into light energy.* This requires that the undulations must oppose a resistance to the advancing wall, and that the mechanical work required to push on the wall is directly transformed into undulatory energy. In fact, let us consider the mechanism of the reflexion. Suppose the displacement in a directly incident wave-train, with velocity of propagation

*Dynamics of moving reflector.* $c$, to be $\xi = a \cos (mx - mct)$; that in the reflected train will be $\xi' = a' \cos (m'x + m'ct)$, where $a'$, $m'$ are determined by the condition that the total displacement is annulled at the advancing reflector, because no disturbance penetrates beyond it; therefore when $x = vt$, where $v$ is its velocity, positive when receding, $\xi + \xi' = 0$.

*Exact form of Doppler principle.* Thus we must have $a' = - a$, and $m' = m \dfrac{c - v}{c + v}$, in agreement with the usual statement of the Doppler effect when $v$ is small compared with $c$. Observe, in fact, that the direct and reflected wave-trains have a system of nodes which travel with velocity $v$, and that the moving reflector coincides with one of them. Now the velocities $d\xi/dt$ and $d\xi'/dt$ in these two trains are not equal. Their mean squares, on which the kinetic energy per unit length depends, are as $m^2$ to $m'^2$.

---

[1] The present form of this argument arose out of some remarks contributed by Professor FitzGerald, and by Mr Alfred Walker of Bradford, to the dis-

*Windmill driven by radiation:* cussion on this paper. Mr Walker points out that by reflecting the radiation from a hot body, situated at the centre of a wheel, by a ring of oblique vanes around its circumference, and then reversing its path by direct reflexion from a ring of fixed vanes outside the wheel, so as to return it into the source, its pressure may be (theoretically) utilized to drive the wheel, and in time to get

*thermo-dynamically.* up a high speed if there is no load: the thermodynamic compensation in this very interesting arrangement lies in the lowering of the temperature of the part of the incident radiation that is not thus utilized.

The potential energies per unit length depend on the means of $(d\xi/dx)^2$ and $(d\xi'/dx)^2$, and are of course in the same ratio. Thus the energies per unit length in the direct and reflected trains are as $m^2$ to $m'^2$, while the lengths of the trains are as $m'$ to $m$; hence their total energies are as $m$ to $m'$; in other words the reflected train has received an accretion of energy equal to $1 - m'/m$ of the incident energy, [positive for an advancing reflector], which can only have come from mechanical work spent in pushing on the reflector with its velocity $v$. The opposing pressure is thus in numerical magnitude the fraction $\left(1 - \dfrac{m'}{m}\right)\dfrac{c}{v}$ of the density of the incident energy, which works out to be $\dfrac{c^2 - v^2}{c^2 + v^2}$ of the intensity of the total undulatory energy, direct and reflected, that is in front of the reflector*.

Law of radiation pressure against advancing or receding reflector.

When $v$ is small compared with $c$, this agrees with Maxwell's law for the pressure of radiation. This case is also theoretically interesting, because in the application to aether waves $\xi$ is the displacement of the aether elements whose velocity $d\xi/dt$ represents the magnetic force; so that here we have an actual case in which this vector $\xi$, hitherto introduced only in the theoretical dynamics of electron-theory, is essential to a bare statement of the facts. Another remark here arises. It has been held that a beam of light is an irreversible agent, because the radiant pressure at the front of the beam has nothing to work against, and its work is therefore degraded. But suppose it had a reflector receding with its own velocity $c$ to work against; our result shows that the pressure vanishes and no work is done. Thus that objection to the thermodynamic treatment of a single ray is not well founded.

The aethereal displacement.

Pressure law consistent for very high speeds.

This generalization of the theory of radiant pressure to all kinds of undulatory motion is based on the conservation of the energy. It remains to consider the mechanical origin of the pressure. In the special case of an unlimited stretched cord carrying transverse waves the advancing reflector may be a lamina, through a small hole in which the cord passes without friction: the cord is straight on one side of the lamina, and inclined on the other side on account of the vibration; and it is easily shown that the resultant of the tensions on the two sides provides a force acting on the lamina which, when averaged, agrees with the general formula. In the case of an extended medium with advancing transverse waves, which are reflected directly,

Analogy of waves swept up on a cord

---

* For details see *infra* "On the Dynamics of Radiation" (1912), where the analogy of waves on a stretched cord is worked out. The factor does not alter with change of sign of $v$, but the energy is piled up in front of an advancing reflector.

the origin of the pressure is not so obvious, because there is not an obvious mechanism for a reflector which would sweep the waves in front of it as it advances. In the aethereal case we can, however, on the basis of electron theory, imagine a constitution for a reflector which will turn back the radiation on the same principle as a metallic mirror totally reflects Hertzian waves, and thus obtain an idea of how the force acts.

Radiation swept back by free electrons.

. The case of direct incidence has here been treated for simplicity; that of oblique incidence easily follows; the expression for the pressure is reduced in the ratio of the square of the cosine of the angle of incidence. If we average up, after Boltzmann, for the natural radiation in an enclosure, which is incident equally at all angles, we find that the pressure exerted is one-third of the total density of radiant energy.

Pressure of natural radiation.

*Adiabatics of an enclosed Mass of Radiation, and resulting General Laws.* Now consider an enclosure of volume $V$ containing radiant energy travelling indifferently in all directions, and of total density $E$; and let its volume be shrunk by $\delta V$. This requires mechanical work $\frac{1}{3} E \delta V$, which is changed into radiant energy: thus

Law of Energy in a shrinking enclosure:

$$EV + \tfrac{1}{3} E \delta V = (E - \delta E)(V - \delta V),$$

where $E - \delta E$ is the new density at volume $V - \delta V$. This gives $\frac{4}{3} E \delta V = V \delta E$, or $E \propto V^{-\frac{4}{3}}$.

As already explained, if the original state has the constitution as regards wave-lengths corresponding to a temperature $T$, the new state must correspond to some other temperature $T - \delta T$. Thus we can gain work by absorbing the radiation into the working substance of a thermal engine at the one temperature, and extracting it at the other; as the process is reversible, we have by Carnot's principle

Carnot's principle gives law of temperature:

$$\tfrac{1}{3} E \delta V / EV = - \delta T / T,$$

so that $T \propto V^{-\frac{1}{3}}$.

Thus $E \propto T^4$, which is Stefan's law for the relation of the aggregate natural radiation to the temperature.

Radiation density and temperature.

Moreover, the Doppler principle has shown us that in the uniform shrinkage of a spherical enclosure the wave-lengths diminish as the linear dimensions, and therefore as $V^{\frac{1}{3}}$, or inversely as $T$ by the above result. Thus in the radiations at different temperatures, if the scale of wave-length is reduced inversely as the temperature the curves of constitution of the radiation become homologous, *i.e.* their ordinates are all in the same ratio. This is Wien's law*.

Wien's correspondence of states.

* Nothing is said about atoms. The laws of Stefan and Wien would thus continue to hold at temperatures so high that the wave-length is a small fraction of the atomic diameter.

The radiation function.

These relations show that the energy of the radiation corresponding to the temperature $T$, which lies between wave-lengths $\lambda$ and $\lambda + \delta\lambda$, is of the form $\lambda^{-5} f(\lambda T)\, \delta\lambda$. The investigation, theoretical (Wien, Planck, Rayleigh, etc.) and experimental (Lummer and Pringsheim, Paschen, etc.) of the form of this function $f$ is perhaps the most fundamental and interesting problem now outstanding in the general theory of the relation of radiation to temperature. The theoretical relations on which this expression is founded have been shown to be in agreement with fact; and it appears [Wien] that the form $c_1 e^{-c_2/\lambda T}$ fairly represents $f(\lambda T)$ over a wide range of temperature*. These relations have been derived, as usual, from a dynamical discussion of the aggregate intensity of radiation belonging to the temperature; it may be shown that the same results, but nothing in addition, will be gained by applying the same principles to each constituent of range $\delta\lambda$ by itself, assigning to each its own temperature†.

*Postscript.* The following letter from *Nature* (Jan. 1925) bears on the history of this subject:

The extremely interesting notes by Sir Arthur Schuster in a recent number of *Nature* (Jan. 17th, 1925, p. 87) may possibly leave with the ordinary reader an impression that Balfour Stewart's contributions to the establishment of the laws of natural radiation were slighter than was actually the case. The considered opinion of the late Lord Rayleigh [with regard to the Stewart-Kirchhoff law of exchanges], set out in *Phil. Mag.* Vol. I, 1901, pp. 98–100, or *Scientific Papers*, Vol. IV, pp. 404–405, can hardly be gainsaid. In Stewart's discussion of radiation in an isolated enclosure containing moving bodies, his expressed conviction, that the second law of thermodynamics is only satisfied through the action of mechanical forces necessary to maintain the motion, is only turning round the other way the considerations employed by Boltzmann and by Wien long after, who by means of these mechanical forces (namely, the reaction of radiation pressure) combined with the second law of thermodynamics, deduced the law of structure of natural radiation. Reference may also be made to footnotes appended to the reprint of Stokes' cognate papers in *Math. and Phys. Papers*, Vol. IV, especially p. 136.

Balfour Stewart's prevision.

---

\* Afterwards improved by Planck to the present standard form, which can also include the Rayleigh law for long waves which asserts proportionality to temperature.

† It will give a verification of the initial postulate that these constituent temperatures remain the same.

# ON THE STATISTICAL DYNAMICS OF GAS THEORY AS ILLUSTRATED BY METEOR SWARMS AND OPTICAL RAYS.

[Report of the British Association: Bradford, 1900.]

IMAGINE a cloud of meteors pursuing an orbit in space under outside attraction—in fact, in any conservative field of force. Let us consider a group of the meteors around a given central one. As they keep together their velocities are nearly the same. When the central meteor has passed into another part of the orbit, the surrounding region containing these same meteors will have altered in shape; it will in fact usually have become much elongated. If we merely count large and small meteors alike, we can define the density of their distribution in space in the neighbourhood of this group: it will be inversely as the volume occupied by them. Now consider their deviations from a mean velocity, say that of the central meteor of the group; we can draw from an origin a vector representing the velocity of each meteor, and the ends of these vectors will mark out a region in the velocity diagram whose shape and volume will repre-

*Spreading of a meteor swarm,* sent the character and range of the deviation. It results from a very general proposition in dynamics that as the central meteor moves along its path the region occupied by the group of its neighbours multiplied by the corresponding region in their velocity diagram remains constant. Or we may say that the density at the group considered, estimated by mere numbers, not by size, varies during its motion proportionally to the extent of the region on the velocity diagram which corresponds to it.

This is true whether mutual attractions of the meteors are sensibly effective or not; in fact, the generalized form of this proposition, *in relation to Action.* together with a set of similar ones relating to the various partial groups of coordinates and velocity components, forms an equivalent of the fundamental law of Action which is the unique basis of dynamical theory.

Now, suppose that the mutual attractions are insensible, and that $W$ is the potential of the conservative field: then for a single meteor of mass $m$ and velocity $v$ we have the energy $\frac{1}{2}mv^2 + mW$ conserved: hence if $\delta v_1$ be the range of velocity at any point in the initial position, and $\delta v_2$ that at the corresponding point in any subsequent position of the group, we have $v_1 \delta v_1 = v_2 \delta v_2$, these positions remaining unvaried and the variation being due to different meteors passing through them. But if $\delta \omega_1$ and $\delta \omega_2$ are the initial and final conical angles of divergence of the velocity vectors, corresponding regions in the velocity diagram are of extents $\delta v_1 . v_1^2 \delta \omega_1$ and $\delta v_2 . v_2^2 \delta \omega_2$: these quantities are, therefore, in all cases proportional to the densities at the group in its two positions. In our present case of mutual attractions insensible, the volume density is thus proportional to $v \delta \omega$, because $v \delta v$ remains constant. Now the number of meteors that cross per unit time per unit area of a plane at right angles to the path of the central meteor is equal to this density multiplied by $v$: thus here it remains proportional to $v^2 \delta \omega$, as the central meteor moves on. In the corpuscular formulation of geometrical optics this result carries the general law that the concentration in cross-section of a beam of light at different points of its path is proportional to the solid angular divergence of the rays multiplied by the square of the refractive index, which is also directly necessitated by thermodynamic principles; as a special case it limits the possible brightness of images in the well-known way.

*Analogy in corpuscular optics of rays:*

*intensity of the stream.*

In the moving stream of particles we have thus a quantity that is conserved in each group—namely, the ratio of the density at a group to the extent of the region or domain on the velocity diagram which corresponds to it; but this ratio may vary in any way from group to group along the stream, while there is no restriction on the velocities of the various groups. If two streams cross or interpenetrate each other, or interfere in other ways, all this will be upset owing to the collisions. Can we assign a statistical law of distribution of velocities that will remain permanent when streams, which can be thus arranged into nearly homogeneous groups, are crossing each other in all directions, so that we pass to a model of a gas? Maxwell showed that if the number of particles each of which has a total energy $E$ is proportional to $e^{-hE}$, where $h$ is some constant (which defines the temperature), while the particles in each group range uniformly, except as regards this factor, with respect to distribution in position and velocity jointly, as above, then this will be the case. In fact, the chance of an encounter for particles of energies $E$ and $E'$ will involve the product $e^{-hE}e^{-hE'}$ or $e^{-h(E+E')}$, and an encounter does not alter this total energy $E + E'$; while the domains or extents of range of two colliding groups each nearly homogeneous and estimated, as above, by deviation from a central particle in position and velocity jointly, will have the same product after the encounter as before by virtue of the Action principle. It follows that the statistical chances of encounter, which depend on this joint product, will be the same in the actual motion as are those of reverse encounter in the same motion

*An invariant:*

*generalized to Maxwell's statistics in gas theory:*

*the argument,*

statistically reversed. But if the motion of a swarm with velocities fortuitously directed can be thus statistically reversed, recovering its previous statistics, its molecular statistics must have become steady; in fact, we have in such a system just the same distribution of encountering groups in one direction as in the reverse direction: thus we have here one steady state. The same argument, indeed, shows that a distribution, such that the number per unit volume, of particles whose velocity deviations correspond to a given region in the velocity diagram, is proportional to the extent of that region without this factor $e^{-hE}$, will also be a steady one. This is the case of equable distribution in each group as regards only the position and velocity diagrams conjointly; but in this case each value of the resultant velocity would occur with a frequency proportional to its square, and a factor such as $e^{-hE}$ is required to keep down very high values. The generalizations by Boltzmann and Maxwell to internal degrees of freedom would lead us too far, the aim here proposed being merely concrete illustration of the very general but purely analytical argument that is fully set forth in the treatises of Watson, Burbury, and Boltzmann.

*implies a steady state.*

*Rôle of the exponential factor.*

# 57

## CAN CONVECTION THROUGH THE AETHER BE DETECTED ELECTRICALLY?

[From *The Scientific Writings of the late G. F. FitzGerald* (1902), pp. 566–569.]

THE considerations explained by Dr Trouton in the preceding paper bear fundamentally on the question whether the convection of matter through the aether sensibly disturbs that medium. The result of the experiment which he proposes to carry out* will either corroborate or modify the results deduced from the Michelson-Morley second-order experiment on optical interference, the interpretation of which has hitherto been somewhat ambiguous because that experiment has stood alone as the only one of its kind.

*Optical relativity.*

We begin by tracing the theoretical consequences of an ideal experimental arrangement. Consider a condenser $AB$ suspended from a point $O$, and charged (or rather its charge altered) from another fixed charged condenser $PQ$ in the same circuit. Separating the plates $P$ and $Q$ increases the charge in $AB$, and *vice versa*. Now charge $AB$ when its plates are perpendicular to the direction of the aether drift; then disconnect it and rotate it through a right angle. Finally, let

*An electric perpetual motion, in the Solar frame,*

---

* This paper relates to a fundamental new departure in the relativity of electrodynamics, which was initiated in Dublin in the autumn of 1900 by FitzGerald and his friend and assistant F. T. Trouton, now also deceased. Hitherto, experimental test had been optical, interpreted in terms of the electric theory of light: there had been no direct electric confirmation. The idea was that when a charged condenser is convected a magnetic field is set up between its plates, of value changing with its orientation, of which the energy has therefore to be traced. It was thought at first from rough analogy that this extra energy might involve a considerable mechanical impulse on charging the condenser. When no such impulse of the amount anticipated was found by Trouton (*Trans. R. Dublin Soc.* April 1902; FitzGerald's *Scientific Writings*, pp. 557–565) they fell back, in the last days of FitzGerald's life, on electrodynamic relativity. There may actually, however, be a smaller translatory impulse equal to the electric momentum excited between the plates, as found in footnote, p. 72.

The amended view, that what was to be looked for was a torque involving the square of the charge, as in the text, originated with Trouton (*loc. cit.*): the experiment was carried through by him along with H. R. Noble at University College, London, where he had become professor of physics, and was reported in *Phil. Trans.* for June 1903, with result wholly negative. The test has recently been repeated by Tomaschek, using modern refinements but with like negative result, near the summit of the Jungfrau, on the idea that altitude might influence an aether drift past the travelling Earth.

*Historical.*

$AB$ be discharged by re-connecting to $PQ$—whose plates have mean-time been moved to the distance that will give them the same difference of potential as $AB$—and then approximating the plates $P$ and $Q$ to their original distance. After this $AB$ may again be rotated through a right angle when it is uncharged, and the cycle

drawing on astronomical energies: analyzed. repeated. The conductors being supposed perfect, this will give a perpetual motion, or rather will use up the energy of the Earth's translatory motion through the aether—unless the potential difference

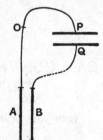

between $A$ and $B$, on which the energy of charge and discharge depends, is the same in the two positions, for a given value of the charge. This potential difference is sustained, in part by the aether strain, in part by the magnetic field between the plates which is produced by the electric convection.

*Electrodynamic relativity for steady systems:* Now a known electrodynamic result may be stated as follows: Compare an electrostatic conducting system at rest in the aether with the identically same system in motion with velocity $v$; suppose the charges to be the same in the two cases: to obtain the potential differences between the conductors in the latter case, find what they would be in the system when at rest, and elongated in the

ratio* $\epsilon^{\frac{1}{2}}$ or $\left(1 - \dfrac{v^2}{c^2}\right)^{-\frac{1}{2}}$ in the direction of the motion, and multiply

the result by $\epsilon^{\frac{1}{2}}$. Let us apply this proposition to the present case.

*applied to a condenser.* (i) When the plane of the condenser is at right angles to the drift, the thickness of the dielectric in the correlative one at rest is increased $\epsilon^{\frac{1}{2}}$ times, thus so is the potential difference for given charge, and the other factor, *supra*, $\epsilon^{\frac{1}{2}}$ makes the increase $\epsilon$ times in all. (ii) When the plane of the condenser is parallel to the drift, the plates are each elongated $\epsilon^{\frac{1}{2}}$ times; hence the density on them is increased $\epsilon^{-\frac{1}{2}}$ times; the dielectric thickness being unaltered, this increases the potential difference $\epsilon^{-\frac{1}{2}}$ times, and the other factor makes the increase in all null.

Thus position (i) of the condenser has $\epsilon$ times as much electric energy as position (ii) for the same charge $Q$; that is, it has more

energy than (ii) by $\dfrac{v^2}{c^2}$ of the total energy of charge in either posi-

tion, this latter energy of charge being $2\pi \dfrac{Q^2}{S} t$, where [$Q$ is in electro-

static measure], $t$ is the thickness of the dielectric, and $S$ the area of either plate.

---

* The first exponent of $\epsilon$ in the text, and in consequence the following ones, is now corrected. Cf. *Aether and Matter* (1900), § 96: also *supra*, p. 39.

In position (ii) there is a magnetic field between the plates, of intensity $4\pi \dfrac{Qv}{Sc}$, having electrokinetic energy $\dfrac{1}{8\pi}\left(4\pi \dfrac{Qv}{Sc}\right)^2 St$, or $\dfrac{v^2}{c^2} 2\pi \dfrac{Q^2}{S} t$, which forms part of the electric energy of the system, the energy of aethereal strain being correspondingly diminished; on rotation of the condenser through a right angle this kinetic energy disappears.

Now the condenser is charged by transferring the charges into the plates against the electric force; the energy required for this operation is half the charge $Q$ multiplied by the potential difference between the plates. As it is charged at one potential difference and discharged at another, there is energy remaining over of the amount estimated above; and as the process is reversible, this energy must be mechanically available. Thus the energy of motion of the Earth through the aether is available for mechanical work to an unlimited extent, unless the potential difference in the condenser [or other electrostatic system] is independent of its orientation; that is, by accepted electrodynamics, unless the FitzGerald-Lorentz shrinkage of moving bodies is a fact.

*Local perpetual motion evaded only by relativity.*

It is to be observed that the proposition that, if this shrinkage occur, the system will in no other respect differ from the state of no motion, is established for uniform translatory motion; so that a torque arising when the motion is rotatory would not be in contradiction to it. But such a torque must depend on the velocity of rotation, vanishing along with the latter, and so does not affect the present argument in which the displacements can be made very slowly.

*Torque for rapid rotation not excluded.*

As the electric energy of the condenser is reduced by turning it from the transverse to the longitudinal position, this motion would, in the absence of the FitzGerald-Lorentz shrinkage, be assisted by a mechanical torque, of amount $\dfrac{v^2}{c^2} W \sin 2\theta$, where $W$ is the energy of charge, and $\theta$ is the angle the plane of the condenser makes with the transverse position[1]; thus the transverse position would be one of unstable equilibrium.

*The stable orientation of the condenser:*

For the case of a single charged conductor, such as may be represented by an ellipsoid, it is not difficult to show* that the stable

*of an elongated charged conductor.*

[1] The torque is equal and opposite to that suggested by the considerations given [by Trouton, *FitzGerald's Papers*, p. 563]. For the condenser there described, suspended vertically at an inclination of 45° to the aether drift, and charged as in the experiments there recorded, the value of this torque would be about 1 c.g.s. unit; while the torsional system as there adjusted was found to respond sensibly to about 7·5 c.g.s. units. [The sensitiveness was made adequate in the later negative experiments by Trouton and Noble, reported in *Phil. Trans.* (June 1903) Vol. ccii, A, pp. 165–182: cf. formula, p. 168.]

*Sensitiveness of Trouton's earlier apparatus.*

* As has been done in detail by M. Abraham, *Wied. Ann.* Oct. 1902, on the same hypothesis of no shrinkage. Cf. also Vol. 1 *supra*, Appendix ix.

orientation is the one in which its longest dimension is along the direction of convection.

The charge and discharge of the condenser involve currents round the condenser $AB$, and round the discharging circuit. So long as the suspending wire $OA$ does not move with the rotation, and also the discharging wire $BQ$, the Earth's magnetic field can produce a torque only by acting on the charging current circulating round $AB$—for charging $A$ positively and $B$ negatively is the same as making a current flow down $A$ and up $B$, if both plates are charged from the top. This torque is negligible in the very slowly effected cycle of the experiment; but any uncertainty can be obviated by reversing its direction each time by an arrangement which first annuls, and then, after half a period, reverses, the charge of the condenser $AB$.

No correction for the Earth's field.

If the condenser $AB$ is held *absolutely fixed* while it is being charged, any impulsive torque there might be could do no work; yet the condenser gets its energy. This seems by itself sufficient to negative the suggestion that the energies of charge and discharge, as distinct from that derived from rotating the condenser, have to do directly with mechanical forces*.

The energies are of aethereal origin.

* The subject may also profitably be approached from another side. For a system isolated electrically, that is subject to no extraneous field of force, the forces it exerts on its supports may be determined (*supra*, p. 72) from a distribution of internal aethereal momentum of density $[DH]$, where $D$ is aethereal displacement. For the present case of a condenser this distribution is uniform over its field between the plates, and amounts to $2W/c^2 . v \cos \theta$ directed along the condenser in the plane of $\theta$, where $\theta$ is the angle between $v$ and its plates. This is for air as dielectric: otherwise there is a factor $K^{-1}$. The forces exerted by the system are the reaction to the time rate of change of this momentum arising from changes of $v$ and of $\theta$. Also a torque is involved, amenable to measurement as above, arising from displacement of the line of application of this momentum by the convection $v$, and thus equal to $W/c^2 . v^2 \sin 2\theta$ round an axis transverse to $v$ in the plane of $\theta$, in agreement with the result in the text. Cf. also Lorentz, "Elektronentheorie," *Ency. Math.* v (1903), p. 259. But the experiments exclude any actual torque of comparable value.

Alternative exposition.

Moreover, if any such torque existed, the energy of the Earth's motion through space could, ideally, be made mechanically available, or "harnessed," as in the text, as Trouton remarked. The reconciliation is that the torque thus determined is relative to a frame in which the system has the velocity $v$, with associated convective deformation: whereas mechanical forces are actually measured by apparatus attached to the observer's local frame for which $v/c$ is of no account, though for the Solar frame it is $10^{-4}$.

Why the effect is null.

# 58

# ON THE ELECTRODYNAMIC AND THERMAL RELATIONS OF ENERGY OF MAGNETIZATION.

[*Proceedings of the Royal Society*, Vol. LXXI, pp. 229–239, Jan. 1903.]

1. There appears to be still some uncertainty as to the principles on which the energy of magnetized iron is to be estimated, and the extent to which that energy is electrodynamically effective*. The following considerations are submitted as a contribution towards definite theoretical views. <small>Available magnetic energy.</small>

The electrokinetic energy of a system of electric currents $\iota_1, \iota_2, \ldots$, flowing in complete linear circuits in free aether, is known to be

$$\tfrac{1}{2}\left(\iota_1 N_1 + \iota_2 N_2 + \ldots\right);$$

<small>Energy of current system,</small>

wherein $N_1$ is the number of tubes of the magnetic force $(\alpha, \beta, \gamma)$ that thread the circuit $\iota_1$, and is thus equal to $\int (l\alpha + m\beta + n\gamma)\, dS$ extended over any barrier surface $S$ which blocks that circuit, $(\alpha, \beta, \gamma)$ being circuital (*i.e.* a stream vector) so that all such barriers give the same result. As under steady circumstances $(\alpha, \beta, \gamma)$ is also derivable from a magnetic potential $V$, which has a cyclic constant $4\pi\iota$ with regard to each current, this energy assumes the form

$$\frac{1}{8\pi}\, \Sigma \int V \left( l\frac{\partial V}{\partial x} + m\frac{\partial V}{\partial y} + n\frac{\partial V}{\partial z} \right) dS,$$

<small>redistributed as magnetic energy in field.</small>

in which the integrals are now extended over both faces of each barrier surface. This is equal by Green's theorem to the volume integral

$$\frac{1}{8\pi} \int (\alpha^2 + \beta^2 + \gamma^2)\, d\tau$$

extended throughout all space. This latter integral is in fact taken in most forms of Maxwell's theory to represent the actual distribution, in all circumstances whether steady or not[1], of the electrokinetic energy among the elements of volume of the aether, in which it is supposed to reside as kinetic energy.

2. The most definite and consistent way to treat magnetism and its energy is to consider it as consisting in molecular electric currents; so that in magnetic media we have the ordinary finite currents, <small>Magnetism as Amperean currents:</small>

* Cf. G. H. Livens, *Roy. Soc. Proc.* Sept. 1916: or *Treatise on Electricity*, Ed. 2.

[1] In the previous electric specification, the fictitious electric currents of aethereal displacement must be introduced when the state is not steady.

combined with molecular currents so numerous and irregularly orientated that we can only average them up into so much polarization per unit volume of the space they occupy. So far in fact as the latter currents are concerned, the only energy that need or can occupy our attention is that connected with some regularity in their orientation, *i.e.* with magnetization, the remaining irregular part being classed with heat. If there were no such molecular currents, the magnetic force ($\alpha$, $\beta$, $\gamma$) in the aether would in steady fields be derived from a potential, cyclic only with regard to the definite number of circuits of the ordinary currents. But when magnetism is present this potential is cyclic also with respect to the indefinitely great number of molecular circuits. The line integral of magnetic force round any circuit is $4\pi$ ($\Sigma\iota + \Sigma\iota'$), where $\Sigma\iota'$ refers to the practically continuous distribution of magnetic molecular currents that the circuit threads. This latter vanishes when these currents are not orientated with some kind of regularity. If we extend the integral from a single line to an average across a filament or tube of uniform cross-section $\delta S$, with that line for axis, we obtain readily the formula, the last integral being the expression of the component along the circuit of the moments of the molecular currents,

$$\delta S \int (\alpha dx + \beta dy + \gamma dz) = \delta S 4\pi\Sigma\iota + 4\pi \int (A dx + B dy + C dz) \, \delta S$$

in which ($A$, $B$, $C$) $\delta\tau$ represents the magnetization in volume $\delta\tau$. Thus, after transposition of the last term, and removal of the factor $\delta S$ after the average has now been taken, we obtain

$$\int \{(\alpha - 4\pi A) \, dx + (\beta - 4\pi B) \, dy + (\gamma - 4\pi C) \, dz\} = 4\pi\Sigma\iota.$$

the cyclic potential: In other words, this new vector ($\alpha - 4\pi A$, $\beta - 4\pi B$, $\gamma - 4\pi C$) is *derived from a potential* which is cyclic in the usual manner with regard to the ordinary currents *alone*.

If we compare this result with the customary magnetic vectors of Kelvin and Maxwell, it appears that ($\alpha$, $\beta$, $\gamma$) must represent the

interpreted: reversed notation. "induction," and so will hereafter be denoted, after Maxwell, by ($a$, $b$, $c$). The new vector, which has a potential cyclic with respect to the finite currents only, represents the "force," and will hereafter be denoted by ($\alpha$, $\beta$, $\gamma$), whose significance is thus changed from henceforth. The "induction" on the other hand has not necessarily a potential, but is, by the constitution of the free aether, always circuital; that is, it satisfies the condition of streaming flow

$$\frac{\partial a}{\partial x} + \frac{\partial b}{\partial y} + \frac{\partial c}{\partial z} = 0.$$

The expression for the energy now includes terms

$$\tfrac{1}{2}\left(\iota_1 N_1 + \iota_2 N_2 + \ldots\right)$$

for the ordinary currents $\iota_1, \iota_2, \ldots$, where $N_1, N_2, \ldots$ are the fluxes, *of magnetic induction*, through their circuits; this transforms as usual into

$$\frac{1}{8\pi}\,\Sigma \int V\,(la + mb + nc)\,dS$$

over both faces of each barrier, which by Green's theorem is equal to

$$\frac{1}{8\pi}\int (a\alpha + b\beta + c\gamma)\,d\tau \qquad\qquad (i)$$

*[margin: Dissection of energy of Amperean magnet:]*

extended throughout all space. But there are also terms

$$\tfrac{1}{2}\left(\iota_1' N_1' + \iota_2' N_2' + \ldots\right)$$

for the molecular currents; now taking $N'$ to be the cross-section of the circuit multiplied by the component of the averaged induction normal to its plane, and remembering that $\iota'$ multiplied by this cross-section is the magnetic moment of this molecular current, it appears that $\iota'N'$ is equal to the magnetic induction multiplied by the component of the magnetic moment in its direction, and therefore $\tfrac{1}{2}\Sigma\iota'N'$ is equal to

$$\tfrac{1}{2}\int (Aa + Bb + Cc)\,d\tau.$$

Thus the magnetic circuits add to the energy the amount[1]

$$\tfrac{1}{2}\int (A\alpha + B\beta + C\gamma)\,d\tau \qquad\qquad (ii)$$

together with

$$2\pi \int (A^2 + B^2 + C^2)\,d\tau. \qquad\qquad (iii)$$

The formula (i) is usually taken, after Maxwell's example, to represent the energy of the electrokinetic field. It here appears that it represents only the part of the energy that is concerned with the currents, arising from their mutual interactions and the interactions *[margin: of currents:]* of the magnets with them: that there exists *in addition* a quantity (ii) which is that taken by Maxwell as the energy of magnetization in *[margin: of magnetization:]* the field $(\alpha, \beta, \gamma)$, and also a quantity (iii), which is purely local and *[margin: constitutive.]* constitutive, of the same general type as energy of crystallization. The question arises whether (iii) is a part of the intrinsic energy of magnetization of different kind from (ii), in that it cannot even *[margin: Availability of latter part?]* partially emerge as mechanical work, or on the contrary the usual formula (ii) must be amended. See §§ 5, 8. In any case the dynamics

---

[1] (These energies as here determined are *kinetic*; if they are (as is customary) to be considered as *potential*, their signs must be changed. Cf. *Phil. Trans.*, A, 1894, p. 806: Vol. I, *supra*, p. 513.)

of the field of currents (when there are no irreversible features) involves only that part of the energy function in which the currents operate, thus excluding both (ii) and (iii).

Ex.: ring coil on iron core:

3. The simplest example is that of a coil of $n$ turns carrying a current $\iota$, wound uniformly on a narrow iron ring core, of cross-section $S$ and length $l$. On the present basis the energy is made up of an electrodynamic part $\frac{1}{2}\iota n \mathbf{B} S$ and a magnetic part $\frac{1}{2}\mathbf{J}\mathbf{B}Sl$; as $4\pi n\iota = \mathbf{H}l$ by the Amperean circuital law, these parts are

$$\frac{\mathrm{I}}{8\pi}\mathbf{B}\mathbf{H}v \quad \text{and} \quad \tfrac{1}{2}\mathbf{B}\mathbf{J}v,$$

where $v$ is the volume of the core; they make up *in all* $\mathbf{B}^2/8\pi$ per unit volume instead of the usual $\mathbf{B}\mathbf{H}/8\pi$. The former part is mechanically available. The question has been raised by Lord Rayleigh[1] whether

Rayleigh's conclusion.

the latter part, which includes the very large term (iii) above, namely $2\pi\mathbf{J}^2v$, in the case of iron, has any considerable mechanical effectiveness; the question can only arise when it belongs in part to *permanent magnetism* whose ultimate annulment can induce a current—when the current vanishes the energy of permanent magnetism, in the present case represented by (iii) alone, is the only part of the energy of the system that remains. The conclusion reached by him is that it cannot be annulled quick enough, when the ring carries a coil, to develop any considerable available electric energy by induction.

Hydraulic analogue:

4. We may form a rough illustration of the mechanical *rôle* of this purely magnetic energy by considering, as the analogy of the currents, a branching system or network of pipes carrying liquids, in one of which a turbine is located, to be driven by the stream, which will be supposed to be an alternating one. The flow will be directed more fully into this particular pipe, and higher pressure will also be attained, after the manner of the hydraulic ram, if it communicates at the side with an expansible reservoir into which the liquid can readily force its way, to be expelled again by the elasticity of its walls when the stream begins to set in the reverse direction. This increase of kinetic pressure on the turbine roughly represents the electromotive pressure on a motor due to the increased magnetic flux, and the energy spent in expanding the reservoir as it fills up represents the energy of magnetization of the iron. If things were perfectly reversible in the reservoir, that is if the iron were perfectly soft, the latter energy would rise and fall concomitantly with the alternations of pressure on the motor, but of course if its temperature remained constant it would contribute nothing to the energy driving the motor,

[1] *Phil. Mag.* 1885: also *Archives néerlandaises*, Vol. II, 1891, p. 6, reprinted in *Phil. Mag.* 1902, and in *Scientific Papers*, Vol. IV, No. 272.

which must be introduced into the system from an extraneous source. But if there are frictional resistances involved in filling the reservoir, the operations will not be perfectly reversible, and mechanical energy will be lost in it by conversion into heat; and moreover on account of the phase of its changes getting out of step—still more by permanent delays such as are classed under hysteresis—it will operate less efficiently in directing the stream of energy towards the turbine. Both these statements have analogical application to the iron in a magnetic circuit.

*two causes of ineffectiveness.*

An example is provided by the ring coil aforesaid. Suppose that when the current has ceased in the coil the core retains permanent magnetism, its energy being the latter term in the formula above. This corresponds to the reservoir becoming temporarily choked, so that it retains its contents after the pressure that drove the liquid into it has been removed. The question arises whether this retained energy is available for mechanical work. The present aspect of the matter appears to lead to the conclusion (Lord Rayleigh's) that it will not be available to any considerable extent unless its pressure in the reservoir is considerable, that is, in the magnetic case, unless the iron is not very receptive of magnetization.

The paradox that energy of residual magnetism, which is outside the electrokinetic system, can on running down affect that system, shows that the circumstances are more general than an analogy of a pure dynamical system of finite number of degrees of freedom can illustrate. In fact the equations of dynamics imply permanent structure of the system; whereas in Professor Ewing's illustrative model of para-magnetization, when the displacement is great enough the structure changes by the component magnets toppling over[1], and after the general disturbance thus set up has subsided with irrecoverable loss of energy into heat, there remains a new structure to deal with. The only way to estimate the available part that may be latent in the great store of energy of residual magnetism of an iron core is thus by the empirical process of detailed experiment. Lord Rayleigh has inferred from the form of the curve of hysteresis for retentive iron in high fields that the fraction that is directly available at the actual temperature must always be small, and he supports the inference by considerations of the nature of the above analogy; in the absence of hysteresis there would be no such direct availability. He derives the practical result that a complete magnetic circuit is deleterious for induction coils in which length of spark is the *desideratum*, the increased total induction attained inside the ring core being more than neutralized by the diminished promptness of magnetic reversal.

*System not purely dynamical.*

*Bearing on sparking induction coil.*

[1] The effective susceptibility $dI/dH$ becoming enormous in the steep part of the characteristic curve.

Ideal most
promising
case:
In fact, if the core, laminated so as to have merely negligible conductivity, is surrounded by a perfectly conducting coil or sheath, and its permanent magnetism is removed at constant rate $- dI/dt$ by an ideal process applied to it, the intensity of induction in the core will diminish at the rate $- 4\pi dI/dt$; and this defect of induction must be made good by the influence of the current thereby induced in the sheath, as otherwise there would be a finite electromotive power in it, which is impossible on account of its perfect conductivity. This restored induction is of the form $H' + 4\pi I'$, where $I'$ is the magnetism induced by the force $H'$ due to the induced current; thus

$$\frac{d}{dt}(H' + 4\pi I') + 4\pi \frac{dI}{dt} = 0,$$

and the actual total rate of fall of magnetization is diminished to $\frac{dI}{dt} - \frac{dI'}{dt}$, which is only the fraction $\frac{1}{4\pi}\frac{dH'}{dt}\Big/\frac{dI'}{dt}$ of the constrained loss of retained magnetism $dI/dt$. In this most favourable case the action of the coil or sheath thus delays the time rate of loss of permanent magnetism in the core in the ratio $(4\pi\kappa')^{-1}$, where $\kappa'$ is the effective permeability for small additional force under the actual circumstances; is unfavourable. that is, the delay in reversal more than compensates the gain in induction.

5. There remains another question, when viscous and other hysteretic effects are practically absent so that the changes of magnetization exactly keep step with those of the currents, and the degree of availability of *residual* magnetic energy thus does not arise;—whether the energy of the magnetization comes from the store of heat of the material and is thus concomitant with a cooling effect when no heat is supplied, or whether it is in part intrinsic inalienable energy of the individual molecules merely temporarily classed as magnetic. So far as it may be the latter, it must for each element of volume depend on the state of that element alone, like the part (iii) of § 2. It has already been seen that no part of (ii) or (iii) can be supplied from the electrodynamic field. This points to the intrinsic energy of paramagnetism, except an unknown fraction of the local part (iii), which depends only on the state of polarization of the element of the medium, being its type derived from purely thermal sources; and the following thermodynamic argument[1] will strengthen this conclusion.

If the value of the magnetic susceptibility $\kappa$ for any material is a function of the temperature, we can perform a Carnot reversible

---

[1] This theoretical deduction of Curie's law has been already given substantially in *Phil. Trans.*, A, Vol. cxc, 1897, p. 287 [: *supra*, p. 116. In connection with §§ 5, 6 see Appendix to this volume.]

The theory of diamagnetism, which assigns it to modification of conforma-

cycle by moving a small portion (say a sphere) of the substance in the permanent field of a system of magnets supposed held rigidly magnetized by constraints. We can move it into a stronger region $H_2$ of this field, of varying strength $H$, maintaining it at the temperature $\theta$ by a supply of heat from outside bodies at that temperature; we can then move it on further, having stopped the supply of heat, until its temperature becomes $\theta - \delta\theta$; we can move it back again isothermally by aid of a sink of heat at this temperature until the stage $H_1$ is reached, when further progress back adiabatically will restore it to its original condition. If $\kappa$ is a function only of the strength of field and of the temperature, this cycle will be reversible. If $E$ is the heat energy supplied at temperature $\theta$, and $W$ is the work done by the sphere on external bodies in the cycle, the principle of Carnot gives the relation

$$\frac{E}{\theta} = \frac{W}{\delta\theta}.$$

Now
$$W = \frac{d}{d\theta}\left(\tfrac{1}{2}I_2 H_2 - \tfrac{1}{2}I_1 H_1\right)$$

$$= -\tfrac{1}{2}\frac{d\kappa}{d\theta}\left(H_2{}^2 - H_1{}^2\right)\delta\theta, \text{ if } \kappa \text{ is small}[1],$$

when the cycle is taken such that the change of $H$ along the adiabatic part of the path is negligible [it is not so restricted as the cycle is expressible by a parallelogram] compared with that along the isothermal part. Thus

$$E = -\tfrac{1}{2}\left(H_2{}^2 - H_1{}^2\right)\theta\frac{d\kappa}{d\theta}.$$

tion in the individual molecule by the inducing field rather than to average spatial orientation of the crowd of molecules, leads to a non-thermal origin as regards that part. The analogous question (*loc. cit.*) as to whether dielectric polarization is mainly an affair of orientation of unaltered molecules like paramagnetism, or one of polarity due to internal deformation of the molecule like diamagnetism, is now answered by the experiments of J. Curie and Compan (*Comptes Rendus*, June 2, 1902). It appears that the dielectric coefficient of glass, for rapid changes, *diminishes*, but not very quickly, with fall of temperature, and that at temperatures below $-70°$ C. duration of charge ceases to have influence on its value. The electric excitation is thus analogous to diamagnetism and has no thermal bearing, its energy being self-contained in the molecule; the signs of the susceptibilities in the two cases are different, because the one is of static, the other of kinetic character. The sharpness of the Zeeman magneto-optic effect has already led (*Aether and Matter*, 1900, p. 351) in this direction, for it requires that the electric polarization in the molecule shall be of isotropic type, so that there may be no axis of maximum susceptibility. [See *supra*, p. 142. Also as regards rapid modern experimental progress, mainly by Debye, on the thermodynamic relations of dielectric polarization, cf. Appendix to this volume.]

[1] This restriction is not necessary for the final result; if $\kappa$ is not small, $W$ and $E$ have both to be multiplied by the same factor.

Law of Curie applied; Now the experiments of Curie on the relation of $\kappa$ to $\theta$ in weakly *paramagnetic* materials make $\kappa$ vary inversely as $\theta$; and this result has more recently been verified down to very low temperatures by Dewar and Fleming. This gives

$$E = \tfrac{1}{2}\kappa \, (H_2{}^2 - H_1{}^2).$$

Thus the movement of the magnetizable material at uniform tempera‑ makes the energy of thermal origin. ture is accompanied by a supply to it of heat, equal to the mechanical work done by it owing to the attraction of the field; and this heat is just what is wanted to be transformed into the additional energy of intrinsic magnetization (ii) of § 2. It is to be observed that in the actual experiments $\kappa$ was small, and the other part (iii) of this energy therefore negligible: so that no conclusion as to the extent to which its source is thermal can be derived from Curie's law.

6. The uncertainties of § 4 do not of course affect the estimation of the loss of motive power arising from cyclic magnetic hysteresis, for we have here to do with the *mutual* energy of the applied field and the magnet, not the intrinsic local energy of the latter by itself. If the applied field is $(\alpha, \beta, \gamma)$, the total energy employed in polarizing the magnetic molecules in volume $\delta\tau$ is

$$(A\alpha + B\beta + C\gamma)\, \delta\tau.$$

So long as the polarization is slowly effected against the resilience Hysteresis loss into heat. of reversible internal elastic forces this is stored as potential energy; but any want of reversibility involves degradation of some of it into heat, while if the field were instantaneously annihilated the molecules would swing back and vibrate, so that ultimately all would go into heat.

Let us pass the magnetic body through a cycle by moving it around a path in a permanent magnetic field $(\alpha, \beta, \gamma)$. An infinitesimal dis‑ placement of the volume $\delta\tau$ from a place where the field is $(\alpha, \beta, \gamma)$ to one where it is $(\alpha + \delta\alpha, \beta + \delta\beta, \gamma + \delta\gamma)$ does mechanical work on its attachments, arising from the magnetic attraction of the field, of amount

$$(A\delta\alpha + B\delta\beta + C\delta\gamma)\, \delta\tau.$$

The integral of this throughout the whole connected system gives the virtual work for that displacement, from which the forces assisting it are derived as usual. Confining attention to the element $\delta\tau$, the work supplied by it from the field, to outside systems which it drives, in traversing any path is thus

$$\delta\tau \int (A\,d\alpha + B\,d\beta + C\,d\gamma),$$

the integral being taken along the path. If $(A, B, C)$ is a function of $(\alpha, \beta, \gamma)$, that is if the magnetism is in part thoroughly permanent, and in part induced without hysteresis, so that the operation is reversible, this work must vanish for a complete cycle [relative to the extraneous mechanical system on which it works]; otherwise energy would inevitably be created either in the direct path or else in the reversed one of the complete system of which $\delta\tau$ is a part. Thus the negation of perpetual motion in that case demands that

$$A d\alpha + B d\beta + C d\gamma = d\phi,$$

General theory (Kelvin) of reversible magnetization.

where $\phi$ is a function of $(\alpha, \beta, \gamma)$, involving only even powers, and practically quadratic for small fields. Its coefficients are then the six magnetic constants for general aeolotropic material, no rotational quality in the magnetization being thus allowable by the doctrine of energy. But if there is hysteresis in the magnetic body, moved in the field, so that the cycle is not reversible,

$$\delta\tau \int (A d\alpha + B d\beta + C d\gamma),$$

Formula for hysteresis loss in cycle:

or in vector product form $\delta\tau \int \mathfrak{J} d\mathfrak{H}$, represents mechanical energy lost in the cycle, degraded into heat in the moving magnetic body. It will not be different, if it is the field that is moving.

In addition to this energy concerned with attraction, the external field imparts mechanical energy in polarizing or orientating the individual molecules against internal forces of the medium, of aggregate amount

$$\delta\tau \int (\alpha dA + \beta dB + \gamma dC).$$

In any case, whatever the hysteresis, the sum of this second part and the first reversed is integrable from one position to another independently of the path, giving

$$\delta\tau \,|\, A\alpha + B\beta + C\gamma \,|,$$

namely, the change in the total energy [supplied to] the element, thus vanishing for a cycle which restores things to their original state as it ought to do. The latter part is purely internal, and of merely thermal value as in § 5. The former part represents the averaged waste of direct mechanical energy in moving the iron armature through the cycle, and accounts for the heat thus evolved. It is the expression of Warburg and of Ewing for magneto-hysteretic waste of mechanical energy in driving electric engines; for a portion of a cycle it represents work partly degraded and partly stored magnetically.

not in a part of the cycle.

7. Reverting to § 5, we may profitably illustrate by working out into detail a suggestion of Lord Rayleigh (*loc. cit.*). Consider a ring coil of $n$ turns with a flexible open core of soft iron of length $l$ and

Open ring coil:

cross-section $S$, whose flat ends are bent round until they face each other at a distance small compared with the diameter of section. We can apply Hopkinson's theory of the open magnetic circuit to trace

the transformation of the energy of attraction between these poles of the core, into electrokinetic energy of the coil, as the poles close up together. As frictional waste is not essential to this question, we can consider the coil to be a perfect conductor which will store all the energy without loss. We need not postulate that the iron is of constant permeability or devoid of hysteresis. When the distance between the pole faces is $x$, very small, let the current in the coil be $\iota$. The total energy is

$$\tfrac{1}{2}n\iota N + \text{energy of magnetization};$$

and part of the latter may remain sub-permanent when $\iota$ vanishes. The principle of the magnetic circuit [cyclic integral of magnetic force] gives, as $BS = N$, the formula

$$\frac{N}{\mu S}l + \frac{N}{S}x = 4\pi n\iota,$$

assuming as usual that the lines of magnetic force are conveyed straight across the air gap between the pole faces. Thus the electromagnetic energy $T$, equal to $\tfrac{1}{2}n\iota N$, is

$$2\pi \frac{n^2\iota^2 S}{x + l/\mu};$$

and when the poles separate by $\delta x$, its increment is approximately

$$- 2\pi \frac{n^2\iota^2 S}{(l/\mu)^2}\delta x + 4\pi \frac{n^2\iota S}{l/\mu}\delta\iota. \tag{i}$$

In this displacement the work gained from attraction between the poles magnetized to intensity $I$, equal to $\kappa N/\mu$, is $- IS2\pi I\delta x$, which is

$$\left(1 - \frac{I}{\mu}\right)^2 2\pi \frac{n^2\iota^2 S}{(l/\mu)^2}\delta x. \tag{ii}$$

Comparing (ii) with (i) it appears that in cases for which $\mu$ is great,

to which alone the principle of the magnetic circuit can be applied, the work of mechanical attraction by which the pole faces can *transfer* potential energy to a spring placed between them, by compressing it, is concomitant with equal increase of the electrokinetic energy if the current do not change. As there is no source of energy, the current must therefore vary, and so that the total change of electrokinetic energy given by (i) and (ii) vanishes; that is, it must alter by $\delta\iota$,

given by $\delta\iota = \tfrac{1}{2}\dfrac{\mu\iota}{l}\left\{1 + \left(1 - \dfrac{I}{\mu}\right)^2\right\}\delta x$, which is practically equivalent to $\dfrac{\delta\iota}{\iota} = \dfrac{\mu}{l}\delta x.$

This result may be immediately verified by the Lagrangian process.
As there is infinitely small resistance, the electric pressure $\dfrac{d}{dt}\dfrac{\partial T}{\partial \iota}$ in

the coil must be infinitely small; hence $\iota \propto x + l/\mu$, so that $\dfrac{\delta \iota}{\delta x} = \dfrac{\mu}{l}$
when $x$ is negligibly small.

The energy of magnetization of the core, which is $\frac{1}{2}\mathfrak{J}\mathfrak{B}lS$, where
$\mathfrak{J} = \kappa\mathfrak{B}/\mu$, and therefore is $\frac{1}{2}\dfrac{\kappa}{\mu}\left(\dfrac{4\pi n\iota}{x + l/\mu}\right)^2 lS$, is not included in this con-

servation. Its increase for a change $\delta x$ is $-\dfrac{\mu - \mathrm{I}}{4\pi}\left(\dfrac{4\pi n\iota}{l/\mu}\right)^2 S\delta x$, which
is large compared with the quantities above. The fraction $\mu^{-1}$ of it,
which is comparable with the other variations, is compensated
*thermally*, by absorption of the heat of the system, and has, therefore, only the limited availability of thermal energy. The remainder
belongs intrinsically to the magnetization, constituting mutual energy
of contiguous molecules; how much of it, as above expressed, is of
thermal origin remains undetermined in the absence of calorimetric
experiment.

8. The main points that it has been sought to bring out are as
follows:

(i) In an electrodynamic field there exists the usual specification
of electrokinetic energy, but also *in addition* the energy of magnetization of magnetic material.

(ii) This energy of magnetization appears as made up of a part
given by the ordinary formula, which (when paramagnetic) is derived
from thermal sources, and so in the absence of hysteresis has the
limited mechanical availability of thermal energy; together with a
local part which is to some extent thus available, but is also in part
permanent intrinsic energy of the molecules, regarded temporarily
as magnetic energy.

(iii) The law of Curie, that the susceptibility of weak paramagnetic
substances is inversely proportional to the absolute temperature, is
involved in these statements.

(iv) The extent of the direct (non-thermal) availability of *retained*
magnetism can be inferred only by empirical procedure, for example,
in general features by inspection of the hysteresis diagram, as pointed
out by Lord Rayleigh.

# 59

# ON THE MATHEMATICAL EXPRESSION OF
# THE PRINCIPLE OF HUYGENS*.

[*Proceedings of the London Mathematical Society*, Ser. 2, Vol. 1 (1903) pp. 1–13.]

*Green's theory of potentials:* 1. In Green's *Essay on the Application of Mathematical Analysis to the Theories of Electricity and Magnetism* (Nottingham, 1828), which laid the foundations of modern analysis as applied to mathematical physics, potentials were first considered from a functional point of view, and the conditions requisite to specify a potential function were laid down; as a particular case of "Green's Theorem," the value *their* of such a function throughout a region of free space was expressed *determinacy for inverse* in terms solely of its own values and those of its gradient normal to *square law.* the boundary of the region. The mathematical development of this *Extended to* general theory was extended by Helmholtz[1], with important and *retarded potentials:* novel applications, to the velocity potential in the theory of the propagation of sound waves in air, which satisfies a differential equa-*Helmholtz,* tion for which an analogous procedure applies. Subsequently the theorem which corresponds to the one quoted above for the gravita-*Kirchhoff.* tional or electric potential was developed in detail by Kirchhoff†, who thus derived a general expression for the vibratory disturbance pro-pagated into a medium in terms of the distribution of the disturbance and its spatial gradient, supposed given throughout the time, over any arbitrary geometrical boundary across which it must have travelled into the region under consideration. Two centuries ago *Huygens'* Huygens explained the general features of the propagation of dis-*general principle,* turbances of elastic type, by considering each differential element of the front of an advancing wave-train to constitute a source of disturbance from which it travels onward into the medium in advance; and in the hands of Young and Fresnel this principle of Huygens furnished not merely an explanation of the geometrical form of the *accounts for* advancing waves, but served, by aid of the principle of interference *optical rays* of undulations, as a basis for the optical principle of rays, and also *(Young),* for the quantitative explanation of the phenomena of diffraction in *and their* which that concept partially fails. In Fresnel's hands the circum-*diffraction:* stances of actual optical diffraction were amenable to mathematical calculation in minute detail, without any knowledge of the law of

---

* Continued and amended in Part II (1919) immediately following.

[1] "Theorie der Luftschwingungen in Röhren mit offenen Enden," Crelle, Vol. LVII, 1859.

† Earlier in unexceptionable manner by Stokes: see Part II, *infra*, p. 263.

propagation of these secondary disturbances from the various elements of the wave-front, except such as could naturally be assumed to represent them in the immediate neighbourhood of the direction of propagation. Yet the formula assumed for this purpose by Fresnel involved some oversights, whose general origin was, however, readily traceable when the dynamical theory of vibrations became more explicitly developed. The occurrence of such discrepancies gave rise to the endeavour to place, if possible, the circumstances of the specification and propagation of secondary disturbances on a precise foundation. This was first essayed, and may indeed be held to have been so fully developed that subsequent discussion has added little except in the way of adaptation to new circumstances, in Sir G. Stokes' memoir on "The Dynamical Theory of Diffraction."[1] The formula obtained in Kirchhoff's memoir, above referred to[2], was, however, advanced by him mainly as an improved mathematical expression of Huygens' principle, and is often taken to be its exact mathematical specification. That such merely analytical specification *à priori*, independently of the mechanics of the propagation as depending on the constitution of the medium, including the relation of stress to strain, does not constitute a definite problem, has been remarked by Lord Rayleigh[3], and illustrated by various allowable laws of propagation from the various elements of the wave-front*.

> Analysis of Fresnel,
>
> leads to a dynamical problem.
>
> Stokes.
>
> Is Huygens' principle definite?

More recently the same subject has been treated by Professor A. E. H. Love[4], and further developments for uniaxal crystalline media were given by Professor A. W. Conway recently in these *Proceedings* (Vol. xxxv).

The usual modes of deduction of the Helmholtz-Kirchhoff formula are not free from analytical complexity[5], and can hardly be said to throw much direct light on the character of the principle which they demonstrate, or on the degree of its determinateness, which is here

[1] *Camb. Phil. Trans.* 1849; or *Mathematical and Physical Papers*, Vol. II [: see especially p. 262 *infra*. Also cf. the final chapters of G. Lamé, *Leçons sur l'Élasticité* (1866)].

[2] "Zur Theorie der Lichtstrahlen," Wied. *Ann.* Vol. xviii, 1883.

[3] *Encyc. Brit.* "Wave Theory," 1888; *Collected Papers*, Vol. iii, No. 148.

* It is however explained in Part ii, *infra*, that what is required is a distribution of sources that will produce the actual transmission of the oncoming wave, outward and without turning any of it back, for no material obstacle is to be supposed on the surface. This condition restricts the localization of the secondary sources to that represented in the Kirchhoff formula. But it is only in the discussion of problems of diffraction, involving only part of the enclosing surface, that this location can matter.

[4] "The Integration of the Equations of Propagation of Electric Waves," *Phil. Trans.* Vol. cxcvii, A, 1901.

[5] Compare, *e.g.* the discussion in Drude's *Optik*, 1900.

*The Mathematical Expression of* [Jan. 8

Its nature, under consideration; there seems, therefore, to be room for the folowing remarks, intended as a contribution towards physical precision, by putting forward a more intuitive view of the content of the
physically formula. It will not here be necessary to attend to all the saving
envisaged. restrictions that are customary in the exact analysis of formal pure
mathematics[1]; such limitations are, in the main, sufficiently obvious, and are satisfied by the nature of the case, in continuous physical analysis, while their detailed exposition would impede and interrupt the main argument.

Case of    2. It will be simplest to begin with the analogous propositions for
instantaneous the more familiar case of the ordinary potential, which is, in fact,
propagation: what the vibrational potential becomes when the speed of propagation is infinite.

Consider, then, a potential specified throughout all space as follows.
leads to It is a function $V$, single-valued and continuous as to itself and its
ordinary gradient, and satisfying Laplace's equation, in all the space outside
potential and
inverse square a boundary $S$, and, as a consequence, diminishing towards infinity,
law. according to the law $r^{-1}$ or a higher inverse power of $r$; it is zero everywhere inside this boundary. What is the distribution of attracting mass to which this potential belongs? This distribution is, as usual, determined, in Green's manner, by the singularities and discontinuities of the potential function. It consists, in fact, of a surface density $\sigma$ over the boundary $S$, and a double sheet of strength $s$ over the same boundary, expressed in terms of the potential and its gradient at the surface by the formulae

Effect is ex-
pressible in
terms of any
$$\sigma = -\frac{1}{4\pi}\frac{dV}{dn}, \quad s = \frac{1}{4\pi}V,$$
surface
enclosing in which $\delta n$ is an element of the normal drawn towards the outside
the sources, of the surface. For it follows, by the usual procedure, that, if $V'$ is the potential of this distribution, then $V - V'$ is a potential function which has no singularities or discontinuities throughout all space, and is, therefore, identically null. Expressed analytically, at any
by an integral point in space the potential is thus
expressing
activity of
sources spread
$$-\frac{1}{4\pi}\int \frac{1}{r}\frac{dV}{dn}dS + \frac{1}{4\pi}\int V\frac{d}{dn}\frac{1}{r}dS.$$
over it.

In other words, this formula gives the value of a potential function $V$ in the free space outside the surface in terms of the values which it and its gradient assume on the surface; it constitutes, in fact, the analytical continuation of the function outward from the surface,

---

[1] A more detailed discussion, in many respects parallel, will be found in Poincaré's *Traité math. de la Lumière*, 1889, Chap. III and IV: cf. *infra*, p. 246, footnote.

while inside the surface the value of this expression is everywhere null. In the case of a closed surface, as well as that of an open sheet, either side may be called the outside for the present purpose. This continuation of the function is necessarily unique and determinate[1], but the form of the integral which expresses it is far from being so. We may, in fact, generalize the formula immediately in Green's manner. Consider a function $V$ which is the potential, throughout space, of any assigned distribution of mass. Draw any surface dividing space into two regions, $A$ and $B$, each of which contains part of the mass, these parts being represented by $M_A$ and $M_B$. What distribution of masses, and of surface densities and normal doublets on the surface $S$, is required to produce a potential equal to $V$ in region $A$, and equal to zero in region $B$? Clearly, $M_A$, together with

Analytical continuation of the potential function.

Determinacy of result:

$$\sigma = -\frac{1}{4\pi}\frac{dV}{dn}, \quad s = \frac{1}{4\pi} V.$$

What distribution is required to make the potential zero in region $A$ and $V$ in region $B$? Clearly, $M_B$, together with the same distribution on the surface but with sign changed if the direction of $\delta n$ remain the same. Thus, a distribution of surface densities and normal doublets is found which exactly cancels the effect of $M_A$ on the other side of the dividing surface; moreover, an infinite number of such distributions can be found, for in determining it $M_B$ is entirely arbitrary. On changing the signs of $s$ and $\sigma$, we infer that a distribution of surface densities and normal doublets can be assigned, in an infinite variety of ways, over any surface surrounding an attracting system, so as to form the exact equivalent of that system as regards the effect transmitted to outside space; and that one of the modes of distribution thus obtained produces null effect in inside space. The problem of electrostatic distribution is, of course, unique in a different manner from this latter; a free electric distribution on a conductor can consist (physically) of simple sources only, and cannot involve doublets. Its specification involves the solution of a problem in which the form of the surface is involved, and cannot be derived from a general continuation formula.

but the distribution of sources not definite,

unless further conditioned:

*e.g.* the electrostatic problem.

Needless to say, an argument similar to the above applies to a function specified by any characteristic partial differential equation of the second order.

3. The same procedure may now be extended to a scalar potential, which is propagated in time, *i.e.* which satisfies a characteristic equation involving the time as a variable.

Extension.

Value throughout any region however small determines a potential.

[1] This is, of course, the Gaussian theorem that, if two systems have the same potential throughout any region, the identity extends to all places that can be reached from that region without crossing attracting matter.

We consider first the simplest case, that of a velocity potential $\phi$,

Application to propagated potential of sound waves in air. say of sound waves in air, which is propagated with velocity $c$ in accordance with the characteristic equation

$$c^{-2}\frac{d^2\phi}{dt^2} = \frac{\partial^2\phi}{\partial x^2} + \frac{\partial^2\phi}{\partial y^2} + \frac{\partial^2\phi}{\partial z^2}.$$

It is a necessary preliminary to ascertain what distribution of sources on a surface $S$ will create given discontinuity in the values of $\phi$, and of its normal gradient, in crossing the surface, it being clear

The discontinuity set up by interfacial sources. that such discontinuities in $\phi$ and in $d\phi/dn$ constitute the most general type involving only first differential coefficients of $\phi$ that can exist.

The velocity potential propagated from a simple point source[1], variable with the time, of strength $f(t)$, is, for this characteristic

The potential propagated from a simple source; equation, $\frac{1}{r}f\left(t - \frac{r}{c}\right)$. The velocity potential propagated from a doublet consisting of simple sources $+f(t)$ and $-f(t)$ separated by an interval $\delta n$ is consequently $\delta n \frac{d}{dn}\left\{\frac{1}{r}f\left(t - \frac{r}{c}\right)\right\}$; thus, for a doublet

from a doublet source. of strength $F(t)$, the equivalent of $f(t)\,\delta n$, it is $\frac{d}{dn}\left\{\frac{1}{r}F\left(t - \frac{r}{c}\right)\right\}$, in which the function $F$ comes under the differentiation. Within a region of such small extent that the functions $f(t)$ and $F(t)$ do not sensibly change in the time required for the disturbance to pass

Simplified local forms for regions small compared with wave-lengths: across it, these velocity potentials are of types

$$\frac{f(t)}{r} \quad \text{and} \quad F(t)\frac{d}{dn}\frac{1}{r},$$

so far as they relate to sources inside the region of which $f(t)$ and $F(t)$ are the strengths at this interval of time; for this modification only neglects lower inverse powers of $r$ than those that are retained. Thus, for such a region enclosing an element $\delta S$ of the surface $S$, and at times for which the functions $f(t)$ and $F(t)$ do not there change abruptly, the velocity potentials, subject to exceptions to be presently encountered in the case of double sheets, take, throughout any time of the order above specified, the form of simple gravitational potentials, the circumstance of propagation not sensibly interfering[2].

[1] In this section (but not in § 4) a source is the analogue of a mass, the flux from unit source being $4\pi$.

Helmholtz-Rayleigh treatment of resonance. [2] The very elegant and fruitful application of this principle, conjoined with the Helmholtz-Kirchhoff continuation formula, by Helmholtz (*loc. cit.*), to trace the vibratory connection established between large regions, through connecting channels of dimensions small compared with the wave-length, including the problem of reflexion from an aperture, may be recalled; cf. the simplified analysis in the chapter on Resonators in Lord Rayleigh's *Theory of Sound.*

We are therefore invited to follow the procedure of Coulomb and Laplace for the ordinary potential, and investigate the discontinuities arising from a surface distribution of simple sources and one of doublets orientated normally to the surface. The discontinuities, on crossing the surface at any point, clearly arise solely from the distribution over a surface element $\delta S$ surrounding that point: for the disturbances propagated from the more distant sources are virtually the same at points on the two sides of the surface, whose distance apart is infinitesimal compared with the linear dimensions of $\delta S$, so that, as regards their effect, no discontinuity can arise.

applicable to analysis of interfacial discontinuities:

Taking first, then, the case of a simple surface density $\sigma(t)$ spread over $\delta S$, which we may take to be uniform all over it at each instant, its effect is to transmit towards both sides a train of plane waves with fronts parallel to $\delta S$, which remain plane until the distance $n$ to which they have travelled becomes comparable with the linear dimensions of $\delta S$. For them the value of $d\phi/dn$ at distance $n$ and time $t$ is $2\pi\sigma\left(t - \dfrac{n}{c}\right)$ but with different sign on the two sides; such a surface distribution $\sigma(t)$ of simple sources thus accounts for a discontinuity in $d\phi/dn$ of amount $4\pi\sigma(t)$, but introduces no discontinuity in $\phi$ itself.

type due to a sheet of simple sources:

This result now assists us to analyse the circumstances of a sheet of normal doublets of strength $s(t)$ per unit area; for we can replace it by two simple parallel sheets of densities $+\sigma_1(t)$ and $-\sigma_1(t)$, at an infinitesimal distance $\delta n$ apart, such that $\sigma_1(t)\,\delta n = s(t)$. At a point at distance $n$ from the sheet of density $+\sigma_1(t)$ the values of $d\phi/dn$ at time $t$ arising from these two sheets are, as above,

doublet sources as limit of two adjacent such sheets.

$$\pm\, 2\pi\sigma_1\left(t - \frac{n}{c}\right) \quad\text{and}\quad \pm\, 2\pi\sigma_1\left(t - \frac{n + \delta n}{c}\right),$$

in which signs are to be determined by the sides of the respective component sheets on which this point lies. If the point is not between the sheets, the signs are opposite, and the sum for both is

$$-\,2\pi\,\frac{\delta n}{c}\,\frac{d}{dt}\,\sigma_1\left(t - \frac{n}{c}\right),$$

which is the value of $d\phi/dn$ due to the element $s\delta S$; but it has *the same sign* on both sides of the double sheet; so that in crossing the double sheet there is no discontinuity in the value of $d\phi/dn$, though the element $s\,dS$ of the double sheet contributes $\dfrac{4\pi}{c}\dfrac{ds}{dt}$ to that quantity on each side. But, between the sheets, the value of $d\phi/dn$ arising from them is of a higher order of magnitude, being a sum instead of a difference, and is $4\pi\sigma_1\left(t - \dfrac{n}{c}\right)$, or simply $4\pi\sigma_1\,(t)$ when $\sigma_1\,(t)$ is not discontinuous; and this value integrated across the interval $\delta n$ gives a discontinuity in $\phi$ itself, on crossing the double sheet, of amount $4\pi\delta n\sigma_1\,(t)$, that is, $4\pi s\,(t)$.

*The general result.* Collecting these results, we see that a discontinuity in $\phi$ over a surface $S$ of amount $\chi\,(x, y, z, t)$ and a discontinuity in $d\phi/dn$ equal to $\psi\,(x, y, z, t)$ over the same surface are accounted for respectively by a double sheet on the surface of strength $s$ equal to $\chi/4\pi$, and a single sheet of density $\sigma$ equal to $\psi/4\pi$.[1]

*Surface distributions of sound sources equivalent for one side.* We are thus in a position to proceed exactly as in § 2. Consider any system of sources, and let $\phi$, a function of $(x, y, z, t)$, be their velocity potential in infinite free space. Assign any surface $S$, dividing space into two regions $A$ and $B$, and let $m_A$ and $m_B$ stand for the sources as divided between the two regions. What distribution of sources would give rise to a velocity potential equal to $\phi$ in region $A$ and equal to zero in region $B$? Clearly the sources $m_A$, together with a distribution $(\sigma, s)$ over $S$ given by $4\pi\sigma = -\dfrac{d\phi}{dn}$, $4\pi s = \phi$; for, if $\phi'$ is the potential arising from this distribution, and $\Phi$ is a function equal to $\phi$ in region $A$ and to zero in region $B$, then $\Phi - \phi'$ will be a potential having no singularities or discontinuities throughout infinite space, and must therefore be null by simple physical intuition, or analytically by the usual type of theorem of determinacy based on

---

[1] These discontinuities are, I find, treated in elegant form by Poincaré (*loc. cit. ante*) for the more restricted case of simple harmonic vibrations of type $e^{\iota pt}$.

*Analytical treatment for a harmonic period:* The characteristic equation then assumes the special form

$$\nabla^2\phi + p^2 c^{-2}\phi = 0,$$

and the potential of a source $m$ is $m\,\dfrac{e^{\iota\kappa r}}{r}$, where $\kappa = p/c$. And, if there is a volume distribution of sources constituting a density $\rho$, the equation is modified, for purposes of continuous analysis, to

$$\nabla^2\phi + p^2 c^{-2}\phi = -4\pi\rho.$$

The application of Green's theorem is direct and simple—in fact in Helmholtz's original manner. As regards discontinuities in crossing surface distributions of sources, the key to the discussion is the remark that the potential

*its device for the local field:* of a source may be written in the form $\dfrac{m}{r} + m\,\dfrac{e^{\iota\kappa r} - 1}{r}$, of which the latter part can never become infinite; so that the discontinuities are just the same as for the gravitational potential, corresponding to the statement in the text.

*now extended to the general problem.* (*Feb.* 14. So in the more general circumstances here treated $r^{-1}f\left(t - \dfrac{r}{c}\right)$ can, under the restrictions specified, be expanded for small values of $r$ by Taylor's theorem in the form $\dfrac{f(t)}{r} - \dfrac{1}{c}f'\left(t - \theta\dfrac{r}{c}\right)$, $(\theta < 1)$, of which only the first term contributes to interfacial discontinuities.)

the energy of disturbance being of necessity essentially positive. As the total effect within the region $B$ is null, we can thus say that $\phi_A$, the part of it arising from sources $m_A$ outside the region, is given at time $t$ by

$$- 4\pi\phi_A = - \int \frac{dS}{r} \left[ \frac{d\phi}{dn} \right] + \int dS \frac{d}{dn} \frac{[\phi]}{r},$$

in which $[\phi]$ and $[d\phi/dn]$ are the values of these quantities for the element $\delta S$ at time $t - r/c$. This formula expresses the vibrational potential due to sources $m_A$ within a surface $S$, throughout the region $B$ outside that surface, as determined by the values which it and its gradient assume at that surface. It is, so to speak, an *analytical continuation* beyond that surface of a function satisfying the aforesaid characteristic differential equation. Such a continuation must be unique, and it is determined by the values assumed by $\phi$ alone on the surface: as therefore $d\phi/dn$ is determined by a know- **Not a** ledge of $\phi$ over the surface, the data for the formula here expressing **practical** the continuation are redundant[1]; if arbitrarily assigned, they will usually be self-contradictory, and the formula thus nugatory[2]. Moreover, the formula determines $\phi_A$ in terms of the surface dis- tribution of $\phi$, equal to $\phi_A + \phi_B$, where $\phi_B$ is due to an entirely arbitrary distribution of sources within the region $B$ to which the formula relates. Thus the quantities integrated in it are very widely **also the dis-** arbitrary, and the element of the integral corresponding to $\delta S$ in **tribution of** no sense represents any influence actually propagated from that part **indeter-** of the surface. The formula is purely analytical, and in no degree **minate.**

**Not a practical solution:**
**also the distribution of sources is indeterminate.**

[1] With regard to the remarks in the text that, for the determination of a field of physical activity in terms of the succession of disturbances over its boundary, the double set of *data* over the boundary are redundant, an important reservation must, however, be made when the field is of limited extent. In that case the field possesses a system of free types of vibration in which either single set of boundary disturbances is null; and when the two sets are of such nature that the energy lost at a surface depends on their product, *i.e.* when they may be characterized generally as *data* of force and of velocity, these free vibrations, when once excited, will go on for ever in the absence of vis- cosity: each type constitutes a spectrum of definite related periods, usually of vast number, when harmonically analysed. Each type thus forms an un- **The modes** determined addition to the disturbance that is determined by the succession **of free** of values in time of that set of boundary *data* which it does not alter, an **vibration in** addition which, however, is made definite by a knowledge of the state of the **space are** whole region at the time from which the analysis begins; if the beginning is **excited in** remote, and slight viscosity of amount otherwise unimportant is assumed, **addition, to** the influence of the initial state of the system dies away and the indeterminacy **an undeter-** is practically non-existent. The nature of the transition to the gravitational **mined degree;** potential, where no such reservations occur, on making the velocity of pro- **but disappear** pagation infinitely great, is readily recognizable. **in time by**

**The modes of free vibration in an enclosed space are excited in addition, to an undetermined degree; but disappear in time by damping.**

[2] In the demonstration this is safeguarded by both being taken as due to **damping.** $m_A$ and $m_B$.

a mathematical formulation of the physical principle of Huygens which relates to propagation of actual disturbance. In fact, if $m_B$ vanishes, the formula represents a distribution of surface disturbances which does not radiate at all into the region $A$; whereas, if the doctrine of rays and of Huygens' (or rather Fresnel's) zones could really be derived from the formula by a mathematical process devoid of limitation or exception, it would appear that [local] radiation ought to reach every internal point $P$ along the directions of the normals drawn from $P$ to the surface $S*$.

4. Similar considerations are applicable to functions determined by an aeolotropic characteristic equation: for example, they apply

*Application to an anisotropic medium:* to the equation of propagation of heat in a crystalline medium, expressible in the form

$$\frac{\partial u}{\partial x} + \frac{\partial v}{\partial y} + \frac{\partial w}{\partial z} = \rho \frac{d\theta}{dt},$$

where $(u, v, w)$ is the current of heat, determined, in terms of the gradient of the temperature $\theta$, by the constitution of the medium in the form

*e.g. thermal conduction,*

$$u = \kappa_1 \frac{\partial \theta}{\partial x} + \kappa_{12} \frac{\partial \theta}{\partial y} + \kappa_{13} \frac{\partial \theta}{\partial z}, \quad v = \kappa_{21} \frac{\partial \theta}{\partial x} + \kappa_2 \frac{\partial \theta}{\partial y} + \kappa_{23} \frac{\partial \theta}{\partial z},$$

$$w = \kappa_{31} \frac{\partial \theta}{\partial x} + \kappa_{32} \frac{\partial \theta}{\partial y} + \kappa_3 \frac{\partial \theta}{\partial z},$$

in which $\kappa_{rs} = \kappa_{sr}$, unless the medium has rotational structure or quality.

If the surface element $\delta S$ has the posture $(\lambda, \mu, \nu)$, and it carries a surface density $\sigma$ of simple sources of heat, they will produce equal and opposite gradients of temperature on its two sides; but opposed tangential gradients would involve discontinuous temperature; hence the gradient on either side due to them must be wholly along the normal, say $(\lambda, \mu, \nu) \, d\theta/dn$. Also

$$\tfrac{1}{2}\sigma = \lambda u + \mu v + \nu w = \kappa_{\lambda\mu\nu} \, d\theta/dn,$$

where

$$\kappa_{\lambda\mu\nu} = \kappa_1 \lambda^2 + \kappa_2 \mu^2 + \kappa_3 \nu^2 + (\kappa_{23} + \kappa_{32})\, \mu\nu + (\kappa_{31} + \kappa_{13})\, \nu\lambda + (\kappa_{12} + \kappa_{21})\, \lambda\mu.$$

Thus a discontinuity of amount $\psi(t)$ in the value of $d\theta/dn$ involves a distribution of sources on the surface of density $\kappa_{\lambda\mu\nu} \psi(t)$. By

---

* These solutions are all equivalent when the sources are spread over a complete closed surface: but in Fresnelian diffraction theory only the local part of the disturbance passing the surface from within is used: if this is to be represented by local sources, their distribution must be definite. The sources will represent the transmitted disturbance completely only if they send no disturbance backwards: thus in this physical sense the Kirchhoff formula, though otherwise arrived at, is the unique solution. See Part II, which follows.

superposition of the effects of two sheets $+\sigma$ and $-\sigma$ at an infinitesimal distance $\tau$ apart, it appears that a double sheet of intensity $s\ (=\sigma\tau)$ does not introduce any discontinuity in $d\theta/dn$ in
crossing it, but only one of amount $\kappa_{\lambda\mu\nu}^{-1}s$ in $\theta$ itself. Thus, when once [*the solution.*]
the law of propagation of temperature, around a point source which
changes with the time, is found, as can be done in finite terms by
the usual Fourier analysis[1], we can express the distribution of temperature throughout any region in terms of the distribution in space
and time of the temperature and its normal gradient along the
boundary of the region.

5. The case of electrodynamic or optical propagation in any [*Case of electric and*]
medium, isotropic or otherwise, can be immediately dealt with, the [*optical pro-*]
nature and source of singularities in the electric vectors being well [*pagation.*]
known[2]. A discontinuity at any surface is completely expressed by
the amounts of the abrupt changes in the tangential electric and
magnetic forces; for the normal components of the forces at the
two sides of the surface are determined from the tangential ones
directly by the fundamental circuital relations of Ampère and
Faraday, and therefore definite discontinuity in them also is in- [*The types of surface*]
volved, being demanded solely by the intrinsic constitution of the [*sources:*]
aether, whatever be the sources that are supplied on the interface.
Now, discontinuity in the tangential magnetic force implies simply
a surface stream of electricity in $\delta S$, at right angles to the direction
of the change in this force, and equal in intensity to it divided by $4\pi$.
The question arises whether all these elements of flux fit together
into a circuital current sheet on the interface, or whether they lead
to accumulation of electricity on it through not satisfying the condition of continuity of flow. If the current sheet were circuital, so [*current sheet:*]
as to produce no accumulation of charge, the tangential magnetic
force would be restricted by being derived from a surface potential.
Such a restriction is not called for; indeed, in the present case of
purely vibratory motions, the slight accumulations of electricity
provide potential energy for the system. If the physical constant

[1] (For a homogeneous medium, see Maxwell, *Electricity*, § 301. The state- [*Ellipsoidal*]
ments above concerning the discontinuities in $\theta$ arising from distributions of [*level surfaces*]
sources may be verified analytically from this expression, or more briefly [*of a point*]
by linear transformation of the space so as to reduce the characteristic equation [*source.*]
to the Laplacian form, in the manner of Stokes, *Camb. and Dub. Math. Journal*,
1851; *Collected Papers*, Vol. III, p. 203.)

[2] Analytically, this case supplies the analysis for the vectorial characteristic
equation

$$c^2\nabla^2\,(P,\,Q,\,R)=\frac{d^2}{dt^2}\,(P,\,Q,\,R) \text{ subject to } \frac{\partial P}{\partial x}+\frac{\partial Q}{\partial y}+\frac{\partial R}{\partial z}=0;$$

and for the aeolotropic generalization thereof.

of the transmitting medium, now aether *plus* matter, were complex, so that conduction currents would exist, there could be more permanent electric accumulation; but in all cases the amount of charge thereby supplied is required to produce the accompanying discontinuity in the normal electric force, which is thus provided for.

Any discontinuity in the tangential electric force, provided it has a surface potential, is accounted for by a distribution of varying electric doublets normal to the surface. These may be considered to constitute a double current sheet, not circuital, but with accompanying double sheet of electric density: of this the former part accounts for the accompanying discontinuity of the normal magnetic force. But when such a surface potential does not exist we must take the sources on $\delta S$ to constitute an element of a "magnetic current sheet," which may be considered to arise from a varying sheet of tangential magnetization, after the analogy of the electric current sheet of the previous case.

*double current sheet,*

*which is a sheet of magnetic current.*

The result then specifies a current sheet of varying intensity, equal at each instant to the discontinuity in the tangential magnetic force divided by $4\pi$, and at right angles to its direction; also a sheet of "magnetic current" of varying intensity, equal at each instant to the discontinuity in the tangential electric force divided by $4\pi$, and at right angles to its direction; these together form the cause of all the discontinuity that can exist at the interface[1].

*This system complete.*

The previous mode of argument is now applicable. Consider the field of activity throughout space of any electrodynamic system; divide space into two regions $A$ and $B$, and let the parts of the sources that are situated in them be denoted by $m_A$ and $m_B$. Consider another field constituted of this one in the region $A$, and zero field in the region $B$; and specify the surface distribution of electric and magnetic sources that is required in addition to $m_A$, in order to constitute it. This superficial system with sign changed must produce an effect in region $B$ the same as that of the $m_A$ sources situated outside that region; and we have thus learned how to effect the continuation of any electrodynamic field beyond a surface up to which it is known.

*The solution:*

As in the previous cases, the data of arbitrarily assigned tangential forces, both electric and magnetic, over the boundary, are redundant,

---

[1] It is possible, theoretically, to construct solvable cases of vibrating systems, in Green's manner, by drawing the lines of electric force, for standing waves due to any known system of sources in infinite space, and transforming one or more of their orthogonal trajectories, where such exist, into perfectly conducting surfaces. This procedure is always available for cylindrical or axial systems; but there does not appear to be any simple surface, other than the obvious one of a plane, which can play the same part as a sphere does in the theory of electrostatic images.

*Construction of solved problems.*

and so, if given independently, will probably be self-contradictory. A knowledge of either tangential force by itself renders the problem physically determinate, and so involves knowledge of the other also.

This analysis holds good for the most general type of aeolotropic medium, because no relations involving the constitution of the material medium have been employed. When the medium is aeolotropic the disturbance emanating from a source of either type will spread in two-sheeted wave-fronts of the Fresnel type*, and its distribution may be obtained in finite terms when the strength of the source varies in a general manner expressed by an arbitrary function. The case of a uniaxal medium has been dealt with in elegant manner by Professor Conway, *loc. cit.*[1]

*extends to anisotropic media.*

(*Feb.* 14. When the medium is free aether the expressions are as follows. If in crossing an element of surface $\delta S$ there is discontinuity $4\pi f(t)$ in the tangential magnetic force, an electric current sheet is required, of intensity $f(t)$, in the direction $\delta s$ at right angles to the direction of this discontinuity. This will produce a magnetic field of intensity $-\sin(r, \delta s)\dfrac{d}{dr}\dfrac{1}{r}f\left(t - \dfrac{r}{c}\right)\delta S$ in circles around $\delta s$; also an electric field in part derived from a potential function

*The factors of the elements of the surface sources determined.*

$$-c^2 \cos(r, \delta s)\frac{d}{dr}\frac{1}{r}f_1\left(t - \frac{r}{c}\right)\delta S,$$

* This requires limitation: though the idea is fundamental in the *aperçu* of the law of refraction of waves in Huygens' manner, by dividing up the incident disturbance into surface elements radiating independently. Each such element is regarded as sending out into the medium a disturbance limited to a spherical shell diverging from it as centre: and, after Huygens, the envelope of these shells is the travelling front. This explanation loses its force for waves travelling in two dimensions; the propagation from an elementary source is not in a compact shell, there being a trailing disturbance left behind it: but the final result still holds, as dynamics in the manner of Green can show. A long time ago (in the final chapters of Lamé's *Leçons de Élasticité*, 1866) it had been shown, and had since been forgotten, that the disturbance emitted from a vibrating centre in a crystalline medium does not advance in compact shells: and the subject has been pursued more recently on general lines of modern analysis by Volterra (1909). An impulse at a point would perhaps emit a thin shell of radiation if the medium were slit along the angle in space between the optic axes drawn from the point. It would seem that the disturbance usually travels out in a stratum bounded externally and internally by the two sheets of an advancing wave-surface. If this be so, it appears, curiously, that the Huygenian explanation is at fault for the very case of double crystalline refraction in which it was practically most successful. The failure was noted for uniaxal crystals by A. W. Conway, *loc. cit., Math. Soc. Proc.* 35 (1902) § 4.

*The two sheets of the wave-surface as the outer and inner boundaries of an expanding pulse.*

[1] For the adaptation to practical optical diffraction formulae, the memoir of Kirchhoff may be referred to, or that of Professor Love.

where $f_1(t)$ represents $\int f(t)\, dt$, and in part of intensity

$$\frac{\mathrm{I}}{r} f'\left(t - \frac{r}{c}\right) \delta S$$

parallel to $\delta s$. If in crossing $\delta S$ there is also discontinuity $4\pi\phi(t)$ in the tangential electric face, a magnetic current sheet is required of intensity $\phi(t)$ in direction $\delta\sigma$ at right angles to it. This will produce a magnetic field in part derived from a potential function

$$-\cos(r, \delta s)\, \frac{d}{dr} \frac{\mathrm{I}}{r} \phi_1\left(t - \frac{r}{c}\right) \delta S$$

and in part of intensity

$$\frac{\mathrm{I}}{c^2 r} \phi'\left(t - \frac{r}{c}\right) \delta S$$

parallel to $\delta\sigma$; also an electric field of intensity

$$\sin(r, \delta s)\, \frac{d}{dr} \frac{\mathrm{I}}{r} \phi\left(t - \frac{r}{c}\right) \delta S$$

Continuation of an electro-dynamic field: but from redundant data. in circles around $\delta\sigma$. These two fields, integrated over the surface $S$, form the continuation of an electrodynamic field which involves tangential magnetic force $4\pi f(t)$ and electric force $4\pi\phi(t)$ at this surface.)

6. This method of analytical continuation of fields of physical activity has now been sufficiently illustrated. It is to be observed that in no case may the bounding surface in these formulae form a The formulae imply absence of obstacles. physical obstacle: the propagation is supposed to be free throughout all space, or, if there are obstacles, the law of propagation of disturbance from the various sources must take them into account and be modified thereby.

Having ascertained ways of distributing sources over any surface surrounding the origin of the disturbance in such manner as to produce the same effects, the way is open for explaining why it is only the parts of these sources that lie near the ray drawn to any external point that are usually effective at that point; and the law Imperfection of the general Huygenian explanations. of their effect is at once expressible in its main features. But the argument must, in the nature of the case, be based on general considerations, rather than exact analysis, as we have seen how to distribute sources in an infinite variety of ways so as to produce no effect at all.

We may finally briefly advert to the more directly dynamical Stokes' mode of treatment by equivalent volume sources: treatment of diffraction theory which was set forth with exhaustive generality by Sir George Stokes in 1849, and in which may be recognized, as Poincaré has remarked (*loc. cit.*), a less ambiguous foundation for Huygens' principle. The mode of operation is to consider the state of the medium as given throughout, at any time, by the strain or displacement and the velocity of each element of its volume, and to compute thence its state at any subsequent time by considering each such element of volume thus disturbed as a source from which disturbance is propagated. The law of this propagation will, of course, depend on the nature and position of any obstacles that may exist in the medium: on the hypothesis of their absence it has been worked out with full generality for elastic vibratory media, and all subsequent applications to more special problems are virtually included in Stokes' analysis. *his analysis covers all types of problem.*

The expression of the disturbance produced at any point at subsequent times can be reduced from a volume integral to a surface integral, which must be some one of the type discussed above; but the process of integration by parts prevents the terms from any longer representing real components derived from corresponding surface sources. The essential element in the theory is that a disturbance in an element of volume lasting for a small time is completely exhausted in sending out a definite spherical pulse of disturbance which travels ahead without leaving any trail behind it. The principle of Poisson for air waves, that the disturbance at any point at time $t$ depends only on the initial disturbance of the medium (including velocity, but *not* any spatial gradient) at the surface of a sphere of radius $ct$ with this point as centre, has therefore general application. This is at the base of the principle of Huygens, for it shows that what happens at this point in the interval $\delta t$ at time $t$ depends on the initial state of a spherical shell of the medium of thickness $c\delta t$ and radius $ct$. If this shell is drawn so as just to meet the region of initial disturbance, it appears that the disturbance at its centre depends only on the initial state of disturbance at this place of contact, and as the exciting disturbance travels on the place of contact travels along the ray. The explanation of Young and Fresnel as to why the point is not disturbed at subsequent times by the more distant elements of the initial disturbance that are off the line of the ray essentially depends on the feature of clean compact propagation without residual effect above adverted to[1]. *Result reducible to a Huygenian surface integral:* *reason therefor.* *Provides alternative explanation of rays.*

An interesting question is suggested by Professor Lamb's recent discussion, in these *Proceedings*, of the propagation of elastic waves in two dimensions, *e.g.* waves on a stretched membrane. Here each pulse leaves a trail, the type of argument above sketched requires modification, and the type of the Fresnel formula for ray propagation *Huygens' principle fails for two dimensions of space:*

[1] Cf. Sir G. Stokes' Wilde Lecture, *Proc. Manchester Lit. and Phil. Soc.* 1897 (or in Vol. IV of *Mathematical and Physical Papers*, now in the press). [See previous footnote, or Part II which follows.]

by diffractional interference is altered. The nature of the difference can be seen by considering the two-dimensional case as the analogue of one of cylindrical propagation in three dimensions of space: the preceding considerations then apply, as each element of disturbance travels in a pulse, but the Fresnel zone that is effective is determined by a sphere of radius $ct$, instead of a cylinder; the rays persist, but the diffraction formula cannot be constructed on the basis of the disturbance from a cylindrical source, and must, in fact, be deduced from the three-dimensional one.

the reason thereof.

# 60

# ON THE MATHEMATICAL EXPRESSION OF THE PRINCIPLE OF HUYGENS.—PART II.

[*Proceedings of the London Mathematical Society*, Ser. 2, Vol. XIX (1919) pp. 169–180.]

THE following considerations are in amplification and further development of a previous paper on this subject, *Proc. Lond. Math. Soc.* Ser. 2, Vol. I (1903) p. I.

In any medium which transmits physical action, the disturbance sent out across any surface $S$, which surrounds the true sources, is wholly determined by knowledge of the stress which is exerted, on the medium beyond, across each element of $S$. It would also be determined by knowledge of the strain which is imparted to the medium outside at each element of $S$. These determinations must agree: and the reason is that the two specifications over the surface $S$, one in terms of stress the other in terms of strain, are not independent, one determines the other.

Let us fix attention on the case of sound waves in a gas, completely determined by a single propagated velocity potential $\phi$. The stress condition is that the change of pressure $-\rho\partial\phi/\partial t$, and therefore the value of $\phi$ as regards sound of given frequency, is given over each element $\delta S$. The strain (in this case velocity) condition is that the value of $\partial\phi/\partial n$, the gradient of $\phi$ along the normal, is everywhere given over $S$. If we had a formula for the effect of a single type of source, namely of an impressed pressure given over an element and *merely absent* over the remainder of the surface, or else of the normal velocity given there and *maintained null* over the other elements of the surface, the total effect would be expressible as an integral taken over the surface.

Symmetry indicates how this is to be effected when the surface $S$ is an infinite plane, say that of $xy$. When the velocity of propagation is infinite, $\phi$ is the ordinary static potential $f(t)/r$, and the respective solutions are $\dfrac{\phi_0}{2\pi}\,\delta\Omega$ and $-\dfrac{1}{2\pi}\left(\dfrac{\partial\phi}{\partial n}\right)_0 \delta S/r$; where $\delta\Omega$ is the element of solid angle subtended, equal to $\delta S\,\dfrac{\partial}{\partial z}\left(\dfrac{1}{r}\right)$ or $\cos(zr)\,\delta S/r^2$, and the subscript zero indicates local value at $\delta S$. The possibility of integration of such formulae rests on the principle that a gradient differentiation of any function of $r$ can be transferred from one end of $r$, the source, to the other, at which the disturbance is to be estimated, by

*Prolongation determined by stress,*

*or by strain.*

*Example of waves of sound.*

*Contrast of cases:*

*propagation across a plane interface.*

mere change of sign. The corresponding forms for a potential propagated with finite speed c are $f\left(t-\frac{r}{c}\right)/r$ and $\frac{\partial}{\partial z}.f\left(t-\frac{r}{c}\right)/r$. Very near the origin these can ·be expanded by Taylor's series into $f(t)/r - c^{-1}f'(t) + \ldots$ and $\partial/\partial z$ of the same. The source or singular region at the origin must be specified completely in the first terms which alone involve inverse powers of $r$: thus as regards its determination the value of c does not enter, and the case is the same as the previous one of the static Laplacian potential. The disturbance

The two solutions: transmitted across the unlimited plane is thus expressed for the point $(r, z)$ by either of the formulae

$$\phi_1 = \int \frac{1}{r} f\left(t-\frac{r}{c}\right) dS, \quad \phi_2 = -\int \frac{\partial}{\partial z} \left\{\frac{1}{r} F\left(t-\frac{r}{c}\right)\right\} dS,$$

in which the surface data are

$$f(t) = -\frac{1}{2\pi}\left(\frac{\partial\phi}{\partial z}\right)_0, \quad F(t) = \frac{1}{2\pi}\phi_0.$$

It is to be noted that in the expression for $\phi_2$, the function $F$ must be kept inside the bracket: otherwise the element of the integral would not be a propagated potential. In fact this second formula is the same as

$$\phi_2 = -\int \frac{\cos(rz)}{cr} F'\left(t-\frac{r}{c}\right) dS - \int \frac{\cos(rz)}{r^2} F\left(t-\frac{r}{c}\right) dS,$$

the local disturbance abstracted. of which it is the first term alone that represents radiation sent away from the system, the second term being, owing to the higher inverse power of $r$, a local alternating disturbance which does not involve steady emission of energy.

The nature of rays. For the special case of radiation transmitted across a wave-front, or surface of constant phase, of any form, the transmission will be by rays of energy, whose paths are determined by the condition that time of transit does not vary sensibly for change to any adjacent path, and therefore that the ray travels by the quickest possible route at each stage, though it may not remain the path of shortest time for the whole course, if the latter is too long. In the case of a uniform medium, to which alone the formulae as here written apply, the rays are straight, and their intensity is expressed in the Fresnel manner in terms of the elements of wave-front from which the disturbance comes. There ought to be agreement as regards the ex-

These solutions identical in detail, pressions $\phi_1$ and $\phi_2$; and that is so in this special case of plane fronts, even for the separate elements, for as regards any function of $t - z/c$, the operator $\partial/\partial z$ is the same as $- c^{-1}\partial/\partial t$.

Here $f$ and $F$ may change with position so that the disturbance varies from place to place on the plane of resolution; but if this plane is not a wave-front the functions do not involve only $z$ and $t$, only in the<br>special case. and the formulae for $\phi_1$ and $\phi_2$ will be altered. The advancing disturbance is, except in the above special case, decomposed into elements at the plane or other surface of resolution in two different ways; these must be equivalent as regards total effect, but we are not as yet entitled to assert that either of them represents a resolution of the source into local equivalent physical parts. And this should not cause surprise: the strain specification expresses assigned local strain combined with constraint everywhere else on the surface $S$: The reason: while the stress specification is local but with no constraining or other condition elsewhere. Thus the latter, being self-contained as regards the element, has better claim to be a true physical resolution  a preference indicated, of the source of disturbance, as will presently be confirmed.

The second of these two special modes of resolution, that by means of stress conditions, is in keeping with the Kirchhoff formula expressing a quantitative formulation of the principle of Huygens as for Kirchhoff's form. regards pressural waves; but the analytical possibility of the first one also was calculated to throw doubt on the definiteness of any such resolution of the disturbance into components emerging from elementary surface sources. It is this point that we are now to consider more closely.

In the previous paper the general problem was approached from the point of view of Green's analysis in his fundamental *Essay on Electricity* of 1828, which formed in fact incidentally the earliest initiation of the theory of continuous analytical functions and their singularities. If the field of activity due to a set of sources inside $S$ and another set outside $S$ is supposed to be mapped out, then it was The previous case for indetermination, established at once on these principles that the outside effect of the sources inside can be represented as due to an equivalent distribution of sources over $S$: and the inside effect of the sources outside is due to the same distribution with sign reversed. It thus appears that there are an infinite number of distributions over $S$ which can represent in the outside region any given set of internal sources: for the external sources of the theorem can be chosen at will. And this applies whether $S$ is a wave-front or not: the resolution of the disturbance actually transmitted into contributions from secondary sources distributed over $S$ is not unique, and the theory of definite rays transmitting the disturbance from the various elements of $S$ seems therefore to be in danger of failure, which ought to be avoidable. endangers the principle of rays.

The proposition now to be advanced is that only one of these resolutions is physically correct. Indeed when we consider the matter, ample reason appears for the problem being definite. A state of stress and strain is continually transmitted up to the surface $S$ from the actual sources inside, and we are to find a distribution of secondary sources that will send it on to the outside as it arrives, without sending anything back. For the sending of a disturbance back into the interior would be an alteration of the physical circumstances, would in fact add to and confuse the effect of the assigned true sources inside. Hence the correct secondary sources are the distribution which produces the effect of the true sources on the side beyond them, but produces no backward effect on their own side. The distribution of such sources was determined in the previous paper by an immediate process indicated above; and this now puts in evidence both its possibility and its uniqueness.

Again it will suffice to illustrate from the case of sound waves which depend on only one potential $\phi$ of scalar type. The value of $\phi$ outside due to the given true sources inside and no sources outside is

$$\phi = \frac{1}{4\pi}\int \frac{1}{r}\left(\frac{\partial\phi}{\partial n}\right)_0 dS - \frac{1}{4\pi}\int \frac{\partial}{\partial n}\left(\frac{\phi_0}{r}\right) dS,$$

in which under the integrals $\phi_0$ and $(\partial\phi/\partial n)_0$ denote the values at the surface element $\delta S$ at time $t - r/c$: cf. *loc. cit.*, [*supra*, p. 247]. Now

$$\frac{\partial}{\partial n}(\phi_0) = -\frac{\cos(rn)}{c}\left(\frac{\partial\phi}{\partial t}\right)_0$$

and *if the surface S be a wave-front*

$$\left(\frac{\partial\phi}{\partial n}\right)_0 = -\frac{1}{c}\left(\frac{\partial\phi}{\partial t}\right)_0.$$

Thus, effecting the differentiation,

$$\phi = \frac{1}{4\pi}\int \frac{1 + \cos(rn)}{r}\left(\frac{\partial\phi}{\partial n}\right)_0 dS + \frac{1}{4\pi}\int \frac{\cos(rn)}{r^2}\phi_0 dS.$$

The latter term involving $r^{-2}$ is merely a local alternation or fluctuation of disturbance; so that it is the former alone that expresses the acoustic radiation transmitted from each secondary source at an element $\delta S$. In the forward direction along $\delta n$ it is $\frac{2}{4\pi r}\left(\frac{\partial\phi}{\partial n}\right)_0 \delta S$,

whence it continuously falls off in sideway directions until in the backward direction it has become null. When the surface $S$ is a wave-front, an actual ray is thus propagated from the group of adjacent local elements $\delta S$, normally in the forward direction, but none in the backward direction. For in dealing with radiation as such, it is implied that it is estimated far enough from the source for the local fluctuating field represented by the other integral term in the formula to be ignored. It is to be remembered also that in this mathematical analysis the surface $S$ is merely a geometrical boundary in the continuous medium, and does not at all act as an obstacle; each secondary source on an element $\delta S$ of it is to be regarded as radiating freely in all directions without any relation to the sources on the other elements.

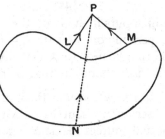

We still consider a surface $S$ which is a wave-front: and it is to be noted that this usually implies a single point source, though with its radiation modified by transit through different media, as in the familiar case of optics. The disturbance that reaches an external point $P$ will travel to it along definite paths which are those of quickest propagation across the region between $P$ and the surface. In a uniform medium they will be straight paths, in fact the normals to the surface, shown as $PL$, $PM$, $PN$ in the diagram: it is only the elements of disturbance crossing $S$ near $L$, $M$ and $N$ that can sensibly affect $P$. But when such secondary sources on $S$ are made determinate by the condition that they produce no backward radiation, the contribution starting from $N$ in the direction along the normal drawn inward ought to be obliterated, and that is in fact secured by the formula above developed: therefore $NP$ is not a physical ray, for no energy travels from $N$ along it. The rays are restricted to be the outward-directed paths of shortest time, or more strictly of quickest propagation. The difficulty of Fresnel's early theory of diffraction, as to why a ray is not also propagated backward, is thus surmounted*.

*(July 10, 1920.* If, more generally, the surface $S$ over which the emerging disturbance is dissected is not a wave-front, let $\nu$ be the direction of the ray at the element $\delta S$ and $(\nu n)$ its inclination. Then

$$\left(\frac{\partial \phi}{\partial n}\right)_0 = \cos (\nu n) \left(\frac{\partial \phi}{\partial \nu}\right)_0 = -\frac{\cos (\nu n)}{c} \left(\frac{\partial \phi}{\partial t}\right)_0,$$

$$\frac{\partial}{\partial n} (\phi_0) = \frac{\cos (rn)}{c} \left(\frac{\partial \phi}{\partial t}\right)_0,$$

and the formula becomes

$$\phi = \frac{1}{4\pi} \int \frac{\cos (rn) - \cos (\nu n)}{cr} \left(\frac{\partial \phi}{\partial t}\right)_0 dS + \frac{1}{4\pi} \int \frac{\cos (rn)}{r^2} \phi_0 dS$$

in which the second term is as before purely local. In this form it applies to any coherent system of rays. The case of a simple pulsating point source of sound is interesting: for it, $r_1$ being the distance of this source from $\delta S$,

$$\phi = \frac{A}{r_1} \cos \left(nt - \frac{r_1}{c}\right),$$

Coherent beam:

its rays:

inward rays excluded.

Fresnel's postulate.

General formula:

---

* Does the existence of rays depend on clean propagation, without leaving a trail? For instance, are there rays in propagation in two dimensions (footnote, p. 261)? The Huygenian argument fails, unless it can be reinforced: for the wave from a secondary source of disturbance on the interface has a trail. But two dimensions is a case of three, with a slit replacing a compact element of area: therefore rays do persist, as indeed the general formulae indicate.

case of a
point source,

thus the transmitted disturbance can be expressed[1] as

$$\phi = -\int \frac{An}{4\pi C} \frac{\cos{(rn)} - \cos{(r_1 n)}}{rr_1} \sin{\left(nt - \frac{r_1 + r}{C}\right)} dS,$$

attenuation with distance, phase, ray path.

which puts in evidence the factor of attenuation $(rr_1)^{-1}$ and the phase depending on the path $r_1 + r_2$, and shows that it is only the elements $\delta S$ that lie near the line joining the source to any point which produce the disturbance at that point.)

But now consider the more general and indeterminate distribution of secondary sources, in which they are derived as representing the actual sources inside $S$ together with any assumed fictitious sources outside. The presence of the latter does not affect the validity of the new system of secondary sources, provided it is taken as a whole,

Energy in rays must not travel backward.

as regards outside space: but that system also sends disturbance into the inside space, rays thus travel and energy is propagated backwards as well as outwards. If the surface $S$ happened to continue to be a wave-front after the fictitious external sources are superposed $NP$ would be a ray; and $LP$, $MP$ would continue to be rays, but of altered intensities such however that the aggregate disturbance at $P$ would remain the same. Usually $S$ would not remain a wave-front, and the rays passing between it and $P$ would

Rays belong only to coherent radiation.

start from it not normally: it is only under very special circumstances that disturbances from different sources could consolidate into one system of rays at all. Anyhow this somewhat artificial illustration suffices to enforce the point, that all these mathematically equivalent distributions of secondary sources are fictitious except the one that happens to have been hit upon originally for this case of pressural waves involving a potential, by Kirchhoff. For the disturbance at each place outside would remain the same whatever fictitious external distribution of sources is superposed, but the paths by which the energy travels across the medium to that place would be different:

Rays the discriminating feature.

as a ray is the definite path of actual transfer of energy, this illustration in terms of rays is specially graphic, indeed is of the essence of the matter. The corresponding general formulae for optical and electrical radiation can now be derived on the lines expounded in the previous paper: for example the expressions for optical radiation may be developed from § 5 [as *infra*]. Enough has been said to illustrate the principles that are involved*.

[1] Cf. Kirchhoff, Wied. *Ann.* Vol. XVIII, 1883, formula (13).

Compact propagation of pulses not essential:

* The argument, to be valid, ought to be independent of whether a point source sends out a compact shell of radiation, or a shell leaving a residual trail of disturbance behind it. Thus consider any set of sources of radiation $S$, and an ideal boundary $B$ separating the region $I$ containing the sources from the region $O$ beyond it. The sources produce a definite surface succession of

*The Law of Oblique Diffraction of Transverse Waves.*

The expression for the electric or optical radiation diffracted from an element of wave-front $\delta S$, which strangely does not seem hitherto to have been determined at all, is perhaps obtained most simply by taking the element at the origin in the plane $xy$, so that the electric force $F$ is along $x$ and the magnetic force (equal to $F/c$ in electrodynamic units) is along $y$, the axis of $z$ being thus the normal to the front. We have to determine the disturbance propagated in the direction of the radius vector $r$ making an angle $\theta$ with the normal

*The Huygenian formula for electric radiation:*

states over the boundary $B$ continuous with definite successions of states in the regions inside it and beyond it. A distribution of ideal sources over the boundary $B$ is here determined which would produce a succession of states of free radiation outside the boundary, continuous with but not disturbing the actual succession of states within it which is determined by the actual sources. The continuity of these two across the boundary then ensures that the outer succession is the free prolongation of the inner one due to the actual sources.

A great variety of other distributions of ideal sources over the boundary would give the same external succession of states, but they would not be continuous with the actual succession when their ideal sources are removed. In consequence such sources would not express for each locality on the boundary the free continuation of the actual succession of states within it. There is nothing here asserted as to whether each source radiates in compact shells or not.

In a two-dimensional medium a point source does not radiate in compact shells: nor does it even in a three-dimensional medium if it is of crystalline quality. See Lamé, *Leçons sur l'Élasticité*, ed. 2 (1866): the problem has been studied later more deeply by Volterra, *Acta Math.* xvi, and *Clark Univ. Lectures* (1909). It would appear that the two sheets of the Fresnel wave-surface, which as Lamé remarks, after Fresnel himself, are analytically inseparable, are the outer and inner boundaries of a disturbance travelling from a point source of limited duration, so that even an instantaneous disturbance at a point produces a shell of radiation widening as it progresses.

*significance of the double-sheeted wave-surface:*

A question may be propounded: if a disturbance is established over one sheet of a Fresnel wave-surface, would it necessarily as it travels out spread towards the other sheet of the widening wave-surface? In fact, for a uniaxial medium, any source of electric type, however complex or extended it be otherwise, provided it is symmetrical around an axis parallel to that of the medium, must emit radiation confined solely to ellipsoidal shells, of whatever thickness: for the magnetic lines of force are by symmetry circles around the axis, so that the field of radiation, for the most general type of such travelling pulse, can be formulated at once after the manner of p. 146 *supra*. In the same way such a source, if of magnetic type, would radiate solely in spherical shells. But for other types of source, even for a simple transverse impulse or vibrating electron, the radiation will be of mixed type as above, and apparently not amenable to any manageable formulation. As however a general pulse travels out, bounded by the two sheets of an expanding wave-surface, it necessarily concentrates toward the outer and inner boundaries as their curvatures diminish.

*illustrated by uniaxal medium:*

Similar considerations apply even for the two sheets, compressional and distortional, but now independent, of the wave-surface for an isotropic elastic solid: near the source the propagation is of mixed type. The case of a spherical source is treated in detail by Kelvin, *Baltimore Lectures*, xiv.

*by isotropic elastic solid.*

to the front and such that the plane of $\theta$ makes an angle $\psi$ with the direction $(x)$ of the electric force in the front. It is easier, and also

components of the secondary source; the local effect rejected: is a variation on the procedure of the previous paper, to resolve the equivalent electric and magnetic oscillators on $\delta S$ into linear components in the directions of the polar elements $\delta r,\ r\delta\theta,\ r\sin\theta.\delta\psi$. When we attend only to the terms of lowest order in $r^{-1}$, representing true radiation, the first component produces no effect in the direction of $r$. The effects of the others involve only the amplitudes of radiation of a linear magnetic vibrator $f'/4\pi$ and a linear electric vibrator $\phi'/4\pi$ (in the previous notation) in directions transverse to their lengths: the former produces in its equatorial plane electric amplitude $f'/4\pi r$ parallel to the element, and magnetic amplitude $f'/4\pi cr$ at right angles, the latter produces magnetic amplitude $\phi'/4\pi c^2 r$ parallel to its direction and electric amplitude $-\phi'/4\pi cr$ at right angles, in each case the argument of the functions $f'$ and $\phi'$ being $t-r/c$.

By combining the contributions from all the components of the equivalent sources on $\delta S$ the result comes out in the simple form electric amplitude:

$$\begin{matrix}\text{along } r\delta\theta & \cos\psi \\ \text{along } r\sin\theta.\delta\psi & \sin\psi\end{matrix}\ (1+\cos\theta)\frac{\delta S}{4\pi r}f'\left(t-\frac{r}{c}\right),$$

magnetic amplitude:

$$\begin{matrix}\text{along } r\delta\theta & \sin\psi \\ \text{along } r\sin\theta.\delta\psi & -\cos\psi\end{matrix}\ (1+\cos\theta)\frac{\delta S}{4\pi cr}f'\left(t-\frac{r}{c}\right),$$

the aggregate that is propagated away from it: in which $f(t)$ is the electric amplitude on the original wave-front of radiation on $\delta S$, which is in the direction of $\psi$ null, and $cf(t)$ is the magnetic amplitude on the original wave-front at right angles.

The resultant electric and magnetic disturbances along any direction $r$ are at right angles, as they ought to be in radiation. The ray diffracted from $\delta S$ along $r$ is thus plane polarized of electric amplitude

described physically: $(1+\cos\theta)\dfrac{\delta S}{4\pi r}f'\left(t-\dfrac{r}{c}\right)$ with the plane of its electric vibration inclined to the plane through the normal to $\delta S$ at an angle equal to the azimuth of $r$ around the normal, the sign of this angle being

its polarization. conveniently recovered by considering directions of $r$ nearly normal to $\delta S$ for which the electric vibration must be nearly parallel to $x$. From this formula, applied for directions near the normal, the doctrine of rays may be developed on the usual lines.

Simple result for diffracted unpolarized light: For ordinary light the result is the same as for equal beams polarized in directions at right angles: in the intensity the azimuth $\psi$ thus disappears, and the obliquity factor $1+\cos\theta$ in the amplitude alone survives squared for the intensity of unpolarized diffracted light.

The resultant ray as integrated from all the elements such as $\delta S$ is normal to the front as usual: and the factor $1+\cos\theta$ ensures, as in the previous case of waves of pressure, that while energy is propagated forward in the ray direction of $\theta$ null, no energy travels backward in the other formally possible ray direction specified by $\theta$ equal to $\pi$.

In the direction of the actual ray the intensity of the electric force propagated from an element of surface is $f'/2\pi r$ parallel to $x$, and of the magnetic force $f'/2\pi cr$ parallel to $y$, in which the argument of $f'$ is $t - r/c$. Integrated over the wave-front for an alternating train, the result should be $f$ and $f/c$: as may be verified by the usual procedure of Fresnel.

(*July* 10, 1920. It should not be forgotten that the diffraction of elastic waves of general type was treated very clearly and completely by Stokes, with results virtually equivalent to those in the text, including the crucial factor $1 + \cos(rn)$, as early as 1849 in Part I, Sec. iii (cf. § 33) of the classical memoir "On the Dynamical Theory of Diffraction."* As however he was chiefly concerned with the direction of polarization of the diffracted light, the formulae were not developed. The disturbance was taken as transmitted by traction, not displacement, of the medium, as was natural: cf. *supra*, p. 257.

But it has not been sufficiently noted that his analysis completely evades any discussion of the mode of transmission. Thus (§ 32) "the disturbance transmitted during the interval $\tau$ ... occupies a film of the medium, of thickness $b\tau$, and consists of a displacement $f(bt')$ and a velocity $bf'(bt')$": and the procedure is to deduce from the general formulae applicable to an unlimited unrestricted medium the way in which this disturbance in these local elements of volume will spread. There is thus no room for any ambiguity in Stokes' mode of analysis, and an appeal to the mode of propagation of energy is not necessary: though he appears to have overlooked (or rather was not concerned with) the result that his formula allows no backward propagation of disturbance into the region within the surface on which the disturbance is thus dissected.

In § 36 he, however, appears to state that for common light the obliquity factor for intensity would involve $1 + \cos^2 \theta$ instead of $(1 + \cos \theta)^2$: this would violate the canon last mentioned, and indeed is not consonant with modern ideas of the nature of ordinary light.

The law of polarization which has emerged above does not seem to agree with the famous law of Stokes, the subject of so much experiment more or less relevant by him and others, namely that the plane of (magnetic) vibration of the diffracted disturbance is parallel to the vibration of the original beam. It would not be surprising that electrodynamic secondary sources should thus give a different result from secondary sources in a mechanical elastic medium, as is in fact

* *Camb. Phil. Trans.*, reprinted in *Math. and Phys. Papers*, Vol. ii.

put in evidence in the different laws for reflexion of waves in the
two cases. The argument of Stokes from symmetry, which is so
effective for diffraction from particles and the polarization of the
blue of the sky, thus does not here seem to apply because the magnetic
vibration is as important as the electric. But further scrutiny of
the discrepancy must be deferred. On this question, and on the
ambiguities resulting from modes of resolution that have produced
embarrassment, cf. Lord Rayleigh's summary, in "Wave Theory."*)

### Curved Wave-Fronts.

We have considered the oblique diffraction from elements $\delta S$ of a
plane wave-front. Curvature, if its radius is very great compared with
the wave-length, may be presumed not to affect sensibly the result.
To examine this question, we may try to express analytically the train
of coherent radiation advancing from a front of finite curvatures.

Indeed before any general formulae for diffraction can be practically
applicable, before we can pass for optical systems of rays much
beyond the elementary modes of argument of Huygens and Fresnel
confined to directions near the ray, it would be necessary to have an
electrodynamic specification of the general optical train on which
the formulae are to operate. In the absence of knowledge of any
previous attempt in this direction, the following sketch is offered
for scrutiny and expansion.

The criterion of a coherent train is that there must be a system of
wave-fronts, surfaces of equal phase. The rays, or paths of the energy,
are for an isotropic medium the curves intersecting this system
of surfaces orthogonally. If the rays are curved the medium must be
of varying density: thus the velocity of propagation $c'$ will be a function of position in the medium, unless the rays are in straight lines.

Let us divide one of the wave-fronts into elementary curvilinear rectangles by drawing on it
the two sets of lines of curvature. The geometrical rays that issue normally to it from each
line of curvature will form a sheet or surface.
The two sets of surfaces thus obtained, and the
set of wave-fronts, form a triply orthogonal
system: therefore by the theorem of Dupin, all
their curves of intersection are lines of curvature
on the surfaces on which they lie.

Let us consider a train of waves such that at
one of the fronts the electric force $F$ is directed along one of the sets
of lines of curvature, of element $\delta s$, and the magnetic force $H$ along
the other set, of element $\delta s'$. Then the circuital relations of Ampère

* *Ency. Brit.* (1888): *Scientific Papers*, Vol. III, p. 165.

and Faraday, when applied around the faces of the curvilinear element of volume $\delta s\delta s'\delta n$, give (with electromagnetic units) six equations of types

$$\delta n \frac{\partial}{\partial n}(H\delta s') = -\frac{\partial}{\partial t}\left(\frac{F}{c'^2}\right)\delta s'\delta n,$$

$$\delta n \frac{\partial}{\partial n}(F\delta s) = -\frac{\partial H}{\partial t}\delta s\delta n,$$

together with the relations that $F\delta s$ and $H\delta s'$ remain constant, but only up to the first order and so only for a short range, in passing to the opposite sides of the rectangle $\delta s\delta s'$—together also with the relations that $F$ and $H$ are along the directions of the lines of curvature on the consecutive wave-front (in the recognized shorthand infinitesimal sense of that term) indicated by the dotted rectangle in the diagram. If this be allowed, the invariance of $F\delta s$ and of $H\delta s'$ along the respective lines of curvature determines the distribution of $F$ and $H$ as regards intensity, over any one selected wave-front, the directions however being fixed. Then on the other wave-fronts their directions will also be along the lines of curvature, and their intensities will be determined by the equations first written. These equations take the form, if $R$, $R'$ are the principal radii of curvature of the front,

*The electric curvilinear relations:*

$$\frac{\partial H}{\partial n}-\frac{H}{R'}=-\frac{1}{c'^2}\frac{\partial F}{\partial t}, \quad \frac{\partial F}{\partial n}-\frac{F}{R}=-\frac{\partial H}{\partial t};$$

we derive

$$\frac{\partial}{\partial n}FH-\left(\frac{1}{R}+\frac{1}{R'}\right)FH=-\tfrac{1}{2}\frac{\partial}{\partial t}\left(\frac{1}{c'^2}F^2+H^2\right).$$

This affords a verification; for it expresses, on Poynting's principle, that the integrated flux of energy (equal in intensity to $FH/4\pi$) out of the element of volume is the rate of diminution of the energy inside it. It expresses the law of intensity along the ray.

*verified by equation of energy.*

On elimination, $F$ and $H$ are determined separately by the two equations

$$\left(\frac{\partial}{\partial n}-\frac{1}{R'}\right)\left(\frac{\partial}{\partial n}-\frac{1}{R}\right)F=\frac{1}{c'^2}\frac{\partial^2 F}{\partial t^2}, \quad \left(\frac{\partial}{\partial n}-\frac{1}{R}\right)\left(\frac{\partial}{\partial n}-\frac{1}{R'}\right)H=\frac{1}{c'^2}\frac{\partial^2 H}{\partial t^2},$$

which are different as regards the second order, such as has been neglected. Radiation of general type may be replaced by a beam polarized along $\delta s$, together with an equal one polarized along $\delta s'$. If the wave-fronts are plane, this is the usual equation of simple transmission of disturbance.

*Special cases:*

The practical important case in which an exact specification for curved fronts is available is that of the radiation from a simple bipolar (or multipolar) source: for from that case of spherical fronts all other systems of rays may be constructed by superposition.

*exact solutions.*

# 61

# ON THE INTENSITY OF THE NATURAL RADIATION FROM MOVING BODIES AND ITS MECHANICAL REACTION.

[From the Boltzmann *Festschrift*, Sept. 1903.]

THE subject of the pressure of radiation, which was first reduced into a definite formula by Maxwell, was placed in new and most fruitful

Boltzmann: Bartoli.

light when Boltzmann showed, by following out an idea of Bartoli, that it stood in intimate relation to the law connecting the radiation

Poynting's cosmical conclusions.

of a body with its temperature. In a recent memoir[1] Poynting has based very remarkable results, as regards cosmical dynamics, on the operation of a retarding force due to the back pressure of its own radiation when the radiating body is in motion. The main object of the present note is to treat this aspect of radiation pressure by more direct methods, and thereby confirm the expression for the mechanical reaction against a moving radiating surface, that has been deduced by Poynting from general considerations, naturally somewhat uncertain, relating to flux of energy.

The pressure exerted by radiation is essentially connected with opacity to it [including perfect opacity which turns it back]. From formulae developed on other occasions[2] it appears that in the case of a medium which may vary in its properties in any manner along the direction of propagation $x$, when it is the seat of electric dis-

Mechanical force in a gradually absorbing medium:

turbances of simple harmonic period $2\pi/n$, polarized so that the electric force is $(0, Q, 0)$ and the magnetic $(0, 0, \gamma)$, the dynamical equations being thus in Maxwell's notation

$$4\pi v = -\frac{d\gamma}{dx}, \quad \frac{dQ}{dx} = -\mu\frac{d\gamma}{dt}, \quad v = \sigma Q + \frac{K}{4\pi c^2}\frac{dQ}{dt};$$

the mechanical force acting on any block or segment of it is representable by pressures of intensity

$$\frac{1}{8\pi}\left(\gamma^2 + \frac{1}{c^2\mu n^2}\frac{dQ^2}{dt^2}\right)$$

applied to the two ends of the segment—these pressures just cancelling each other, as they ought, when the segment consists

---

[1] *Roy. Soc. Proc.* 1903: *Phil. Trans.* 1903. [Poynting's *Scientific Papers* (1920), p. 304: also editorial note, p. 754, reprinted *infra*.]

[2] *Phil. Trans.* 1897, A [*supra*, p. 117]; or more fully in *Aether and Matter* (1900), pp. 130–3.

of free aether without matter. The mean value of this end pressure is

$$\frac{1}{16\pi}\left(\gamma_0{}^2 + \frac{Q_0{}^2}{c^2\mu}\right),$$

where $\gamma_0$ and $Q_0$ represent the amplitudes of $\gamma$ and $Q$ [: when $K$, $\mu$ are each unity it is the energy density of the radiation].

When the amplitudes are diminished owing to absorption [so gradual that there is no reflexion] as the disturbance travels onward, there is this steady mechanical force exerted in the medium in the direction of propagation. When the electric disturbance is incident on a transparent reflector there is no resultant force on the [geometrical] reflecting surface itself, because $\gamma$ and $Q$ both remain continuous in crossing it. When however the reflector is nearly perfectly opaque, the electric forces in front of it in the incident and reflected disturbances almost cancel each other, while the magnetic force just outside is doubled by its presence: there must be disturbance of the nature of alternating electric flux in the skin layer of the reflector such as will annul this magnetic field in its interior, and it is the electrodynamic forces acting on this layer of current that constitute the aggregate electric pressure, which can be shown (*loc. cit.*) to agree with Maxwell's formula.

From this way of considering the mechanical force, it is readily verified that when the incidence on the reflector is oblique, Poynting is right in taking the incident and reflected wave-trains each to exert their full oblique thrust on the reflector along their directions of propagation.

For radiation to exert steady non-alternating [resultant] pressure on a small body, it must (*loc. cit.*) be of opaque* material. A dielectric mass constituted of perfectly elastic elementary vibrators should not be repelled by radiation. In illustration, consider the simplest type of vibrator, an electric doublet consisting of charges $+ e$ and $- e$ separated by a varying distance $l$, parallel to $x$, so that its moment $M$ is $el$. When it is subjected to a simple wave-train travelling along $x$ with electric force $(0, 0, A \cos pt)$ and therefore magnetic force $- c^{-1} (0, A \cos pt, 0)$, the equation of its forced vibration is

$$\frac{d^2M}{dt^2} + \kappa^2 M = eA \cos pt,$$

so that

$$M = \frac{eA}{\kappa^2 - p^2} \cos pt;$$

and, the vibrator constituting a current element $dM/dt$, the magnetic field pushes it along $z$ with a mechanical force $\beta\, dM/dt$, which is

$$p\, \frac{cA^2 c^{-1}}{\kappa^2 - p^2} \cos pt \sin pt.$$

* The paradox of the two following paragraphs has been explored in Vol. 1, *supra*, Appendix VII.

This electromagnetic force is however purely alternating and so adds up in time to nothing: the only way to obtain steady mechanical pressure on the vibrator is to put the forced vibration out of phase with the exciting field, by the introduction of a frictional term into the equation of vibration, which will correspond to opacity [, or by the secondary radiation from the matter].

Condition for the pressure to get a grip.

In the theory of exchanges of radiation it is customary to represent a perfect reflector as a body of very high electric conductivity. Any body across which the radiation cannot penetrate is as already stated subject to a pressure from the radiation just outside it, determined by Maxwell's formula.

The absorption in re-flexion from a good conductor,

It is worth while to verify explicitly that the absorbing quality thus associated with this pressure does not act so as to vitiate the perfection of the reflexion by degrading the energy. This is of course readily done. The equations of wave-propagation already formulated lead to

$$\frac{d^2Q}{dx^2} = 4\pi\mu\sigma \frac{dQ}{dt} + K\mu c^{-2} \frac{d^2Q}{dt^2}.$$

Writing $\qquad Q = A e^{int} e^{-ipx}$

this gives $\qquad p^2 = K\mu c^{-2} n^2 + 4\pi\mu\sigma ni.$

Thus if the conductivity $\sigma$ is largely preponderant we may write

$$p = (2\pi\mu n\sigma)^{\frac{1}{2}} (1 + i), \text{ say } = r (1 + i).$$

Taking for the transmitted train the real part[1]

$$Q = A e^{int} e^{-rx} \cos rx$$

the heat developed per second comes out to be

$$H = \sigma \int_0^\infty \tfrac{1}{2} Q^2 dx = \frac{3\sigma A^2}{16r} = \frac{3A^2}{32\pi} \left(\frac{\lambda\sigma}{\mu c}\right)^{\frac{1}{2}}.$$

Now if $A_1$ is the coefficient for the wave-train directly incident from the free aether on this medium and $A'$ that for the wave-train reflected back, the continuity of $Q$ and of $\mu^{-1}dQ/dx$ across the surface gives

$$A_1 + A' = A, \quad \frac{n}{c}(A_1 - A') = \frac{p}{\mu} A,$$

so that $\qquad A_1 = \tfrac{1}{2}\left(1 + \frac{p}{\mu n}c\right) A, \quad p = r(1 + i);$

[1] But this is for stationary waves; it should have been for progressive waves $Q = A e^{-rx} \cos(nt - rx)$, giving

$$H = \frac{A^2}{4r} = \frac{A^2}{4}\left(\frac{\lambda\sigma}{\mu c}\right)^{\frac{1}{2}}.$$

and passing again to real parts by taking moduli, the amplitude of the incident train is approximately

$$\tfrac{1}{2} (\mu^{-1} \lambda \sigma c)^{\frac{1}{2}} A.$$

The energy incident per second is thus

$$\frac{1}{8\pi c^2} c \cdot \tfrac{1}{4} \mu^{-1} \lambda \sigma c A^2 \text{ or } \frac{A^2}{32\pi\mu} \lambda\sigma, \qquad \text{is negligible:}$$

of which the part $H$ that is degraded thus forms a negligible fraction inversely proportional to the square root of the conductivity $\sigma$.

The waves are thus turned back without sensible loss by degradation, because, for an ideal good conductor, the surface layer is at a node of the electric force. There is superficial current in the conductor which gives rise to the Maxwellian repulsion by the agency of the magnetic field; while there is no sensible electric resistance, the small as was to be expected: electric force near the node establishing the necessary current without production of heat. [Cf. Vol. I, *supra*, Appendix v.]

The conditions which here obtain for very high conductivity and short waves also hold for lower conductivity and longer waves. For long heat waves the proportionality of the absorbing powers of metals to the square roots of their specific resistances has, as is well known, been discovered by Hagen and Rubens, and explained in the conductance attaining its normal value for heat waves. advance by Drude and afterwards by Planck; this observation carries the interesting result that the resistance coefficients are nearly the same for such heat waves as for ordinary steady currents.

Any doubt that may be entertained as to whether radiation exerts a back pressure on the body that emits it, may be diminished by considerations of the kind here employed. The emitting body being opaque, the source of the radiation is vibratory disturbance of The electron mechanism of the pressure. electrons in its surface layer; these constitute a self-damped current sheet which is pushed back by the magnetic field it produces, precisely as happens for the corresponding current sheet at a conducting surface on which radiation is incident as above. [Cf. *loc. cit.*]

We now proceed to our problem of the radiation from a moving body. Consider an enclosure, with ideal perfectly reflecting walls, at a uniform temperature throughout and thus pervaded by the steady natural radiation corresponding to that temperature. The principle of Carnot requires that we cannot by cycles of slow move- The Stewart-Kirchhoff principle now involved in that of Carnot. ment of the bodies in the enclosure transform any of this energy at uniform temperature into mechanical effect through the agency of the pressure of radiation. There must therefore be a unique state of density of the total enclosed radiation, independent of the nature of the surfaces of the bodies in movement; for otherwise direct movement with one kind of surface combined with the reverse movement with another kind would constitute a working cycle. The steady aggregate density of radiant energy in the enclosure is therefore not affected by the motion of the bodies; indeed if this were not so, by opening and closing a window in the enclosure while it is moving at different speeds, cycles could be established which would violate Carnot's principle. Now compare a moving perfectly reflecting surface, which reflects back all the incident radiant energy, with the same moving surface rendered perfectly absorbing; this is allowable, the analogous change from conducting to non-conducting being contemplated in elementary thermal reasoning about Carnot's principle.

*The equilibrium of exchanges involves law of radiation from moving matter.* It follows from the theory of exchanges, that in the state of equilibrium the radiation that is returned must be the same as regards constitution and intensity in both cases. Now the solution of the electrodynamic problem of reflexion from a moving perfect reflector is known[1]; therefore the law of the radiation from a perfect radiator in motion is determined in complete detail. When the reflector is advancing in a stationary enclosure, the energy density of the reflected radiation is greater than that of the incident, and the excess is a fraction of the latter equal to four times the ratio of the velocity of the reflector in its direction to the velocity of light[2]. Thus when *Law of free radiation in a moving enclosure:* the enclosure is moving as well as the reflector, the energy of the incident stream coming from its receding walls is in defect by twice the ratio of these velocities and that of the reflected stream is in excess by twice the same ratio. This latter factor therefore also expresses the excess in the volume density of natural radiation coming from a perfect radiator that is produced by its own advancing motion; but in a detailed specification of this radiation the modification of the wave-lengths in accordance with the Doppler principle is also to be borne in mind.

A different and generalized mode of treatment may also be adopted, based on the Lorentz transformation for passing from the field of activity of a stationary electrodynamic material system to that of *verified by the Lorentz correspondence,* one moving with uniform velocity of translation through the aether. If $(f, g, h)$ and $(a, b, c)$ represent the field of a material system at rest in the aether, then to the first order of $v/c$,

$$\left(f,\ g + \frac{v}{4\pi c^2} c,\ h - \frac{v}{4\pi c^2} b\right)$$

and
$$(a,\ b - 4\pi v h,\ c + 4\pi v g),$$

[1] Cf. Larmor, *British Association Report*, 1900 [*supra*, p. 218]: *Encyclopaedia Britannica*, Article "Radiation," Vol. XXXII, 1903.

[2] The Maxwellian formula for the pressure of radiation may be based (*loc. cit.*) on this result, in connection with the conservation of the energy; or conversely, the value of that pressure being assumed on other grounds, this result for the intensity of the reflexion may be based upon it.

represent the values of the same vectors, say $(f_1, g_1, h_1)$ and $(a_1, b_1, c_1)$, for a system in motion parallel to $x$ with velocity $v$; and the positions and magnitudes, and therefore relative velocities, of the electrons which produce these fields in the surrounding aether in the two cases are identical at each instant, so that the fields belong to the same material system[1]. quoted, to first order.

An enclosing material boundary is supposed to form part of the system, so as to retain the radiant energy at uniform density. Let us compare the densities $E$ and $E_1$ of energy in the two cases of rest and translation, as given by Maxwell's formula

$$E = 2\pi c^2 \left(f^2 + g^2 + h^2\right) + \frac{1}{8\pi} \left(a^2 + b^2 + c^2\right).$$

We obtain, neglecting $(v/c)^2$ as before,

$$E_1 = E + 2v \left(gc - hb\right).$$

Now the flux of energy in the aether is by Poynting's rule the vector

$$- c^2 \left(hb - gc, \ fc - ha, \ ga - fb\right),$$

so that the last term in $E_1$ is $2c^{-2}$ times the scalar product of this flux and the translatory velocity of the system.

Thus the density of the radiation that is travelling in the enclosure in directions inclined towards $v$ is increased; but in the opposite directions it is diminished by equal amount, so that the aggregate density is unaltered as already seen. Radiation becomes anisotropic by the convection:

Taking a particular case, for a plane wave-train represented by $(f, g, h)$ and $(a, b, c)$, forming part of the steady radiation, which thus travels in the direction perpendicular to both these vectors, the flux of energy per unit time is increased for the moving material system by a fraction of itself equal to twice the component of $v$ along its direction of propagation divided by the velocity of light. There is diminution in the flux for waves coming from the receding parts of the boundary of the enclosure, and an equal increase for those reflected back, giving in all the factor four previously obtained for the change of volume intensity on reflexion. It may be remarked that this mode of selected orientation of the steady radiation in the moving enclosure clearly satisfies the necessary condition that, when an aperture has been made anywhere into an outer region of steady radiation, the radiation that issues through it is the same as had been previously sent back from the wall at that place. as does the energy flux in the waves: undisturbed by a small aperture.

The same results for the change in the energy flux in any direction may be obtained directly from the flux formula of Poynting, when

[1] Cf. *Aether and Matter*, p. 169. (The change to "local time" merely introduces [accounts for] the Doppler effect.)

the modified values of the vectors in the moving system are inserted. The connection between the two methods rests on the remark that for a plane progressive wave the flux per unit time is the density multiplied by the velocity of propagation, when there is no dispersion.

The volume density of radiation emitted from a perfect radiator in any direction thus involves a factor $1 + 2k$, where $k$ is the ratio of the velocity of the radiator in that direction to the velocity of light; and the pressure of this ray, exerted directly backward, is altered accordingly, with consequences considered by Poynting in the memoir already referred to.

*The cosmic effects.*

This result is in fact what clearly obtains if on an ultimate dynamical theory the energies of the vibratory motions of the radiating sources are not affected by the uniform translation, but depend only on the temperature or other physical cause, as Carnot's principle requires; for the amplitude of the vibration communicated to aether then remains the same, but owing to the shortening of the waves, the velocity in this vibration is changed, and therefore the volume density of vibratory energy in the aether is modified as above. And the Lorentz transformation has shown us what is not so immediately obvious, that also on the electric view which considers the sources to be constituted of vibrating electrons, though their relative motions are not affected by the uniform translation as again Carnot's principle demands, yet the vibratory energy emitted from them is modified in the manner here described.

*Final result demanded by Carnot's principle.*

(*Note added Dec.* 26. As the intensity of the pressure of radiation depends on the instantaneous state of the adjacent medium, it may be expected to remain equal to the energy per unit volume, as above assumed, whether the body that it acts on is at rest or in motion.

*Law of pressure as determined by local state of the medium:*

We may verify in detail for a plane-polarized wave-train with electric force $(0, Q, 0)$, current $(0, v, 0)$ and magnetic force $(0, 0, \gamma)$, incident directly on an absorbing face perpendicular to $x$. Then[1] the mechanical force in the absorber per unit volume is

$$X = \left(v - \frac{dg}{dt}\right)\gamma,$$

where

$$4\pi v = -\frac{d\gamma}{dx}, \quad \frac{dQ}{dx} = -\mu\left(\frac{d}{dt} + v\frac{d}{dx}\right)\gamma$$

and

$$g = \frac{Q}{4\pi c^2},$$

[1] *Aether and Matter*, § 65.

$v$ being the velocity of the material medium with which the axes of coordinates travel. Thus

$$\int_{x_1}^{x_2} X dx = - \left| \frac{\gamma^2}{8\pi} \right|_{x_1}^{x_2} - \frac{1}{4\pi c^2} \int_{x_1}^{x_2} \gamma \frac{dQ}{dt}\, dx.$$

Let the slice between $x_1$ and $x_2$ be an indefinitely thin one containing the absorbing interface; as $Q$ is continuous across it, $dQ/dt$ is very small outside it; thus $\gamma$ being finite, the last term is negligible, and the mechanical force acting on the slice is equal to the value of $\gamma^2/8\pi$, just outside it where $Q$ is null; thus it is equal to the energy density just outside, whether the absorber is in motion or not.  verified.

From the way of considering the origin of this mechanical force above, as acting on the interfacial current sheet, it is not difficult to verify that when the incidence is oblique, the incident, reflected, and refracted wave-trains exert independently on the reflecting surface their full oblique thrusts in their own directions of propagation, as is implied in Professor Poynting's calculations referred to at the beginning.  The thrusts<br>are along the<br>rays:

The result here verified, that motion of a material body does not affect the pressure exerted on it by the ambient radiation, has been rejected by Professor Poynting* in a later postscript added to the memoir above referred to, on the ground that radiation escaping from a radiator $A$ into a moving absorber $B$ would, according to it, alter the store of momentum of the two bodies.  inapplicable<br>to a convected<br>radiator?

It may be noticed, in connection with p. 272 *supra*, that for the same amplitude of ionic excursions in the vibrating molecule, as determined by its maximum electric moment, and for the same periodic time, it follows from Hertz's formulae for a simple radiator, and may be generalized by the theory of dimensions, that the radiation emitted per unit time is proportional to the refractive index of the surrounding medium, and therefore the equilibrium density of the radiation in that medium is proportional to the square of the same index, in accordance with Balfour Stewart's law derived from the doctrine of equilibrium of exchanges between sources at uniform temperature.)  The law of<br>dependence<br>of free radia-<br>tion on re-<br>fractive index<br>of the<br>medium.

* This subject is disentangled, and Professor Poynting so far justified, in a note added to the posthumous *Collected Papers* (1920), p. 754, now reprinted *infra*. The system is convected with velocity $v$, and $Q$ is the electric force in it: but the aethereal force which induces $g$ is $Q'$ equal to $Q + v\gamma$, not $Q$ as above: this adds a term to the stress, into which $v$ now enters.

Both the momentum and the energy of radiation are relative to the frame of reference. It is only the Action that is invariant for uniform translatory convection, involving an invariant stress-momentum-energy field tensor in which the atoms are mechanically imbedded.

# 62

## ON THE ASCERTAINED ABSENCE OF EFFECTS OF MOTION THROUGH THE AETHER, IN RELATION TO THE CONSTITUTION OF MATTER, AND ON THE FITZGERALD-LORENTZ HYPOTHESIS.

[*Proceedings of the Physical Society of London*, Vol. XIX.; also *Phil. Mag.* June 1904.]

IN a recent paper by Professor D. B. Brace (*Phil. Mag.* April 1904, p. 318) the author removes by very refined experimenting all trace of doubt from Lord Rayleigh's conclusion that motion of transparent solids through aether does not induce any double refraction, even to the second order of the ratio of the velocity of the translation to that of radiation; but he infers from this the non-existence of the second-order deformation of the solid due to its translation, suggested by FitzGerald and by H. A. Lorentz to account for Michelson's earlier demonstrated absence of effect on optical interferences over long paths in free aether. As he remarks, it had previously been suggested by Lord Rayleigh that such an inference might possibly follow from this result. The object of this note is to explain that the inference in question is the opposite to that which I still hold to be the natural result of the theory of the motion of molecular aggregates through aether, as hitherto developed[1].

*Misinterpretation.*

The argument of Professor Brace proceeds on the basis that the *whole* effect of the convection through the aether is to *introduce new forces* between the molecules, causing the shrinkage aforesaid along the direction of convection; and it can be readily granted that if this were all, double refraction must result. But both the line of argument suggested as probable by Lorentz[2], and the molecular analysis offered by me some years later[3], proceed by comparing a system shrunk in the FitzGerald-Lorentz manner and convected through the aether, with the *same system* unshrunk and at rest, and finding a complete correspondence between them as regards the states and activities of

*Not an affair of forces.*

*Material identity in the two frames:*

[1] *Aether and Matter*, Camb. Univ. Press, 1900, Chapter XI. [Compare the cognate discussions by Professor Lorentz, *Theory of Electrons* (1906), §§ 169, 177, 186, building on his exact correspondence for the field in *Proc. Amsterdam Acad.* Dec. 1904.]

[2] *Versuch einer Theorie...*, 1895, §§ 91–2, translated in part in *Aether and Matter*, p. 186.

[3] *Loc. cit.*

the individual molecules. As the argument is somewhat complex and has been misunderstood, a brief re-statement of the result may prove useful.

We are to compare the field of physical activity of a system of molecules at rest, with the field of the identically same configuration of molecules in uniform translatory motion through aether. If small quantities of the order of the square of the ratio of the velocity of convection to that of radiation ($v/c$) are neglected, the Maxwellian physical equations for the second system, referred of course to axes of coordinates moving along with it, can be reduced to the form belonging to the same system at rest, by the transformation first developed by Lorentz: namely, each point in space is to have its own origin from which time is measured, its "local time" in Lorentz's phraseology, and then the values of the electric and magnetic vectors

$$(f, g, h) \quad \text{and} \quad (a, b, c),$$

at all points in the aether between the molecules in the system at rest, are the same as those of the vectors

$$\left( f, \ g - \frac{v}{4\pi c^2} a, \ h + \frac{v}{4\pi c^2} b \right) \quad \text{and} \quad (a, \ b + 4\pi v h, \ c - 4\pi v g)$$

at the corresponding points in the convected system at the same local times. This correspondence can, in fact, be shown to locate the electrons at corresponding points in the two systems, and to make them equal; if, then, they are held in rigid connection, or more generally if their states of orbital motion in the molecules are conserved, the effect of translatory motion of the system with velocity $v$ is to transform the aethereal field around them and between them as here specified. The fields of aethereal activity are *not identical*, but where one vanishes at any point so does the other at the same point. This conclusion was reached by Lorentz, who pointed out that it carried with it a null result for all recognizable optical tests of convection in the system, up to the first order, with the one exception of the Doppler effect which is involved in the "local" time measurements, and which is only a partial exception because it refers to radiation coming from outside the system.

Does, however, the system of electrons need to be constrained in order to prevent change of configuration when being convected? The force acting on an individual electron $e$ is thereby changed from

$$4\pi c^2 e \left( f, \ g - \frac{v}{4\pi c^2} c, \ h + \frac{v}{4\pi c^2} b \right) \quad \text{to} \quad 4\pi c^2 e \ (f, g, h).$$

If there is a magnetic field $(a, b, c)$ there will thus be alteration: if there is no sensible average magnetic field, even among the molecules, we may perhaps fairly assume, with Lorentz, that no constraint is

*[margin note: affected however by local times.]*

*[margin note: The fields correlated.]*

*[margin note: The effect of the local time.]*

needed in order to prevent change in molecular configuration in the
The constitu-
tive moduli. system due to convection. Anyhow, the absence of recognizable
optical result to the first order is certain, as the physical constants
of the system in bulk must be unaltered to that order.

But the brilliant experimenting of Michelson and Morley had
A second-
order cor-
relation
demanded: already led to the recognition of absence of optical result up to the
second order of the ratio of the velocities. Thus the question was
suggested whether the above correspondence between the resting
and convected systems can be effectively extended up to the second
order. It is, in fact, found that the Maxwellian circuital equations
is secured by
shrinkages
in the frame
in space and
time: of aethereal activity, in the ambient aether, referred to axes moving
along with the uniform velocity of convection $v$, can be reduced to
the same form as for axes at rest, up to and including $(v/c)^2$, but not
$(v/c)^3$, by adopting a local time $\epsilon^{-\frac{1}{2}} (t - vx/c^2)$ as before, but with a
new unit $\epsilon^{-\frac{1}{2}}$, and also a reduced unit of length parallel to $x$ equal
to $\epsilon^{-\frac{1}{2}}$, where here and in what follows $\epsilon$ represents $1 + v^2/c^2$, the
the material
system being
unchanged, units of length along $y$ and $z$ remaining unaltered. It is found that
for two aether fields, one referred to fixed axes and the other to
moving axes, standing in this mutual correlation, the electrons, or
poles, in approaching which the aethereal electric vector becomes
infinite as $er^{-1}$, are situated at corresponding points and are of equal
values: the relation, exact to the second order, is now that

and the fields
correlated.
$$(f, g, h) \quad \text{and} \quad (a, b, c)$$

in the field belonging to the fixed system of poles correspond to

$$\epsilon^{\frac{1}{2}} \left( \epsilon^{-\frac{1}{2}} f, \, g - \frac{v}{4\pi c^2} c, \, h + \frac{v}{4\pi c^2} b \right)$$

and
$$\epsilon^{\frac{1}{2}} \left( \epsilon^{-\frac{1}{2}} a, \, b + 4\pi vh, \, c - 4\pi vg \right)$$

for the field belonging to the convected system; where $\epsilon$ is $1 + v^2/c^2$,
as above, the factor $\epsilon^{\frac{1}{2}}$ being needed to make corresponding poles
equal in value instead of merely proportional*.

If each pole or electron is connected with a molecule possessing
extraneous mass, and it may be having an extraneous field of gravitational and other force of its own, and thereby interacting with other
molecules, we shall want to know the forces exerted on that molecule

---

* The observation by Lorentz (Dec. 1904) that this transformation is exact
as regards the field, not merely true to the second order, where $\epsilon$ is $(1 - v^2/c^2)^{-1}$,
is the foundation of modern theories of relativity. The influence of the size
of the electron relative to the atom, however, comes in with the third order,
and the validity for atomic phenomena has to be assumed not proved: which
was the reason for not probing beyond the second order in the present discussion
of date previous to 1900.

by the surrounding aether, in order to form its own equations of
motion, which must be combined with those of the aether field
around it in order to constitute a complete system. But if such other
forces are molecularly insignificant, or better, if the electron is a
mere passive pole—nucleus of beknottedness in some way—in the
aether, conditioned and controlled entirely by the aether around it,
just as a vortex ring is conditioned by the fluid in which it subsists
and is also carried along thereby, then, as in the familiar hydrodynamics of vortices, the motion of the aether determines the motion
of the entirely passive electrons, and the idea of force acting between
them and the aether is dispensed with.

*The aether wholly controls locally its passive electrons: as mobile knots, or like vortices in a fluid. The idea of force thus transcended.*

If, then, matter is for physical purposes a purely aethereal system,
if it is constituted of simple polar singularities or electrons, positive
and negative, in the Maxwellian aether, the nuclei of which may be
either practically points or else small regions of aether with internal
connections of pure constraint, the propositions above stated for the
first order are extended to the second order of $v/c$, with the single
addition of the FitzGerald-Lorentz shrinkage in the scale of space,
and an equal one in the scale of time, which, being isotropic, is
unrecognizable.

*Result valid on the hypothesis that matter is an electric structure: nature of electrons.*

On such a theory as this the criticism presents itself, and was in
fact at once made*, that one hypothesis is needed to annul optical
effects to the first order; that when these were found to be actually
null to the second order another hypothesis had to be added; and
that another hypothesis would be required for the third order, while
in fact there was no reason to believe that they were not exactly
null to all orders. Such a train of remarks indicates that the
nature of the hypotheses has been overlooked. And if indeed it
could be proved that the optical effect is null up to the third order,
that circumstance would not demolish the theory, but would rather
point to some finer adjustments† than it provides for: needless to
say the attempt would indefinitely transcend existing experimental
possibilities.

*Destructive criticism of the correspondence: unfounded.*

As, then, the theory contains no further power of immediate
adaptation, what are the hypotheses on which it rests, and how far
are they gratuitous hypotheses introduced for this purpose alone?
Up to the first order the electron hypothesis, that electricity is
atomic, suffices by itself, as Lorentz was the first to show. Yet, even
if the nature of the particles of the cathode discharge had never
been made out, and the Zeeman effect had never been discovered,
the facts known to Ampère and Faraday were sufficient to *demonstrate*

*To first order, theory is absolutely secure.*

---

* By Poincaré, as often since quoted to its detriment.
† Showing themselves in the structure of the electron as an extended system.

that no other conception of electricity than the atomic one is logically self-consistent[1].

*To second order, matter has to be electric:*

Up to the second order the hypothesis that matter is constituted electrically—of electrons—is required in addition. For this there is no independent evidence except perhaps the general simplicity of the correlations of physical law. The circumstance that positive electrons

*including the positive ions and their inertia.*

have not yet been isolated naturally counts considerably on the other side; yet the theory puts no limit to the size and inertia and complexity of an electron, it only prescribes that it must be a collocation of aether poles connected together by some sort of pure constraint, but with no extraneous activities.

Any rival theory must on the threshold give an account of the Michelson null optical result, of Trouton's null electric result for convection of a charged condenser[2], and of Rayleigh's absence of double refraction now rendered thoroughly secure by Brace[3].

*The aether fundamental in physical activities.*

As electrons are already held to be a reality on various grounds, theoretical and experimental, it would appear therefore that there is much to be said for a benevolent attitude to the proposition that all the interactions of matter, so far as the laws of physics and chemistry extend, are to be described as phenomena occurring in and through the aether, and thus differentiated from the more recondite world of vital growth and change which they make manifest to our senses. This principle does not yet, so far as one can see, stand in the way of any other branch of physical science, while it accounts for the very remarkable absence of influence of the Earth's motion through space on the most sensitive phenomena, and is almost led up to thereby.

*Maxwellian scheme resumed.*

[*Sept.* 1926. The argument from correspondence, when reduced to the bare essentials, seems to be expressible as follows. Following Maxwell, an electrodynamic nuclear system has a field in free space. The forces transmitted to the system and acting on it are to be derived from the aggregate field determined by this system and the systems affecting it, by constructing after Maxwell but with correction, a stress tensor and a distribution of momentum in this field, which may represent a reality in the free aether or may be a purely formal construct: that is, the force on the part of the system within any boundary is the resultant of this stress over the boundary diminished by what is needed for change of this aethereal momentum inside it.

[1] Cf. *Aether and Matter*, p. 337.

[2] *Phil. Trans.* 1903. [See FitzGerald's *Scientific Writings* (1902), p. 566, or *supra*, p. 225.]

[3] The null influence on optical rotation, observed by Rayleigh, counts here as a first-order effect.

How is this affected when the system is convected?  The field can be transformed into a new convected field, referred to shrunken scales of space and time, and certainly the electrons for this new field are in corresponding positions: and the field can be adjusted so that their strengths are not changed.  May we therefore go further and assert that *this modified electrodynamic field belongs to the same material system as before but now convected?*  That is the sole hypothesis.  If it is admitted, and no alternative could be simpler, the forces on the convected system are to be calculated through the stress and momentum, from this modified field.  When that is done (cf. *Math. Congress*, 1912, as *infra*) the only difference that presents itself is due to an increased inertia of the moving system.  All this is strictly Maxwellian transformation: there is no essential extension involved, unless it is a mere formal synthesis of all the frames of the permitted group.  Use has been made of the stress and momentum solely as an intermediary: that need not imply that they are anything more than formal. *Complete correspondence is identity.*

Whether the space in which the stars are set so obviously is to be regarded as a privileged one, and so belong to an aether, depends from this point of view on whether this postulate of relativity is asserted up to the second power of $v/c$ or absolutely.  What is known of atomic structure favours the first alternative: the second is pure hypothesis. *The space of the stars.*

In strict epistemology, the transformed system, in which each nucleus point corresponds to one nucleus in the original, cannot differ from that original if the mutual relations correspond.  For we can know only the relations between the permanent things, how they disturb one another: and here they bear no deciding feature of change except mere change of the frame in which they are expressed.  The recent schemes of relativity would in self-consistency stress this criterion of identity.  But if the transformation is of cyclic type instead of pure one to one, so that each nucleus transforms into a cyclic set of nuclei, it may be said to create a new world.] *Transformation into new universes.*

It is pertinent to the present subject to refer to Mr Sutherland's recent remarks (*Phil. Mag.* April, p. 406) on the magnetic effect of electric convection, in relation to the mysterious action of a dielectric varnish that has been announced by Crémieu and Pender.  The discrepancy in the conservation of energy, there described, applied to the domain of electric polarization, is too startling to have been overlooked by the current theory[1]; and accordingly closer considera-

---

[1] Cf. *Phil. Trans.* 1897, A, p. 248 [*supra*, p. 65], and *Aether and Matter*, 1900, Appendix A.

tion gets rid of the difficulty. When an electron $e$ is transferred in
an electric field from a place where the potential is $V_1$ to a place
where it is $V_2$, the force acting on it, being $e$ multiplied by the
gradient of $V$, does work equal to $e\,(V_1 - V_2)$. When, however, the
electron is embedded in a piece of dielectric matter which is so transferred, the force acting on the electron itself is diminished by the
presence of the surrounding polarized matter, and so the work done
on the electron is less than before: but now the electric polarization
induced by the electron in this surrounding matter is also acted on
by the electric field, and if we add the work done on it during the
movement, we shall get the same total work as before for the system
that is moved, and there will be no discrepancy to be otherwise
explained.

Convection of
an electron
embedded
in matter:

balance of
interactions.

# 63

## JOSIAH WILLARD GIBBS*. (1839–1903.)

[Obituary notice: from *Proceedings of the Royal Society*,
1904, pp. 1–17.]

JOSIAH WILLARD GIBBS was the son of Josiah Willard Gibbs (Y.C. 1809), the distinguished Professor of Sacred Literature in the University from 1822 to 1861, and of Mary Anna (Van Cleve) Gibbs. He was born in New Haven, Conn., on February 11, 1839, and died on April 28, 1903. He was prepared for college at the Hopkins Grammar School, New Haven, and entered the class July 24, 1854. In his College course he won the Berkeley Premium for Latin Composition; 1857, Bristed Scholarship; 3rd Prize Latin Examination, 2nd term Junior year; Berkeley Premium for Latin Composition; 1858, 1st De Forest Mathematical Prize; Clark Scholarship; Latin Oration.

He occupied the first five years after graduation in 1858 in mathematical and other studies in New Haven. In the autumn of 1863 he became tutor in Yale, and was engaged with the duties of that position until August, 1866, when he went to Europe. Biographical

The winter of 1866–67 he spent in Paris, and the winter of 1867–68 and the following summer, in Berlin, studying especially physics, but devoting a part of his time to mathematics. The winter of 1868–69 he passed in Heidelberg, and the next spring in France, reaching home in June, 1869. In July, 1871, he was elected Professor of Mathematical Physics in Yale.

This synopsis of the early life of the distinguished American Natural Philosopher, whose decease occurred on April 28, 1903, is quoted from the *Yale Alumni Weekly*, of May 6[1]. Not the least of the lessons to be learned from the careers of distinguished men is conveyed by a recital of their early education and activities. It has been remarked that the present case is one of many in which some training in literary studies has helped to mould and brace a mind, destined to contribute materially to the unravelment of the simple fundamental principles that regulate the complex phenomena of Nature.

The prevailing interests revealed in Gibbs' work are those of a mathematician; though his facility in algebra was perhaps slight, and he was most successful when casting his arguments into graphical form. His mind was always straining towards complete general His powers geometrical:

* Cf. the Appendix on thermodynamic principles, at the end of this volume.
[1] Reference should also be made to the very interesting account by his colleague Professor H. A. Bumstead, in the *American Journal of Science*, Vol. XVI, pp. 187–202, which appeared after the present notice was completed [: now republished as introduction to *The Scientific Works*].

views. His direct geometrical or graphical bent is shown by the attraction which vectorial modes of notation in physical analysis exerted over him, as they had done in a more moderate degree over Maxwell, the interpreter of Faraday in this respect; his generalizing tendency was illustrated in the formal address, in which he expounded to the American Association the fascinations of the mathematical notations and operations appropriate to this subject, where he could not reach finality until his treatment had got into $n$ dimen-

*bent towards generalization.* sions of space. This bent towards exhaustive survey of his subject probably served Gibbs in good stead, by driving him to mathematical completeness in his exposition of thermodynamics, where others would have stopped short with the fragment of the theory that covered the physical applications then prominent or likely to arise. But his tendency to wind up the exposition and regard the account as closed when the logical fabric has been welded together, and to assign a subsidiary place to the details of such particular physical illustrations as then existed—from restraint, be it noted, not from lack of knowledge—retarded for many years the application of his methods by experimenters, to whom the behaviour of actual things is of more interest than the perfection of an abstract formulation of their relations.

### History of Thermodynamic Science.

The achievement by which Professor Gibbs will chiefly be remembered is his development, into their full scope, of the fundamental principles which regulate equilibrium and the trend of transformation in inanimate matter in bulk, that is, in general chemical and physical phenomena.

*Conserved mechanical work:* The laws of mechanical equilibrium of material systems, when constitution and physical state remain unaltered, have received attention ever since the ancient beginnings of the science of Statics. Their coordination was fruitfully considered by Galileo, who fixed his attention on the indications then current of a general law of compensation in mechanical transformations, revealing itself in the rule that what is gained in force is usually lost in the distance over which it operates. This principle of conservation of work, for all

*its final formulation by Newton:* possible virtual displacements, in balanced frictionless systems, was afterwards briefly noted in a form independent of special systems, by Newton; it appeared as a generalization of the third of his *Axiomata*, or fundamental unresolvable laws of inertia and motion, which asserted that to every intrinsic dynamical action of one part of a system on another there is an equivalent (or compensating) reaction of the second part on the first. After this idea of compensation in mechanical work had become more familiar, in connection with special problems of increasing complexity, during the succeeding century, the thread of Newton's brief generalization was again picked up by Lagrange, who, in the *Mécanique Analytique*, made this principle of work the unique foundation of all statics—that is, of the dynamics of steady systems—thus, as he expressed it, eliminating from the general principles of the science all accidental considerations peculiar to special types of mechanism. In the wider ramifications of modern abstract dynamics, in which, following the lead of Kelvin, Maxwell, and Helmholtz, its generalized principles are now extended to the elucidation of electrical and other recondite physical manifestations, the query is often put to specify the reaction, to show that Newton's Third Law of Motion is not violated; and the convenient answer is to appeal to Newton's own *Scholium*, which practically asserts that, wherever the analysis is based on an energy function, the compensation of action and reaction, considered as work of intrinsic forces, is secured in advance; and we may thus claim to have obtained permanent footing without the necessity of exploration of the concealed working of the system in order to trace out the exact mode of occurrence of this compensation.  *the Lagrangian virtual work. Newton's principle of reaction: may be latent.*

The general consideration of the motions generated in unbalanced mechanical systems—a subject the details of which Newton expressly excluded from the *Principia*—led directly, in the hands of Lebniz, Huygens, the Bernoullis, and others, to the recognition that, in the absence of frictional (*i.e.* irreversible) resistances, work that went uncompensated reappeared as *vis viva* of the motion in corresponding amount; so that, by including the kinetic energy, the principle of compensation or conservation still maintains its validity, unless friction is present. We have Helmholtz's own statement that it was early familiarity with these mathematical investigations of the previous century* that prompted him to extend the law of conservation into a far-reaching conspectus of all physical phenomena, in his famous essay on the "Conservation of Force" of 1847—waste energy being assumed to take the unorganized kinetic or thermal form—about the same time that the irrefragable evidence for such a universal principle of transmutation of energy had been supplied, unknown to him, by the experiments of Joule.  *Conservation of energy: historical: thermal waste.*

The whole of the statics of reversible (*i.e.* frictionless) mechanical systems had thus been condensed into the law that the states of equilibrium are defined by the energy being stationary, the criterion of stability being that it is a minimum. A corresponding generalization was now required for the wider science in which chemical and  *The energy foundation for statics:*

* A special place belongs to the development contained in a manuscript of Cavendish, only recently published: *Scientific Papers*, Vol. II, pp. 415–430.

other non-mechanical sources of power are in operation. Here again the theory will be exhaustive only in its application to reversible types of change. In fact, if the complex of possible transformation

*extended to reversible thermo-dynamic systems.* is to be reducible to a theory involving analytical functions, reversibility must be an essential feature; without it the courses of individual transformations may be traced, but their features cannot be interlaced into a scheme of relations. The essay of Sadi Carnot, "Sur la

*Engine cycles developed by Carnot:* Puissance Motrice de Feu," of date 1824, had already pointed out the way, by abstract reasoning based on the idea of complete cycles of processes such as all mechanical engines execute. Such cycles were, in fact, afterwards appealed to in mechanical statics, by the originators of the doctrine of conservation of energy, to establish the existence of a definite energy function, on the ground that otherwise work could be obtained in unlimited amount out of nothing. In the hands of Carnot, so far back as 1824, they had been already employed in a wider scope to prove that all reversible heat engines are mechanically equivalent if they work between the same temperatures, on the ground that otherwise work in unlimited amount would be derivable—as he

*and enforced long after by W. Thomson.* then thought—from nothing. In 1848, W. Thomson, whose notice was attracted to this theory by Clapeyron's graphical exposition of a part of it, seized upon it as affording a new and purely dynamical conception of the notion of temperature—hitherto unconnected with other physical ideas—as a function specifying the capacity of heat for doing work. In the following year he published an exposition of Carnot's doctrine, still reposing, provisionally, on the dogma of the

*Difficulties in thermal theory:* indestructibility of heat; he then professed himself at a loss[1] to reconcile Carnot's argument with Joule's doctrine of the transmutation of heat into work, so much so as to lead to an unguarded expression of his belief that no case of this had yet been made out, though further experiment was urgently necessary. The difficulties that presented themselves to the pioneers in this new realm are necessarily hard to appreciate by us, to whom so much has, through their labours, become obvious. Yet one weighty circumstance may be recalled.

*did not delay applications.* Before Clausius' attention was attracted to the subject, the doctrine of Carnot made, in the hands of James Thomson, the first of its long series of predictions in the entirely new domain of change of physical state[2]; and no delay ensued before his propositions were verified by

[1] In his memoir of Joule (*Manchester Memoirs*, 1892) Osborne Reynolds connects this difficulty with the recognition that something is lost when heat merely diffuses; he considers that it was overcome only when the principle of dissipation of energy clearly emerged, which asserted that while the energy is conserved its mechanical availability is in part destroyed.

[2] Clapeyron had already deduced the corresponding formula for change of boiling point from Carnot's principles.

his brother's experiments on the lowering of the freezing point of water by pressure. It is in Clausius' great memoir of February, 1850, written from a knowledge of Clapeyron and Thomson alone, that the reconciliation of Joule's and Carnot's principles is effected; so simple yet significant is it, merely the change of the words "derivable from nothing" above to "derivable from completely diffused heat," that in the reprint of James Thomson's memoir on the lowering of the freezing point, in 1851, in the *Camb. and Dub. Math. Journal*, only one sentence in the argument had to be altered. The doctrine of Carnot, including most of the results he derived from it, might in fact have been quite well developed in a form which would leave open the question as to Joule's principle. Carnot himself before his death in 1832, long previous to this time, had already advanced further; continued reflexion on the phenomena and properties of gases had impelled him to recognize the law of conservation in total energy, as appears very clearly in his manuscript remains published to the world in 1882.

*Elucidation.*

*Clausius.*

*Nature of heat can remain open: Carnot's further progress.*

The first suggestion that a dynamical scale of temperature more convenient than Thomson's original one would be gained by taking Carnot's function to be the reciprocal of the temperature, measured from absolute zero by the expansion of any ideal gas, came to him from Joule himself in 1848, who, in reasoning on this subject, presumably attended to his own principle[1]. In the procedure of Clausius' memoir, the value of Carnot's function was derived from the discussion of an ideal gas, including in its definition Joule's property of absence of internal work, or rather the indications thereof deducible from general considerations and from Regnault's results. It was shown to be the reciprocal of absolute temperature as measured by this ideal body as thermometric substance; and he thence advanced at once to the fundamental simple formula for the efficiency with finite range of temperature. But the precise practical determination of this ideal absolute scale in terms of practical scales based on known actual thermometric substances was a great effort of genius, entirely the work of Joule and Thomson, which amply atoned for Thomson's earlier slip.

*Temperature as absolute:*

*Joule's proposal:*

*fundamental scale determined by Thomson-Joule.*

The final general form toward which Carnot's principle was to crystallize was first adumbrated by W. Thomson, in 1852, in the statement that the trend of spontaneous change is towards the dissipation or irrecoverable diffusion of the sources of mechanical power. In 1854

*Principle of dissipation of conserved energy.*

---

[1] Lord Kelvin, under date February 19, 1852 (Joule, *Collected Papers*, Vol. 1, p. 353), describes Joule's letter to him of December 9, 1848, as giving this result "as the expression of Mayer's hypothesis in terms of the notation of my account of Carnot's theory."

Clausius contributed to placing the matter on a simple and quantitative basis by introducing the scalar quantity which he afterwards
called the entropy of a system; and in 1865[1] the law of trend of
changes took the final form that in spontaneous change in a system*
with given energy the entropy must tend always to increase. In this
form, viz., expressed by means of entropy, the proposition, quoted
from Clausius, is placed by Gibbs at the head of his great memoir,
as virtually constituting the culmination of the creative ideas in this
subject, what remained being the application of the principle, re-
stated by him in far more manageable form†, in the ramifications of
change of state in actual matter.

*Entropy discovered: by Clausius.*

The trend of Thomson's ideas had led him away towards other
problems, such as the cosmical results of his principle, and (1853) the
amount of mechanical energy that can be restored from an unequally
heated system of bodies in equalizing their temperatures. By taking
one of the bodies to be of infinite thermal capacity and of temperature
$\theta_0$, the energy so restorable would have been determined immediately
as $H - \theta_0 \int dH/\theta$; as it is necessarily the same for all reversible paths of
transformation, this formula leads directly to Clausius' fundamental
proposition about entropy.

*Thomsonian availability:*

*its relation to entropy.*

That function had, in fact, been introduced first of all, at a very
early period, by Rankine, who approached it through a special theory
of molecules and their atmospheres, and afterwards generalized the
argument, in his *Treatise on the Steam Engine*, into a chain of abstract
propositions, which have never been satisfactorily interpreted or
understood.

*Adumbrated by Rankine.*

This sketch of the history of the first stage of the development of
abstract thermodynamics is outlined, for the sake of comparison with
the next stage of progress. It had been marked by the activity of two
great minds working contemporaneously at the same problems;
questions of priority have naturally arisen; but we are now far
enough removed from the period to appreciate that the merit of each
is not interfered with by the work of the other.

Any want of balance in the account above given will be counteracted by an extract from Willard Gibbs' obituary notice (1889) of
Clausius, whom he seems to have regarded as in a special sense his

*Gibbs' obituary tribute to Clausius:*

---

[1] The passages may be readily found in Hirst's translation of Clausius'
earlier Memoirs, published by Van Voorst, 1867.

* In Clausius' formulation the system, which has to be isolated, is the
universe. Cf. the remarks of Planck, *Thermodynamik*, § 135.

† That is, entropy becomes at once a universal coordinate for the material
system, while Thomson's available energy was restricted primarily to uniform
temperature. The equation $A = E + \theta \partial A/\partial\theta$, often named after Helmholtz,
which connects available energy $A$ with total energy $E$ through temperature $\theta$,
was, however, obtained early by Thomron: cf. p. 292.

master; the extract is appropriate here, as conveying the attitude of mind of the writer towards the history of the science which he had himself reduced into its canonical form.

But it was with questions of quite another order of magnitude that his [Clausius'] name was destined to be associated. The fundamental questions concerning the relation of heat to mechanical effect, which had been raised by Rumford, Carnot, and others, to meet with little response, were now everywhere pressing to the front. "For more than twelve years," said Regnault, in 1853, "I have been engaged in historical: collecting the materials for the solution of this question: Given a certain quantity of heat, what is, theoretically, the amount of mechanical effect which can be obtained by applying the heat to evaporation, or the expansion of elastic fluids, in the various circumstances which can be realized in practice?" The twenty-first volume of the *Memoirs of the Academy of Paris*, describing the first part of the magnificent series of researches which the liberality of the French Government enabled him to carry out for the solution of this question, was published in 1847. In the same year appeared Helmholtz's celebrated memoir, *Ueber die Erhaltung der Kraft*. For some years Joule had been making those experiments which were to associate his name with one of the fundamental laws of thermodynamics and one of the principal constants of Nature. In 1849 he made that determination of the mechanical equivalent of heat by the stirring of water, which for nearly thirty years remained the unquestioned standard. In 1848 and 1849, Sir William Thomson was engaged in developing the consequences of Carnot's theory of the motive power of heat; while Professor James Thomson, in demonstrating the effect of pressure on the freezing point of water by a Carnot's cycle, showed the flexibility and the fruitfulness of a mode of demonstration which was to become canonical in thermodynamics. Meantime, Rankine was attacking the problem in his own way, with one of those marvellous creations of the imagination, of which it is so difficult to estimate the precise value.

Such was the state of the question when Clausius published his first memoir on thermodynamics, *Ueber die bewegende Kraft der Wärme, und die Gesetze, welche sich daraus für die Wärmelehre selbst ableiten lassen*. This memoir marks an epoch in the history of physics. If we say, in the words used by Maxwell some years ago, that thermodynamics is "a science with secure foundations, clear definitions, and distinct boundaries," and ask when those foundations were laid, those definitions fixed, and those boundaries traced, there can be but one answer; certainly not before the publication of that memoir. The synthesis by Clausius: materials indeed existed for such a science, as Clausius showed by constructing it from such materials, substantially, as had for years been the common property of physicists. But truth and error were in a confusing state of mixture. Neither in France, nor in Germany, nor in Great Britain, can we find the answer to the question quoted from Regnault. The case was worse than this, for wrong answers were confidently urged by the highest authorities. That question was completely answered, on its theoretical side, in the memoir of Clausius, and the science of thermodynamics came into existence.

And, as Maxwell said in 1878*, so it might have been said at any time since the publication of that memoir, that the foundations of the science were secure, its definitions clear, and its boundaries distinct. The constructive power thus exhibited, this ability to bring order out of confusion, this breadth of view which could apprehend one truth without losing sight of another, this nice discrimination to separate truth from error—these are qualities which place the possessor in the first rank of scientific men.

his relation to W. Thomson.

In the development of the various consequences of the fundamental propositions of thermodynamics, as applied to all kinds of physical phenomena, Clausius was rivalled, perhaps surpassed, in activity and versatility by Sir William Thomson. His attention, indeed, seems to have been less directed toward the development of the subject in extension than toward the nature of the molecular phenomena of which the laws of thermodynamics are the sensible expression. He seems to have very early felt the conviction that behind the second law of thermodynamics, which relates to the heat absorbed or given out by the body, and therefore capable of direct measurement, there was another law of similar form, but relating to the quantities of heat (*i.e.* molecular *vis viva*), absorbed in the performance of work, external or internal.

(The laws of partition of molecular energy are then compared with Clausius' hypothesis of the Disgregation; cf. his 9th Memoir, § 14.)

### Generalized Thermodynamics.

The principle of dissipation of energy:

In its second stage, with which Gibbs is fundamentally associated, the science of thermodynamics widened out into the tracing of the consequences of applying the Carnot-Thomson-Clausius principle of dissipation to natural changes of all kinds: only such changes will be spontaneously possible as do not involve, so to speak, negative or reversed dissipation. From the beginning, the question of the range of the validity of this principle attracted attention.

its wide range.

Lord Kelvin, in his earliest exposition of the "Dissipation of Energy" (April, 1852), already feels justified in the assertion that "restoration of energy is probably never effected by means of organized matter, either endowed with vegetable life or subjected to the will of an animated creature." At a much later date (1882) Helmholtz still considers it an open question whether such a possibility could arise in the very delicate arrangements of the animal organization. Nowadays, probably, the presumption would be in favour of Lord Kelvin's view; violation of it would, in fact, necessitate the recognition of "vital force" in a very unambiguous and conspicuous form.

Living organisms.

The cosmical results of the dissipation of energy mainly occupied

* This challenging phrase occurs in a review of the progress of thermodynamics, of critical historical character, written in reference to Tait's *Thermodynamics*, in *Nature*, Vol. XVII, quoted here textually as in Maxwell's *Scientific Papers*, Vol. II, pp. 660–671.

Thomson's attention; he has nowhere, any more than Clausius, Cosmical applications. essayed the development of the theory into a doctrine of chemical and physical statics of matter in bulk. The first to emphasize in Physical. general terms, by aid of examples, the rich possibilities in this field, was Lord Rayleigh, in 1875, in a lecture on "The Dissipation of Rayleigh. Energy"[1]. Later in the same year he made the first quantitative application in this direction, by calculating the dissipation intrinsically involved in the mixture of different gases[2]. This discussion revealed various clues towards general procedure. The idea of reversible chemical absorption of one of the gases (of carbonic acid for example by quicklime) at once solves the main question proposed, though it is not stated whether this novel and direct mode of argument was suggested by the extreme simplicity of the result already obtained through more complex considerations. This investigation, expounded later by Maxwell[3] in terms of entropy, went far towards completing the doctrine of thermodynamics for interacting gaseous systems, and its results are naturally referred to by Gibbs in that connection.

The investigations of Gibbs began in 1873, with two papers on the Gibbs: graphical expression of thermodynamic relations, energy and entropy his graphical appearing explicitly as variables; of these papers, the later one scheme, based on entropy: described the well-known thermodynamic surface, afterwards constructed to scale by Maxwell for water substance, and gracefully presented by him to its discoverer. It would seem as if it was the study of these surfaces that directed Gibbs towards his general thermodynamic functions, whose minima at any given temperature determine the phases of stable equilibrium of a physical or chemical system at that temperature. Next appeared his great memoir, or leads to rather treatise, published in instalments in the *Transactions of the* analytic. *Connecticut Academy*, in 1875 and the three following years, in which this property of minimum in these functions is applied to the most general material systems—going, in his theoretical deductions, far ahead of any experimental procedure that was then contemplated. Thus, advancing beyond Rayleigh's postulated separator of gaseous Osmotics mixtures by reversible chemical absorption, he introduces into anticipated. physics, by a stroke of pure theory, the so-called semi-permeable membrane*, and the osmotic pressure against it, which has more recently played so fundamental a part in theory, and to some extent

[1] *Roy. Inst. Proc.* 1875; *Scientific Papers*, Vol. I, p. 238.
[2] *Phil. Mag.* Vol. XLIX, 1875, p. 311; *loc. cit.* Vol. I, p. 242.
[3] *Ency. Brit.* Article "Diffusion."
* The earliest artificial semi-permeable membrane, but of very different type, was perhaps the wire gauze of Sir Humphry Davy's safety lamp for miners.

in practice, as a mode of expression of the reversible energy relations between solutions.

The specification of a definite formula for the available energy in a thermodynamic system, in Lord Kelvin's original order of ideas, proved to be a subject not devoid of perplexity. It has been noticed above that the addition to the system of a condenser, or sink of heat, of unlimited capacity at a fixed temperature $\theta_0$, leads to a simple expression for the energy which can be available in this composite *Motivity relative:* system in the process of reversibly reducing all its heat to this temperature, namely $E - \theta_0\phi$, the quantity afterwards termed "motivity" by Lord Kelvin; and that this involves as a corollary that $\phi$, the entropy of Clausius, is a definite analytical function of the actual state of the system, which cannot spontaneously increase provided *entropy* the system is isolated. A hasty confusion between entropy and *absolute.* availability*, which occurred in the first edition of Maxwell's *Theory of Heat*, was corrected by Gibbs in his early memoir on the thermodynamic surface; and it is, perhaps, not unlikely that he was thereby led to reflect on the question as to the true measure of available energy of the system considered by itself.

If we adhere to the Kelvin order of ideas, we can reason as follows. The dissipation of energy in any material system must be relative to some standard state of the system, for with regard to the absolute zero of temperature all energy is mechanically available. Any group of states of the system, which are mutually convertible by reversible adiabatic processes, are on the same plane as regards dissipation, and can serve as equivalent standard states, those namely of equal entropy $\phi_0$. In any working process, for each infinitesimal amount of heat $\delta H$ or $\theta\delta\phi$, which is acquired in a specified state of the system at temperature $\theta$, let the portion $\delta H_0$ ultimately arrive at a selected *The Kelvin* standard state, say the one whose temperature is $\theta_0$; this process *Dissipation:* involves somehow waste, relative to this standard state†, of a part of this energy originally existing at temperature $\theta$, of amount

$$\delta H - \theta\delta H_0/\theta_0, \text{ that is } \theta\delta\phi - \theta\delta\phi_0 \text{ or } \delta(\theta\phi - \theta\phi_0) - (\phi - \phi_0)\delta\theta.$$

Thus the total unutilized energy in the given state, relative to this standard, is

$$\theta\phi - \theta\phi_0 - \int_{\theta_0}^{\theta} (\phi - \phi_0)\, d\theta,$$

where $\phi_0$ is constant. In this expression the integral is perfectly definite; in it $\phi$ is to be expressed as a function of $\theta$ and the constitu-

---

* Maxwell does not admit more than a misunderstanding.

† Kelvin's problem of the amount of thermal energy potentially available in an isolated system not at the same temperature throughout, is of course different.

tion of the system, and $\theta_0$ is the temperature at which with this constitution the value of $\phi$ is $\phi_0$. This dissipation is readily represented on the temperature-entropy diagram for a given constitution of the system.  graphical scheme.

For isothermal change, in which $\theta$ is not altered by the addition of the heat $\delta H$, the dissipation for an infinitesimal change is the increment of $\theta\phi - \theta\phi_0$; thus $E - \theta\phi + \theta\phi_0$, of which the last item is now a constant, serves as a function representing the mechanically available energy for changes of state conducted isothermally: its gradient is therefore downward in spontaneous change.  Isothermal free energy:

This last result may be reached more directly, following Planck's mode of exposition, by including in the system the surrounding medium in thermal equilibrium with it. The change of the total entropy is now $\delta\phi - \delta E/\theta$, for heat of amount $\delta E$ has been lost from the surroundings; this must be positive; thus as before it is $\phi - E/\theta$ that tends to a maximum in a system maintained at constant temperature.  alternatively.

But neither of these ways of arriving at the adaptation of the principle of dissipation of energy, or of maximum entropy, to use in the practical case of slow reaction proceeding at uniform temperature, is as direct or comprehensive as Gibbs' original analytical statement. As Maxwell remarked, the key to his advance on Kirchhoff and other writers who had previously treated some cases of physico-chemical change in more complex manner, was in definitely introducing the functions of thermodynamics $E$ and $\phi$ as generalized coordinates of the system. This involved the existence for the system of a unique characteristic relation of total differentials, which might well be called the equation of Gibbs; namely,  Gibbs' variables.

$$\delta E = \theta\delta\phi - p\delta v + \mu_1\delta m_1 + \mu_2\delta m_2 + \dots,$$  Characteristic equation.

where $\theta\delta\phi$ represents $\delta H$, which is not itself the differential of any analytical function, and the other terms represent non-thermal energy of expansion, of chemical change, and of other types, the coefficients $\theta, p, \mu_1, \dots$ being intensities which depend only on the state of the system, not on its mass*. The principle of Clausius is that, for self-contained systems in which $E$ does not vary, the trend of $\phi$ must be upwards in spontaneous change; thus at constant temperature the value of $- p\delta v + \mu_1\delta m_1 + \mu_2\delta m_2 + \dots$ must be negative; therefore at constant temperature the trend of $E - \theta\phi$ must be downwards, contrasting with the case of constant energy when that of $- \phi$ is downwards. The stable equilibria at constant temperature are at the

* This is for a fluid system for which the only mechanical force is pressure: cf. Appendix at end of this volume.

minima of this function. If there is the additional restriction that the reactions occur also under constant extraneous atmospheric pressure, so that $p$ is constant as well as $\theta$, it is the modified available energy $E - \theta\phi + pv$ that tends to a minimum; and a similar simplification obtains if any other force on the system is supposed to be maintained invariable. The variation of this part of the energy, $A$ or $E - \theta\phi$, is wholly available mechanically between states at the same tempera-ture $\theta$; for each temperature it constitutes a potential energy function of the forces, and was in fact afterwards named on that account the free energy by Helmholtz. When states of the system at different temperatures are compared, the availability must be referred to some standard entropy, as in the discussion above, and the idea becomes complex. Free energies at different temperatures must thus not be compared as available relatively to each other; but $\dfrac{\partial A}{\partial \theta}$ is

equal to the entropy $\phi$, and $\theta\delta\dfrac{\partial A}{\partial \theta}$ is equal to $\delta H$, so that the heat

capacities of the system are determined directly in terms of $A$. It had, indeed, been pointed out long before by Massieu, as Gibbs re-marked, that this expression $E - \theta\phi$ constituted a single function fully characteristic by itself of the system, over the range of purely physical changes with which thermodynamics was then concerned, in that all the thermal quantities belonging to the system are involved in it and derivable from it by differentiation.

An important development, in this new science of chemical ener-getics, was the discussion of the number of different states or phases that could exist alongside each other with given materials, as depend-ing on the number of constituent substances that are actually inter-acting; one result of this was the simple and invaluable phase rule, giving a limit to the number of independent substances actually present in terms of the number of phases that can coexist, and conversely. In fact, the great service of Gibbs to general chemistry was his definite marking out of the channels within which a scheme of reactions can proceed, by aid of a discussion of the relations of coexisting states or phases of the material, of so general and abstract a character that it could only be carried through effectively by his graphical regional representations, which have shown their power in the application of the method to special problems ever since.

In subsequent parts of the memoir the phenomena of interfacial films were considered in relation to the physical and chemical con-stitution of the films, leading for the first time to an insight into the causes of their permanence, which has since been expanded, mainly by Lord Rayleigh. Here his essential problem was to try to formulate

*Side notes:*
Available energy at constant temperature: dissipation at constant energy. Other modified functions.

Limitations.

Massieu's function.

Concomitant phases:

their number limited.

Practical importance.

Permanence of soap films.

conditions for interaction, such as would initiate new substances at the interface where two systems are in contact. Not to recur again to the complete establishment of osmotic principles almost before the phenomena had attracted quantitative attention, and to delicate mathematical applications to the effect of stress on chemical equilibrium of crystalline solids in their mother liquid, which make precise James Thomson's early ideas* but have hardly yet borne fruit, the memoir ends, in 1878, by the fundamental application to voltaic phenomena. The electromotive force in a reversible cell turns out to be the available energy, not the total energy, liberated per electrochemical equivalent of decomposition, and it is shown by actual cases that this distinction may play an essential part. One brilliant illustration may be quoted; it cannot matter to the available energy at the temperature of liquefaction whether a substance is solid or fused, for the transfer of the latent heat at the uniform temperature can produce no mechanical effect; accordingly electromotive forces should not alter abruptly owing to solidification or fusion, thus explaining a striking fact that had already been experimentally noted.

This electrical application of the theory had been developed independently, by a less general analysis but with extensive experimental confirmations, by Helmholtz in 1882, in ignorance of Gibbs' work. It was not till an even later time that the essential bearings of the theory on general chemistry were thrown into practical light and developed experimentally by van der Waals, van't Hoff, and the other great physical chemists of the Dutch school. In England the work was earlier known. Towards the end of his career (1876) Maxwell contributed an enthusiastic exposition to the Cambridge Philosophical Society, confined to the first part of the memoir, the only part then published, and conveying the opinion that by utilizing fully the fundamental concepts of energy and entropy, the author seemed to him "to throw a new light on thermodynamics"; and, in the early eighties, copies of the memoir were there highly valued, but procured with difficulty. In the abstract of Maxwell's discourse, published in *Proc. Camb. Phil. Soc.* Vol. IV, p. 427†, the theory is illustrated from F. Guthrie's published experiments on selective precipitation by the introduction of solid substances into a mixed solution, and by observations recently made, perhaps for this purpose, by P. T. Main, on the phases of the triple system composed of chloroform, alcohol,

*Marginal notes:* Emergence of new phases. Chemical influence of stress. Voltaic chemical cells: change by fusion inessential. Helmholtz: the Dutch school: Maxwell: at Cambridge, special applications.

---

* Also Thomson's ideas on coexistent phases of a simple substance such as water or carbonic acid, and its triple point, probably stimulated by the experimental investigations of his friend Andrews, seem to have been a stimulus to Gibbs' general theories.

† Reprinted in an Appendix to this volume.

and water; while the simple and natural exposition of the nature of catalytic action was singled out for remark.

It is characteristic of Professor Gibbs' extreme care for completeness and perfect elaboration, that probably the most interesting and instructive account in existence of this great memoir of over 300 pages *His own* is the abstract of 18 pages which he contributed himself to the *abstract.* *American Journal of Science* for December, 1878.

*Historical.* The nineteenth century will be remembered as much for the establishment of the dynamical theory of heat at the very foundation of general physics, as for the unravelment of the nature of radiation and of electricity, or the advance of molecular science. In the first of these subjects the name of Carnot has a place by itself; in the completion of its earlier physical stage the names of Joule and Clausius and Kelvin stand out by common consent; it is, perhaps, not too much to say that, by the final adaptation of its ideas to all reversible natural operations, the name of Gibbs takes a place alongside theirs.

### *Formal Development of Anisotropic Optical Theory.*

Afterwards Gibbs turned his attention to the electrical theory of light, then in the tentative stage of development, and published in 1882 three papers in which the electrical relations forming the foundation of Maxwell's theory were expounded on the most general formal *General* basis. The medium is taken to be heterogeneous (molecular) in its *electric* *formulation* smallest parts, but of an averaged homogeneous structure as regards *for optics:* elementary regions of dimensions comparable with a wave-length. General linear relations of a formal type between the Maxwellian vectors are assigned, involving the case of rotational media when they are not self-conjugate. The precise part of the electrical basis utilized was the universally admitted general type of formula (Neumann-Maxwell) for the kinetic energy of the (circuital) displacement currents in the field. The object of the papers was to point out how naturally the laws of optical reflexion, and of double refraction including its disperson, flow from the electrical ideas as contrasted with the mechanical theory of Cauchy and Green. Thus the analysis is in some respects open to the remark that the electrical foundation is refined and generalized until there is but little distinctively electrical that is left, except the frame; there remains a very general scheme of *its very* formal analytical relations, not unlike Hertz's later manner of con*abstract* ceiving the Maxwellian electrical equations, which has to become *quality,* more restricted and particularized for practical use. In a subsequent paper he contrasted Lord Kelvin's remarkable labile mechanical aether, then recently announced, with electric theory in the same general manner, again laying stress on the formal optical fitness of the system of equations which are the expression of the latter. In all this work we recognize the same penetration and skill, in the formulation and expression of the utmost generality of outlook, which he showed in pure mathematics by his partiality for the study of generalized algebras and vectorial analysis, and which in thermodynamics has largely constituted the strength of his work, though at the same time it has retarded its absorption into the general body of scientific doctrine.

is characteristic.

### Statistical Foundations of Thermodynamics.

After a period in which Professor Gibbs' work was much interrupted by ill health, he again appeared before the scientific world early in 1902 as the author of a notable treatise of 207 pages octavo, entitled *Elementary Principles in Statistical Mechanics, developed with especial reference to the Rational Foundation of Thermodynamics*, which was published in connection with the Bicentenary of the University of Yale. Having had a principal share in evolving the ultimate form of the principles that govern physical and chemical equilibrium, when matter in bulk is considered in terms of its observable properties alone, and having taken care to state them free of all vestige of molecular theory, it was natural that his thoughts should have turned to the less definite problems opened out by the molecular hypothesis of the constitution of matter, which provides a rational base for the axioms on which Carnot, Clausius, and Thomson originally built. While, however, Maxwell and Boltzmann, the creators of the subject here called statistical dynamics, had treated of molecules of matter directly, it is characteristic of Professor Gibbs that his exposition relates primarily to the statistics of a definite vast aggregate of ideal similar mechanical systems of types completely defined beforehand*, and then compares the precise results reached in this ideal discussion with the principles of thermodynamics, already ascertained in the semi-empirical manner. This reversal of order can only profitably be made, as he remarks, after the pioneers of statistical molecular theory have cleared the ground and defined the scope of the relations that are to be explored; but, nevertheless, he holds that the interests of precision invite such a paraphrase of their results.

Systematic statistical theory:

statistics of states of the system as a whole,

Moreover, we avoid the gravest difficulties when, giving up the attempt to frame hypotheses concerning the constitution of material bodies, we pursue statistical enquiries as a branch of rational mechanics. In the present state of science it seems hardly possible to frame a dynamic theory of molecular action which shall embrace

---

* The method originates, as he states, with Boltzmann (1871) and Maxwell (1879): see Appendix to this volume.

the phenomena of thermodynamics, of radiation, and of the electrical manifestations which accompany the union of atoms. Yet any theory is obviously inadequate which does not take account of all these

*evading structural hypotheses,*

phenomena....Certainly one is building on an insecure foundation who rests his work on hypotheses concerning the constitution of matter.

These remarks, and others of the same tenour, in the preface to the treatise above mentioned, coincide with a general tendency in scientific

*if that be desirable,*

exposition which is now prominent \*. But it may be doubted whether there are many who have pondered much on special physical theories, who have not perforce realized the vastness of Nature, so as not to require any reminder that they are merely following out analogies in one aspect of its immense but rational scheme, or improving in one direction our mental outlook on its operations. If we are dissuaded from framing any dynamic theory of molecular action, we shall certainly not progress in welding together the regions of phenomena which Professor Gibbs enumerates; while to establish, or at any rate trace, their inter-connections, the unattainable ideal of complete knowledge is not required, provided our hypotheses are held in a state of suspense so as not to force us permanently in a wrong direction†. And, moreover, recent indications hardly bear out the hopelessness of direct physical speculation in this very subject. The course of progress has rather been usually an evolution from the special simplified theory to the general formal scheme of relationship; an analogy with simple phenomena already understood suggests a special type of hypothesis—so to speak, a calculus adopting that theory as its notation or mode of expression; this is worked out and tested on the facts; it is thus improved and generalized by dropping

*except in the final stages.*

unessential restrictions—it may ultimately become strong and vivid enough to drop all analogies, and emerge as a purified scheme of

---

*Scope of the statistical method.*

\* It may perhaps be held that the power of the statistical method consists in its being able to conduct its operations with respect to known general laws of the system, such as the conservation of energy, averaging out the effects of other laws that are unknown. As such laws come to be discovered or recognized, the effect is not to invalidate the previous results, but to add others more special.

*An anthropomorphic setting not irrelevant.*

† The so-called Gibbs' paradox shows up the anthropomorphic element in available energy, the "homo mensura," which, as we are constituted relative to Nature, cannot be banished from science even if we would. For example, with Rayleigh, the fall of available energy by isothermal mixture of two gases is the same whatever be their natures, the same even if their molecules differ only infinitesimally from one another, though infinite time would then be required; yet there could be no such loss if they were absolutely identical. The paradox cannot be transcended except by probing deeper into Nature in the special case, and finding by closer tests, such as could utilize the fall of available energy of mixture, whether the individual molecules are really not identical.

relations embracing the sensible phenomena without requiring any forms of expression that are not directly inherent in themselves. Such a scheme of general relations is an improvement on the special analogical theory, provided it does not become too abstract and intangible; but experience may be held to suggest that essential relations restricting the excessive number of the variables in purely descriptive mathematical formulations of experience are, in the first instance, suggested through the analogies of simpler systems. This other point of view may in fact also be put in Professor Gibbs' own words relating to molecular theory, again quoting from his notice (1889) of the work of Clausius.

The origin of the kinetic theory of gases is lost in remote antiquity, and its completion the most sanguine cannot hope to see. But a single generation has seen it advance from the stage of vague surmises to an extensive and well-established body of doctrine. This is mainly the work of three men—Clausius, Maxwell, and Boltzmann, of which Clausius was the earliest in the field, and has been called by Maxwell the principal founder of the science. *Gas theory:*

In the meantime, Maxwell and Boltzmann had entered the field. Maxwell's first paper, "On the Motions and Collisions of perfectly Elastic Spheres," was characterized by a new manner of proposing the problems of molecular science. Clausius was concerned with the mean values of various quantities, which vary enormously in the smallest time or space which we can appreciate. Maxwell occupied himself with the relative frequency of the various values which these quantities have. In this he was followed by Boltzmann. In reading Clausius we seem to be reading mechanics; in reading Maxwell, and in much of Boltzmann's most valuable work, we seem rather to be reading in the theory of probabilities. There is no doubt that the larger manner in which Maxwell and Boltzmann proposed the problems of molecular science enabled them in some cases to get a more satisfactory and complete answer, even for those questions which do not at first sight seem to require so broad a treatment. *Maxwell's statistical method,* *which imparted precision.*

Boltzmann's first work, however (1866), *Ueber die mechanische Bedeutung des zweiten Hauptsatzes der Wärmetheorie,* was in a line in which no one had preceded him, although he was followed by some of the most distinguished names among his contemporaries. Somewhat later (1870), Clausius, whose attention had not been called to Boltzmann's work, wrote his paper, "Ueber die Zuruckführung des zweiten Hauptsatzes der mechanischen Wärmetheorie auf allgemeine mechanische Principien." The point of departure of these investigations, and others to which they gave rise, is the consideration of the mean values of the force-function and of the *vis viva* of a system in which the motions are periodic, and of the variations of these mean values when the external influences are changed. The theorems developed belong to the same general category as the principle of least action, and the principle or principles known as Hamilton's, which have to do, explicitly or implicitly, with the variations of these mean values.... *The method of Action described.*

Atomic
theory
essential for
practical
physics.

The first problem of molecular science is to derive from the observed properties of bodies as accurate a notion as possible of their molecular constitution. The knowledge we may gain of their molecular constitution may then be utilized in the search for formulas to represent their observable properties....

Personal.

Professor Gibbs, during his lifetime, was invited to honorary membership of most of the leading learned societies and academies of both hemispheres that pursue physics and mathematics. In particular, he became a Foreign Member of the Royal Society in 1897, and was awarded its crowning distinction, the Copley Medal, in 1901. His thermodynamic writings are accessible in German and in part in French; the curious fatality which has rendered them almost unprocurable, in the language in which they were written, seems happily to be about to cease, through the publication of a memorial edition of his works. We may apply to them his own reflexion on one of his peers:

Such work as that of Clausius is not measured by counting titles or pages. His true monument lies not on the shelves of libraries, but in the thoughts of men, and in the history of more than one science.

This notice may fittingly be brought to a close by another quotation, expressing the sentiments of the University in which Professor Gibbs passed his life.

It was not given to laymen to appreciate his services, but all who thought at all of what was being done at this University knew that the roll of Yale teachers was illuminated by a great name; that one of the men who passed in and out so quietly among his colleagues and his students, bearing in his face and forehead such unusual marks of the scholar, was familiarly known in his works wherever in the world the highest scholarship was pursued, and was frequently followed with admiration in new paths of learning. The very presence of such a man as Professor Gibbs is an asset to a University whose value is beyond measure as an influence upon its members of which they are often unconscious, but by which they are powerfully affected to their good.

# 64

## ON THE DYNAMICAL SIGNIFICANCE OF KUNDT'S LAW OF SELECTIVE DISPERSION, IN CONNECTION WITH THE TRANSMISSION OF THE ENERGY OF TRAINS OF DISPERSIVE WAVES[1].

[*Proceedings of the Cambridge Philosophical Society,*
Vol. XIII, pp. 21–24, 1904.]

1. This subject has been treated by the writer, *Phil. Trans.* 1897, A, p. 243 [*supra*, p. 57]. If we are content with a naïve theory of an independent vibrator for each line of the spectrum, or even with a conception of a molecule as a connected dynamical system vibrating about a position [configuration] *of rest*, it is readily inferred that the index of refraction always trends upward with increasing frequency, so as to be abnormally great on the lower side of an absorption band and abnormally small on the upper side of it, as was originally remarked by Kundt to be actually the case in anomalously dispersive media. But when the molecule is considered, as it must be, notwithstanding analytical difficulties, to be a dynamical system vibrating about a permanent state of steady cyclic motion, the gyrostatic terms in the equations of its vibrations render a theoretical discussion of Kundt's law difficult, and it was not then completely effected; though it was easy to see that the law should be connected with the necessarily positive quality of the vibratory energy. A recent paper by Professor Lamb, in *Proc. Math. Soc.*, consequent upon a remark by Professor Schuster, seems to afford a key to the matter.

*Kundt's law not demanded if the atoms are kinetic systems.*

When we speak in optics of wave-length $\lambda_0$, we can only mean light of wave-lengths comprehended within a small interval $\delta\lambda$ around $\lambda_0$—that is a train of undulations of wave-length uncertain from point to point within this interval $\delta\lambda$, which is itself a measure of the defect of purity of the beam. There is no such thing as an absolutely homogeneous train of wave-length exactly $\lambda_0$. Now Sir George Stokes has explained* (Smith's Prize Examination Questions, 1876) how in such an undamped train, nearly homogeneous, the velocity of propagation, considered as the rate at which the actual

*The velocity of a wave-group:*

---

[1] The first section of this note was communicated to the British Association at the meeting in Cambridge last August.

* The subject was considered long before, on an atomic foundation, by Hamilton, *Proc. R. Irish Acad.* 1839: also exhaustively by Rayleigh in 1877, *loc. cit. infra*, following O. Reynolds.

disturbance travels onwards, is in a dispersive medium not the wave-length $\lambda$ of the wave-form divided by its periodic time $\tau$, but is[*]

$$d \frac{1}{\tau} \Big/ d \frac{1}{\lambda}.$$

And Lord Rayleigh has verified (*Proc. Math. Soc.* 1877: "Theory of Sound," Vol. 1, Appendix) by a direct analysis of high generality that the energy of the undulations is in fact conveyed with this velocity[†]. Thus if the medium could be such that the frequency

is that of the diminished with diminishing wave-length, the energy would travel
energy: backwards, namely in a direction opposite to that in which its wave-form is propagated. The question considered by Schuster and Lamb is as to whether this is actually possible. Examples are given by Professor Lamb of infinitely extended trains in which it occurs. But such a train requires to be fed with energy, so to speak, at both

which cannot ends. With a train of light advancing across a dispersive medium
travel from one side only, the case cannot occur, for the energy cannot
backwards. become negative[1].

The curve of dispersion is usually constructed with the frequency $\nu$, equal to $\tau^{-1}$, as abscissa, and the index of refraction $\mu$ as ordinate. If $c$ is the velocity of light in a vacuum

$$\lambda = \frac{c}{\mu} \Big/ \nu.$$

Thus the group velocity $U$ is $c \, d\nu/d \, (\mu\nu)$. Hence

$$\frac{c}{U} = \frac{d \, (\mu\nu)}{d\nu} = \mu + \nu \frac{d\mu}{d\nu}.$$

This must be positive; thus $d\mu/d\nu$ can indeed be negative, but the greatest allowable negative value at any point is $-\mu/\nu$. So far then

[*] The simplest and sufficient illustration is that of two harmonic wave-trains of equal amplitudes superposed: then

$$y = A \cos (mx - nt) + A \cos \{(m + \delta m) \, x - (n + \delta n) \, t + a\}.$$
$$= 2A \cos (\tfrac{1}{2}\delta m . x - \tfrac{1}{2}\delta n . t + \tfrac{1}{2} a) \cos \{(m + \tfrac{1}{2}\delta m) \, x - (n + \tfrac{1}{2}\delta n) \, t + \tfrac{1}{2}a\}.$$

The combined result thus appears as a simple wave-train, but one of fluctuating amplitude expressed by the first factor. The groups of prominent waves, of large amplitude, thus travel with the velocity

$$\frac{dn}{dm}, \text{ or } d \frac{1}{\tau} \Big/ d \frac{1}{\lambda}, \text{ or } \nu - \lambda \frac{d\nu}{d\lambda}, \text{ as } \tau \text{ is } \frac{\lambda}{\nu},$$

where $\nu$ is the velocity of propagation of phase which is $n/m$.

It is to be observed that the length of one of the groups of waves is $\pi/\tfrac{1}{2}\delta m$. Thus, generally, when the wave-motion becomes more pure so that $\delta\lambda$ the range of wave-length is smaller, the groups become longer, without limit, the length being of the order of $\lambda/\delta\lambda$ wave-lengths.

[†] Cf. the *quantum* theories of L. de Broglie and E. Schrödinger, in which material mass is identified with such concentrated undulatory energy.

[1] This argument would be evaded if we could suppose that the wave-form travels backward: but in the case of a train which is not endless, the wave-form is of necessity propagated forwards, that is, in the direction in which the front of the train advances.

as this present condition goes, the dispersion curve *may* trend downwards, but to a limited extent, outside a band of absorption, whereas upward trend is unrestricted. The condition that a source of radiation must be emitting, not absorbing, energy thus allows the dispersion curve to be an undulating line in a region not containing absorption bands, but with a limit to the steepness of the downward steps of the curve. In a region of absorption this restriction is not imposed; hence the trend will usually be downward. Thus the character of the curve of dispersion would be in the main that remarked by Kundt, though the present consideration by itself does not require an upward trend in transparent regions to be invariable. Trend of the dispersion curve thus limited, in the main as observed.

2. The general analytical verification of the rate of propagation of energy by Lord Rayleigh, above mentioned, prompted by a more special discussion by Professor O. Reynolds, involves use of the proposition that the mean energy, in stationary or progressive vibratory systems, not affected by viscosity, is on the average half potential and half kinetic. The ultimate foundation of this important general principle seems to be adequately contained in the following remark, and more abstruse considerations are hardly necessary. The characteristic of all kinds of steady standing vibrations or undulations in a non-dissipative system, however complicated in its structure, is that each particle of the system describes an orbit in simple harmonic motion. For each particle therefore the two kinds of energy, kinetic and potential, are equal on the average; and this includes their equality in the aggregate. For uniform trains of progressive waves similar considerations remain applicable. The argument extends also to the most complex types of dispersive media, in cases where absorption is negligible: it is to be noticed that the energy that is "propagated" with the group velocity is there the energy in the aether, together with that of the vibratory disturbance of the stationary molecules—which latter is on the theory of cyclic systems of Kelvin and Routh a perfectly definite quantity added on to the intrinsic kinetic energy of each molecule. Reason for the balance of potential and kinetic energies. The energy that travels.

3. The principle involved in propagation by groups may be somewhat generalized, thus affording insight into its essential character. An equation Group velocity generalized:

$$y = Af(nx - pt)$$

represents a disturbance whose profile is of the form $y = Af(nx)$, propagated onward with uniform velocity $p/n$. When the form is a periodically undulating one the equation represents a progressive wave-train of definite wave-length. The problem is to specify the general features, if any, presented by an aggregate of related progressive forms, all differing but slightly from a central type

$y \propto f (n_0 x - p_0 t)$ but otherwise undetermined. We may represent one of these component forms by $\delta y = \delta A . f \{(n_0 + \delta n) x - (p_0 + \delta p) t\}$, where $\delta n$ and $\delta p$ are treated as infinitesimal. The definiteness of type of the forms under consideration implies that $n$ is given as some function of $p$, so that $\delta n / \delta p$ is here a function of $p_0$ and so constant, say $U_0$. The application of Taylor's theorem allows us in general to express the component form above mentioned as

$$\delta y = \delta A f (n_0 x - p_0 t) + (\delta n . x - \delta p . t) \, \delta A f' (n_0 x - p_0 t).$$

This expansion holds so long as $\delta n . x - \delta p . t$ is small: that is, it is valid for the neighbourhood of a point travelling with the definite velocity $\delta p / \delta n$, say $U_0$. Under this restriction, the aggregate of a crowd of such closely related progressive forms is therefore represented by an equation

two aspects of a crowd of similar pulses.

$$y = \phi (x - U_0 t) \, F (n_0 x - p_0 t).$$

Thus, to an observer travelling with velocity $U_0$, so that $x - U_0 t$ remains constant, the aggregate disturbance *in his neighbourhood* has the features of a simple disturbance of type $y = CF (n_0 x - p_0 t)$, namely of form $y = CF (n_0 x)$ travelling with velocity $p_0 / n_0$. This resultant form $y = CF (n_0 x)$ is not the same as the original standard form $y = Af (n_0 x)$, belonging to the components which make it up. But if the component form is represented by a sine or cosine, the resultant is also of that type; and an observer travelling with velocity $U_0$ then keeps in touch with a steady aggregate disturbance of the standard type, which travels as a group in company with him.

Local features progress through the pulses:

The generalized result is the proposition that the local features of a train formed of disturbances of nearly identical types travel with a definite velocity, different from the velocity of propagation of the individual waves, and therefore travel through the waves; but the

but generally not like them.

travelling group is of the same type as the individual disturbances, which in the aggregate give rise to it, only when these latter are simple harmonic trains represented by sines or cosines.

Thus, for example, we may for an instant imagine a radiation, the aggregate of an indefinitely great number of similar discrete radiant pulses shot out from the radiating molecules, each pulse being propagated without change of type with a velocity determined by the scale of its dimensions in space and time: in the aggregate they would present the appearance of disturbances travelling onward with the group velocity determined above, but there would be no similarity between these groups and the original type of pulse. This illustration is however purely ideal, for the dynamics of propagation require that it is only single harmonic wave-trains that can travel unchanged in a dispersive medium.

# 65

## NOTE ON THE MECHANICS OF THE ASCENT
## OF SAP IN TREES*.

[*Proceedings of the Royal Society*, B, Vol. LXXVI, pp. 460–463, 1905.]

THE following remarks, relating to one of the most powerful and universal of the mechanical operations of organic nature, are based mainly on the numerous experimental results reported in Dr A. J. Ewart's recent memoir[1]. Their chief object is to assert the view that we are not compelled to suppose the sap, in the column of vessels through which it rises, to be subject to the great actual pressure, amounting in high trees to many atmospheres, that is sometimes postulated. It is hardly necessary to remark that the problem of the rise of sap is one of mechanics, in so far as concerns the mode of the flow and the propelling power. {Very intense pressures not required:}

Contrary to the view above referred to, it seems not unreasonable to consider that the weight of the sap in each vessel is sustained in the main by the walls and base of that vessel, instead of being transmitted through its osmotically porous base to the vessels beneath it, and thus accumulated as hydrostatic pressure. {local support along the column:}

We could in fact imagine, diagrammatically (as happens in ordinary osmotic arrangements), a vertical column of vessels, each provided, say, with a short vertical side tube communicating with the open air, in which the pressure is adjusted from moment to moment, and yet such that the sap slowly travels by transpiration from each vessel to the one next above, through the porous partitions between them; provided there is an upward osmotic gradient, *i.e.* if the dissolved substances are maintained in greater concentration in the higher vessels[2]. This difference of density must be great enough, {like an osmotic column.}

---

\* For the botanical view, which apparently does not differ essentially, cf. Professor H. H. Dixon's book on transpiration (1914). The facts as to the paths of the circulation seem still to be obscure.

[1] *Roy. Proc. Soc.* Vol. LXXIV, p. 554; *Phil. Trans.* B, Vol. CXCVIII, p. 41.

[2] Thus in an ordinary osmotic experiment with a U-tube, the percolation of water through the plug gradually *produces* a difference of hydrostatic pressure on its two faces, which is *sustained* by the fixity of the plug itself, but would be at once neutralized if the plug were free to slide in the tube. This increase of volume of the salt solution, by the percolation of pure water into it, is on the van't Hoff analogy correlated with the free expansion of the molecules constituting a gas. It goes on with diminished speed under opposing pressure, until a definite neutralizing pressure is reached, inaptly called the osmotic pressure of the molecules of the solute, which just stops it, while higher pressures would reverse it. The stoppage is due to the establishment of a balance between the amounts of water percolating one way under osmotic {Partitions must be rigid.}

between adjacent vessels, to introduce osmotic pressure in excess of that required to balance the head of fluid in the length of the upper *with concen-* one, into which the water has to force its way. Thus, in comparing *tration of* vessels at different levels, the sap must be more concentrated in the *solute* *upwards,* upper ones by amounts corresponding to osmotic pressure more than counteracting the total head due to difference of levels, in order that it may be able to rise. As osmotic pressure is comparable with gaseous pressure for the same density of the molecules of the dissolved *but not great.* substance, the concentration required on this view is considerable, though not very great.

Such a steady gradient of concentration could apparently, on the whole, become self-adjusting, through assistance from the vital *Adjustment* stimuli of the plant; for concentration in the upper vessels is promoted *by vital* by evaporation. Yet pressures in excess or defect of the normal *stimuli:* atmospheric amount might at times accumulate locally, the latter *not perfect.* giving rise to the bubbles observed in the vessels, through release of dissolved gases.

*Experimental* It may be that this assumes too much concentration of dissolved *test.* material in the sap, *as it exists inside the vessels of the stem*, to agree *Capillary* with fact. In that case the capillary suction exerted from the nearest *suction from* leaf surface might be brought into requisition, after the manner *the leaf may* *help.* of Dixon and Joly, to assist in drawing off the excess of water from the vessels. The aim proposed in this note is not to explain how things happen, which is a matter for observation and experiment, but merely to support the position that nothing abnormal from the passive mechanical point of view need be involved in this or other vital phenomena.

As regards estimating the amount of flow, at first sight it may not appear obvious, *à priori*, that the transpiration through a porous *Osmotic flow* partition or membrane, due to osmotic gradient, is equal or even *as mechanical* *not molecular.* comparable in amount to what would be produced, with pure water, by a hydrostatic pressure head equal to the difference of the osmotic pressures on the two faces of the partition. But more exact consideration shows that on the contrary osmotic pressure is *defined* by this very equality[1]; it is that pressure difference which would produce such an opposite percolation of water as would just balance the direct percolation due to the osmotic attraction of the salt solution.

*The pressure* attraction, and the opposite way under hydrostatic pressure. The pressure *as reaction to* established, *e.g.* in an organic cell immersed in salt solution, is thus really *osmotic* *affinity, the* the reaction which is set up against the osmotic process. That process itself *molecular* is perhaps more directly and intelligibly described as the play of osmotic *effort to* affinity or attraction, even though it must be counted as of the same nature *uniformity.* as the affinity of a gas for a vacuum. Cf. *Proc. Camb. Phil. Soc.* Jan. 1897, or Whetham's *Theory of Solution*, p. 109.          [1] See preceding footnote.

It would, however, appear that the great resistance to flow offered by what botanists call Jamin tubes, viz. thin liquid columns containing and carrying along numerous broad air bubbles, is conditioned mainly by the viscosity of the fluid, and involves only indirectly the surface tension of the bubbles. In fact the resistance to flow may be expected to remain much the same if each bubble were replaced by a flat solid disc, nearly but not quite fitting the tube. Its high value arises from the circumstance that the mass of liquid between two discs moves on nearly as a solid block when the flow is steady, so that the viscous sliding has to take place in a thin layer close to the wall of the tube, and is on that account the more intense, and the friction against the tube the greater. The increased curvature of the upper capillary meniscus of the bubble is thus merely a gauge of the greater intensity of the viscous resistance instead of its cause; and modification of the surface tension cannot be involved as a propelling power. The experimental numbers given by Dr Ewart show that, even where the vessels are largely occupied by bubbles, the greater part of the resistance to active transpiration still resides in the partitions between them.

If the osmotic gradient, assisted possibly by capillary pull at the leaf orifices, is insufficient to direct a current of transpiration upward, *capillary* alterations inside the vessels, arising from vitally controlled emission and absorption of material from the walls, cannot be invoked to assist: rather it must be *osmotic* alterations from one vessel to the next, of, so to speak, a peristaltic character, that might thus come into play. But any such alteration (of either kind) will involve local supply of energy. Is there a sufficient fund of energy, latent in the stem, to provide permanently the motive power for the elevation of the sap? In what form could this energy get transported there? The energies of the plant economy come from the sunlight absorbed by the leaves. The natural view would appear to be that the work required to lift the sap is exerted at the place where the energy is received, and that it operates through extrusion of water by evaporative processes working against the osmotic attraction of the dissolved salts; while the maintenance of equilibrium along the vessels of the balanced osmotic column, with its semi-permeable partitions, demands that an equal amount of water must rise spontaneously to take the place of what is thus removed.

The subject might, perhaps, be further elucidated by observation of the manner in which the flow is first established at the beginning of the season, or possibly by experiments on the rate at which water would be absorbed by a wounded stem high above the ground.

# 66

# ON THE CONSTITUTION OF NATURAL RADIATION*.

[*Philosophical Magazine*, Nov. 1905.]

THE recent paper by Lord Rayleigh on "The Origin of the Prismatic Colours"[1] recurs to fundamental and delicate points in the philosophy of Optics, first effectively expounded by himself in 1881 and the following years[2], on which I desire to offer some observations; especially as the mode of exposition of the dispersive action of a prism which was adopted by me several years ago[3] has been the subject of criticism in the papers by Schuster and Ames, to which Lord Rayleigh refers at the beginning of his paper.

The first part of the following remarks would doubtless bear condensation, as the full force of Lord Rayleigh's comparison of the dispersion problem with that of a travelling maintained source had not been grasped when they were written; but in a subject which so largely turns on the mode of expression, condensation might involve obscurity. There does not, however, seem to be in what follows anything that stands in definite contradiction with Lord Rayleigh's published views†.

That there is room for still further precision of terminology in this subject is indeed suggested by the beginning of Lord Rayleigh's paper. It would seem that a definite choice can be made between the two modes of exposition, both of which he considered to be allowable. According to the first, "The assertion that Newton's experiments prove the colours to be already existent in white light is usually made in too unqualified a form." On a first impression this remark might be imagined to strike at the roots of all the various instrumental methods that have been elaborated for analysing complex radiations; for if the analysis brings out features that are not already existent in the radiation, two different methods of analysis (*e.g.* by a grating and by a prism) can hardly be expected to give concordant results.

*In what sense is the spectrum pre-existent in natural light?*

---

* See Appendix at the end of this volume.

[1] *Phil. Mag.* p. 401 (Oct. 1905).

[2] Cf. especially "Wave Theory," § 7, *Ency. Brit.* 1888; *Scientific Papers*, Vol. III.

[3] *Aether and Matter*, Chap. xv (1900).

† This, as the writer found in personal discussion, is rather doubtful. See his further remarks in *Phil. Mag.* 1905; *Scientific Papers*, Vol. v, pp. 279–282.

The alternative mode of exposition is to say that each complex type of radiation is constituted definitely of those colours (simple trains of various wave-lengths), into which the Fourier mathematical analysis would divide its vibration curve: and that various analysing instruments (gratings, prisms, etc.) are capable of revealing this constitution with different amounts of precision, the outstanding differences between these analyses being treated as due to imperfections of the instruments as regards the purpose in question. This (the usual) point of view is claimed as a valid alternative in Lord Rayleigh's third paragraph: to hold, as is done *infra*, that in some cases the resulting analysis is so imperfect as to be valueless, need not disturb the general validity of this point of view.

So far, the matter is one as to the most suitable mode of theoretical description or formulation. But we presently reach questions on which opinions may perhaps differ as to physical fact. In the spectral analysis of ordinary continuous radiation, the prism and the grating give consistent results, when well understood corrections and adaptations are applied before making the comparison. If these instruments are applied to radiation consisting of a system of sharp, entirely uncoordinated, discrete pulses, such as the Röntgen rays are [were] supposed to be, will this general agreement continue? It is clear that the grating (if of ideal perfectly reflecting quality) will draw out each pulse into a spectrum, and thus will analyse the radiation. It seems open to question whether a prism will not merely gradually dissipate it by scattering, however wide the pulses may be, even if they are of breadth comparable with the wave-lengths of visible light. If this be so, the prism is very badly suited for the analysis of this type of radiation, and no amount of adaptation of the result will bring the prism into conformity with the ideal grating.

The Fourier mathematical process, as also the ideal grating which reflects back the disturbance in *échelon* so to say, operates by simply selecting and piecing together elements existing in the original radiant disturbance, so as to isolate periodic wave-trains that on superposition would reproduce the form of the original vibration curve.

On the other hand, what the prism may do to a given isolated pulse would seem to depend on its own constitution. The customary mode of investigation would be to replace the pulse by the equivalent infinite system of component Fourier wave-trains, to find the effect produced on each of them by substituting its expression in the differential dynamical equations of the dispersive medium, and to add the results thus found. Though the original Fourier expansion of the pulse is always analytically legitimate and definite, it is not always allowable, without scrutiny as to convergency, thus to operate

*or produced by the analyser?*

*Test case of pulses.*

*Analysis as redistribution.*

*Nature of analysis by refraction:*

on its separate terms and add. Indeed, the particular component waves whose period is a free period of the medium would increase infinitely in importance in the result: thus it must be ascertained whether this infinity of intensity is more than compensated by infinite smallness of the element of period over which it ranges, before the procedure which includes it can be accepted as mathematically legitimate*. The insertion of a very small frictional term in the dynamical equation will, however, secure that the component vibration remains finite though great at this critical period, and the analysis then becomes entirely valid. But the problem is only shifted; it has now to be ascertained whether the limit which this solution approaches as the friction is reduced indefinitely is the same as the solution previously arrived at in the absolute absence of friction,— whether in fact there exists a definite limit. That there is in many cases no definite limit is merely another way of expressing the theory of anomalous or selective dispersion, in which the final steady result depends essentially on the magnitude of the small viscous term that must be introduced in order to evade infinities, while the mode of the gradual establishment of that result is likewise undetermined. The question thus arises, whether the proportion of the energy of an incident isolated pulse that goes into this selective vibration is capable of determination by operating analytically upon its Fourier analysis in this way—whether, in fact, a different line of attack would not be required in order to determine it. At any rate, such questions of mathematical validity can arise only in regard to the presence of selective or anomalous dispersion[1].

The argument of Professor Schuster (*Phil. Mag.* p. 6, Jan. 1904[2]) arrives at the conclusion that a single pulse is split up regularly into a spectrum by a prism. It appears to start from an implied hypothesis that even an abrupt pulse travels unchanged across the dispersive medium, with the velocity appropriate to a group of waves [*supra*, p. 300]. If for a pulse is substituted a train of waves with wave-

*Marginal notes:*
may cut out a range of periods altogether.

Pulses are double.

* Cf. *Aether and Matter*, § 162 (1900), on white light: also an address "On the Fourier analysis," *Proc. Lond. Math. Soc.* (1916) as *infra*, where it is explained how light must be of uniform structure if it is to be analysable by a Fourier integral.

[1] I fear that I have on previous occasions orally assigned to them a wider importance. It is only the fate of the constituent wave-trains that are near the free period that is undetermined. [The practical answer is that the molecules of the prism drink up and scatter this portion of the light: the calculation may be feasible.]

[2] The paper, as its title indicates, is concerned mainly with a brilliant application of groups of undulations to the instantaneous explanation of Fox Talbot's interference bands. In connection with § 6 it may be remarked that the dynamical relations require that a limited disturbance, travelling in a transparent medium, must consist of compensating positive and negative parts.

lengths variable within the narrow limits $\lambda$ and $\lambda + \delta\lambda$, so that the train is very nearly simple harmonic, this statement will be sensibly exact except near the beginning and end of the train: and Professor Schuster's representation of the emergent radiation, as consisting of groups of waves, most concentrated in the neighbourhood of surfaces which are oblique to the wave-fronts, then affords an instructive view of the process of dispersion, whether prismatic or diffractive. But if this argument is to be pressed so as to include a single sharp pulse, what value of $\lambda$ are we to take as applicable to it? The theorem of definite group velocity is demonstrated only for the compound disturbance arising from the superposition of simpler trains of some common type but with slightly differing parameters—the trains being unlimited and of simple harmonic type in the usual Stokes-Rayleigh theory [*supra* p. 302]. A single pulse will thus not have any definite group velocity with which it can travel; or, what comes to the same, the different parts of it will travel in the dispersive medium with widely different velocities, so that it will spread out and be dissipated. An argument which assigns a definite velocity to a complex disturbance can thus be applicable only to very special types of disturbance[1]: for them it must of necessity lead to the same law of prismatic dispersive power as holds by the same argument for a disturbance consisting of uniform trains of simple waves, if the average wave-length of the latter corresponds to this group velocity. *(marginal note: Graph of a dispersed wave-group.)*

As these considerations relating to the mode of propagation of pulses apply to both Professor Schuster's and Professor Ames' arguments, it will not be superfluous to fortify them by the following quotation from Lord Kelvin[2], written in connection with the features exhibited near the beginning and end of a regular gravitational train of surface waves travelling on deep water:

Our present solution shows how rapidly the initial sinusoidality of the head and front of a one-sided infinite procession, travelling right-wards, is disturbed in virtue of the hydrokinetic circumstances of a procession invading still water. Our solution, and the item towards it represented in Figs. 6 and 7, and in Fig. 2 of § 6 above, show how rapidly fresh crests are formed. The whole investigation shows how very far from finding any definite "group-velocity" we are, in any initially given group of two, three, four, or any number, however great, of waves. I hope...to return to this subject in connection with the energy principle set forth by Osborne Reynolds, *(marginal notes: Kelvin on the mode of travel of the front of a wave-train, in a dispersive medium:)*

---

[1] Formation of the differential equation for the forms of disturbances that are propagated without change of type shows that, when simple wave-trains possess this property, there are in general no other solutions; the existence of a wave-group in fact implies the existence of the wave-train through which it travels. *(marginal note: Wave-groups imply extended trains.)*

[2] Kelvin, "On the Front and Rear of a Free Procession of Waves in Deep Water," *Phil. Mag.* Vol. VIII, p. 468 (1904).

and the interferential theory of Stokes and Rayleigh giving an absolutely definite group-velocity in their case of an infinite number of mutually supporting groups.

*limitations.*

It would appear then that there is no certain ground, on the basis of the ideas pertaining to group velocity, for concluding that a prism is competent to disperse any isolated aethereal pulse, or any series of pulses with absolutely irregular statistics, into a series of simple wave-trains, in a regular manner[1], as an ideal grating could do, the number of undulations in each train being in that case the number of rulings in the grating or a sub-multiple thereof.

*Grating constructs a regular train from a pulse.*

But Lord Rayleigh in his recent paper has thrown fresh light on the subject of the general action of dispersive media, by examining the disturbance that follows an impressed travelling aperiodic pulse, *maintained at constant intensity*, and showing that such a pulse imitates in some respects closely the behaviour of a wave-train[2]. He in fact points out the analogy with the surface waves produced by a boat travelling with uniform velocity on a smooth lake, which, as everybody has observed, are dispersed into simple wave-trains each travelling in its own appropriate direction. It would seem indeed that this illustration bears more closely on the action of a travelling *source* impressed on the medium than on the fate of an unsupported pulse travelling across it spontaneously. In default of a constant supply of power to the boat, to be spent in making new waves, it would soon lose its velocity unless it had a store of kinetic energy great for its size. Thus this close analogy with ordinary dispersion, which is afforded by the dispersed wave-trains excited by a pulse, impressed and maintained from outside, appears to leave where it was the question of the fate of an isolated unsupported pulse, propagated into the dispersive medium and then left to itself. The steadiness which in the ordinary dispersion theory arises from the succession of fresh waves of the train, is obtained in the illustration above by the maintenance of the energy of the pressural source, with results in close analogy in the two cases.

*A travelling boat creates trains of surface waves,*

*and maintains them.*

It thus still seems difficult to evade the force of the argument of Sir George Stokes[3]:

[1] What applies to a prism would probably also apply to colour perception by the eye. [Even for the grating, a thin incident pulse gives rise to a single series of thin secondary pulses issuing from its elements, and these have only one common tangent plane, the front of the one thin pulse which alone the grating produces.]

[2] A more direct investigation than that quoted from Lord Kelvin's note of 1877 is given in Professor Lamb's *Hydrodynamics*, § 227 (1895).

[3] Wilde Lecture, "On the Nature of the Röntgen Rays," 1897, in *Math. and Phys. Papers*, Vol. v, p. 271. [But on the other side is to be put the statistical consideration by Lord Rayleigh, *Scientific Papers*, Vol. v, pp. 279–282, which has the advantage of being expressed in direct graphical detail.]

When you let a ray of light fall on a refracting medium such as glass, motions begin to take place in the molecules forming the medium. The motion is at first more or less irregular; but the vibrations ultimately settle down into a system of such a kind that the regular joint vibrations of the molecules and of the ether are such as correspond to a definite periodic time, namely that of the light before incidence on the medium. That particular kind of vibration among the molecules is kept up, while the others die away, so that after a prolonged time—the time occupied by, we will say, ten thousand vibrations, which is only about the forty thousand millionth part of a second—the motion of the molecules of the glass has gradually got up until you have the molecules of the glass and the ether vibrating harmoniously together. But in the case of the Röntgen rays, if the nature of them be what I have explained, you have a constant succession of pulses independent of one another. Consequently there is no chance to get up harmony between the vibrations of the ether and the vibrations of the body. Stokes on establishment of wave-trains: pulses not refracted.

The distinction may perhaps be put more definitely. White light from an incandescent solid is made up of a vast number of pulses arising from the molecular shocks incessantly occurring in the hampered spaces to which the molecules are confined. On the other hand, the Röntgen rays are [were supposed to be] made up of the independent sporadic shocks transmitted through the aether from the impacts of the separate and independent cathode particles. Both kinds of disturbance are resolvable by Fourier's principle into trains of simple waves. But if we consider the constituent train having wave-length variable between $\lambda$ and $\lambda + \delta\lambda$, *i.e.* varying irregularly from part to part of the train within these limits, a difference exists between the two cases. In the case of the white light the vibration curve of this approximately simple train is in appearance steady: it is a curve of practically constant amplitude, but of wave-length slightly erratic within the limits $\delta\lambda$ and therefore of phase at each point entirely erratic*. In the Fourier analysis of the Röntgen radiation the amplitude is not regular, but on the contrary may be as erratic as the phase. The origin of this difference is that the body radiating the white light is presumably so far in a steady state that each element of it has a definite temperature at each instant, which implies a statistical uniformity in the vibratory disturbance that is emitted. Or, approaching the subject from the side of the thermodynamics of radiation, each elementary constituent wave-train, say that corresponding to the interval $\delta\lambda$ above, has its own temperature Statistical smoothness in white light.

* That is, it is analysable only as regards its energy, cf. *Proc. Lond. Math. Soc.* 1916 as *infra*, and that only when it is statistically uniform. The *x*-rays were supposed to be pulses too sparse to admit of statistics continuous within the range of the length of a pulse: cf. *loc. cit. infra*, on the intricacy of the Fourier analysis of two identical pulses far apart.

which it carries permanently along with it, the same as the temperature of its source supposed a perfect radiator. This temperature is at each point of it a function of the energy density, and therefore

of the amplitude of the radiation. If the amplitude were different along two reaches of the train, the reach of higher amplitude could be in equilibrium of spontaneous exchanges of energy with a perfect radiator of higher temperature than its own source, and ideal automatic arrangements involving intensification of the energy would be possible, in opposition to Lord Kelvin's fundamental principle

of degradation. In the internal equilibrium to which a material system nearly instantaneously settles down, in acquiring a definite temperature for each element of its mass, such differences of amplitude must thus have disappeared: the state of uniform amplitude is, in fact, the most probable one[1].

In the distinction which is here suggested, the average degree of suddenness of the Röntgen pulses is not involved. That would still be capable of estimation by experiments on diffraction of rays travelling in free aether, in the manner of Haga and Wind. But no physical *rationale* of prismatic dispersion[2] except the influence of the vibrations excited in the material system seems now to be entertained; and this appears to impose a limit to the types of disturbances that are subject to regular dispersion.

Even in the case of wave-trains excited on the surface of water by a travelling source, where, as Lord Rayleigh remarks, there is no structural periodicity[3], the presence of the wave-trains travelling in any direction does, at any rate, depend on the persistence of free periodic trains of waves. If we imagined the water replaced by a medium in which no free wave-trains could travel with velocities within certain limits, then we would expect a gap in the wave-pattern formed by the travelling source, corresponding to those limits. The passing remark of Sir George Stokes, which likens the synchronous optical vibration

of a transparent solid body to the sonorous vibration of the sounding board of a pianoforte, thus assigning it to regions of the material medium in bulk rather than to its individual molecules[4], would bring the optical effect somewhat nearer to the water-wave phenomenon.

[1] It is hoped to pursue this idea in another connection.

[2] Propagation in limited systems such as bars is not to the point.

[3] Professor Schuster (*Phil. Mag.* Jan. 1904) puts the point as follows: "As we may imagine continuous media of such elastic properties as to give dispersion, the true explanation must be independent of the sympathetic vibrations" on which I had relied in *Aether and Matter*, p. 248. The force of this is obvious; yet, when friction is ruled out, what can there be, as a matter of fact, except conspiring periodicities in time (free periods) or space, to modify simple elastic waves, which travel without change, into the dispersive type?

[4] In the case of a vapour, the molecules, being isolated, must however operate independently.

But the feature in the case of the water waves to which Lord Rayleigh doubtless intends to draw special attention, is the absence of gradual initiation and delay of effects; as soon as the source begins to move with uniform velocity, the wave-pattern begins to travel out from it, and as soon as motion of the source stops, so does the formation of the wave-pattern. By a legitimate application of the principle of group velocity, the number of undulations formed is shown to be connected with the distance over which the source has moved; just as the number of waves formed from a single freely travelling pulse by a grating is determined by the number of its lines over which the pulse has travelled.

*Instantaneous development of regular wave-pattern on water.*

At first sight, as above stated, it is difficult to detect sufficient similarity to the optical case. But if we consider (with Lord Rayleigh, p. 404) a thin plane pulse incident from free aether obliquely on an infinitely extended plane face of a refracting medium, the intersection of the pulse with the face will be just such a maintained disturbance, travelling along the face, as we require. If the medium has the dispersive quality fully developed, for all disturbances however sharp, *i.e.* if the differential equation determining dispersive vibration has no limits to its full application to such disturbances, the resolution into trains of waves must be granted as a necessary consequence of the analysis for a steady travelling source.

*A broad pulse refracted as a travelling source.*

This *rationale* of the dispersive refraction, at a plane surface, of an obliquely incident thin plane pulse, is one of those obvious things that when once grasped form a permanent addition to our stock of physical imagery[1]. One can picture its application to water waves. A tract of water may be imagined, of small uniform depth $h$, in which therefore all waves travel with the same velocity $\sqrt{gh}$, separated from a region of deep water, in which the velocity of a train depends on its wave-length, by a straight boundary. A disturbance consisting of a thin plane ridge can advance obliquely towards the deep water without change of form; the successive parts of the ridge reach the boundary in the manner of a maintained local disturbance running along the boundary with uniform speed, of which a definite fraction is transmitted across into the deep water, the rest being reflected back. The mode of this transmission is, by symmetry, the same as if that part of the disturbance were doubled and the deep water were unlimited on both sides: regular wave-trains are shed off dispersively in the different directions, as in the case of the boat described

*Analogue of a ridge advancing obliquely to deeper water:*

[1] Lord Rayleigh considers this explanation of the *refraction* of a pulse into a dispersive medium to be less simple than his first case of the propagation of a plane pulse in such a medium; but owing to the difficulty described above, regarding the maintenance of such a pulse, I have failed to appreciate the argument in that case.

above, with wave-lengths such that their velocity can just keep up with the travelling source, while the distribution of intensities between the various directions depends on the character of the moving source, *i.e.* of the incident travelling ridge of water. The waves must, in fact, form a steady pattern travelling with the source: thus the velocity of free propagation of the component train travelling in any direction must be the component in that direction of the *details.* velocity of the source. But as the travelling source has finite size, this component train, though nearly homogeneous in wave-length, is not quite so; being nearly homogeneous, the dispersive quality of the medium will make its waves travel in groups, which progress in the known manner with only half the velocity of their component waves. Thus after the train is well formed, the groups of disturbance will recede from the source with a velocity equal to the difference *Group* of these two velocities, and Lord Rayleigh's determination of the *velocity* length of disturbance emitted in a given time ensues—subject, how-*the rapidity* ever, to the reservation quoted above from Lord Kelvin as regards *of establish-* the extreme head of the train. The account of the process, which is *ment of the* *train.* indicated by Lord Rayleigh, seems in its essentials to be fully verified.

Yet the quotation above made from Sir George Stokes, in which the imagery is optical instead of hydrodynamical, appears to show a different aspect of the picture which we are bound to follow out; though Lord Rayleigh has guarded himself against it by his reserva-tion, "so long at least as we are content to take for granted the character of the dispersive medium—the relation of velocity to wave-length—without inquiring further as to its constitution." The *Differential* postulate thus indicated is that the partial differential equation of *equation to be* propagation is to hold true without limitation. This implies that the *of unlimited* *scope:* dispersive medium must be homogeneous in space; if it had minute alternating structure, then this differential equation could not of course be applied without modification to waves of length comparable with the dimensions of that structure—a circumstance on which Cauchy reared his original attempt at an explanation of optical dispersion. But it requires also that the medium should, so to speak, be homogeneous in *time*. An optical dispersive medium is made up of elements which have periods of free vibrations of their own, that are more or less durable; the differential equation will not hold for disturbances whose scale of duration is so small as to be of the same order as the time of the natural subsidence of free disturbance among the elements of the medium. In connection (originally) with the dynamical theory of viscosity in gases, Maxwell introduced the term *time of relaxation* to express the time, roughly assignable, that it would take for a local derangement of the molecules of the medium to smooth itself out. In optics it is the time needed for the free irregular vibrations of an element of the medium, produced by a local shock or other disturbance, to die out by dissipation into surrounding elements. The theory of regular dispersion of a disturbance into wave-trains caused by refraction, re-stated above for the hydrodynamic case of waves on water, cannot be applied in the optical case unless the scale of duration of the disturbance is long compared with the *time of optical relaxation* of the dispersive medium. In Sir George Stokes' illustration, taken at a venture, the time of relaxation would be ten thousand times the period of a light wave; if so, regular refraction and dispersion would hardly be established for sequences of less than ten thousand similar waves. Perhaps the only means of even roughly guessing at the time of optical relaxation is by the time lag in such phenomena as fluorescence, which are connected in part with free internal vibrations excited in the elements of the medium. Stated in the present form, the criterion that a Röntgen aether pulse should be regularly refracted and dispersed into wave-trains, according to a process of which Lord Rayleigh's *rationale* has been paraphrased above, is that its duration should be long compared with the time of optical relaxation of the dispersing medium. In the hydrodynamic illustration the restriction does not arise, for the time of molecular relaxation is far beneath the period of any observable surface waves.

*[margin: time of relaxation is to vanish.]*

*[margin: A criterion.]*

To sum up, it now seems clear that Lord Rayleigh's application of the phenomena of a *maintained moving source* gives an adequate picture of the *modus operandi* of the dispersion of an incident aperiodic disturbance into regular wave-trains by refraction, for all types of disturbance that are slow compared with the period of natural molecular relaxation of the refracting medium—provided, however, anomalous dispersion, which cannot be included unless a *quasi*-frictional term is *assumed* in the analysis, plays a part which is unimportant*. But it is still held to be unlikely that aethereal pulses of the type of the Röntgen rays come as a rule within this limit. If this be so, white light, such as can be regularly dispersed by a prism, cannot consist of wholly irregular aethereal disturbance; each Fourier component, comprised within say the infinitesimal range of wave-length between $\lambda$ and $\lambda + \delta\lambda$, must have sequences of regularity [uniformity] in its amplitude, of duration comparable with the time of optical relaxation of the dispersing medium.

---

* As *supra* it merely cuts out the local range of periods by scattering. Also as *supra* the analysis of natural light by dispersion can be only a statistical one, that is, an analysis of its energy.

# 67

# THE IRREGULAR MOVEMENT OF THE EARTH'S AXIS OF ROTATION: A CONTRIBUTION TOWARDS THE ANALYSIS OF ITS CAUSES*.

[*Monthly Notices of the Roy. Astron. Soc.* Vol. LXVII, pp. 22–34, Nov. 1906.]

The problem: MUCH material, defining with increasing accuracy the irregular wanderings of the Earth's axis of rotation, has now been accumulating for a long series of years. The attempt to decompose the movement into regular harmonic components excited interest some six or eight years ago. Since that time the more systematic data obtained and analysed by the International Organization have given greater precision to the path of the Pole; and, while a definite astronomical discussion must rest with the experts, curiosity as to the general physical causes of the phenomenon is legitimate. It has long been recognized that displacement of material on the Earth's surface due to meteorological changes (melting of polar ice, long-period barometric fluctuations, etc.) must be a prominent agent, and may indeed be taken to be the main one; while Newcomb has pointed out that the free Eulerian oscillatory period must be very different, for a nearly spherically balanced Earth, if it is elastically deformable under the centrifugal force, from what it would be if it were rigid, thus accounting for the unexpected value of the Chandler period.

to determine displacements of terrestrial material. It is shown below that, without making any hypothesis except the natural one, that this free precessional period is fixed in duration and determines the average duration of the revolutions of the Pole of rotation, it is easy by a graphical process to deduce from the path of the pole a map of the varying torque which must be acting in order to produce that path, and thence to infer as to the character of the displacements of terrestrial material that must be taking place in order to originate that torque on the Earth as a whole. It has seemed worth while to carry this out in a preliminary way. It has also been thought worth while to set down various dynamical considerations which may prove useful in a systematic analysis of the observational results.

Let $\omega_1$, $\omega_2$, $\Omega$ be the component angular velocities of the Earth referred to axes moving with itself, the latter being around the axis

* By J. Larmor and Major E. H. Hills, R.E. Read in part at the British Association, August 1906.

of figure. Thus $\omega_1/\Omega$, $\omega_2/\Omega$ are the angular coordinates of the Pole of rotation measured on the Earth's surface, and represent directly the change of latitude; they are different from the absolute coordinates of the Pole on the celestial sphere, in the order of the ratio $C/(C-A)$, as appears from the Poinsot geometrical representation of the free precessional motion, $C$ and $A$ being the *effective* polar and equatorial moments of inertia[1]. The dynamical equations of the Earth's free precession are, with sufficient accuracy, in terms of the angular momentum $(h_1, h_2, H)$,

$$\frac{dh_1}{dt} - h_2\Omega + H\omega_2 = 0,$$

$$\frac{dh_2}{dt} - H\omega_1 + h_1\Omega = 0,$$

where, $D, E, F$ being products of inertia,

$$h_1 = A\omega_1 - F\omega_2 - E\Omega, \quad h_2 = B\omega_2 - D\Omega - F\omega_1, \quad H = C\Omega - D\omega_2 - E\omega_1.$$

Thus   $A\dot\omega_1 - (C-A)\,\Omega\omega_2 = -L'$,   $A\dot\omega_2 + (C-A)\,\Omega\omega_1 = -M'$,

where $\Omega$ may be taken constant, because its variation would multiply in these equations two factors, each small of the first order; and where $L'$, $M'$ include, in addition to the kinetic forcives due to the location of mobile material attached to the Earth, the terms

$$\dot{A}\omega_1 - \dot{F}\omega_2 - \dot{E}\Omega, \quad \dot{B}\omega_2 - \dot{D}\Omega - \dot{F}\omega_1.$$

Here the terms on the left would be rates of change of angular momentum if the configuration remained constant. The terms on the right include the reversed rates of change of the same, taking the angular velocity constant but the configuration altering. To determine these latter terms, as arising from the reaction of mobile ter- restrial material, we can consider them directly as representing the centrifugal forces of this loose material, together with the reversed gradient of the angular momentum, due to the Earth's rotation, of

---

[1] Cf. Routh, *Dynamics*, Vol. II, §§ 180–2, 533. On account of the largeness of this factor $C/(C-A)$ the axis of rotation is practically fixed *in space*, except in so far as the extraneous attractions of the Sun and Moon operate in causing forced precession, so that the effect here in question is simply a change of latitude. Cf. Maxwell ["On a Dynamical Top...with some Suggestions as to the Earth's Motion," *Trans. R. S. Edin.* 1857; *Scientific Papers*, Vol. I (pp. 259–261)], who appears to have been the first to apply the Eulerian theory precisely to the Earth, and to examine the Greenwich observations in search of a 306-day period in latitude. According to Professor Newcomb, *Monthly Notices*, Vol. LII, 1892, pp. 336–341, searches for terms in the latitude of this period, instituted at Pulkowa by Peters, and at Washington, in 1862–7, had already yielded negative results. In Newcomb's paper, which pointed out and estimated roughly the increase of free period produced by elastic yielding, the influence of irregular meteorological displacements was also considered.

new negative material in the original position and of new positive material in the altered position.

**The method.** The difference between this procedure and the usual one adopted by previous writers[1] is that the investigation of the slight changes of position of the Earth's principal axes of inertia arising from displacement of material is evaded, by considering an unchanging Earth, with effective moments of inertia $(A, A, C)$, which is subject to force arising from the kinetic reaction exerted on it by this additional and independent material, moving over it and at the same time maintained by it in diurnal rotation.

**Effect of elastic axial yielding:** The effect of centrifugal force, in flattening elastically the terrestrial spheroid, simply modifies its effective or dynamical moments of inertia according to the principle arrived at in a previous discussion[2], viz. that the moments of inertia which would exist in the absence of diurnal rotation, but on the assumption that the Earth's form when

[1] For an account with references, cf. Routh, *Dynamics*, Vol. II, §§ 24, 533–5.
[2] "On the Earth's Free Eulerian Precession," *Proc. Camb. Phil. Soc.* May 25, 1896, p. 186 [: *supra*, p. 1].

**deduced.** The argument there employed, briefly stated in more analytic form, is as follows. Let $A'$, $B'$, $C'$ be the principal moments of inertia of the Earth when unstrained by centrifugal force; and let $I$ be the change of moment of inertia round its own axis, due to the equatorial protuberance raised by that force. This axis is in the direction of the resultant angular velocity $(\omega_1, \omega_2, \omega_3)$; and it is implied that it is very near to the principal axis of greatest moment $C'$, so that $\omega_3 (= \Omega)$ is practically constant and great compared with $\omega_1$ and $\omega_2$. It is involved also in this restriction that $I$ is a constant up to the first power of $\omega_1/\Omega$ or $\omega_2/\Omega$. Referred to the principal axes, the total component angular momenta are

$$h_1 = A'\omega_1 + I\omega_1, \quad h_2 = B'\omega_2 + I\omega_2, \quad h_3 = C'\omega_3 + I\omega_3.$$

The equations of motion referred to the rotating axes are of the well-known vector type

$$h_1 - h_2\omega_3 + h_3\omega_2 = L.$$

When $A$ and $B$ are equal, the third of them is

$$\frac{d}{dt}(C\omega_3) = N,$$

where $C$ is the effective moment of inertia $C' + I$: when $N$ is null $\omega_3$ is thus constant, say $\Omega$, up to the first order. The other two equations are

$$\frac{d}{dt} \cdot (A' + I)\,\omega_1 + (C' - B')\,\Omega\omega = L_2,$$

$$\frac{d}{dt} \cdot (B' + I)\,\omega_2 - (C' - A')\,\Omega\omega_1 = M,$$

which in the case of approximate symmetry involve a free period

$$2\pi\,(A' + I)/(C' - A')\,\Omega,$$

and similarly in the general case, thus depending only on $A'$, $B'$, $C'$ when $I$ is small.

The present procedure absorbs the effect of this regular change of form due to strain into modified moments of inertia, while it sets out the effect of erratic displacement of (additional) material on the rotating Earth as a kinetic forcive.

centrifugal force is thus removed is determined by a linear law of elasticity, are to be employed in dynamical investigations which take account of the elastic yielding of the Earth. There would be consequently an increase in the free precessional period (actually as observed it is from 306 to about 428 days) in the manner first pointed out by Newcomb.

Refer now the problem thus formulated to axes of $\omega_1$ and $\omega_2$ rotating with angular velocity $\Omega (C - A)/A$, that of the undisturbed free Eulerian precession, viz. about 428 days. The equations of movement assume the form

$$A\dot{\omega}_1 = - L', \; A\dot{\omega}_2 = - M',$$

Rotating frame of reference.

the same as if the axes were fixed and there were no diurnal rotation; that is, the polar axis moves in the Earth, relative to these rotating axes of coordinates, along the direction of the reversed resultant of the torques $L'$ and $M'$.

It seems useful, therefore, to plot the course of the Pole relative to coordinate axes rotating with the mean Chandler period, marking, at intervals along the curve, both the time, and the longitude of one of the revolving axes of coordinates at that time; for the velocity along the curve at any instant will then give the direction and magnitude of that part of the rate of change of $(L', M')$, the transverse component of the centrifugal torque of the loose material and the time gradient of the angular momentum arising from transport of this material, when the velocity of rotation is imagined unaltered. Such change can therefore be partially located, as *infra*; if it is mainly due to displacements of surface material, of thermal or meteorological type, it should show seasonal recurrences, and may prove to be in part due to slight change in oceanic or barometric levels.

Graphical solution:

the main disturbing causes.

Aggregate rough estimates of mere order of magnitude are easiest made directly, without use of these rotating axes. Thus a surface depression of 1 foot over a square mile, extending down in gradually diminishing amount to 30 miles, would involve an effective displacement of a layer 1 foot thick through 15 miles downward. In latitude 45°, where the effect in this respect would be greatest, this displacement would change the resultant transverse component of angular momentum of the Earth by $4000 \, \Omega \, \dfrac{2\frac{3}{4} \cdot 15}{5280} \cos^2 45°$, the whole angular momentum of the Earth being in the same units

Earthquakes unimportant.

$$\Omega \cdot 5\tfrac{1}{2} \cdot \tfrac{4}{3}\pi \, (4000)^3 \cdot \tfrac{2}{5} \, (4000)^2;$$

this takes the density of surface material to be $2\frac{3}{4}$ and that of the whole Earth $5\frac{1}{2}$. The polar axis would thereby be displaced through an angle equal in absolute measure to the ratio of these quantities;

in seconds of arc it would be about $3 \cdot 10^{-13}$. Thus local displacements by earthquakes can have no sensible direct effect on motion of the Pole. But more important is the centrifugal effect representable by displacement of the axis of inertia of the Earth, round which the free precession of the polar axis is taking place. This angular displacement is $h/(C - A)$, when $h$ is the product of inertia thus introduced: this is of $C/(C - A)$ times (about 300 times) the order of the direct effect on the Pole of rotation, so that in the present way of viewing the matter the changing origin of precession is practically everything[1]. It has been found, however, by Milne (Bakerian Lecture, *Roy. Soc. Proc.* 1906), that, for the small range of time (two years) then investigated by him, sharp curvatures in the polar movement appeared to be on the whole concomitant with earthquakes; the latter may be promoted perhaps by the changes of superficial or internal loading along meridians, that are the main cause of the irregular motion of the Pole and are greatest when the curvature of its path is sharpest. The procedure indicated in this note would locate to some extent this displacement or change of loading, and thus test that theory[2].

The effect of transfer of water from the Poles towards mean latitudes, arising from melting of Arctic ice, may be estimated either by considering an added layer, of thickness positive or negative according to the locality, and of null aggregate amount, spread over the whole ocean, or by estimating directly as above the change of angular momentum involved in the displacement of each portion of the material. A displacement of the Pole of rotation in the Earth in a given direction, when referred to the rotating axes as above, would imply alteration of intrinsic angular momentum of surface load in the neighbourhood of the meridian circle containing that direction, which would be a defect in the northern quadrant in front of that direction or an increase in the northern quadrant behind it, and *vice*

*Marginal notes:*
Even indirectly:
contrasted with the records.
Melting of Arctic ice important,

---

[1] Sir G. H. Darwin has recently expressed the opinion that on the whole the effect of earthquakes may be to bring the axis of rotation nearer to the axis of figure, and thus damp the polar movement.

[2] The case here considered is 15 cubic miles of material displaced vertically 1 foot. Professor Milne informs us that the result of an actual earthquake might be 10,000,000 cubic miles displaced vertically or horizontally through 10 feet. This would multiply the figure in the text by $7 \cdot 10^6$, thus giving $2 \cdot 10^{-6}$ seconds of arc. After this sudden shift of the axis of rotation in the Earth, the free precession would continue, but it would be around a new principal axis of inertia displaced from the original one by a quantity of the same small order of magnitude multiplied by $C/(C - A)$, that is by 300, giving a result of the order $10^{-3}$ seconds.

Sir G. H. Darwin has estimated (*Phil. Trans.* 1876) that $\frac{1}{100}$ of the area of Africa rising or falling *in situ* through 10 feet would produce 0·2 second of change, the rising or sinking being presumably taken as not merely superficial.

*Marginal notes:*
Maximum effect of shift of material.

*versa* for the southern quadrants. Thus water rapidly moved from the Poles, where it has little angular momentum, so as to cover to a depth of 1 foot a region 4000 miles square, in middle latitudes, would displace the Pole of rotation in the Earth by something of the order of 2 seconds of arc; for it would involve a new transverse angular momentum $\Omega \cos^2 45°\ 4000^5/5280$ in the same units as above. It is readily seen that the principal axis of inertia, about which the free precession would continue, would be displaced in the opposite direction through an angle of the same order. in displacing the Pole in the Earth, and the axis of the free precession.

In reducing the International Observations of change of latitude at the selected observatories extending round the Earth, it has been found necessary to include a change common to all longitudes, which at first sight could only arise from change of form of the spheroid which represents the terrestrial sea-level. This might be in part due to gravitational influence of the displaced material; yet the removal by melting of 10 feet of ice over a polar area 500 miles in diameter would produce a change of attraction which could not, in middle latitudes, raise the Pole by more than 1/1000 of the order of magnitude required. As the effect seems to be real, some indirect, perhaps seasonal instrumental, cause must apparently be sought for. The Kimura change of latitudes: not real.

In trying, as here, to separate out the meteorological displacements of the Pole from the true free precession which would in their absence be a regular circular motion, by referring the whole to rotating axes, the essential point is to assign as correctly as possible the period of this precession, for that determines the velocity of rotation to be given to the axes of coordinates. Elastic yielding of the Earth will prolong the period beyond the 306 days that would belong to a rigid solid. Hough has shown[1] that an average modulus of rigidity of a solid Earth even so great as that of steel would involve the prolongation of the period to the above value, 428 days, which represents the periodicity of the observed path of the Pole. Now whatever be the cause, elastic or fluid displacement, or both, that thus alters the effective dynamical moments of inertia of the Earth, it may be presumed that it alters them to a constant extent over a fairly long period of time. Thus the true free, or Eulerian, precession would maintain a rotation of the Pole fairly constant, while the meteorological disturbance superposed on it would in the long run have no rotational quality one way round or the other. In applying the method of analysis of the complex motion that is here proposed, the angular velocity of true precession may therefore be obtained by An estimate of actual free period essential: how obtainable.

[1] *Phil. Trans.* 1895. It is (as above) the near approach to sphericity that makes so slight a yielding to the changes of centrifugal and other stresses effective in this manner.

taking the mean of the observed times of revolution of the Pole, as Chandler originally pointed out.

The Eulerian principal axes fixed in the Earth, or still better the axes rotating as specified above, are appropriate to the analysis of the effects of moments or torques, which arise from change of loading and are therefore themselves revolving on the whole with the Earth. On the other hand, torques which only slowly change in direction in *Contrast with* space, such as those arising from the attraction of the Sun and Moon *the forced* on the protuberant parts of the oblate terrestrial spheroid, are most *precessions:* amenable to dynamical analysis when referred to fixed axes. They produce mainly the ordinary forced astronomical precessions and nutations, on which the varying elastic yielding of the Earth has no *the data* sensible kinetic influence; while the intensity of the torque of attrac- *derivable* tion depends on the instantaneous geometrical value of $C - A$, as *therefrom.* determined by distribution alone. The distribution of mass thus governs the solar and lunar precessions[1]. These forced precessions depend on internal fluidity, etc., only in so far as it modifies the effective inertia moment $C$ or $A$; while, on the other hand, the free precession depends on the effective or kinetic value of the small difference $(C - A)/C$.

The tides would be about the same at antipodal points if contours of the land were symmetrical, and the two opposite tidal protuberances would be additive in their effect on the free precessional moment. At first sight, it might appear that the estimate given above for a local overflow of a foot of water from the polar regions would involve that, even apart from the irregularity of form of the oceans, the features of the various tidal components travelling round the Earth must be reproduced to some extent in an exact diagram of the torque, owing to the tidal flow demanding alteration of the angular momentum of the water relative to the Earth's rotation, and to its centrifugal *Effect of* force. But in so far as it is the Earth that turns round under the *inertia of the* nearly stationary tide, the direct dynamical effects of the tidal move- *tides not* ments are very slight and belong to the astronomical class, and are *great:* in fact merged in the lunar and solar nutations. But if the Earth's surface were divided by meridional barriers, so that the tidal flow would be in the main north and south instead of around the Earth,

[1] There is one case, however, in which, as the equations of motion given *Condition for* above show, a small extraneous forcive would cause a wandering of the Pole *effective* in the Earth itself, much greater than its change of direction in space, *i.e.* than *extraneous* the astronomical precession thereby caused, that, namely, of a forcive having *causes.* a period in longitude relative to the Earth's rotation, of the order of the free period of 428 days. Such a term in the forcive would cause a forced oscillation in latitude of its own period, which would be superposed on the free oscillation of a nearly equal period. thus producing an alternation in its amplitude.

we might expect a tidal aberration in the latitude of amount not entirely insensible.

The magnitude of this tidal torque would then compare even with that of the precessional couple of the Moon's attraction on the protuberant parts of the terrestrial spheroid. The latter is

$$\tfrac{2}{3}Mr^{-3}\,(C - A)\,\sin 2\delta,$$

so that its amplitude is about $\tfrac{2}{3}\,(0''{\cdot}1\,g/a)\,10^{-2}\,C$ while the Earth's angular momentum is $C\omega$. If this torque were to rotate with the Earth for six hours it would produce an angular displacement of the Pole of amount $0''{\cdot}001\,g\,.\,6h/\omega a$, where $\omega^2 a/g = \tfrac{1}{289}$, that is, of amount $0''{\cdot}001 \times 289\,.\,\tfrac{1}{2}\pi$ or $0''{\cdot}5$; in contrast with the actual lunar fortnightly nutation of amplitude one or two seconds. It has just been seen that a partial tidal overflow from the Poles covering $(4000)^2$ square miles to the depth of a foot, and carried on with the Earth's rotation, could in an extreme case account for a displacement of the Pole as much as $2''$: and the antipodal high waters reinforce each other. It is true that if the Earth were covered symmetrically with water, the Sun and Moon travelling in the equator would produce no effect. But the obliquity of their paths and the irregular distribution of the oceans must lead indirectly to nutations of the Pole of the short periods of the various tidal components, which are not insensible in the present connection. Thus, for example, a forced nutation of this kind might introduce discrepancies into observations around a parallel of latitude, of character in part systematic owing to the progressive semi-diurnal change of phase, which would be eliminated in smoothing out the observations for each observing station of the International chain of longitude.

Possibilities in this direction are perhaps worth bearing in mind. For example, in addition to the difference of phase at different stations mentioned above, if the same stars were observed at all the stations, a solar tidal nutation might partially simulate a change of latitude with a yearly period, common to all the stations, such as appears in the reduced observations. The deviation of the vertical by the attraction of the tidal water is in all cases very small compared with this deviation of the Pole (cf. Thomson and Tait's *Natural Philosophy*, ed. 2), and is in fact here negligible.

The influence of a nutational torque of fairly short period could be laid off by aid of the diagram referred to rotating axes; but even in this case it can be effected rather more easily in the usual astronomical manner, on account of the relatively slow change of its direction in space.

But whether these tidal influences are sensible or not, assuming 428 days to be the period of free precession, we can transform the

*Margin notes:* estimated: compared with lunar nutation and with melting of polar ice: tidal nutations, simulating a Kimura term.

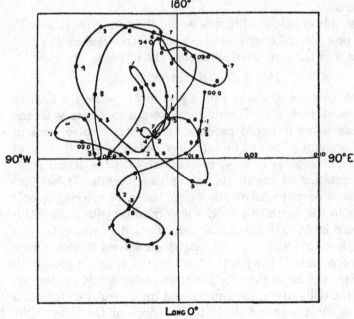

1. Albrecht, 1900-1905.

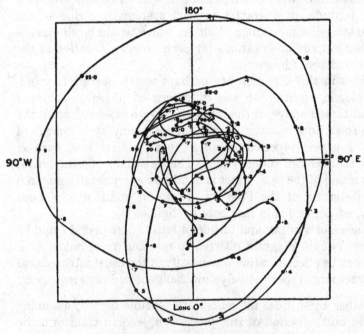

2. Albrecht, 1890-1897.

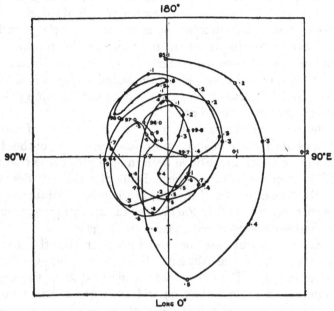

3. Chandler, 1895–1899.

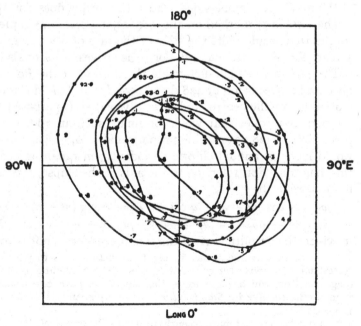

4. Chandler, 1890–1897.

curve of wandering of the Pole as above to axes of coordinates rotating with this period: and the hodograph of this new curve, when referred back again to axes connected with the Earth, will represent the distribution in direction and time of the torque arising from displacement of terrestrial material, which is continually modifying the motion of the Pole. From the point of view of geophysics, this

*Geophysical aspect.* curve would appear to be worth setting out, and might be expected to show seasonal recurrences.

*The diagrams of torque.* The annexed diagram (1) gives, from the path of the Pole since 1900, as officially published by Professor Albrecht in the *Ast. Nachr.*[1], the torque which must have been in action to cause that motion. It will be observed that the expected annual periodicity does not appear, but that the direction of the axis of the torque points preponderantly towards the side of the Pacific Ocean, an extraneous feature which we shall show how to eliminate later (p. 328).

Then follow, but based on more imperfect data, the torque diagram (2) derived from the Albrecht diagram of polar wandering for the period 1890–97. The other two are derived from the Chandler diagrams for the periods 1895–99 and 1890–97. It will be noticed that for the first one and a half years of the overlapping period 1895–97 the torque diagram is much the same in both, but not for the later part. The appearance of all these diagrams is very different from that of the first, *e.g.* the one-sided bias in the torque does not appear.

The unit (0·1) marked on the axes of the diagrams represents the torque that would shift the Pole at the rate of $0''\cdot05$ in one-tenth of a year; the dates are marked along the curves in decimals of a year.

The analysis of Chandler made the motion of the Pole consist of his circular precession of 428 days' period, with an additional elliptic

*Annual motion superposed on precession.* motion about the centre, and of yearly period, superposed on it. The component torques that would originate such an addition to a free precession of 428 days, being proportional to $\dot{\omega}_1 - n\omega_2$ and $\dot{\omega}_2 + n\omega_1$, would have an annual elliptic periodicity, in rough agreement with this diagram. For, referring to the axis of the ellipse, this theoretical motion would be

$$\omega_1 = A \cos (nt + \alpha) + a \cos pt, \quad \omega_2 = A \sin (nt + \alpha) + b \sin pt,$$

so that $\dot{\omega}_1 - n\omega_2 = - (ap + bn) \sin pt, \dot{\omega}_2 + n\omega_1 = (an + bp) \cos pt,$ in which the amplitude $A$ of the free precession is not involved*.

[1] The most convenient way to use these diagrams is to place them on a terrestrial globe, with the origin at the Pole. The Albrecht and Chandler path diagrams from which these force diagrams have been constructed may be found in *Bericht über den Stand...der Breitvariation*, and *Ast. Journal*, Vol. XIX, respectively.

* It has been found more recently by Sir F. Dyson (*M.N. Roy. Astron. Soc.* 1918, pp. 452–462) that the actual wandering of the Pole is well expressed by

We have now to inquire into the kind of information that these diagrams can convey. On our plan of analysis, on the basis of a definite elastic Earth on which additional matter can be displaced, the forcive necessary to supply new angular momentum to material that has come into a position of greater diurnal velocity has to be supplied by this Earth, while it has also to sustain the centrifugal force of this material. The other ways in which the mobile material reacts on the motion of the Earth on which it is superposed are negligible in comparison with these two[1].

*Influence of moving material:*

For $m$ in co-latitude $\theta$ the centrifugal torque is $m\Omega^2 r^2 \sin\theta \cos\theta$ around an axis at right angles to the meridian of $m$. The aggregate torque corresponding to its angular momentum in its present position

combining a circular motion in about 14 months with another annual one, of which only the amplitudes would thus remain amenable to secular change. The former term would represent the free precession as affected, after Newcomb, by elastic yielding, with amplitude changing owing to accidental disturbances. The latter term would require, in order to maintain it, a regular torque of harmonic type precessing round the axis in an annual period, presumably due to seasonal meteorological effect of the Sun's radiation. The amplitude of this disturbing torque would change slowly from year to year, involving transfer of energy into the free vibration, as is necessary to account for its own persistence with changing amplitude. Any disturbing cause not strictly annual and harmonic must, so far as it is sensible, upset this balance of interacting circular motions. Cf. also p. 335 *infra*.

*Annual component artificial:*

This direct analysis of the record is however of statistical rather than dynamical type. It is analogous to extracting the annual period, which must be there, out of the meteorological records. Thus it does not aim at penetrating to causes. Indeed any special disturbances, such as we have here been trying to locate, would remain in the residue: which therefore might be well worth scrutiny.

The most direct way to extract an annual term, if such exist, *of constant amplitude*, is to take the mean positions of the Pole at each month of the year: the mean curve of their plot will be the annual part, and will not be symmetrical round the centre if harmonics are present in it. A similar averaging for times spread uniformly over an amended Eulerian free period will take out the precessional part: as it is dynamical, harmonics would hardly be expected in it, so that the symmetry of the graph may perhaps be a test of its purity and thus of correctness of the period that had been assumed. When both these parts are taken out the residue of the graph of the polar wanderings would show what fraction of its statistics is sporadic: it might also point out exceptional shocks, which are not to be expected, or rapid changes, such as would be due to convection of ice over the ocean, or of air as H. Jeffreys has indicated. The Kimura term would be in the residue. Cf. p. 335.

*how to be extracted:*

*nature of residue.*

But an annual period of amplitude which is varying is not a pure period. For example a term $\cos mt$, with amplitude $A + B\cos kt$, is compounded out of three pure periods: the general case of amplitude $f(t)$ may be very complex.

[1] This is readily seen by the procedure of the note, p. 318, if we introduce the exact formula $h_1 = A'\omega_1 - E\omega_3 - F\omega_2 + I\omega_1$, where for an additional mass $m$ at $xyz$, $F = mxy$, $E = mxz$; we are in fact merely neglecting $\Omega\omega_1$ and $\Omega\omega_2$ compared with $\Omega^2$.

its centri-
fugal

and trans-
verse forces.

is $m\Omega r^2 \sin\theta\cos\theta$, as regards the equatorial component which is in the meridian of $m$. The former operates as a whole as a forcive, but only the time gradient of the latter thus acts, which is $2m\Omega rv\cos 2\theta$, where $v$ is the velocity of $m$ along the meridian. The former preponderates, in the ratio $\tfrac{1}{2}\Omega r$ to $v$, therefore usually very much so, as $\tfrac{1}{2}\Omega r$ is of the order of 300 miles per hour. On the other hand, the centrifugal force of a new steady local load merely makes the steady precession occur about a new axis of inertia, thus is not progressive or cumulative. This is readily verified by reversing the graphical pro-

Result.

cedure; a steady torque, represented by the end of a radius vector, becomes represented by an arc of a circle when referred to axes rotating with the free precessional velocity, and this corresponds to a velocity of circular precession of the Pole while it lasts.

This double mode of action of transported material, through centrifugal force and through change of momentum of diurnal motion, renders interpretation of the torque diagram to some extent indefinite.

But, neglecting the effect of change of intrinsic angular momentum, which may be as much as one-tenth of the whole, the torque will be due to the centrifugal force of the distribution of the mobile load at each instant, and will thus indicate the general features of that distribution; while the rate of change of the torque, *i.e.* the velocity in this torque diagram, will give an indication of the movement of the load. The radius vector $OP$ of the torque diagram at any instant

Interpreta-
tion.

will imply a proportional accumulation of materials in middle latitudes on the meridian at right angles to $OP$, so that antipodal accumulations reinforce, but adjacent ones, north and south of the equator, counteract each other. The marked tendency of the torque diagram for the period 1900–05 towards the side of the Pacific Ocean might thus be due to simultaneous accumulation of load not on the side of the Pacific, but in the neighbourhood of the perpendicular meridian. There is, however, a possible alternative to be kept in view, as follows.

It has been suggested to us by Professor Turner that the position of the origin to which the curve of wandering of the Pole of rotation

Uncertainty
as to centre of
graph:

is referred, is subject to considerable uncertainty. The observations and their reduction do not, however, seem to be at fault specially in this direction; unless the unexplained constant (Kimura) term may be taken to indicate a radius of uncertainty due to seasonal instrumental changes. This term, which recent discussion has confined to a smaller amplitude and to an annual period, was referred roughly to a displacement (mainly N. and S.) of the Earth's centre of gravity: we have verified above, however, that no likely meridional transfer due to seasonal change of temperature could produce an effect so great.

There does not, in fact, seem to be any ground, apart from mere uncertainties, for taking the origin to which the wanderings of the Pole of rotation are referred to be other than a fixed point on the Earth. But, on the other hand, this fixed point may not be the Pole of inertia of the solid Earth. We can, however, make it so, by separating from the solid Earth a thin superficial sheet, and counting this with the mobile material. The centrifugal forcive due to this sheet will then constitute a torque invariable with respect to the Earth: and we have merely to subtract this from the torque diagram referred to the Earth in order to obtain the torque due to the loose material alone. This subtraction of a constant vector term amounts simply to a change of origin. Thus on the final torque diagram the origin is uncertain, and would naturally be placed in as central a position as possible. *effect of change of centre.*

The process here carried out graphically may be compared with the procedure by successive steps as employed by Newcomb, in which free precession occurs for an infinitesimal time $\delta t$ round an axis of inertia $O$ supposed fixed in the Earth, then $O$ is moved on to $O_1$ as the result of the change of the mobile load during that time, then free precession takes place round $O_1$ for a time $\delta t_1$, then $O_1$ is moved on to $O_2$, etc. Our method has virtually amounted to the elimination of the free precessional motion, with its constant angular velocity, thus leaving the causes which displace the Pole of inertia on the Earth's surface open to inspection—the constrained shift of the axis of rotation in space being neglected for the nearly spherical Earth, as *infra*, p. 331. *Newcomb's procedure.*

How far the results may throw light on their causes depends largely on a comparison of the diagrams with the displacements of matter on the Earth's surface that are known to meteorology and ocean-ography. It may be that not much certain information may yet be derivable; but, considering the long time that observations of the wandering of the Pole have been accumulating, it can hardly be said that it is too soon to prepare for their preliminary discussion from the geophysical point of view. *Geophysical.*

The mode of reduction on which this paper is founded gives a force diagram which exhibits the torque sustained by the rest of the Earth owing to the displacement in and over it of the movable masses treated as independent bodies. It remains valid, however rapid the free precession may be. In the case of the actual Earth the latter is slow, being $C/(C - A)$ sidereal days, in which the difference of moments of inertia has its effective or dynamical value, thus lengthening the free period from 306 to 428 solar days. In this case various

Case of a body
inertially
nearly
isotropic. features assume simple forms, as appears in the Poinsot representation by a rolling ellipsoid. When this ellipsoid is very nearly spherical,
the axis of rotation, drawn from its fixed centre to the point of contact
with the plane on which it rolls, is at a small inclination to the
invariable direction of resultant momentum (which is normal to that
plane) compared with its inclination to the axis of inertia; thus the
axis of rotation is practically fixed in direction in space—except as
regards the superposed luni-solar precession. When the ellipsoid of
inertia is nearly of revolution, as in the case of the Earth, the Pole of
inertia thus revolves in space with the uniform free precessional
velocity, around the fixed direction of the Pole of rotation, while at
the same time it is undergoing such shifts as the redistribution of
material geometrically requires. Our procedure in the above has
been to eliminate the uniform precession, and the residue is a graphical
representation of the irregular shifts, or rather of the torques which
produce them.

The discussion of the question whether in past geological history
the Pole of rotation has wandered extensively in the Earth seems also
Effect of past
geological
changes on
migration of
the Pole: capable of being based on simple graphic representation [*]. We shall
assume, as before, that the dynamics of the Earth's rotation are based
at each instant on a simple kinetic energy $T$ given by

$$2T = A\omega_1^2 + B\omega_2^2 + C\omega_3^2,$$

where $A$, $B$, $C$ are effective moments of inertia; it follows that the
angular momentum $(L, M, N)$ is given by the formula

$$(L, M, N) = (A\omega_1, B\omega_2, C\omega_3).$$

During the history of the Earth $(L, M, N)$, resultant $G$, must have
remained constant, while $T$ will probably have diminished through
frictional agency. Abstraction is here made of the solar and lunar
forced precessions, which compensate the torque of extraneous
attractions without affecting the position of the axis of rotation in
the Earth. The free motion is represented, after Poinsot, by the
angular motion of a momental ellipsoid, say

$$Ax^2 + By^2 + Cz^2 = K.$$

The direction of the axis of instantaneous rotation intersects its
surface at the point $(xyz)$ such that

$$\frac{x}{\omega_1} = \frac{y}{\omega_2} = \frac{z}{\omega_3}, \text{ therefore } = \left(\frac{K}{2T}\right)^{\frac{1}{2}}.$$

---

[*] The earliest concise graphical discussion of the Eulerian free precession and
its astronomical implications occurs in the Cavendish manuscripts: see *Scientific
Papers*, Vol. II, p. 411.

The distance $p$ of the tangent plane at this point $(xyz)$ from the centre is given by

$$\frac{1}{p^2} = \frac{A^2}{K^2} x^2 + \frac{B^2}{K^2} y^2 + \frac{C^2}{K^2} z^2$$

$$= \frac{G^2}{2KT}.$$

Thus if $K^{-1} = 2T$, we have $p = G^{-1} = $ constant; this tangent plane is then fixed as regards distance from the centre, as well as regards its direction, which is perpendicular to the invariable momentum $(L, M, N)$. Thus it is entirely fixed.

Thus the free precessional motion of the slowly changing Earth is represented by the varying ellipsoid *[represented by a modified Poinsot rolling ellipsoid:]*

$$Ax^2 + By^2 + Cz^2 = (2T)^{-1},$$

rolling with centre fixed, so as to keep in contact with this fixed plane whose distance from the centre is $G^{-1}$.

If the kinetic energy keeps constant, this is simply Poinsot's representation. The axis of rotation will circulate in the body around the axis of greatest moment of inertia, and in space around the axis of resultant angular momentum. The amplitude of this free precession *[amplitude kept down by friction.]* will be kept small by internal friction, so that the axis of rotation will always be near the principal axis, and can never wander further from its original position than the latter does. It will require a good deal of change of distribution of mass to move this principal axis very far: to move it into the equator the radius of the Earth must shrink to the order of 40 miles along the equator near the new Poles, and expand to about an equal extent near the original Poles.

If the Earth is shrinking uniformly, the moments of inertia vary *[Uniform shrinkage ineffective.]* as $l^2$, where $l$ represents linear dimensions; thus the angular velocities vary as $l^{-2}$ and the kinetic energy varies as $l^{-2}$. Thus the dimensions of this rolling ellipsoid remain unaltered. The distance of the plane on which it rolls also keeps fixed. Hence uniform shrinkage without frictional loss of energy would not affect the amplitude of the free precession, or cause the Poles to migrate.

Diminution of the energy of rotation through internal friction, without change of $(A, B, C)$, would increase the dimensions of the *[Dissipation of the energy of rotation makes for stability.]* rolling ellipsoid (in the proportion of $T^{-\frac{1}{2}}$), and so would make strongly for stability of the axis of rotation, as above remarked. Such increase can only proceed to a limited extent, determined by the ellipsoid just failing to intersect, and so touching the fixed plane at the extremity of its axis.

A discussion of the geological problem of displacement of the polar axis in the Earth must take account of considerations such as these.

# 68

# THE IRREGULAR MOVEMENT OF THE EARTH'S AXIS OF ROTATION: A CONTRIBUTION TOWARDS THE ANALYSIS OF ITS CAUSES. PART II*.

[*Monthly Notices of the Roy. Astron. Soc.* Vol. LXXV, pp. 518–521, May 1915.]

IN *Monthly Notices, R.A.S.* Nov. 1906 [as *supra*, p. 316], we considered the graph of the minute wanderings of the Earth's axis of rotation over its surface, as published from year to year under the auspices of the International Polar Commission, with a view to ascertaining something of the terrestrial origins of the irregularities[1]. An undisturbed free precession would make the Pole trace on the Earth a small circle (actually a few yards in radius) uniformly in about 428 days. The principle of our procedure was that, if the actual graph of the polar motion with respect to the Earth's surface is changed to a graph with respect to axes of coordinates travelling round, over the surface of the Earth, uniformly with the velocity of this free precession, then the velocity of description, by the Pole, of the graph so obtained, multiplied by the Earth's moment of inertia, represents in direction and magnitude the disturbing torque acting on the solid Earth at that instant. Such a torque could arise from change of position and from rate of motion of loose material, supported by the solid Earth, but for the purpose of formation of dynamical equations not counted in its inertia. The succession of torques exerted by the loose material is thus represented by the hodograph of the path of the Pole relative to the new axes moving round with the precession: and the form of that curve was accordingly traced.

Determination of the disturbing torque:

its sources:

is represented on the hodograph of the Pole.

At each observing station it is the course of *change* in latitude that is determined; and that information, by comparison for the various stations, determines the course of *movement* of the pole of rotation. But the position of the Earth's Pole of inertia is not determined on the diagram. It is presumably a nearly central point, chosen to make the precession as nearly circular as possible. Changing it to another point is equivalent to merging a certain shell of the solid Earth with the mobile material, that shell, namely, which must be shaved off

The Pole of inertia assigned on trial:

effect of an error,

---

* By J. Larmor and Colonel E. H. Hills, R.E., F.R.S.

[1] Cf. also *Proc. Roy. Soc.* Vol. LXXXII (1909), p. 89; *Monthly Notices, R.A.S.* Jan. 1915; *Proc. Lond. Math. Soc.* April 1915.

to make the remaining Earth dynamically symmetrical around the new axis thus postulated. The difference produced in the torque diagram will be that the torque arising from this shell will now be included in it; but this torque is invariable with respect to the Earth, so that its inclusion amounts merely to changing the origin in the *merely refers* vector diagram of torque. Thus the origin in that diagram also is *the hodo-graph to* indeterminate, depending on what surface deposits we choose to *another origin,*

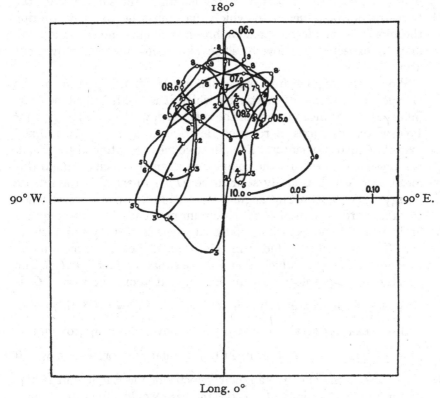

Torque diagram of the Earth's axis, years 1905–1910.

include with the solid Earth, as we had previously seen [p. 328, *supra*]; and the natural course is to place it as centrally as possible.     *as is natural.*

The diagram of torque for the period 1905–10, here annexed, is constructed on the same principles as the former ones (*Monthly* *Continuation* *Notices*, Vol. LXVII, p. 29). Owing to the revision of the deduced *of earlier* positions of the Pole since the diagram for the period 1900–05 was *hodograph.* constructed, the last two points on the earlier diagram, *i.e.* those for 1904·8 and 1904·9, must now be erased, and the new positions as given on the present diagram substituted.

It will be seen that an inspection of the directions and magnitudes of the torques during 1907 does not lend support to the view advanced by H. J. Swiers (*Proc. R. Acad. Amsterdam*, Vol. XIV, p. 211) that there was then some special cause of perturbation of the regular motion of the Pole. The disturbing torques during that year were not of any exceptional magnitude, nor were they changing rapidly. To judge by the diagram, 1907 was a much quieter year than 1905 or

Abrupt change in sidereal day not indicated. 1909. But if recent abnormal discrepancies in the Moon's mean motion really mean a change in the length of the sidereal day, the terrestrial displacements to which that must be due, by altering the moments of inertia, presumably in a manner not symmetrical, may be expected to show to some extent in the polar movement also.

Would the regular free precession by itself, irrespective of the changes of moments of inertia by which it is excited, have sensible influence on the mean length of the sidereal day? In the most general Poinsot case the axis of rotation describes a cone round the fixed

Effect of a free precession on rotational period: axis of angular momentum. The effective mean rotation of the Earth is represented closely by the component angular velocity around this mean axis, which is easily seen to be $2T/G$, where $T$ is the kinetic energy and $G$ the constant angular momentum[1]. A disturbance, when the axis of rotation is that of maximum moment of inertia, requires an increase of energy, while frictional agencies steadily diminish it: the effect of each is included in the formula. Let us follow out the case of symmetry, in which $A$ and $B$ are equal; and let $G = C\Omega$. The effective average rotation of the Earth, that about the axis of $G$, is

$$\omega_3 + \omega_1 \frac{A\omega_1}{C\omega_3},$$ approximately, where $C^2\omega_3{}^2 + A^2\omega_1{}^2 = C^2\Omega^2$; this agrees

with the exact general expression $2T/G$ above, and is approximately

$\left(1 + \frac{1}{2}\frac{\omega_1{}^2}{\Omega^2}\right)\Omega.$ The establishment of a regular free precession would

negligible. thus *increase* the effective angular velocity of the Earth; but if it is of amplitude $n$ seconds of arc the increase would integrate to only $n^2/30$ seconds of time in a century.

If the Earth's magnetism is due to its rotation, We may remark in passing that the energy of the Earth's magnetic distribution is usually supposed to have come somehow out of the Earth's rotation. If it were wholly accounted for by work done by the rotating Earth against some extraneous cosmical resistance, the

Effect of gradual loss of energy on the free rotation. [1] As follows: The motion is the Poinsot rolling of the momental ellipsoid $Ax^2 + By^2 + Cz^2 = 1$ on a fixed plane; if so, $\Omega = (\omega_1, \omega_2, \omega_3) = \theta\,(x, y, z)$, where $(x, y, z)$ is its point of contact with the plane; hence $T = \frac{1}{2}\theta^2$ and $G = \theta/p$, where $p$ is the distance of the centre from this plane, which is thus constant while its direction is that of $G$, the fixity of the plane of rolling being thereby in fact verified. The component angular velocity around $p$ is $\Omega p/r$, which is $2T/G$ as in the text.

consequent change in the diurnal angular velocity $\omega$ would be given very roughly by an equation

$$\frac{1}{8\pi} H^2 \text{ vol.} = -\delta \left(\tfrac{1}{2}Ek^2\omega^2\right),$$

in which the Earth is supposed uniformly magnetized so that $H$ is uniform throughout. Taking $H = \cdot 5$ c.g.s. and $k^2 = \tfrac{1}{3}a^2$, this gives

$$\frac{\delta\omega}{\omega} = -25\cdot 10^{-10},$$

a change which would make the Earth lag by 8 seconds of time in a century. If the fraction $1/n$ of the terrestrial magnetic energy were thus renewed each century, there would be a secular retardation of the Earth's rotation such as would produce a lag of $4/n$ seconds in one century, an amount which, when the rapid motion of the magnetic poles is considered, is perhaps not an absolutely negligible contribution to the lag arising from other causes, especially as the loss of purely magnetic energy would not be the whole loss concerned. *length of the day must be influenced, perhaps sensibly.*

We have recently discovered that the diagram of observed polar displacement has been analysed in a different way in the treatise by F. Klein and A. Sommerfeld, *Theorie des Kreisels*, Heft III (1903), p. 676. Their procedure is to eliminate the 428-day periodic part by graphing the difference between the position of the Pole at each instant and the position 428 days earlier. This is, of course, done immediately by plotting the line joining the two positions, with regard to a new origin. The result which they thus obtained showed traces of an annual period; and when that also was removed by similar treatment, the residue appeared to be quite irregular. The question suggests itself—can any physical interpretation be attached to this process and to the derived type of graph? It would appear from our previous discussion that in it each radius represents the vectorial time aggregate of the torques that have been acting during the preceding 428 days, and so hardly admits of comparison with the course of disturbances; indeed that interpretation is obvious, for it represents the total change of angular momentum during that time, as the free precession has completed its period. The second derived graph does not appear to admit of any direct dynamical meaning. *Procedure by successive removal of periods: imperfect.*

# 69

# ON THE RANGE OF FREEDOM OF ELECTRONS IN METALS[1].

*[Philosophical Magazine, August 1907.]*

The establishment of conduction in metals:

IT has been ascertained that complete metallic conduction is established in a small fraction of the period of low ultra-red radiation. This is proved by the experimental result of Hagen and Rubens that radiation of about ten or more times the period of light is reflected from all metals in proportions determined by their ohmic conductivities alone. For that type of radiation, the square of the *quasi*-index

is complete for radiation of low period.

of refraction is therefore a pure imaginary quantity.

Contrast of high period:

The opposite extreme case, that of radiation of period rapid compared with the times of undisturbed motion of the electrons, is also of interest. It will appear that, if the effective electrons are free, the square of the *quasi*-index must be a real negative quantity. The optical determinations of Drude indicate that this is not very far from being true with light waves for some of the nobler metals; in the case of white metals such as silver, the property, moreover, persists

Historical.

Results.

[1] This brief discussion was drawn up in forgetfulness of Professor Schuster's estimate of the number of free electrons in metals (*Phil. Mag.* VII, 1904). His method was, I find, criticized and enlarged on in an interesting way later in the same year by Drude (*Ann. der Phys.* XIV, p. 936), with a view to greater precision; but the widely uncertain assumption that the optical frictional term is represented for visible radiation by the full ohmic resistance seems to enter fundamentally. The values there obtained are about 10 times larger than the admittedly very rough estimate in the text. Later in the same paper (p. 956) Drude estimates the length of free path of the electrons, obtaining results ranging in order of magnitude from $10^{-6}$ cm. for the nobler metals to $10^{-8}$ cm. for bismuth; these were intended to replace estimates (of almost the same order for the nobler metals) deduced by J. J. Thomson and J. Patterson from magnetic influence on the resistance of thin sheets. The lower estimate, [a small multiple of] $10^{-8}$ cm. for most metals, deduced in the text directly from the time necessary for establishment of complete conduction, is the main point of this paper; the result might possibly be stretched toward $10^{-7}$, but hardly further, if the velocities of the free electrons are really determined by the gas law. If velocities so small as these are retained, the negative conclusion that hardly any portions of the paths by which the electrons travel are free, and that therefore estimates of number made in this way are uncertain, seems not unnatural.

[The subject of free electrons in metals had been systematized in relation to conduction, under the methods of gas theory, by Professor Lorentz early in 1905, *Proc. Amsterdam Acad.*; it has now been passed into the statistical domain of quantum theories.]

over a considerable range of period, though at length it fails. Thus, both on this ground and by reason of the shorter period, over a rather wide range of optical period the electrons are perhaps not far removed from being free. Moreover, theories of ordinary complete metallic conduction have been developed with some promise which ascribe it to electrons entirely free.

*free electrons indicated:*

When the electrons in the optical problem are supposed to be virtually free, then for each of them, of inertia $m$ and charge $e$, under electric force $(P, Q, R)$,

*influence on wave-velocity.*

$$m\ (\ddot{x}, \ddot{y}, \ddot{z}) = e\ (P, Q, R);$$

thus if there are $N'$ of them per unit volume, and $(u', v', w')$ is the current of conduction,

$$\frac{d}{dt}\ (u', v', w') = N'\ \frac{e^2}{m}\ (P, Q, R),$$

the sign of the charge $e$ not entering if $m$ is the same for all.

Consider a plane-polarized wave-train travelling along $z$, with electric vector $(P)$ along $x$ and magnetic vector $(\beta)$ along $y$. Its equations of propagation are, by the circuital relations of Ampère and Faraday,

$$\frac{\partial \beta}{\partial z} = -\ 4\pi u, \quad \frac{\partial P}{\partial z} = -\ \frac{d\beta}{dt},$$

where*

$$4\pi u = 4\pi u' + Kc^{-2} \frac{dP}{dt},$$

[part of] the last term representing the *quasi*-current of aether strain, which will prove to be here negligible. Thus

$$\frac{\partial^2 P}{\partial z^2} = 4\pi\ \frac{du}{dt}$$

$$= 4\pi\ N'\ \frac{e^2}{m}\ P + Kc^{-2} \frac{d^2 P}{dt^2},$$

which determines the mode of propagation.

In a simple wave-train of period $2\pi/p$,

$$P = P_0 e^{\iota \frac{p}{c'} z - \iota p t}$$

which yields on substitution

$$-\frac{p^2}{c'^2} = 4\pi N'\ \frac{e^2}{m} - K \frac{p^2}{c^2},$$

thus giving for the square of the *quasi*-index $\mu\ (= c/c')$ the value

*The refractive index:*

$$\mu^2 = -\ 4\pi N'\ \frac{e^2}{m} \frac{c^2}{p^2} + K.$$

* The factor $K$ has now been introduced, to include dielectric polarization.

For radiation ($D$ line) of wave-length $6 \cdot 10^{-5}$ cm., $2\pi/p$ is $2 \cdot 10^{-15}$, so that

$$\mu^2 = -\frac{N'}{\pi}\frac{e}{m} 10^{-20} (9 \cdot 10^{20})(4 \cdot 10^{-30}) + K$$

$$= -N'\frac{e}{m} \cdot \frac{9}{8} \cdot 10^{-29} + K \text{ approximately.}$$

If the ions are all free electrons, $e/m$ is $17 \cdot 10^6$, thus [if the effect of collisions is neglected]

$$\mu^2 = -N' \cdot 2 \cdot 10^{-22} + K.$$

estimates.    For the nobler metals, in red and yellow light, the real (and five times preponderant for gold, silver, magnesium) part of $\mu^2$ is of the order $-10$; this would give $N' = \frac{1}{2} \cdot 10^{23}$,[1] so that the number of free electrons taking part would be of somewhere about the same order as that of the molecules of the metal*.

Drude's theory of conduction.    A view of metallic electric conduction, which fits in with its parallelism to thermal conduction, is that it takes place by free electrons, whose velocities are prescribed by collisions with the molecules and so are taken as determined by the law of equality of mean energies in gas theory. For hydrogen under standard conditions, the average velocity in the free path is $2 \cdot 10^5$ cm./sec.; as the electrons are [$2 \times 1850$] times less massive, their average velocity would thus be about $12 \cdot 10^6$. And we know that metallic conduction is fully es-

Optical facts require very short free paths:    tablished in a small fraction of the time $2 \cdot 10^{-14}$ sec., which is the period of ultra-red radiation of ten times the wave-length of light. If then, as theories involving free electrons require, the establishment of conduction is intimately concerned with the duration of the free paths[2], the mean free path of the electrons must [be small compared

---

[1] If undamped resonance of adjacent free molecular periods contributes sensibly to the real part of $-\mu^2$, otherwise than by shaking out more electrons into freedom, this would be an over-estimate.

* It is too great a value, as J. J. Thomson pointed out later for the Drude theory *infra*, if there is to be equipartition of velocities including the electrons, for it would influence specific heat unduly.

[2] Because the mean additional velocity imposed upon the electron by the electric field is proportional to the duration of the free path.

This view of conduction, as stated by Drude, seems to require tacitly that the average velocity is in most instances restored by collision at the end of each free path. Moreover the conductivity varies inversely as the temperature for pure metals: thus $N \times$ free path $\times$ velocity must be constant. These conditions are difficult to interpret, unless each molecule may be taken to emit electrons at a constant rate and absorb a definite proportion of those that en-
Possible balance of electronic exchanges.    counter it, the same at all temperatures, thus establishing an equilibrium in which $N'$ varies inversely as the velocity. Such increase of $N'$ at low temperatures would make $\mu^2$, and also the absorbing power, for those shorter wave-lengths for which $-\mu^2$ is large, tend to fall off inversely as the square root of the temperature.

with 25 . $10^{-8}$] cm., so that their excursions must in fact be largely confined to the spaces between each molecule and the next. If their free paths were larger than this, complete conduction could not be established throughout so short a periodic time.

If a free path longer than this were demanded, then the optical incipient conduction would have to be ascribed mainly to the gradual deflexion of the path by the electric field without introducing limitation due to interruption of the path by collision with the molecules; and the theory of propagation which has been developed above would be the one applicable to ultra-red radiation. The results of Hagen and Rubens [correlating it with ohmic resistance] seem emphatically to preclude that type of hypothesis. *the alternative excluded.*

If then metallic conduction is due to free electrons, their freedom is spatially very much restricted, almost in fact within molecular limits. *Freedoms only incipient:*

While if it is inferred from the approximation of $\mu^2$ to a real negative value, that the *incipient* conduction, which is the main agent in the optical phenomena of the nobler metals, must be largely due in the above manner to deflexion of paths of electrons, effectively free for times exceeding the period of the vibration, it would appear that the number of them that are concerned is roughly of about the same order as the number of the metallic molecules. *numbers if wholly free.*

The value of the mean velocity of the free electrons, above employed, about 12 . $10^6$, is involved in Drude's form of the theory (*Ann. der Phys.* 1, 1900, p. 577), in order to get the right (universal) ratio of electric to thermal conduction, on the assumption that both of them are effected through the agency of the free electrons. If, however, their velocity were of the order 3 . $10^9$, like the electrons from radioactive substances, then the free path might be as much as 5 . $10^{-4}$ cm. without vitiating the conditions laid down. The work of O. W. Richardson, on the escape of ions from incandescent solids, which he finds to agree with the requirements of the gas law, supports the former view, in fact seems to be its main experimental support; he concludes, moreover (*Phil. Trans.* 1906), that the number of free electrons in platinum ranges somewhere about $10^{21}$, which is not discordant, under the circumstances, with the general estimate $\frac{1}{2} . 10^{23}$ made above. *Thermionic support for equipartition of energy.*

The conclusion reached above, that in fact the free electron is deviated or entangled by each molecule which it meets or traverses, also points to the smaller velocity, as the rapid electrons are known to pass straight through thin sheets of metal. But the free path now seems hardly long or definite enough to substantiate, except very *Uncertainties.*

roughly, Drude's formula for metallic conduction as effected entirely by the steady deflexion of the free paths by the electric field. If the rough general estimates of order of magnitude, made above, for the number of free electrons and their velocities and free paths in the best conducting metals, are substituted in Drude's formula (*loc. cit.*) they give a conductivity of the order $12 \cdot 10^{15}$, which is about one-fiftieth of that of silver, and errs on the wrong side, though hardly so far as to cause surprise when the necessary vagueness of the data is kept in view.

# 70

# NOTE ON PRESSURE DISPLACEMENT OF SPECTRAL LINES.

[*Astrophysical Journal*, Vol. XXVI, 1907, pp. 120–122.]

THE important paper by Mr Humphreys (this *Journal*, July 1907) gives *data* for somewhat closer scrutiny of the origin of the pressure shift of lines in the spectrum. The change must be connected with electric properties of the surrounding gas; mechanical pressure arises merely from the translatory motions of the molecules, and these are so slow as hardly to count in connection with radiation periods. Thus the phenomenon is probably more strictly describable as a density effect. Electrically, the effect of increase of density is to increase the inductive capacity of the medium, that is, to diminish the effective aethereal elasticity which propagates the radiation. This is the averaged result; each molecule individually, through the agency of its plastic field of force or aether strain, provides a yielding region in the aether in which the effective stiffness is diminished*. The elastic energy which maintains the free vibrations of a radiator is located in its field of force in the adjacent aether; and by dynamical principles any loosening of the constraint in that field such as an adjacent molecule would produce, which would itself be somewhat intensified by equality of period, must in general tend toward increasing the free period, involving displacement of the radiation toward longer wavelength.

*The pressure shift in spectra,*

*as due to the increase of density,*

*which diminishes the effective electric elasticity around the vibrators:*

*intensified by resonance.*

By known dynamical principles[1], the change in free period due to slight change of constitution of the vibrating system can be estimated by calculating the altered kinetic and potential energies of the type of vibration under consideration on the assumption that the type remains unaltered.

*Rayleigh's principle.*

In the present case some light may be thrown on the amount of effect to be expected, by supposing the vibrating molecule to be situated in the centre of a sphere of free aether, beyond which the molecularly constituted gas is taken to be smoothed out into a uniform medium having the inductive capacity $K$ of the gas. The type of vibration being retained as before, its kinetic (magnetic) energy is not thereby affected from what it would be in a vacuum, but its elastic

*Rough estimate of pressure effect:*

---

* In marked contrast to the influence of an electron or ion.
[1] On this subject see Lord Rayleigh's *Theory of Sound*, § 88.

(electric) energy is altered in the ratio $K^{-1}$, wherever $K$ is different
effective<br>spherical<br>cavity: from unity. To obtain a rough estimate, suppose the vibrating aether field to be that outside a concentric spherical surface of radius $a$, and suppose the electric field to fall off with distance according to the law $r^{-n}$, multiplied of course by a function of direction. The static energy in it, measured from outside up to a concentric spherical surface of radius $c$, will be proportional to

$$\int_c^\infty r^{-2n} 4\pi r^2 \, dr, \text{ or } \frac{4\pi}{2n-3} c^{-2n+3}.$$

change of<br>elastic energy, If therefore the region beyond a distance $c$, equal say to $ka$, is filled with material of inductive capacity $K$, not much different from unity, the total static energy of the vibration is thus altered in the ratio of $a^{-2n+3} - c^{-2n+3} (1 - K^{-1})$ to $a^{-2n+3}$, that is, of $1 - k^{-2n+3}(1 - K^{-1})$ to $1$.
and so of<br>period. The frequency of the vibration is increased as the square root of this. For air at pressure of one atmosphere $K = 1 \cdot 0006$, and Mr Humphreys gives (p. 31) the proportionate change of wave-length as about $10^{-6}$. Thus $k^{-2n+3} \times 0 \cdot 006$ is about equal to $10^{-6}$. If the vibrator operates as
Estimate for<br>air at normal<br>density: a simple Hertzian doublet, $n = 3$; the other term, which constitutes the radiation, not being of account close to the vibrator. This would make $k^{-3}$ of the order $\frac{1}{600}$, so that $k$ would be about $8\frac{1}{2}$. In a gas at pressure of one atmosphere the molecules are spaced at a mean distance of very roughly 10 times the molecular diameter; and if $n = 3$, only about $\frac{1}{27}$ of the energy of one of them is beyond three molecular radii from its centre. Thus it is not unreasonable to replace the influence of the discrete distribution of gas molecules by that of a uniform averaged medium extending inward to about eight molecular radii from the centre of the vibrator. But these *data* are of course far too vague to justify more than the mere statement that the dielectric
of the right<br>order. influence of the neighbouring molecules is a *vera causa* of the right order of magnitude. For the next higher type of possible vibration, $n = 4$, the value of $k$ would be about $3\frac{1}{2}$, which may be just barely permissible; but for $n = 5$, the value of $k$ would be slightly over $2$, which would be ruled out. There is thus some presumption that the free vibration corresponding to each line of the spectrum is (except of course close up to the nucleus) of the simple type of that of a Hertzian doublet source*. Moreover if $n$ were not 3, the effect would not be proportional to the density of the gas. We have been estimating the average effect, on which a general broadening of the band due to irregular nearer approaches of molecules is superposed.
Displacement<br>absent in<br>bands.     The shift has not been observed in band spectra. The vibrator would then presumably be a molecule; and it may not be fanciful to

* As of course the Rayleigh analysis of the blue sky requires.

suppose that this circumstance may point to its field of energy being more concentrated into the region between its [oppositely charged] atoms; a higher value of $n$ would make the difference.

The remarkable one-sided broadening of absorption bands of pure mercury vapour, and its abolition by the admixture of a foreign gas, reported by Professor Wood (this *Journal*, July 1907) may perhaps have suggested similar considerations. The tendency to condensation in the pure vapour may proceed to an equilibrium*, when the formation of loose molecular aggregates by what may be called adhesion would be balanced by their destruction by collisions. The molecules in such loose aggregates would, owing to their (slight) mutual influence, vibrate in longer periods, and give rise to the displaced part of the band: but the average amount of this incipient aggregation would be much diminished by the admixture of a neutral gas.

Effect of admixture.

* The result would now be connected with statistical equilibrium between the "excited states" of the molecules on the Bohr theories.

# 71

# THE PHYSICAL ASPECT OF THE ATOMIC THEORY.

[*The Wilde Lecture: Memoirs of the Manchester Lit. and Phil. Soc.*
Vol. LII, 1908, No. 10.]

WHEN Descartes proceeded methodically to shake himself free from the trammels of the scholastic philosophy, and to reconstruct the content of his knowledge on what is essentially the modern basis, he found among other things—some of them now fantastic—that it was unintelligible to suppose that matter could act where it was not, *i.e.* could produce an influence in regions with which it was not in continuous structural connection. So powerfully did this feeling dominate his thought, that he appears to have been unable to form any conception of mere empty space as distinct from some mode of occupation of it: to him, space was a *plenum*, the seat of the processes of communication between the sensible objects which it contains. Direct interaction of these objects across distances, with mere nothingness between, was not a satisfying account of the relations between the apparently discrete masses which constitute our sensible universe.

Space a *plenum*:

The modern idea of an aethereal medium, as a means of transmission of physical influences from mass to mass, receives its first systematic exposition in his physical writings. The Sun is for him the centre of a great aethereal vortex, by which the planets are swept round in their orbits; light consists of impulsion or pressure propagated through the fluid *plenum*. An attempt is carried through to reconstruct the phenomena of physics and physiology on a basis of mutual connection—thereby starting afresh the aspiration which is fundamental to all scientific instinct, the effort to push to the utmost the unravelling of rational foundations of the scheme of things in which we subsist. And, whereas in Descartes' time there was little to go upon, except the imagination applied to the common facts of experience, now there are the vast and growing accumulations of ascertained experimental knowledge in the various Sciences, affording a most urgent stimulus towards the continued cultivation and improvement of general syntheses; these in turn react as the ever present incitement to the further pursuit of scientific experiment into regions economically devoid of profit.

Descartes.

The Cartesian system of celestial vortexes had been absorbed into common modes of thought, as a natural and intelligible feature in the cosmogony, when the precise observations of [Tycho Brake as utilized by] Kepler and the deductions of Newton came to replace this obvious mode of representation by new but exact principles whose foundations were entirely concealed from view. It is not easy for us now to imagine how strange must have been the idea that the planets were drawn inward to the Sun by a direct pull across space, depending on nothing but their distance apart—by a force which was postulated to act quite irrespective of whatever obstacles might intervene between them. To prepare for the unimpeded operation of direct forces across space such as Newtonian gravitation, the aether of Descartes—resisting medium so-called in this connection—which had to carry round the planets in its vortex, must be rigorously abolished: and space appears again as empty. Even the intellect of Huygens—whose vast range of achievements has been largely masked by the contemporary presence of Newton—seemed unable to accept the new doctrine in its full scope. To a mind like his there could remain no question about the chief Newtonian deductions: the evidence was conclusive that the interactions through the aether of bodies far apart in comparison with their magnitudes, and with nothing between, must somehow adjust themselves into the gravitational law of attraction. But he was unable to understand how the complex mutual influence of masses near together could possibly, in all cases without exception, resolve itself into a result so uniform and so simple. And to this day we remain largely in the position of Huygens with regard to this subject. The evidence, which in its beginnings enabled the genius of Newton to detect and develop his cosmical system, has of course long ago become overwhelming. Yet why is the gravitational attraction of a particle of matter sunk at the centre of the Earth entirely unaffected by all the intervening mass? We do not know, any more than Newton did, how this action is transmitted. We may take refuge in the idea that the nuclei in the aethereal medium, which constitute the cores of the fields of activity known to us as material atoms, must even in the densest matter occupy a space absolutely infinitesimal compared with the whole region of aether, and so not obstruct or modify the transmission of the gravitative influence. To get play for rational conceptions, we are thus thrown back on the atomic theory of matter, and that in its more modern physical aspect, to which it is now time to pass on.

It does not appear that Descartes was able to penetrate to any idea of the relation of the cosmical vortex to the atoms of the material bodies which it carried round in its grasp; they were merely like extraneous dust or mist whirled in the wind. For an adequately exact type of unifying conception of the relation between matter and aether,

Kepler:

Newtonian direct attraction:

the demur of Huygens,

its palliation.

Atoms.

of their structural connection, science had to wait until the middle of the nineteenth century. The profound analysis of Helmholtz had revealed the unexpectedly simple scope of the principles determining the interactions of vortexes in fluid—one of the most brilliant of the achievements of mathematical reasoning, whose highest function must always be to condense the unmanageable mass of relevant particulars into the practicable limits of general principles. His main result was that, in the entire absence of friction in the fluid, each vortex ring would be a permanent state of motion, capable of temporary modification (distortion, vibration, etc.), though interaction across the fluid with other vortexes which come within its range, but always in the end receiving its original condition, and thus retaining its individuality through unlimited time. We can well imagine the keen interest excited by Lord Kelvin's rapid *aperçu* that such vortexes may represent in essential features the atoms of matter. For here we have a type of atom that is not something foreign to the aether, merely immersed in it and pushed about by it, but a permanent located whirl or stable state of motion which subsists in the aether itself, and is of its very essence.

Here was suggested a mode of unification of duality previously unresolvable; and we can appreciate the zeal with which the problem of the investigation of this vivid image of one of the fundamental modes of interconnection in the scheme of nature was attacked by mathematical physicists. Its development, in which alongside the name of Helmholtz that of Lord Kelvin will ever stand, has constituted, as we all know in this place, a new science, that of abstract hydrodynamics, which analyses the interaction of uniform inertia and simple fluid pressure, and is in itself one of the most elegant and perfect constructions of modern mathematics. As a result of these thorough investigations, we are now more familiar with the limitations of the so-called vortex theory of atomic constitution, than with the initial successes of this mode of representation of physical reality. But limitation ought not to be taken to imply failure. Human reason is finite in its potentialities: it is not competent to frame a picture of the activities of the *cosmos*, of which it is itself a part, such as can bear comparison with actuality throughout the whole range of phenomena. It is of course absurd, as we are often reminded with much insistence, especially in recent years, to imagine that a material atom is merely a vortex ring in fluid; but, on the other hand, we can never know any object, even an atom intrinsically, but only through its relations with other objects, and the picture of atoms as motional configurations subsisting in some way in the universal *plenum*, and not merely objects foreign to it, is possibly the greatest expansion

*[Margin notes:]*
Helmholtz's vortex rings.

Kelvin's vortex imagery for atoms:

stimulated a new branch of science.

All models necessarily limited.

which our modes of thought on these matters have received since Descartes; and it has come to us, or has at any rate been made definite, through the vortex illustration.

This procedure, the study of the relations of the universe by the construction of working models, easily apprehended as a whole, is not restricted to external nature. A considerable part of present activity in abstract mathematics seems to consist in the construction of schemes of representation, such as will elucidate the inner scope and connection of the processes of mathematical thought, involving analysis of the ideas of number and magnitude, limit, infinity, etc. It is significant that in both cases some kind of atomism has been a mental necessity, as it was in the earliest Greek inquiries. *Mathematical thought and atomism.*

We have recalled that, in order to make way for the principle of gravitation, Newtonians were compelled to clear the celestial spaces of the "resisting medium" which constituted the aether of Descartes. But that by no means implied any belief that gravitation did not require a medium for its transmission. Towards the end of his life, Newton allowed himself to set down formally in the famous series of "Questions" appended to the second edition (1717) of his *Opticks*[1], his speculations on this and related subjects concerning the constitution of matter, pervaded as they were by constant suggestion of the vibratory motions which constitute heat, the radiation which these motions excite, and their close relation to chemical change. His ideas (Query 17 *seq.*), of essentially modern type, involved a medium infinitely more rare than air and of infinitely stronger elasticity—*aether* is his own name for it—amidst the waves of which the atoms of matter and the corpuscles which he took to constitute light were agitated like logs in a sea; of such a medium the elastic pressure, weaker on the adjacent sides of bodies, might, as he thought, in some way represent gravitational attraction, while its dead resistance to planets moving through it would be, owing to its small mass, quite negligible. *Newton's speculations on matter:* *his rarefied aether:*

At the end of a prolonged physico-chemical discussion he sums up his atomic view of the constitution of matter (*loc. cit.* p. 375) in archaic terms with deep modern significance, that have often been quoted:

All these things being consider'd, it seems probable to me that God in the Beginning form'd Matter in solid, massy, hard, impenetrable, moveable Particles, of such Sizes and Figures, and with such other Properties, and in such Proportion to Space, as more conduced to the End for which he form'd them; and that these primitive Particles being Solids, are incomparably harder than any porous Bodies compounded of them; even so very hard as never to wear or break in *his hard atoms:*

[1] *Opticks*, ed. (3), pp. 313–382.

pieces: No ordinary power being able to divide what God himself
imperishable: made one in the first Creation. While the Particles continue entire,
they may compose Bodies of one and the same Nature and Texture in
all Ages: But should they wear away, or break in pieces, the Nature
of Things depending on them would be changed. Water and Earth
unchangeable: composed of old worn Particles and Fragments of Particles, would
not be of the same Nature and Texture now, with Water and Earth
composed of entire Particles in the Beginning. And therefore that
Nature may be lasting, the Changes of corporeal Things are to be
placed only in the various Separations and new Associations and
Motions of these permanent Particles: compound Bodies being apt
to break, not in the midst of solid Particles, but where those Particles
are laid together and only touch in a few Points.

It seems to me farther, that these Particles have not only a *Vis
inertiae*, accompanied with such passive Laws of Motion as naturally
result from that Force, but also that they are moved by certain active
Principles [Energies], such as that of Gravity, and that which causes
Fermentation, and the Cohesion of Bodies. These Principles I consider not as occult Qualities, supposed to result from the specifick
Forms of Things, but as general Laws of Nature, by which the Things
laws of themselves are form'd; their Truth appearing to us by Phænomena,
combination. though their Causes be not yet discover'd. For these are manifest
Qualities, and their Causes only are occult....

This survey carries us about as near as purely physical speculation,
based on the broad simple principles of universal dynamics that
Newton was the first definitely to codify, could approach towards an
atomic theory. And it is a considerable advance. The uninstructed
tendency, judging from one's own early recollection, is to assume
that the apparently continuous substances around us are divisible
without limit, and afterwards to wonder what sort of evidence it is
The main that has suggested the contrary conclusion. Such evidence must have
appeal is to
chemistry. an essentially chemical flavour: its gist has been strikingly expressed
by Newton in the argument of which the conclusion has been quoted.
It is true that the mathematical physicists of the beginning of the
nineteenth century were accustomed to conduct their investigations
in terms of atoms or particles of bodies: but for their purposes, these
terms were hardly much more than the embodiment of the fact that
exact reasoning requires numerical expression, and therefore, the
mathematical resolution of the media with which it is concerned into
infinitesimal geometrical parts[1]. On a rather higher plane must how-

Young's      [1] An exception must be made at any rate in the case of Young, as will appear
atomic    in connection with optical dispersion, and, as Lord Rayleigh has remarked, in
theory:   connection with capillarity. In a letter to Arago (Jan. 12, 1817), he reports:
"I have been reconsidering the theory of capillary attraction and have at last
fully satisfied myself with respect to the fundamental demonstration of the
to explain general law of superficial contraction, which I have deduced in a manner at once
capillarity: simple and conclusive from the action of a cohesive force extending to a con-

ever be placed the (subsequent) electric atoms of W. Weber, which
may be held to have been somewhat unduly discredited by the
destructive criticism of Helmholtz.

Weber's
electrons.

### The Daltonian Atoms.

Early in the nineteenth century the time had come for the translation of these dim physical perceptions into secure experimental
knowledge. Perhaps the new feature developed by Dalton is at
bottom describable as the principle of the essential homogeneity of
each pure substance, that it is composed of molecules of only one
type, absolutely alike. Once it is postulated that only one kind of
aggregation into molecules occurs, *e.g.* that in water there is only one
way in which the hydrogen attaches itself to the oxygen, the laws of
definite and multiple proportions are self-evident. The only way to
ascertain the truth of this hypothesis was to test the consequences
experimentally. In the hands of Lavoisier it had become clear that
in chemical transformation mass does not to a sensible extent ever
disappear or re-appear, that chemical operations are not attended by
dissipation or destruction of matter. In the hands of Dalton it became clear that each type of substance is characterized by its own
specific type of aggregation of constituent atoms, by its own molecule.

Dalton's
principle:

Lavoisier's.

At that time neither principle could have stood out in the full light
in which we are accustomed to view it now. Physical ideas had
retrograded since Newton's day. The heat which to his view seemed
so obviously to be vibratory motion, due to the clash of atoms under
their specific energies, had come to be regarded since Stahl, aided
perhaps by a misreading of Black's doctrine of specific and latent
heat, as a substance combined in various proportions with different
bodies, on the idea that it is only something material that could be
conserved.

The caloric
doctrine.

And, moreover, as in all fundamental advances, the result attained
was not so much the vindication of any inflexible experimental fact,
as the introduction of an abstract guiding principle into the Science,
fortified of course by experimental support. For it is still a legitimate
aim of experiment to try whether any detectable change, either in
mass or in gravity, is produced by that re-aggregation of atoms to
form new molecules, which constitutes chemical reaction. Some day
the ascertained fact, that such influence of the close partial super-

Are masses
additive?

siderable number of particles within a given invisible distance. This solution
has very unexpectedly led me to form an estimate, something more than merely
conjectural, though not fully demonstrative, of the magnitude of the ultimate
atoms of bodies; in water for instance, about" $10^{-9}$ cm. in diameter (modern
estimates being nearer $10^{-8}$ cm.). Young's *Works*, ed. Peacock, Vol. I, p. 382;
also article "Cohesion," *loc. cit.* p. 462.

his estimate
of atomic
magnitudes.

position of the fields of energy of the adjacent atoms in the molecule is at any rate almost infinitesimally small, may play a part in the The reason as suggested by gravitation. elucidation of the mode of operation of gravitation. The essentially cognate fact, that no intervening obstacle can modify sensibly the gravitation of two masses, has already drawn us towards the position that the nuclei of the fields of stress, which constitute the physical aspect of atomic forces, are excessively small compared with the distances apart of these interacting nuclei in the fields of activity which are the atoms.

Mobile chemical equilibrium: Berthollet: its implications. In the same way the Daltonian principle, of a definite molecule for each substance, now stands in intimate connection with the Berthollet idea of statistical or mobile equilibrium, which requires that in the active interchange that is always going on among ultimate constituents of a substance, all the possible types of molecules which have any degree of stability must be present in some amount, though in most cases practically infinitesimal. It has also to take cognizance of fundamental considerations of a biological character, which will be referred to later.

The *rôle* of theory. While theory is aimless and impotent without experimental check, experiment is dead without some theory, passing beyond the limits of ascertained knowledge, to control it. Here as in all parts of natural knowledge, the immediate presumption is strongly in favour of the simplest hypothesis; the main support, the unfailing clue, of physical science is the principle that, Nature being a rational *cosmos*, phenomena are related on the whole in the manner that reason would anticipate.

### Radiation.

In sketching the progress of the purely physical notion of atomic structure there is also another fundamental order of ideas, arising from the phenomena of radiation, which must be included—the Atom as a vibrator. conception of an atom or molecule as a vibrating system of some sort, complete in itself and reacting by resonance with such waves of radiation as have periodic times adjacent to the periods of its own free vibration.

In the famous memoir "On the Theory of Light and Colours" read by Thomas Young before the Royal Society on November 12th, 1801, which, in the form of a mass of brief and pregnant suggestions, lays the foundation of modern physical optics, the view of the refraction of light, as due to the reaction of natural free vibrations of the constituent parts of the refracting medium, had already been advanced. The passage perhaps demands quotation[1]. After giving a correct

[1] *Lectures on Nat. Phil.*, quarto edition, Vol. II, p. 623.

*aperçu* of the mechanism of total reflexion, as involving and being supported by surface waves in the rarer medium, he proceeds as follows:

*Proposition VII.* If equidistant undulations be supposed to pass through a medium, of which the parts are susceptible of permanent vibrations somewhat slower than the undulations, their velocity will be somewhat lessened by this vibratory tendency; and, in the same medium, the more, as the undulations are more frequent.

For as often as the state of the undulation requires a change in the actual motion of the particle which transmits it, that change will be retarded by the propensity of the particle to continue its motion somewhat longer; and this retardation will be more frequent and more considerable as the difference between the periods of the undulation and of the natural vibration is greater.

It is hardly possible to extract definite meaning from this cryptic explanation*: indeed, the dynamics of a compound vibrating system is nowhere treated by Young; but, at any rate, he would have been quite prepared to predict the modern phenomena of anomalous optical dispersion as arising from the sympathetic vibrations of the molecules of the material medium. Later (1817), on again taking up the subject of Physical Optics, in his article in the *Encyclopaedia Britannica* on "Chromatics," he seems to have been dominated by the new puzzles connected with his principle of polarization due to transverse vibration, and this mode of explaining dispersion is dropped.

What are these parts of bodies—solid, liquid, or gaseous—which are thus taken to be susceptible of permanent vibrations of their own? At any rate, the hypothesis implies a thoroughly discrete structure of matter. And it is perhaps remarkable that Young was not tempted to ascribe the slowing of the period belonging to a given wave-length to the mere loading of the aether by inert structureless particles of matter, in the manner of the explanation of dispersion which Cauchy and Poisson afterwards imposed on optics; he went to the root of the question, in the Newtonian manner (as he remarks in the appended Corollary and Scholium), by ascribing to these particles free periods of intrinsic vibration, and therefore definite and identical structures.

Nothing further is heard of this point of view until the origin of the lines of the spectrum, and the mechanism of the production of the dark Fraunhofer lines, were discussed between Stokes and W. Thomson in 1854[1]. In 1860, in introducing the practical discovery of

*Marginal notes:* Young's view of total reflection of light: his guess at dispersion through atomic vibrations, far in advance of later theories,

---

* It may reasonably be read as ascribing dispersion of the normal type to absorption in the ultra-red, whereas the exact dynamics of nearly a century later would locate it in the ultra-violet.

[1] G. G. Stokes, *Math. and Phys. Papers*, Vol. IV, 1904, Appendix.

until Stokes' theory of absorption spectra: spectrum analysis by Kirchoff and Bunsen to the notice of the British public, Stokes based his dynamical explanation of selective absorption of radiation on the simple remark that, when the waves passing through the medium are closely attuned to it, so as to induce strong sympathetic vibration in the parts of the medium which they traverse, the radiant energy emitted by these active vibrators must be supplied from that of the exciting train of waves.

But it appears that, in following out this chain of ideas on the nature of optical absorption, it had not occurred to either Stokes or Thomson to consider the reaction exerted by the vibrating absorber on the train of waves; and the elucidation of the dispersion of colours in light as due to the induced vibrations of the molecules, which had naturally presented itself to Young at the beginning of the century, as utilized by his successors, remained to be enforced by Maxwell, Rayleigh[1], Ketteler, and especially Sellmeier, having been as it seems definitely perceived and especially Sellmeier and Helmholtz. sought for experimentally, in the form of anomalous dispersion near a region of absorption, by the latter, as early as 1866. As the matter Rayleigh's model. presented itself in luminous brevity to Lord Rayleigh, the medium is capable of free standing vibrations after the manner of a plucked string, executed in a periodic time which is determined for each type of vibration, *i.e.* wave-length, by its elasticity and inertia alone, provided it contains no independent internal vibrators that could be excited cumulatively by this motion: but where it contains structures that can set themselves vibrating in sympathy, the circumstances are analogous to those of an oscillating string on which light free pendulums with periods of their own, near those under consideration, are hung: the regular swing of the system now takes place in a modified time, and the velocity of propagation, determined by the ratio of wave-length to periodic time, must be altered, precisely in the manner that optical observation confirms.

The atomic constitution of matter is thus involved, in a highly refined manner, in the group of phenomena connected with the chromatic dispersion and absorption of radiation. But to gain conditions of ideal simplicity, we must attend to the case of gases where each molecule is isolated and free from encounter with others during the period required for thousands of its optical vibrations, where in fact it can be treated as vibrating free. The sympathetic vibration and resulting absorption along the spectrum, in liquid and solid bodies, are influenced in complex ways by the transient combinations

Maxwell. [1] See an early paper by Lord Rayleigh, *Scientific Papers*, Vol. I, pp. 141–6 (1872)—prior to the publication of Sellmeier's demonstration, but, as he now thinks, possibly written with recollections of Maxwell's ideas—in which the practical inadequacy of mere differences of passive optical density is insisted on.

of the molecules into groups; though the general relations elucidated by experiment between specific mean refraction and chemical structure suggest something often approximating to mere simple aggregation.

*Chemical equivalents in refraction.*

### The Function of Conceptual Models.

In the case of every successful scientific theory, the time must come when its first easy triumphs become exhausted, and what prominently confront the investigator are its outstanding defects and difficulties. When this stage arrives, one way of saving appearances is to purify the theory by banishing all terms which have an illustrative or analogical connotation, expressing its verified relations alone by new words which represent simply general types of mathematical quantity —vectors, scalars, rotors, etc. When such a state of crystallization has fully set in, further progress in general views is hardly to be hoped for—the sources of invention are dried up—though details in such a restricted abstract scheme will continue to be filled in, while new phenomena will probably suggest arbitrary unexplained additions to its content.

*Models soon reach their limits:*

*the results are then abstracted and crystallized.*

Even in chemical philosophy it has at times been a matter of concern that, for example, water is described as containing oxygen and hydrogen, whereas really it retains precisely none of the properties of either of these substances. Though it be admitted that it is constituted of molecules, yet the molecule of water is something different from its constituents; and it is held to be a crude or even unwarranted image that suggests that in it an oxygen atom and two hydrogen atoms lie alongside either at rest or in orbital motions. Criticism like this attaches to all inferences that cannot be tested by direct sensual perception. What we can know in any direct manner about chemical combination is expressed merely in the laws of definite and multiple proportions. Such a revision of the mode of expression of our knowledge as this criticism suggests may be useful occasionally as a stocktaking; but the misconceptions which it guards against are seldom real, and indeed it makes little, if any, permanent appeal on the physical side of the science. Here almost everything has been constructed on the basis of dynamical ideas—those fundamental Newtonian ideas of force and inertia which constitute the simplest formal scheme that admits of permanence of free motions—applied to conceptual models; such a theoretical representation is never perfect or complete, but it is vivid and illuminating, and historically it has been progressive; to give it up would be to replace a growing system by a collection of fragments of knowledge. The physicist in his own range is never likely to forget that any simple piece of matter is a vast interlacing, interdependent complex, which he can never hope completely to disen-

*Nature of chemical synthesis:*

*the appeal to dynamics.*

tangle or resolve: he is certain that matter is of grained structure, but to him the grains are very far from being mutually isolated things— each of them is actively influenced by all the others around it. Yet he has no alternative but to hold that each ultimate grain is itself a self-existing cosmos, of complexity probably beyond any complete analysis on our part, which may indeed to appearance merge itself in combination with another atom or molecule, but is always recoverable unaltered—that there is no degradation of matter. He holds probably that it is necessary to believe that in the same pure substance the molecules are all exactly alike, or, at any rate, that they are as nearly alike as individuals of a very sharply defined species in the organic world; though he knows no natural reason which would compel them to be so constituted, except in so far as they may represent the limited number of types of dynamical structures that can be built up from simpler identical primordial elements. It is vastly more suggestive to accept this wonderful inference, which constitutes the Daltonian theory, as our working hypothesis, than to try to refrain altogether from analogical reasoning about unseen molecules: moreover, this procedure is almost imposed *à priori* by the general principle already alluded to, that the simplest theory is probably the most fruitful representation of reality.

There is one branch of actual observational knowledge in which this identity of the molecules of a substance asserts itself with special strength: if the molecular theory had not been introduced on the evidence of the laws of definite and multiple proportions in chemical compounds, it must have demanded recognition as a result of a study of the crystalline structure of bodies. We call to mind that correspondences are now coming to light by which it is becoming possible to reason regarding the type of the molecule, and the geometrical grouping of its constituent atoms, from measurement of the crystalline aggregate: in such cases the single molecule would itself be the ultimate formative crystalline element. Where an atom has a higher valency, it must, according to any formula of spatial chemical constitution, aggregate more atoms around it and in touch with it in the molecule: it must, on that account alone, itself occupy or exist in a larger central space. In this way greater atomic volume would be in general a result of greater valency, while the atomic volume will always be nearly the same in similar surroundings: the very striking recent investigations to ascertain how far the structure of the crystal is determined by the arrangement of the atoms in its molecule on the basis that equivalent atoms require about the same atomic volume, are known to all of us here[1].

*The atom as a cosmos,*

*of unlimited stability,*

*identical within each of the types,*

*narrowly limited in numbers.*

*Emphatic evidence of crystals.*

*Atomic volumes.*

[1] Barlow and Pope: cf. *Trans. Chem. Soc.* 1906.

The contrast has recently been sketched by Professor Voigt in eloquent terms between this domain of the properties of crystals, where all is definite, orderly arrangement, and that of liquids and gases where physical properties are merely average values which belong in the statistical sense to crowds of jostling molecules. But even here of course the regularity is limited; the molecules become confined more or less securely in definite positions by the mutual forces of cohesion, but not so firmly as to prevent them from taking part in the conduction of heat and other modes of equalization or dissipation of energy; the very bonds of cohesion are themselves functions of the temperature. *Precision of crystal-physics:* *has limits.*

The tendency of most physicists would still probably be to take comfort from a remark of Helmholtz, published in one of his letters, to the effect that organic chemistry progresses steadily and surely, but in a manner which, from the physical standpoint, appears not to be describable as quite rational. Yet as time goes on it becomes increasingly difficult to resist the direct evidence for the simple view that, in many cases, chemical combination is not so much a fusion or intermingling of the combining atomic structures, as rather an arrangement of them alongside each other under steady cohesive affinity, the properties of each being somewhat modified, though not essentially, by the attachment of the others; and that the space formulae of chemistry have therefore more than analogical significance. The many instances, thermal capacity, refractive index, etc., in which the physical properties of the compound molecule can be calculated additively with tolerable approximation from those of its constituent atoms, are difficult to explain otherwise. The crystallographic evidence has already been referred to. *Organic chemistry largely schematic.* *Architecture of organic molecules:* *additive qualities.*

### The Spectrum.

Yet the spectrum, which the physicist is accustomed to regard as the most complete (though largely undeciphered) index of the structure of the molecule, is totally different, at any rate in the simpler combinations as compared with single atoms, unlimited groups of lines (forming bands) taking the place for the molecule of the single lines of atomic spectra. It may be permissible to believe—it is now in fact widely accepted—that no stimulation of an atom, less violent than complete disruption of some molecule in which it exists, can suffice to excite sensibly its atomic line spectrum. But there seems to be more correspondence between the absorption spectra of complex molecules and those of the molecules or radicles of which they are built up. The difference is fundamental between the firm, almost unalterable structures which are the atoms, and the molecules, con- *Contrast of the spectrum.* *Origin disruptive.*

sidered as intimate definite aggregations of atoms capable of definite disruption; it ought perhaps to involve the corollary that sensible
*Enhanced lines.* internal vibratory disturbance in the former is far more difficult to excite than in the latter. In the case of molecules of easily condensable gases, between which, therefore, strong mutual forces come into play, there seems to be evidence that the natural thermal collisions alone can excite selective radiation, as indicated by the appearance of
*Band spectra.* band spectra under conditions of high temperature without chemical or electric action: in the case of atomic line spectra the negative results of Pringsheim and other observers seem consonant with what
*Possibilities of internal equilibrium.* was to be expected. Line spectra are of very great luminous intensity compared with any natural continuous spectrum on which they can be superposed, unless the temperature of the latter is extremely high, and therefore the molecular collisions very violent. This seems to afford sufficient reason why they cannot be considered as in any kind of energy equilibrium with the surrounding continuous radiations.

An adequate interpretation of the master clue to dynamical molecular structure afforded by the spectrum is still lacking. The researches of Liveing and Dewar, Balmer, Rydberg, Kayser and Runge, Rayleigh, Schuster, and others, have led to the division of the simpler line
*The spectral series:* spectra into correlated series of lines, with the successive vibration frequencies in each series, after the first one or two, determined by simple approximate formulae, obviously the asymptotic forms of more complex exact relations which remain to be discovered; but very little progress has yet been made towards the dynamical interpretation of this ordered system. The radiations from electrons involve their accelerations, while those from ordinary material vibrators, as, for example, in the case of sound waves, depend only on velocities; thus,
*of special mode of origin.* as Lord Rayleigh has remarked, it is hardly surprising that the law connecting the overtones (so to speak) with the fundamental in each spectral series is of a type that is not met with in ordinary dynamics. A probably easier problem, as yet unravelled, is the mode of genesis
*Band spectra:* of banded spectra; here the law connecting the frequencies of the series of lines which constitute a band is of type not unfamiliar[1], but the known conditions in which these relations occur seem rather complex for an ordinary molecule. The facts that increased density of the
*their physical properties.* surrounding medium does not shift the bands, and that the Zeeman magnetic effect is absent in bands, are very pertinent to the problem: it has been thought that these facts are somehow correlated: it may well be that the former indicates close concentration of the steady aethereal vibration into the interatomic spaces in the molecule[2].

[1] Cf. *Ency. Brit.* ed. IX, supplement (1900), art. "Radiation."
[2] Cf. *Astrophysical Journal,* 1907, p. 120 [*supra*, p. 341].

But though the problem of the dynamics of the spectrum has not hitherto yielded much under the accumulation of knowledge, the primal property of the spectrum as an analytical agent remains unimpaired. It is still true that the occurrence of a definite line marks the presence of a definite substance. With variation of the conditions of excitation that substance may or may not emit the line in question: but wherever the line is seen the inference backward is still valid, though high dispersion may of course be requisite to distinguish it from closely adjacent lines. The inference as to the presence of a substance is easy; but it would be far more difficult to establish its absence. *Excitation of lines.*

Among the remarkable results of recent research in this field are those of R. W. Wood on the fluorescent spectrum of sodium vapour, and the conditions of its stimulation—for instance, the fact that excitation by a homogeneous vibration of period close to one of the free periods of the vapour excites the emission of the line of that period, accompanied by a series of lines, equidistant in frequency, ranged on both sides of it. The similarity to the circumstances of G. W. Hill's (and Adams') dynamical analysis of the lunar perturbations, first transferred to problems in other branches of physics by Lord Rayleigh, has been remarked more than once[1]. On the general view that a molecule is a structure such as a vortex ring or an orbital system of electrons, it has intrinsic cyclic motions of its own, which must, however, be so balanced as to avoid the unlimited draining away of its constitutive energy by radiation; we should then anticipate that each of these structural cyclic periods would interact with the periodic disturbance exciting a natural vibration, and give rise to the analogues of summation and difference tones, if we may use the acoustic terminology. On this view the common difference of the vibration frequencies, in one of Wood's series, would be the frequency of an intrinsic cyclic motion of the kinetic system which constitutes the molecule. Here again Wood has found that many of the lines which exhibit these phenomena are specially sensitive to magnetic influence. *R. W. Wood.* *Complex stimulation by a pure period:* *analogue in the lunar oscillations.* *Interaction with cyclic latent periods in the molecule.*

Recent investigation seems still to confirm that in a given spectrum the lines that shine out brightly are determined by the temperature, without concomitant chemical change: lines which appear conspicuously at one range of temperatures do not show sensibly at another, where others of the same system show up instead*. We are thus required to imagine a way in which the mode of electric or other excitation of the same molecule may be different at different tempera- *Effect of temperature.*

[1] *E.g.*, A. Stephenson, *Phil. Mag.* July 1907, p. 115.
* The theory of mobile equilibrium of electronic combination advanced by Saha now illuminates these questions.

tures, just as among the tones that belong to a bell the particular ones that ring out from it depend on the mode in which it is struck. The increased translatory motion at higher temperatures can hardly make a difference in the vibration; there remains the increased spinning motion accompanying it, which at first sight offers some promise. It might even be asked whether, in the dissociative equilibrium which arises in some gases at high temperatures, it may not be just as likely that the atoms should slip apart gradually as the result of the increased whirling motion of the molecule at the higher temperature, as **Changes must be abrupt,** that they are broken apart at the time of collision. But if ultimate separation can thus gradually arise, the earlier stage of merely modified configuration would change the periods of the spectral lines or bands of the molecules in which it occurs, and if there were enough of them, it would show as a widening of the lines of the spectrum. Thus, whatever may be the case for the far smaller numbers of degrading radioactive molecules, for ordinary gases no way is open in this direction: the change must be abrupt, and the manner in which the molecule is struck, the nature of the collision, must somehow supply the cause of the variation in the intensity of the lines. Or it may be that the dissociation which accompanies the production of a **and transient.** line spectrum takes place in successive transient stages (cf. p. 364) and that the durations of these stages have an influence on the relative brightness of the lines.

### Electric Phenomena.

A survey of the general features of the atomic theory would be far from adequate which omitted the fundamental atomic properties **Faraday:** announced by Faraday in 1834, in following the path opened up by Davy's electrochemical work, and carefully formulated by him under the new terminology of electric ions. It must have presented itself from the first to the mind of any atomist who was daring enough, in defiance of Faraday's own express caution, to transcend the limits of **atoms of electric charge,** experimental observation, that here we have to do somehow with electric atoms, and that the essence of chemical change is involved in the passage of these entities across from molecule to molecule. Maxwell contemplated but shrank from making this plunge, being in fact fully occupied in a more accessible and equally fundamental subject, the mode of transmission and propagation of electrical influence; it **enforced by Helmholtz.** was Helmholtz, further removed and thus not so much under Faraday's direct influence, who in his Faraday lecture to the Chemical Society of London in 1881[1] and elsewhere, first marshalled sys-

[1] "On the modern development of Faraday's conception of Electricity," *Wissen. Abhandl.* ii, pp. 52–87. Cf. Faraday, *Experimental Researches in Electricity*, Vol. i, Nos. 852, 871.

tematically the evidence that electricity must be atomic, and that the main energies of chemical affinity are, as Faraday held, of electric type. More recently the electron, or electric atom, has become a necessary idea for electrodynamic theory, if that is to include the origin of electric disturbance as well as its mode of propagation. At first the view was natural that the electron could be transferred only during the intimate encounter of molecules, and so could hardly exist free: though Helmholtz had enforced the idea that electric excitation by friction consisted in the rending of the molecules into material ions by the mutual forces arising on the intimate contact between dissimilar substances thereby produced, and later investigations have been concerned with traces of the same kind of phenomenon appearing in chemical reactions such as ionization of air by phosphorus. The great experimental discoveries of the last dozen years (J. J. Thomson, Schuster, etc.) have, however, shown that in rarefied gases, where the molecules are far too widely separated for any appreciable amount of immediate interchange, disruptive electric discharge establishes itself by driving the electron into the open, where its intense and extended field of energy has made it far more amenable to physical scrutiny than the more self-contained neutral material molecules could ever themselves have been. The enormous speed with which it travels, approximating often towards that of radiation, which is the maximum conceivable in an aether not subject to rupture, can perhaps be ascribed only to a comparable orbital velocity when the electron is in the molecule, from which it occasionally glides away through some kind of overbalancing of the internal kinetic adjustments, either enforced by disruptive electric excitation or spontaneous as in the case of radioactive substances. It appears significant that in the atomic disintegration involved in the later case the emission is at much greater speed, almost up to the limit of what is possible.

The electron:

now found free,

and conspicuous.

Source of intrinsic high speeds.

Considerations of this kind concur with the very precise magnetic subdivision of the lines of the spectrum discovered by Zeeman and Lorentz. Since Maxwell's coordinating analysis we are certain that light and radiation generally are phenomena of electric disturbance: or more precisely, as he put it, we know that electrostatic phenomena are manifestations of strain, and electromagnetic phenomena are manifestations of interaction of strain and inertia, in the same medium whose undulations constitute radiation. As soon as a definite structural picture could be formed for an atom of electricity as a region of strain abutting on a central nucleus at which it is locked together, even though the free mobility of the nucleus has to be assumed rather than understood, the suggestion of an electric theory of matter on the analogy of the vortex illustration lay open to dynamical development.

Magnetic influence on radiating atom.

Possible purely electric type of atoms:

Such a picture must, on principles clear ever since the time of Ampère, give positive and negative electrons, which differ essentially only as a system differs from its mirror image, or as a right hand differs from a left hand: and the suggestion was obvious, to pass from the Daltonian principle of the identity of all atoms of the same substance, to the *and their* hypothesis of the equality of all electrons, except as regards this *ultimate* distinction into positive and negative. We are thus invited to discuss *electrons.* how far progress is possible towards a mode of representation of physical nature on this foundation alone. It will help towards wider *Scope must* synthesis; but we know in advance that it will hardly avail us further *be limited.* than the interactions between molecules across intervening aether.

The early extension of Dalton's principles which arose from Gay-Lussac's laws of multiple combining volumes in gases, had already *The* proceeded in this direction. To Avogadro the Daltonian atom itself *molecule.* appeared as a more ultimate constituent in the molecule, which is the actually subsisting discrete element of matter[1].

## *Limitations.*

The fundamental limitation of any conceivable atomic theory, *Atomic theory* which was emphasized by Maxwell in the seventies[2], seems to have lost *transcended* none of its force in the years which have since elapsed. The trans-*in biology.* mission of qualities from ancestors to descendants in organic nature, now studied under the name of heredity, is of far too subtle a cha-racter to be managed by aggregations composed of discrete atoms of matter as large as we know the Daltonian atoms to be. The identity of the atoms is a safe foundation in chemistry and physics; but if an attempt is made to carry it over into biology we get lost in two directions, both in the enormous complexity of the organic chemical molecule, and in the infinite delicacy of transmission of characteristics in organic life, the latter being in sharp contrast with the rapidity of *The rapid* the spread of its lower forms, once inoculation has taken place under *spread of life.* suitable conditions as to food. We are entitled to conclude, not that there is here any essential contradiction with the atomic theory, but rather that the complexities of the phenomena transcend our powers

*From Dalton*   [1] The working of the presumption in favour of the simplicity of Nature is *to Avogadro.* illustrated by the relations of Avogadro to Dalton. For want of the simplifying idea of the molecule Dalton was logically compelled to reject the view, after-wards associated with Avogadro, that all gases contain the same number of particles: but he fell back on the next simplest hypothesis, by asserting that the particles of all gases are accompanied by the same amount of caloric in their "heat atmospheres." On the other hand, Avogadro demurred to this hypothesis, as really irrelevant to the particular question at issue. Yet Dalton's instinct for simplicity proved to be right, when interpreted into energy, though the re-pelling heat atmospheres of the period were wrong.

[2] *Encyc. Brit.* art. "Atom"; *Collected Papers*, Vol. II, p. 461.

of mental analysis. Will they always do so? Every new physico-chemical explanation in physiology is fresh evidence that the processes, so far as they can be extricated, are all rational; wherever the complex of phenomena can be partially disentangled, we find order. The only method of progress is to seize the salient or large-scale manifestations of order as they present themselves, and by correlating them and fitting on new regularities thereby discovered, to go on improving the working scheme of representation. The model is not erroneous, because it is incomplete; explanation is usually worth more than criticism: an imperfect physical representation which has stood the test of substantial prediction can be improved, but it is not often absolutely refuted. Whatever be the world of reality behind and within the Daltonian atoms, of which at present we can form no idea, we are entitled by all experience to push the consequences of their physical interaction as far as possible, without fear of meeting irresolvable contradiction. *[margin: Extrication of principles.]* *[margin: Method.]*

This digression, and the related postulate that the simplest representation of Nature is the most probable and effective one, receive illustration from the outstanding puzzle of the electron theory, already referred to. Why do positive electrons not exist? Put in this crude way the question is absurd. They do exist, and they are equal quantitatively to the negative electrons; otherwise ordinary molecules would not be neutral. But why are they apparently so different in structure, whereas we should expect them to be mirror images of the negative ones? The phenomena in this domain recall the other question, why have the albumens and most other vital products evolved on this planet left-handed optical structure? It may be that the union of positive with negative electricity is something much more intimate than mere orbital motion; for the structure of the nuclei of electrons is as yet totally unconceived. It may thus be that in the one or more types of molecule that happen to be practically amenable to the expulsion of an electron into the open, it is the negative that can be readily driven out. If the central nucleus which knots together the field of force constituting a positive electron were large, such as, *e.g.*, might be represented by a sphere of electrification of atomic dimensions, the intensity of its field of strain would be small throughout, its energy and inertia would thus be slight, and its effects would not be prominent except as a mere centre of attraction. All electric inertia would then belong to the negative electrons. Is there any other inertia belonging to the atom besides electric inertia? This is the same as to ask, Is the atom something foreign to the aether and self-existent, which, however, can attract and hold certain aethereal nuclei called electrons, constituting a connection with the aether *[margin: The positive electron: why absent.]* *[margin: An analogous question.]* *[margin: Selectively.]* *[margin: Large positive would be nearly devoid of inertia:]* *[margin: implies inertia that is not electric.]*

through which atoms, which would otherwise be isolated worlds, are in physical relation with one another to form a *cosmos*? Such an extension of our conceptions may as we have seen be welcome in biological science, involving as it would that atoms in intimate entanglement may interact differently from the simple and regular methods which obtain when the aether is their only mode of inter-communication. But all who have appreciated the course of evolution of the principles of modern physical explanation through Newton, Lagrange, Young, Fresnel, Faraday, Stokes, Kelvin, Maxwell, Helmholtz, Hertz, to mention only a few names of the past, will still hold that even if

*The more physical atomic interaction must be electric.*

the atom itself is intrinsically unfathomed, yet the interaction between atoms separated in space is in its larger features understood and has its seat in their sub-electric connections with the aethereal medium.

### Chemical Reaction in Gases*.

The laws of chemical equilibrium in the rarefied state of matter, with free molecules, which belongs to gases, ought to be the most amenable to the direct indications of molecular theory: but in practice it is here that complexities show most prominently.

*Willard Gibbs' studies in dissociation of vapours.*

The earliest and one of the most complete numerical discussions of the facts of the simpler phenomena, those of gaseous dissociation, was made by Willard Gibbs himself in 1879[1]. In the four cases treated he satisfied himself that the observations of pressure were consistent, on his own principles, with an equilibrium involving simple binary dissociation.

But subsequent detailed investigations of gaseous reactions, mainly by van't Hoff and his school, showed that often, especially when more than two molecules were concerned in the change, the results were abnormal both as regards speed of reaction and ultimate equilibrium

*Effect of foreign bodies.*

—and in fact indicated that the main part of the reaction occurred not in the volume of the gaseous mixture, but in contact with the walls of the containing vessel.

A discussion of these difficulties in detail is hardly to be ventured in this place, where so much of our knowledge as to the circumstances promoting or inhibiting reaction in pure gases has been acquired. But just for that reason it appears desirable to embrace the opportunity to refer to a principle, which so far as my reading goes, seems to have

* The subjects here glanced at have now expanded into vast range and experimental complexity: reference may be made to C. N. Hinshelwood's recent book *The Kinetics of Chemical Change in Gaseous Systems*, Oxford, 1926.

[1] "On the vapour densities of peroxide of nitrogen, formic acid, acetic acid, and perchloride of phosphorus." *Silliman's Journal*, (3) Vol. XVII: *Scientific Papers*, Vol. I, pp. 372–403.

escaped attention[1]. In my copy of the English edition of van 't Hoff's *Studies in Chemical Dynamics* (1896) I find the following memorandum of ten years ago. "It would be a corollary from this [estimate of chances of encounter of molecules] that in gaseous reactions where more than two molecules are concerned, such as $2H_2 + O_2$ and $2CO + O_2$, the chance of them all being in contiguity at the same instant is extremely small compared with that for two, and the reaction can therefore only proceed very slowly, unless it proceeds by intermediate binary reactions such as the formation of $H_2O_2$ in the first case or reaction with contained water vapour in the second case. This superior velocity of binary combination has possibly a bearing on the specific catalytic action of traces of certain foreign vapours in facilitating explosive combination, as determined by Dixon." I see no reason to abandon the conclusion thus expressed. It is quite sound to reason, in the statistical manner introduced by Guldberg and Waage, that the relative number of direct diad combinations between molecules of types $A$ and $B$ is $k_{AB} . n_A . n_B$, where the factors of type $n$ are the respective numbers of these molecules, and that the number of direct triad combinations between types $A$, $B$ and $C$ is $k_{ABC} . n_A . n_B . n_C$; but it would appear that the coefficient $k_{ABC}$ of the latter combination must be almost infinitesimally small compared with $k_{AB}$. Thus if we imagine the scale of magnitude of a gas at a pressure of one atmosphere to be magnified so that the diameter of each moving molecule becomes about one inch, there will be in the model roughly one molecule in each cubic foot, and a molecule will have to travel about a hundred feet before it encounters another one. Such binary encounters will thus happen with some frequency, and from some of them combination may ensue. But the chance of three molecules coming together simultaneously is negligible: the only way in which a trimolecular combination can arise is by one of the molecules

*Marginal notes:* Ternary reactions proceed in stages. — The scheme of Guldberg and Waage.

---

[1] I find that Professor H. B. Dixon has reasoned from the extreme rarity of trimolecular encounters, in a paper on "The Mode of Formation of Carbonic Acid in the Burning of Carbon Compounds," *Trans. Chem. Soc.* 1896 (cf. p. 777), which also brings to a focus many of the peculiarities of gaseous reaction. In the same paper it was announced that the bimolecular reaction of CO with $N_2O$ is not excited by the electric spark when the gases are well dried, though the mixture becomes explosive on the addition of water vapour. The argument in the text need not imply that every bimolecular reaction proceeds spontaneously; in such a case as the one quoted the ordinary explanation is appropriate, which regards a catalyst as opening a path of transformation around an obstructing ridge on the surface of available energy, such as would present a barrier to direct combination. As in electric phenomena, so in chemical change, the apparent anomalies of reaction in pure gases seem destined to provide the clue towards deeper insight, as, in fact, Lord Rayleigh pointed out long ago.

*Marginal notes:* H. B. Dixon on trimolecular encounters. — Thermodynamics of catalysts.

attaching to itself another, in a manner perhaps relatively transient, and this pair going off together to meet the third, each acting, so to speak, as a carrier for the one united with it. Without some such *Else tri-* intermediate transient stage of combination a dissociation of a com-*molecular* *dissociations* plex into three molecules must proceed to an end, for the chance of *must be* an equilibrium through recombination would be negligible. Where an *complete.* equilibrium is found to become established, either the reaction must occur in binary stages, or else it must take place in contact with solid *Adsorption.* or liquid boundaries where the molecules form a denser layer in which each is always in relation with others. This consideration reinforces the importance of the study of reaction in pure gases, as a means of disentangling the intermediate stages of chemical combination and the durations of the products formed in them. The fundamental importance of this kind of knowledge for the adequate interpretation of banded spectra has already been alluded to. It appears, indeed, to be *The actual* commonly recognized that direct trimolecular combinations occur *facts:* seldom: the inference from the present line of argument is that in gaseous reactions they do not occur at all. Recently I have learned *Mendeléef.* that Mendeléef had always maintained that gaseous reaction occurs in monomolecular or bimolecular stages in all cases: there seems to be strong presumption in favour of such a view.

It would require the instincts of a chemist to venture on any attempt to apply this principle to special cases, to discuss why, for *Catalysts:* example, the presence of a foreign substance sometimes promotes the occurrence of the necessary intermediate binary reaction, and in *and poisons.* other cases presumably destroys its product, and so inhibits the final transformation. The recent results obtained by Bone and Edmunds for the thermal dissociation of $H_2O$ agree with the conclusions drawn in 1884 by Dixon with regard to the explosion of a mixture of CO and $O_2$, in assigning the presence of hydrogen free or combined as a potent stimulus to ternary reaction[1].

### Ionization and Solution: Available Energy and Berthelot's rule.

At first sight it might appear that the principles of statistical equilibrium in dilute systems would afford a criterion as to the intimate process of dissociation, whether into ions or molecules. Thus, to *Partial* fix the ideas, the two reactions, $HCl = H + Cl$ and $2HCl = H_2 + Cl_2$ *equilibria* *fuse into a* would come to different equilibria, of the types $n = k \cdot n_1 n_2$ and *consistent* $n^2 = k' \cdot n'_1 n'_2$. But in reality the subsequent aggregation of H and *aggregate.* H into $H_2$ itself involves an equilibrium $n_1 = \sqrt{(\kappa n'_1)}$, so that the

[1] I have ventured to add an abstract discussion on the formal possibilities that are open, which was drawn up about a year ago, as an Appendix [p. 373].

discrimination is not possible on these considerations alone. Nor is it ever possible in this way on the thermodynamic theory, which can be seen (cf. Appendix) to be consistent with separate independent equilibria as regards every type of reaction that is formally possible in the system. Principle of separate equilibria.

The process of ionization in a liquid solvent is obviously very different from ordinary gaseous dissociation. The view that some such special type of dissociation is required in order to form a coherent mental picture of Faraday's electrolytic results, must really in strict logic go back to Clausius' ideas of about fifty years ago. It is true that he did not venture to suggest more than extremely slight ionic dissociation. But once the mere possibility is granted, there is no ultimate escape from the permanent ionic separation of Arrhenius: for it is only a question of making the solution more and more dilute in order to diminish indefinitely the chance of any two ions ever meeting again to unite, as compared with the unaltered chance of any remaining whole molecules becoming divided into ions. Complete ionization must ultimately arrive; and there is only the question remaining over as to the degree of dilution at which it is practically attained. Electrolytic ionization: Clausius: must be complete at ultimate dilutions.

It will be observed that we are in this argument applying the principles of mobile equilibrium to the ionization of the dissolved substance. Here an appeal to experiment becomes feasible. It was by Ostwald that this test of Arrhenius' view was first applied. As is well known, he found that for acids and salts that are but slightly ionized, *e.g.*, acetic acid, the mathematical relation expressive of equilibrium of simple dissociation is satisfied; but for highly ionized substances it is widely departed from. The verification in the former case appears to be sufficient by itself to confirm the general point of view. For it seems natural to suppose that high ionization indicates the presence of some type of direct affinity with the solvent, which is too powerful to be altogether omitted from the equation which expresses the ultimate equilibrium. How such an influence should be included is one of the main unsolved problems in this subject. At higher concentrations either the tendency to re-combination of the ions is resisted, or else the tendency to ionization of the molecules is promoted. Thus if increased ionization were a result of collision of molecules with the existing ions, in analogy with known effects of collision of (rapidly moving) ions with the molecules of gases in promoting further ionization, the equation of equilibrium would be altered from $n_1^2 = kn$ to $n_1^2 = kn + k'nn_1$, in which $k'/k$ would be sensible only for easily ionizable substances [cf. however footnote, p. 379]. The solution of this problem, if there is any simple colligatory principle, must, however, be a matter of experimental scrutiny. Ostwald's law for high dilution. Affinity: how it may operate.

The acceptance of the idea of ionic dissociation by solution has been opposed by scruples of a more fundamental kind, not altogether unlike the difficulties once attaching to the Berthollet doctrine that the extent to which a reaction proceeds depends on the relative amounts of reacting substance. It is a fundamental postulate that a molecule is a self-existent aggregate, whose intrinsic binding affinities are independent of temperature: as we have seen, one of the main *à posteriori* reasons for this conclusion is that each (sharp) line of the spectrum is characteristic of the molecule, alterable in position only very slightly, or not at all, by any change of physical conditions. And this view agrees with the kinetic theory which connects temperature with the average translatory (and concomitant rotary) motions of the molecules in space, and sometimes with partial dissociation, but not with any intrinsic change in structure in the molecules that are present. Thus the bonds of atomic affinity which have to be overcome, say in the ordinary non-ionic dissociation of a gas, are the same at high temperatures as at low. But occasionally a collision with another molecule may be well-directed towards breaking these affinities, like the sharp impact of a mason's trowel on a brick or tile, and as a rule it will be the more effective the higher the temperature. The verification of the theoretical law of equilibrium, in ordinary gaseous dissociation, enables us to assert that strong affinities are in fact occasionally thus shattered; that high affinity is to be measured not by entire absence of dissociation but by its relative rarity, though the products of disruption can be accumulated when the opportunities for recombination are removed. The dissociation into ions in a solvent is perhaps a more fundamental change than the ordinary dissociation of a gas: yet the ions may accumulate notwithstanding, for the obstacles to recombination presented by the dense molecular aggregation of the liquid in which they are entangled are also enormous in comparison with any that are present in the gas. Thus the circumstance that self-ionization is hardly detectable in a gas is not conclusive evidence that it never occurs without the assistance of a liquid solvent medium.

It seems worth while to follow up these relations somewhat further in the light of Faraday's conclusion, so emphatically enforced by Helmholtz in 1881, that the strongest forces of chemical affinity are of electric type. It would almost seem as if we must adopt the view that the active atom in ordinary chemical change is the ion, with its large intrinsic electric charge as an essential feature. No permanent state such as we associate with a simple dense material substance can be reached until these enormously active positive and negative bodies have become paired; until, in fact, their domains of activity,

*Marginal notes:*

Nature of affinity:

strong affinities no protection against dissociation.

Recombination obstructed in a denser medium.

Final state, one of neutral molecules.

in place of being the widely ramifying fields of force of free ions, are changed to the more concentrated and individualized fields of mole-cules[1], in which the lines of force instead of spreading out far into space simply pass across in more or less curved lines from one ion to its adjacent conjugate. No substance could exist completely ionized in free space for a moment, nor with any considerable excess of ions of one sign: ionization is, however, continually occurring in substances to a small extent, spontaneously and so to speak by accident, *i.e.*, in a manner not controllable, the duration of separate existence of the ions depending on the obstacles to their finding new partners. In the case of gases in which the formation of ions is stimulated by electric shock (Röntgen rays, etc.) knowledge is now highly developed (J. J. Thomson, Townsend, etc.) regarding their rate of formation, and its equilibrium with recombination or with their extraction from the region by an electric field: the processes in solutions are much more rapid, and only resulting equilibria can be directly investigated. The operations of inorganic chemistry consist largely in presenting to these ions of solutions the possibility of taking on new partners, either by simple admixture, or by pulling them away into a new environ-ment by electric agency. The energy required for the guidance of chemical processes is expended in this latter way, perhaps none of it goes (except very indirectly through thermal interchange) to the pulling asunder of the ions in the atom; that separation takes place sporadically, with purely local adjustment of energy, to an extent dependent, however, on the environment and in this way modifiable. *(margin: Stages traceable in a gas.)* *(margin: The rôle of energy in chemical change.)*

Any intense motional disturbance liberated in such ionization lapses into mere thermal energy. But by means of guiding control (constraints in bulk) little of it may be left to this fate—*e.g.* in the Daniell voltaic cell, as Lord Kelvin discovered half a century ago. Here again we recall the emphasis placed by Helmholtz, in 1881, on the inference that in a voltaic cell there is but slight expendi-ture of energy in getting the current across the solution, merely, in fact, the Joulean heat, while the large amounts of energy which become available or stored have their origin at the electrodes, in that process (often, as will be seen, nearly statical) of passing over the Faraday unitary electric charge from atom to atom, which is, therefore, the essence of the change in the state of chemical combination. *(margin: Thermo-dynamic efficiency can be nearly complete:)* *(margin: so reversible.)*

[1] In the molecular groups of solvent media, which have abnormally high dielectric capacities, the two conjugate ionic poles are so far apart that these groups may be held to occupy a position intermediate between ordinary mole-cules and active ions. Cf. next page. *(margin: Solvent effect of high dielectric capacity.)*

The source of the energy required for the natural degree of ionization which occurs when, say, K . HO is dissolved in water, has been a standing problem in this department. The current mode of explanation seems so far sound, that when a dilute base is neutralized by an acid, say K . HO by H . Cl, the resulting heat is derived from the free ions H and HO clashing together at high speed generated by their strong electric attraction, as $H_2O$ does not exist sensibly ionized— that therefore the heat evolved is about the same per chemical equivalent in all such cases. But when K . HO is added to water, whence comes the supply of energy demanded for the pulling apart of the K and the HO, into separated ions, against their mutual attraction? The concomitant absorption of heat is far too slight to account for it[1]. We are tempted to conclude that internal potential energy is released owing to the ions falling into relations of closer affinity with the solvent, and that the process is nearly a self-contained interchange of energy, reversible as regards each molecule separately, being a steady static drawing apart of the ions unaccompanied by the generation of violent subsidiary electronic motions whose energy would escape into the general store of heat. The fact above alluded to, that in voltaic batteries so large a proportion of the chemical energy is usually mechanically available, also seems to point this way; it shows, too, that the fundamental interchanges of electrons at the electrodes, which are the sources of the transformation of energy, are intrinsically of the same not merely reversible but almost static type, accompanied by but little energy of intense agitation such as would be partially dissipated into the surroundings. It may thus be the close quarters at which these operations are developed, in the liquid environment, that limit both the occurrence and the diffusion of irregular intense disturbance such as would pass away into sensible heat. In ordinary dissociation of the nearly free molecules of gases, accompanied by comparatively large changes of volume or pressure, or in the cognate phenomena involving osmotic expansion of dilute solutions already formed, a relatively greater degradation of energy is involved, and is indicated by the greater variation of the equilibrium on change of the temperature.

The history of this problem may be recalled. When Lord Kelvin opened up the subject of availability of chemical energy in 1851, he found by experiment that in a Daniell cell nearly the whole of the energy of chemical combination was available for mechanical work. Later Gibbs in 1878 and Helmholtz in 1882 pointed out that the change of electromotive force with temperature gave a measure of the

*Marginal notes:*
Constancy of heats of neutralization.

Ionization a nearly static balanced process,

as contrasted with dissociation in gases.

Kelvin's empirical law for voltaic cells,

amended by Gibbs and by Helmholtz.

[1] Cf. G. F. FitzGerald: Helmholtz Memorial Lecture, *Trans. Chem. Soc.* 1896: *Scientific Writings*, p. 363 *seq.*, also p. 521.

proportion of the energy that is not thus available: and the accumulating cases of discrepancy with Kelvin's principle thus became rational. Now, the problem is rather why the unavailable part proves to be often so slight, as compared with other chemical processes more thermal in character. Not merely can the operations be conducted in a nearly reversible manner, so that all the available energy is utilized, but in addition nearly all the energy of the chemical change is often actually available so that there is but slight evolution of heat where it occurs. It will be remembered that Professor Nernst devoted his recent Silliman Lectures to this subject. "To enable us to proceed it is necessary to find the conditions under which the principle of Berthelot comes nearest to expressing the true relation between [available] chemical energy and heat, or, what amounts to the same, between the magnitudes $A$ and $Q$. In this direction we can show that in reactions between solids, liquids, or concentrated solutions, the values of $A$ and $Q$ approach each other very closely, while on the other hand in dilute solutions or with gases we actually find large differences between the two qualities;...[1]" The case of galvanic cells operating even by dilute solutions is included in the generalization for the reason given above. Obviously also on the present view the unavailable part of the energy should become steadily less at lower temperatures, as Nernst concludes. The principle of Gibbs, that the fraction of the energy of chemical combination that is unavailable is equal to the ratio of the actual temperature of reaction to the temperature of dissociation, provided correction can be made for work of expansion and heat on change of state, etc., is seldom effective on account of this latter complication.

Nernst's condensed systems as non-dissipative;

Gibbs' references to the temperature of dissociation.

The Faraday unitary charges have now a specific name, the electrons. Their unchanging magnitudes were strong presumption from the first of their intrinsic atomic existence: the Zeeman-Lorentz effect has almost exhibited them to us in action in the molecule, as the agents of radiation through their combined vibratory motions, in the now familiar manner foreshadowed by the Maxwell-Hertz theory of radiation. But the most far-reaching of recent discoveries has been that not merely can they pass at close quarters from molecule to molecule in some hitherto inscrutable way, according to the Faraday law, and also reveal their vibrations inside the molecule through its spectrum and its magnetic modifications, but that they can be drawn out into the open by electric shock, and securely manipulated (J. J. Thomson, Lenard, etc.) as atoms of pure disembodied electricity in those cathode streams across highly rarefied

Free electrons.

[1] Nernst, *Applications of Thermodynamics to Chemistry*, 1907, p. 43 *seq.*, where extensive examples are given.

gases which Sir W. Crookes long ago insisted on calling a fourth state of matter.

In the electrolysis of an acid we are to imagine the negatively charged massive hydrogen ions, which happen sporadically to be free in the solution, as being slowly drawn towards the negative electrode by the electric field pervading the medium, as accumulating there with *Polarization* production of polarization reacting against this impressed electric *of electrodes:* field, until they are so crowded together by the constraint that some kind of instability arises, whereby one of them takes over, but without violent disturbance such as could diffuse away, two positive unitary *suggests* charges from the electrode. Surely these charges must in ultimate *transfer of* analysis be positive electrons, or else the power of losing negative *positive* *electrons.* ones must be unlimited. At any rate, in virtue of them, the ion can associate with another and become released as a free self-contained molecule of hydrogen, a very different thing from the mutually constrained ions that gave rise to it—the energy required for the electrolysis being expended mainly, as Helmholtz insisted, and, as we have seen, usually without much necessary waste, in this requisition of the two positive charges.

The relevance of the mode of operation of the Grove gas battery will here occur to mind; the finely divided or porous platinum surface *Reversible* promotes ionization of the gas alongside it through the opportunities *reaction in* arising from intimate contact in confined spaces (cf. p. 368), and so *confined* *spaces:* manages to utilize much of the available energy of gaseous combination, which in gases is very different from the heat of combination on account of the change occurring in the energy of expansion. Fortu- *gives scope to* nately for exact knowledge, the principles of thermodynamics give *dynamical* in all such cases the means of estimating the final result, without *methods.* requiring hypothesis as to the nature of the process that is involved, provided only it be reversible; in dilute systems the argument can be expressed in terms of the available energies, or thermodynamic potentials, of the constituents, interpretable in the simpler cases by partial osmotic pressures—in extension of which the idea of solution pressure of an ion can be employed as a graphic mode of expression, without implying that the processes involved can be placed in effective analogy with those of evaporation or solution. The very remarkable quantitative connection of the electric potential gradient between solutions with the diffusions of their ions, established by Nernst, must ever remain one of the solid foundations in this subject.

*Electric view* The modern expression of Helmholtz's provisional generalization, *of affinity.* that chemical affinity arises from intrinsic attraction of the different kinds of matter for electricity, may roughly be that every active atom is an ion, and that saturated inert molecules are welded into unity by each constituent atom keeping hold, through the aethereal agency of electric attraction, of the (perhaps interpenetrating) electrons belonging to the other. The electrical view provides a reason for the ordinary saturated inactive molecule of elementary bodies being often polyatomic, a fact otherwise of undiscovered import. The exceptions afforded by monatomic gases and metals indeed suggest themselves at once: but spectrum analysis shows that these molecules are intrinsically just as complex in sub-electric structure as the others, while the physical test of monatomicity perhaps only verifies that the components of the molecule are somehow so closely compacted that the thermal collisions do not induce sensible internal commotion.

*Atomic electrons held in common.*

But it is time to conclude this discursive survey. We have recognized that the Daltonian molecular theory is still the indispensable guide, if we wish to continue constructive efforts in the physical elucidation of Nature, and are not content to take down our scaffoldings for the sake of logical symmetry, and, in the future, make the most of the edifice as it now stands. While we are certain with Dalton that molecules are very definite, identical, structures, it has been seen that, when we inquire into the detail of their constitution, though many guiding principles mainly of electric and spectroscopic types have been made secure, yet we have not much more than their distant analogy with familiar dynamical systems to aid us. But for many branches of the science knowledge of detailed molecular structure is not required. The pioneering example of this kind was the kinetic theory of gases. The domain of electrodynamics is now securely founded on the displacements and movements of electrons, each of which may be considered merely as a point at which the unitary electric charge is concentrated, so small are the unknown nuclei of the electrons compared with their distances apart. In the same way the wide domain including the course and equilibrium of reactions in dilute systems can be studied by pure numerical statistics in the manner of Guldberg and Waage, or by the more generalized but fundamentally equivalent thermodynamic methods associated mainly with Willard Gibbs. But the aim of structural chemistry must go much deeper; and we have found it difficult, on the physical evidence, to gainsay the conclusion that the molecular architecture represented by stereochemical formulae has a significance which passes beyond merely analogical representation, and that our dynamical views must so far as possible be adapted to it. We have recognized that the interaction of atoms at a distance apart, which is necessary to a *cosmos*, is provided for by a very special mechanism, consisting in the activity through the aether of the electrons that are attached to them. The

*Daltonian model atoms still necessary:*

*and in some problems sufficient.*

*Point-electrons suffice for atomic models.*

*Reactions studied by statistics:*

*or by thermo-dynamics.*

*Static architecture of the molecules.*

*Electric nature of affinity.*

artificial aspect of this arrangement would be relieved if we could assume these entities to be of the essence of atomic structure; we are justified in following out this hypothesis as far as it can carry us; and the totally unexpected phenomena of disintegration of complex atoms, *The clue of radioactive disgregation.* very definitely detected, even in part predicted, by Rutherford and his colleagues and successors (Soddy, Ramsay, etc.), itself arising from Becquerel's and the Curies' discovery of spontaneous radioactivity, may ultimately lead us far. But there remains the question whether the facts of biology demand an underlying complexity in the atoms vaster than could be embraced in any definite physical scheme.

*The foundation of the physical cosmos:* Our conviction of an orderly connection between things constitutes the conception of a *cosmos.* We have placed the foundation of this in the existence of a uniform medium, the aether, the physical groundwork of interstellar space, through which the actions between material bodies are established and transmitted. The idea of such a medium, when analyzed mathematically, almost demands that matter should consist of discrete atoms involving nuclei each of which binds together into permanence some mode of local disturbance in the medium. The illimitable complexity of the phenomena resides in matter; but our grasp of the physical relations to which all its manifestations are subject arises from their being such as can be established through the aether. The Daltonian principle of identity of all atoms of the same substance—they are the same to the remotest limits of our universe, as Huggins demonstrated—may well arise from these atoms being the *explaining the restrictions of atomic types.* limited number of definite intimate types of structure into which more ultimate atoms can arrange themselves. These ultimate atoms would be limited as regards their relations at a distance, for they would in this respect involve only the few fundamental types of strain centres which are capable of subsisting in the simple aether. The keystone of such a physical scheme is the aether: and the only ground for postulating the presence of this medium is the extreme simplicity and uniformity of the constitution which suffices for its functions. Needless to say, there remain many unresolved features, some still obscure, *The aether.* but hardly contradictory. But should it ever prove to be necessary to assign to the aether as complex a structure as matter is known to possess, then it might as well be abolished from our scheme of thought altogether. We would then fall back on simple phenomenalism; proximate relations would be traced, but we need not any longer oppress our thought by any regard to a common setting for them; the various branches of physical science would cultivate with empirical success independent modes of explanation of their own, checked only by the mutual conservation of the available energy, while the springs of their orderly connection would be out of reach. That time, however, is not yet.

# APPENDIX (1908)

## *On the Possible Types of Direct Chemical Combination*[1].

The question must often have arisen why chemical combinations and decompositions take place on a simple plan which can be represented as the addition to the molecule, or removal from it, of definite blocks or complexes of atoms, named radicles, which are therefore assumed to have a transient corporate existence; so much so that in a class of cases their existence has to be introduced explicitly in order to assist as carriers in the reactions. The physical analogy of an atom as a kinetic system of orbitally moving sub-atoms or electrons, such as is pointed to by its very definite intrinsic spectrum, would lead us to expect that, as a rule, tampering with the structure of a molecule by slicing off a block would lead to its total dissolution: while, on the other hand, the practically effective conceptions of organic chemistry suggest architecture rather than dynamics. *Existence of radicles:*

Some light may perhaps be thrown on this subject by the consideration that it is only those structures that do fall to pieces in successive stages, and are thus capable of definite experimental dissection, that can have a chance of being produced in quantity, and of becoming segregated, in the clash of molecules amidst which all things chemical have their origin. *not anomalous.*

In illustration, imagine a substance, say gaseous for simplicity, formed by the immediate instantaneous combination of three gaseous components $A$, $B$, $C$. When these gases are mixed, the chances are very remote of the occurrence of the simultaneous triple encounter of an $A$, a $B$, and a $C$, which would be necessary to the immediate formation of an $ABC$; whereas, if ever formed, it would be liable to the normal chance of dissociating by collisions; it would thus practically be non-existent in the statistical sense. But if an intermediate combination $AB$ could exist, very transiently, though long enough to cover a considerable fraction of the mean free path of the molecules, this will readily be formed by ordinary binary encounters of $A$ and $B$, and another binary encounter of $AB$ with $C$ will now form the triple compound $ABC$ in quantity. The cognate subject of the dynamics of gas theory illustrates the point, in fact is closely implicated: that theory proceeds by aid of statistics of encounters, yet in its analysis *Transient interactions in a ternary system:*

---

[1] This Appendix was included substantially in a course of lectures at Columbia University, New York, in March 1907.

as in gas theory. triple and multiple encounters are left aside as negligible in number*.

The principle thus suggested, that immediate molecular combinations and dissociations are practically all binary, may have a wider application. It would tend to explain, as above indicated, how it is Dissection of that in organic chemistry only those types of molecules—perhaps organic molecules. very few compared with what are in totality possible—have any chance of being isolated amid the chances of natural reactions, involving no control whatever of individual molecules, which can of themselves divide into parts or radicles of appreciable persistence, some of them replaceable by other such blocks or parts without allowing time for the dissolution of the whole. That science, in fact, proceeds by searching out and classifying the ways in which complex molecules may be thus definitely dissected and reconstructed.

There is involved, on this view, the proposition that in the ultimate type of chemical interaction each molecule is divided into two parts at most, which it may interchange with another molecule in the proResolution cess of double decomposition; that in fact any transformation of more into binary complex type than this must be expected to occur in successive reactions: stages. Thus the simplest case of a triple dissociation, say of a molecule $ABC$ into $A$ and $B$ and $C$, may be held to occur in two stages, the first stage being such as a change into $AB$ and $C$: the reason may be repeated, that without an intermediate diad stage the velocity of association of such molecules would be extremely slow compared with the velocity of dissociation of those already formed, so that in equilibrium the triad compound would practically not exist[1].

It appears indeed from the facts that chemical equilibria involving processes more complex than double decomposition occur but rarely. Where they occur at all, it is here suggested that there must be an intermediate stage, perhaps very transient; and the question arises whether its existence may modify the usually accepted deductions from the Guldberg-Waage statistical representation of the chemical

* This seems to vitiate to some degree any application of thermodynamic formula to ionized constituents of media, whether stellar atmospheres or fluid solutions, as the mutual inverse-square forces are not entirely local to the binary encounters. Cf. *supra*, p. 365. The discrepancy will fall away at high temperatures.

[1] Considerations such as these must often have occurred to molecular theorists. It is only recently (a year after this note was composed) that, in an obituary notice of Mendeléef by Dr E. C. Edgar (*Manchester Memoirs* 51, 1907) a remark about "his persistent devotion to the Mendeléef-Gerhardt law, that gases combine only in equal volumes" has prompted a reference to the section on Atoms and Molecules in the "Principles of Chemistry," where views essentially equivalent to the above are powerfully supported on purely chemical grounds. (See footnote, p. 363, *supra*.)

equilibrium, or the application of thermodynamics of which that *its influence on chemical equilibrium:* theory forms one aspect.

The matter will, however, assume a sufficiently complicated form, at any rate for initial consideration, in the very simplest example. *considered in a simple case.* Let us then examine the chemical equilibrium of a dissociating gaseous substance $ABC$, mixed with its components. Let this symbol $ABC$ denote quantity of the substance $ABC$, measured in chemical equivalents, say by number of molecules per unit volume, with similar meanings assigned for the other symbols. There can be present in the *The possible constituents:* interaction seven substances, viz. $ABC$, $AB$, $BC$, $CA$, $A$, $B$, $C$, some of them perhaps of such slight permanence that they are not apparent.

Of the amount $ABC$, suppose the quantity $k_1 . ABC$ changes into *their interchanges formulated.* $BC$ and $A$ per unit time, $k_2 . ABC$ into $CA$ and $B$, $k_3 . ABC$ into $AB$ and $C$: suppose the quantity $a . BC$ dissociates into $B$ and $C$ per unit time, and so on: suppose $f . B . C$ is the quantity which associates into $BC$ per unit time, and so on: suppose $l . BC . A$ is the quantity which associates into $ABC$ per unit time from $BC$ and $A$, and so on: the coefficients $a$, $b$, $c$, $k$, ... being proper fractions. Thus we have a scheme of formally possible transformations

$$(k_1 + k_2 + k_3) ABC$$
$$a . BC, b . CA, c . AB$$
$$f . B . C, g . C . A, h . A . B$$
$$l . BC . A, m . CA . B, n . AB . C.$$

If any of these intermediate substances (say $AB$) is so transient as practically not to occur, the corresponding association factor ($h$) must be very small compared with the dissociation factor ($c$).

When chemical equilibrium is attained, the dissociations and *Conditions for steadiness:* associations continually going on do not alter the amount of the substance $A$; therefore

$$k_1 . ABC + b . AC + c . AB - g . A . C - h . A . B - l . BC . A = 0,$$

and there are two similar equations.

In the same way the constancy of the amount of the substance $BC$ requires

$$- k_1 . ABC + a . BC - f . B . C + l . BC . A = 0,$$

and there are two similar equations.

And the constancy of the amount of the substance $ABC$ requires

$$(k_1 + k_2 + k_3) ABC - l . BC . A - m . CA . B - n . AB . C = 0.$$

There are seven equations in all, but only four can be independent; *reduced in number.* for the total amount of $A$ that is present, free and combined, cannot change, therefore

$$A + AB + AC + ABC$$

is constant, and there are two other such relations. This reduction is verified; for adding any one of the first group of equations to the corresponding one of the second group gives the same result, viz.,

$$a \cdot BC + b \cdot CA + c \cdot AB = f \cdot B \cdot C + g \cdot C \cdot A + h \cdot A \cdot B,$$

and subtracting the sum of the first group from the last equation also gives this result.

*Intermediate formations eliminated.* We shall take the second group and the last equation as the independent relations. To eliminate the intermediate substances $AB$, $BC$, $CA$, we have from the former

$$BC = \frac{k_1 \cdot ABC + f \cdot B \cdot C}{a + l \cdot A};$$

and substituting in the last equation the value of $l \cdot BC \cdot A$ thus derived,

$$\frac{lf \cdot A \cdot B \cdot C - ak_1 \cdot ABC}{a + l \cdot A} + \frac{mg \cdot A \cdot B \cdot C - bk_2 \cdot ABC}{b + m \cdot B}$$
$$+ \frac{nh \cdot A \cdot B \cdot C - ck_3 \cdot ABC}{c + n \cdot C} = 0,$$

*Law of equilibrium.* a complicated relation which, in conjunction with the expressions for binary combinations such as $BC$ above, and the total atomic amounts of interacting material, determines the equilibrium.

*Case of all intermediates transient:* If $l/a$, $m/b$, $n/c$ are very small, and $f$, $g$, $h$ correspondingly large, so that the intermediate compounds $AB$, $BC$, $CA$ are all very transient, we have very approximately

$$\left( \frac{lf}{a} + \frac{mg}{b} + \frac{nh}{c} \right) A \cdot B \cdot C = (k_1 + k_2 + k_3) ABC,$$

*yields usual simple result.* which is the type of formula usually assigned for the equilibrium of a triple dissociation. When the amounts of interacting materials are given, this formula determines their distribution.

*Otherwise result is more complex,* If $BC$ and $CA$ are very transient compared with $AB$, we have $l/a$ and $m/b$ very small compared with $n/c$, while $f$ and $g$ are large compared with $h$ if they take a sensible part in the equilibrium; then

$$\left( \frac{lf}{a} + \frac{mg}{b} + \frac{nh}{c + n \cdot C} \right) A \cdot B \cdot C = \left( k_1 + k_2 + k_3 \frac{c}{c + n \cdot C} \right) ABC.$$

*unless in special cases.* But if we suppose only five substances sensibly operative, say $ABC$, $AB$, $A$, $B$, $C$, the equations will be the (usual) binary ones,

$$c \cdot AB = h \cdot A \cdot B, \quad k \cdot ABC = n \cdot AB \cdot C,$$

yielding

$$kc \cdot ABC = hn \cdot A \cdot B \cdot C,$$

which is a different law of equilibrium, being the same as if $AB$ also did not occur.

If $A$ and $B$ and $C$ are identical, this latter law will hold universally: if only $A$ and $B$ are identical, it need not do so. Generally, the conditions for its validity are that $l$, $m$, $n$ should be very small, or else $l/a = m/b = n/c$.

The *thermodynamic* condition of equilibrium employs conceptions and physical constants different from those pertaining to this statistical view of Guldberg and Waage, but at bottom connected and in ordinary cases leading to the same results. If $m_1$, $m_2$, $m_3$, $m_{12}$, ... denote the quantities of the different simple and compound substances that are present in any phase, and $A$ the available energy, The thermo-<br>dynamic<br>method<br>(Gibbs):

$$\delta A = \ldots\ldots + \mu_1 \delta m_1 + \mu_2 \delta m_2 + \ldots\ldots + \mu_{12} \delta m_{12} + \ldots\ldots$$

And as the available energy tends to a minimum, under the appropriate conditions, including constancy of temperature, any slight reactive change that can occur in the phase must leave $A$ sensibly unaltered, provided equilibrium has arrived. Thus the thermodynamic potentials

$$\mu_1, \mu_2, \mu_{12}, \ldots$$

in the phase must satisfy a number of relations indicating the equilibrium of each possible partial reaction that can occur in it, *e.g.*, as there is a reaction possible of type

$$m_1 + m_2 \rightleftarrows m_{12}$$

we must have

$$\mu_1 + \mu_2 = \mu_{12}.$$

The thermodynamic potentials of all compound substances in the phase are thus found in terms of those of the simple (or other) independent constituents

$$m_1, m_2, \ldots\ldots m_r;$$

that is, the system will settle down to an equilibrium in which they have the values thus determined.

If two phases coexist in contact, $\mu_1, \mu_2, \ldots\ldots \mu_r$ must moreover have the same values in both of them.

If $P$ phases can coexist, there are thus $r(P - 1)$ conditions to be satisfied: and each phase has a characteristic equation of state connecting $m$, $v$, and the temperature $T$—thus making up in all $P$ conditions. Now there are independent variables $rP$ in number, together with $T$, and the total volume $v$—the portions of the volume occupied by the various phases being determined by their characteristic equations. The system will be wholly determined if $P = r + 2$. Phase rule.

This is, in fact, Willard Gibbs' theory limiting the number of phases that can coexist in given material, and conversely. The thermodynamic potentials that are here involved must be functions of the velocities of interaction of the previous analysis. But are they always

Comparison with unrestricted statistics. consistent with the previous statistical view, without restriction? In dilute solution $\mu \propto \log (m/v)$, thus we should have a relation $m_1 m_2 / m_{12} =$ constant. Thus all the thermodynamic equations of equilibrium will take the form of the constancy of simple factorial ratios. Does this imply more than mere statistics of chance encounters can provide? It involves a further principle, that the reaction between $m_1$, $m_2$ and $m_{12}$ is in equilibrium *by itself*, just as if the other com-

Isolation of component equilibria in dilute systems: ponents containing the same elements were prevented by constraint from changing. If this principle of isolation of the equilibria of the component reactions is warranted, it produces extensive simplification not inherent in the customary statistical point of view: it can for instance specify at once the proportions of the intermediate compounds that are present, replacing a system of complex linear equations by constancy of simple ratios.

If we may apply it to the problem on page 375, the equations there deduced will be replaced by

$$k_1 . ABC = l . BC . A, \ldots\ldots f . B . C = a . BC, \ldots\ldots$$

so that

$$k_1 . ABC = \frac{lf}{a} A . B . C;$$

involving

$$k_1 \Big/ \frac{lf}{a} = k_2 \Big/ \frac{mg}{b} = k_3 \Big/ \frac{nh}{c};$$

involving conditioned statistics: and also expressing the relative frequencies in which $ABC$ splits up into different intermediate compounds.

In further exemplification of the simplification thus introduced, consider the system $N_2O_4$, $NO_2$, N, O; if we are sure, experimentally, that N and O are infinitesimal in a partial system $NO_2$, N, O, then they are so in the wider system, for by the equilibrium of the partial system N and O are determined.

equivalent to only binary interactions, *On this thermodynamic view it is in fact by trains of single or double decomposition that substances are formed.* For if we do not admit this postulate, then the equations of statistical equilibrium will contain more than two terms as exemplified above; and that aspect of chemical equilibrium will be at variance with the usual thermodynamic theory, which expresses an independent equilibrium for every type of reaction that is formally possible. And the reason has been already indicated, viz., the usual expressions for the thermodynamic entropy

for dilute systems. and available energy of gaseous systems, and through them of dilute solutions, involve implication that only binary molecular encounters need be considered. The two points of view will agree only if all reactions take place in binary stages; and it becomes a question whether this is a universal rule under all circumstances, or only one prevalent in the prominent cases which are naturally those governed by simple recognizable relations.

An actual case in which these distinctions may make theoretically a difference is worked out from the thermodynamic side in Planck's *Thermodynamics*, § 247, under the heading of graded dissociation, viz., that of hydriodic acid HI into $H_2$, $I_2$, and I. | An actual case.

Another question in which such considerations may have scope is that of Ostwald's law of equilibrium of ionization. If only two ions can arise, they must be equal in number; thus if $c'$ is their concentration (dilute) and $c$ that of the non-ionized part, $c'^2/c$ may be expected to be constant at each temperature, the ionization proportional to $c$ being balanced by the recombination proportional to $c'^2$. But this assumes that all the ionization is spontaneous, whereas in the cognate phenomena of gases the encounter of an ion (in rapid motion) with a molecule has been shown by Townsend to be a potent cause of further ionization. This suggests the question whether $c$ should not be replaced by $c + kcc'$ or $c(1 + kc')$, which may make a difference in the direction actually occurring, whenever the concentration of ions $c'$ is considerable. Moreover, the spheres of mutual electric influence of ions are far greater than those of molecules, which may also make a difference\*. | Ostwald's ionic equilibrium, may require correction.

It is, however, to be noticed that in the discussion above of the ordinary association of a substance $ABC$, this type of action, in which, *e.g.*, a component $C$ acts in a special manner in breaking up a component $AB$, has been excluded. The presence of such actions in which more than one cause contributes to the result would seem hard to adapt to the usual thermodynamic theory involving as we have seen independent binary stages of reaction. | Other alternative processes also excluded.

\* If $a$ is the fraction of the molecules thus dissociated into ions, then for a gram-mol of the substance $c' = a/v$, $c = (1 - a)/v$: thus this relation would give

$$K = \frac{c'^2}{c} = \frac{a^2}{1 - a} \frac{1}{v}.$$

For an electrolyte $a$ is obtained as the conductance per molecule divided by its value at infinite dilution: the formula is found to be verified for weak ionization, but for strong electrolytes for which $a$ is greater the dilution $v$ becomes replaced by $\sqrt{v}$, for some reason not yet fully understood. The alternative in the text would only replace $v$ in the formula by $v + ka$.

The relation has recently had tentative application to the degree of ionization in stellar atmospheres.

# 72

# THE RELATION OF THE EARTH'S FREE PRECESSIONAL NUTATION TO ITS RESISTANCE AGAINST TIDAL DEFORMATION.

[*Proceedings of the Royal Society*, A, Vol. LXXXII, Dec. 18, 1908, pp. 89–96.]

THE modern investigation of the wandering of the Earth's axis of rotation, considered as a physical problem relating to the actual non-rigid Earth, may be said to have been initiated in Lord Kelvin's

Historical: address to the Physical Section of the British Association in 1876. After referring[1] to the scrutiny of the recorded observations of change of latitudes, conducted by Peters in 1841 and independently by Maxwell in 1851, in search of the regular Eulerian free period of 306 days which would belong to a rigid Earth, with negative results, he insisted that the irregular motions brought out in these analyses are not merely due to instrumental imperfections, but represent true motions of the Pole, due to displacement of terrestrial material. For

effect of sudden shifts of material: example, he estimates that existing shifts of material, of meteorological type, are competent to produce displacements of the axis of rotation ranging from $\frac{1}{2}$ to $\frac{1}{20}$ of a second of arc. A sudden shift of material on the Earth will not at once affect the axis of rotation, but will start it into motion round the altered axis of inertia, with a period of 306 days if the Earth were rigid, which will go on displacing the Pole until it is damped out by the frictional effects of the tidal

starts an ocean tide: motions thus originated. A radius of rotation of 1 second of arc would raise an ocean tide of the same period as the rotation, having as much as 11 cm. of maximum rise and fall. Thus the motion of the Pole is to be considered as continually renewed by meteorological and other

damped off. displacements, as it is damped off by tidal and elastic friction; it was therefore, perhaps, not to be expected that it would show much periodicity, though the movements were eminently worthy of close investigation. Their nature was examined more closely by Newcomb at Kelvin's request; but not much more had been done regarding

Chandler's period. their cause when Chandler announced that the records of changes of latitude did actually indicate a period of precession—of 427 days, however, instead of the Eulerian period of 306 days, which, if any, had previously been taken for granted. Soon after, in 1890, observations were organized systematically by the International Geodetic

[1] Reprint in *Popular Lectures and Addresses*, Vol. II, see pp. 262–272.

Union on the motion of Professor Foerster, of Berlin; and already, in 1891, he was able to inform Lord Kelvin that a comparison of European observations with synchronous ones made at Honolulu gave direct proof of his conclusion of 1876 (*supra*), "that irregular movements of the Earth's axis to the extent of half a second may be produced by the temporary changes of sea level due to meteorological causes[1]." *Direct evidence of irregular causes.*

In the following year the synchronous observations had already indicated periodicity, apparently in about 385 days, considerably less than Chandler's estimate, which, however, longer observation has since confirmed substantially. Lord Kelvin remarks in his next annual address as follows[2]: "Newcomb, in a letter which I received from him last December, gave what seems to me undoubtedly the true explanation of this apparent discrepance from dynamical theory, attributing it to elastic yielding of the Earth as a whole. He added a suggestion, especially interesting to myself, that investigation of the periodic variations of latitude may prove to be the best means of determining approximately the rigidity of the Earth. As it is, we have now for the first time what seems to be a quite decisive demonstration of elastic yielding of the Earth as a whole, under the influence of a deforming force, whether of centrifugal force round a varying axis, as in the present case, or of tide-generating influences of the Sun and Moon, with reference to which I first raised the question of elastic yielding of the Earth's material many years ago." But "when we consider how much water falls on Europe and Asia during a month or two of rainy season, and how many weeks or months must pass before it gets to the sea, and where it has been in the interval, and what has become of the air from which it fell, we need not wonder" that the amplitudes of the polar wanderings "should often vary by 5 or 10 metres in the course of a few weeks or months." *Very early estimate of longer period, suggests (1877) to Newcomb elastic yielding of the Earth: irregular part, causes.*

It will be recalled that the main object of the original calculations of Lord Kelvin, which assigns to the Earth as a whole an effective rigidity of the same order as that of steel, was to combat the view then prevalent which assumed for the Earth a fluid interior. Even a solid shell of very considerable thickness, enclosing a fluid core, was thus ruled out, unless its materials were preternaturally rigid; and it is clear that placing a solid core in the middle of the fluid interior cannot affect this conclusion so long as an equilibrium theory is applicable, *i.e.*, so long as the layer of fluid material is not so thin or viscous as to *The Earth solid:*

[1] *Presidential Addresses R.S.*, Nov. 30, 1891; *Popular Lectures...*, Vol. II, p. 504. Lord Kelvin's investigations up to 1876 are collected in *Math. and Phys. Papers*, Vol. III, especially pp. 312–350.

[2] *Presidential Address R.S.*, Nov. 30, 1892; *loc. cit.* p. 525.

prevent its adjusting itself immediately by flow to the alternating
tidal stresses impressed upon it from its solid walls. By passing to the
other limit, and thus taking it so thin that the outer shell practically
rides on the solid nucleus, but without effective tangential stress-
connection, we obtain a hypothesis to which this objection does not
apply.

*except for a thin layer of isostasy.*

In a brief note in *Monthly Notices R.A.S.* that year (1892), Newcomb
showed, by a general estimate, that the effect of elastic yielding is
competent to prolong the free period to about the amount required
by observation. A formal mathematical discussion on the bases of
calculation of the elastic deformation of a homogeneous sphere was
first given by Mr S. S. Hough, now H.M. Astronomer at the Cape of
Good Hope, in a memoir on "The Rotation of an Elastic Spheroid,"
in *Phil. Trans.* 1896.

*Newcomb follows up his estimate:*

*also S. S. Hough.*

He concluded that the Chandler free period required an effective
rigidity of the whole Earth of the order of that of steel, agreeing with
Lord Kelvin's previous estimates from tidal phenomena; and his
result seems to have been substantially confirmed by more recent
calculations, giving for the average effective rigidity estimates derived from various possible hypotheses and simplifying assumptions
ranging between extreme values $17 \times 10^{11}$ and $4\cdot4 \times 10^{11}$, while
Hough's estimate was put at $8\cdot98 \times 10^{11}$. This shows an even
striking degree of agreement in calculations necessarily vague on
account of the unknown constitution of the Earth's interior, especially
in so far as observations of the equilibrium tides of long periods,
and of the deviation of sea-level due to tidal attraction which is
essentially the same thing, lead to results of the same order as those
of free precessional rotation[1]. It indeed suggests, as we shall actually
recognize, that this internal terrestrial constitution really is not involved in these various phenomena, except in the common feature of
determining the surface effects arising from a given tidal or rotational
stress[2]. The key to the matter, from the general point of view, is contained in the remark of Hough that the free precession of the yielding
Earth is the same as that of a rigid one of the shape that would result
when the bulging arising from the centrifugal force of diurnal rotation
is removed. It is not difficult to show, from geometrical considerations
regarding momentum[3], that this result is general, and extends to an
Earth of any degree of heterogeneity or plasticity. The argument may

*Earth as rigid as steel.*

*Confirma-tions.*

*The actual problem.*

*Hough's conclusion:*

*verified as a general principle.*

---

[1] Cf. Professor A. E. H. Love, *Roy. Soc. Proc.* Vol. LXXXII, p. 73. To this
paper I am indebted for information as to results of recent calculations.

[2] The identity of these two types runs through the discussions in Thomson
and Tait's *Natural Philosophy*.

[3] *Proc. Camb. Phil. Soc.* May 1896, p. 185.

be reproduced in analytical form and rather wider scope, from another place[1] (with definition of $I$ rewritten), as follows.

Let $\omega$ be the angular velocity of the Earth about the instantaneous axis, $\omega_1$, $\omega_2$, $\omega_3$ its components referred to the principal axes in the configuration that the Earth would have if the motion were steady. The Earth is deformed from this configuration by the inequality of centrifugal force due to the deviation of the instantaneous axis from the principal axis, with which it would coincide if the motion were steady. This deforming force is the resultant of the centrifugal force, directed outwards from the instantaneous axis, and the reversed centrifugal force, directed inwards towards the principal axis in question. A linear law of elasticity applies to the small resultant of these two forces. If the same law applied to the two forces separately, the reversed centrifugal force would change the moments of inertia $A$, $B$, $C$ to certain values $A'$, $B'$, $C'$, which might, under simplifying hypotheses, be calculated from the theory of the deformation of an elastic sphere; and the centrifugal force directed outwards from the instantaneous axis would produce a certain change of density at each internal point, and would raise a certain protuberance on the surface, which might be calculated by the same theory. Let $I$ denote the moment of inertia (about the instantaneous axis) of a mass arranged as specified by this change of density and this protuberance. The instantaneous axis is a principal axis of this mass, and therefore the contributions of this mass to the components of moment of momentum are $I\omega_1$, $I\omega_2$, $I\omega_3$. The complete expressions for the components $h_1$, $h_2$, $h_3$ of moment of momentum are therefore

*(margin note: Mobile centrifugal protuberances treated separately.)*

$$h_1 = A'\omega_1 + I\omega_1, \quad h_2 = B'\omega_2 + I\omega_2, \quad h_3 = C'\omega_3 + I\omega_3.$$

The equations of motion referred to the rotating axes are of the well-known vector type,

$$dh_1/dt - h_2\omega_3 + h_3\omega_2 = L.$$

When $A$ and $B$ are equal, the third of them is

$$\frac{d}{dt}(C\omega_3) = N,$$

where $C$ is the effective moment of inertia $C' + I$; when $N$ is null $\omega_3$ is thus constant, say $\Omega$, up to the first order. The other two equations are

$$\frac{d}{dt}\cdot(A' + I)\,\omega_1 + (C' - B')\,\Omega\omega = L,$$

$$\frac{d}{dt}\cdot(B' + I)\,\omega_2 - (C' - A')\,\Omega\omega_1 = M,$$

[1] E. H. Hills and J. Larmor, "The Irregular Movement of the Earth's Axis of Rotation," *Monthly Notices R.A.S.* Nov. 1906, p. 24: *supra*, p. 318.

which in the case of approximate symmetry involve a free period $2\pi (A' + I)/(C' - A') \Omega$, and similarly in the general case, thus depending only on $A'$, $B'$, $C'$ when $I$ is small.

The result is that the period of the free precession is not $C/(C - A)$ days, as it would be for a rigid Earth, but approximately $C/(C - A')$, where the denominator is that difference of principal moments of

*Result.* inertia which would remain after the imposition of a bodily forcive having as potential

$$W = - \tfrac{1}{2}\omega^2 r^2 \sin^2 \theta = - \tfrac{1}{3}\omega^2 r^2 (\mathrm{I} - P_2),$$

namely, that of the centrifugal force reversed, $P_2$ representing the zonal harmonic $\tfrac{1}{2} (3 \cos^2 \theta - \mathrm{I})$.

The first part of $W$, the term $- \tfrac{1}{3}\omega^2 r^2$, corresponds to slight contraction of volume, which is immaterial as regards the desired quantity $C' - A'$. The other part, $\tfrac{1}{3}\omega^2 r^2 P_2$, will produce an extension of the same harmonic type as itself, around the polar axis, which will in turn alter the potential of the Earth's attraction at its own surface by $k \cdot \tfrac{1}{3}\omega^2 r^2 P_2$, where the value of $k$ depends on its effective resistance to deformation. Moreover the Earth's potential is at distant points, by Laplace's formula,

$$V = \gamma \left( \frac{E}{r} + \frac{A + B + C - 3I}{2r^3} + ... \right),$$

which gives
$$V = \gamma \left( \frac{E}{r} - \frac{C - A}{r^3} P_2 + ... \right)$$

in the present special case; and if, as in the actual circumstances, further harmonics do not occur to sensible amount, this expression holds right up to the Earth's surface. The free surface, of ellipticity $\epsilon$, is

$$r = a_0 (\mathrm{I} + \epsilon \sin^2 \theta)$$
$$= a (\mathrm{I} - \tfrac{2}{3} \epsilon P_2),$$

where $a = a_0 (\mathrm{I} + \tfrac{2}{3}\epsilon)$. The value of $\epsilon$ is determined by the constancy over the ocean surface of the total potential $V - W$, as $- W$ is the potential of the centrifugal force, viz., of

$$\gamma \left\{ \frac{E}{a} (\mathrm{I} + \tfrac{2}{3}\epsilon P_2) - \frac{C - A}{a^3} P_2 \right\} + \tfrac{1}{3}\omega^2 a^2 (\mathrm{I} - P_2);$$

whence, equating to zero the coefficient of $P_2$,

$$\tfrac{2}{3}ga \left( \epsilon - \frac{\omega^2 a}{2g} \right) - \frac{\gamma}{a^3} (C - A) = 0,$$

*Confirmation for astronomical precession:* thus deriving from data of the distribution of gravity, or of the form of the Earth's surface, the value of $C - A$, which determines the astronomical precession. Again, if taking off the centrifugal force would change $C - A$ to $C' - A'$, it would alter $V$ by

$$\gamma r^{-3} \{(C - A) - (C' - A')\} P_2,$$

which must, according to the above specification of $k$, be equal to $k \cdot \frac{1}{3}\omega^2 r^2 P_2$. Thus

$$\frac{C' - A'}{C - A} = 1 - \frac{\frac{1}{3}k\omega^2 a^5 \gamma^{-1}}{C - A}$$

$$= 1 - \frac{k\omega^2 a / 2g}{\epsilon - \omega^2 a / 2g}.$$

Hence, if $\tau$ is the periodic time of actual free precession and $\tau_0$ is what it would be if the Earth were rigid,

and elastic
increase of
its period,

$$1 - \frac{\tau_0}{\tau} = k \frac{\omega^2 a}{2g} \bigg/ \left( \epsilon - \frac{\omega^2 a}{2g} \right).$$

This is the formula (15) in Professor Love's paper before referred to; it is there deduced from a hypothesis of concentric spheroidal stratification of the Earth's interior, after the manner of Laplace. We have found that, like Clairaut's formula for gravity, this relation is independent of any hypothesis as to the Earth's internal structure, except such as is involved in the definition and value of $k$.

depends on
$k$ alone:

As $\omega^2 a/g$ is $1/289$ and $\tau$ is found to be $428$ days, and $\tau_0$ is $306$ days, this relation makes $k$ equal to $4/15$.

its value.

The values of $k$ corresponding to various moduli of rigidity and compressibility of the Earth considered as a homogeneous globe might perhaps be deduced and tabulated for comparison, from Lord Kelvin's and similar elastic analysis.

Further
procedure.

The height of the long-period equilibrium tides provides different data; corresponding to an extraneous tide-producing potential $W_2$ of this type, the absolute rise of the water is $(1 + k) W_2/g$, from which has to be subtracted $hW_2/g$ for the rise of the solid Earth due to this tide-producing potential, thus leaving a factor $1 + k - h$ for the relative tide which alone can be the subject of observations. The reductions of tidal data for the Indian Ocean gave Kelvin and G. H. Darwin the value $\frac{2}{3}$ for this factor, which is confirmed by more recent discussions: the observations of Hecker with a horizontal pendulum at the bottom of a well, which obviously determine the same thing, viz. the change of level due to tide-producing potential, concur in a remarkable manner*. Thus $h = \frac{3}{5}$.

Long-period
tides:

with
determined
part in solid
Earth:

These values of $k$ and $h$, as defined in the last paragraph, would not be independent for a homogeneous incompressible globe: they would, in general, require for their consistency both elasticity of volume and of form. The phenomena of free precession give the value of $k$, but

* There are now also available Michelson's observations at Chicago on alternations of level in long pipes, reported in *Astrophys. J.*, perhaps showing an influence of Atlantic tides.

with reference to compression along the polar axis; those of tidal change of level give the value of $h - k$, or rather its mean value, with reference to compression along axes in the neighbourhood of the equator[1]. This statement is the purest and simplest expression of the information relating to the solid Earth's resistance to deforming forces that the data of periodic change of latitude and of equilibrium (*i.e.* long-period) tides can supply, prior to any hypothesis regarding the internal distribution and the effective elasticity or plasticity of its materials.

*extent of tidal results.*

*Feb.* 2. It has been remarked above, after Lord Kelvin, that a sudden shift of material from one part of the Earth's surface to another would alter the position of the principal axis of inertia round which the free precession of the Earth's axis of rotation takes place, and thus cause a sharp bend in the path of the Pole. If the shift were merely local, such as an earthquake may be expected to produce, the effect would be inappreciable. The connection of sharp curvature in the path of the Pole with seismic disturbance, if it really exists, would thus be indirect, the earthquake being itself started possibly by the slight changes, meteorological or other, of distribution of surface load, which are indicated by the disturbance of the free precession.

*Earthquakes ineffective,*

*effect not cause:*

But it is to be noticed that a *submarine* seismic subsidence, if uncompensated by adjacent elevation, or *vice versa*, would be competent to produce sensible direct disturbance of the path of the Pole; for water would have to flow, in part from distant regions, to fill up the defect of level thus produced. The same would be true for earthquake subsidence near coast lines, if it is compensated by rise of the land. In reply to an inquiry on this subject, Professor Milne writes as follows: "When a very large earthquake occurs on land, we find vertical and lateral displacements of, let us say, 20 feet, along lines which may be one or two hundred miles in length. The majority of big earthquakes, however, are sub-oceanic in their origin, along lines parallel to mountain ridges, as, for example, at the bottom of the trough which runs parallel to the Andes. The mass movement appears to result in the deepening of the trough and the rise of the coast line. We have measurements where depth has increased as much as 200 fathoms: see *Brit. Assoc. Seismic Report*, 1897, for a number of these measurements."

*unless suboceanic.*

*Amounts of their actual subsidences (J. Milne),*

*Analysis by Love.*

[1] Cf. Professor Love, *loc. cit.* p. 81, to whom this proposition is substantially due, having been reached by him through analysis appropriate to a centrically stratified body. The quantities $h$ and $k$, in other notation, enter essentially into the tidal discussions by Kelvin and Darwin in Thomson and Tait's *Nat. Phil.*

An estimate of the effect of such displacements is easily made. Thus, an uncompensated subsidence of the ocean floor, of volume corresponding to a fall of one foot over a thousand miles square, in middle latitudes, would produce[1] a direct shift in the Pole of rotation amounting to about one-eighth of a second of arc; and at the same time the Pole of the principal axis of inertia, round which the 428-day precession of the axis of rotation takes place, would be displaced in the opposite direction through an angle of the same order of magnitude.

*might shift the Pole by ⅛″.*

In connection with the possibility of irregularity in the Earth's diurnal rotation due to causes of this kind, similar considerations arise[2]. A slight subsidence, due to shrinkage around the equator, unless it extended downward a long way toward the Earth's centre, would have negligible direct effect on the moment of inertia and, therefore, on the length of the day; but if we were under sea it would involve transference of water from regions nearer the Earth's axis, in order to make up the deficiency, and if the equatorial regions were all under water, a contraction of 50 cm. in equatorial radius would in this way alter the length of the year by an amount of the order of half a second of time, which would be astronomically of high importance.

*Notable effect in length of day.*

[1] *Loc. cit., Monthly Notices R.A.S.* Nov. 1906, p. 26 [: *supra*, p. 328. It appears that the Pole has been drifting towards Greenwich at the rate of about one-tenth of a second of arc per year for the last ten years or so. See also H. Spencer Jones, *M.N.R. Astron. Soc.* March 1926].

[2] Lord Kelvin, *loc. cit.* § 38 [: also *supra*, p. 320, from *M.N.R. Astron. Soc.* 1906].

# 73

## THE KINETIC IMAGE OF A CONVECTED ELEC-TRIC SYSTEM FORMED IN A CONDUCTING PLANE SHEET.

[*Proceedings of the London Mathematical Society*,
Ser. 2, Vol. VIII, 1909, pp. 1–9.]

*Maxwell's receding trail of images:*

THE mathematical problems associated with Arago's rotating disc led in Maxwell's hands[1] to his beautiful theory of a receding trail of images, as representing the reflexion, from an infinite plane conducting sheet, of the disturbance due to a magnetic pole or other magnetic system travelling in front of it[*]. In these days, when electrons at rest or in motion are recognized as the source of all electric

*extended to moving electrons.*

and magnetic phenomena, it is natural to try to develop his solution in the direction of moving electric charges. It turns out that the trail of images can in both the electric and magnetic cases, and indeed for any convected system, readily be reduced to a simple and direct synthetical form. The velocities of convection are not to reach the order of magnitude of the speed of light; otherwise radiation would become sensible, and the present "equilibrium theory" would not suffice.

*A moving conductor can screen off electro-dynamic influence:*

The screening action due to motion of a conducting layer has been treated in a previous paper[2]. The direct synthetical analysis of that problem, and of the one here treated, have recently acquired new physical interest from G. W. Hale's observations of the Zeeman

*magnetic field in sunspots confined to a thin layer:*

magnetic effect in the spectrum at spots in the Sun. Observations appear at present to show that the magnetic field is confined to a thin layer in the Solar atmosphere; and this would seem to require that the natural extension of the field due to the electric flow in the layer

*perhaps by such screening.*

must be cut off by some kind of screening action from the regions above and below this layer.

If $\phi$ is the stream function of the currents $(u, v, 0)$ induced in the sheet, itself situated in the plane of $xy$, and

*A screening sheet of currents:*

$$\chi = \int \phi^{-1} \, dS,$$

[1] *Proc. Roy. Soc.* February 1872; *Elec. and Mag.* §§ 656–669.
[*] The theory is extended to cylindrical and spherical sheets by G. H. Bryan, *Phil. Mag.* May 1898, pp. 381–397.
[2] "Electromagnetic Induction in Conducting Sheets and Solid Bodies," *Phil. Mag.* January 1884, cf. pp. 21–23 [: *supra*, Vol. I, pp. 20, 26].

then the magnetic potential due to them is $-\partial\chi/\partial z$, and its vector potential $(F, G, H)$ is $(\partial\chi/\partial y, -\partial\chi/\partial x, 0)$. The equations of electric flow in the sheet, of specific resistance $R$, are

$$Ru = -\frac{\partial F}{\partial t} - \frac{\partial\psi}{\partial x}, \quad Rv = -\frac{\partial G}{\partial t} - \frac{\partial\psi}{\partial y},$$

which are both satisfied by

$$R\phi = -\partial\chi/\partial t,$$

without requiring any potential $\psi$ arising from accumulated free electricity.

Also at the sheet

$$-\partial\chi/\partial z = 2\pi\phi.$$

Thus at the sheet
$$\frac{R}{2\pi}\frac{\partial\chi}{\partial z} = \frac{\partial\chi}{\partial t}.$$
its mode of change:

This means, provided the sheet be *plane*, that the magnetic potential due to a decaying current sheet alters, just as if the system to which interpretation. it belongs were receding, without other change, with velocity $R/2\pi$. For, on introducing the relative coordinate $z + Rt/2\pi$ in place of $z$, the relation becomes $\partial\chi/\partial t = 0$, showing that at the sheet and therefore beyond it $\chi$ is of the form $F(x, y, z + Rt/2\pi)$.

We first apply this principle of Maxwell's to the case of an electric point charge $e$, moving with velocity $v$ in front of the sheet. This con-  Electron tinuous motion of the charge may be replaced by a series of instan-  travelling in front of taneous jerks between the positions which it occupies after successive  resting sheet: infinitesimal intervals of time $\tau$. Each such displacement of its position is equivalent to the creation of a doublet of moment $ev\tau$. The initial (magnetic) effect of the currents induced in the sheet is to annul the field of the doublet thus instantaneously created, as regards the region beyond the sheet: thus the action of the currents in the sheet is on either side of it equivalent initially to that of this doublet with sign changed, supposed to be situated on the other side. The currents thus impulsively induced in succession gradually die away by resistance $(R)$ in the sheet, and in so doing each of them exerts the same influence (Maxwell) as if the equivalent doublet on the other side of the sheet moved steadily away with velocity $(U)$ equal to $R/2\pi$. The fields of force persisting from successive past intervals of time $\tau$ are thus expressed as due to the sources graphically represented in the diagram; in it the image system is inserted on the remote side of the sheet, the conjugate image for the near side being, of course, its  its trail of reflexion in the sheet, with sign changed if it is a magnetic, but not if  receding images. it is an electric pole. Each step between a receding doublet image and the next is of length $U\tau$. At the instant when the moving charge $+ e$

has reached the point marked as the end of the path, the aggregate image system consists of the doublets represented by the short continuous lines, each of which is a persisting travelling reflexion of the instantaneous effect of the jerk representing a previous element of the path of $+ e$.

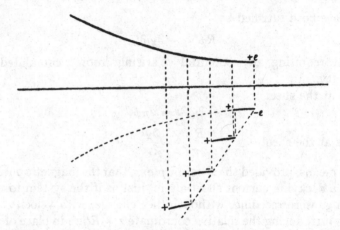

It is the change, occurring per unit time, in the conformation of this image system of doublets, that represents convection of electric charges and so produces magnetic effect. It is, in fact, only while complementary charges are being separated, so as to create a doublet, that a current element exists, of moment equal to charge multiplied by velocity: when the doublet has once been established, and merely continues to exist, the magnetic flux around it ceases, though there is a latent accumulation of such flux which persists without further electrodynamic effect, and would be undone if the doublet were again absorbed by its poles moving into coincidence.

Latent magnetic flux.

Thus the magnetic effect of the induced currents in the sheet is the same as that of the electric flux in the image system; and the aggregate flux in the time $\tau$ amounts to the distribution of vertical doublets, each of moment $eU\tau$, as marked by the $+$ and $-$ in the diagram, each receding from the sheet a distance $U\tau$, and in addition the creation of the fresh isolated image $- e$ at the end of the series. It is the limiting form assumed by this convection as $\tau$ vanishes with which we are concerned.

If the inducing charge $e$ is at rest, this image system amounts to nothing, as of course it ought. If the inducing charge is moving directly towards the sheet with velocity $u$, the image system again assumes a simple form; for if each doublet $- e \,.\, U\tau$, where $U$ is $R/2\pi$, is replaced by an equivalent but longer one $- eU/(U + u) \,.\, (U + u) \,\tau$, the doublets will form here also a continuous line in which adjacent poles cancel, and will thus represent in the aggregate the removal in time $\tau$ of an image charge $-eU/(U + u)$ to infinite distance, or rather to the distance appropriate to the duration of the motion. Thus, of the instantaneous image $-e$, there remains in position, after subtracting the part thus removed, a residue $-eu/(U + u)$, moving towards the sheet with velocity $u$. This reduced geometrical image $-eu/(U + u)$ constitutes in the present case the entire image system, as regards magnetic field due to the disturbance excited in the sheet; but, as regards electric field, the effect of the sheet is the same as that of the complete electrostatic image $-e$. For a receding charge the sign of $u$ would be changed.

This result may, of course, be generalized. When any rigid electric system is approaching directly with uniform velocity, from a distance towards an infinite sheet of specific resistance $R$, the magnetic effect on its own side of the sheet, due to the currents induced in the sheet, is the same as that of the moving optical image of the system with all charges altered in the ratio $-(1 + R/2\pi u)^{-1}$. *(margin: Reduced image of an approaching charged system:)*

Obviously the argument here applied to a moving charge applies equally to a moving magnet pole[1]; but in that case the image must be positive, instead of negative as here, in order to annul the normal component of the magnetic field at the sheet: thus the last result can be generalized so that the system may include magnets as well as electric charges. As circuital electric currents can be represented by magnetic polarity, the principle also extends so as to include such currents. *(margin: extended generally.)*

The representation by reduced geometrical images thus applies to any steady electromagnetic system whatever, which is approaching directly to the conducting sheet or thin layer; and it is easy to extend it to a changing system.

When a moving charge $e$ describes a closed orbit, the receding doublet images are distributed along curves drawn on a cylinder. Thus when it describes a circle parallel to the sheet with uniform velocity the image is a distribution of doublets, each of them parallel to the sheet, located along a spiral. *(margin: Image for an orbital current:)*

Here also we can arrange cases so that these doublets will fit together end to end. They will do so if a large number of equidistant discrete charges are attached along the circle so that each moves round it with the same velocity $v$. The image system is then a uniform distribution of tangential doublets spread around the image cylinder, each perpendicular to its axis. They can be enlarged in length so as to fit end to end, and thus form an area of continuous cyclic electric displacement around the cylinder, spreading downwards along it with velocity *(margin: as a spiral of electric displacement:)*

---

[1] Maxwell, *loc. cit.* § 655, gives these special results.

$U$. The intensity of this sheet of electric displacement is obtained by noting that, in the time of a revolution $2\pi r/v$, the images have spread down the cylinder a length $2\pi r U/v$, and that the whole charge $E$ on the circle has in this time crossed a line of this length: thus the intensity of the sheet is $E/2\pi r \,.\, U/u$ or $\lambda u/U$, if $\lambda$ is the electric line density on the ring. The increment of this sheet of cyclic electric displacement, accruing at its remote end, is equivalent magnetically to a circular current there, of amount equal to the intensity of the sheet multiplied by $U$, that is, to $\lambda v$, or to $i$, the measure of the current in the ring. This image current is too remote to have sensible effect, except at the very beginning of the steady motion. But, if the inducing ring is approaching the sheet with velocity $w$, the cylindrical image sheet of electric displacement will now be expanding at both ends; there would be an image current $iw/(U+w)$ in the position of the optical image of the ring, and another one $iw/(U-w)$ at the remote end of the cylindrical sheet. These results, generalized to a closed current system of any form, directly approaching the conducting sheet, may be at once verified from the magnetic image when the cyclic currents are replaced by equivalent magnetic shells.

*worked out for a circle.*

*The magnetic equivalent.*

We now revert to the case of a single electron $+e$ travelling in a curved path, as represented by the diagram. In the differential interval of time $\tau$ next before the time exhibited, each of the vertical doublets in the diagram will have receded a distance $U\tau$, while a new image $-e$ will have been instituted opposite to the charge, or rather transferred there from the further end of the image system. To determine the magnetic field due to convection of these parallel doublets, we make use of the vector potential, which for a current element $eu$ is parallel to it and equal to $eu/r$. The operation of moving an electric doublet of moment $\mu$ along its own direction (that of $z$) with velocity $-U$ thus produces a vector potential $U\partial/\partial z\,(\mu/r)$. In the present case $eU \,.\, \delta t$ is the aggregate moment of these doublets, all normal to the sheet, which correspond to an element $\delta t$ of the time of motion. Thus the aggregate vector potential is parallel to $z$ and equal to

*Vector potential of the receding image system:*

$$U\int \frac{\partial}{\partial z}\left(\frac{eU}{r}\right)dt, \text{ or } U^2 \int \frac{\partial}{\partial z}\left(\frac{e}{v'r}\right)ds';$$

it is thus $(R/2\pi)^2$ multiplied by the component along $z$ of static attraction (toward the sheet) of a line density equal to $e/v'$, distributed along the (elongated) image curve, where $v'$ is the velocity along that curve corresponding to the motion of the inducing charge. If $(0, 0, H)$ denote this vector potential, the magnetic field is parallel to the sheet and equal to $(\partial H/\partial y, -H/\partial x, 0)$; in fact, $H$ is its stream function. The whole magnetic influence reflected from the conducting sheet is this magnetic field, together with the electric and magnetic fields of an image $-e$ existing and travelling in the position of the optical image of the inducing charge: as the latter is the instantaneous shielding image, the former part represents the trail due to imperfect conductance of the sheet, which prevents the currents from adapting themselves instantaneously into the shielding distribution. expresses defect of shielding due to imperfect conductance.

If the inducing charge travels uniformly in an oblique straight line, the distribution of which $H$ is the Newtonian potential is a uniform straight linear one; thus $H$ is expressible in logarithmic form, and the solution is at once completed in simple finite terms, which it is needless to express at length. An example.

The reaction of a conducting plane on any moving electric system can in this way be synthetically set out. Owing to the trail arising from imperfect conductance, each moving charge experiences a force transverse to its velocity equal to the component of the magnetic field in that direction multiplied by $v$, and a force towards the sheet equal to $edH/dt$, where $H$ is determined as above; while, in addition, there is the forcive on the moving charge due to its moving reversed optical image. It is implied throughout that the velocity of convection does not approach within, say, one-tenth of that of radiation, as if it did so an "equilibrium" theory which neglects the propagation of waves in the aether could not effectively apply. Reaction of the screen.

In the case of an inducing magnetic pole $+ m$, it is the actual poles of the image system that produce the magnetic field, in contrast with the convection which alone operates in the electric case above. Thus, in addition to the direct instantaneous image $+ m$ (and its complement $- m$ at the other end of the trail), there is a trail of decaying previous disturbance having a magnetic potential

$$- U \frac{\partial}{\partial z} \int \frac{m}{v' r} \, ds',$$

wherein the integral is the potential of a line density $m/v'$ distributed along the receding image path in the diagram[1].

Similar conclusions apply if the screening sheet itself is in motion, everything being determined by the relative motion of the inducing system with regard to it.

Thus the specification of the disturbance reflected from the infinite sheet for any convected system containing charges, magnets, and currents is formally complete. The conducting sheet will cut off the direct action of the moving system from the other side, replacing it by a decaying trail represented by the receding images here investigated. Transmitted disturbance.

[1] Cf. Maxwell, *loc. cit.* § 664, where the case of uniform rectilinear motion is worked out into detail analytically, in rather different form.

But the advantages of this separation between shielding and decaying currents are in the main analytical, unless the velocity is large; it is only for velocities comparable with $R/2\pi$ that the field of force on the other side is much affected by them, and even for copper 1 cm. thick $R$ is about 1600 cm./sec.

*Only high velocities influential.*

### The Mode of Decay of a Plane Sheet of Currents.

It has been seen that, when an electromagnetic system is suddenly established in front of an unbounded plane sheet, of specific resistance $R$, it originates a current system induced in the sheet, which at first entirely cuts off the magnetic disturbance from the region beyond it; and that in time such a current system gradually dies away, so that its magnetic field is always the same as would be due to the inducing electromagnetic system, as thus suddenly established but with algebraic sign changed, itself moving away unchanged with velocity $R/2\pi$. Or, in other words, the magnetic field, belonging to an infinite plane current sheet, simply moves up into the sheet as it decays, without change of form or intensity, and with this velocity.

*Statement of shielding influence.*

*Very simple graph of subsidence of field of a plane sheet.*

The character of the electric flow in this masking current sheet and the mode of its decay are amenable to simple graphical expression. They are to be such as would produce, in the region beyond the sheet, the field of magnetic potential belonging to the receding reversed equivalent of the inducing system, as above specified. Now, if $\phi$ is this magnetic potential close to the sheet, the currents in the sheet are derived from a stream function $\phi/2\pi$; that is, the current crossing an element of arc $\delta s$ is

*The currents in the sheet:*

$$\frac{1}{2\pi}\frac{d\phi}{ds}\,\delta s.$$

Thus, if the magnetic equipotential surfaces of the inducing system are constructed, their sections by the plane of the conducting sheet are the initial lines of flow (with direction of course to be reversed) of the currents induced in it. The stages of subsequent decay of these currents in the sheet are determined similarly in terms of the sections of this initial system of equipotential surfaces by a parallel plane, moving away with velocity $R/2\pi$. In this representation, the current flowing in the strip of the sheet included between the same two equipotential surfaces remains the same throughout the decay, now represented by the receding of the sheet. Thus, on this mode of expression, the currents instantaneously established in the sheet may be said to decay (the sheet being infinite) simply by their lines of flow contracting in this manner without alteration of the flux along each, the central filaments of flow disappearing in succession by shrinking to nothing.

*graphically expressed:*

*also as regards decay,*

*thus by shrinkage of their circuits.*

For example, the case may be taken of a current suddenly established in a circular wire in front of the sheet, whether its plane be parallel or oblique to the sheet. We have to draw the magnetic equipotentials of such a circular current. These are the orthogonals of the stream lines of a circular vortex filament as drawn by Maxwell in *Elec. and Mag.*, after Lord Kelvin, *Trans. R. S. Edin.* They are a system of surfaces abutting on the current, and if they are constructed so as to spring from it equally spaced at angles $2\pi/n$, the strip between two successive surfaces will, in the specification above, carry one $n$th of the current flowing in the wire. The features of the current sheet and its mode of decay are thus open to direct inspection on Lord Kelvin's diagram aforesaid, or rather on a set of parallel sections of a solid model of it. Moreover, in cases of induction not merely instantaneous or impulsive, the lines of electric flow in the sheet at each instant are still the sections by the sheet of the magnetic equipotentials of the image system. *[Reference to relevant graphs for a circular current. Generalized.]*

For the case of an electric system convected in front of the sheet, it has been seen (p. 392) that the effect of the currents induced is that of the instantaneous image (the same as if the sheet were a perfect conductor) together with a vector potential $(0, 0, H)$ which is readily calculated, or graphically expressed, in the simpler cases. The value of $H$ represents the effect of that part of the currents which is due to resistance in the sheet. This part, in fact, involves a magnetic field $(\partial H/\partial y, -\partial H/\partial x, 0)$ at the sheet; thus it belongs to a distribution of currents in the sheet expressed by $(2\pi)^{-1} (\partial H/\partial x, \partial H/\partial y, 0)$; and the part of the electric flow which arises from resistance, the only part which is not cancelled on the remote side of the sheet, is thus represented by the gradient of the function $H/2\pi$. For an inducing electron moving uniformly in a straight line, it is easy to see that these outstanding resistance currents flow along systems of ellipses which are parallel plane sections of the confocal level surfaces of a uniform line distribution. *[The sheet of currents: a simple case.]*

# 74

## THE STATISTICAL AND THERMODYNAMICAL RELATIONS OF RADIANT ENERGY*.

[Bakerian Lecture. *Proceedings of the Royal Society*,
A, Vol. LXXXIII (1909) pp. 82–95.]

The atoms:

IT was surmised by a prominent school of thought in ancient times, it became absorbed into the mechanical philosophy of Newton, and it was at length established by the experiments and reasonings of Dalton and his contemporaries, that matter is not divisible without limit, but is constituted of an aggregate of discrete entities, all alike for the same homogeneous substance, and naturally extremely minute compared with our powers of direct perception. The smallest portion of matter which we can manipulate (at any rate until very recently) consists of a vast assemblage of molecules, of independent self-existing systems which exert dynamical influences on each other. The direct knowledge of matter that mankind can acquire is a knowledge of the average behaviour and relations of the crowd of molecules. To a sentient intelligence with perceptions of space and time minute enough to examine the individual molecules, each of them would

each a cosmos.

probably appear as a *cosmos* in itself, influencing and influenced by others—not unlike stars in a firmament.

Knowledge statistical: yet exact.

The observed laws of Nature are thus laws of averages—are statistical relations. Yet they are for practical purposes exact. To illustrate this in a way that will presently be of use let us imagine a row of urns whose apertures are of different areas, and let us consider how $N$ objects will be distributed at random among them, assuming that the chance of an object getting into an urn is proportional to the area of its aperture, and is otherwise indifferent as regards them all. If the number of objects is not very large in comparison with the number of urns, no direct law of numbers emerges in this random

Events and their opportunities to occur.

distribution: though by the doctrine of probabilities we may calculate definitely the relative numbers of times that the various distributions will occur in a vast total number of cases, and this will represent the chances of recurrence of these distributions. The most likely arrangements are those ranging close around the equable distribution, in which the contents of the urns are proportional to their apertures.

* This subject now of course ramifies far beyond the simple preliminary stage in its history that is represented in the present paper.

Those far removed therefrom are much less likely. When the number $N$ is very great, a relatively small deviation from the equable distribution has an almost negligible chance of occurring. Equable distribution then assumes the aspect of a rigid law; nevertheless occasionally in an aeon it will be widely departed from. The abstract laws governing the extent and distribution of the various kinds of deviations from the mean distribution constitute an important part of the theory of statistics, first explored and developed by James Bernoulli[1].

Relative frequency of occurrence of the various collocations among all the possibilities, that are equally open subject to the restrictions essential to the constitution of the system, is the sole criterion of statistical law: and when the numbers concerned are vast, it assumes a precise and absolute form.

This procedure for determining the natural distribution of objects must be applicable to every subject that has to do with indefinitely numerous assemblages of similar molecules. It is more fundamental than the test of the ultimate state employed by Maxwell and his successors in the theory of gases, that the steady state is the one having the property that it is not disturbed (departed from) owing to the mutual collisions or encounters of the molecules. Indeed, we can regard this latter procedure as being itself the test for the state of maximum probability, viz., that a slight derangement does not alter its constitution. The application of the present method seems to have been first attempted in gas theory by O. E. Meyer, but was soon after more effectively realized by Boltzmann (1877); in his lectures on gas theory, however, only a few pages are devoted to this aspect of the subject[2]. *(The steady statistical state: is that of maximum probability.)*

It may be, and has been, objected that this mode of procedure in molecular physics is ambiguous and unreliable; that different ways of calculating the chances lead to entirely different results, and that there is no certainty as to which method is to be preferred. Such discrepancies can, however, arise only because the imposed conditions of physical necessity (dynamical laws) restricting the distribution that would otherwise be entirely fortuitous must have been chosen differently in the cases compared. But there can be only one set of conditions that are right: the correct and tolerably complete formulation of the physical problem may be difficult, only to be attempted in simple cases, but the method is sound and, in general, no other *(Precision imperfect: results of improvement:)*

---

[1] Cf., *e.g.*, Lord Rayleigh, "On James Bernoulli's Theorem in Probabilities," *Phil. Mag.* 1899; *Collected Papers*, Vol. IV, p. 370 [: cf. the applications in gas theory by Smulochowski and Einstein: cf. also Appendices III and V *infra*].

[2] *Vorlesungen über Gastheorie*, 1895, pp. 39–42, referring back to a longer and less definite discussion in the original paper.

method seems to be open. In the very simplest physical aggregates of molecules that we can consider, such as a gas where they are practically independent of each other, the restricting conditions are, of course, far more complex than in the ideally simple illustration of urns and balls above. But if we have got hold of the main necessary conditions restricting fortuitous distribution, even though a whole series of minor ones are ignored, we shall be able to deduce approximately the actual statistical relations: although every new condition that is recognized and introduced into the analysis will alter to some extent the whole circumstances. There is no rigorous finality; and this, in fact, corresponds to Nature, where we never know all about an actual system of bodies, but have to refine it down into an ideal simple *tested by* system capable of more exact specification, and then compare the *experience.* results with the phenomena of the natural system of which it is a limited representation, but one that we can learn gradually to make more and more complete as knowledge advances.

The method of statistical probability thus corresponds closely to *Knowledge is* the features of the experimental exploration of Nature. We never get *of the surface* to complete knowledge of anything, however simple; but we seize *of Nature.* upon prominent groups of uniformities or correlations, more or less isolated, and thus capable of analysis more or less approximately by themselves with subsequent introduction of corrections representing the influence of the other groups.

There would be little utility in following out these statistical methods if they were confined to systems of molecules that have already settled down into steady distribution—to systems forming homogeneous bodies in thermal equilibrium. The methods of dynamics of solid and fluid systems have been fully elaborated, and practically cover that field. It is as regards the residuum of uncorrelated movements of the molecules that the statistical method is essential—those *Residual or* which are put aside in ordinary dynamics, yet are available and power- *thermal* ful to modify mechanical forces in bulk when there is difference of *energy:* temperature, owing to the combined effects of their striving towards mobile equilibrium of distribution, of which uniformity of temperature is the test. The laws of transformation of residual (thermal) energy must find their justification in this connection. The fundamental thermodynamic principle that thermal energy tends towards the equilibrium distribution, never by any chance spontaneously in *falls away* the opposite direction, is itself a direct result of the discrete statistical *irreversibly.* point of view, and cannot otherwise be justified. The degradation of mechanically available energy is the progress towards its equable molecular distribution, a progress which cannot be undone by any operations on matter which do not rearrange the molecules [cf. p. 411];

and the measure of the amount of degradation must rest ultimately on statistical principles.

The most salient aspect, in Boltzmann's hands, of the method of pure statistics in the theory of molecular distribution in gases is the expression of this fact. In Clausius' treatment of the doctrine of degradation of the energy, this principle becomes the principle of continual increase of the entropy, that being the name given to a definite thermodynamic function of the state of the system, specified statistically in terms of ordinary physical *data*. The proposition, that a molecular system always tends in spontaneous change towards states of greater entropy, runs parallel to the principle that it always tends towards molecular distributions of greater probability subject to the restraints and laws that are inevitable. The entropy ($\eta$) is thus in some way a measure of the probability of the molecular collocation. By testing this idea on the known gas theory, Boltzmann arrived at the conclusion that $\eta = k \log \Pi$, where $\Pi$ is the probability of the molecular collocation, and $k$ is a universal constant to be suitably chosen in terms of fundamental units. And the present train of ideas is here corroborated; for this logarithmic law is just as it ought to be, in order to satisfy the condition that the entropies of independent systems are additive, the conjoint chance $\Pi$ being obtained by multiplication of the independent partial chances.

Moreover, by thus defining $\eta$ for molecular distributions in terms of $\Pi$, the idea of entropy is extended to systems of which the differential element of mass is not itself in statistical equilibrium, that is to systems in which the idea of temperature at a point is inapplicable. Such systems, for example, may be made up of superposed systems of different temperatures, which interpenetrate yet keep their energies separate. Radiations of different wave-lengths in a perfectly reflecting enclosure form a case in point: their energies remain isolated, and become averaged independently, unless some absorbing molecular matter is present which opens a path for interchange. It is, in fact, in the theory of natural radiation, where it was first applied by Planck, that the statistical method has been effective as suggesting inferences beyond what formal thermodynamics, involving the existence of temperature, could supply.

What follows may be regarded as an expansion and generalization of ideas implied in Planck's analysis for the case of natural radiation; it will involve, however, a modification of Boltzmann's application to gas theory, such as is apparently necessary to a direct general view of the molecular significance of temperature. This quality, temperature, which we know directly through a special sense, has, in fact, so universal a *rôle* in physical equilibrium that a definition of it in terms

*Margin notes:*

The measure of degradation:

found in additive entropies:

interpreted as log probabilities.

Entropy a wider idea than temperature:

*e.g.* in superposed radiations:

so transcends thermodynamics of steady states.

Planck's radiation statistics.

Temperature is sensually given to us.

of complex analytical formulas, such as is obtained in gas theory, can hardly represent the fundamental aspect of the subject. And the thermodynamic scale of Lord Kelvin, though providing a universal criterion, does so rather by utilizing a dynamical property than by elucidating the essence of the quality that is measured.

*Its physical meaning?*

In the case of radiation, some of the main features inviting explanation may be set out in stronger light as follows. Consider a homogeneous solid body with an internal cavity, all being in thermal equilibrium. The question was put early in the history of thermodynamics, whether the radiation from its walls, across this cavity, may not institute inequalities of temperature, in contradiction to Carnot's principle. The necessary negation was provided in a very elegant paper by Clausius, in which he explained how, on the Hamiltonian theory of rays, there is an exact balance of exchanges; that each element of surface must radiate to any other element just as much energy as it receives back from the radiation of the latter, provided their intrinsic brightnesses are equal. Let us now suppose that just beneath the radiating skin of the cavity there is an interface adiabatic to radiation and conduction, having two small apertures which can be imagined to be opened, say by sliding shutters, thus establishing communication with the mass of the solid body. When they are opened, transmission of energy among the molecules of the body, including heat conduction, comes into play, conjointly with radiation across the cavity. Why is it that when the two elements of surface are of equal brightness as regards radiation, they are also adapted for an equilibrium of exchanges by the other route, across the mass of the body—because, if they were not, there could be no state of equilibrium? Again, to take another case: if we have three different bodies, $A$, $B$, $C$, in a row, so that $B$ is in contact with each of the others, and in equilibrium of exchange of thermal energy with both of them, why is it that when $B$ is removed, and the other two are moved up into contact, there is no disturbance of the equilibrium of thermal exchange? Facts such as these suggest a settling down of the wandering elements of energy into an equilibrium, in which each body, or each region, takes its share according to its capacity: and the task at present attempted is simply the expansion of this idea of distribution into quantitative form.

*Temperature adjustable by exchanges of radiation:*

*or by molecular transfer of heat:*

*why with identical results?*

*Independent of chemical structure:*

The general thesis of which a development is here attempted is thus the molecular statistics of distributions of energy; and the method that will be followed may be foreshadowed in a general way as follows. Our complex physical system is capable of containing energy in various forms, with paths of transformation open between them. We may apportion the system which is the seat of the energy into

*thus an affair of energy,*

elementary receptacles of energy as regards each form assumed by it, which we may call *cells*: and we may establish a relation between the extents (in a generalized sense) of these cells by the condition that they shall be of equal opportunity, that the element of disturbance possessing the element of energy under consideration is as likely in its travels to occupy any one of them as any other. Our test of equality of opportunity is the Euclidean test of equality of extent, namely, so-called superposition; when an element of disturbance of one type is transformed so as to be of another type and its course is traced during the transformation, the regions which it occupies in its progress are regions of equal opportunity for that type, or, say, cells of equal extent. The term "region" is here used in the sense of geometry of many dimensions, which is familiar in generalized gas theory. The energy itself, belonging to the same element of disturbance whose progress is traced, may be different in two states of it; for the transformation from the one to the other may involve the addition of energy, either abstracted from the surrounding region or entering by work applied from without. Thus we may make a mathematical formulation in which the energy is made up of differential elements, which are of different amounts, $\epsilon_1, \epsilon_2, \ldots \epsilon_r$, for the $r$ possible states of the energy, in which it is distributed among equivalent cells of numbers $N_1, N_2, \ldots N_r$; and the problem is to find $n_1, n_2, \ldots n_r$, which are the numbers of elements of energy that, in the ultimate state of statistical equilibrium, will reside in these groups of cells respectively, subject to conservation of its total amount $E$. The chance of the distribution just expressed, being made up of independent chances, is proportional to the number of ways $\Pi$ in which this statistical distribution can be realized, which is of the form

*and its material and aethereal receptacles.*

*Cells of equal opportunity:*

*the test.*

*Elements of energy, changeable in transit.*

*Formulation of problem of distribution.*

*Chance of assigned distribution:*

$$\Pi = F\,(N_1,\, n_1)\; F\,(N_2,\, n_2)\ldots F\,(N_r,\, n_r);$$

and in the natural distribution this product must be a maximum subject to the restriction*

*maximum subject to energy conserved,*

$$\epsilon_1 n_1 + \epsilon_2 n_2 + \ldots + \epsilon_r n_r = E, \text{ constant.}$$

This gives
$$\delta \log \Pi = \frac{F'\,(N_1,\, n_1)}{F\,(N_1,\, n_1)}\, \delta n_1 + \ldots;$$

which, therefore, must vanish at the maximum, subject to the condition
$$0 = \epsilon_1 \delta n_1 + \epsilon_2 \delta n_2 + \ldots .$$

---

* In the recent new statistic (1925) of Bose, Einstein, and de Broglie, the cells are not fixed but only their total number $N$ is assigned, against the spirit of the present discussion: this loosens the problem by introducing instead of fixity merely another equation of restriction $\Sigma n_r = N$, thus replacing $e^{\epsilon\lambda}$ in the final Planck formula by $e^{\epsilon\lambda+\phi}$. Two specifications of thermal state $\vartheta, \phi$, would moreover be involved, instead of one temperature $\vartheta$.

This requires that $$\frac{F'(N_s, n_s)}{F(N_s, n_s)} = \vartheta\epsilon_s$$

for all values of $s$, where $\vartheta$ is an undetermined quantity, the same for all the sets of *cells*. We have thus been led to a quantity $\vartheta$ which, in the state of equilibrium of exchanges of molecular energy, is the same throughout all parts of the structure of the system. It can be none
other than temperature; or rather, the scale and length of degrees being still undefined, it is some function of temperature as measured on any convenient scale. [Cf. p. 410.] The value of $F(N, n)$, the number of modes of distribution of $n$ like objects in $N$ like compartments is,
following Planck, $(N + n - 1)!/(N - 1)!\,n!$, which, when $N$ and $n$ are very large numbers, by application of Stirling's approximation $n! = n^n e^{-n}\sqrt{2\pi n}$, assumes the form

$$F(N, n) = (N + n)^{N+n}/N^N n^n,$$

so that $\log F(N, n) = (N + n)\log(N + n) - N\log N - n\log n,$

and therefore $$\epsilon\vartheta = \frac{d}{dn}\log(F, n) = \log\frac{N + n}{n},$$

and, finally, $$n\epsilon = \frac{N\epsilon}{e^{\epsilon\vartheta} - 1}.$$

This* becomes Planck's formula for natural radiation, with $\vartheta$ representing temperature on some scale, when the values of $N$ and $\epsilon$ are substituted in terms of $\lambda$, and when another factor, $\lambda^{-1}$, is adjoined to reduce the value to the same unit range of wave-length in all cases, as *infra*.

On this skeleton of a general theory of temperature various remarks are to be made. In the first place, there are many dynamical principles to be satisfied throughout the progress of the elements of disturbances, in addition to the conservation of energy; as witness the elaborate dynamical discussions in gas theory. For example, linear and angular momenta must be conserved during the migrations of the elements.
It may be recognized, however, that all such principles will be required and will be used up in determining the equivalence of the various sets of cells, among which the wandering elements of disturbance are distributed at random. Of this, in fact, gas theory offers
a convenient illustration; there the dynamical principles governing encounters are not ignored, but are fully used up in the determination of the relative extents of equivalent cells of different types ($\delta u\,\delta v\,\delta w$), so that no conditions are outstanding except the constancy of the total energy and the constancy of the total number of molecules. The energies, together with the paths followed by the elements of the disturbance, which latter are needed to determine the cells, make everything determinate, and therefore cover the whole dynamical

* $\epsilon\vartheta$ is $\epsilon_q/R\tau$, where $\epsilon_q$ is the quantum energy and $\tau$ is standard temperature.

field. It is only in a few of the simpler cases that this procedure for mapping out cells can yet be carried out, notably gas theory and the theory of natural radiation; but the point gained is that they here fit into a general scheme of distribution of the uncorrelated elements of energy instead of being isolated fragments of molecular theory. Our specification of cells must remain, however, an imperfect and merely approximate one, to be modified and improved by each fresh addition to our knowledge of the system; but this is as it should be, for the available energy of thermodynamics is itself provisional in the sense that new discoveries may and do reveal stores previously inaccessible for mechanical purposes.

*Method general:*

*adaptable to new knowledge.*

Again, the foundation here assigned to the universal concept of temperature rests on the implied hypothesis that the distributions of the various types of energy in the various sets of similar cells are independent chances, so that the total probability is their product. If that were not so, each type of energy could not have a temperature of its own—equal to that of every other type with which a path of interchange is open, when the system is in equilibrium. On the other hand, if this be so, without limitation, each range of wave-length in an adiabatically isolated region of radiation, such as the radiation in the range between $\lambda$ and $\lambda + \delta\lambda$, acquires its own temperature when it has become fortuitously distributed by reflexions at the boundaries; while to equalize the temperatures of the different ranges of wave-length the presence of some absorbing matter is required to act as a means of exchange. We have here an illustration in which the *same* region is occupied by distributions of energy of different temperatures; even if this extension is demurred to, the principles will still apply to the contents of *adjacent* regions.

*Free exchanges necessary for a unique temperature.*

Again, the necessity for associating different amounts of energy with the same unitary element of disturbance, as it passes into regions of cells of different types, may be illustrated from the circumstance that reversible physical operations in bulk, such as the adiabatic compression of a mass of gas, cannot alter the statistics of distribution of the energy: thus indicating that the energy put into the system by compression is distributed among the said unitary elements of disturbance.

The principle of independent chances as regards the distributions of the various types of energy requires, as already mentioned, a restatement, in altered form, of Boltzmann's statistical determination of entropy for gas theory. Under natural conditions, however, concerning the relation of the amount of the element of energy to the extent of a cell, the final results are not sensibly affected; thus the validity of the ordinary principles of gas theory will not come into question. As regards the general underlying idea, it seems reasonable that the distribution of things of different types (*e.g.*, molecular groups within various velocity ranges $\delta u \; \delta v \; \delta w$ in gas theory), so far as concerns their manner of mere mixture among one another in space,

The cells are the frame of reference necessary for the statistic.

has not direct physical significance, which was what Boltzmann assumed. What matters is rather the opposite, the extent of the region of cells within which the elements of each type of disturbance have play to spread out, in view of their freedom to adjust their relative quantities by transformation into and from other types, subject only to the condition of constancy of the total energy.

The procedure of Planck for the elucidation of the statistical relations of natural radiation, which we have attempted to generalize,

Planck's energy element large:

depends essentially, as Lorentz remarked, on the assumption of a discrete or atomic constitution of energy; and the indivisible element of energy, as estimated from the constants of the formula for natural radiation, proves to be of considerable amount, even compared with the energy of a molecule of a gas. A somewhat similar implication survives in the present development; but it now appears in the form that the ratio of the energy element to the extent of the standard unit cell is an absolute physical quantity determined similarly by the observations on natural radiation. A theory of the present type, as Mr Jeans has recently remarked, has more likelihood of being true in the limit, if there be one, in which the element of energy becomes indefinitely small, after the manner of an ordinary differential; thus,

modified form.

here it would be the limiting differential ratio of energy element to extent of cell that is somehow predetermined, but now without any implication that energy is itself constituted on an atomic basis*.

In short, what is formulated in the cells, into which the system is

Closer specification of problem:

divided, is equality of opportunity, not for elements of energy, but for elements of disturbance, which pass into different types during their progress while retaining their identity, and in so doing receive or lose

implications:

energy. This gain or loss we must assume to occur according to definite laws, independent of the path of transition of the disturbance between the configurations under comparison. In self-contained dyna-

justified by Hamiltonian dynamics,

mical systems this is secured as a corollary from the Hamiltonian principle of variation of the Action, after the manner of Liouville's theorem of differential invariance. In its more general aspect it may

or more generally by cyclic considerations.

perhaps be justified by a cyclical argument, after the manner of Carnot's principle in the classical thermodynamics.

The inquiry obviously presents itself: What is the relation of the order of ideas here sketched to the Maxwell-Boltzmann principle of equipartition of the energy, on the average, among the various

---

* The idea is that the statistics hold for infinitesimal elements of energy and therefore infinitesimal extents of cell: and that their ratio tends to some limiting value.

degrees of freedom of each dynamical system, with which it in effect is in contradiction?  The answer must be that in this order of ideas the different degrees of freedom of the system represent and are replaced by cells, or receptacles for differential elements of energy\*, which are of *equal opportunity* or *extent* as regards an element of disturbance which retains its identity, because they enter absolutely alike and without any mutual influence into the dynamical relations.  Whether the amount of the element of energy is the same for them all can be determined by following the course of an element of disturbance from one into the other, and ascertaining whether the amount of its energy has to change.  If there were no collisions and no aether, different types of energy would be entirely isolated—there would be no transition; and even collisions operate not by contact, but involve fundamentally the aether as an elastic buffer.  If then, we may take it that vibratory disturbance passes from one molecule to another, or from one free period to another, by interchange across the aether during mutual encounter, a consideration of the process of emission and subsequent absorption of radiation ought to reveal how the energy has to change in the transition.  When this transition is effected, through the agency of the work of the pressure of radiation as the mode of change in free aether from one free period to another, it fits consistently into the above general scheme; and the energy elements of the statistics are, in fact, different while the opportunities are equal in all the modes of vibratory molecular freedom[1].

*Statistics wider than dynamics:*

*paths must be opened up.*

It remains to give some account of the physical bases of the argument as applied to the special case of radiant energy†.  A ray or filament of light, even sensibly homogeneous and therefore of infinitesimal range $(\delta\lambda)$ of wave-length, is not a train of uniform waves;

*A ray is a filament of light:*

* The averaged energy in an element of volume can be formulated as a spectrum: but no such representation is possible as regards the disturbance itself or its phase. Cf. the Petzval theorem in Fourier analysis, as *infra*, p. 549. Thus free radiant energy has of necessity to be treated on statistical lines running parallel with gaseous thermodynamic energy.

*Restriction to energy.*

The statistical interchange of elements of radiation between a set of dynamical cells, as here developed, now goes by the name of quantification of the phase extension. To bring it into correlation with molecular thermodynamics of gases the laws of the interchange must be extended so as to include exchange, by absorption and emission, with the moving atoms: this has been formally carried through (Einstein, 1917) on the basis of the Bohr atomic radiative scheme.

*Phase extension here quantified.*

[1] The investigation by Lorentz, in which the natural radiation of metals is deduced (for an ideal metal) as due to the deflections of free electrons, moving with the velocities of gas theory, by collision with the molecules, is one of the ways in which radiation may be connected up with translatory molecular motions.

*Electrons may share in equipartition of energy.*

† The statistics of free radiation have recently undergone various types of formulation, of which a concise account is given by L. Brillouin, *Journ. de Physique*, 1927.

but it is the aggregate of a vast complex of trains of limited lengths, coming from the various molecules that take part in the radiation. The ray is thus a statistical aggregate; and the statistical relations of molecular equilibrium that are implied in the existence of a temperature in the radiating element of mass will be transmitted in some form into the constitution of the ray, and will thus limit its generality if it is to belong to natural radiation[1]. An early way of representing this constitution was to consider the ray as a system of separate impulses or elements following one another; and the problem took the form of determining the limitations imposed on their character and succession in natural light in consequence of the molecular equilibrium in each element of the source. This problem seems to have been first essayed by W. Michelson in 1887; it was soon after developed by Lord Rayleigh[2] in various papers. We propose, in this way, to divide the natural ray or filament of light into a set of elements, one of which may be, *e.g.*, a short train of simple undulations; each of these elements of disturbance travels so that its content, measured by $\delta S\, \delta\omega$ per unit length ($l$) where $\delta S$ is cross-section and $\delta\omega$ solid angle of divergence at it, remains unchanged, in a manner that however involves a rather close interpretation of the idea of differential invariance[3]. We may take $\delta S\, \delta\omega \,.\, l$ to be the measure of the extent of an aether cell for radiation of the given period. To compare the extent of cells for different periods we must follow the transition of a filament of disturbance from one period to another, say a higher one. The change can be imagined to be effected gradually, by shortening $l$ by compression exerted by perfect reflectors working against the pressure of the radiation. It appears that energy is added so that the amount of energy in the filament of disturbance is always inversely as the wave-length $\lambda$*; while the extent of the cell containing it varies directly as $\lambda^3$. We assume that these relations, obtained in this special manner, are universal, relying for the evidence thereof in the simpler cases on dynamical theory and on special verification, but generally on a principle analogous to that of Carnot, as above. Then a statistical

*Margin notes:*
not homogeneous along its length:

early hypothesis of short elements.

A radiation-cell:

nature of equivalence of cells for different periods determined,

from a special case, *infra*,

---

[1] Cf. "On the Constitution of Natural Radiation," *Phil. Mag.* Nov. 1905: also recent investigations in the *Annalen der Physik* by Planck, Laue, and others.

[2] Cf. "On the Character of the Complete Radiation at a given Temperature," *Phil. Mag.* vol. xxvii, 1889; *Scientific Papers*, vol. iii, p. 268.

[3] Cf. "On the Statistical Dynamics of Gas Theory as illustrated by Meteor Swarms and Optical Rays," *Brit. Assoc. Report*, 1900; or *Nature*, Dec. 27, 1900[: *supra*, p. 222]. But the analogy does not extend to reflection from a moving mirror, $\delta S\, \delta\omega$ not retaining its value for the reflected swarm of particles.

* The radiation *quantum* postulated by Einstein from photo-electric considerations in 1905, and much discussed later, appears to remain of constant energy, thus not changing its frequency, until later developments by Compton and Debye.

procedure on the lines sketched above gives, $\tau$ being $\vartheta^{-1}$, for the densities of energy in corresponding infinitesimal ranges of wave-length the expression $C\lambda^{-4}/(e^{c_2/\lambda\tau} - 1)$. But the extents of corresponding ranges $(\delta\lambda)$ are proportional to the mean wave-lengths $(\lambda)$; thus the density of energy per unit range involves an additional factor $\lambda^{-1}$, and the constitution of natural radiation is given by the expression $E_\lambda\delta\lambda$, where

$$E_\lambda = C_1\lambda^{-5}/(e^{c_2/\lambda\tau} - 1),$$

which is Planck's well-known formula.

*leads to Planck's consolidated formula.*

The demonstration of these relations between the extents of cells occupied by the same filament of disturbance when it is transferred to different wave-lengths, and its energies, and its ranges $(\delta\lambda)$, is obtained by following out the circumstances of the oblique reflexion of a filament from an advancing perfect reflector. The relations thus obtained then form a basis, leading by pure statistical procedure, without admixture of extraneous thermodynamics, to the Planck law for natural radiation*. This law of course includes the Stefan-Boltzmann law for the total radiant energy, and the Wien displacement law pertaining to the form of the energy curve at different temperatures, which marked the limit of the knowledge derivable from the classical thermodynamics.

*How cells are compared.*

It may be recalled that the original demonstration of Wien's law involved the compression of a slice of natural radiation by approach of its totally reflecting parallel bounding walls, and a rather complex statistical re-classification of its elementary parts as thus altered. By Carnot's principle these parts must constitute the natural radiation at some other temperature: which leads to Wien's inference. The demonstration is reduced to the simplest and most direct terms by arguing from the special case of a volume of natural radiation enclosed in a spherical boundary which is gradually and uniformly shrunk. Here the wave-length of *each* elementary filament is altered in proportion to the radius of the boundary, and its energy in the inverse ratio; thus the statistical averages are not disturbed, and no process of estimating them afresh is now required[1]. If the general statements above made hold good, not only do these relations hold for the spherical enclosure, but also each filament of radiation contained in it must occupy similar positions with regard to the boundary, at the end and at the beginning. This is readily proved independently, and

*Wien's displacement correspondence;*

*it holds for each element in a shrinking spherical enclosure:*

*otherwise it involves statistical rearrangement.*

---

* The values of $C_1$ and $c_2$ determine the extent of the cells. In Planck's later general formulation for a quantified phase space the extent of the cell in the sixfold, combined momentum and position, comes out to be $h^3$, obviously of the right dimensions.

[1] Cf. "On the Relations of Radiation to Temperature," *Brit. Assoc. Report*, 1900; or *Nature*, Dec. 27, 1900[: *supra*, p. 220].

it constitutes a confirmation of the preceding statements; for $r$ being the radius, $\delta S$ varies as $r^2$, and $l$ varies as $r$, while $\delta\omega$ is as before, so that the extent $\delta S\,\delta\omega\,.\,l$ varies as $r^3$, that is as $\lambda^3$, while the energy involved varies as $\lambda^{-1}$.

We have now to consider how far it is feasible to extend these statistics of radiation so as to include the material molecules and their **Relative** types of free vibration in the spectrum. If the type of vibration for a **capacity of a** free period is that of a simple Hertzian bipole, it has been shown by **bipole** Planck[1] that in a field of natural radiation specified per unit volume by **radiator in** given field of radiation: $E_\lambda\delta\lambda$ (or by $cE_\lambda n^{-2}\delta n$, where $n$ is the frequency) the average vibratory energy of a molecule for a free period of wave-length $\lambda$ is $E_\lambda/8\pi\lambda^{-4}$. We may thus say that the capacity or extent of the single molecular period is equal to that of a volume $(8\pi\lambda^{-4})^{-1}$ as regards the radiation, per unit range, near that wave-length. But this volume of free radiation specified by $E_\lambda$ is equivalent in extent, by the transformation by shrinkage above, to a volume $(8\pi\lambda_1^{-4})^{-1}$ of radiation specified by $E_{\lambda 1}$, which is again equivalent in extent to the energy in a free **independent** molecular period corresponding to $\lambda_1$. Thus the capacities, or oppor- **of period:** tunities, or extents, whichever term we prefer, are the same for all **but elements** free molecular periods; but the elements of energy, namely those of **of radiation** the same disturbance as transformed from one into the other, are **different.** inversely as the wave-length.

In fact, different free periods are of equal opportunity for another **Interpreta-** reason, that of gas theory, because their vibrations are entirely inde- **tion:** pendent and superposable; so that there is nothing to determine whether an element of disturbance locates itself in one or other of them. But in the process, spontaneous or not, of getting from one to another receptacle or cell, the disturbance has to absorb energy from the surrounding elements of disturbance, or reject it to them, in such manner that the unitary energy elements are different for the two cells. It is the disturbance, here the element of radiation, that is an entity, identifiable in its wanderings, but the energy belonging to it **of equal** may change as it passes from one location to another, just as in gas **opportunity.** theory it is the molecule that can be traced from one group ($\delta u\ \delta v\ \delta w$) to another, but its energy changes in the transition.

For very long wave-lengths the analysis shows that the elements of **Equipartition** energy may be taken as practically of the same amount in all types of **holds for** cells, so that the energy is distributed among the cells in proportion **long waves:** to their extents or opportunities,—equipartition holds good. The reiterated opinion of Lord Rayleigh may here be recalled, that the weakness of the usual argument for equipartition of the energy seems

[1] Cf. also H. Lamb, *Trans. Camb. Phil. Soc.*, Stokes Jubilee Volume, vol. XVIII, 1899, p. 349.

to lie in the unsatisfactory treatment of the potential energy residing fails for lack of convergence of its formulae. in very powerful (*i.e.* almost rigid) restraints, such, namely, as involve very short times of relaxation. [Cf. p. 411.]

Probably the most remarkable incident in the long discussions to which this subject has given rise has been a brief indication by Lord Rayleigh of how he would expect the energy to be partitioned—one which would have satisfied him completely were it not that it led to frittering away into higher periods without any final equilibrium. In this brief note[1] he obtained the correct value for long waves, Rayleigh's theoretical law: practically by isolating in thought a rectangular block of aether, and considering all its possible modes of free vibration as excited with equipartition of energy, postulating that the natural radiation is made up of them all taken together—that is, made up as regards distribution of energy but by no means as regards configuration[2].

In what relation does this very remarkable and successful *aperçu* verified for long waves. stand to the ideas here sketched? Its exposition and development have been essayed recently in several very interesting papers by Mr Jeans[3], and have been continued in this direction in Professor Lorentz's Address on the present subject delivered to the last Mathematical Congress at Rome. The great obstacle *à posteriori* is, as aforesaid, the frittering away of the energy into the unlimited number of very high free vibrational periods. On the other side, the remark Avenues of transfer in dynamics. made above seems worthy of emphasis, that in such an illustrative system the energy does not distribute itself at all: it remains as it was originally, unless avenues for transformation are admitted; whereas if means are so provided for one of the partial disturbances to alter its wave-length, it will alter its energy too, at the expense of its neighbours.

The problem which we have here essayed to discuss is, on one of its sides, of long standing. For more than thirty years, ever since Boltzmann and Maxwell generalized the principle of equipartition of Historical: molecular energy, the resulting paradox has been urgently in need of unravelment. Lord Kelvin resolved it by denying that the analysis Kelvin: put forward involved any real contact with the principle it sought to establish, and afterwards spent much time in destructive criticism of

---

[1] *Phil. Mag.* vol. XLIX, 1900, pp. 139, 140; *Scientific Papers*, vol. IV, pp. 483–485, with note of date 1902 indicating the subsequent experimental confirmation of his result for long waves. Planck subsequently pointed out that not merely the form, but also the numerical coefficient, as corrected by Jeans, was experimentally correct, agreeing with his own later formula; for the procedure requires that the energy of each vibrational mode must be put equal to that of each translational mode of a molecule in a gas. Cf. Planck, *Vorlesungen über die Theorie der Wärmestrahlung*, 1906, § 154.

[2] It may be noted that uniform compression of this very specially coordinated system alters all the energies in the same inverse ratio of the linear dimensions.          [3] Cf. *Phil. Mag.* June, July, 1909.

Rayleigh. special cases[1]. Lord Rayleigh has played throughout the part of umpire, warding off unfounded objections to the principle, while at the same time keeping an open mind regarding its validity: indeed hardly anything has been urged on the side of equipartition of energy that is not to be found in his two papers of 1892 and 1900[2].

The motive of this present discussion is the conviction expressed at the beginning, that the statistical method, in Boltzmann's form, must in some way hold the key of the position, no other mode of treatment sufficiently general being available. The writer has held to this belief, with only partial means of justification, ever since the appearance in 1902 of Planck's early paper extending that method to radiation. In the *British Association Report*, 1902 [as *infra*], there is a brief abstract of a communication "On the Application of the Method of Entropy to Radiant Energy," in which it was essayed to replace

Can the vibrators be eliminated? Planck's statistics of bipolar vibrators by statistics of elements of radiant disturbance. "It was explained that various difficulties attending this procedure are evaded and the same result obtained, by discarding the vibrators, and considering the random distribution of the permanent elements of the radiations itself, among the differential elements of volume of the enclosure, somewhat on the analogy of the Newtonian corpuscular theory of optics" (cf. *Brit. Assoc. Report*, 1900, as *supra*, p. 223).

Since that time the present point of view has been presented and discussed at various times in University lectures[3]. The writer has recently been encouraged, from various quarters, to offer the type of

[1] Cf. his *Baltimore Lectures, passim.*

[2] "Remarks on Maxwell's Investigation regarding Boltzmann's Theorem," *Phil. Mag.* vol. XXXIII, 1892; *Scientific Papers*, vol. III, pp. 554–557. "On the Law of Partition of Kinetic Energy," *Phil. Mag.* vol. XLIX, 1900; *Scientific Papers*, vol. IV, pp. 433–451.

[3] Among others, at Columbia University, New York, in March 1907.

The deduction of a temperature. [If there is more than one undetermined multiplier, thus safeguarding more than one restriction on the maximum, the direct necessity (p. 402) for a universal temperature will be vitiated. Thus in the Boltzmann application to a gas, now familiar, there are two independent restrictions, to constancy of energy and of number of molecules: but there it is only the ratio of the two parameters that counts. While in an extension of what is practically the present scheme to gases, by Einstein, Berlin *Berichte*, 1927, the resulting distribution formula involves two independent natural constants instead of the one of Planck: the temperature has to be introduced as the value of $\partial E/\partial \eta$, where $\eta$ is entropy, and the proof that it is universal, the same for each component of the system, would presumably be supplied in some indirect way.]

The necessity for cells. The fixation of the relative extents of the cells on dynamical principles as in p. 408 is an essential feature: indeed this is how the structure of the system controls the statistics, in themselves purely numerical. The determination of a set of cells of equal opportunity supplies the backbone of every statistical discussion.

argument here very briefly, and perhaps obscurely, outlined, to the consideration of a wider audience, including the experts in this complex subject, which is so fundamental in molecular physics.

### (1928) *Note on Anthropomorphism and its Quantification.*

The outstanding historic figures in the incipient thermodynamics are Carnot, Kelvin, Clausius. The paradox encountered by Kelvin on the threshold, as to how energy could be conserved yet only partially convertible into mechanical work, was after much perplexity resolved by his introduction of the concept of available energy and its dissipation. His phrase "available to man for mechanical effect" became in time a challenge to the elimination of any such anthropomorphic feature, which, it was argued, must be irrelevant to the scheme of external nature; and the entropy of Clausius, as defined in absolute manner by the Boltzmann formula, was the most promising resource for such purpose. Maxwell illuminated early implications of dissipation by the introduction of his ideal very small demons, operating under their own intellectual guidance, who could transcend the limitations that were insuperable for the grosser faculties of mankind. The degree of availability of energy would then depend on the fineness of the physical faculties of this race of operators. One way to formulate the problem would be to specify the smallest magnitude of the cells, in the uniform dynamical extensional framework of coordinates and momenta, that the demon could control: the classical thermodynamics, as "available to man," would constitute the asymptotic limit approached for the coarser races of demons when each cell can contain very many atoms or other relevant units of the statistics.

*[marginal note: The early thermo-dynamic paradox,]*

*[marginal note: resolved by introducing quantification]*

*[marginal note: essential for a Maxwellian demon-theory;]*

Here the anthropomorphism appears as essential to the nature of the problem, and by no means an irrelevant importation. The statistical problem is a double one, involving division of the physical field into a set of equivalent practically effective cells, expressing in fact the degree of coarseness of the framing in which nature has to be set, and the distributions of the statistical units among them. The scheme for natural radiation, above·sketched, is a direct example of such a scheme. In its formula (cf. pp. 402, 414) only a ratio survives, that of extent of standard cell to size of the effective unit, along with the constant of gas theory. What determines the extent of the cells is the problem of quantification that remains, which is now perhaps being thrown back on uniformities in atomic structure, in its relation to radiation as an exploring instrument.

*[marginal note: actually in operation as here for natural radiation.]*

This notion of the classical world as a limit is now familiar: cf. the principle of correspondence of N. Bohr. The wide interlockings

Surprising correlations. of trends of representation, apparently most remote from one another, are illustrated by the remark (Eddington, *Stars and Atoms*) that the Einstein displacement in spectra, derived from relativity, may also be classed as a kind of Compton interaction (in which energy and momenta are conserved but only in *quanta*) of the light, with gravitational energy taking the place of energy of encounters with atoms.

From the statistical side, the problems of formulation in general thermodynamics are to identify the physical units; the relations of their permutations among the cells of the frame (much developed in theory recently by C. G. Darwin and R. H. Fowler) are then reduced to abstract combinatory analysis. But the infinite variety of Nature can be only partially enchained in numbers.

A cognate train of ideas has recently emerged. Instead of considering the opportunities for arrangement of the independent dynamical entities as above, we may fix attention on the measure of statistical uncertainty inherent in the specification of the members of the set of cells. The Liouville invariant may be taken as expressing that the coarseness of grain of configurational position multiplied by that of the state of the system as regards momentum remains a constant; and cognate invariants for partial sets of the

An inherent limitation of analysis. variables enlarge the scope of this statement. The most recent speculation seems to recognize in this prescribed extent of cell a definite *quantum* of unavoidable uncertainty: which seems to be none other than the formulation advanced tentatively in the text. Cf. N. Bohr in *Nature*, April 1928; also an Appendix to this volume "On the Doctrine of Molecular Scattering of Radiation."

The map of human knowledge. As in the case of the correlations of the acquired science of a postulated *corps* of observers throughout the universe, in relation to their space and time (cf. final Appendix), so here also, the physical limitations of the faculties of the intercommunicating group of observers in face of the group of external facts might well prove to be of the essentials as regards the character of their map of knowledge.

# 75

# ON THE STATISTICAL THEORY OF RADIATION

*[Philosophical Magazine, August 1910.]*

In the *Philosophical Magazine* for July (p. 122) Professor H. A. Wilson, in a valuable review of my recent paper on the statistical theory of natural radiation[1], concludes that its procedure does not really evade the main difficulty, that an atomic constitution of energy must be implied in such investigations. One of the positions advanced in the paper was that the magnitude of the element of energy needed for the statistics might be chosen at will, provided the size of the elementary cell was chosen in a fixed proportion to it. Though such a theory has, and must have on the most favourable view, imperfect and provisional features, it does not appear to me that Professor Wilson has established this formidable addition to their number, and for the following reason.

Using his notation, the heads of the argument there set out, perhaps too briefly, were as follows. If $S$ is entropy and $W$ is the number of ways in which the system can be arranged in the actual state, then

$$S = k \log W$$
$$= k \log W_1 W_2 \ldots W_n,$$

*Boltzmann's formulation of entropy:*

where $W_1, W_2, \ldots$ are the numbers for the parts of the system. If the first part contains $n_1$ elements of energy each of amount $\epsilon_1$, contained in $N_1$ cells, and similar for the others, then the total energy is

$$E = \epsilon_1 n_1 + \epsilon_2 n_2 + \ldots \epsilon_r n_r.$$

The natural state of an isolated system is the one that makes $S$ maximum subject to $E$ remaining constant. This requires

*combined with conservation of energy,*

$$\frac{1}{\epsilon_1}\frac{\partial S}{\partial n_1} = \frac{1}{\epsilon_2}\frac{\partial S}{\partial n_2} = \ldots = \vartheta,$$

where $\epsilon_1 \delta n_1 = \delta E_1, \ldots$ . Thus $\vartheta$ is a quantity the same for all the parts of a system which is in equilibrium of exchanges of energy: in fact if absolute temperature $T$ is defined by the Clausius formula $dS = dE/T$, then $\vartheta$ is $T^{-1}$. Also the working out of the actual value of $W_1$ leads [*supra*, p. 402] to

*demands a temperature.*

$$\frac{\partial S}{\partial n_1} = k \log \left(1 + \frac{N_1}{n_1}\right).$$

[1] *Roy. Soc. Proc.* 1909, vol. LXXXIII, A, pp. 82–95[: *supra*, p. 396].

Thus, finally, for the distribution of energy among the parts of the system we have the formula (Planck's)

$$E_1 = n_1\epsilon_1 = \frac{N_1\epsilon_1}{e^{k\epsilon_1/T} - 1}.$$

The criticism is that $E_1$ ($= n_1\epsilon_1$) as thus determined cannot be independent of the size of the energy element $\epsilon_1$, because $\epsilon_1$ is the only variable that enters except $N_1$, which measures the extent of the system, so that any change of $\epsilon_1$ must change the value of $E_1$, even though $\epsilon_1 N_1$ is kept constant: for example, if $\epsilon_1$ is taken very small, the formula becomes

$$E_1 = N_1 k^{-1} T,$$

which represents the law of equipartition. But this unwelcome conclusion is evaded simply by recognizing that the value of $k$ must be some function of the size of the energy element which is taken as the basis of the statistics; it would indeed be strange if it were otherwise. If $k\epsilon_1$ remains finite as $\epsilon_1$ diminishes, the equipartition is not attained unless $T$ is very great. We shall find that it is $k\epsilon_1$ that is to be taken as constant when $\epsilon_1$, the statistical element for any given type of energy, is changed.

The *two* independent constants in the formula are in fact $N_1\epsilon_1$ and $k\epsilon_1$. Their ratio $N_1 k^{-1}$ is equal to the gas constant. That universal
quantity, and $N_1\epsilon_1$ (say $a$) which is the ratio of the energy element to the extent of a cell, are what affect the distribution and are thus of predetermined values; but there seems to be nothing that demands a definite magnitude of the energy element itself.

On the Boltzmann form of the theory of probability of distributions of energy among the molecules of gases, $k$ turned out indeed to be the
gas constant. On the present form of theory, which involves distribution of elements of disturbance with their appropriate energies in the containing system as mapped out into cells*, instead of mere collocation of elements with regard to one another, this conclusion need not hold. We may probe this point further. It is known as a fact that, under ideal conditions, equable partition is very nearly attained as regards the translatory and rotatory parts of the energy of the molecules of a gas. This requires that, if $\epsilon_r$ is the value of $\epsilon$ corresponding to each of the translatory or rotatory types of freedom, it must prove to be so small compared with $\epsilon_1, \epsilon_2, \ldots$ that the exponent $k\epsilon_r/T$ is also small; for that is needed in order to lead to this law of approximately equable partition in the atoms, except at very low temperatures, which takes the form

$$E_r = n_r\epsilon_r = N_r k^{-1} T.$$

In this special result the value of the element of energy $\epsilon_r$ has be-

* Cf. Einstein, after Bose, Berlin *Berichte*, 1927: also *supra*, p. 410, footnote.

come eliminated. Also $N_r k^{-1}$ must be the gas constant $R$; and since $N_r \epsilon_r$ must be $\alpha$, another universal constant, we have $k \epsilon_r = \alpha/R$. Hence in this case of simple gas theory the value of $k$ should be inversely as the scale of magnitude of the elements of energy chosen; and the size of a standard cell should be directly as that element. And this result must be universal. *A ratio emerges.*

Thus the conclusion is, briefly, that to render the entropy independent of the scale of minuteness of subdivision of the statistics, as is natural, we have only to define it as $k \log W$, where the value of $k$ (if we decide to retain it in the formulas) must vary directly as this amount of subdivision, or inversely as the scale of sizes of the elements of energy that are employed in the analysis. But, on the other hand, if $k$ had the same value whatever be the scale of the statistics that is adopted, conclusions such as those of Professor Wilson regarding the magnitude of the ultimate element of energy would necessarily follow. *The scale of entropy depends on scale of statistics of the energy:*

To connect formally the values of $\epsilon$, thus demanded by experimental knowledge for gas theory, with those that obtain for the types of radiant energy, would involve a rather long argument. But the present type of theory works out for the domain of radiation as above, and it is readily seen that it works out for the domain of gas theory on the ordinary lines as indicated in the paper referred to; while a bridge can be constructed between the two, as there suggested, by noting that both for translatory and rotatory motions in gas theory and for radiation of long wave-length, the principle of equipartition is practically effective, so that we may take advantage of Professor Lorentz's train of ideas connecting these equipartitions by a calculation of the amount of the natural radiation from a thin metallic plate, considered as arising from the collisions of the moving free electrons that are required by its electric conductivity. *transition to radiation, may be possible through radiation from electrons in a metal.*

The existence of another universal physical constant $(\alpha)$, in addition to that of gas theory, has been postulated without any explanation as yet. But its existence is independent of these statistical theories; and it thus seems to have come to stay in some form or other. In fact it was early pointed out by Wien and by Thiesen that the value $\lambda_m T$, where $\lambda_m$ is the wave-length of maximum radiation at temperature $T$, and which is by Wien's displacement law a universal constant, suffices and is required, in conjunction with the other recognized universal constants of Nature, to establish an absolute system of fundamental units of mass, length, and time independent of special kinds of atoms; its dimensions are therefore not expressible in terms of those of other universal constants, and it must have an independent existence of its own. *A second universal constant required for radiation: and not more.*

# 76

## WILLIAM THOMSON (LORD KELVIN): *MATHEMATICAL AND PHYSICAL PAPERS*, Vol. IV, HYDRODYNAMICS AND GENERAL DYNAMICS.

[*Scientia*, Vol. VIII (1910).]

DE son vivant, lord Kelvin a à peu près complété la réédition de ses mémoires scientifiques, en publiant trois volumes de *Mathematical and physical papers* et un volume de *Baltimore lectures*, le tout suivi de trois volumes d'un genre moins abstrait, sous le titre: *Popular lectures and addresses*. L'achèvement de cette tâche par la publication de ce IV<sup>ème</sup> volume de la première série et du dernier et V<sup>ème</sup> volume, actuellement sous presse, fut confié, après sa mort survenue il y a deux ans, à d'autres mains.

La plus grande partie du présent volume renferme des mémoires sur l'hydrodynamique. L'étude du mouvement des liquides avait préoccupé lord Kelvin toute sa vie durant, et cette préoccupation se manifeste soit dans la théorie difficile, mais séduisante, des vagues et des véhicules se déplaçant sur l'eau, théorie qui lui a été sans doute suggérée par Stokes pendant ses années d'études, soit dans les idées plus générales sur la dynamique dans lesquelles son esprit rigoureusement concret s'était toujours complu, celles par exemple sur les différentes manifestations de l'inertie des liquides et sur la nature de la résistance qu'ils opposent aux solides qui se déplacent à travers leur milieu. La théorie des vagues tire beaucoup de son intérêt de cette circonstance qu'il suffit, pour l'élucidation complète et élégante d'une importante catégorie de phénomènes naturels, de considérer un liquide idéal exempt de friction. Mais, d'un autre côté, le contraste qui se manifeste entre les types de résistance dynamiquement simples des liquides sans friction et les phénomènes irréguliers et complexes que produit la friction des liquides naturels a privé la deuxième partie de l'hydrodynamique de tout contact avec les besoins de la technique pratique, laissant pour seule ressource la combinaison des idées physiques générales avec les indications fournies par l'expérimentation.

C'est ainsi qu'en ce qui concerne l'étude détaillée des vagues des canaux, les pénétrantes recherches expérimentales de Scott Russell ont servi de point de départ à une théorie, et lord Kelvin a discuté ailleurs les améliorations de la navigation sur canaux qui est née,

*Sources of inspiration.*

*Ideal fluid of wave-motion:*

*in other domains not practicable.*

*Canal traffic.*

disait-il, de la découverte faite par un cheval que lorsqu'il précipitait ses pas de façon à maintenir constamment sa barque sur les sommets des vagues, la force nécessaire pour la traction se trouvait considérablement diminuée. C'est ainsi encore que plus tard, lorsque la résistance que les navires rencontraient en pleine mer était devenue le problème pratique, ce furent les expériences de W. Froude sur de petits modèles dans des réservoirs et les conclusions tirées des rapports entre leurs différentes dimensions qui ont fourni à la théorie ses principaux éléments. Les constructions navales peuvent aujourd'hui conclure directement et en toute confiance des observations faites sur le modèle aux conditions exigées par le navire. Mais cette solution empirique n'eût jamais pu être obtenue sans cette interaction entre l'expérience et la théorie, dans laquelle lord Kelvin a joué un rôle prépondérant, placé comme il était à Glasgow, au centre même des constructions navales. Et si l'on veut obtenir des progrès ultérieurs, on n'a qu'à suivre le même procédé—qui sera sans doute partiel et aura besoin, lorsqu'il faillira à la tâche, d'être complété par l'expérience—, à savoir l'étude théorique des facteurs physiques dont dépend la résistance à vaincre.

*Ship resistance: theory of models.*

*Interaction of theory and practice.*

La moitié environ du présent volume est consacré au problème du mouvement des vagues. Et la belle explication (p. 413) des principaux caractères des modèles de vagues qui suivent les navires est un exemple de l'intérêt à la fois théorique et pratique que présente le sujet. Plus haut (pp. 76–92), se trouvent réimprimées les notes brèves et bien connues, dans lesquelles la distinction entre vagues et rides est illustrée par quelques exemples pratiques, et qui renferment des recherches sur l'instabilité de la surface de séparation de deux courants, ainsi que sur celle d'une surface d'eau exposée au vent. Lord Kelvin en a vu lui-même les applications météorologiques, par exemple en ce qui concerne l'explication des bandes de nuages dans un "ciel couleur maquereau," sujet auquel Helmholtz a donné plus tard un développement complet.

*Ship waves.*

*Ripples.*

*Effects of wind:*

*instabilities.*

Mais c'est peut-être la partie la plus abstraite du livre, celle où se trouvent réimprimés les mémoires sur le mouvement en tourbillon, qui doit être considérée comme la plus puissante et la plus belle. La preuve fournie par Helmholtz de la permanence d'anneaux de tourbillon dans les liquides parfaits constitue la découverte la plus étonnante des temps modernes en mathématique appliquée, celle qui a exercé l'influence la plus suggestive sur l'évolution des idées dans la physique mathématique. Son grand mémoire de 1857 a posé d'une façon complète les principes fondamentaux du sujet. Mais les vues romantiques et stimulantes formulées dix années plus tard par lord Kelvin, relativement à la représentation des propriétés des atomes de

*Vortex motion.*

*Fundamental as physical ⌡ analogue:*

la matière par analogie avec les anneaux des tourbillons, imprimèrent à la question une nouvelle impulsion. Il fallait se rendre compte des influences réciproques que les tourbillons exercent les uns sur les autres, et pour atteindre ce but il était nécessaire d'inventer des méthodes intuitives et géométriques, destinées à remplacer les équations différentielles difficiles à manier. Il fallait également déterminer les périodes de leurs divers modes de vibration. Plus importantes encore étaient peut-être les discussions relatives à la permanence ou stabilité des anneaux (p. 166 et suiv.), moins à cause de ce qui a été accompli dans ce domaine spécial et de la perspicacité et de l'ingéniosité qui lui ont été consacrées, que parce que ces discussions ouvrent accès aux théories générales de la stabilité dynamique appliquée aux phénomènes physiques; ces théories renferment, entre autres, les principes de l'énergie maxima ou minima pour des moments donnés, ainsi que la distinction entre la stabilité ordinaire qui ne se maintient qu'en l'absence de friction et la stabilité séculaire que la friction ne peut plus modifier et qui apparaît avec évidence dans les problèmes séculaires de l'astronomie physique.

C'est en connexion avec ces données que naquit l'idée fondamentale de l'impulsion propre au mouvement d'un solide dans un milieu liquide, cette idée étant nécessaire si l'on veut réaliser des progrès au point de vue des principes généraux dans les cas où les intégraux exprimant le moment sont divergents et perdent ainsi leur signification directe. Cette méthode a été en effet transcrite (p. 70), avec toutes ses applications dynamiques, d'un carnet de notes remontant à 1858, mais sans les détails logiques développés dix ans plus tard dans le mémoire sur les tourbillons. L'idée hardie d'appliquer la dynamique de Lagrange aux mouvements des solides dans des milieux liquides, idée qui a formé un autre nouveau point de départ dans la physique mathématique, a été émise pour la première fois vers la même époque dans *Natural philosophy* de Thomson et Tait.

Ce volume renferme également (pp. 458–531) une collection de recherches brèves, mais souvent très élégantes, sur la dynamique analytique. Ces recherches proviennent de sources très dispersées, et la plupart d'entre elles font apparaître, à l'arrière-plan, leur application physique ou astronomique directe. Ces discussions et d'autres encore, par exemple les mémoires sur le mouvement en tourbillons, ont été jugées, à cause de leur forme inaccoutumée, comme étant d'un abord difficile pour les mathématiciens professionnels. Elles s'inspirent plutôt des intuitions de l'ingénieur qui est habitué à sentir les forces, au lieu de les calculer; et l'on peut considérer la fraîcheur des aperçus sur l'actualité qui les caractérise comme étant justement

*Marginalia:*
interactions of vortices:

vibrations:

stabilities:

relation to general dynamics.

Theory of the impulse:

its origin.

Great extension of general dynamics.

The intuitions of the engineer.

nécessaire pour empêcher la physique mathématique de dégénérer en une forme peu intéressante d'analyse algébrique.

Le volume se termine par une liste des mémoires sur la question générale de la propagation élastique dans ses aspects matériel et optique, mémoires dont la plupart ont déjà paru ailleurs.

Le dernier volume, actuellement sous presse, renfermera l'ouvrage le plus récent de lord Kelvin sur la thermodynamique, les recherches et spéculations sur la physique cosmique et géologique, à laquelle il s'est intéressé toute sa vie durant, et le reste de ses mémoires sur l'électrodynamique et l'électrolyse. Il y aura aussi une partie considérable consacrée à la géométrie et à la dynamique des structures cristallines et à la dynamique moléculaire, comme par exemple la théorie sur la couleur bleue du ciel, théorie à laquelle il a collaboré avec des physiciens italiens. Il renfermera enfin les recherches expérimentales sur la radio-activité, dans laquelle il a été pionnier sur beaucoup de points, comme par exemple dans la question de la conductibilité électrique induite des gaz. *Cosmical and geological physics. Molecular dynamics: crystals. The blue sky. Radio-activity.*

Avec la publication de ce dernier volume, la tâche sera achevée de dresser l'inventaire d'une vie de travail d'un chercheur infatigable dont l'activité a exercé une des plus grandes influences sur le progrès scientifique du XIX$^{\text{ème}}$ siècle. Cet inventaire occupera à tout jamais une place d'honneur parmi les œuvres complètes des autres grands fondateurs de la physique moderne.

# 77

# ON THE DYNAMICS OF RADIATION.

[*Lecture at the International Congress of Mathematicians:*
Cambridge, August 1912: from *Report*, Vol. II, pp. 1–20.]

THE subject of this title is coextensive with the whole range of the physics of imponderable agencies. For if it is correct to say with Maxwell that all radiation is an electrodynamic phenomenon, it is equally correct to say with him that all electrodynamic relations between material bodies are established by the operation, on the molecules of those bodies, of fields of force which are propagated in free space as radiation and in accordance with the laws of radiation, from the one body to the other. It is not intended to add to the number of recent general surveys of this great domain. The remarks here offered follow up some special points: they are in part in illustration of the general principle just stated: and in part they discuss, by way of analogy with cognate phenomena now better understood, the still obscure problem of the mode of establishment of the mechanical forces between electric systems.

*Forces propagated as radiation.*

*Mode of establishment,*

The essential characteristic of an electrodynamic system is the existence of the correlated fields, electric and magnetic, which occupy the space surrounding the central body, and which are an essential part of the system; to the presence of this pervading aethereal field, intrinsic to the system, all other systems situated in that space have to adapt themselves. When a material electric system is disturbed, its electrodynamic field becomes modified, by a process which consists in propagation of change outward, after the manner of radiation, from the disturbance of electrons that is occurring in the core. When however we are dealing with electric changes which are, in duration, slow compared with the time that radiation would require to travel across a distance of the order of the greatest diameter of the system—in fact in all electric manifestations except those bearing directly on optical or radiant phenomena—complexities arising from the finite rate of propagation of the fields of force across space are not sensibly involved: the adjustment of the field surrounding the interacting systems can be taken as virtually instantaneous, so that the operative fields of force, though in essence propagated, are sensibly statical fields. The practical problems of electrodynamics are of this nature— how does the modified field of force, transmitted through the aether

*of field of activity:*

*for changes not extremely rapid,*

*an instant adjustment.*

from a disturbed electric system, and thus established in the space around and alongside the neighbouring conductors which alone are amenable to our observation, penetrate into these conductors and thereby set up electric disturbance in them also? and how does the field emitted in turn by these new disturbances interact with the original exciting field and with its core?  For example, if we are dealing with a circuit of good conducting quality and finite cross-section, situated in an alternating field of fairly rapid frequency, we know that the penetration of the arriving field into the conductor is counteracted by the mobility of its electrons, whose motion, by obeying the force, in so far annuls it by Newtonian kinetic reaction; so that instead of being propagated, the field soaks in by diffusion, and it does not get very deep even when adjustment is delayed by the friction of the vast numbers of ions which it starts into motion, and which have to push their way through the crowd of material molecules; and the phenomena of surface currents thus arise. If (by a figure of speech) we abolish the aether in which both the generating circuit and the secondary circuit which it excites are immersed, in which they in fact subsist, the changing phases of the generator could not thus establish, from instant to instant, by almost instantaneous radiant transmission, their changing fields of force in the ambient region extending across to the secondary circuit, and the ions in and along that circuit would remain undisturbed, having no stimulus to respond to. The aethereal phenomenon, viz., the radiant propagation of the fields of force, and the material phenomenon, viz., the response of the ions of material bodies to those fields, involving the establishment of currents with new fields of their own, are the two interacting factors. The excitation of an alternating current in a wire, and the mode of distribution of the current across its section, depend on the continued establishment in the region around the wire, by processes of the nature of radiation, of the changing electromagnetic field that seizes hold on the ions and so excites the current; and the question how deep this influence can soak into the wire is the object of investigation. The aspect of the subject which is thus illustrated finds in the surrounding region, in the aether, the seat of all electrodynamic action, and in the motions of electrons its exciting cause. The energies required to propel the ions, and so establish an induced current, are radiant energies which penetrate into the conductor from its sides, being transmitted there elastically through the aether; and these energies are thereby ultimately in part degraded into the heat arising from fortuitous ionic motions, and in part transformed to available energy of mechanical forces between the conductors. The idea—introduced by Faraday, developed into precision by Maxwell, expounded and illustrated in

*Side notes:*

Penetration into material bodies: reaction.

Alternating field soaks in:

not deep.

Connection by aethereal propagation essential.

Elastic transmission

is damped by free ions. various ways by Heaviside, Poynting, Hertz—of radiant fields of force, in which all the material electric circuits are immersed, and by which all currents and electric distributions are dominated, is the root of the modern exact analysis of all electric activity.

The elementary phenomena of steady currents, including Ohm's law and all the rest of the relations which are so easily formulated directly, are the simple synthesis to which this scheme of activity leads, when the changes of the controlling fields are slow enough to be considered as derived from statical potentials. In the electric force integrated round a circuit, viz., the electromotive force so called, the undetermined portion of the electric field, that arises from electric distributions which adjust the current to flow "full-bore," after the manner of a stream, is eliminated by the integration; and the fully developed current, thus adjusted to be the same at all sections of the wire, is of necessity proportional to the impressed electromotive force, as Ohm postulated.

*Rationale* of Ohm's circuit laws.

The principle of the controlling influence of the activities in the intervening aether emerges in the strongest light when we recall the mechanical illustrations of the Kelvin period; we may say that the disturbance of the field of aether exerts influence on the conductors in the same general kind of way as the pressures involved in the inertia of moving fluid control the motion of the vortices or solid bodies which are immersed in the fluid.

Analogies of mechanical media.

As regards the mode of establishment of these fields of aethereal activity, we here merely recall that the phenomena of free propagation in space are covered by Hertz's brilliant analysis for the case of a simple electric vibrator or dipole, that being the type of element out of which all more complex sources of radiation, including the radiation of moving ions, may be built up by superposition. More remarkable, on account of its sharp contrast with the familiar phenomena of light and sound, is the guidance of electric radiation by a wire, which was explored experimentally by Hertz, in full touch with the mathematical theory, but with so much trouble and vexation arising from the influence of casual conductors serving as the "return circuits" on which the issuing lines of electric force must find their terminations. It was not long however until the conditions were made manageable, first by Lecher, by the simple device of introducing a parallel wire, on which as return circuit the lines of force from the other wire could again converge, thus restricting the propagated field of force to the region extending between the original and this return circuit; this arrangement, by preventing lateral spreading of the radiation, thus avoided all effects both of moderate curvature of the guiding conductor and of disturbance by neighbouring bodies. The problem of

Elementary dipole vibrator.

Waves guided along conjugated wires,

conducting free radiant energy to a distance in the open space around without loss. the guiding conductor, without lateral loss any more than there is when electric undulations travel along the interior dielectric region between the coatings of a submarine cable, thus became securely realized and understood.

The equations of propagation of radiant effects in free space are, as we have seen, the foundation of all electric phenomena, whether static or kinetic. In the case of slow changes, both the electric field $(PQR)$ and the magnetic field $(\alpha\beta\gamma)$[1] have their potentials; namely we have

$$\text{curl }(PQR) = 0, \quad \text{curl }(\alpha\beta\gamma) = 0.$$

Curl and divergence relations:

In the case of rapid alternations these relations have to become adapted to the transmission of undulations; which we know in advance must be of necessity transverse, owing to the absence of divergence of the vectors concerned in them, as expressed by the equations

$$\text{div }(PQR) = 0, \quad \text{div }(\alpha\beta\gamma) = 0.$$

When this order of ideas is pursued, as it was in a way by Hertz, the appropriate kind of modification of the statical equations can hardly be missed: it must in fact be expressed by relations of the type

lead to modifications of static potential properties.

$$\text{curl }(PQR) = -A\frac{d}{dt}(\alpha\beta\gamma), \quad \text{curl }(\alpha\beta\gamma) = A\frac{d}{dt}(PQR).$$

We are bound to recall here that precisely equivalent relations had been laid down, in brilliant fashion, as the result of a tentative process of adaptation of analytical theory to optical phenomena, by Mac-Cullagh[2] as far back as 1838, as a scheme consistently covering the whole ground of physical optics: and especially that their form was elucidated, and their evidence fortified, by him in the following year, by showing that they fitted into the Lagrangian algorithm of "Least Action," which was thus already recognized in physics as the generalized compact expression and criterion of the relations appropriate to a dynamical system. The analysis proceeded on the basis of the wider range of indications afforded by the study of optical phenomena in crystals, and accordingly the result was reached in the generalized form appropriate to aeolotropic media. But MacCullagh could not construct any model of the dynamical operation of his analytical equations, on the lines of the properties of ordinary material bodies, with which alone we are familiar through experiment—a task which indeed is now widely recognized to be an unreasonable one, though at that time it largely dominated all problems of physical interpretation.

MacCullagh's optical scheme expanded.

Least action:

procedure by adaptation.

Aether is not matter.

[1] A British writer will be pardoned for retaining the commodious and classical notation of Maxwell, in which he has been educated, until there is some consensus of opinion as to what other notation, if any, is to replace it.

[2] *Collected Works of James MacCullagh* (1880), p. 145.

## Guidance of electric radiation by wires.

*Mode of influence of wires: arrival of the field:* In the arrival of a magnetic field and its related electric field, in the space around the conducting circuit, the field being transmitted there by radiant processes, we have recognized the essential cause of the excitation of an alternating current in the circuit and of its location, when the changes are rapid, in the outer part. An electric field so transmitted through space will not, of course, be along the length of the wire; but the component that is oblique will be at once compen-*transverse part compensated:* sated and annulled statically by the electric separation (of amount in other respects negligible) which it produces across the section of the conductor. The longitudinal force that remains may be treated as the "induced electromotive force per unit length" at the place in question. When this adjustment of the electric field has been effected by excita-*thus effective force is along wire.* tion of free charge, the total force becomes longitudinal, and its distribution in the cross-section of the wire is necessarily restricted to those solutions of its cylindrical harmonic characteristic equation which remain finite across the section; in the simplest and usual case of axial symmetry the distribution is represented as *infra* by the Bessel function of complex argument and zero order.

*This cylindric problem treated:* It is perhaps worth while formally to set out the analysis from this point of view[1], when this longitudinal electric field, arising from the transmitted field and the transverse electric distribution induced by it—which is the field propelling the current—is a uniform force $R_0 e^{i p t}$, operating on a long conducting cylinder from outside and all along its surface.

The equations of the field are, in Maxwell's notations,

$$-\frac{da}{dt} = \frac{dR}{dy}, \quad \frac{db}{dt} = \frac{dR}{dx}, \quad \frac{dc}{dt} = 0,$$

where $\quad 4\pi w = \frac{d\beta}{dx} - \frac{d\alpha}{dy}, \quad$ where $w = \left(\frac{K}{4\pi c^2}\frac{d}{dt} + \sigma\right) R.$

Thus $\quad \dfrac{d^2R}{dx^2} + \dfrac{d^2R}{dy^2} = -m^2 R,$ and $w = -\dfrac{m^2}{4\pi\mu p}\iota R,$

where $\quad -m^2 = -\dfrac{K\mu}{c^2}p + 4\pi\sigma p\mu\iota, \quad \iota p = \dfrac{d}{dt}.$

*leading to the Fourier-Bessel equation:* Thus, for symmetrical distribution in a circular section,

$$\frac{d^2R}{dr^2} + \frac{1}{r}\frac{dR}{dr} + m^2 R = 0;$$

[1] Cf. the rather different analysis in Maxwell, *Phil. Trans.* 1865, *Elec. and Mag.* Vol. ii, § 690, developed by Rayleigh, *Phil. Mag.* Vol. xxi (1886), p. 387.

so that inside the section the distribution, converging towards the origin, is determined as

$$R_{\text{in}} = A J_0 (mr),$$

$m$ being a complex quantity; representing inward travelling waves, thus, while outside the section, in free space, it must be a form converging with distance, so that

$$R_{\text{out}} = R_0 + B I_0 \left(\frac{p}{c} r\right);$$

where[1], for small values of $x$, if $S_n = 1 + \frac{1}{2} + \frac{1}{3} + \ldots + \frac{1}{n}$, and $\gamma$ represents Euler's constant $\cdot 577$,

and Stokes-Rayleigh solutions.

$$I_0 (x) = (\gamma + \log - \tfrac{1}{2} \iota x) \left(1 - \frac{x^2}{2^2} + \frac{x^4}{2^2 \cdot 4^2} - \ldots\right)$$

$$+ \frac{x^2}{2^2} S_1 - \frac{x^4}{2^2 \cdot 4^2} S_2 + \frac{x^6}{2^2 \cdot 4^2 \cdot 6^2} S - \ldots.$$

Also   $J_0 (x) = 1 - \frac{x^2}{2^2} + \frac{x^4}{2^2 \cdot 4^2} - \ldots.$

The continuity of the field across the boundary of the conductor requires that $R$ and $\frac{1}{\mu} \frac{dR}{dr}$ are there continuous. Thus for a wire of radius $a$,

$$A J_0 (ma) = R_0 + B I_0 \left(\frac{p}{c} a\right) \quad \ldots\ldots\ldots\ldots\ldots (1),$$

$$A \frac{m}{\mu} J_0' (ma) = B \frac{p}{c} I_0' \left(\frac{p}{c} a\right) \quad \ldots\ldots\ldots\ldots\ldots (2).$$

Subject to interfacial continuities.

Also, the total current $C$ is determined by

$$C = \int_0^a w \cdot 2\pi r dr = - \frac{m^2}{2\mu p} \iota A \int_0^a J_0 (mr) \, r dr$$

$$= \frac{1}{2\mu p} \iota A \left| mr J_0' (mr) \right|_0^a,$$

giving finally          $C = \frac{ma}{2\mu p} \iota A J_0' (ma) \ldots\ldots\ldots\ldots\ldots\ldots (3).$

The current:

The elimination of the undetermined constants $A$ and $B$ between the equations (1), (2), (3) will lead to the circumstances of propagation of this total current $C$ excited in the infinite cylinder.

As $pa/c$ is usually very small, we may write approximately

$$I_0 (x) = \gamma + \log \tfrac{1}{2} \iota x, \quad I_0' (x) = \frac{1}{x};$$

---

[1] Cf. *e.g.* Rayleigh's *Theory of Sound*, Vol. II, § 341; referring back to cognate physical applications by Stokes, *Phil. Trans.* 1868, or *Collected Papers*, Vol. IV, p. 321. [The explanations have been amplified in this reprint.]

and after some reductions we have*

equation of
induced
current for
ordinary cases:

$$\frac{R_0}{C} = - \frac{2\mu p}{ma} \iota \frac{J_0\,(ma)}{J_0{}'\,(ma)} - 2p\iota \left(\gamma + \tfrac{1}{2}\pi + \log \frac{p}{2C}\,a\right);$$

and finally, eliminating the complex character of this equation by writing $dC/dt$ for $pC\iota$, bearing in mind that $m$ is complex, we arrive at the usual form of relation

$$R_0 = L \frac{dC}{dt} + \rho C,$$

---

\* Correction of the value of $\log \iota$ from $\tfrac{1}{2}\pi$ to $\tfrac{1}{2}\pi\iota$ is here required. This will add $\pi/p$ to the effective resistance $\rho$ per unit length and delete the term $-\pi$ *But radiation* in the inductance $L$. But this addition to resistance would be very great for *from the wire* high periods: and it turns out that the form involving $I_0$ which it implies for *is excluded,* the external field belongs to waves travelling outward from the wire (cf. Rayleigh, *Theory of Sound*, Vol. II, p. 305, following Stokes, *Phil. Trans.* 1868, p. 461) which thus radiates its energy rapidly into space.

For a field of radial waves converging on the wire the value of $\rho$ is diminished by $\pi/p$ instead of increased. In both cases the change, being radiative, is naturally independent of the material of the wire.

Actually the circuit must be completed, so its return part is an essential feature. The interaction of the two parts sets up waves between them that are almost stationary, with radiation not indeed null but negligibly slight unless $p$ is great. This corresponds to a different form for the outside field. Practically the superposition of outward and inward waves gives standing undulations: *so flux estab-* then the two additions to the resistance that arise from them cancel out, so *lished by* that the result in the text then stands, with the term $-\pi$ omitted from the *standing* expression on next page for the inductance $L$. *waves,*

At first glance this result may imply that the standing waves are practically symmetrical at all distances around the wire, therefore that the return circuit is provided by adjacent conductors around it, strictly by a concentric cylinder. But if the wire is narrow compared with the distance of the return conductors, *abutting on* the lines of force are practically radial near it and uniformly distributed, and *a return* they belong to an oscillating but, as above, practically not radiating field. This is *circuit.* the justification of the radial formulae in the text, which originated with Maxwell (*Phil. Trans.* 1864; *Scientific Papers*, Vol. I, p. 592) as a correction in his measurements of standard inductances, and acquired a fundamental application much later, under Kelvin's lead, in the electrotechnics of alternating currents.

*Correction* It appears from the formulae of Stokes (*loc. cit.*) that to avoid radiation *as regards* $I_0\,(x)$ must be modified by replacing the constant $\gamma$ by $\gamma - \tfrac{1}{2}\pi\,(1 + \iota)$, thus *establish-* changing $\gamma + \tfrac{1}{2}\pi\iota$ into $\gamma - \tfrac{1}{2}\pi$ in the corrected expression in the text for $R_0/C$: *ment:* it is this that cancels the term $-\pi$ in the inductance factor later in the text without affecting the resistance factor. The function of the return circuit is here solely to choke off radiation.

*the simple* Practically, when the period of alternation is large compared with the time *practical* of light across the system, the procedure is simple: (1) the total current has the *case.* same value at each instant across all sections of whatever form; (2) its distribution over any section is determined as due to a uniform lengthwise electric field penetrating from outside; (3) the strength of this field is determined by the assigned value of the total current; (4) its integral around the circuit must compensate the applied electromotive force.

and thus obtain expressions for the inductance $L$ and the resistance $\rho$ of a wire per unit length when it is so nearly straight that the radius of curvature is many times its diameter, and when no disturbing conductor is near.

When the conductivity $\sigma$ is very large, and the frequency $p/2\pi$ is not excessive so that the second term in the expression for $m^2$ is preponderant, we have

$$m = (2\pi\sigma\mu p)^{\frac{1}{2}} (1 - \iota).$$

Also when, as [for very high periods], the real part of the argument $z$ is very great and positive,

$$J_0(z)/J_0'(z) \text{ becomes equal to } \iota.$$

Thus in this case

$$\frac{R_0}{C} = \left(\frac{\mu p}{2\pi\sigma}\right)^{\frac{1}{2}} - \iota p \cdot 2 \left\{\gamma + \tfrac{1}{2}\pi + \log\frac{pa}{2c} - \tfrac{1}{2}\left(\frac{\mu}{2\pi p\sigma}\right)^{\frac{1}{2}}\right\},$$

viz., $$R_0 = \left(\frac{\mu p}{2\pi\sigma}\right)^{\frac{1}{2}} C + \left\{2\log\frac{2c}{pa} - \pi - 2\gamma + \left(\frac{\mu}{2\pi p\sigma}\right)^{\frac{1}{2}}\right\}\frac{dC}{dt}.$$

reduced form<br>for large<br>conductance.

As $R_0$ is the electric force impressed along the conductor just outside it, the coefficient of $C$ is the effective resistance $\rho$ of the conductor per unit length, and that of $dC/dt$ is the effective inductance $L$ per unit length, in this limiting case. These expressions agree with Rayleigh's results[1], except that in this mode of formulation the value of $L$ is definite and does not retain any undetermined constant.

The transmission, along a wire of circular section, of electric waves which maintain and propagate their own field subject to the inevitable damping, involves a different point of view; as the impressed force $R_0$ is now absent, a [frequency which can persist for given wave-length] is determined by equating the values of $A/B$ derived from equations (1) and (2).

Definite speed<br>of free electric<br>waves on a<br>wire.

### General theory of pressure exerted by waves.

If a perfectly reflecting structure has the property of being able to advance through an elastic medium, the seat of free undulations, without producing disturbance of structure in that medium, then it follows from the principle of energy alone that these waves must exert forces against such a reflector, constituting a pressure equal in intensity at each point to the energy of the waves per unit volume. Cf. p. 434, *infra*. The only hypothesis, required in order to justify this general result, is that the velocity of the undulations in the medium must be independent of their wave-length; viz., the medium is to be non-dispersive, as is the free aether of space.

Pressure<br>of waves:<br>in non-<br>dispersive<br>medium:

[1] Cf. *Theory of Sound*, ed. 2, Vol. 1, § 235 v.

deduced by
considerations
of energy.
This proposition, being derived solely from consideration of conservation of the energy, must hold good whatever be the character
of the mechanism of propagation that is concerned in the waves. But
the elucidation of the nature of the pressure of the waves, of its mode
of operation, is of course concerned with the constitution of the
medium. The way to enlarge ideas on such matters is by study of
special cases: and the simplest cases will be the most instructive.

*Transverse
waves on
a cord:*
Let us consider then transverse undulations travelling on a cord of
linear density $\rho_0$, which is stretched to tension $T_0$. Waves of all
lengths will travel with the same velocity, namely $c = (T_0/\rho_0)^{\frac{1}{2}}$, so
that the condition of absence of dispersion is satisfied. A solitary
wave of limited length, in its transmission along the cord, deflects
each straight portion of it in succession into a curved arc. This process
implies increase in length, and therefore increased tension, at first
locally. But we adhere for the present to the simplest case, where the
cord is inextensible or rather the elastic modulus of extension is in-
*tension is
constant
because
inextensible:*
definitely great. The very beginnings of a local disturbance of tension
will then be equalized along the cord with speed practically infinite;
and we may therefore take it that at each instant the tension stands
adjusted to be the same $(T_0)$ all along it. The pressure or pull of the
undulations at any point is concerned only with the component of
this tension in the direction of the cord; this is

$$T_0 \left( 1 + \frac{d\eta^2}{dx^2} \right)^{-\frac{1}{2}},$$

where $\eta$ is the transverse displacement of the part of the cord at distance $x$ measured along it; thus, up to the second order of approximation, the pull of the cord is

*diminished
pull along the
cord when
vibrating,*
$$T_0 - \tfrac{1}{2}T_0 \left( \frac{d\eta}{dx} \right)^2.$$

The tension of the cord therefore gives rise statically to an undulation pressure

$$\tfrac{1}{2}T_0 \left( \frac{d\eta}{dx} \right)^2, \text{ or } \tfrac{1}{2}T_0 c^{-2} \left( \frac{d\eta}{dt} \right)^2, \text{ or } \tfrac{1}{2}\rho_0 \left( \frac{d\eta}{dt} \right)^2.$$

*interpreted.*
The first of these three equivalent expressions can be interpreted as
the potential energy per unit length arising from the gathering up of
the extra length in the curved arc of the cord, against the operation of
the tension $T_0$; the last of them represents the kinetic energy per unit
length of the undulations. Thus there is a pressure in the wave,
arising from this statical cause, which is at each point equal to half
its total energy per unit length.

There is the other half of the total pressure still to be accounted for.
That part has a very different origin. As the tension is instantaneously

adjusted to the same value all along, because the cord is taken to be Add kinetic inextensible, there must be extra mass gathered up into the curved part due to extra travel- segment which travels along it as the undulation. The mass in this ling mass, arc is

$$\int \rho_0 \left(1 + \frac{d\eta^2}{dx^2}\right)^{\frac{1}{2}} dx,$$

or to the second order is approximately

$$\rho_0 l + \int \tfrac{1}{2}\rho_0 \left(\frac{d\eta}{dx}\right)^2 dx.$$

In the element $\delta x$ there is extra mass of amount

$$\tfrac{1}{2}\rho_0 \left(\frac{d\eta}{dx}\right)^2 \delta x,$$

which is carried along with the velocity c of the undulatory propagation. This implies momentum associated with the undulation, and of

amount at each point equal to $\tfrac{1}{2}\rho_0 \mathrm{c} \left(\frac{d\eta}{dx}\right)^2$ per unit length. Another through the transmission of its portion of the undulation pressure is here revealed, equal to the rate momentum. at which the momentum is transmitted past a given point of the cord;

this part is represented by $\tfrac{1}{2}\rho_0 \mathrm{c}^2 \left(\frac{d\eta}{dx}\right)^2$ or $\tfrac{1}{2}\rho_0 \left(\frac{d\eta}{dt}\right)^2$, and so is equal to

the component previously determined.

In our case of undulations travelling on a stretched cord, the pressure exerted by the waves arises therefore as to one half from transmitted intrinsic stress and as to the other half from transmitted momentum.

The kinetic energy of the cord can be considered either to be energy

belonging to the transverse vibration, viz., $\int \tfrac{1}{2}\rho \left(\frac{d\eta}{dt}\right)^2 ds$, or to be the

energy of the convected excess of mass moving with the velocity of

propagation c*, viz., $\int \tfrac{1}{2}\rho \left(\frac{d\eta}{dx}\right)^2 \mathrm{c}^2 dx$; for these quantities are equal by

virtue of the condition of steady propagation $\frac{d\eta}{dt} = \mathrm{c}\frac{d\eta}{dx}$.

On the other hand the momentum that propagates the waves is The wave-pressure an

transverse of amount $\rho \frac{d\eta}{dt}$ per unit length; it is the rate of change of indirect effect,

this momentum that appears in the equation of propagation

$$\frac{d}{dt}\left(\rho \frac{d\eta}{dt}\right) = \frac{d}{dx}\left(T \frac{d\eta}{dx}\right).$$

But the longitudinal momentum with which we have been here

---

* This is *half* the total energy; but it is only a shadow, not the energy of an actual moving mass.

specially concerned is $\frac{1}{2}\rho \left(\frac{d\eta}{dx}\right)^2 c$ per unit length, which is $\frac{1}{2}\frac{d\eta}{dx} \cdot \rho \frac{d\eta}{dt}$.

*small of the second order.*  Its ratio to the transverse momentum is very small, being $\frac{1}{2}\frac{d\eta}{dx}$; it is a second-order phenomenon and is not essential to the propagation of the waves. It is in fact a special feature, and there are types of wave-motion in which it does not occur. The criterion for its presence is that the medium must be such that the reflector on which the *Criterion:* pressure is exerted can advance through it, sweeping the radiation along in front of it, but not disturbing the structure; possibly intrinsic *a feature.* strain, typified by the tension of the cord, may be an essential feature in the structure of such a medium.

*This wave-pressure not in evidence in an analysis by Action.*  If we derive the dynamical equation of propagation along the cord from the Principle of Action

$$\delta \int (T - W)\, dt = 0, \text{ where } T = \int \tfrac{1}{2}\rho \left(\frac{d\eta}{dt}\right)^2 ds \text{ and } W = \int \tfrac{1}{2}T_0 \left(\frac{d\eta}{dx}\right)^2 dx,$$

the existence of the pressure of the undulations escapes our analysis. A corresponding remark applies to the deduction of the equations of the electrodynamic field from the Principle of Action[1]. In that mode of analysis the forces constituting the pressure of radiation are not *Nor is radiation pressure:* in evidence throughout the medium; they are revealed only at the place where the field of the waves affects the electrons belonging to the reflector. Problems connected with the Faraday-Maxwell stress lie deeper; they involve the structure of the medium to a degree which the propagation of disturbance by radiation does not by itself give us means to determine*.

We therefore proceed to look into that problem more closely. We now postulate Maxwell's statical stress system; also Maxwell's magnetic stress system, which is, presumably, to be taken as of the *which is due wholly to a secondary travelling momentum,* nature of a kinetic reaction. But when we assert the existence of these stresses, there remain over uncompensated terms in the mechanical forcive on the electrons which may be interpreted as due to a distribution of momentum in the medium[2]. The pressure of a train of radiation is, on this hypothetical synthesis of stress and momentum, due entirely (p. 432) to the advancing momentum that is absorbed by *in contrast with the cord.* the surface pressed, for here also the momentum travels with the waves. This is in contrast with the case of the cord analysed above, in which [cf. also p. 447] only half of the pressure is due to that momentum.

[1] Cf. Larmor, *Trans. Camb. Phil. Soc.* Vol. XVIII (1900), as *supra*, p. 164; or *Aether and Matter*, Chap. VI.

* But see, with reference to this and the next page, pp. 135-7, *supra*.

[2] For the extension to the most general case of material media cf. *Phil. Trans.* Vol. CXC (1897), p. 253 [: *supra*, p. 71].

The pressure of radiation against a material body, of amount given by the law specified by Maxwell for free space, is demonstrably included in the Maxwellian scheme of electrodynamics, when that scheme is expanded so as to recognize the electrons with their fields of force as the link of communication between aether and matter. But the illustration of the stretched cord may be held to indicate that it is not yet secure to travel further along with Maxwell, and accept as realities the Faraday-Maxwell stress in the electric field, and the momentum which necessarily accompanies it; it shows that other dynamical possibilities of explanation are not yet excluded. And, viewing the subject from the other side, we recognize how important have been the experimental verifications of the law of pressure of radiation which we owe to Lebedew, too early lost to science, to Nichols and Hull, and to Poynting and Barlow. The law of radiation pressure in free space is not a necessary one for all types of wave-motion; on the other hand if it had not been verified in fact, the theory of electrons could not have stood without modification.

*(margin: It appears, however, as pressure against the oscillating electrons.)*

*(margin: Formally necessary field of a fourfold stress tensor.)*

*(margin: Its experimental verification necessary and fundamental.)*

The pressure of radiation, according to Maxwell's law, enters fundamentally in the Bartoli-Boltzmann deduction of the fourth power law of connection between total radiation in an enclosure and temperature. Thus in this domain also, when we pass beyond the generalities of thermodynamics, we may expect to find that the laws of distribution of natural radiant energy depend on structure which is deeper seated than anything expressed in the Maxwell equations of propagation. The other definitely secure relation in this field, the displacement theorem of Wien, involves nothing additional as regards structure, except the principle that operations of compression of a field of natural radiation in free space are reversible. The most pressing present problem of mathematical physics is to ascertain whether we can evade this further investigation into aethereal structure, for purposes of determination of average distribution of radiant energy, by help of the Boltzmann-Planck expansion of thermodynamic principles, which proceeds by comparison of the probabilities of the various distributions of energy that are formally conceivable among the parts of the material system which is its receptacle.

*(margin: Radiation pressure essential to thermo-dynamics.)*

*(margin: Expanded entropy theory.)*

### Momentum intrinsically associated with Radiation.

We will now follow up, after Poynting[1], the hypothesis thus implied in modern statements of the Maxwellian formula for electric stress, namely that the pressure of radiation arises wholly from

*(margin: Pressure as involved in travelling momentum.)*

[1] Cf. *Phil. Trans.* Vol. ccii, A (1903).

momentum carried along by the waves. Consider an isolated beam of definite length emitted obliquely from a definite area of surface $A$ and absorbed completely by another area $B$. The automatic arrangements that are necessary to ensure this operation are easily specified, and need not detain us. In fact by drawing aside an impervious screen from $A$ we can fill a chamber $AA'$ with radiation; and then closing $A$ and opening $A'$, it can emerge and travel along to $B$, where it can be absorbed without other disturbance, by aid of a pair of screens $B$ and $B'$ in like manner. Let the emitting surface $A$ be travelling in any direction while the absorber $B$ is at rest. What is emitted by $A$ is wholly gained by $B$, for the surrounding aether is quiescent both before and after the operation. Also, the system is not subject to external influences; therefore its total momentum must be conserved, what is lost by $A$ being transferred ultimately to $B$, but by the special hypothesis now under consideration, existing meantime as momentum

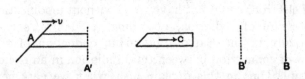

in the beam of radiation as it travels across. If $v$ be the component of the velocity of $A$ in the direction of the beam, the duration of emission of the beam from $A$ is $(1 - v/c)^{-1}$ times the duration of its absorption by the fixed absorber $B$. Hence the intensity of pressure of a beam of issuing radiation on the moving radiator must be affected by a factor $(1 - v/c)$ multiplying its density of energy; for pressure multiplied by time is the momentum which is transferred unchanged by the beam to the absorber for which $v$ is null. We can verify readily that the pressure of a beam against a moving absorber involves the same factor $(1 - v/c)$. If the emitter were advancing with the velocity of light this factor would make the pressure vanish, because the emitter would keep permanently in touch with the beam: if the absorber were receding with the velocity of light there would be no pressure on it, because it would just keep ahead of the beam.

*Back pressure on moving radiator:*

*moving absorber.*

There seems to be no manner other than these two, by altered intrinsic stress or by convected momentum, in which a beam of limited length can exert pressure while it remains in contact with the obstacle and no longer. In the illustration of the stretched cord the intrinsic stress is transmitted and adjusted by tensional waves which travel with velocity assumed to be practically infinite. If we look closer into the mode of this adjustment of tension, it proves to be by the transmission of longitudinal momentum; though in order that the pressure

*More general case of waves on cord.*

may keep in step, the momentum must travel with a much greater velocity, proper to tensional waves. In fact longitudinal stress cannot be altered except by fulfilling itself through the transfer of momentum, and it is merely a question of what speeds of transference come into operation.

In the general problem of aethereal propagation, the analogy of the cord suggests that we must be careful to avoid undue restriction of ideas, so as, for example, not to exclude the operation, in a way similar to this adjustment of tension by longitudinal propagation, of the immense but unknown speed of propagation of gravitation. We shall find presently that the phenomena of absorption lead to another complication. Various possibilities.

So long, however, as we hold to the theory of Maxwellian electric stress with associated momentum, there can be no doubt as to the validity of Poynting's modification of the pressure formula for a moving reflector, from which he has derived such interesting consequences in cosmical astronomy. To confirm this, we have only to contemplate a beam of radiation of finite length $l$ advancing upon an obstacle $A$ in which it is absorbed. The rear of it moves on with velocity $c$; hence if the body $A$ is in motion with velocity whose component along the beam is $v$, the beam will be absorbed or passed on, at any rate removed, in a time $l/(c - v)$. But by electron theory the beam possesses a distribution of at any rate *quasi*-momentum equal to $c^{-1} \times$ the distribution of its energy, and this has disappeared or has passed on in this time. There must therefore be a thrust on the obstructing body, directed along the beam and equal to $\epsilon (1 - v/c)$, where $\epsilon$ is the energy of the beam per unit length [: and there will be similar back thrusts from the reflected and transmitted beams]. Thrust in a ray.

The back pressure on a radiating body travelling through free space, which is exerted by a given stream of radiation, is by this formula smaller on its front than on its rear; so that if its radiation were unaffected by its motion, the body would be subject to acceleration at the expense of its internal thermal energy. This of course could not be the actual case. Apparent paradox:

The modifying feature is that the intensity of radiation, which corresponds to a given temperature, is greater in front than in rear. The temperature determines the amplitude and velocity of the ionic motions in the radiator, which are the same whether it be in rest or in uniform motion: thus it determines the amplitude of the oscillation in the waves of aethereal radiation that are excited by them and but intensity of internal radiation in moving body depends on direction.

travel out from them. Of this oscillation the intensity of the magnetic field represents the velocity. If the radiator is advancing with velocity $v$ in a direction inclined at an angle $\theta$ to an emitted ray, the wavelength in free aether is shortened in the ratio $1 - \dfrac{v}{c}\cos\theta$; thus the period of the radiation is shortened in the same ratio; thus the velocity of vibration, which represents the magnetic field, is altered in the inverse ratio, and the energy per unit volume in the square of that ratio, viz., that energy is now $\epsilon\left(1 - \dfrac{v}{c}\cos\theta\right)^{-2}$, cf. p. 444, *infra*; and the back pressure it exerts involves a further factor $1 - \dfrac{v}{c}\cos\theta$ owing to the convection; so that that pressure is $\epsilon\left(1 - \dfrac{v}{c}\cos\theta\right)^{-1}$, where $\epsilon$ is the energy per unit volume of the natural radiation emitted from the body when at rest. The pressural reaction on the source is in fact $E'/c$, where $E'$ is the actual energy emitted in the ray per unit time.

### Limitation of the analogy of a stretched cord.

In the case of the inextensible stretched cord, the extra length due to the curved arc in the undulation is proportional to the energy of the motion. The loss of energy by absorption would imply slackening of the tension; and the propositions as to pressure of the waves, including Poynting's modification for a moving source, would not hold good unless there were some device at the fixed ends of the cord for restoring the tension. The hypothesis of convected momentum would imply something of the same kind in electron structure.

The analogy of sweeping up waves on a cord.

It is therefore worth while to verify directly that the modified formula for pressure against a moving total reflector holds good in the case of the cord, when there is no absorption so that the reflexion is total. This analysis will also contain the proof of the generalization of the formula for radiant pressure enunciated *supra*[1].

[1] See *Brit. Assoc. Report*, 1900 [*supra*, p. 219]. The statement that follows here is too brief, unless reference is made back to the original, especially as a *minus* sign had fallen out on the right of the third formula below. The reflector consists of a disc with a small hole in it through which the cord passes; this disc can move along the cord sweeping the waves in front of it while the cord and its tension remain continuous through the hole—the condition of reflexion being thus $\eta_1 + \eta_2 = 0$ when $x = vt$. In like manner a material perfect reflector sweeps the radiation in front of it, but its molecular constitution is to be such that it allows the aether and its structure to penetrate across it unchanged. For a fuller statement, see *Encyclopaedia Britannica*, ed. IX or X, article Radiation.

Let the wave-train advancing to the reflector and the reflected wave-train be represented respectively by

$$\dot\eta_1 = A_1 \cos m_1 (x + ct),$$
$$\dot\eta_2 = A_2 \cos m_2 (x - ct).$$

At the reflector, where $x = vt$, [there is no transverse displacement so] we must have

$$\int \dot\eta_1 dt = -\int \dot\eta_2 dt;$$

this involves two conditions,

$$\frac{A_1}{m_1} = \frac{A_2}{m_2} \text{ and } m_1 (c + v) = m_2 (c - v).$$

Now the energies *per unit length* in these two simple wave-trains are

$$\tfrac12 \rho A_1{}^2 \text{ and } \tfrac12 \rho A_2{}^2;$$

thus the gain of energy *per unit time* due to the reflexion is

$$\delta E = (c - v) \tfrac12 \rho A_2{}^2 - (c + v) \tfrac12 \rho A_1{}^2$$
$$= \tfrac12 \rho A_1{}^2 \left\{ (c - v) \left( \frac{c + v}{c - v} \right)^2 - (c + v) \right\}$$
$$= \tfrac12 \rho A_1{}^2 \cdot 2 \frac{c + v}{c - v} v.$$

This change of energy must arise as the work of a pressure $P$ exerted by the moving reflector, namely it is $Pv$; hence

$$P = \tfrac12 \rho A_1{}^2 \cdot 2 \frac{c + v}{c - v}.$$

The total energy per unit length, incident and reflected, existing in front of the reflector is

$$E_1 + E_2 = \tfrac12 \rho A_1{}^2 + \tfrac12 \rho A_2{}^2$$
$$= \tfrac12 \rho A_1{}^2 \cdot 2 \frac{c^2 + v^2}{(c - v)^2}.$$

Hence finally

$$P = (E_1 + E_2) \frac{c^2 - v^2}{c^2 + v^2},$$

becoming equal to the total density of energy $E_1 + E_2$, in accordance with Maxwell's law, when $v$ is small*.

If we assume Poynting's modified formula for the pressure of a wave-train against a travelling obstacle (p. 434), the value ought to be

$$P = E_1 \left( 1 + \frac{v}{c} \right) + E_2 \left( 1 - \frac{v}{c} \right);$$

and the truth of this is readily verified.

* One may remark the contrast between the limiting cases $v = c$, when both pressure and energy density are infinite, and $v = -c$ when the pressure is null because the wave does not catch up the receding reflector. Cf. p. 464, *infra*.

It may be remarked that, if the relation connecting strain with stress contained quadratic terms, pressural forces such as we are examining would arise in a simple wave-train[1]. But such a medium would be dispersive, so that a simple train of waves would not travel without change, [and the energy would travel in wave-groups with a different speed,] in contrast to what we know of transmission by the aether of space.

*Elasticity must be linear.*

[A discussion follows* of the frictional resistance to the motion through space of a radiating body, whose mere existence, as is pointed out, had been predicted by Balfour Stewart as early as 1871. Estimates are made for [reaction of issuing ray alone], including one for the sphere which by itself verifies Poynting's formula in *Phil. Trans.* 1893. The important applications to cosmical astronomy which Poynting has there developed do not seem to have yet received the attention they deserve.

Then it is recalled that if we assume the real existence of the Maxwell stress in the aether, suitably modified for modern ideas, as the source of all mechanical interactions between electric systems, and we retain the ascertained mechanical electrodynamic forces as part of its effect, yet another phenomenon is required to make up the complete result, and this can be represented as a distribution of momentum in the aether of density equal to the vector product of aethereal displacement and magnetic induction. In the case of trains of waves, the latter agrees with Poynting's momentum of radiation. As regards the resultant momentum and forces for any self-contained system, the Maxwell stress is eliminated, and no hypothesis as to its reality is involved.

But for such a complete system, free from external disturbance, we require to compare this outstanding force, visualized as rate of change of some kind of latent momentum, when the system is convected with uniform velocity $v$, with what it would be for the same system at rest in the aether; for although the system remains the same the convection modifies the electrodynamic field around each electron which it contains, and thus may modify the effective electromagnetic mass of that electron as well as the distribution of latent momentum. When this comparison is made by aid of the classical correlation first employed by H. A. Lorentz, it turns out that the forcive acting on the convected system exceeds that acting on the same system when stationary in the aether, by the effect of a convection of latent momentum specified exactly as before, together with a force equal to $v \frac{d}{dt}\left(\frac{E}{c^2}\right)$, where $E$ is the energy in the system. On the principle that force is expressed as $\frac{d}{dt}(mv)$ we can infer that an increase $\delta E$ of the electrodynamic energy of a system increases the effective mass of the

[1] Cf. Poynting, *Roy. Soc. Proc.* Vol. LXXXVI, A (1912), pp. 534–562, where the pressure exerted by torsional waves in an elastic medium, such as steel, is exhaustively investigated on both the experimental and the mathematical side.

* This enlarged synopsis of the rest of the present paper was printed in Poynting's *Collected Scientific Papers* (1920), p. 433.

system by $\delta E/c^2$. This additional result, as well as the momentum result, is necessitated beyond cavil by the ascertained laws of electro-dynamics, which however are themselves established only when $(v/c)^2$ is negligible[1]: extension of its validity beyond that limit requires new postulates of "relativity." In astronomical applications such as Poynting's, the effect of any change of mass due to cooling is totally insignificant compared with the results which he derives from the latent momentum.]

### Momentum in convected aethereal fields.

If any transfer of momentum, analogous to what has been here described for the case of a stretched cord, is operative in free aether, the concentration of inertia on which it depends must, as in that case, be determined by and involved in the nature of the strain system. Now this strain is expressed by the electric field, and therefore by the tubes of electric force. Thus we have to consider in what cases the change of the electric field can be supposed to be produced by con-vection of the tubes of force[2].

*[margin: Electric convection: how far representable by moving tubes of force.]*

Let the scheme of tubes of force be in motion with velocity varying from point to point, equal at the point $(xyz)$ to $(pqr)$, but without other change. If $N$ be the number of electric tubes enclosed by a fixed circuit, then by Ampère's circuital electrodynamic relation

$$\frac{1}{c^2}\frac{dN}{dt} = \int (\alpha dx + \beta dy + \gamma dz),$$

*[margin: Amperean circuital relation:]*

for the left-hand side is equal to $4\pi$ times the total (in this case aethereal) current through the circuit. But if the tubes of the current $(uvw)$ all enter the circuit by cutting across its contour with the velocity $(pqr)$, i.e. if none of them originate *de novo* during the operation, the rate of gain of total current $(uvw)$ is expressed kinematically by

*[margin: in terms of such flux of tubes:]*

$$-\int \{(qw - rv)\,dx + (ru - pw)\,dy + (pv - qu)\,dz\}.$$

And as the current is here wholly aethereal, $(u, v, w) = (\dot{f}, \dot{g}, \dot{h})$.

The equivalence of these two line integrals, as it holds good for all circuits, requires that

$$-\frac{\alpha}{4\pi} = qh - rg - \frac{d\psi}{dx},$$

$$-\frac{\beta}{4\pi} = rf - ph - \frac{d\psi}{dy},$$

$$-\frac{\gamma}{4\pi} = pg - qf - \frac{d\psi}{dz}.$$

[1] Cf. Larmor, *Aether and Matter*, 1900.
[2] The considerations advanced in this section were suggested by the study of a passage in J. J. Thomson's *Recent Researches* (1893), § 9 *seq.*

gives laws of convection of tubes of electric displacement: These relations must in fact be satisfied for every field of aethereal strain $(fgh)$ whose changes occur by pure convection.

If permanent magnets are absent, the potential $\psi$ will not enter, and we have then the relations

$$\alpha f + \beta g + \gamma h = 0 \quad \text{and} \quad p\alpha + q\beta + r\gamma = 0.$$

which is a very special type of field. Thus in a convected field the magnetic vector must be at right angles to the electric vector and to the velocity of convection, is in fact $4\pi$ times their vector product.

In the case of the stretched cord, the kinetic energy is expressible as that of the convected concentrated mass on the cord. Following that analogy, the kinetic energy is here to be expressed in terms of the velocity of convection $(pqr)$ and the electric field; viz.

Field energy for this type expressed in terms of the convection:

$$T = \frac{1}{8\pi} \int (\alpha^2 + \beta^2 + \gamma^2) . d\tau$$

$$= 2\pi \int \{(qh - rg)^2 + (rf - ph)^2 + (pg - qf)^2\} \, d\tau.$$

Now generally in a dynamical change which occurs impulsively, so that the position of the system is not sensibly altered during the change, if $\Phi$ is the component of the impulse corresponding to the coordinate $\phi$, the corresponding applied force is $\dot{\Phi}$, and the increase of kinetic energy is equal to the work of this force, viz., to

$$\Sigma \int \dot{\Phi} d\phi = \Sigma \int \dot{\Phi} \dot{\phi} dt = \Sigma \int \dot{\phi} d\Phi ;$$

thus
$$\delta T_\Phi = \Sigma \dot{\phi} \delta \Phi.$$

But $T_{\dot{\phi}}$ is a quadratic function of the velocities: thus $T_\Phi$ must be a quadratic function of the momenta, and therefore

$$2T = \Sigma \Phi \frac{dT}{d\Phi} = \Sigma \Phi \dot{\phi}.$$

Hence
$$\delta T_{\dot{\phi}} = \Sigma \Phi \delta \dot{\phi}.$$

This argument applies for example in the field of hydrodynamics. In the present case the Cartesian components of momentum would leading to a distribution of impulse agreeing with the Maxwellian: therefore be $\dfrac{dT}{dp}, \dfrac{dT}{dq}, \dfrac{dT}{dr}$; so that in volume $\delta\tau$ they are

$$\begin{vmatrix} f & g & h \\ a & b & c \end{vmatrix} \delta\tau.$$

We thus arrive at the same distribution of momentum as the one that has to be associated with the Maxwellian stress system. In this but reducible for this special case: case of supposed pure convection, that momentum is of type having for its $x$ component

$$4\pi p \, (f^2 + g^2 + h^2) - 4\pi f \, (pf + qg + rh).$$

If the motion of the tube of force is wholly transverse, as symmetry would seem to demand, we have the condition

$$pf + qg + rh = 0;$$

thus the momentum, as also the kinetic energy, is now along the direction of the motion $(pqr)$ and belongs to a simple travelling inertia $4\pi (f^2 + g^2 + h^2)$ per unit volume. The convected inertia here suggested is equal to twice the strain energy multiplied by $c^{-2}$, double what it was in the case of the stretched cord. [The kinetic energy $T$, thereby envisaged, is only the fraction $v^2/c^2$ of the electric strain energy.]

The two conditions above introduced

$$pf + qg + rh = 0 \quad \text{and} \quad af + \beta g + \gamma h = 0$$

are equivalent to

$$\frac{f}{q\gamma - r\beta} = \frac{g}{ra - p\gamma} = \frac{h}{p\beta - qa}$$

which, as we shall see presently, may be treated as an indication that the magnetic tubes are convected as well as the electric tubes. Under these circumstances of complete convection, electric and magnetic, it is thus suggested that there is a mechanical momentum in the field, which arises from convection of inertia of amount equal to twice the energy of strain multiplied by $c^{-2}$. But here again restrictions will arise.

Meantime we have to supply the condition that the magnetic lines are simply convected. If $n$ denote the number of unit magnetic tubes that are enclosed by a circuit, then by Faraday's circuital law

$$-\frac{dn}{dt} = \int (Pdx + Qdy + Rdz),$$

where $(PQR)$ is at each point the force exerted per unit charge on an electric particle moving along with the circuit.

Now when the change of the tubes is due to convection solely we must have

$$-\frac{dn}{dt} = \int \{(qc - rb)\, dx + (ra - pc)\, dy + (pb - qa)\, dz\}.$$

These line integrals are therefore equivalent for all circuits fixed in the aether: hence

$$P = qc - rb - \frac{dV}{dx},$$

$$Q = ra - pc - \frac{dV}{dy},$$

$$R = pb - qa - \frac{dV}{dz}.$$

*Margin notes:*

suggests a varying travelling mass,

contrast with the electric amount.

But the convection should be complete:

the criterion of convection of the magnetic tubes:

If there are no free charges and the potential $V$ is thus null,

$$4\pi c^2 (fgh) = \begin{vmatrix} p & q & r \\ a & b & c \end{vmatrix}.$$

Thus when the system is completely convected, as regards its electromagnetic activity, and it has no charges, the velocity of convection is at each point at right angles to both the electric and magnetic lines which are at right angles to each other: this velocity is everywhere c, that of radiation, and the ratio of magnetic to electric induction is c⁻¹. But these are precisely the characteristics of a field of pure radiation, to which alone therefore the preceding argument can possibly apply: and the general argument even in this case is destroyed by the circumstance that $(pqr)$ is restricted to the constant value c, and so is not amenable to variation. But as already seen, if we accept on other grounds this convection of momentum by radiation, the validity of the Maxwell stress will follow.

*restricts to the tubes of pure radiation in free space:*

In the wider case when charges are operative, both these circuital relations for the convective system are satisfied, if $v^2 = p^2 + q^2 + r^2$, by

*or, when charges are present, to their steady convected fields.*

$$\left(1 - \frac{v^2}{c^2}\right)(P, Q, R) = -\left(\frac{d}{dx}, \frac{d}{dy}, \frac{d}{dz}\right) V,$$

$$(\alpha, \beta, \gamma) = -4\pi\,(qh - rg,\ rf - ph,\ pg - qf).$$

When $(pqr)$ is constant, this is the well-known field of a uniformly convected electrostatic system specified by the potential $V$. But to no other system is it applicable unless it can satisfy the necessary conditions of absence of divergence in $(PQR)$ and $(\alpha\beta\gamma)$.

### Frictional resistance to the motion of a radiating body[1].

We proceed to examine the retarding force exerted on a body translated through the aether with uniform velocity $v$, arising from its own radiation. A ray transmitting energy $E'$ per unit time pushes backward the moving source with a force $E'/c$. Consider first by themselves the sheaf of nearly parallel rays of natural radiation emitted from all parts of the surface whose directions are included within the same cone of infinitesimal angle $\delta\Omega$. Their energy $(E')$, emitted per unit time, as regards the part issuing from an element of surface $\delta S$ is

*Ray pressure as emerging momentum:*

*Early thermodynamic aperçu.*

[1] It appears to have escaped notice that Balfour Stewart definitely established, by a qualitative thermodynamic argument based on the disturbance of compensation in the exchanges, as early as 1871, that a moving body must be subject to retardation owing to its own radiation; and that we should "expect some loss of visible energy in the case of cosmical bodies approaching or receding from one another." See *Brit. Assoc. Report*, 1871, p. 45.

(p. 434), for a perfect radiator, if $\epsilon$ *now* is taken to represent natural radiation per unit time,

$$\epsilon\delta\Omega \left(1 - \frac{v}{c}\cos\theta\right)^{-1} \times \text{projection of } \delta S \text{ on } \delta\Omega;$$

and for the whole surface, taking the front and rear parts separately, they give a back pressure along the rays

$$\frac{\epsilon\delta\Omega}{c}\left(1 - \frac{v}{c}\cos\theta\right)^{-1} S_\theta - \frac{\epsilon\delta\Omega}{c}\left(1 + \frac{v}{c}\cos\theta\right)^{-1} S_\theta,$$

where $S_\theta$ is the projection of the surface on the plane perpendicular to the rays: neglecting $(v/c)^2$ this is

$$2\delta\Omega \frac{\epsilon v}{c^2}\cos\theta . S_\theta.$$

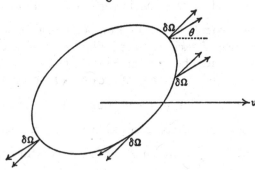

If the body is symmetrical around the direction of its translatory convection $v$, so that $S_\theta$ is independent of azimuthal angle, we have $\delta\Omega = -2\pi . d\cos\theta$; and the aggregate backward [ray-]pressure opposing the motion of the body is

$$\int_0^1 2\frac{\epsilon v}{c^2}\cos^2\theta . 2\pi . d\cos\theta . S_\theta,$$

that is

$$4\pi\frac{v}{c^2}\epsilon\int_0^1 S_\theta \cos^2\theta . d\cos\theta.$$

its amount from the rays alone:

If the travelling perfect radiator is a sphere of radius $a$, we have $S_\theta = \pi a^2$, and the force resisting its motion is

$$\tfrac{4}{3}\pi\epsilon . \frac{v}{c^2} . 4\pi a^2 .[1]$$

for a sphere:

For a plane radiating disc of area $S$ advancing broadside on, the resisting force is

$$4\pi\frac{v}{c^2}\epsilon\int S\cos^3\theta \, d\cos\theta, \text{ which is } \pi\frac{v}{c^2}\epsilon S,$$

for a plane disc.

namely is $\tfrac{3}{4}$ of the value for a sphere of the same radius.

[1] Agreeing with Poynting, *Phil. Trans.* Vol. ccii, A (1903), p. 551, where important applications in cosmical astronomy are developed. [But these results are finally amended, *infra*, p. 448, by dropping a factor $\tfrac{4}{3}$. The sure way to evade intricacies at moving and radiating boundaries, and effects of changes in the field, is to take advantage of the convective invariance of the Maxwellian stress, as on p. 443.]

The natural radiation is more usually defined by $R$ the total radiation per unit area in all directions per unit time: then

$$R = \int \epsilon \cos \phi \, d\Omega \text{ where } d\Omega = -2\pi d \cos \phi,$$

so that

$$\epsilon = \pi^{-1} R.$$

If the radiating body is in an enclosed region whose walls are also convected with the same uniform velocity $v$, the radiation contained in the region will attain to a steady state. Then the density travelling in the region in each direction ($\theta$) will be equal to that emitted in that direction from a complete radiator; thus it will involve a factor $\left(1 + \dfrac{v}{c} \cos \theta\right)^{-2}$, and so be an aeolotropic distribution.

The separate elements of surface of a perfect radiator will not now maintain a balance of exchanges of radiant energy in their emission and absorption. It may be calculated[1] that the extra pressure, due to its own radiation, on an element of area $\delta S$ whose normal makes an angle $\beta$ with the direction of convection $v$, is $\delta S \dfrac{v}{c^2} \epsilon \tfrac{1}{4} \pi \left(1 + \cos^2 \beta\right)$, agreeing with the two special cases above. Also the extra radiation emitted by it is $\delta S \cdot \epsilon \dfrac{v}{c} \pi \cos \beta$ and the extra radiation absorbed by it is $-\delta S \cdot \epsilon \dfrac{v}{c} \pi \cos \beta$; the equilibrium of exchanges is thus vitiated [for the perfect radiator] and there is either a compensating flux of heat in the radiator from rear to front, or, if an adiabatic partition is inserted, there is a diminished temperature of the part in front. The same statements apply as regards the front and rear walls of the enclosure itself.

---

[1] As follows: The extra pressure resolved along $v$ is

$$\delta S \iint \epsilon \frac{v}{c^2} \cos \phi \, (dA \cdot d \cos \theta) \cos \phi,$$

and the extra radiation of the surface is

$$\delta S \iint \epsilon \frac{v}{c} \cos \phi \, (dA \cdot d \cos \theta),$$

where $A$ is the azimuth of the ray, and

$$dA \cdot d \cos \theta$$

is an element of its solid angle referred to the normal $n$ to $\delta S$, while

$$\cos \phi = \cos \theta \cos \beta + \sin \theta \sin \beta \cos A.$$

Integrating for $A$ from 0 to $2\pi$ and for $\cos \theta$ from 0 to 1, the expressions in the text are obtained.

### Generalization to forcive on any convected system.

It is of interest to attempt to extend this analysis so as to include the resistance to the motion through aether of any electrodynamic system whatever. The transformation of Lorentz is appropriate to effect this generalization. When that transformation is extended to include the second power of $v/c$,[1] the phenomena in any system of electrons at rest or in motion, uniformly convected with velocity $v$ parallel to $x$, are the same as for that system unconvected, provided only that (in electromagnetic units)

$$(f, g, h) \text{ and } (a, b, c) \dots\dots\dots\dots\dots(1)$$

Effect of convection, referred to shrunk frame.

for the convected system are put equal to the values of

$$\epsilon^{\frac{1}{2}}\left(\epsilon^{-\frac{1}{2}}f, g - \frac{v}{4\pi c^2}c, h + \frac{v}{4\pi c^2}b\right)$$

$$\text{and } \epsilon^{\frac{1}{2}}\left(\epsilon^{-\frac{1}{2}}a, b + 4\pi vh, c - 4\pi vg\right) \dots\dots(2)$$

for the stationary system, and this change is accompanied by a shrinkage of space and local time involving $\epsilon^{\frac{1}{2}}$ and so of the second order, where $\epsilon = (1 - v^2/c^2)^{-1}[$, but also a local origin of time expressed by $t - vx/c^2*]$.

Now the force exerted on any system of electrons is determined by and statically equivalent to the Maxwell *quasi*-stress over a boundary enclosing the system, diminished by what would be needed to maintain in the region a distribution of momentum of density $\begin{vmatrix} f & g & h \\ a & b & c \end{vmatrix}$.

Formal stress specification of forces, for any frame.

In free space $(abc)$ and $(\alpha\beta\gamma)$ are the same. The $x$ component of this stress yields a force

Force on system calculated therefrom:

$$2\pi c^2 \int \{(f^2 - g^2 - h^2)\, l + 2fg\,.\,m + 2fh\,.\,n\}\, dS$$

$$+ \frac{1}{8\pi} \int \{(\alpha^2 - \beta^2 - \gamma^2)\, l + 2\alpha\beta\,.\,m + 2\alpha\gamma\,.\,n\}\, dS.$$

---

[1] See *Aether and Matter* (1900), Chap. XI, p. 176; or *Phil. Mag.* June 1904 [*supra*, p. 276]. That the transformation, with $\epsilon$ thus inserted in it, so as to include the second order, is in fact exact *so far as regards free space*, not merely to the second order which is as far as experiment can go, but to *all* orders of $v/c$, was pointed out by Lorentz later, thus opening a way for the recent discussions on absolute relativity, an idea which involves of course, on the very threshold, complete negation of any aethereal medium.

* While spatial gradients are altered by it to the first order, the inertial effect of this first order local epoch of time, is, up to the first order inclusive, that in the convected system the accelerations of all masses are to be measured in its own moving frame of space and time. When we conclude that mass is diminished by radiation, it is intrinsic mass relative to the convected system itself as its own inertial frame. See final Appendix.

how the convection affects it, referred now to unconvected axes:

The effect on this force of the translatory velocity is found by substituting (2) in place of (1). When we neglect as usual terms in $(v/c)^2$, it is (expressed thus in terms of the stationary system as the frame of reference)

$$v \int \{((g\gamma - h\beta) \, l - f\gamma \cdot m + f\beta \cdot n\} \, dS$$

$$+ v \int \{((g\gamma - h\beta) \, l + h\alpha \cdot m - g\alpha \cdot n\} \, dS,$$

that is,

$$v \int \{2 \, (g\gamma - h\beta) \, l + (h\alpha - f\gamma) \, m + (f\beta - g\alpha) \, n\} \, dS. [1]$$

Now the rate of increase of the $x$ component of the *quasi*-momentum inside this fixed surface, which represents the effect of the pressure of radiation on the system supposed self-contained, is, so far as it arises from convection only,

allowing for increased internal momentum,

$$v \int (g\gamma - h\beta) \, l \, dS.$$

When this is subtracted there still remains, for this case of a steady convected system, a force of the same order of magnitude parallel to $v$ of amount

there remains a part which expresses

$$\frac{v}{c^2} R,$$

where $R$ is the loss of energy per unit time by flux across this fixed surface, which in our present problem can arise only from excess of radiation emitted over radiation received.

diminution of mass by radiation.

It may perhaps be suggested that this excess of the total force on the system, beyond pressure connected with the radiation, arises from change of effective mass ($\delta m$) of the source itself, owing to loss of radiant energy from it; for this would involve an increase of momentum $v\delta m$, to be supplied by impressed force inside the radiator itself, if the velocity is to be maintained. A loss of energy $\delta E$ would thus increase [decrease] the effective mass* of the system by

Mass depends on store of energy.

---

[1] It may be shown that owing to the steady convection the Poynting flux of energy in the field is modified by the following additions: twice the total density of energy is carried on with velocity $v$, from which is subtracted twice the electrostatic energy $W$ carried along the electric field with velocity equal to the component of $v$ in that direction, and twice the magnetic energy $T$ carried along the magnetic field with like velocity. And similarly the *quasi*-momentum is altered by that of a mass $2E/c$ convected with the system, of a mass $- 2W/c$ convected along the electric field with velocity the component of $v$, and of a mass $- 2T/c$ convected along the magnetic field with like velocity.

Flux of energy in moving frame:

and momentum.

* It is a *decrease* of mass because this aethereal momentum $v\delta m$ must in the absence of extraneous impressed force be taken from the mechanical store of momentum $vm$ of the material body.

$\delta E/c^2$,[1] owing presumably to resulting change in minute internal configuration. But Poynting's astronomical calculation as to the time
of clearance of cosmical dust from celestial spaces would stand, as the
changes of thermal content are balanced and so cannot introduce any
sensible change of effective inertia[2].

But these second order structural phenomena appear to be still
obscure.

In final illustration of the principle that *quasi*-momentum is somehow transmitted along the rays, I take this opportunity formally to
correct two statements occurring in a paper "On the Intensity of the
Natural Radiation from Moving Bodies and its Mechanical Reaction";
Boltzmann *Festschrift*, 1903, p. 591, as *supra*, p. 273.

It was shown there that when radiation is incident directly on a   Tractions
reflector it exerts no force on the surface layer, unless indeed the   on an
conductivity is so great that we can regard this layer as containing   interface
a current sheet. But this does not prove absence of radiation pressure   balanced.
where there is no conduction. The argument from a beam of limited
length, on p. 433, is decisive on that point, so long as the principles
of the theory of electrons are maintained. The correct inference is that
the radiation pressure in a parallel beam is *transmitted* without change,
unless where the existence of conductivity gives rise to electric flow.
Thus it is the Amperean force acting on the current sheet induced in
the surface of a good conductor that *equilibrates* the radiant pressure
advancing with the incident beam [as modified by the reflected one],
and prevents its transmission beyond that surface.

The attempted disproof, in a postscript, of Poynting's modification
of the law of pressure of radiation for a moving body, is also at fault,
and must be replaced by the general discussion just given (cf. p. 447).

[1] The writer is reminded (by a summary in W. Schottky, *Zur Relativtheo-
retischen Energetik und Dynamik, Berlin Dissertation*, 1912, p. 6) that this   Historical on
relation, obtained by Einstein in 1905, expanded in 1907 by Planck, Hasenöhrl,   varying mass.
Laue, etc., is [universal] in the [fourfold world of] relativity theory. It may be
permitted to mention that the analysis in the text is taken, modified by various
corrections, from private correspondence with Poynting relating to his *Phil.
Trans.* memoir of 1903 on resistance due to radiation, which has been already
referred to. [To make agreement with Vol. I, Appendix IX, *supra*, this loss of
energy by radiation would fall on the potential energy of the radiating system.]

[2] On similar principles, the pressure of solar radiation on the Earth would
produce, owing to its orbital motion, an increase of the secular retardation of   Influence on
the Earth's diurnal rotation, which however proves to be only about a ten   Earth's
thousandth part of the amount astronomically in question.   rotation.

# APPENDIX.

## *Retardation by Radiation Pressure: A Correction.*

[Postscript added in Poynting's *Scientific Papers* (1920), pp. 754–7.]

THERE is a discrepancy, noticed also recently by Professor Schuster, between the result of Poynting and the one above obtained by me (p. 434) on Maxwellian principles for the retardation of a body arising from its own radiation, which demands consideration. The subject, if it is to be reduced to terms of simple statement of physical principle, without intricate algebra, requires methodical exposition, as the synopsis now submitted may show.

Poynting's analysis of reaction on travelling source.     The foundation of Poynting's argument was the postulate that the effect of convection $v$ on a radiating system is, up to the first order of $v/c$, (i) to retain unaltered the state of the system as determined by the positions and motions of its electrons, (ii) thus to conserve the amplitudes of the transverse optical vibration along all the rays, but to alter the wave-length or period on account of the following-on of the source by the Doppler-Fizeau principle, and therefore to alter the energy density inversely as the square of the wave-length. This statement proves to be correct: it can be verified by an application of the Lorentz correspondence analogous to the one mentioned *supra* [, p. 443]. The radiation thus altered thrusts back on its source on account of the momentum it carries away: the thrust of each ray being altered by convection in the same ratio as the density of momentum, or of energy multiplied by c, as Poynting asserted in the text—but also altered in the opposite direction on account of the shortening of the ray along which the momentum is emitted, in accordance with his correction made in an appendix which reduced the result to one-half.

A faulty application,     But the thrust deduced by him on these principles for a spherical body, assumed to be a full radiator if that be possible, is only one-third of the value deduced by a complete formal analysis on Maxwellian lines (*supra*, p. 444), namely $- Rv/c^2$ where $R$ is the rate of radiation from the body. Some element must have been overlooked.

    In the case of an isolated radiator nothing that could happen can alter the resultant momentum of the entire system, that belonging to the body and that belonging to the radiation that has left it. The momentum in the radiation that is moving out from it over a distant boundary fixed in aether is equal and opposite to its reaction, which is the time integral of the back thrust of its rays on that boundary. corrected.  There is also the change of the momentum in the region within the boundary and within the source itself. The latter part fortunately can be determined, and an equation of conservation is established; as follows.

It is easy to verify, by application of the Lorentz transformation in its exact form, that the effect of convection on the electrodynamic energy distributed within any regions that correspond in the two states, convected and unconvected, of the system[, all being referred to the unconvected frame as *supra*, Vol. I, p. 673,] is expressed by

$$E = E_0 \left(1 - \frac{v^2}{c^2}\right)^{-\frac{1}{2}}$$

$$= E_0 + \tfrac{1}{2}\frac{E_0}{c^2} v^2 + \ldots,$$

where $E_0$ is the energy in the unconvected state of the system. Thus when $(v/c)^4$ is neglected, the change of energy arising from convection is described by the ascription to the system of an extra inertia $E_0/c^2$, where $E_0$ is its energy of electrodynamic type. In this interpretation the boundary is supposed to be so far away that the field energy that is attached to the system itself and gives rise to its electric inertia is practically all included within that boundary; there is also energy of free radiation included within it, but the momentum of that part will come separately into the account.

The momentum attached to the moving source is of form $mv$, and the force required to alter its motion is $\frac{d}{dt}(mv)$, which is $m\frac{dv}{dt} + \frac{dm}{dt}v$. The latter term in the force, involving the velocity as a factor, is in this case $\frac{d}{dt}\left(\frac{E}{c^2}\right)v$, here $-Rv/c^2$ where $R$ is the radiation per unit time, in agreement with the value as directly determined in the discussion reproduced above.

When these items are all collected together the account is balanced, the total momentum not changing with time: this is in accordance with expectation and gives confidence that no element has been Conservation. omitted. But the change of sign hastily suggested near the end in a footnote appended in the Math. Congress volumes[1] must be withdrawn, the original text being correct.

To sum up: the total change of momentum in the field of any convected radiator is made up of

(i) the backward thrust of the radiation travelling out across some definite boundary fixed in the aether, which it is desirable to choose so far away that there is sensibly no energy in the aether adjacent to it except free radiation:

---

[1] Vol. I, p. 19, "On the Dynamics of Radiation" [: now adjusted, p. 444].

(ii) the change of momentum of the free radiation inside this boundary:

(iii) the effect as regards momentum of the change of the inertia of the source owing to the energy of electrodynamic type which it has lost by its own radiation.

The analysis above referred to shows that these parts cancel: and that the force required to change the velocity of the source whose mass is diminishing owing to its radiation, as measured by the rate of change of momentum due to the activity of that force, is simply

*Isolated radiating body not retarded,* $m\,\dfrac{dv}{dt}$, where $m$ is the mass the system happens to have at the instant.

This force has been here determined from the rate of change of momentum over the whole field, which implies that when the source is disturbed the field attached to it has time to attain sensibly to a steady distribution at each instant, and thus to transmit its reactions to the source.

*though its mass diminishes:* Thus far for an isolated radiator: the loss of radiant energy affects its mass $m$, but at each instant the effect of an applied force is determined as $m\,\dfrac{dv}{dt}$, where $m$ is the mass at that instant.

*but the opposite when its energy is maintained.* But for Poynting's particle describing a planetary orbit the radiation from the Sun comes in, which restores the energy $\delta E$ lost by radiation from the particle, and so establishes again the retarding force $-\dfrac{1}{c^2}\dfrac{dE}{dt}\,v$, which is the reaction of the increased inertia. It does more; the thrust of the momentum of Solar radiation $\epsilon$ scattered and in part absorbed by the particle produces a repulsion from the Sun equal to

*Gravitation to radiating body diminished.* $c\,\dfrac{d\epsilon}{dt}$, which is inversely as the square of the distance: but this merely modifies the effective modulus of gravitation, producing once for all a slight permanent change in the orbit.

But there is another interesting effect, which proves to be important. The radiation $\epsilon$ from the Sun, considered relative to the orbital motion of the particle, is subject to the astronomical aberration of light: therefore so is the momentum which it carries: coming on to the particle obliquely along with the rays this momentum has a

*Aberration of the rays doubles the retardation.* tangential component equal to $\dfrac{v}{c}$ of the whole, that is to $\dfrac{v}{c^2}\dfrac{d\epsilon}{dt}$. Here $\dfrac{d\epsilon}{dt}$ represents the Solar radiation absorbed and also scattered, of which the former part is balanced by the radiation $-\dfrac{dE}{dt}$ of the particle.

Thus the aberrational effect doubles the previous result, so that for a planetary particle which is a complete absorber the total retarding force is $2Rv/c^2$, which is six times Poynting's result, or three times that result as corrected in his Appendix. However, in the astronomical application, to explain the transparency of the interplanetary spaces, which was his chief aim, it is only the order of magnitude that is essential: and that remains practically as it was.

Final result.

But the remarkable result seems to be established that an isolated body cooling in the depths of space would not change its velocity through the aether, the retardation due to the back thrust of the radiation issuing from it being just compensated by increase of velocity due to momentum conserved with diminished mass: it will move on with constant velocity, but with diminishing momentum so long as it has energy to radiate.

Isolated radiator conforms to relativity.

The cause of inadequacy of Poynting's form of argument from momentum has yet to be specified. He calculates the part (i) above, the back thrust of the radiation, for a spherical body assumed to be a perfect radiator, by integration over a boundary which coincides with the surface of the body. This procedure is at fault; for so close up to the source there are the electrodynamic fields of the adjacent vibrators as well as the field of the free radiation, and the stress over the boundary due to the former fields is operative and should have been included. He would have attained to the same result for back thrust of the rays alone, more simply, by integrating over a very distant boundary; but in that case the changing momentum of the disturbance inside this fixed boundary will also contribute as in (ii), and necessarily to the same degree that the neglected fields of the convected vibrators would have done on the other plan which he adopted.

Discrepancy is due to field of momentum.

The root principles of the thermodynamics of [free] radiant energy, the transmission of momentum to a distance by radiation, and the alteration of its amount by convection of the source or of a reflector, thus stand firm without doubt or ambiguity.

# 78

## THE ELECTROMAGNETIC FORCE ON A MOVING CHARGE IN RELATION TO THE ENERGY OF THE FIELD.

[*Proceedings of the London Mathematical Society,*
Ser. 2, Vol. XIII, 1913, pp. 50–56.]

THE basis of the modern theory of the electric field consists in the conjugate circuital relations of Maxwell[1]; these relate to the electric and magnetic forces in the aether, the medium that establishes connection between material bodies at a distance apart through the agency of the electrons which they contain. It is a direct consequence of these equations of the field that an electron $e$, considered simply as a mobile pole of the electric force, continually generates, while in motion, a field both electric and magnetic, which spreads out from it with the velocity of radiation[2]: that in particular, while it is moving with uniform velocity $v$, it thus establishes and convects with it a steady magnetic field, arranged in circular lines of force around its line of motion, and with distribution of intensity following the Amperean law $evr^{-2}\sin\theta$, where $e$ is in electromagnetic units.

In order to complete the foundation on which the theory of ordinary electric currents rests, it is usual, in purely formal discussions which do not attempt to probe into the dynamical connections of phenomena, to add another principle—namely, that an electron $e$, moving with velocity $v$, in a magnetic field $H$, is subject to electromagnetic force $e\,[vH]$, where $[vH]$ represents in magnitude and direction the vector product of $v$ and $H$.

This principle, which determines all the mechanical phenomena of electric currents, and so dominates electrical technology, must however be, on the face of it, in intimate relation to the derived result described above, which gives the magnetic field belonging to a moving electron. It ought, therefore, to be deducible as a consequence of the circuital relations and the value of the energy; and it would not, in a completed logical exposition, persist under the guise of an independent relation.

*Field of moving charge.*

*Law of electromagnetic force on a charge is usually postulated:*

*should be deducible:*

[1] Cf. J. Clerk Maxwell, *Phil. Trans.* 1868: *Collected Papers*, Vol. II, pp. 137–143.

[2] Cf. for example *Phil. Mag.* December 1897 [*supra*, p. 145]: *Aether and Matter*, pp. 221–234.

Its complete demonstration, by means of the formulation of the activity in the electric field under the general dynamical principle of Least Action, is a somewhat complex piece of mathematical analysis, involving abstruse points of interpretation[1]. It is desirable, however, that a connection so fundamental should be amenable to elucidation in simple and direct physical terms, and that it should be viewed from various angles. The brief discussion which follows is a contribution in this direction, deducing the law of the mechanical force from consideration of a special simple system of moving charges.

*is inherent in the Action analysis.*

*A direct elementary deduction now offered.*

We have to deal with electrons moving in a magnetic field; and the fundamental dynamical quantity is the energy concerned in the motion. If we take the field to be uniform, and unlimited in extent, the mutual part of the energy of a single electron and the field is found to involve divergent integrals. Thus, if we considered one electron by itself, we would have to introduce the origin of the field, which is, of course, in actuality not unlimited in extent, and so have to discuss the interaction between the existing magnets and the electron. This, however, would carry us back into the complexities of the general discussion. But all electric charge arises, in fact, from electric separation: and we may, therefore, rather inquire whether the infinities can be evaded by dealing with a pair of conjugate electrons, positive and negative, moving together in the uniform magnetic field.

*Divergent integrals involved: interpreted: and evaded.*

When we proceed on this track, mathematical simplicity will be gained if, instead of two point charges, we consider the positive and negative charges uniformly spread and insulated on the faces of a flat condenser, having a thin uniform dielectric space of thickness $t$, and in motion with velocity $v$ through a uniform magnetic field $H$. We can calculate the forces acting on this double sheet of electric charge from the principle that each element of charge $\delta E$ (in electromagnetic units) constitutes an Amperean element of current $\delta E \cdot v$, and is therefore acted on by a force $\delta E\,[vH]$ as above postulated. We resolve all such forces into their components transverse to the condenser and along its plane: the former components cancel as regards opposite conjugate elements of charge $+ \delta E$ and $- \delta E$: the latter sum up in all to a torque or couple, in the plane containing the pole of $[vH]$ and $n$ the normal to the condenser, and of moment equal to

*Forcive on charged condenser convected in a magnetic field:*

*obtained from Amperean formula:*

$$Et\,[vH] \sin (vH \cdot n),$$

where $(vH \cdot n)$ represents the angle between these directions.

Now let us examine whether the same result as is here deduced by postulating the law of magnetic force on a moving charge can be

*and now to be verified.*

[1] Cf. *Aether and Matter*, pp. 82 *seq.*: or *Camb. Phil. Trans.*, Stokes Commemoration volume, 1899[, *supra*, p. 165.]

obtained directly, in this simple case, from the energy in the field. If so, it will follow, since it remains true for a condenser of any form of contour and any thickness of dielectric, that the forcive on a pair of conjugate elements of charge $+ \delta E$ and $- \delta E$ forms an elementary torque, the same as if each charge were acted on as would be the equivalent Amperean current element $[\delta E \cdot v]$. The essential dependence between these electromagnetic forcives sustained by moving charges and the electromotive properties of the field will then have been confirmed in a simple way. And if, further, we can obtain a like confirmation for the component, at right angles to its plane, of the mechanical force on one of the charges by itself, we shall have deduced the complete law of mechanical force from the relations of the field.

Now the moving charges of the condenser will produce a magnetic Its own magnetic field: field, here to be determined, which is superposed on the uniform field $H$. The Amperean element $\delta E \cdot v$ can be replaced by its components, $\delta E \cdot v \cos \theta$ normal to the condenser and $\delta E \cdot v \sin \theta$ in its plane, where $\theta$ is the angle $vn$. In the steady motion of the system the magnetic field of the former is annulled by that arising from the opposite negative element $- \delta E$. The magnetic fields of the latter components aggregate, for all the elements of the system, to a field of uniform intensity $4\pi\sigma v \sin \theta$ in a direction along the dielectric and at right angles to $v$, where $\sigma$ is the surface density of charge; while they give *no field whatever* outside the plates of the condenser.

Generally, if a magnetic field $(\alpha_0, \beta_0, \gamma_0)$ is superposed on an existing field $(\alpha, \beta, \gamma)$, the kinetic energy

$$(8\pi)^{-1} \int (\alpha^2 + \beta^2 + \gamma^2)\, d\tau$$

is increased by

is superposed:
$$\frac{1}{4\pi} \int (\alpha\alpha_0 + \beta\beta_0 + \gamma\gamma_0)\, d\tau + \frac{1}{8\pi} \int (\alpha_0{}^2 + \beta_0{}^2 + \gamma_0{}^2)\, d\tau.$$

In the present case the superposed field exists only between the plates, and there

$$\alpha\alpha_0 + \beta\beta_0 + \gamma\gamma_0 = H \cdot 4\pi\sigma v \sin \theta \cos (H \cdot vn),$$
$$\alpha_0{}^2 + \beta_0{}^2 + \gamma_0{}^2 = (4\pi\sigma v \sin \theta)^2.$$

the mutual part of the energy. Thus the mutual or interacting term on the magnetic energy is, integrated over the volume of dielectric,

$$H \cdot Evt \sin \theta \cos (H \cdot vn).$$

Now on the spherical projection in the diagram

$$\cos (H \cdot vn) = \sin (Hv) \sin \psi,$$

and
$$\sin \theta \sin \psi = \cos \chi,$$

where $\psi$ is $(vH \cdot vn)$, and $\chi$ is $(n \cdot vH)$ viz. the inclination of $n$, the normal to the condenser, to the fixed direction of the pole of the vector $[vH]$.

The mutual term $(\tau)$ in the magnetic energy is therefore

$$H \cdot Evt \sin (Hv) \cos \chi.$$

On varying $\chi$, viz. the angle $(Hv \cdot n)$, by rotation of the condenser, this local store of energy is altered; and if this is done not too fast, it might be thought that there would be no gain or loss of energy by radiation or otherwise. Then the increment of energy would represent work that must be put into the system consisting of the charged condenser and its surroundings, by an extraneous torque assisting the variation of $\chi$: the reacting forcive exerted by the charged condenser itself on external bodies is the opposed torque, whose moment in the direction of $\chi$ increasing would therefore be

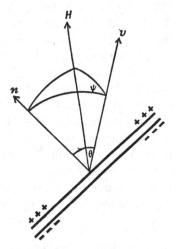

At first sight energy not conserved.

$$H \cdot Evt \sin (Hv) \sin \chi,$$

measured in the direction of $\chi$ increasing. This is the same as the value $Et\,[vH] \cos (vH \cdot n)$ already obtained: but it has the opposite sign[1], so the hypothesis of conservation of the energy from which it has been derived would not be borne out unless there is compensation.

The momentum along $v$, in the system, is here $T/v$, which is $H \cdot Et \sin (Hv) \cos (Hv \cdot n)$, equal to $H \cdot Et \sin (nH) \cos (v \cdot nH)$, for here and *infra* the coordinate of which $v$ is the velocity is not involved in $T$: thus showing that the resultant momentum in the system, of which this is a component, is $H \cdot Et \sin (nH)$ directed towards the pole of $(nH)$. This momentum arises from the interaction of the uniform magnetic field $H$ with the uniform electric field of the charged condenser. It is thus to be regarded as uniformly distributed throughout the dielectric space, and so of density equal to the vector product of magnetic force and aethereal displacement—in accordance with the well-known formula*: there are in addition the intrinsic momenta of the individual electrons of the charge. This derivation of it is exact

The electrokinetic momentum in the field.

[1] I am indebted to Dr Bromwich for pointing out this mistake in the original draft of the paper, and its source in the kinetic character of the energy, the force being thus $+ dT/dv$.

\* This direct derivation of the field momentum seems to be noteworthy, as it usually appears merely as a residue of the Maxwell stress formulation: as *supra*, p. 71, *A Dynamical Theory...*, Part III, § 39.

when $v$ is diminished indefinitely, *i.e.* in all ordinary systems: the modification required for exceedingly rapid convection does not now concern us.

What is the source of the energy that is expended by the system when the convected condenser does work on external bodies? As the working force exerted by the system is $+ \partial T/\partial \phi$, it appears at first sight to gain in the end energy of the same amount $+ \delta T$ as it has expended in work, which would involve a supply $2\delta T$ from outside.

Compensation of energy to be sought. This could only come from some of the energy of the extraneous field being transmitted by radiation across the intervening region, during the time that the condenser is undergoing change of orientation. But this change can be as slow as we please; and, as the rate of receipt of radiation must be proportional to the square of the rate of change, while the duration of receipt is inversely as that rate, the total receipt can be diminished indefinitely. Thus the conservation of energy cannot be secured in this way; though conservation of momentum can thus be effected.

Mutual kinetics of condenser and magnets. We therefore revert to the general theory of the kinetics of a system, made up of two parts, which exert interaction expressed by mixed terms in the energy, say of type

$$T = \ldots + L v_1 v_2 + \ldots,$$

where $L$ depends on the coordinates of position, of type $\phi$ only, of the two parts of the system. Let us suppose that the second part is maintained steady in its coordinates and velocities: it will however suffice if its field of influence is maintained steady throughout the region in which the other part is situated, which covers the case of the magnets producing the extraneous field of our present system. The first part of the system then exerts on external bodies a force $- v_2 \cdot dL/dt$ in the direction which works on $v_1$. But it also exerts in the direction which works on $\phi$ a force $v_1 v_2 \partial L/\partial \phi$. The total rate of working of this part on external bodies is thus

$$- v_2 \frac{dL}{dt} v_1 + \Sigma v_1 v_2 \frac{\partial L}{\partial \phi} \phi,$$

which is

$$- v_1 v_2 \frac{\partial L}{\partial t}.$$

Thus when $L$ does not contain the time explicitly, *i.e.* when the structure of the first part of the system is not changing with the time, Condenser retains its energy. and the field arising from the second part is steady, this rate vanishes and there is no loss of energy from the first part of the system to external bodies, the work expended through one coordinate being recovered through the others. Interesting hydrodynamic examples can readily be stated.

In the present case, when the condenser system acquires energy through work gained in rotating itself, this gain is merely a transfer from the energy of convection: for that convection is opposed by a force, extremely small, which appears as the rate of loss of intrinsic momentum of the portion of the system under consideration.

The torque on the condenser vanishes when its plane is parallel to $H$ and $v$. Under the continued free operation of this forcive the condenser ought, in fact, finally to assume the position in which it cannot increase the local magnetic energy further, *i.e.* the position in which it is maximum, which is that in which the field due to the moving charges is in the same direction as the extraneous field and as great as possible. This agrees with the result above stated.

This calculation of the torque from the energy of the field confirms, but does not prove, the expression for the force on a moving electron. We can now go further. We may displace one face of the moving condenser away from the other. Variation of the energy with regard to $t$ should thus give the component, in the direction of $t$, of the force acting on the moving charge $\delta E$. The value thus obtained is easily seen to agree with the Amperean formula. As then the torque for the pair $+ \delta E$, $- \delta E$ agrees, and also one component of the force on a single charge, it is easy to see that the formula for the mechanical force on a charge moving in a magnetic field is derivable completely from the electromotive relations of the field and its energy.

The intrinsic electrokinetic energy of the moving charged condenser is the remaining term, viz. $(8\pi)^{-1}(4\pi\sigma v \sin\theta)^2$ per unit volume of the dielectric, or in all $2\pi E\sigma v^2 t \sin^2\theta$, where $\theta$ is the inclination of the normal to the plane of the condenser to the direction of its motion. Thus the condenser sustains a torque $2\pi E\sigma v^2 t \sin 2\theta$ tending to set it transverse to that direction. With these small quantities we are, however, in the debatable domain of the principle of relativity. It can be shown[1] that by means of such second order forcives the energy of the absolute motion of the Earth through aether could be drawn upon for terrestrial work, unless motion of bodies through the aether is accompanied by the FitzGerald-Lorentz shrinkage, which would annul, up to the second order inclusive, both this and all other methods of testing the presence of absolute motion through the aether by operations within the system.

The forces above considered arise from the motional, or electromagnetic, energy of the system. Are there also forces of electrostatic origin? The electrostatic field of a uniformly convected system is not altered, up to the first order (cf. *Aether and Matter*, p. 154), the electric

[1] Cf. Note appended in *Scientific Writings of G. F. FitzGerald* (Dublin, 1902), pp. 566–8[, as *supra*, p. 228].

An electric potential: in convected system: unaffected to order $v/c$.

force of the field being defined relative to the moving charges; thus the electric potentials exist as before, and there are no new first-order forces of this type.

Reference has just been made to annulments in a convected system by adjustment up to the second order. We have, however, obtained in this discussion a force of the first order: thus, as the impressed magnetic field may be itself due to magnets themselves forming part of the convected system, a contradiction is suggested. The explanation is to remember that a moving magnet produces, on Amperean principles, a field of aethereal force, measurable as force exerted on stationary ions: this gives a force equal and opposite at each point to the force experienced by an ion on account of its motion with the system, so that, on the whole, a convected ion is undisturbed. Thus, if the magnets producing the impressed field are moving with the system, then, even though that field be absolutely uniform, the state of motion of its source is revealed by the presence of an aethereal force which would disturb a fixed ion, but for a moving ion convected with the magnets would just annul the other force on it arising from its motion. Further elucidation as to how this comes about would be interesting, but not now relevant.

Effect of the system's own magnetic field, how compensated.

# 79

# PROTECTION FROM LIGHTNING AND THE RANGE OF PROTECTION AFFORDED BY LIGHTNING RODS*.

[*Proceedings of the Royal Society*, A, Vol. XC, 1914, pp. 312–318.]

THE elasticity of the aether is perfect; and a perfect vacuum is infinitely strong, so that no electric force, however great, can produce disruptive discharge through it. The electric weakness of a rarefied gas is now known to be connected (J. S. Townsend, J. J. Thomson) with the long free paths of whatever ions may exist in it; their length allows time for the ions to attain abnormally high velocities, under the acceleration produced by the forces of an electric field, before they are deflected by collisions, and thus enables them to produce more ions in geometrical progression, by shattering impact as they come into collision with the molecules of the gas. Discharge conditioned by presence of gas molecules, yet facilitated at low densities.

At ordinary densities, with their short free paths, it requires much stronger fields to get up the velocities requisite to induce such an explosively increasing accumulation of ions. In free space, away from disturbing solid walls, each gas, in fact, is recognized to have definite electric strengths corresponding mainly to the different values of its density. Electric strength of a gas.

In the strong fields generated by atmospheric accumulations of electricity, we can thus conceive that, owing to displacements of the charged masses by wind and their mutual attractions, the limit of strength is reached at some point of most intense force. We can think of a small region surrounding that point as breaking down, so that positive ions are drawn away from its interior, and accumulated towards its boundary following the direction of the line of force, the conjugate negative ions accumulating on the boundary in the opposite direction. These aggregations constitute an electric polarity of this labile region, which disturbs the surrounding field that produced them, intensifying it along the line of force (three times for the case of a sphere) and reducing it in sideways directions. Thus the disruption, which there began, will tend to spread along the line of force both forward and backward. The electric discharge is therefore a rupture of the gas *along a line*, not along a surface like rupture of a solid body due to disruptive waves or elastic overstrain; and the line of rupture Rupture initiated where field is greatest: and proceeds by boring a linear path: contrasts with tearing.

* By Sir J. Larmor and J. S. B. Larmor.

tends to spread in both directions along the line of electric force which passes through that point of maximum force at which the discharge originated. It might be asked whether this tendency would be modified by the electric force induced kinetically by the discharge; but that also is in the same general direction as the other part.

Direction of a discharge along path.
When it is borne in mind that discharge is due to electric rupture of the gaseous medium, which is propagated in time, the question of the direction in which a discharge strikes acquires a meaning. The initial rupture is to be expected at a place of maximum force; and it spreads in both directions, though with different characteristics, along the line of force*. For example, in the case of a lightning rod the discharge would start at the summit of the rod, which is the place of most intense strain, and strike away from the rod instead of towards it.

Zig-zag lightning.
The zigzag character of many discharges also becomes more directly interpretable. For discharge from one cloud mass to another alters the charge in each, and so alters profoundly the fields of force between them and other adjacent masses, thus inciting immediate new discharges between those bodies. Something of this kind is perhaps indicated by the discontinuous crackle of a sharp discharge. It is hardly necessary to mention that the duration of thunder, so far as

Reverberation.
it is not of the nature of reverberation by reflexion, is due to the longer time that the sound takes to arrive from the more distant parts of the flash, and thus gives a sort of analysis of its progress†.

When once an ionized path has been opened up in the gas there will be a strong tendency for subsequent discharges to take place along it.

Multiple flashes.
Meantime the line of ionized gases, forming the open path, will be blown into a new position by the wind; and this may not inconceivably have something to do with the origin of the frequently photographed parallel flashes.

Brush.
Such considerations, as regards the propagation of disruptive discharge, may have also to do with the establishment of the beautiful brush discharges which are propagated into the gas, and are of different types for positive and negative ions. The branching character of a sharp negative spark suggests obstacles to the natural direction

Direction of discharge.
* Cf. more recent discussions on lightning by C. T. R. Wilson and by G. C Simpson. The latter investigator concludes from the evidence of statistics of lightning (*Roy. Soc. Proc.* 1926) that the discharge is opened up (loses itself in branches) only in the positive direction.

Data of thunder-clouds.
† According to the experience of C. T. R. Wilson and his associates the field under a thunder-cloud may be $10^4$ volts per metre, falling off to $10^3$ at a distance of 10 kilometres and to $10^2$ at 20 kilometres. They consider the cloud to be charged negatively below, positively above, to an electric moment of 30 coulomb-kilometres, discharging 20 coulombs in a single flash: in the cloud there may be $10^6$ volts before discharge and an electric energy $10^7$ ergs. There are usually two thousand storms going on: this upward flux of charge may balance the usual downward flow in fair weather.

of travel along the line of force; the positive discharge, being con- Contrast of
ditioned by different and perhaps more massive ions, will be of more negative and
positive
sluggish type. The nature of the gas is known to affect the exquisite flashes.
patterns made by discharge along photographic plates.

Considerations like the above suggest that we may estimate the Region of
region of protection of a lightning rod, by inspection of a diagram of protection
the modification of the field of force which is established around the
rod, before discharge, owing to its influence*. We can render the
problem definite by supposing that a single vertical rod or column by earthed
connected to a horizontal earth is situated in an atmospheric field of vertical wire
vertical electric force of uniform intensity. A discharge striking into
the modified field surrounding the structure will tend to follow the
lines of force, except so far as its initial electric momentum delays
that tendency; and, to an extent to be now determined, it will be
guided to the upper part of the structure instead of striking downward to the earth as it would do in the absence of that conductor.

This special form of the problem of determining a disturbed field which modi-
of force has a ready solution for the case of a semi-ellipsoidal con- fies a vertical
field,
ductor $(a, b, c)$ standing on the ground; and a rod (of suitably varying
section) may be illustrated by the special case in which the semi-axes
$a$ and $b$ are small, while $c$ is large. In fact, if the undisturbed vertical
atmospheric field is $F$, the modified potential

$$V = - Fz + Az \int_{\epsilon}^{\infty} \frac{d\lambda}{(a^2 + \lambda)^{\frac{1}{2}} (b^2 + \lambda)^{\frac{1}{2}} (c^2 + \lambda)^{\frac{3}{2}}}$$

* Copious illustrative natural and experimental results are given by F. W.
Peek, *Journ. Franklin Institute*, 1924–25, reprinted in *Smithsonian Report*, 1925.
It appears that a vertical rod protects a radius of four or five times its length.
These field results, and those of Norinder (*loc. cit.* 1927), contemplate charges
of one sign in the cloud, but do not seem to disagree generally with Wilson's
orders of magnitude: cf. A. W. Simon, *Proc. Amer. Nat. Acad.* April 1928, p. 458,
who reasons from the maximum possible field in the atmosphere, taken as
$3 . 10^6$ volts per metre.

Maxwell's opinion in 1876 (*Nature*, vol. xiv, *Scientific Papers*, ii, p. 538) was Maxwell.
that a lightning rod is effective rather for the relief of the clouds than the protection of the building, and that a thin copper wire carried round the edges of the
building would be an effective safeguard.

In Maxwell's introduction to the *Cavendish Papers* there is a history of the
proceedings of the Committee of the Royal Society which examined for the
Government the question of protection of powder magazines from lightning.
The recommendations of the final committee are given very systematically Historical
without reasons assigned in *Phil. Trans.* 1778, p. 313. They prescribe a system views.
of pointed elevated rods rising about 10 feet above the roof, connected below
by metal bands, in agreement with the view arrived at *infra*, a single elevated
conductor having been shown by experience to be ineffective; and they conclude as follows: "We give these directions, being persuaded that elevated rods
are preferable to low conductors terminated in rounded ends, knobs, or balls
of metal; and conceiving that the experiments and reasons made and alleged
to the contrary by Mr Wilson are inconclusive."

will be null over the ground, and also null over the ellipsoid $(a, b, c)$, provided

$$\frac{F}{A} = \int_0^\infty \frac{d\lambda}{(a^2 + \lambda)^{\frac{1}{2}} (b^2 + \lambda)^{\frac{1}{2}} (c^2 + \lambda)^{\frac{3}{2}}}.$$

For our special case of a thin symmetrical semi-ellipsoid of height $c$, this gives

$$V = - Fz + Az \int_\epsilon^\infty \frac{d\lambda}{\lambda (c^2 + \lambda)^{\frac{1}{2}}}.$$

The value of this integral, however, increases indefinitely towards its lower limit as $\epsilon$ falls to zero, when $a$ and $b$ are null. Thus as the semi-ellipsoid becomes thinner the value of $A$ diminishes without limit: that is, the modification of the field of force by a very thin rod **but only near** is negligible along its sides unless close to it. A thin isolated rod thus **its summit.** draws the discharge hardly at all unless in the region around its summit.

Our special problem is, in fact, the same as if the earth were abolished, and we had to examine the influence of the presence of a complete ellipsoidal conductor, uncharged, on a uniform field of force, in this case parallel to its longest axis. As the ellipsoid (or solid of other form) becomes very thin, the moment of the charge induced on it, depending as it does on considerations of volume, becomes negligible, the actual value for any rod depending on the form and mode of variation of the cross-section; cf. Green's problem of the magnetization induced in an iron cylinder in a uniform magnetic field, treated in his Essay, and referred to in Maxwell, *Electricity and Magnetism*, Vol. II, § 439.

We are to take it, then, as a general principle, that it is the building **The building** to which the lightning rod is attached that modifies the electric field, **draws the** and so draws the discharge, by virtue of its conducting materials and **discharge,** on account of its breadth being substantial compared with its height[1]. **the projecting** The structure protects the region around its base, by directing dis- **wire diverts it.** charge to its own upper parts, which therefore need protection by conductors adequate to draw off this discharge to earth; and vertical rods, joining together if need be lower down, but rising from the corners of the structure to a height which need not be more than **The ideal** about half its breadth, will lift up the field of concentrated electric **protection.** force from the region directly above the building to the region above their summits, and thus will themselves take the discharge instead of

---

[1] Mr S. G. Brown informs us that instances of damage of radiotelegraphic stations by lightning have occurred, and it has been the subject of remark that the discharge does not come down the antennae. On the other hand, conducting stay-wires tend to transfer an antenna from the character of a pole to that of a cone: the mast of a ship may come under this case.

the upper part of the structure. It would thus appear to be the top of a building that needs protection by rods of some height studded over it, and in direct easy connection with earth. For convenience and security of communication the rods may rise from a network spread over the roof, but the essential feature appears to be their lifting the field of intenser force to the region in which their free ends are immersed.

This mode of analysis can be illustrated more widely by the diagram annexed, of the lines of force belonging to a semi-ellipsoid of finite breadth standing on the ground. For the influence of a semi-ellipsoidal conducting structure on the field is the same as the influence of a larger confocal semi-ellipsoid on a field less intense by a factor which can be readily estimated. In this way, by combining the same ellipsoidal field with uniform exciting fields of various intensities, we can realize a wide range of cases. *Diagram for wire protector* *applicable also to thick confocal ellipsoids.*

The mode in which a uniform vertical field will be fenced off from a structure by an arrangement of projecting rods may be illustrated by Plate XI in Maxwell's *Electricity and Magnetism*, Vol. 1, which represents the protection afforded by a series of parallel plates running inward from the left of the page, the lines of equal potential being the vertical curves. The field of force is promptly annulled, being reduced to half its intensity before it penetrates within the points to a quarter of the distance between adjacent ones. *Set of projecting wires pushes up the field.*

It is easy to see that a lightning rod is most effective when it lies along the lines of force. If it is at right angles to them it is hardly effective at all. It is commonly held that a powder magazine would be best protected if it were enclosed in a network of rods. The horizontal rods would be of use, however, only when the electric discharge is oblique. If we could assume that the lines of force in a storm are always vertical, a system of vertical rods would serve as well as a network. *Network most effective protection,*

When a discharge has once become established, by the opening up of an ionized conducting path, a lightning rod can do little to deflect its course. What it can do, and still better a series of rods rising over the whole structure can do, is to provide the easiest and therefore most probable path for such discharge as may be attracted by the structure. It seems open to question whether merely encasing a building in a wire network will always protect it from a disruptive discharge striking down upon it, unless the network is so fine as to approximate to a complete metallic covering; a spread of connected metallic points some height above the building would appear to be more effective, and might even by themselves suffice to take up and guide away any likely stroke. *or wires if elevated above the structure.*

The considerations above set out have rested on the hypothesis that

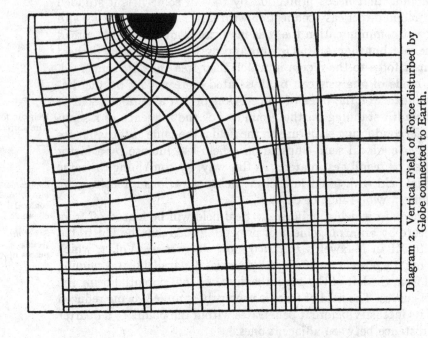

Diagram 2. Vertical Field of Force disturbed by Globe connected to Earth.

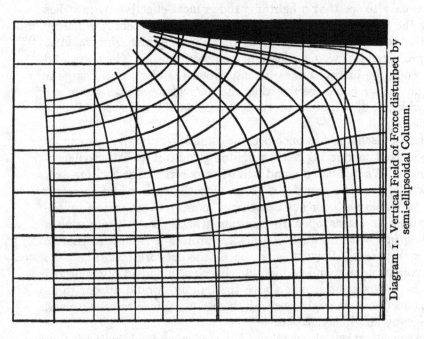

Diagram 1. Vertical Field of Force disturbed by semi-ellipsoidal Column.

there is no discharge from the rod into the surrounding field. But if it is sharply pointed, the very intense field set up close to the point establishes a quiet discharge of ions which, so far as they are not blown away, will form a conducting region around the point, in which the field is relieved. The influence of the rod is then to be likened to that of a conducting globe at its summit, connected to earth by a thin wire. Accordingly, it is of some interest to exhibit the modification which is thus imposed on a uniform vertical field of force, $F$, say, by a globe of radius $a$ with its centre at height $c$ and earthed. The globe acquires such a charge $e$ as reduces its potential to zero: thus $-Fc + e/a$ must vanish, so that its charge is $Fca$; the potential of the field therefore becomes

*The silent discharge in still air.*

*Illustrated by diagram.*

$$V = -Fz + Fca \left( \frac{1}{r_1} - \frac{1}{r_2} \right),$$

where $r_1$, $r_2$ represent distances from the centres of the globe and its image. If the lines of equal potential which pass underneath the globe are continued down the length of the rod to meet those coming in from the side, the circumstances of the rod with conducting region at its summit will be roughly represented.

The diagram 1 represents a uniform field with vertical lines of force and horizontal equipotential lines overhead, as disturbed by a thin conductor of the form of a semi-ellipsoid standing on the ground; the left-hand half of the field is drawn.

The diagram 2 represents a similar field overhead disturbed by the presence of a conducting sphere connected to earth by an infinitely thin wire*.

---

* Other cases have been worked out somewhat later into graphical expression by Prof. C. H. Lees, in *Roy. Soc. Proc.*

# 80

# THE REFLEXION OF ELECTROMAGNETIC WAVES BY A MOVING PERFECT REFLECTOR, AND THEIR MECHANICAL REACTION.

[*Philosophical Magazine*, Vol. xxviii, Nov. 1914, pp. 702–7.]

I FIND that I cannot accept the criticism of my views on this subject which is contained in Mr Edser's paper in the present volume of the *Phil. Mag.* pp. 508–527.

In a train of radiation, the magnetic and the electric force intensities must be of the types of velocity and strain in the transmitting medium, if we are to liken the radiation to a train of dynamical waves at all. The amplitude with which we are naturally concerned is the amplitude of the swaying displacement or shift of that medium, a shift of which the magnetic force intensity represents in some way the time gradient or velocity. In reflexion of a wave-train directly incident, the amplitude of this displacement undergoes no change; but when the reflecting surface is receding, the amplitude of its velocity, here the magnetic force, is no longer thereby preserved.

*Condition at receding reflector:*

If the reflector is receding with velocity $v$, equal to $c$, that of the radiation, it just keeps ahead and there is no reflexion, and so there ought to be no pressure. The argument in the quotation made by Mr Edser [cf. *supra*, p. 435] leads in fact to the result that the pressure on the reflector is equal to the energy density in front of it multiplied by a reducing factor $(c^2 - v^2)/(c^2 + v^2)$. Mr Edser challenges this result on p. 513. But it appears from his discussion that he is considering a reflector moving up into the radiation instead of one receding from it; in that case the reflexion piles up the radiation [which is the other factor in the pressure] as it advances, and to an infinite degree when it advances with the velocity of radiation.

*radiation pressure on moving reflector.*

I have been for long aware that the general problem of reflexion from a moving reflector, with which Mr Edser's paper is mainly concerned, involves considerations as to the relation of aether to moving matter, which it would be interesting to disentangle. But as the modification introduced in the law of radiation pressure, which is the main object of the theory, seemed to be of the order $(v/c)^2$, nothing substantial appeared to be obtainable without much complication: for it would thus rank with the other second-order phenomena, which either require the introduction into the analysis of the FitzGerald-Lorentz deformation of material media produced by their motion through aether or, alternatively, involve the complexities of the relativity theories.

It will assist and test ideas, however, to pursue to some degree the problem of oblique reflexion. It will suffice for present illustration to restrict it, following Mr Edser, to a reflector moving with only translational velocity, and that at right angles to its own plane.

The law of reflexion of the rays, and the law of the change in their wave-length, are a purely kinematic affair, independent of dynamics of propagation. Thus if, following Mr Edser's equations (13) and (14), the properties of the incident wave-train, travelling now towards $x$ positive, are defined as functions of $m$ $(x \cos \theta + y \sin \theta - ct)$, and those of the reflected as functions of $m'$ $(x \cos \theta' + y \sin \theta' + ct)$, the quantities involved in these arguments must satisfy the condition that the waves pass along the reflecting surface $x = vt$ in step. Now when $vt$ is put for $x$, these arguments become

$$m \ (v \cos \theta - c) \ t + my \sin \theta \quad \text{and} \quad m' \ (v \cos \theta' + c) \ t + m'y \sin \theta';$$

and these must therefore be equal.  Hence

$$-\frac{c + v \cos \theta'}{c - v \cos \theta} = \frac{\sin \theta'}{\sin \theta} = \frac{m}{m'}. \tag{A}$$

To determine the relation between the amplitudes of the waves, the structural interfacial conditions must enter. If $v/c$ is small, we can apply general electric principles. For up to the first power of that quantity, the Maxwellian expression for the electric force in a moving body (viz. the intensity of force acting on electrons connected with the body) is deducible from fundamental principles; and its expression thus obtained satisfies the Faraday circuital relation expressing the electromotive effect round a circuit moving with the medium[1]. Apply then this principle to a very narrow circuit made up of two very near parallel lines, one on each side of the reflecting interface, both travelling with it, and connected at their ends: as the area enclosed here tends to nothing, the electric force related to the moving medium (as distinguished from the aethereal force, which is what would act on an electron at rest) must be the same along both lines.

Again, the magnetic flux must be continuous. In our present problem, on which there is no field of force transmitted into the reflector, but only a shielding current sheet, we have thus tangential electric force and normal magnetic force continuous. But as only one relation between the intensities of the incident and reflected wave-trains is to be determined, these relations ought to be identical; and in fact the one is involved in the other through the Faraday circuital

*Reflexion from receding surface.*

*Surface conditions:*

[1] Cf. *Aether and Matter*, pp. 98, 113.

relation. As already stated, they hold good only when $(v/c)^2$ is
*limitation.* neglected; if we go beyond that, the nature of the linkage of the
electron with the aether may have to come into consideration and we
get into unexplored regions.

We consider now the problem of a simple harmonic wave-train.
If the aethereal (not electric) forces $E$ and $E'$ in the incident and
reflected waves are perpendicular to the plane of reflexion, say if
they are

$$E = a \cos m \, (x \cos \theta + y \sin \theta - ct),$$

$$E' = a' \cos m' \, (x \cos \theta' + y \sin \theta' + ct),$$

then the magnetic forces are $- E/c$ and $E'/c$, and are in the plane of
reflexion, and the vanishing of their total component along the normal
gives

$$\frac{E}{c} \sin \theta - \frac{E'}{c} \sin \theta' = o,$$

that is

$$a \sin \theta = a' \sin \theta'. \tag{B}$$

*Result.* This equation, in conjunction with the kinematic relations (A), solves
the problem.

Again, if the radiation is polarized perpendicular to the plane of
incidence, it is the magnetic forces that are given as above; say

$$H = b \cos m \, (x \cos \theta + y \sin \theta - ct),$$

$$H' = b' \cos m' \, (x \cos \theta' + y \sin \theta' + ct).$$

The aethereal forces are $Hc$ and $- H'c$ in the plane of incidence, so
that the total tangential component of the electric (not aethereal)
force in the reflector is

$$Hc \cos \theta - H'c \cos \theta' + v \, (H + H').$$

This must vanish: so that

$$b \, (c \cos \theta + v) = b' \, (c \cos \theta - v'). \tag{C}$$

Now

$$(c \cos \theta + v)^2 = (c + v \cos \theta)^2 - (c^2 - v^2) \sin^2 \theta;$$

hence each term of (A) is equal to $- \dfrac{c \cos \theta' + v}{c \cos \theta - v}$, the sign being fixed
by continuity. It readily follows that $b'/b$ in (C) has the same value
as $a'/a$ in (B).

*No*
*polarization*    This agrees with Mr Edser, and is the interesting result of his in-
*produced.* vestigation. It means that a moving total reflector cannot operate so
as to partially polarize an incident beam of ordinary light.

This result is in accordance with anticipation; inasmuch as the thrust
of radiation in the parallel beam is simply equal to the density of its
energy, and does not even depend on $v/c$ up to the first order, while

the work done by the combined thrusts on the receding reflector must be expected thermodynamically to be the same, whatever be the nature of the polarization of the beam.

The question presents itself: What is the vector, corresponding to $\xi$ in the investigation for direct reflexion, which represents the actual shift in the aether? It appears now that it is not in the general problem the time integral of the magnetic force; for we have taken it that it is the normal component of the magnetic force itself that is continuous across the surface.

As regards the pressure of the radiation, the uncertainty in this way of estimating may be obviated by the more general and precise procedure based on the Maxwell stress theory. Consider the system consisting of an incident beam of restricted breadth, the reflected beam, and the receding reflector. Draw a boundary, fixed in aether, represented by the dotted curve of the diagram [omitted]. Then the Maxwell stress exerted from outside over this boundary, together with the force represented by the time rate of increase of the actual momentum throughout the volume, is equal to the mechanical force exerted by the electric field on the bodies inside the boundary. In time $\delta t$ this momentum is increased by that contained in the layer of breadth $v\delta t$ which is added to the radiation space during that time. Now the Maxwell stress incident is a thrust along the beam equal to its energy density (say $\epsilon$); reflected it is a like thrust. The momentum density in each beam is along its direction of propagation and equal to its energy density divided by $c$; hence the normal components of the increment of momentum cancel, and the tangential components make up $2\epsilon/c \cdot v\delta t \cdot \sin\theta$. The production of this momentum is to be assigned to the mechanical forces. Thus, up to the first order of $v/c$, the forces exerted on the reflector are made up of the thrust of the incident beam, the back thrust of the reflected beam, and a tangential traction on the surface in the direction of travel of the radiation and equal to $2\epsilon v/c \cdot \sin\theta$, which may in a general statement be apportioned between the incident and reflected beams. [Cf. p. 443.]

This tangential traction was missed in the previous estimate. It is easy to see in a general way whence it comes, in the case of polarization perpendicular to incidence. The electric density $E\sin\theta/4\pi c^2$, receding with velocity $v$, then constitutes a current which is acted on by the tangential magnetic field $E/c$. Again, the thrusts along the two beams may not balance tangentially. But further scrutiny in detail must be deferred until a happier time.

*Derivation from stress and momentum.*

# 81

# IRREGULARITIES IN THE EARTH'S ROTATION, IN RELATION TO THE OUTSTANDING DISCREPANCIES IN THE ORBITAL MOTION OF THE MOON.

[*Monthly Notices of R. Astron. Soc.* LXXV, Jan. 1915, pp. 211–219.]

1. The results of the application of the law of gravitation to the lunar motion have now been summed up in magisterial manner by Professor E. W. Brown[1], as the culmination of his classical investigation on the Lunar Theory. The circumstance that there remains an outstanding irregularity in the orbital motion, composed roughly of *Outstanding lunar orbital fluctuations:* a fluctuation of 13 seconds of arc on each side of a mean in a half-period of about 140 years, combined with a like fluctuation of about 3 seconds of arc in the shorter half-period of about 35 years, has been felt to create an intolerable discrepancy, which demands every effort of the gravitational astronomer to resolve. No higher tribute than this could be paid to the extreme refinement and exactness of the gravitational explanation of the celestial motions, a department of knowledge which leaves far behind the highest amounts of precision that the presence of intractable disturbing agencies allows to us in other branches of physical science.

The object of this note is to examine such grounds as may exist for *ascribed to Earth's rotation,* attributing all or part of these discrepancies, whose rate of change amounts to the order of 15 to 20 seconds of arc per century of time, to irregular changes in the Earth's velocity of rotation on its axis, due to rearrangement of surface masses. For such changes would introduce fluctuating uncertainties into the fundamental reckoning of time* that would be, in astronomical procedure, thrown on to the celestial motions which that time is employed to specify. This suggestion towards a possible explanation can claim classical authority *by Kelvin and Newcomb.* of long standing. In his address to the Mathematical and Physical Section of the British Association at Glasgow in 1876[2], Lord Kelvin referred, towards the end, to these fluctuations of the lunar motion, of which he had just received private information from Newcomb;

[1] *British Association Report*, Australia, Sept. 1914, "Address on Cosmical Physics."

* The aim of Professor de Sitter's recent formulation, as *infra*, p. 480, is towards a practical fixation of pure Newtonian time.

[2] Reprinted in *Popular Lectures and Addresses*, Vol. II; see p. 271. This explanation is developed numerically in 1888 in a later paper on Polar Ice-caps, *loc. cit.* p. 319; while the effect of tidal friction in slowing the Earth's rotation was estimated in 1868, in a memoir on Geological Time, *loc. cit.* p. 10.

and he agreed with Newcomb in ascribing them to the cause here described. The subject of the diminution of the Earth's axial motion by tidal friction had occupied his attention before 1867[1].

The actual astronomical evidence that has been brought forward is of two types. First, the times of day are recorded at which some ancient eclipses occurred: this will be affected the same way whether there be a retardation of the Earth's motion of rotation, or an equal acceleration of the Moon's motion, causing it to gain faster on the Sun. Second, observations during modern times have revealed the irregularities, above stated, in the Moon's position among the fixed stars, which prove to be the same whether that position is determined directly by occultations, or indirectly by meridian transits with reference to a clock which is itself rated by transits of fixed stars; in either case, if the Earth has been rotating too slow, the actual position of the Moon will be referred to a time behind the actual absolute time, involving apparent acceleration of its motion. *[side note: Evidence from ancient eclipses: from modern occultations of stars:]*

Such an irregularity in the practical measure of time, viz. the Earth's rotation, as is here contemplated, would be a universal cause, introducing similar discrepancies into all periodic phenomena in the solar system*. Although there is no other feature sensitive enough to form a definite test of so small an effect, there are, it appears, two in which the fluctuations might just show, namely, the motions in longitude of the Earth and Mercury. The very interesting comparative diagrams printed near the end of Professor Brown's recent address (*loc. cit.*) possibly show some trace of correlation of this kind: at any rate, we may take it that they contain all that can be made out in this regard. *[side note: other astronomical tests.]*

It is proposed here to estimate what possibilities exist in this field. It will appear that there ought to be a tidal secular apparent acceleration of the Moon's motion, which may conceivably be of about the order usually thought to be outstanding, and that the fluctuating irregularities which remain unexplained are of an order not inconsistent with their being due, in substantial part, to terrestrial movement of material. *[side note: A general tidal influence not inadequate.]*

[1] Cf. Thomson and Tait, *Nat. Phil.*, ed. 1, 1867, § 276; also § 830, where an estimate of the ratio of this effect to its rotational reaction on the body producing the tides is quoted from a private communication made by J. C. Adams, which will be referred to and confirmed later.

* There have been recent important investigations of this subject. In *Yale Observatory Transactions* 3, VI, 1926, pp. 209–235, E. W. Brown published a close discussion of the deviations of the Sun and Moon from their gravitational orbits from the point of view of changes in the Earth's rotation, with positive results. In *Monthly Notices R.A.S.* Nov. 1926, pp. 4–31, H. Spencer Jones has compared the irregularities in longitude of the various planets; he considers that they show sufficient agreement with that of the Moon to justify ascribing them, all except the slow fluctuations, to irregularities in the Earth's rotation. Cf. also de Sitter, as *infra*, p. 480. *[side note: Recent discussions.]*

2. Irregular changes in the Earth's velocity of rotation are to be ascribed to transfers of terrestrial material, of accidental, for example meteorological, character. Merely local transfer, for example disturbance due to an earthquake, will be practically ineffective. To alter sensibly the moment of inertia of the Earth the transfer must be from places near the axis of rotation to places remote from it, or conversely. Such change can occur through displacement of the oceanic surface waters. But nothing except local movements can be expected inside the solid surface of the Earth, either from earthquakes or other causes: the unchangeableness of the solid Earth, as a whole, is one of the remarkable results of the astronomical data.

*Local terrestrial displacements inoperative:*

*remarkably:*

For example, the necessary conditions are satisfied by water melting off from an elevated antarctic ice-cap and spreading uniformly over the whole ocean surface. This case was first discussed by Dr J. Croll in connection with the origin of changes of sea-level and of climate: and it has been considered at length under the present aspect by Lord Kelvin in 1888[1].

*not so oceanic, by melting of arctic ice,*

The necessary conditions are also satisfied by a sudden or gradual local rise or subsidence of an ocean floor; for this must cause compensating rearrangement of water over the whole ocean. According to the late Dr J. Milne, the effect of an earthquake may be a displacement over a large area, vertically or horizontally, through as much as 10 feet. Such a fall of the sea bottom, 10 feet, say over 400 miles square, if it is balanced by the rise of a land surface and not by an adjacent rise of the ocean floor, would lower the level all over the ocean by about $\frac{1}{4}$ of an inch, supposing the ocean to cover $\frac{2}{3}$ of the Earth's surface. If it occur in latitude $\lambda$, and the mass of water thus displaced is $M$, it will alter its angular momentum from $\omega M a^2 \cos^2 \lambda$ to $\frac{2}{3}\omega M a^2$. Inserting the figures of our illustration, this event would diminish the period of the Earth's rotation, as the total angular momentum must remain constant, by the fraction $8 \cdot 10^{-10} \left(\frac{2}{3} - \cos^2 \lambda\right)$ of itself, taking its mean density to be $5\frac{1}{2}$; this amounts to $25 \cdot 10^{-3} \left(\frac{2}{3} - \cos^2 \lambda\right)$ seconds of time in a year, or at the rate of about $2\frac{1}{2} \left(\frac{2}{3} - \cos^2 \lambda\right)$ seconds in a century. If all this discrepancy in the time standard were thrown on to the Moon's motion, an apparent disturbance at a constant rate of 20″ of arc per century in the Moon's position, which is equivalent to 37 seconds of time, would require a rise or fall of the ocean floor of 420 feet over 400 miles square, if it is near the equator,

*or by change in ocean floor:*

*possible effect estimated.*

*Estimate as a lunar effect:*

[1] *Loc. cit. ante.* [But the South Polar region is now known to be a high table-land.

The efficacy of the Croll-Kelvin cause has been resumed by W. D. Lambert, *Proc. Washington Acad.* Mar. 1927, with results of the same order as here. Especially he strongly suggests correlation between the lunar irregularity and the wandering of the terrestrial Pole.]

or half that amount if it is in the polar regions[1]. The general ocean level would thereby be changed by 6 inches, or by half that amount, in the two cases respectively[2]. This, or a proportionally smaller rise or fall of the ocean bed over a larger area, thus seems too large an amount to postulate in the course of a century; though tracts in the Pacific near the western coast of South America have been known to drop by 1000 feet[3]. But the present type of explanation of the outstanding lunar irregularities seems to be at any rate partially adequate. *and change of sea-level.*

Lord Kelvin verified, at an early stage of his consideration of these subjects, that the Earth's gradual shrinkage, owing to its known rate of loss of heat, is far too small to produce any appreciable effect of this kind. It may readily be verified that the slight wanderings of the Pole cannot be sensibly effective. The most remarkable physical alteration that is going on in the Earth is perhaps the rapid change of its magnetic moment: but the change of dimensions thus brought about, owing to magnetic change of form, is also far too small to be sensible in this way. *Thermal shrinkage not effective: nor change of latitudes nor change of magnetic moment, directly.*

3*. The lunar tidal drag as the Earth may be estimated roughly as the torque exercised by the Moon on the tidal spheroidal elevation. Assuming at first the ocean to cover the whole Earth, this is a sheet of water, of surface density $\sigma$ on the terrestrial sphere given roughly by $\sigma = h \cos^2 \theta$ or more conveniently by $h (\cos^2 \theta - \frac{1}{3})$, that is $\frac{2}{3}hP_2$, where $\theta$ is measured from the axis of the spheroid, which lags behind the moon by an angle $\theta_0$. The mutual energy $W$ of this sheet and the moon is $MV$, where $V$ the potential of the sheet at the Moon is given outside and inside it by *Lunar torque on equilibrium tide.*

$$V_{\text{out}} = A \frac{a^3}{r^3} P_2, \quad V_{\text{in}} = A \frac{r^2}{a^2} P_2,$$

so that

$$4\pi\gamma\sigma = 5AP_2.$$

---

[1] Cf. in E. H. Hills and J. Larmor, "The Irregular Movement of the Earth's Axis of Rotation," *Monthly Notices R.A.S.* Nov. 1906, p. 26, [*supra*, p. 320] there too it is shown that a displacement of the order in the text must be distributed in longitude if it existed, for if localized it would affect latitudes by 2″, which is far too much.

[2] The quicker fluctuation by itself would even demand a rise and fall of this order about every seventy years, which is rather beyond probability.

[3] Cf. *Proc. Roy. Soc.* 1908 [*supra*, p. 385], "On the Relation between the Earth's Free Precessional Rotation and the Resistance to Tidal Deformation." (The large drop above mentioned is, I am informed, now considered doubtful.)

* This section has been rewritten in more convenient form, and with result doubled, Oct. 1926.

The mutual torque is therefore

$$G = -M\frac{\partial V}{\partial \theta} = M\tfrac{6}{5}\pi\,\frac{a^4}{R^3}\,\gamma h\sin 2\theta_0$$

$$= M\tfrac{6}{5}\,\frac{g}{\rho}\,\frac{a^3}{R^3}\,h\sin 2\theta_0,$$

as $E/M = 82$, $\gamma E/a^2 = g$, $R/a = 60$, while the Earth's mean density $\rho$ is $\tfrac{11}{2}$.

Thus as the Earth's retardation $-d\omega/dt$ is given by

$$-\tfrac{1}{3}a^2 E\frac{d\omega}{dt} = G,$$

we have

$$-\frac{d\omega}{dt} = \tfrac{2}{3}\frac{M}{E}\frac{gah}{R^3}\sin 2\theta_0 = \tfrac{2}{3}\tfrac{1}{82}\frac{1}{60^3}\frac{gh}{a^2}\sin 2\theta_0.$$

In foot-sec. units this negative angular acceleration makes the Earth fall back in a century ($\tau$) by an angle $\tfrac{1}{2}\tau^2 d\omega/dt$, which works out, when changed into seconds of time by the factor 13700, to be $180h\sin 2\theta_0$ seconds in a century. This rough estimate compares apparently with Lord Kelvin's $137h\sin 2\theta_0$.[1]

*(margin: Kelvin's estimate.)*

If the attraction of the tidal water could be neglected, the value of $H$ between high and low water for the ideal lunar equilibrium tide would be given by $gH = \dfrac{Ma^2}{R^3}$, by balancing the difference of terrestrial potential at the pole and equator of the lunar tidal spheroid against that of the system Moon and anti-Moon; this would make $H$, which is $\tfrac{3}{5}h$, equal to about $\tfrac{7}{8}$ feet. The augmentation due to self-attraction of the tide is small.[2]

4. The amount of its lag $\theta$ depends on the retarding friction that is in operation. For the actual tide, to which, however, this equilibrium theory is not applicable, except most roughly, some light on its value is thrown by a remark of Sir George Airy, which has been elaborated by Lord Kelvin, namely, that spring tides occur more than 12 hours after the conjunction of Sun and Moon.[3]

*(margin: Airy's clue to general tidal lag)*

If we might assume that the tidal displacement is that of a funda-

---

[1] Rede Lecture, Cambridge, May 23, 1866, "On the Dissipation of Energy"; *Popular Lectures*, Vol. II, p. 65, where the calculation is made differently for a special numerical example. The value given, for a viscous liquid spheroid, by Sir G. H. Darwin (Thomson and Tait, ed. II, Appendix G, *a*) is difficult to compare.

[2] *Loc. cit., Popular Lectures*, Vol. II, p. 26.

[3] Cf. Thomson and Tait, §§ 804, 817.

mental free oscillation of the ocean, its coordinate $\phi$ would be given worked out. by an equation of the form

$$\frac{d^2\phi}{dt^2} + \kappa \frac{d\phi}{dt} + n^2\phi = F \cos pt,$$

where, when the friction is not prominent and $\kappa$ is therefore small, the period of the transient free oscillation is $2\pi/n$ and the period of the forced permanent tide is $2\pi/p$. This tide is expressed by the particular solution

$$\phi = \frac{F \cos (pt + a)}{\{(n^2 - p^2)^2 + \kappa^2 p^2\}^{\frac{1}{2}}},$$

where

$$\tan a = \frac{\kappa p}{n^2 - p^2}.$$

Thus, neglecting $\kappa^2$, the tide is *reversed*, so that low water is under the Moon, and the phase is *retarded*, if $n < p$, *i.e.* if the period of free oscillation exceeds the tidal period, as is actually the case; and the lunar tide, being nearer the free period, would be magnified by the sympathy of periods[1]. As $\kappa$ is the same for both solar and lunar periods, the lag, which is $a/p$ in time, is inversely as $n^2 - p^2$, and increases as the periods approach, up to a limit near which this reasoning becomes inapplicable.

Observation, as above quoted, shows that the difference of the lags for the lunar and solar tides exceeds $6°$. If we thus took $10°$ as the lag for one of them, and took $h$ to be $\frac{7}{8}$ of a foot [instead of $\frac{2}{3}$ of this now as above], the lag in time due to tidal friction in a century, as reckoned above, would amount to 36 seconds, of which $\frac{1}{2}$ (see § 5), giving $10''$ of arc, would appear as advance in the Moon's position. This, as it happens, is rather more than the value which has been assigned to the unexplained acceleration of the Moon's motion on the basis of Airy's interpretation of ancient eclipses[2]. Approximate coincidence would be an accident: but the order of magnitude is evidence that, so far from tidal friction not being an effective *vera causa* for an apparent secular acceleration of the motion of the Moon, it seems to be competent to account for an effect of the same order as the actual amount which has been assumed to be outstanding*.

---

[1] The other case, tide direct and phase accelerated, of course also places full tide after the Moon has passed the meridian, and so gives also a torque in retarding direction on the Earth.

[2] Cf. Thomson and Tait, part II, § 830, where the estimate is $6''$, plus $1\frac{1}{2}''$ on account of thermodynamic atmospheric acceleration. (But the $10''$ in the text may have to be reduced to $7''$: see next footnote.)

* See *infra*, p. 498, on estimates from turbulent degradation of energy in shallow seas.

The problem
solvable from
tidal records.
If we knew from observation the actual distribution of the tide over the Earth, the actual torque could be calculated with safety by quadratures, and a definite result obtained. Perhaps even with modern tidal knowledge this would not be feasible: but it ought not to be difficult from general inspection of the co-tidal lines on the charts to recognize in a general way whether the retarding torque on the Earth, and consequent permanent lunar apparent acceleration, is actually much smaller than the value (itself barely sufficient)[1] here deduced from an equilibrium theory[2].

5. In this analysis of tidal friction we have tried to estimate the retarding influence on the Earth. In addition to this influence in causing lunar apparent orbital acceleration by delaying the time, Reaction of
tide on actual
lunar motion there is still the reaction on the true motion of the Moon to be considered. For simplicity in estimating its order of magnitude, imagine an Earth-Moon system isolated, the two bodies rotating about parallel axes so that the equator and ecliptic coincide. The angular momentum of the system will be conserved, as there is no dynamical interference from outside, though there may be thermal. Thus if $a$, $E$, $\omega$ are the radius, mass, and angular velocity of the Earth on its axis, $a'$, $M$, $\omega'$ the corresponding quantities for the Moon, $\Omega$ its orbital angular velocity, $R$ its orbital radius with respect to the centre of mass of the Earth and Moon, or say with respect to the centre of the Earth, determined by
conservation
of total
angular
momentum: as we can imagine that point to be fixed and take moments round it, without excessive error, then

$$Ek^2\omega + Mk'^2\omega' + MR^2\Omega = \text{constant},$$

where, assuming the Laplacian law of internal density, $k^2 = \frac{1}{3}a^2$. In this equation of conservation, the coefficients of $\omega$, $\omega'$, and $\Omega$ are roughly as $1$, $\frac{1}{1000}$, $140$. As $\omega'$ is equal to $\Omega$ we see that, in a rough estimate of exchanges of angular momentum, the second term will not sensibly influence matters. Hence simultaneous changes in $\omega$ and $\Omega$ are about as $-140$ to $1$ absolutely, or relative to the lengths of the month and day they are as about $-5$ to $1$.

This estimate is, however, on the supposition that the moments of inertia remain constant, which requires that $R$ be kept constant by constraint. Actually, diminished velocity of the Moon must increase

[1] Too small by $\frac{1}{3}$ when the continents are allowed for.

[2] Professor Eddington points out to me that the tide in the solid Earth ought also to come in. If, as seems very likely, that tide is almost purely Earth not
viscous. elastic, it cannot lag in phase and so can contribute nothing to the retardation. This solid tide is $\frac{1}{3}$ of the total ideal equilibrium tide (cf. *Proc. Roy. Soc.* Dec. 1908, p. 94, as *supra*, p. 385): thus $\frac{1}{3}$ of the moment estimated in the text must be deducted from the result of the quadratures, which is a very substantial reduction.

$R$, in order that the centrifugal reaction may continue to balance the gravitational attraction, viz. in order to maintain the relation

$$\Omega^2 R = (E + M)/R^2.$$

Thus $\Omega^2 R^3$ remains constant: therefore

$$\delta\,(MR^2\Omega) \propto \delta\Omega^{-\frac{1}{3}}, \text{ and so } = -\tfrac{1}{3}MR^2\delta\Omega,$$

so that the effective relative coefficient of $\Omega$ is not $-140$, but $\frac{1}{3}$ of that number with sign changed, viz. $+47$; and the proportional effects on the month and the day are about as 5 to 3, both in the direction of increase. *amended.*

To estimate how the frictional loss of energy is divided between them we note that $\delta\,(\tfrac{1}{2}MR^2\Omega^2) \propto \delta\Omega^{\frac{2}{3}}$, and so $= \tfrac{2}{3}\tfrac{1}{2}MR^2\Omega\delta\Omega$. Thus the ratio of the changes of energy, both losses, is that of the changes *Loss of energy* of angular momentum, which is unity, multiplied by the ratio of $\omega$ to *falls mainly* $2\Omega$, so that the Earth's rotation sustains $\frac{14}{15}$ of the whole loss of energy. *on Earth's rotation.*

But hitherto we have been considering only mutual action between Earth and Moon, and have neglected to take in the change of $\omega$ due *Solar in-* to the solar tide. The solar tide-generating force is roughly $\frac{2}{5}$ of the *fluence added.* lunar, hence the solar torque is roughly $(\frac{2}{5})^2$ of the lunar. Thus in all, we have instead of 47 a ratio $(1 + \frac{4}{25})$ 47, or 54. Hence an accumulated lag, due to tides, of $1''$ of arc on the Moon's true position among the stars corresponds to a deficiency of $54''$ of arc or $3\frac{3}{5}$ seconds of *Final result:* time in the Earth's rotational position[1], so that it has gone slow by this number of seconds, making the Moon appear to be nearly $2''$ in advance. The total apparent advance of the Moon's position, being the difference of these, is thus half of that due to lag of the terrestrial time alone[2].

The tidal result agrees with that communicated by J. C. Adams for *same as* the first edition of Thomson and Tait (1867, § 830). His argument, *Adams',* which doubtless followed these lines, does not seem to have ever been published. In Thomson and Tait, ed. II (Appendix G, *a*), Sir George Darwin, from the special numerical data for the problem of viscous *verified for* liquid spheroids calculated in his paper, *Phil. Trans.* 1879, verified *special case* that Adams' ratio held good. *by Darwin.*

In 1874 the same method of momentum was employed by J. Purser *J. Purser's* (*Brit. Assoc. Report*), much as above, to answer Airy's unsolved query *investigation.* as to how the loss of energy is divided between the Earth and Moon; and in 1879 the method gave, by graphical treatment in Sir George *Darwin's* Darwin's hands, his famous analysis of the past history of the Earth- *lunar* Moon system (see Thomson and Tait, ed. II, Appendix G, *b*). *evolution.*

[1] The neglected obliquity of the ecliptic will diminish this estimate by something like $\frac{1}{18}$ of itself, an amount in other ways uncertain.

[2] The necessity for this factor was pointed out to me by Professor Eddington.

*Added Jan.* 28, 1915. The maximum estimate in § 4 *supra* for the effect of tidal friction is an apparent *advance* of 5″ of arc in a century of the Moon's motion, *i.e.* an *acceleration* of twice that amount. But this value must be reduced by about one-third (see footnote), on account of the tide being in part an elastic Earth-tide; this correction, as it happens, brings it to half the amount usually assumed as outstanding. There would also be involved necessarily in it an apparent advance of the Sun's motion among the stars, amounting to $\frac{1}{13}$ of the Moon's 7″ in a century, such as Professor Brown has tried to find traces of*.

*Estimates lowered by the elastic solid tides.*

Professor Eddington has drawn my attention to a remarkable synthesis of these matters, avowedly tentative and apparently independent of the discussions of previous writers, by Dr Cowell (*Monthly Notices*, vol. LXVI (1906), p. 354), and has assisted me to understand the exposition, which I am not alone in having found difficult. My paraphrase and criticism of the argument, which, if correct, may be of use to others who do not think astronomically, is as follows.

From his discussion of ancient eclipses, demurred to, however, by Newcomb, Dr Cowell feels compelled to ascribe an outstanding apparent advance of 5″ in a century to the Moon's motion and of 4″ in a century to the Sun's motion among the stars †, or accelerations of 10″ and 8″ respectively. If these effects are both ascribed to slowing of the Earth's rotation due to tidal action, the latter figure will give its amount, for there is no other source of solar acceleration[1]; there would have to be a delay of $365 \times 4″$ of arc in the Earth's rotation ‡, so that it would go slow by about 100 seconds of time in a century. This is more than four times the extreme higher limit found for possible tidal effect in § 4 *supra*, as corrected in this note for elastic solid tide; and, as Dr Cowell remarks, it is greatly in excess of current estimates. As he also points out, an apparent sidereal advance of the solar motion, of 4″ of arc in a century, would be too great for the *Nautical Almanac* to ignore, though the value $2 \times \frac{7}{13}″$, or say 1″, *supra*, might perhaps not be so.

*Cowell's much increased value.*

If, however, Dr Cowell's value for the slowing of the Earth's rotation is for the moment adopted, the conservation of angular momentum in the Earth-Moon system will determine the concomitant effect on the Moon's motion. Dr Cowell refines on previous estimates, such as the one discussed *supra*, § 5, by including the eccentricity *e*

* These numbers, if the elastic tide be left out, are half, as it happens [for *h* was $\frac{2}{3}$ too big], of Dr Fotheringham's estimates from ancient eclipses, *infra*.

† Now confirmed as *infra*, p. 479, but apparently with 40 seconds of time per century in place of 100 for the Sun, and 9 for the Moon.

[1] Tidal influence is entirely negligible.

‡ Viz. 4″ of arc as the Sun's displacement *among the stars* involves for the same time $365 \times 4″$ as the Earth's displacement *around its axis*, either of which is reduced to the unambiguous specification, that of time elapsed, with the result above.

óf the lunar orbit. The equation of angular momentum, neglecting as before that of the Moon's rotation about its axis, now takes the form, inclination being neglected for the present rough purposes, $R$ being now mean distance so that $n^2R^3$ is constant, viz. $\gamma\,(E + M)$,

$$Ek^2N + MR^2n\,(\mathrm{I} - e^2)^{\frac{1}{2}} = \text{constant,}$$

Applied in argument from angular momentum,

where $N$, $n$ are the mean motions of terrestrial rotation and lunar orbit (represented *supra* by $\omega$, $\Omega$), so that for example the Moon's longitude is $\int n\,dt + \ldots$[1]. If we take for $k^2$ the Laplacian value $\frac{1}{3}a^2$ instead of Dr Cowell's rougher $\frac{2}{5}a^2$, the equation of variation comes out

$$\frac{\delta n}{n} - \tfrac{23}{36}\frac{\delta N}{N} - \frac{3e}{\mathrm{I} - e^2}\,\delta e = 0,$$

the second coefficient being not far from $\frac{2}{3}$, as before. For the Earth's rotation $\delta N/N$ is $-\,6 \cdot 10^{-8}$, as required by the assumed solar acceleration. If the term in $\delta e$ is neglected this would give for $\delta n/n$ the value $-\,4 \cdot 10^{-8}$, and for the Moon's advance of tidal origin (viz. a real part due to tidal dynamical reaction and an apparent part owing to the lag of astronomical time behind absolute dynamical time) one-half of

$$\delta n - \frac{n}{N}\,\delta N,$$ which is $\mathrm{I} \cdot 10^{-8}n$ or $16''$ of arc in a century, instead of

including effect of orbital change,

$3 \cdot 10^{-9}n$ or $5''$ of arc. This excessive amount Dr Cowell proposes to reduce towards the latter figure, which his view requires, by aid of the term involving $\delta e$: complete identity would demand, with the Laplacian value of $k^2$, for the change of $e$ per century the value $60 \cdot 10^{-9}$ in place of Dr Cowell's $54 \cdot 10^{-9}$ [, amounts totally unrecognizable otherwise].

Apparently the legitimacy of this course is amenable to test. We have to compare the magnitudes of the disturbances $\delta n/n$ and $e\delta e$, terms in the equation of variation, which arise in the lunar orbit from the retarding attraction of the tide. As the orbit is not far from circular, we may employ the known equations[2] giving the effect of a tangential retarding force $T$, viz. $r$ being focal distance and $a$ mean distance in an elliptic orbit,

$$\frac{\mathrm{I}}{a}\frac{da}{dt} = -\,\tfrac{2}{3}\frac{\mathrm{I}}{n}\frac{dn}{dt} = -\,\frac{2a}{\mu}\,Tv,$$

$$\frac{\mathrm{I}}{e}\frac{de}{dt} = -\,2\,\frac{\mathrm{I} - e^2}{e^2}\left(\frac{a}{r} - \mathrm{I}\right)\frac{T}{v},$$

where

$$v^2 = \mu\left(\frac{2}{r} - \frac{\mathrm{I}}{a}\right).$$

---

[1] That is, $\int\left(n_0 + \dfrac{dn}{dt}t + \ldots\right)dt + \ldots$, so that the *advance* in unit time due to a constant part $f$ in the *acceleration dn/dt* is $\frac{1}{2}f$. The astronomical usage, which Dr Cowell follows, is to give to this advance the name secular acceleration.

[2] Adapted from Cheyne, *Planetary Theory*, §§ 120–1, as the source nearest at hand.

It appears from them that $e\delta e$ is to $\delta n/n$ in the ratio

$$\tfrac{2}{3}\frac{n^2 a^2}{v^2}(1-e^2)\left(\frac{a}{r}-1\right), \text{ or roughly } \tfrac{2}{3}\left(\frac{a}{r}-1\right).$$

**small** The time average of $a/r - 1$ round an elliptic orbit works out to be
**as regards** zero. Thus for an orbit not far from circular, tidal change in eccen-
**eccentricity,** tricity seems to be so small as not to affect sensibly at all the orbital
acceleration*, and so is unavailable to effect a reconciliation†.

[This conclusion is now modified in the footnote. Thus, after Cowell
followed by de Sitter, there is no presumption that the tidal pull on
**but adequate.** the Moon is not partly used up in infinitesimal change of form of the
orbit without affecting the mean motion.]

A more recent revision of ancient eclipses by Fotheringham
(*Monthly Notices*, 1919) appears to lead to a lunar half-acceleration
10·5″ per century, along with an apparent Solar one 1·0″.

With the latter value $\delta N/N$ is $-\tfrac{2}{3}10^{-8}$; so $\delta n/n$ is $10^{-8}$: the Moon's
apparent advance would be 4″ in a century comparing with the previous
5″. Thus of Dr Fotheringham's value $10\tfrac{1}{2}$″ a part $6\tfrac{1}{2}$″ would arise from
$\delta e$, giving for $3e\delta e/(1-e^2)$ the value $\tfrac{13}{8}10^{-8}$. This value would conserve
the angular momentum of the Earth-Moon system: but its ratio to
$\delta n/n$ is [not] far too great for the dynamical estimate above given.

POSTSCRIPT (1928).

*The Standard of Newtonian Astronomical Time.*

The problems of astronomical geology, and of the Earth as a time-
keeper, have exercised an attraction, which was much stimulated by
Lord Kelvin‡, following on Laplace, J. C. Adams, and James Croll,

**Correction.** * This forgets the extreme smallness of the value of $\delta e$ that is required, of the
order $10^{-8}$. Pushed to the second order the ratio $e\delta e$ to $\delta a/a$ seems to come out
about $-\tfrac{1}{2}e^2$: and $n^2 = \mu/a^3$ so that $\delta n/n = -\tfrac{3}{2}\delta a/a$ not $+\delta a/a$: this indicates
a value of $\delta e$ of the right sign, and too small only by a factor of order 10.

† The change of energy of the system is expressed by

$$\delta \epsilon = \tfrac{22}{33}E k^2 N (\delta n - \tfrac{222}{333}\delta N).$$

If the eccentricity does not change the conservation of momentum requires

$$\frac{\delta n}{\delta N} = \tfrac{22}{33}\frac{n}{N} = \tfrac{1}{44},$$

therefore a loss of energy of the system, which must arise from frictional causes,
falls almost entirely on the Earth, the Moon gaining by about $\tfrac{1}{30}$ of the
Earth's loss.

If there were no loss of energy by friction

$$n = \tfrac{2}{3}\delta N, \qquad \frac{3e}{1-e^2}\delta e = \left(\frac{2}{3n}-\frac{2}{3N}\right)\delta N.$$

Thus the two mean motions – $n$ and $N$ would either both increase or both
diminish, but $e$ would have to change strongly in the opposite sense. On a
uniform ocean there would be no lag or prime, when friction is neglected, thus no
change of $e$ would be required to balance the energy account.

‡ *Math. and Phys. Papers*, Vol. III.

and followed by Newcomb and G. H. Darwin, more especially by the publication of Thomson and Tait's *Natural Philosophy* in 1867. Prompted by Kelvin*, we are permitted to imagine that if the Earth had been solid and rigid, with no oceanic tides to lag, its actual rotation might have been accumulated, in quite a brief time cosmically, by the thermodynamic action of the Sun on its atmosphere alone†. Also why should not a tide on an elastic Earth lag or *prime* to some extent: though Laplace showed that on a uniform ocean it would not? Several papers in this volume resulted from the interest in these subjects thus aroused. Very recently the subject has had the advantage of a brief critical pronouncement from the practical side by a high authority on dynamical astronomy‡, from which the *data* and opinions that follow are abstracted.

*[margin: A thermal cause of planetary rotations.]*

*[margin: Is friction essential to lag?]*

It appears that the Hanoverian astronomer Tobias Mayer, who worked on the lunar tables for the British Admiralty, had suggested tidal retardation in the middle of the eighteenth century‖, but it was only after Laplace's reconciliation of the Moon's mean motion had been shown to be only partial by Adams in 1853, that the subject claimed modern attention. The detection by P. H. Cowell (1906) of a secular acceleration of the Sun's motion has now been confirmed by Mercury, being about 4·4 times as much in time as the residual unexplained acceleration of the Moon. It was E. W. Brown (1914) who first advanced strong evidence that the irregular long-period fluctuations in the lunar motion are repeated with a factor about $\frac{4}{5}$ in those of the Sun and planets. The inner satellites of Jupiter are, it appears, for some reason not available as reliable time-keepers. If the Sun and planets are made the standard of time, as Professor de Sitter finally proposes, the Earth's rotation and the Moon's motion may be regarded as uniformly retarded in time in the ratio of 1 to 0·77,

*[margin: Historical.]*

---

* *Loc. cit.*, or *Proc. R. S. Edin.* 1881. Application to double stars may not be negligible. But the almost inconceivable vastness of the changes going on, in brief periods, in giant variable (and double) stars is illustrated by the recent evidence of pulsations of visual radius, even up to half its value, in Betelgeuse: cf. Spencer Jones' discussion in *M.N.R. Astron. Soc.* June 1928.

*[margin: Vastness of stellar phenomena.]*

† As the Earth rotates underneath the Sun, the expansion by solar heat produces a barometric tide, mainly semidiurnal, whose apex is behind the radius to the Sun, whereas the apex of the semidiurnal tide of attraction of the ocean is in front of it. According to Kelvin's careful estimates as reprinted in *Math. and Phys. Papers*, Vol. III, p. 341, the Earth gains 2·7 seconds of time in a century from this cause, which is as much as one-eighth of Adams' estimate of the nett actual loss, so there would have to be more oceanic tidal drag in this proportion.

*[margin: Atmosphere as a thermal engine.]*

‡ W. de Sitter, "On the Rotation of the Earth and Astronomical Time," Supplement to *Nature*, Jan. 21, 1928, pp. 99–106.

‖ The date of Kant's tract was 1754. Delaunay seems to have revived the idea in 1866, after Adams had made his estimate (referred to *supra*, p. 469), which was revealed in Thomson and Tait, *Nat. Phil.* § 830, in 1867.

but subject to superposed fluctuations which are in the ratio of 1 to 0·2.

The causes that may slow the Earth's rotation are as discussed *supra*, p. 469. Professor de Sitter accepts, after Taylor and Jeffreys, the influence of tidal flow in the narrow seas, usually misnamed

**Tidal drag in shallow seas:** retardation by friction, which being an internal force must be inoperative except indirectly\*: he suggests, in contrast with p. 472 *supra*,

**question as to its origin:** that sufficient reaction on the Moon is to be found "in the mutual attraction of the Moon and the small secondary waves set up from the areas of dissipation as centres": and he sees in this a sufficient reason why the drag may change widely in amount, as his view requires, remarking that it cannot by itself explain more than half the fluctuations, which are not drag but alternation of drag and acceleration.

**already included in argument from momentum.** He finds not of course an additional but an equivalent cause in the tidal dynamical interaction between Earth and Moon, as on p. 477 *supra*, but helped out by assuming, on probable grounds, that the lunar orbit may become eccentric to such degree as may be required. On that view there would be something wrong on the tentative discussion *supra*, p. 477†.

**A graphical interpretation of the secular data:** The most valuable feature is his synthesis of the relevant astronomical evidence, exhibited graphically in curves after the manner of that of E. W. Brown. He decides for rates of acceleration practically uniform for intervals of time separated by a number of cataclysms, the chief one (1918) being of amount corresponding to a uniform shrinkage of the whole Earth by 5 inches in radius, which would perhaps correspond (as on p. 470) to removal of more than a foot of water from the oceans into polar ice. As regards the superposed

---

\* There is a note by H. Hertz (Berlin Phys. Soc. 1883, *Miscellaneous Papers*, pp. 207–210) on the friction required to transfer the Moon's oblique pull on the Ocean to the solid Earth. It appears that for a uniform shell of ocean smooth viscosity could not do it. For bounded oceans the water becomes piled up against their western coasts to the order of a foot, with restoring eastward currents.

† The radial component of the tidal force has there been ignored, which might perhaps vitiate the result. For if there were no change of eccentricity, conservation of momentum along with lunar drag would demand loss of energy: on the other hand it appears to be conceivable that there can be lag in the tide and therefore a drag on the Moon, even if there is no viscosity and so no loss of energy. Change of eccentricity as well as of mean distance can reconcile these positions.

Professor de Sitter, quoting from Darwin and Kelvin, finds a plausible cause in the fluctuating intensity of the tidal drag at spring and neap tides: this seems to be a second-order influence, but its period is near that of the lunar orbit. But the whole effect is of almost infinitesimal order except in its geophysical aspects, and the requisite estimate is now attempted in footnote, p. 478.

fluctuations the conclusion is "that $\frac{3}{4}$ of them are produced by changes of the size and shape of the Earth and $\frac{1}{4}$ by variability of the coefficient of tidal friction: $\frac{1}{2}$ of this $\frac{1}{4}$, or $\frac{1}{8}$ of the whole, is then transferred to the Moon." — *superposed fluctuations*

The supposed terrestrial cataclysms are taken to be wholly internal and out of sight, as are indeed the causes of the Earth's very great magnetic changes. The question how far such a cataclysm would show itself on the free path of the Pole has been discussed *supra*, p. 320. The variation of latitude along any meridian is the projection on it of the angle between the axis of instantaneous rotation of the Earth and the principal axis of inertia. The former is practically fixed in space (cf. p. 1), the latter depends on the Earth's configuration. The result of a cataclysm will suddenly shift this axis of inertia, which is, at any time, the Pole on the Earth round which, relative to the terrestrial astronomers, the axis of rotation revolves*. Did then the mean centre of the convolutions of this latter polar path show any sign of shifting about the year 1918? — *May be tested by path of the Pole on the Earth:*

But actually the amplitude of the fluctuation of latitude is of the order of 0·2" of arc. The displacement of the axis of rotation in the Earth by this amount would, by diminution of the inertial moment, prolong the length of the day by an amount some thousands of times smaller than what is observationally in evidence. Thus for a cataclysm to be at all effective, it would have to be great enough to displace the Pole by very many times the total observed amount, unless the off-chance of symmetry around the axis is fulfilled. — *result prohibitive.*

Finally, the interesting judgment may be quoted that it looks as if the Greenwich trials have already shown that the clocks with free pendulum, recently elaborated by Mr Shortt with counsel from Professor Sampson, could with further experience "be depended on to keep time within a few hundredths of a second for a period measured in years instead of weeks," thus giving an unexpectedly favourable answer to Lord Kelvin's proposal, perhaps half serious, of half a century ago, that the Earth as a time-keeper ought to be checked by a clock. — *A clock to correct the Earth.*

* The angle between the Earth's axis of figure and axis of rotation is involved in determination of latitude because it enters with different signs into the two meridian observations of a circumpolar star: the axis of rotation is almost coincident with the fixed direction of the angular momentum, while the axis of figure, meaning here the Polar axis of inertia which is the one attached to the Earth's configuration, revolves round it in disturbed precessional manner.

# 82

# THE INFLUENCE OF LOCAL ATMOSPHERIC COOLING ON ASTRONOMICAL REFRACTION.

[*Monthly Notices of the Roy. Astron. Soc.* Vol. LXXV,
pp. 205–210, Jan. 1915.]

*Apparent symmetrical variation of latitude.* 1. The following considerations, which can hardly be altogether new, were prompted by reading the note on latitude variation in the Annual Report for 1913 of the R.A.S.[1] A question is there broached, whether the residual Kimura annual variation in latitude, symmetrical around the Earth's axis and amounting in amplitude to about $0''\cdot06$, is of such magnitude as can reasonably be referred to changes of refraction. As the term attains its extreme values at midsummer and midwinter, vanishing at the intermediate equinoxes, such an inquiry is encouraged.

In this connection we need not trouble about the curvature of the Earth. We can take the atmosphere to be a flat stratified plate of air; and then, as is well known, the determination of the refraction becomes very simple and direct. For if we imagine a thin horizontal layer of vacuum, interrupting the atmosphere just above the level of the observer, a ray will emerge from above into this crevice along a path parallel to its original direction, because it has passed across a flat stratified plate: the ray will thus be incident in its original direction on the lower horizontal surface of the crevice, to be there refracted into its actual direction as presented to the observer, the crevice being, by the principles of wave-transmission, without effect on the final direction of propagation. If then $z$ is the true zenith distance of a star, $z_0$ the apparent, and $\mu$ the refractive index at the level of the observer, the refraction $r$, equal to $z - z_0$, is given by the law of sines

$$\sin z = \mu \sin (z - r),$$

*The law of astronomical refraction.* leading, when $r$ is small, to the usual formula applicable to large altitudes,

$$r = (\mu - 1) \tan z.$$

For smaller altitudes a correction for terrestrial curvature must be added.

*Meteorological correction.* The value of $\mu$ depends on the density of the air at the place, as determined from its pressure and temperature; for changes in each

---

[1] *Monthly Notices*, Vol. LXXIV, p. 370. [See, however, *infra*, p. 488.]

of these data a correction is in practice duly made. It is the density at the position of the object-glass that here enters, the presumption being that the metallic frame of the observing telescope ensures, by its conducting quality, constancy of temperature, for it must ensure absence of flexure from thermal causes throughout its structure.

But such considerations apply also to the dome in which the telescope is enclosed; being itself at a temperature more or less uniform, it will disturb the horizontal stratification of the air above it. It is true that in usual circumstances the temperature in the free air falls only about $1°$ C. in a rise of 300 feet. But the phenomenon of the deposition of dew involves much steeper fall close to the wetted surface. If this surface is dome-shaped instead of flat, the effect is doubtless much mitigated by the sliding of the cooled air down the sloping surface; yet local difference of temperature is still maintained by radiation into open space.

Disturbance of refraction by the dome:

We are thus prompted to examine the effect on the general refraction that is produced by a local elevation of the strata of equal density, elsewhere flat, in passing over the dome, whether such lifting arise from a sharp temperature gradient through a few feet over the dome, or from a more gradual one extending through a greater height.

strata curved over it.

2. To attain this end, we can establish our imagined horizontal crevice in the atmosphere just above this region of disturbance, say, at a height $h$ from the object-glass, and trace out what happens to the ray in its path beneath that level. The value of the refraction at the level of the crevice is obtained by subtracting from the usual refraction $(\mu - 1) \tan z + \ldots$, calculated for the position of the object-glass, an amount $-\dfrac{\partial \mu}{\partial h} h \tan z$; viz. we have to add $\dfrac{\partial \mu}{\partial h} h \tan z$, or say $(\mu_2 - \mu_1) \tan z$. We have then further to add the actual additional refraction due to the ray crossing the curved strata of equal index down to the telescope, as indicated in the diagram. To obtain its value we compare the ray with a parallel ray at very small distance $a$ from it, say below it. As $\Sigma \mu \delta s$ is the same for all rays between the same two wave-fronts, the difference of its values for these two rays, when divided by $a$, is equal to the angle the wave-front has been slewed round, and is therefore the additional refraction in this part of the path.

At any point on the ray, let $\theta$ be the slope of the stratum of constant index: then as $\delta n$, along the normal to this stratum, is in the direction of the gradient of $\mu$, we have, $\delta s$ being an element of the ray, as before, measured upwards let us say to agree with $\delta h$, and $\delta s'$ at right angles to it,

$$\frac{\partial \mu}{\partial s} = \frac{\partial \mu}{\partial n} \cos (z - \theta), \qquad \frac{\partial \mu}{\partial s'} = \frac{\partial \mu}{\partial n} \cos (z - \theta + \tfrac{1}{2}\pi).$$

The value of $\mu$ at the nearest point on the parallel ray being $\mu + \dfrac{\partial \mu}{\partial s'}\, a$,

is thus $\mu - \dfrac{\partial \mu}{\partial s}\, a \tan (z - \theta)$.

The additional refraction due to this part of the path is therefore

$$- \int \frac{\partial \mu}{\partial s}\, ds \tan (z - \theta),$$

integrated upward, which is equal to $- (\mu_2 - \mu_1)$ multiplied by the mean value of $\tan (z - \theta)$ along the path. To get the whole disturbance due to the building or dome we have to add to this $(\mu_2 - \mu_1) \tan z$, as above, where $\mu_2$ refers to the upper end, and $\mu_1$ to the lower end, of the curved part of the ray.

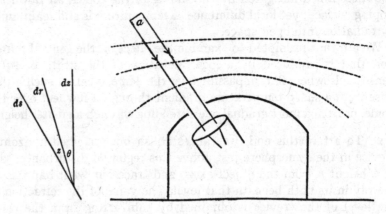

The effect of the crowding down of the temperature gradient on to the dome, where the strata are oblique, is thus equal to

$$(\mu_2 - \mu_1) \{\tan z - \text{mean} \tan (z - \theta)\}, \tag{A}$$

and is usually, as it clearly ought to be, a diminution of the refraction,

Result as apparent increase of z. d.; seasonal. $\mu_2 - \mu_1$ being negative; it would count as an apparent increase of zenith distance. It may well be sensibly different in winter from what it is in summer. If it is smaller in summer, the zenith distances as observed are then in defect; for example, the latitude, as determined from a northern star alone, would be in excess. We proceed to attempt an estimate of the possible discrepancy.

Estimated effect, It is some moderate fraction of $\mu_2 - \mu_1$, the difference of indices at the ends of this terminal curved part of the ray, say usually of the order of one-fifth, while it may be much higher. If we take the local differences in $\mu$ as roughly due to temperature $\tau$ alone, measured from absolute zero, then locally $\mu - 1$ is proportional to $\tau^{-1}$, so that

$$\mu_2 - \mu_1 = \delta\mu = - \frac{\mu - 1}{\tau}\, \delta\tau.$$

Here $\mu - 1$ is about ·0003, that being the value given by the simple formula $r = (\mu - 1)\tan z$, from the knowledge that the refraction is about 58″ $\tan z$, and $\tau$ is say 273°. Thus $\delta\mu$ is of the order $10^{-6}\delta\tau$: and an abnormal fall of 1° C. in a layer, whose thickness is of minor importance, surrounding the dome, would diminish the refraction by the order of one-fifth of 0″·2. For example, an effect equal to the whole Kimura latitude difference 0″·06 between summer and winter would need a local upward change of temperature amounting to two or three degrees centigrade*. compared with the Kimura term.

But such estimates of local error of refraction as are here found for a single star (here taken north of the zenith) do not apply to latitude determinations made from several stars, some north and some south; for these additional refractions would presumably compensate on the whole, though they would appear if the two groups of observations north and south of the zenith were analysed separately. The case would be different if the strata of the atmosphere were tilted as a whole; but it may be estimated from formula (A) that a horizontal relative pressure gradient of 1 in $n$ in an atmosphere of homogeneous height $H$ (actually about five miles) would shift the zenith by about 0″·0003 $n \cos^2 z$, so that for an error of this type to approach the value 0″·06 the mean horizontal barometric gradient would have to be at least $\frac{1}{200}$ of the total pressure, per mile, which is dynamically a quite impossible amount. But compensation in an average of N. and S. stars, unless strata are tilted, to an impossible extent.

The question whether considerations of barometric gradient could have any actual bearing on the puzzling Kimura seasonal effect on latitude would be put to one kind of test by comparing the values derived from the two hemispheres, where the seasons are opposite. If these values agreed symmetrically in phase, in the sense that the variation of north latitude is positive when the variation of south latitude is positive, then such a suggestion could have no place. As a fact, the observations give a deflection in the same direction in both hemispheres at the same time of year, for they are found to be such as an imagined axial displacement of the Earth's centre of mass would account for. Considerations of local refraction are, however, excluded, unless indeed it be unsymmetrical, as the Talcott procedure† involves observations at the same altitude on both sides of the zenith. A tilt in the stratification of the atmosphere, swaying between summer and winter, would serve; but the formula indicates for it, as above, some- Advantage of zenith stars.

* More recent measures seem to ascribe the Kimura effect to instrumental causes: see p. 488.

† This procedure insures against erratic astronomical refraction: to evade local refraction, as here estimated, observations must not be made through a sloping roof, a precaution that has been observed, see p. 488.

thing of the order of several degrees of angle, which is an impossible amount to contemplate as spreading over any considerable distance.

In any case, this general estimate of the possible magnitude of these irregularities in the refraction may not be superfluous. Whether they ever affect meridian observations to any sensible degree is for Practical test. practical astronomers to decide; indeed, the question can readily be tested by attaching for a time an ordinary recording thermograph at the object-glass of a working transit circle, another a few yards above the dome, and another at some distance on the level.

3. The formula (A) expressing the result of local accidental refraction may be verified directly from the simple consideration that, instead of a ray crossing the atmospheric strata at an angle $z$, we have one crossing them at a mean angle $z - \theta$. More generally it may Varying readily be shown, by considering a series of differential refractions, optical medium: that the total deviation of a ray, incident on the strata at angle $z$, where the index is $\mu$, is equal, for any arc of its path, exactly to

$$\int \tan z \, \frac{\partial}{\partial s} \log \mu \, . \, ds, \quad \text{or} \quad \int \tan z \, . \, d \log \mu,$$

which is, in fact, a more directly evaluatable form of the well-known

formula for deviation $\int \frac{\partial \log \mu}{\partial n} \, ds$. For example, in the present case, $z$ is nearly constant, say is $z_0 + \delta z$, and the deviation in any arc becomes

$$\int (\tan z_0 + \sec^2 z_0 \, . \, \delta z) \, d \log \mu,$$

or $$\tan z_0 \, . \, \log \frac{\mu_2}{\mu_1} + \sec^2 z_0 \int \delta z \, . \, d \log \mu,$$

integrated where $\log \mu$ may be replaced by $\mu - 1$. If the strata between indices deviation along ray: $\mu_2$ and $\mu_1$ are tilted through the same small angle $\epsilon$, then $\delta z = \epsilon_0$, and the effect is

$$\sec^2 z_0 \, . \, \epsilon \, (\mu_2 - \mu_1),$$

being independent of the thickness of the stratum.

The effect of the curvature of the Earth on the deviation can be determined from the same formula, with the appropriate value of $\delta z$, derived from

$$(a + h) \sin (z_0 + \delta z) = a \sin z_0:$$

applied to but it is to be noted that owing to the changing direction of the vereffect of Earth's tical, deviation, which is the same as refraction, is not now the same curvature. as $\int dz$. In this way Laplace's expression for the refraction for any arc of path is readily obtained[1].

[1] As, in fact, in Lord Rayleigh's important paper "On the Theory of Stellar Scintillation," *Phil. Mag.* Vol. XXVI, 1893; *Scientific Papers*, Vol. IV, pp. 66–72, especially p. 71.

4. After the above was written, it was found in looking through The problem Henry Cavendish's extensive MS. calculations[1] relating to the plan- ning by the Royal Society, between 1772 and 1774, of the Schehallien experiment for the determination of the Earth's mean density, in which Cavendish took a very prominent part, that he had then con- sidered the problem of accidental refraction error due to the mountain, substantially as here. If the summit is colder than it is at the same level directly above the foot of the mountain, the strata of equal atmospheric density tilt upward towards the summit; and the apparent position of a star, chosen near the zenith to avoid ordinary refraction, will be deflected away from the mountain, thus making the observed effect of attraction of the vertical towards the mountain too small. Cavendish calculated roughly that for a defect of tempera- ture of $12\frac{1}{2}°$ F. at the summit, which he took as an extreme estimate, and for a hill sloping at $21°$ to the horizon, the error would be about $0''\!\cdot\!6$ on each side of the hill, depending, as we have recognized, only on the temperature difference and not on the height. The formula (A) gives for a local temperature change of this total amount, from a mountain of that slope to the air above it, a value of about $0''\!\cdot\!5$ for $z$ small, if the strata are taken parallel to the slope, whereas Cavendish made them inclined to the slope so that the temperature fell along it. In the Schehallien observations the effect of the attraction of the mountain came out as $11''\!\cdot\!6$, which Cavendish's extreme estimate for refraction would increase by 10 per cent., thus, as it happens, leading to consistency with modern determinations. There does not appear to have been any attempt to apply such a correction, either to Maskelyne's Schehallien observations, or to the later series by Sir Henry James at Arthur's Seat, where the result was 3 or 4 per cent. too small, while owing to flatter conditions the refraction error would be much less: see the account in Poynting's Adams Prize Essay of 1894 on *The Mean Density of the Earth*, pp. 15–22.

The problem already treated by Cavendish.

Estimate for deviation of ray on the slope of Schehallien,

as a correc- tion to the deflection of gravity there: important.

ADDITION: from Professor Kimura: *Sept.* 29, 1926.

I have the honour to accept your kind letter containing a suggestion of an anomaly of refraction due to the slope of the observing roof. Such anomaly may exist especially when the opening of the roof is very narrow compared to the size of the observing room. But in the case of the international observatory, the opening is wide enough (more than twice the breath of the pier of the instrument), and the size of the room is comparatively small. Moreover, the operation of opening is daily made soon after sunset, while the observation begins

Precautions for latitude variation:

[1] Now under examination with a view to a second completing volume of Collected Papers [recently published by the Cambridge University Press].

21 o'clock at earliest time; and also the objective of the telescope when erected vertically is just over the height of the wall of the room. I believe, therefore, the effect of the roof would be negligible. On the contrary, the temperature difference at the upper and lower parts of the telescope is usually pretty considerable, which is due to the radiation and the bodily heating of the observer. Therefore, it is probable that anomaly of refraction might occur in the interior of the telescope.

the anomaly discussed by Kimura.

Meanwhile, from the empirical research on $Z$, it is certainly known that the annual part of it is almost the same in its amplitude and in its phase in the observatories lying near the latitude $+ 40°$, provided the observations are made by the chain method having their middle hours near midnight. It is noted those observations were made in several forms of the observing house and their methods, the instruments used, were different.

Thus it may be concluded that the annual part of $Z$ is due to a common cause in that latitude. Whether such variation comes out from the seasonal change of the atmospheric condition near the surface or near the room and the instrument, or the diurnal change of the plumb line due to the deformation of the Earth's crust, nobody can decide now. I believe the principal cause of the systematic variation of $Z$ would be the latter; though the effect of the former is occasionally considerable it would be generally irregular.

# 83

# THE INFLUENCE OF THE OCEANIC WATERS ON THE LAW OF VARIATION OF LATITUDES.

*[Proceedings of the London Mathematical Society,*
Ser. 2, Vol. XIV (1915) pp. 440–449.]

IF the Earth's permanent axis of steady rotation—which must be a principal axis of inertia of its mass, and to ensure *secular*[1] stability must be that of maximum moment of inertia, say $C$—is slightly disturbed, then according to the Eulerian analysis afterwards transformed into geometrical shape by Poinsot, the Pole of the axis of terrestrial rotation ought to describe on the Earth an ellipse around the Pole of inertia in $\sqrt{\dfrac{AB}{(C-A)(C-B)}}$ sidereal days. One of the significant facts relating to the Earth's constitution and origin is that $A$ and $B$ are equal, so far as investigation has detected: thus the period in this free Eulerian polar orbit, now circular when undisturbed, and in the same direction as the diurnal motion by Poinsot's representation, is $A/(C-A)$ sidereal days. The value of this quantity is directly determined by the amount of the ordinary forced Precession of the Equinoxes arising from the known solar and lunar attractions. The free period thus precisely estimated is 304 sidereal days: and accordingly the records of the slight observed fluctuations of the latitudes of observatories were examined long ago, by Peters and by Maxwell, in search for a recurring component, of this period, but without success. In 1892, however, S. C. Chandler announced an actual period of about 428 days: and following on this Newcomb showed briefly[2] by general reasoning that the elastic yielding of the Earth to centrifugal force, as it changes with axis of rotation, would, in fact, lengthen the period, and conceivably to the order of magnitude thus announced, while, of course, irregular meteorological or other transfers of mass would disturb the movement[3]. A mathematical

*Marginal notes:* Stability of rotation. Path of Pole on the Earth: its axial dynamic symmetry perfect. Period of free nutation for a rigid Earth known, and sought for: but an altered period found, explained by elastic yielding.

[1] Cf. *Monthly Notices Roy. Astron. Soc.* 1906: as *supra*, p. 330. The Poinsot representation is an ellipsoid $Ax^2 + By^2 + Cz^2 = (2T)^{-1}$ rolling on a plane at distance $G^{-1}$ from its fixed centre, where $G$ is the angular momentum which is invariable, while $T$ is the energy which is subject to dissipation and to renewal by disturbances but always lies between $G^2/2A$ and $G^2/2C$.

[2] *Monthly Notices Roy. Astron. Soc.* 1892.

[3] Cf. also Lord Kelvin, *Brit. Assoc. Address*, 1876, in "Popular Lectures...," Vol. II, p. 262.

*Marginal note:* Drift of secular stability when energy is being dissipated.

investigation by Hough made shortly afterwards[1], involving a determination of the yielding of an elastic sphere, taken as incompressible, to centrifugal or tidal forces, gave definite formulas which would agree with this conclusion if the Earth were about as rigid as steel.

Hough's principle:    He remarked the result, on his special investigation, that the effect of this varying centrifugal force on the free precession is the same as if the centrifugal elastic bulging were annulled by removing the diurnal rotation, and the resulting form of the Earth were treated as invariable and absolutely rigid. Actually, it is small changes in the total centrifugal protuberance that are concerned, arising from change of axis of rotation, so that the calculation which thus estimates the whole of it is to be effected on the hypothesis of perfect linear elasticity.

generalized.    This result was proved by the writer[2] to express a general dynamical principle, whatever be the nature of the centrifugal deformation, elastic or otherwise, to which the Earth, however heterogeneous, may be subject. A method was later worked out, and applied in

Analysis of causes of actual disturbance of Pole.    conjunction with E. H. Hills[3], for deducing and mapping from the observed path of the Pole, very irregular as it is and far from circular, the character of the meteorological and other disturbances, of the nature of shift of surface terrestrial material, that mask the circularity of the free precession. Finally, generalizing the result of an important investigation by A. E. H. Love, it was shown[4] that the

Effect of elasticity of Earth on tides is expressed by two moduli:    amount of the change in period of the free precession determines the factor ($k$) by which the potential of a tidal or centrifugal bodily force, of the type of a harmonic of the second order, is to be multiplied in order to give the change in the potential of the Earth's gravitational attraction resulting from it—or, what is the same thing, $1 + k$ is the factor by which the total height of the statical oceanic tide is increased by the yielding of the Earth as a whole; and that the further knowledge of the ratio in which the observed part of the long-period equilibrium oceanic tides is reduced owing to elastic solid tides elevating the coasts of the Earth, which is $1 + k - h$, allows us to determine roughly the amount ($h$) of this elastic tidal yielding over the surface relative to the true oceanic equilibrium tide on an invariable Earth, viz. the height of the solid tide, and so to compare

[1] *Phil. Trans.* 1896.

[2] *Proc. Camb. Phil. Soc.* 1896 [*supra*, p. 1]: cf. also *Monthly Notices Roy. Astron. Soc.* 1906 [*supra*, p. 316]; *Proc. Roy. Soc.* 1908 [*supra*, p. 380]. The extension of Hough's principle is also inferred, as I find, by a somewhat complex argument, in Klein and Sommerfeld, *Theorie des Kreisels*, Heft 3, 1903, p. 607; where pp. 663–730 are devoted to a dynamical and historical discussion of the whole subject. [See also historical preliminary paragraphs, *Roy. Soc. Proc.* 1908 as *supra*, p. 380.]

[3] *Monthly Notices Roy. Astron. Soc.* 1906.    [4] See note 1 on p. 487.

the average or effective elastic quality of the solid Earth with that of a globe, say, of glass or steel. If the distribution of mass and elastic quality in the Earth were known, one of these data would suffice to determine the other: but as the constitution of the Earth's interior remains unknown, the two data together suffice without further knowledge to determine the information here recounted—just as, for example, the form of the sea-level suffices by itself, without any reference to the constitution of the Earth, to determine the distribution of gravity over its surface. One of the aspects in which mathematical analysis of natural phenomena appears most to advantage is in its power of suppression and adaptation, through general revision, of its own unnecessarily restricted initial hypotheses and computations.

*independent, as internal constitution is not known.*

*Cognate principles.*

A part of the displacement of terrestrial material which disturbs the free precession of the Pole is that of the waters of the oceans; and it is easy to see, from the formula for $x$ determining the ocean level *infra*, that an oscillation of the axis of rotation, of amplitude one second of arc, would produce a concomitant tidal fluctuation with rise and fall of the order of two inches, such as was detected early by Bakhuysen in the Dutch records, and also recently extricated by General Madsen from the Danish Survey Observations of variation of coast level. But this oceanic disturbance is synchronous with the oscillation of the Pole, and so its effect cannot be estimated by the methods appropriate to irregular disturbing influences. As Clairaut was compelled to recognize in the Lunar Theory, following Newton, the effect in such cases is not merely to disturb the existing oscillation, but to alter its period, thus producing fundamental change in the type of motion. Lord Rayleigh has shown how the cognate refined lunar investigations of J. C. Adams and G. W. Hill are applicable to a general class of prominent physical phenomena[1], *e.g.* to trace the profound influence, on the persistence of acoustical or other vibrations, that may arise from forces that are operative only indirectly by slightly changing the elastic or other constants of the system, provided this alteration occurs in a periodic time half that of the free vibrations in question.

*Oceanic disturbance of the free precession.*

*Amplitude of precessional tide:*

*its reaction on precession is complex.*

*Cognate problems.*

The argument above referred to, for estimating change of free period of terrestrial precession, implying the presence of disturbances harmonic only of the second order, is valid effectively for elastic deformation of the Earth, surface irregularities practically not counting; but it does not directly include these oceanic displacements owing to the irregular form of the oceans. We are thus prompted to estimate independently how much of the lengthening of the period of the free precession the oceanic movement will account for: if it proves to be

*Oceanic influence on free precession:*

[1] *Phil. Mag.* 1887; *Collected Papers*, Vol. III, p. 1.

a substantial proportion, only what remains after it is deducted can be used in the geophysical argument directed towards the Earth's degree of internal rigidity. The general principle which determines directly its influence on the phenomena is that the momentum of the centrifugal protuberance is allowed time to follow closely the axis of rotation. Thus, if $I$ is the moment of inertia of the centrifugal effect around the axis of moment $C$, which remains close to the wandering axis of the Earth's rotation—the latter with components $\omega_1$, $\omega_2$, and $\omega_3$, with resultant the practically constant axial velocity $\Omega$—then the angular momentum of the system has components*

*its nature:*

$$A'\omega_1 + I\omega_1, \quad B'\omega_2 + I\omega_2, \quad C'\omega_3 + I\omega_3.$$

*the effective moments of inertia:* In other words, the effective principal moments of inertia are $A' + I$, $B' + I, C' + I$, where $A', B', C'$ are what they would be if the rotation were stopped and the centrifugal change thus removed. These are sensibly constant when the free precession is of small amplitude; and $I = C - C'$, where $C$ is the actual principal moment of inertia with which the Precession of the Equinoxes is concerned.

But in our present case, owing to the continental masses, the change in the distribution of oceanic waters is not symmetrical around the axes of rotation. The momentum belonging to oceanic shift has three components, specified in terms of its moments and products of inertia; one around the axis of rotation as above, which possesses a permanent configuration; and two others around equatorial axes which travel round with the diurnal motion, and, in so far as they are unsymmetrical with respect to the polar axis, have no sensible influence on phenomena of period long in comparison with a sidereal day. It is thus the change of $C$ and the averaged change of $A$, produced by the heaping up of oceanic waters by the Earth's rotation, that we have to estimate.

*how to be estimated:*

The form of the ocean level is given by

$$E\gamma r^{-1} + \tfrac{1}{2}\omega^2 r^2 \sin^2 \theta = \text{constant};$$

so that, if $r = a + x$, we have

$$x = \frac{\omega^2 a^4}{2E\gamma} \sin^2 \theta + \text{constant}$$

$$= \tfrac{1}{2 \cdot 289} a \sin^2 \theta + \text{constant},$$

on substituting

$$g = \frac{E\gamma}{a^2}, \quad \frac{\omega^2 a}{g} = \tfrac{1}{289}.$$

The moments of inertia $C_0$, $A_0$ of a spheroid of water $(a, a + h)$ are, $\tfrac{1}{2}$ being the Earth's mean density,

$$C_0 = \tfrac{1}{5} \cdot \tfrac{2}{11} E \cdot 2 (a+h)^2, \quad A_0 = \tfrac{1}{5} \cdot \tfrac{2}{11} E \{a^2 + (a+h)^2\}.$$

* Because, as also *supra*, p. 328, the bulge follows the axis of rotation and so has momentum $I\Omega$ of which $I (\omega_1, \omega_2, \omega_3)$ are the components.

Thus for a centrifugal protuberance of water covering the whole contribution of the tidal protuberance.
Earth

$$\frac{C_0 - A_0}{C} = \frac{\frac{1}{5} \cdot \frac{2}{11} E \cdot 4ha}{E \cdot \frac{2}{5} a^2} = \frac{4}{11} \frac{h}{a} = \frac{4}{11} \frac{1}{2 \cdot 289} = \frac{1}{1590}.$$

If we take it that the ocean covers $\frac{2}{3}$ of the Earth's surface in low latitudes, we may apply roughly the factor $\frac{2}{3}$;[1] and we have to remember that $\frac{1}{3}$ of the protuberance belongs to the elastic Earth, requiring another factor $\frac{2}{3}$. In all we obtain for $(C_0 - A_0)/A$ a value $\frac{1}{3600}$. If some possible free oceanic tide of similar type had a period in the neighbourhood of 428 days, this estimate would, of course, be exceeded owing to resonance.

The actual value of $(C - A)/A$ is $\frac{1}{304}$. The influence of the mobile oceanic waters on the free precessional oscillation of the Poles is the same as if $(C_0 - A_0)/A$ were subtracted from it, which by itself would reduce the effect by about $\frac{1}{12}$. Thus the period of the free precession for Precessional free period for rigid Earth with mobile ocean, compared with actual. the Earth supposed rigid but covered by the actual ocean would be about $\frac{12}{11}$ of 304, which is 332 sidereal days. The actual period disentangled from the observations of latitudes is about 428 days: and this remaining difference is attributable to elastic yielding of the solid Earth. Of the total increase from 304 days, about $\frac{24}{31}$ is thus ascribable to elastic yielding: thus we have $k = \frac{24}{31} \cdot \frac{4}{15} = \frac{1}{5}$ say. And observations of tides of long period give $1 + k - h = \frac{2}{3}$. Hence roughly

$$h = \frac{8}{15}, \quad k = \frac{1}{5},$$

Values of the two tidal moduli.

in place of the values[2]     $h = \frac{2}{3}, \quad k = \frac{4}{15},$

deduced by Professor Love from considerations of the elastic centrifugal deformation alone[3]. The elastic terrestrial tide is thus of height Conclusions: $\frac{8}{15}$ of that to which the tidal force would draw the ocean if it lay on an invariable Earth: and the yielding of the Earth increases the absolute tidal protuberance by $\frac{1}{5}$: similar statements holding as regards centrifugal forces arising from axial rotation.

The inferences to be drawn as regards the measure of the Earth's resistance to stress are thus not very substantially altered by oceanic mobility.

It may be observed that, according to the principle which forms the basis of these estimates, a planet without an ocean could not rotate permanently about a principal axis, in face of internal friction, if when the centrifugal strain of the rotation is supposed taken off, the

[1] But see *addendum infra*, which computes the actual value $\frac{1}{3}$, so that the modified. oceanic effects which follow in the next paragraph must be halved, with result as on p. 497.

[2] Cf. *Roy. Soc. Proc.* Dec. 1908, p. 94 [: *supra*, p. 385].

[3] *Proc. Roy. Soc.* Dec. 1908; or *Problems in Geodynamics*, § 58.

planet were no longer dynamically oblate, *i.e.* if that axis ceased to be the axis of maximum moment of inertia. This conclusion harmonizes with what we would in fact have been compelled to anticipate, from comparison with other possible axes. But we are required also to infer that, when there are oceanic areas, secular axial permanence requires that, after the strain arising from rotation is taken off, the planet shall still be dynamically oblate to an extent sufficient to compensate the influence of the free waters. For the Eulerian equations subsist in modified form, and so does their Poinsot representation. Thus, if the planet were dynamically nearly spherical, but with oceanic waters, there would appear to be no axis about which rotation could persist secularly; once it is disturbed, the Pole of rotation would wander over the surface of the planet, keeping however the same direction in space, so that glacial periods of the resulting type and other climatic changes would be normal events, though at long intervals of time[1]. What effective ellipticity would be required to prevent the imperfectly anchored material from producing this loosening of the position of the axis of rotation, in the case of the Earth? One-third of the present ellipticity of $\frac{1}{300}$, and therefore roughly of the existing value of $(C - A)/A$, is due to rotational strain, and we have seen that an additional part in the value of $(C - A)/A$, equal to $\frac{1}{2} \cdot \frac{1}{3600}$, is required to compensate the mobility of the actual oceanic waters. The atmosphere contributes a part depending on the mean difference of the weight of air on the equator and elsewhere, for this is mainly a temperature effect attached to the axis of rotation: but it is practically nothing, and in any case it would be largely compensated by barometric influence on ocean level. Adding these components we find that a value of $(C - A)/A$ at least equal to $\frac{1}{810}$ is needed in order to prevent the Pole of rotation wandering over the Earth's surface, whereas the actual value of this measure of dynamical oblateness is as much as $\frac{1}{304}$. In other planets it might readily be different: but the path towards the explanation of glacial local climatic changes on the Earth in this way still remains firmly closed. The estimate may be expressed otherwise by the statement that with the actual ellipticities, in order for wide polar mobility to be attained the density of the ocean would have to be as much as 16*, instead of 1: while we know that if it were more than $5\frac{1}{2}$ the Earth must emerge and

*Margin notes:*
Instability of Pole of a nearly spherical Earth:

resulting glacial epochs:

shown to be at present excluded for the actual Earth:

not perhaps for other planets.

[1] The Poinsot representation by a rolling ellipsoid teaches that, in this case, when a position of secular stability does exist, the falling away towards steady motion around the axis of greatest effective inertia, and its dissipation of energy, would if unchecked be far from a gentle process: it must involve free precessions of wider and wider radius and corresponding ocean deluges, ultimately again attaining quiescence with rotation around the axis of greatest effective inertia.

Violent tides.

* This may have some slight application in recent theories which allege that the continents are drifting.

so to speak float gravitationally on the ocean, the water being heaped up on one side as Laplace's formula showed.

We can follow graphically the first stages of secular falling away from steadiness. For so long as the displacement of the axis of rotation remains small enough for $A'$, $B'$, $C'$ to be treated as constants, the Eulerian equations are of type

$$(A' + I)\,\dot{\omega}_1 + (B' - C')\,\omega_2\omega_3 = 0,$$

leading readily to

$$(A' + I)\,\omega_1{}^2 + (B' + I)\,\omega_2{}^2 + (C' + I)\,\omega_3{}^2 = 2T,$$
$$(A' + I)^2\,\omega_1{}^2 + (B' + I)^2\,\omega_2{}^2 + (C' + I)^2\,\omega_3{}^2 = G^2,$$

where $T$ is the kinetic energy and $G$ the constant angular momentum of the system. Thus, as in footnote, p. 489, as $T$ is slowly dissipated, the motion continues to be represented by an invariable ellipsoid

$$(A' + I)\,x^2 + (B' + I)\,y^2 + (C' + I)\,z^2 = 1,$$

rolling with centre fixed, on a plane whose distance from the centre is $(2T)^{\frac{1}{2}}/G$ and so is slowly diminishing.

*The Poinsot rolling model adjusted to dissipation of the energy.*

*May* 20, 1915 [: see also *Postscript*]. In the above estimate the factor $\frac{2}{3}$ has been taken to represent the effect of the limited extent of the ocean, that friction having been presumed to be as exact as the factor $\frac{2}{3}$ which reduces the accumulation of oceanic waters, when the tidal rise of the bottom is deducted and separately allowed for by a calculation of elastic deformation. As remarked above, if the solid Earth were rigid, the influence of the ocean could be estimated exactly, in so far as an equilibrium theory applies so that its surface assumes always the statical form corresponding to the changing instantaneous axis of the diurnal rotation. It seems worth while to record here the actual terrestrial data and the result of such an estimate, for which I am indebted to my brother J. S. B. Larmor.

*Improved estimates of the two tidal constants.*

The elevation of the ocean in latitude $\lambda$ due to the whole centrifugal effect of the diurnal rotation is

$$h = ka^2 \cos^2 \lambda - H,$$

where $k = \omega^2/2g$, and $H$ is a uniform fall required to maintain the constancy of volume and so is given by

$$H \int dS = ka^2 \int \cos^2 \lambda \, dS.$$

The changes thereby produced in the polar and other moments of inertia are given by

$$\delta C = \int ha^2 \cos^2 \lambda \, dS,$$

$$\delta A - \tfrac{1}{2}\delta C = \int ha^2 \sin^2 \lambda \, dS.$$

*Estimates of inertial change:*

Thus $\qquad \delta\,(C - A) = \int ha^2\,(\tfrac{1}{2}\cos^2 \lambda - \sin^2 \lambda)\,dS.$

*by quadratures over actual ocean.*

To evaluate this we require the values of $\int dS$, $\int \cos^2 \lambda \, dS$, $\int \cos^4 \lambda \, dS$, integrated over the ocean surface, from which the values of $\int \sin^2 \lambda \, dS$ and $\int \sin^2 \lambda \cos^2 \lambda \, dS$ are readily inferred. These quantities have been estimated by quadratures in the following table, which is determined from the number of parts of water out of 24 parts in all representing each zone of 10° breadth, as taken off from a Mercator map of the world. The entries on the table have to be multiplied by $2\pi a^2$ to obtain absolute values.

### Northern Hemisphere.

| Latitude | Water (24) | $\delta S$ (area) | $\cos^2 \lambda \, \delta S$ | $\cos^4 \lambda \, \delta S$ |
|---|---|---|---|---|
| 0°–10° | 19 | ·137 | ·1367 | ·1356 |
| 10°–20° | 17 | ·119 | ·1110 | ·1035 |
| 20°–30° | 15 | ·098 | ·0810 | ·0665 |
| 30°–40° | 13½ | ·080 | ·0539 | ·0362 |
| 40°–50° | 11 | ·056 | ·0281 | ·0140 |
| 50°–60° | 10 | ·041 | ·0137 | ·0045 |
| 60°–70° | 6 | ·018 | ·0033 | ·0006 |
| 70°–80° | 18 | ·034 | ·0022 | ·0001 |
| 80°–90° | 0 | — | — | — |
| | | ·583 | ·4299 | ·3610 |

### Southern Hemisphere.

| Latitude | Water (24) | $\delta S$ (area) | $\cos^2 \lambda \, \delta S$ | $\cos^4 \lambda \, \delta S$ |
|---|---|---|---|---|
| 0°–10° | 18 | ·130 | ·1295 | ·1285 |
| 10°–20° | 18½ | ·129 | ·1207 | ·1126 |
| 20°–30° | 18½ | ·122 | ·0999 | ·0820 |
| 30°–40° | 21 | ·125 | ·0839 | ·0563 |
| 40°–50° | 23 | ·118 | ·0589 | ·0294 |
| 50°–60° | 24 | ·100 | ·0330 | ·0108 |
| 60°–70° | 24 | ·074 | ·0132 | ·0023 |
| 70°–80° | 24 | ·045 | ·0030 | ·0002 |
| 80°–90° | 0 | — | — | — |
| | | ·843 | ·5421 | ·4221 |
| half sum for both | | ·713 | ·486 | ·391 |
| if all submerged | | 1·000 | ·666 | ·533 |

the last line corresponding to an ocean covering the whole Earth.

It may be noted in passing that the preponderance of water over *Effect of increase of arctic ice on latitudes.* the Southern hemisphere involves greater abstraction from it when the ocean level for any reason falls, and in consequence some displacement of the Earth's centre of mass. But such an effect would not be sensible astronomically: for to alter latitudes to the order of one-tenth of a second of arc the shift would have to be of the order of 15 feet, so that changes of ocean level of the order of 50 feet would be needed.

From the table above we obtain

$$H = ka^2 \tfrac{972}{1426};$$

and therefore for the actual limited ocean

$$\delta\,(C - A) = 2\pi a^6 k \int (\cos^2 \lambda - \tfrac{979}{1496})\,(\tfrac{3}{2}\cos^2 \lambda - 1)\,dS,$$

where
$$k = \frac{\omega^2}{2g} = \frac{1}{2.289}\frac{1}{a},$$

while
$$C = \tfrac{4}{3}\pi a^3 \cdot \tfrac{11}{2} \cdot \tfrac{2}{5}a^2,$$

leading by aid of the tabular data to

$$\frac{\delta\,(C - A)}{C} = \tfrac{1}{4700}.$$

If the ocean covered the whole Earth, $H$ would not count in $\delta\,(C - A)$, and we would have for that case as above

$$\frac{\delta\,(C - A)}{C} = \tfrac{4}{11}ka = \tfrac{1}{1890}.$$

Thus the value of this quantity is reduced owing to limitation of oceanic area by a factor slightly over $\tfrac{1}{3}$, instead of $\tfrac{2}{3}$ as was too hastily assumed above.

This correction amends the final result (with the previous meanings of $h$ and $k$) to

$$h - k = \tfrac{1}{3}, \quad k = \tfrac{3}{13};$$

so that, for example, for tidal action of long period, the absolute tidal deformation of the solid Earth is the fraction $h/(1 + k)$, here $\tfrac{11}{24}$, of the observed oceanic tidal elevation, and would not be far from $\tfrac{1}{2}$ for any reasonable value of $k$. Improved results:

*June* 9, 1915. The effect has to be reduced still further. For the oceanic tidal load itself compresses the solid Earth, and so diminishes its momental difference $C - A$, to a degree that may be calculated from the formulae in Professor Love's *Treatise on Elasticity*, § 177. It is easy to recognize that this reduction is considerable; for a vertical bar of steel of the length of the Earth's radius, with $2 . 10^{12}$ c.g.s. for Young's modulus, would be compressed $\tfrac{1}{4}H$ by an end load of water of depth $H$. Bearing in mind that the Earth's mean density is $5\tfrac{1}{2}$, it would seem that the reduction in the oceanic precessional influence arising on this account might well, if the Earth were like steel, be 20 or 30 per cent. Also $h - k$ might well be changed substantially, even from $\tfrac{1}{3}$ to $\tfrac{1}{2}$, when allowance is made for this sinking of the ocean bed owing to the tidal load, while $k$ would remain at about $\tfrac{1}{4}$. further amended, for compression, to final estimate.

## POSTSCRIPT (1927).—*Tidal Turbulence.*

What appeared to be a new practical departure was initiated by R. O. Street (*Roy. Soc. Proc.* 1918) by a theoretical dynamical analysis, in close touch with the data of the actual very de-tailed tidal records for the Irish Sea, of the tidal phenomena in that basin, partly to explore the influence of the Earth's rotation on the tidal flow. The viscosity of the water was introduced into the analysis: and partly on the instigation of the present writer, the paper concluded with an estimate of the rate of dissipation of tidal energy by such friction; for it was known that most of the energy thus dissipated must come ultimately from the Earth's rotation, so that the result would be relevant to the slowing of the period of diurnal rotation. This application at once stimulated G. I. Taylor to repeat the calculations, replacing, from his vivid practical nautical experience in tidal races, the normal viscosity by a large-scale viscosity of eddy-motion which he had found to be effective to explain the phenomena of surface winds and their change with height, and later in aeronautic applications\*. The value of the *quasi*-viscosity of turbulence was estimated by extrapolation from the data for air; and the Irish Sea was judged to be in the main shallow enough to necessitate turbulent flow instead of the smooth oceanic tidal heave, on the basis of the general dimensional criterion for turbulent motion formulated long previously by Osborne Reynolds. The re-sulting dissipation of energy was naturally hundreds of times greater than Street's value. Soon after, following up the subject on the same lines, H. Jeffreys found the surprising result that in the shallow seas alone there is enough dissipation, or more than enough, to account for all the outstanding slowing of the Earth's rotation.

Such an influence of the dissipation of energy must however be indirect. The slowing is actually caused by the backward pull of the Moon on the tidal elevation which persists after the parts of the Earth have passed on, in its diurnal rotation, from below the Moon. This is the familiar lag of the tides, and some notion of its effects was essayed in Thomson and Tait's *Nat. Phil.* (1867): the discussion in the paper next preceding (p. 472) at any rate suggested that the lag of the oceanic tide might be a sufficient cause by itself, thus without any special reference to shallow seas, for the drag on the Earth's rotation. Its effect could be estimated with certainty by quadratures, as Lord Kelvin suggested, if the tidal records over the Earth were adequate:

*Marginal notes:* Effect of Earth's rotation on tides in a confined basin. / Estimate of dissipation of tidal energy: / motion smooth or turbulent. / Preponderant drag in shallow seas: / involves difficulties.

\* Cf. Rayleigh, *Sci. Papers*, Vol. VI, p. 602; Lamb, *Hydrodynamics* (1924), §§ 366 *c*, 371.

but the writer is informed by Mr Street that they are still not complete or exact enough for this purpose.

The consideration then arises, that if the shallow seas are really predominant, while a conspicuous lagging tidal elevation over the whole ocean is no more than sufficient*, it must be because a delayed tidal protuberance of exceptional magnitude is raised in their neighbourhood on which the Moon gets a horizontal grip: but no such local elevations seem to be in evidence. On the other hand the writer is informed by Mr Street that his own more recent estimate of the excess of energy flowing into the Irish Sea at its southern end and that flowing out at its northern end does, as Mr Taylor first verified, agree with the higher estimate of dissipation by friction. One may note that this excess is correctly calculated relative to the shores, as if the Earth were not rotating, for the "geostrophic" force on the water is transverse to its flow. The subject thus seems to remain not free from obscurity†. In any case the importance would attach presumably only to shallow seas that are confined but have a sufficient inlet, as a shallow region in the open ocean would act largely after the manner of an island and deflect the tidal flow round its sides. For a general summary cf. H. Jeffreys' treatise, "The Earth...," pp. 218 *seq.* Turbulence arising from reflection from the shelving shores has perhaps not yet been considered.

> The problem one of relative energy.

* Cf. also Kelvin's *Math. and Phys. Papers*, Vol. III, p. 337.
† Cf. *supra*, p. 480, in relation to astronomical pure time.

# 84

# VISCOSITY IN RELATION TO THE EARTH'S FREE PRECESSION.

[*Monthly Notices of the Royal Astronomical Society*, Vol. LXXVI (1915).]

ONE of the results of Mr H. Jeffreys' interesting paper on this subject (*Monthly Notices*, June 1915, p. 648) is so surprising that some way out of it seems to be necessary: as, in fact, he fully recognizes. He calculates that an internal viscosity in the Earth sufficient to account, by the lag of the solid tides, for the secular acceleration of the Moon's mean motion, would be sufficient to stop down any free precession of the Pole in a few days. Assuming the calculations to be correct, are his physical hypotheses stretched beyond their range of validity?

The law of viscosity that is assumed consists, following G. H. Darwin, in replacing the rigidity $n$ of the material by

$$n \left( 1 + 1 \Big/ t \, \frac{d}{dt} \right)^{-1}$$

where $t$ may be called time of relaxation of the substance. This hypothesis probably gives a fair representation of the effects of internal friction for oscillations whose period is very short compared with $t$. It also embraces, as was doubtless intended, the facts that *Influence cannot be great.* pitch has fluid properties for disturbances of very slow period, while it can be moulded into a tuning-fork for sound vibrations: but ordinary solids do not at all show this fluidity for very slow changes of stress, at any rate for stresses so small, due to a tidal head of a few feet of water, as we are here concerned with[1].

On the basis of analysis adapted to the case of a complex $n$ from Thomson and Tait's *Nat. Phil.*, Mr Jeffreys finds (p. 651) that $t$ must be so small as about a day and a half, in order that the lag of the solid terrestrial tide may produce sufficient drag on the Moon. But the period of the tide is half a day, which is probably much too near to

---

[1] If the elasticity of the core were perfect, the lunar drag on the Earth's rotation must arise from the lag of the oceanic tides alone. It has not yet been proved that this cause is insufficient to account for the lunar acceleration: cf. *Monthly Notices*, Jan. 1915, [*supra*, p. 473], and on the other side Poincaré, *Théorie des Marées*. [The expectation that an elastic tide would not lag does not however appear to be always justified, unless its period were long.]

But a very slow absorption of energy of tidal strain in the core is certainly not precluded, especially if it is of the type of the elastic fatigue investigated for the case of metals by Kohlrausch, Kelvin, and others.

this value of $t$ for the assumed law of viscous elasticity to have any validity.

Moreover, it is then applied to the free precession of the Pole, where the period involved is over 400 days; and this will appear on reflexion to be very far beyond its possible range—whether we assume fluidity for very slow stresses, and consequent precise isostasy throughout the Earth's mass, or not.  For example, in the case of a maintained vibration of plane shear in a slab of the material, everything would vary as $e^{\iota p t}$ where here $2\pi/p = 400$ days: thus the effective value of the rigidity would be $n\,(1 - 42\iota)^{-1}$, or $\dfrac{n}{42^2} + \dfrac{n}{42}\,\iota$, where $\iota = \sqrt{-1}$, for a period of 400 days, as contrasted with the full value $n$ for sound vibrations—an entirely improbable extrapolation.

Formula at fault

as illustrated.

# 85

# NOTE ON THE SOLUTION OF HILL'S EQUATION.

[*Monthly Notices of the Royal Astronomical Society*, Vol. LXXV (1915).]

THE differential equation treated by G. W. Hill for the purposes of the Lunar Theory is so important also for other branches of mathematical physics[1] that it is desirable to bring into view in what respects, if any, the modern abstract theory of linear equations with periodic coefficients can add to or modify Hill's original analysis. The exposition by Mr Lindsay Ince in the *Monthly Notices* for March (p. 436) allows perhaps, owing to its succinctness, an immediate answer to be given to this question.

The type of solution worked out by Mr Ince is

$$y = e^{\mu z} u$$

[for the equation with $\phi$ as variable instead of $y$, on next page], where

$$u = \sin (z - \sigma) + a_3 \cos (3z - \sigma) + a_5 \cos (5z - \sigma) + \ldots$$
$$+ b_3 \sin (3z - \sigma) + b_5 \sin (5z - \sigma) + \ldots$$

wherein $\mu$ and $\sigma$ are constants.

This is the real part of another solution, the coefficients in the differential equation being real,

$$y = e^{\mu z} \{ - \iota e^{\iota (z-\sigma)} + a_3 e^{\iota (3z-\sigma)} + a_5 e^{\iota (5z-\sigma)} + \ldots \},$$

or, employing different complex constants and writing $w$ for $\mu + \iota$, in which, however, there seems to be no reason why $\mu$ should be always real,

$$y = A_0 e^{\iota \nu z} + A_1 e^{\iota (\nu+2) z} + A_2 e^{\iota (\nu+4) z} + \ldots.$$

Rayleigh's application to physical problems. The solution developed by Hill, as summarized in view of physical applications by Rayleigh (*loc. cit.*), is of type

$$y = b_0 e^{\iota \nu z} + b_1 e^{\iota (\nu+2) z} + b_2 e^{\iota (\nu+4) z} + \ldots$$
$$+ b_{-1} e^{\iota (\nu-2) z} + b_{-2} e^{\iota (\nu-4) z} + \ldots,$$

in which, so long as $\nu$ comes out real, the coefficients are all real quantities.

This latter form leads to a system of linear equations determining the coefficients, and a corresponding determinant determining $\nu$, which are infinite both ways. The previous form, employed by Mr Ince, with the value $- \iota e^{-\iota \sigma}$ for $A_0$, leads to a system infinite only towards the positive side, in which, however, the coefficients are always complex. This scheme, which Mr Ince works out in trigonometrical form, may be advantageous for the determination of the

---

[1] Cf. Rayleigh, *Phil. Mag.* Vol. XXIV, 1887; *Collected Papers*, Vol. III, p. 1.

coefficients in non-astronomical applications for which $\nu$ is usually complex. The expansion set out by Mr Ince for real values of $\mu$ and $\sigma$ belongs to the special case when the root $\nu$ of the determinantal equation comes out of the form $1 - \iota\mu$.

*Postscript* (1928). The structure, so to say, of a differential equation with periodic coefficients, of this general type, is exhibited intuitively in its application, after Rayleigh, to undulatory transmission in a medium of periodic quality. A modern example is a train of X-rays traversing a crystal. Thus if the dielectric modulus $K$ of the medium is given by

$$K/c^2 = a_0 + a_1 \cos (qz + \alpha_1) + a_2 \cos (2qz + \alpha_2) + \ldots$$

the equation of propagation of trains of waves with fronts parallel to $y$ is of form

$$\frac{\partial^2 F}{\partial x^2} + \frac{\partial^2 F}{\partial z^2} = \frac{K}{c^2} \frac{\partial^2 F}{\partial t^2}$$

for the case when they are polarized so that the electric vector $F$ is parallel to $y$ so that $K$ escapes differentiation. The type of undulation being

$$F = e^{\iota mx} e^{\iota pt} \phi$$

the equation becomes

$$\frac{\partial^2 \phi}{\partial z^2} + \{m^2 - a_0 p^2 - a_1 p^2 \cos (qz + \alpha_1) - a_2 p^2 \cos (2qz + \alpha_2) - \ldots\}\phi = 0.$$

The solution of this equation for $\phi$, with coefficients of the present periodic type, must conform to the facts of X-ray spectroscopy, as elucidated by the familiar original *aperçu* of W. L. Bragg in terms of general ideas of diffraction, namely that there are small ranges of incidence of the rays, therefore of values of $m$, for which they cannot penetrate on account of reinforcing reflexions, separating wide regions for which the partial reflexions, being out of step, cannot cumulate in energy. Cf. the systematic treatment by R. Schlapp, *Phil. Mag.* Jan. 1925, utilizing the forms of solution worked out mainly by Whittaker and his school at Edinburgh as now abstracted in Whittaker and Watson's *Modern Analysis*.

The remark may be added, that in accordance with recent developments in X-ray analysis, there can be configurations involving several molecules in the crystalline cell owing to their not being mere points, so that each reflecting stratum may be a layer several molecules in depth instead of being a plane surface. Thus this general type of procedure, which takes $K$ to be a periodic function resolvable into a Fourier series, would appear to be necessary in such cases to adequate theoretical discussion of the crystalline diffraction in detail.

Structure intuitively visualized:

by application to X-ray spectra in crystals:

for which Hill's equation is essential.

# 86

# THE TRANSITION FROM VAPOUR TO LIQUID WHEN THE RANGE OF THE MOLECULAR ATTRACTION IS SENSIBLE.

[*Proceedings of the London Mathematical Society*,
Ser. 2, Vol. xv (1916) pp. 182–191.]

THE physical properties of fluid media, as regards change of state, and also as regards capillary phenomena, have been closely illustrated in theory by consideration of a model medium, subject to internal expansive pressure, of kinetic or other origin, which is counteracted by the contractive effect of mutual attraction between its molecules—the latter force extending through much the greater*, though for ordinary purposes still insensible, range. For liquids, the difference between these two much larger quantities, of different types, constitutes the transmitted hydrostatic pressure†.

*Contact pressure equilibrating molecular attraction.*

Suppose the potential (reckoned as energy) of forces arising from a plane layer of molecules of density $\rho$ and thickness $\delta x$ is, at distance $x$ from it, equal to $f(x)\,\rho\delta x$, so that the attraction is $f'(x)\,\rho\delta x$. Let the homogeneous medium of uniform density $\rho_0$ be disturbed so that $x$ becomes $x + \xi$, and therefore $\rho_0$ becomes $\rho_0 \left[ 1 - \dfrac{d\xi}{dx} + \left(\dfrac{d\xi}{dx}\right)^2 + \ldots \right]$.

*Disturbed potential function of molecular forces:*

---

* The nature of cohesion between solids, across liquid films, has been recently explored, mainly on the initiative of Sir W. B. Hardy: a general account is given by F. H. Rolt and H. Barrell, "Contact of Flat Surfaces," *Proc. Roy. Soc.* Oct. 1927, pp. 422 *seq.* The thickness of films of separation is of the order of $5 \cdot 10^{-7}$ cm., about the same as that of black free liquid films. The limiting case of a surface layer of foreign matter reduced by spreading to the monomolecular type, is of course excluded from smoothed out discussions like the present one. The present state of this attractive subject is summarized by N. K. Adam, one of the principal workers in it, in *Science Progress*, Jan. and April, 1927.

*Films of separation:*

*monomolecular.*

† If a cylinder of length $l$ and density $\rho$ strikes with velocity $v$ an immovable obstacle, its momentum $l\rho v$ per unit cross-section is destroyed in the time $2l/V$ in which a shock wave travels up and down the cylinder, where $V$ is the velocity of sound in it, provided the limits of perfect elasticity are not passed: thus the average pressure on the obstacle is $\frac{1}{2}\rho v V$. A result of the same order (doubled) is arrived at by S. S. Cook by equation of kinetic energy lost to compressive energy gained for a *single* element of the cylinder. The pressures thus produced are the basis of explanation of erosion by impinging water drops or collapsing bubbles. In the latter case the pressure is intensely concentrated at the centre of the collapse: as also on impact of two spheres it is so concentrated, cf. Hertz's calculations, *Miscellaneous Papers*, English Edition, p. 162.

*Pressure in impact:*

*different rough estimates not in discord:*

*erosion by water drops.*

The potential of the whole mass taken at the origin of measurement of $x$ is then

$$V = \int_{-\infty}^{\infty} f\left(x + \xi - \xi_0\right) dx \rho_0 \left[ 1 - \frac{d\xi}{dx} + \left(\frac{d\xi}{dx}\right)^2 + \ldots \right],$$

wherein, the suffixes denoting values at the origin of $x$,

$$f\left(x + \xi - \xi_0\right) = f(x) + (\xi - \xi_0) f'(x) + \tfrac{1}{2}(\xi - \xi_0)^2 f''(x) + \ldots,$$

while

$$\xi - \xi_0 = x\left(\frac{d\xi}{dx}\right)_0 + \tfrac{1}{2}x^2 \left(\frac{d^2\xi}{dx^2}\right)_0 + \ldots,$$

and

$$\frac{d\xi}{dx} = \left(\frac{d\xi}{dx}\right)_0 + x\left(\frac{d^2\xi}{dx^2}\right)_0 + \ldots.$$

The displacement $\xi$ from the condition of uniform density, and its gradients, can be regarded as very small, so that, for the purpose of forming equations of disturbance from the potential energy, terms higher than squares or products of such quantities are negligible. On this understanding,

$$V = \int_{-\infty}^{\infty} f(x)\, dx\rho_0 \left[ 1 - \frac{d\xi}{dx} + \left(\frac{d\xi}{dx}\right)^2 - \frac{d^2\xi}{dx^2}x - \tfrac{1}{2}\frac{d^3\xi}{dx^3}x^2 - \ldots \right]$$
$$+ \int_{-\infty}^{\infty} \left\{ f'(x)\,\rho_0 dx \left[ \frac{d\xi}{dx}x - \left(\frac{d\xi}{dx}\right)^2 x + \tfrac{1}{2}\frac{d^2\xi}{dx^2}x^2 + \ldots \right]\right.$$
$$\left. + f''(x)\,\rho_0 dx \left(\frac{d\xi}{dx}\right)^2 x^2 \right\},$$

wherein the gradients are now to be understood as having the constant values belonging to the origin of $x$.

The second part of this expression can be integrated by parts. The terms at the limits vanish, so far as we shall require to go, that being, in fact, a condition necessary to our system being defined at each point by a local constitution. This part thus becomes

$$- \int_{-\infty}^{\infty} f(x)\,\rho_0 dx \left[ \frac{d\xi}{dx} - 2\left(\frac{d\xi}{dx}\right)^2 + \frac{d^2\xi}{dx^2}x + \tfrac{1}{2}\frac{d^3\xi}{dx^3}x^2 + \ldots \right].$$

Hence

$$V = \int_{-\infty}^{\infty} f(x)\,\rho_0 dx \left[ 1 - 2\frac{d\xi}{dx} + 3\left(\frac{d\xi}{dx}\right)^2 - 2\frac{d^2\xi}{dx^2}x - \frac{d^3\xi}{dx^3}x^2 - \ldots \right]. \qquad \text{rearranged by integration by parts:}$$

If then

$$A_0 = \int_{-\infty}^{\infty} f(x)\, dx, \ldots, A_r = \int_{-\infty}^{\infty} x^r f(x)\, dx, \ldots,$$

so that $A_r$ is null by symmetry when $r$ is odd, we have for any point $x$ of the medium

$$V = A_0\rho_0 \left[ 1 - 2\frac{d\xi}{dx} + 3\left(\frac{d\xi}{dx}\right)^2 \right] - A_2\rho_0 \frac{d^3\xi}{dx^3}\left(1 - \frac{d\xi}{dx}\right) - \ldots. \qquad \text{final form.}$$

This expresses the potential of the attractions in terms of the strain $d\xi/dx$ and its gradients, with coefficients that are physical constants of the medium.

To obtain, from this value of the potential, the force tending to increase the displacement $\xi$ (which is not the gradient of $V$), we must first form in the Lagrangian manner the potential energy of the system

$$W = \tfrac{1}{2}\int V\rho\,dx, \quad \text{where} \quad \rho = \rho_0\left[1 - \frac{d\xi}{dx} + \left(\frac{d\xi}{dx}\right)^2 + \ldots\right].$$

**Potential energy of system:** Thus
$$W = \tfrac{1}{2}A_0\rho_0^2\int\left\{1 - 3\frac{d\xi}{dx} + 6\left(\frac{d\xi}{dx}\right)^2\right\}dx$$
$$- \tfrac{1}{2}A_2\rho_0^2\int\frac{d^3\xi}{dx^3}\left(1 - 2\frac{d\xi}{dx} + \ldots\right)dx + \ldots$$
$$= 3A_0\rho_0^2\int\left(\frac{d\xi}{dx}\right)^2 dx - A_2\rho_0^2\int\left(\frac{d^2\xi}{dx^2}\right)^2 dx + \ldots,$$

**rearranged:** on omitting all perfect differentials, which affect only the dynamical equilibria at boundaries. Thus

**its variation,**
$$\delta W = 6A_0\rho_0^2\int\frac{d\xi}{dx}\frac{d\delta\xi}{dx}dx - 2A_2\rho_0^2\int\frac{d^2\xi}{dx^2}\frac{d^2\delta\xi}{dx^2}dx + \ldots$$
$$= [\ldots] - 6A_0\rho_0^2\int\frac{d^2\xi}{dx^2}\delta\xi\,dx - 2A_2\rho_0^2\int\frac{d^4\xi}{dx^4}\delta\xi\,dx + \ldots.$$

**gives distribution of transverse force.** The force acting on the element at $dx$ to alter its displacement $\xi$ is therefore $Xdx$, where
$$X = 6A_0\rho_0^2\frac{d^2\xi}{dx^2} + 2A_2\rho_0^2\frac{d^4\xi}{dx^4} + \ldots.$$

**Combined with intrinsic pressure:** If $\varpi$ represent internal pressure arising from sources other than these attractions, and we assume that it is determined by a relation $\varpi = F(\rho)$, the equation of slight disturbance of the elements of the medium is

**the dynamical equation,**
$$\rho\frac{d^2\xi}{dt^2} = -\frac{d\varpi}{dx} + X.$$

Now, here
$$-\frac{1}{\rho}\frac{d\varpi}{dx} = -\frac{1}{\rho}F'(\rho)\frac{d\rho}{dx},$$

where
$$\rho = \rho_0\left(1 - \frac{d\xi}{dx} + \ldots\right).$$

The equation of motion therefore becomes
$$\frac{d^2\xi}{dt^2} = F'(\rho_0)\frac{d^2\xi}{dx^2} + 6A_0\rho_0\frac{d^2\xi}{dx^2} + 2A_2\rho_0\frac{d^4\xi}{dx^4} + \ldots,$$

or, now omitting subscripts, it is

**in reduced form.**
$$\frac{d^2\xi}{dt^2} = \frac{d}{d\rho}\{F(\rho) + 3A_0\rho^2\}\frac{d^2\xi}{dx^2} + 2A_2\rho\frac{d^4\xi}{dx^4} + \ldots.$$

For the type of simple undulatory disturbance represented by
$$\xi \propto e^{-i\kappa t - imx},$$

we have therefore $\quad \kappa^2 = \dfrac{dp_0}{d\rho}m^2 - 2A_2\rho m^4 + \ldots,$

where
$$p_0 = \varpi + 3A_0\rho^2.$$

Thus very short wave-lengths ($m$ great) will exhibit instability unless $-A_2$, and other such quantities, are positive constants. We must stop in this series for $\kappa^2$ at or before $A_{2r}\left(=\int x^{2r} f(x)\, dx\right)$, the last of them that is a convergent integral, and, as such, depends only on the state of affairs around the origin to which it belongs. Thus if $A_4$ is to be included, $f(x)$ would have to fall away with distance after the manner of $x^{-5}$ or a higher inverse power. The higher divergent constants of this series are to be regarded as excluded from operation by the ultimate molecular character of the medium, which they do not sufficiently represent.

Long waves are propagated with velocity $(dp_0/d\rho)^{\frac{1}{2}}$, where

$$p_0 = \varpi + 3A_0\rho^2;$$

shorter ones with greater velocity $(dp_0/d\rho - 2A_2\rho m^2)^{\frac{1}{2}}$. As $dp_0/d\rho$ diminishes, stability will decrease; and the medium of uniform density $\rho$ will become unstable at the value of $\rho$ for which $dp_0/d\rho$ is about to change sign by vanishing.

If the internal forces of the medium are effectively cohesive and not repulsive, as they must be in an ordinary liquid, $A_0\left(=\int f(x)\, dx\right)$ must be negative. As the potential energy must be taken to be zero at infinite distance, this requires that the graph of $f(x)$ should be in the main below the axis of $x$.

If the range of the attraction is considerable, disturbances of very short wave-length can be contemplated in this continuous analysis; and these, as we have seen, would be unstable unless $A_2\left(=\int x^2 f(x)\, dx\right)$ is negative. Thus if the graph of $f(x)$ is undulating, so that the force (its gradient) may change from attractive to repulsive with change of distance, the regions of repulsion must be in the main for small values of $x$, and they must be limited by this condition. We might introduce such repulsion of small range into the argument instead of the kinetic momentum pressure: and we could thus contemplate, after the manner of Young, the relations of surface tension, and of change of state as well, as established for a medium, liable to variation of density, with its properties determined solely by a single law of attraction of this type, viz. an attraction on the whole, counteracted by repulsion at shorter range.

We have now to determine what is the value of the pressure transmitted into the medium, in compression experiments such as the classical investigations of Andrews on change of state. In the first place, let us imagine a plane of separation in the medium, at the origin of $x$, and calculate the total force per unit area exerted by the medium situated to the left of this plane on the medium situated to the right. If distance $x$ is measured to the right of the plane and distance $X$ to the left, this force is equal to $\varpi$ diminished by mutual attractions which amount to

$$P = \int_0^\infty \rho_x dx \int_0^\infty \rho_X dX f' (X + x),$$

where, when the disturbance of $\rho$ is very gradual,

$$\rho_x = \rho + \frac{d\rho}{dx} x + \tfrac{1}{2}\frac{d^2\rho}{dx^2} x^2 + \ldots,$$

$$\rho_X = \rho - \frac{d\rho}{dx} X + \tfrac{1}{2}\frac{d^2\rho}{dx^2} X^2 - \ldots,$$

the values of $\rho$ and its gradients after substitution being here the constant values belonging to the origin. This gives

$$P = \int_0^\infty \rho_x dx \left\{ \rho \int_0^\infty f' (X + x) \, dX - \frac{d\rho}{dx} \int_0^\infty Xf' (X + x) \, dX \right\} + \ldots$$

$$= \int_0^\infty \left( \rho + \frac{d\rho}{dx} x + \ldots \right) dx \left\{ - \rho f (x) - \frac{d\rho}{dx} f_1 (x) \right\} + \ldots$$

$$= - \rho^2 \int f (x) \, dx - \rho \frac{d\rho}{dx} \int \{xf (x) + f_1 (x)\} \, dx - \left(\frac{d\rho}{dx}\right)^2 \int xf_1 (x) \, dx + \ldots$$

$$= - \rho^2 . \tfrac{1}{2}A_0 - \rho \frac{d\rho}{dx} \{\tfrac{1}{2}A_1 - f_{11} (0)\} + \left(\frac{d\rho}{dx}\right)^2 f_{111} (0) + \ldots,$$

where $A_0 = - f_1 (0)$, and $A_1$ is null.

If the interface is situated at a place where the medium is uniform, the pressural force $\varpi - P$ transmitted across it is thus of intensity

*now determined.*

$$\varpi + \tfrac{1}{2}A_0\rho^2,$$

where, as we have seen, $A_0$ must be negative for attractive forces.

Now consider the equilibrium of a finite composite slice bounded on one side by this interface and on the other side by an interface situated in the solid piston which applies the pressure. A transition layer, of very rapid variation of density, between the fluid and the solid, will be included in the slice, but the forces in it are compensating mutual forces which will not affect the equilibrium of the whole. Thus $p$, equal to $\varpi + \tfrac{1}{2}A_0\rho^2$, is equal to the pressure transmitted into the fluid as measured on a hydrostatic piston[1].

If $dp/d\rho$ were negative, compression would diminish the statical

[1] In an important modern paper by Rayleigh on these topics, especially those of surface tension (*Phil. Mag.* Vol. XXXIII, 1892; *Collected Papers*, Vol. III, p. 514), this result is obtained through the equation

$$p = \int \rho dV = | \rho V |_0 - \int V d\rho.$$

*Solid surface tension.* [Mathematically it seems hardly to be possible, on the lines of *Roy. Soc. Proc.* (1921) as *infra*, p. 620, that a substance uniformly polarized on the electric model, right up to the interface, thus with sharp transition, such as a crystal might be, could exhibit ordinary surface tension, *i.e.* uniformly distributed surface energy, at all.]

applied pressure; that is, the fluid of uniform density $\rho$ would be an unstable system for applied disturbances even when its density is constrained to remain uniform.

But if our previous calculation is correct, instability for internal disturbance—*i.e.* owing to irregularity in density—sets in earlier than this, viz. when $dp_0/d\rho$ vanishes, where $p_0 = p + \frac{1}{2}A_0\rho^2$. For $\frac{1}{2}A_0$ is negative, as above noticed: it is in fact $- a$ if the equation of van der Waals is adopted. Generally, $p = \varpi - P$, where $\varpi$ represents the repulsive effect of molecular motion, and $P$ the attractive effect $- \frac{1}{2}A_0\rho^2$ of molecular forces. Thus the limit of internal stability is when $dp_0/d\rho$ vanishes, not $dp/d\rho$; where $p_0 = p - 5P$, in which $p$ is Hastened instability; the hydrostatic pressure, and $P$ the internal cohesion of the mutual molecular attractions[1]. It is naturally to be expected that these limits will be narrower than would obtain when the density is kept uniform.

Below the critical temperature, on the Andrews-Thomson $(p, v)$ diagram of isotherms representing the vapour-liquid state, this distinction is not practically important, because at such temperatures gaseous and liquid phases definitely separate out, at a level on the diagram which is determined by the Maxwell-Clausius rule of equal areas cut off above and below by the horizontal line representing vapour and liquid in contact. But above the critical temperature there will be a region in which the single phase will become internally unstable, while yet no such vapour-liquid line can be drawn. For, as $\rho = v^{-1}$, and so $p_0 = p + \frac{1}{2}A_0v^{-2}$, where $A_0$ is negative, the point its range $(dp_0/dv$ null) at which internal instability sets in is determined by extended even $dp/dv = 5A_0v^{-3}$; viz. it begins when the downward slope with increasing critical point, $v$ is less than the positive quantity $- 5A_0v^{-3}$, and thus it obviously obtains even in a region above the critical point where the $(p, v)$ curve continually trends downwards so that no liquid phase can separate off*.

If we sketched in along with the van der Waals graph of $p = \dfrac{RT}{v-b} - \dfrac{a}{v^2}$,

the hyperbola $\varpi = \dfrac{RT}{v-b}$, then $p_0 = 6p - 5\varpi$, and the place at which

---

[1] This quantity $p_0$, which plays the part of pressure as regards internal kinetic disturbances, will usually over large ranges be negative.

* The Andrews-Thomson diagram has recently been explored by N. K. Adam for the case of the two-dimensional pressure in a surface layer, which shows itself as diminution of surface tension. If the molecules are far apart so as to be free it ought to obey the laws of gas pressure with the standard gas constant, as regards expansion of surface: the suppression of one of the three modes of Condensed freedom affects the change of volume to surface and of volume pressure to mono- surface pressure in the same way. The characteristic curve when undulating molecular must involve lines of abrupt transition in the film, just as in three dimensions. films: Such tendency of a foreign film to assemble itself into separate regions of condensed *quasi*-liquefied phase may stand in close relation to the limitation and catalysis. of surface catalytic action.

and possibly below the liquid line. $dp_0/dv$ vanishes would be seen on inspection to fall in general well below the horizontal vapour-liquid line. In case this did not obtain, there would be a region of instability before liquefaction, of the same kind as is here suggested on isothermals above the critical point.

Whether our scheme sufficiently represents actual fluid matter or not, we can certainly form an image of a medium possessing its properties. For such a medium there can be no separation of phases unless the straight line representing the constant pressure of liquid in presence of vapour can be drawn to cut off equal areas from the characteristic curve above and below it. This criterion assumes, on the Maxwell-Clausius principle[1], that the work done in a cycle conducted at constant temperature, *LMNPL* in the diagram, is null,

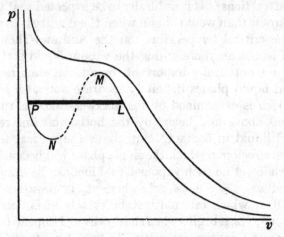

even when part of the cycle is conducted through unstable, though possible states, those represented by the arc of dots in the diagram. The unstable arc can, in fact, be imagined, for the purposes of the argument, as maintained against very slight disturbance by very slight constraints; but these must be such as the internal motion constituting the constant temperature does not vitiate.

*Legitimate to reason across instabilities.*

What then can happen to such an unstable medium if liquefaction is precluded? It has appeared that any slight change of density in the interior of the medium will be augmented in unstable manner when the downward slope of the characteristic $(p, v)$ curve becomes less than a definite limit, and that thus there can be internal instability at temperatures extending to a considerable range above the critical temperature, which is that for which $dp/dv$ vanishes at a point on the curve and is the highest at which liquid can appear[2].

*Region of instability without new phases arising:*

[1] Cf. Maxwell's *Theory of Heat*.

[2] When the constraint to plane strata that is the basis of this analysis is released, the range of internal instability will naturally be greater than that here determined.

If the precision and definiteness attained in the measures of critical temperatures are held to indicate that such a state of matters does not occur sensibly in actual fluids[1], this result might possibly arise from the range of molecular attraction extending over so few molecules that they cannot be treated by the present continuous methods. Or it may be due to the intrinsic pressure, arising from the momentum of the molecules, having itself a range (which is of the order of the free path in gases) comparable with that of molecular attraction; though it might seem that the greater freedom of a long free path would rather increase instability. In either case, the application of the Young-Laplace principles in illustration of capillary phenomena would in actuality be considerably limited. But we can face an alternative that is still open. We may assert that on a $(p, v)$ isothermal not too far above the critical point there is an arc, in the neighbourhood of the critical volume, along which the state of physical aggregation must change as the volume increases, recovering the previous or more probably some other molecular constitution again after this arc, which would otherwise be unstable, has been passed. We would thus hold that there is a region of instability in the $(p, v)$ field in crossing which the substance gradually changes its molecular state of adhesion, but without any tendency to divide into two phases, in which therefore our present $(p, v)$ formula can no longer obtain. There seems to be nothing in the experimental facts to disprove such a view. Indeed the actual phenomena would not be very different: without an $S$ bend in the $(p, v)$ curve the phenomenon of abrupt liquefaction cannot occur; and a definite sharp critical point must in any case be exhibited where the bend vanishes. Notwithstanding the aptness of the simple van der Waals formula, the gradual transition from the molecular freedom of perfect gas to the high molecular crowding of liquid can hardly be claimed to be accomplished without some sort of gradual change of aggregation, especially where it is intensified just above the critical point[2]. The very remarkable flickering striae described and figured by Andrews[3], when a liquid is heated in a confined space at a few degrees

*[margin note: possible explanations.]*

*[margin note: Gradual structural change.]*

---

[1] There has, I find, been much discussion as to whether the phenomena of the critical point are really perfectly definite: an account with references is given by Graetz in Winkelmann's *Physik*, under the section "Wärme": cf. also Poynting, in Poynting and Thomson's *Heat*, p. 192: also for a complete account, Onnes and Keesom, "Die Zustandsgleichung," *Encyc. Math.* or *Leyden Reports*, 1912, cf. § 50.

*[margin note: Critical points definite.]*

[2] It is even to be expected, on any reasonable view, that the physical cohesion of encountering molecules, which at lower temperatures is frequent enough to establish change of state, should still occur less persistently at temperatures somewhat above the critical point, and in fact be a main cause of the large changes of volume that there accompany slight differences of pressure.

*[margin note: Sluggish cohesion.]*

[3] *Scientific Papers*, p. 337. There is nothing to contradict this in the more recent explorations of the critical isotherm.

above the critical point, might be due partly to such a cause; they are ascribed by him to the very great changes of density there produced by small changes of pressure or temperature, and the sluggish recovery from disturbance of density that is involved; but that need not be

Confirmed by<br>Andrews'<br>observations. their sole cause. "They are always a clear proof that the matter in the tube is homogeneous, and that we have not liquid and gas in presence of one another"; the latter state of affairs being indicated, as he states, by a fog or cloud in the tube, instead of by striation.

In the above the propagation of internal disturbance has been investigated on the basis of constant temperature. It may well be that adiabatic conditions would be more in keeping with the rapid

Constant<br>temperature. periods, as in the propagation of waves of sound in gases. But the conclusions as to internal stability would hardly be affected; for, in accordance with Andrews' remark just quoted, when instability is approached the oscillations become very sluggish.

It is well known that a liquid like water can sustain very large negative pressure, or tension, on the hydrostatic piston, provided it fills the reservoir without the presence of any bubble of vapour or other gas. This corresponds to the first descending arm ($PN$) of the isothermal ($p$, $v$) curve. If the temperature is far enough below the critical point, the lowest point $N$ may fall below the axis of zero pressure; but the vapour-liquid line ($PL$) must, of course, be above the axis in all cases. Actually, as shown above, internal stability breaks up before $N$ is reached; thus the condition of $N$ situated on the axis gives too high a value to the limiting temperature below which negative pressure in the fluid is possible.

If we could employ the simple van der Waals equation

$$p = \frac{RT}{v-b} - \frac{a}{v^2},$$

Negative<br>pressure<br>impossible<br>at high<br>temperatures. where $- a$ represents the $\frac{1}{2}A_0$ above, so far from the critical point $T_1$, there would be two real values of $v$ for which $p$ vanishes so long as

$$RT < \frac{a}{4b}.$$

Only at temperatures below this limit (say $T_0$) could the pressure become negative.

The usual determination of the critical point $T_1$ by the conditions $dp/dv = 0$ and $d^2p/dv^2 = 0$ gives

$$RT_1 = \frac{8a}{27b}.$$

Thus we would have        $T_0 = \frac{27}{32}T_1.$

for all substances. For example, the critical temperature for water is Case of pure water. given as 365 [or 638 absolute]; thus the equation would indicate that water could not sustain negative pressure at temperatures above [538] absolute, or [265°] C., and the conclusions drawn above would show that this limit is in excess. [This temperature is too high to interfere with] certain theories of the suspension of sap in high trees[1], and other phenomena in which intervention of liquid tension has been invoked.

[1] Cf. J. Joly and H. H. Dixon, *Phil. Trans.* 1895 B; also, Worthington, "On the Continuity of the Phenomena from Positive into Negative Pressures," *Phil. Trans.* 1892 A. [Cf. the osmotic view of transpiration of sap as *supra*, p. 304.]

# MUTUAL REPULSION OF SPECTRAL LINES AND OTHER SOLAR EFFECTS CONCERNED WITH ANOMALOUS DISPERSION[1].

[*Astrophysical Journal*, Vol. XLIV, pp. 265–272, 1916.]

IN *Philosophical Transactions of the Royal Society of London*, Vol. CLXXXIX, p. 240, 1897, which I believe was the earliest attempt at detailed discussion of such matters, the refractive index of a gas in the neighbourhood of an *ideally simple* line is given by

Classical law of optical dispersion in a gas.

$$\frac{\mu^2 - 1}{\mu^2 + 2} = \frac{ng_1}{p_1{}^2 - p^2}$$

where
$$p_1 = \frac{2\pi c}{\lambda_1}, \quad p = \frac{2\pi c}{\lambda};$$

$n$ is the number of molecules per unit volume, $g_1$ a molecular constant for the line $(p_1)$.

The absorption line is black *absolutely* from

Nature of absorption band near a free period.

$$p^2 = p_1{}^2 - ng_1 \quad \text{to} \quad p^2 = p_1{}^2 + 2ng_1;$$

thus its breadth $\delta p \left( = -\frac{2\pi c}{\lambda_1{}^2} \delta\lambda \right)$ is given by

$$\frac{\delta\lambda}{\lambda} \left( = -\frac{\delta p}{p} \right) = \frac{3ng_1}{2p_1{}^2}.$$

It is asymmetric with respect to the emission line by one-sixth of its breadth toward the violet end.

How far from this absorption line is the extra refractive power of the medium, due to the material that produces the line, recognizable? 

Pressure effect. The pressure effect in a gas is substantially, or largely, an effect of density operating through the increased value of the dielectric

---

[1] From letter to G. E. Hale, of date April 1916, prompted by S. Albrecht's paper [*Astrophys. J.*]; revised August 26.

Solar ionization. [This theoretical discussion would lose its solar and stellar application on the recent view that the pressure in regions accessible to observation is very small compared to an atmosphere. Yet the general magnetic field of the Sun was found by Hale and his associates (1918) to be cut off very rapidly with increase of height: on possible cause for such an effect, applicable to a highly electronized atmosphere, cf. Vol. I, *supra*, p. 13; also pp. 644, 650 *infra* on atmospheric electric waves, where it appears that a slight distribution of electrons or ions, provided their free path is long, can prevent any electromagnetic effect, and therefore any change of magnetic field, from being transmitted through.]

coefficient $K$ or $\mu^2$. The train of ideas in *Astrophysical Journal*, Vol. XXVI, p. 120, 1907 [*supra*, p. 341], which of necessity is most rough, only indicating the order of magnitude to be expected, shows that change of $\mu$ from 1 to 1·003 in the surrounding gas may be expected to alter the wave-length of a vibration of a molecule immersed in that gas by an amount of the order given by

$$\frac{\delta\lambda}{\lambda}\left(=-\frac{\delta p}{p}\right)=10^{-6}.$$

This uses the data of Humphreys' original observations. *A change in $\lambda$ of* 0·001Å. *means* $\delta p/p = \frac{1}{3}.10^{-6}$, *say, for yellow light, and to produce it would thus require the change in $\mu$ to be of the order* $\frac{1}{2}.10^{-3}$. Using this value $\frac{1}{2}.10^{-3}$ for $\delta\mu$, put then for the gaseous atmosphere there

$$\mu^2 = 1 + \epsilon + 10^{-3},$$

where $1 + \epsilon$ is ordinary, owing to bands at a distance, and $10^{-3}$ is local, selective, or anomalous (the value of the constant of astronomical refraction makes $\epsilon$ for air at the bottom of the atmosphere $3 \cdot 10^{-4}$) and the dispersion formula gives, as $\mu^2 + 2$ is practically 3,

$$\frac{10^{-3}}{3} = \frac{ng_1}{p_1^2 - p^2};$$

while the breadth of the absolutely black part of the line is

$$(\delta\lambda)_0 = \frac{3\lambda_1}{2p_1^2}\,ng_1.$$

This gives $\qquad p_1^2 - p^2 = 3\cdot 10^3 ng_1 = 2\cdot 10^3 \dfrac{p_1^2}{\lambda_1}(\delta\lambda)_0$

or $\qquad\qquad 1 - \dfrac{\lambda_1^2}{\lambda^2} = 2\cdot 10^3 \dfrac{(\delta\lambda)_0}{\lambda_1},$

giving $\qquad\qquad \dfrac{\lambda\delta\lambda}{\lambda^2} = 10^3 \dfrac{(\delta\lambda)_0}{\lambda_1},$

or say $\qquad\qquad \delta\lambda = 10^3 (\delta\lambda)_0.$

Thus $\mu$ is altered in the gas by the assumed amount $\frac{1}{2}.10^{-3}$ for radiation at a distance $\delta\lambda$ from the line $\lambda_1$ causing the alteration, which is given by $\delta\lambda = 10^3 (\delta\lambda)_0$; viz. if $(\delta\lambda)_0$ is the breadth of the part of an absorption line of *simplest type that is absolutely black, the wave-length of an adjacent independent line will be affected to the order* 0·001Å. *if that line is at a distance from it of* 1000 $(\delta\lambda)_0$. The molecules of the gas are supposed to be stationary in this deduction: thus really $(\delta\lambda)_0$ is the very small residue that remains when the Doppler-Fizeau effect of molecular translatory motion is subtracted.

Estimate of this repulsion of adjacent lines:

This is very rough and incomplete, and we have had to introduce the unknown $(\delta\lambda)_0$. But Becquerel made observations on the sodium $D$

can be improved.

lines long ago, from which the anomalous change $\delta\mu$ in their neighbourhood can be measured directly without doubt: in *Comptes Rendus*, Vol. CXXVII, p. 899, 1898; Vol. CXXVIII, p. 146, 1899, quoted by Kelvin in *Philosophical Magazine* (5), Vol. XLVI, p. 494, 1898, or *Baltimore Lectures*, p. 176, where, however, the investigation is imperfect (except for a gas) and should be replaced by the above. R. W. Wood has, I think, measured the deviation more precisely since. It is of course easy to measure it with adequate apparatus. Using such direct data, the only remaining question is whether the rough hypothesis quoted above from the *Astrophysical Journal* gives the right order of magnitude; it errs, I think, as an under-estimate.

I *now see* that King's flexured spectra give pertinent information.

Test by prisms of vapour.

The flexure is due to the light passing across a horizontal trough of vapour, something like a prism, which involves deviations of the order $\mu - 1$ as the angle of the prism is about half a right angle or more. If this angular deviation is about $\frac{1}{2}.10^{-3}$ (which would give $\delta\lambda$ about 0·001 of the black breadth of the absorption line), it must give rise to a transverse displacement in the spectrum of $\frac{1}{8}$ inch, as it appears on the photograph, if the plate was at a distance of 1000/4 inches, say 20 feet, from the vapour prism, lenses in the path being allowed for. Even in that experiment it is thus likely that the density of the vapour was below the limit that could be expected to show a dynamical influence on the wave-length of an adjacent line (separated by 1000 $(\delta\lambda)_0$ from it.

I have now looked up R. W. Wood's paper (*Philosophical Magazine*, Vol. VIII, p. 295, 1906) and that of Julius (*Astrophysical Journal*, Vol. XXV, p. 95, 1907). The latter gives on p. 99 a table of densities of saturated sodium vapour; at temperature 420° C. the density is

Effect very slight for metallic vapours.

only $\left(\dfrac{0\cdot0013}{0\cdot000007}\right)^{-1}$ or 1/200 of that of atmospheric air. Such figures seem to give no chance of affecting sensibly the wave-length of an adjacent line.

Julius speaks of $\mu - 1$ being so very great as 0·36 at 0·4 of an angstrom from a $D$ line, quoting Wood. This would imply deviation of that radiation (0·4 from $D$) of the order $\mu - 1$ radians in getting refracted obliquely out of such a region of sodium vapour. He says nothing about the breadth of Wood's $D$ line, probably far removed from the ideal simple type with which we have been dealing; if it came anywhere near to 0·4Å., such adjacent light just outside the margin of the line cannot travel far without absorption or scattering in the sodium vapour. In any case its elimination would only produce a blurring of the $D$ line, what, I think, Julius calls a "dispersion band." This does not exist sensibly in the solar spectrum, though it does in Wood's vapour—thus indicating that the density of vapour in the Sun is much less than Wood's, which makes it to me un- Low density in Sun obviates expectedly, but of course not surprisingly, small and obliterates anomalous influences—extending even to the case of the arc when its effects of anomalous dispersion: lines are sharp, for which the Pasadena laboratory has now a negative result, even in an electric furnace. If one had thought of all the above, the experiment on mutual repulsion of furnace lines might have been unnecessary.

As to the suggested distortion of the form of solar prominences by anomalous refraction: Consider a volume of glowing vapour as represented in Fig. 1. Light emitted from a point $A$ in it is refracted to the observer as if it came from $A'$. This crude representation would be absent even exaggerates the effect. Even if the anomalous part of $\mu - 1$ were for very 0·36, according to Wood's measurement, which is absurd in the high density, present circumstances, the angle $ACA'$ would be only about 20°, so that the abnormal displacement (of $A$ to $A'$) would be less than one-third of the depth $AC$ of the highly refracting vapours, and that in the most favourable case; but under such conditions the path of the on account of short path: light from $A$ to $C$ could not be more than a few miles before extinction arrives, so that the distortion of position ($A$ to $A'$) could not possibly be so much as one-third of this. On the other hand, if the vapour density is much smaller, so is the angle $ACA'$, to an extent compensating the longer possible path $AC$.

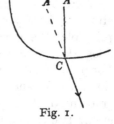

Fig. 1.

The stratification of the vapour must be on the whole parallel to the Sun's surface. If then distortions existed, they would be more prevalent and conspicuous and the prominence lines would appear more ragged nearer the limb, where the light issues obliquely, than around the centre of the disc.

The abnormal flare of the Sherman $F$ line of 1872, figured in Abbot's *The Sun*, p. 163, is surely broadening owing to the condensation or other affection of the gas which is revealed by the concurrent continuous spectrum.

If anomalously refracted light from $A$ thus appears to come from $A'$, $x$ miles distant, while the other parts of the light appear to come from the true source $A$, the spectroscope will receive the former at an inclination $x/D$ to the true line of sight, where $D$ is the Sun's distance. If $AA'$ were as much as 10 miles, the inclination would be 1/42 of a second of arc. This would show itself as an apparent change in the angular dispersion for that constituent of the light, and therefore as an apparent proportionate change to $\lambda$, of the same order

an extreme estimate.

when a high-dispersion grating is used; it would then amount to about $\frac{1}{2} . 10^{-3}$ of an angstrom in the visible spectrum[1]. On the other hand, the first estimate made above refers to a true change of wave-length (or period) of the source, arising from the influence of neighbouring sources of adjacent wave-length. The considerations advanced by W. H. Julius (*Astrophysical Journal*, Vol. XLIII, p. 43, 1916) refer to deviation of rays in the Solar atmosphere by anomalous refraction without change of period; these apparently are subject to the extreme estimate just now made, and thus seem to be beyond present instrumental means[2].

The considerations *supra* as to change of period hold good also for influence of an adjacent line of the *same* substance, provided this line arises from a separate independent vibration. A very searching test

Satellites and independent lines.

would be afforded by the behaviour of the satellites of a line, when the vapour density of the source is increased; but a negative result can be taken as an indication that the satellite is an essential part of the vibration of the main line, and not due to a different source which can be varied independently of it; cf. Evershed's observations, recently reported in *The Observatory*, Vol. XXXIX, p. 59, 1916, and the suggestion on these lines that followed in connection therewith.

The case of rays part of whose path is tangential to the gaseous strata of density needs further consideration. By James Thomson's principle, the curvature of a ray in a heterogeneous medium is

Case of rays along the strata:

$d \log \mu / dn$, where $dn$ is measured along its normal drawn inward. For a gas this is practically $d\mu/dn$; and if the ray travels for a considerable distance nearly tangential to a stratum across which the gradient of density is very steep, its total deviation will be large, giving rise to possibly hundreds of times the $\frac{1}{2} . 10^{-3}$ of an angstrom mentioned above. But would it be shown by the spectroscope?

Suppose the curves of the diagram (Fig. 2) to represent the stratification of density in the mass of gas. The light from the different point sources, in or behind the mass, which is caught up by the spectroscope, is a narrow ray from each point, all parallel. Where the ray is tangential to a surface of stratification it is drawn around toward it, as regards that constituent of frequency susceptible to anomalous refraction by the gas, while the rest passes on; but elsewhere the paths are practically straight. If in its path the ray

[1] See footnote *infra*.

criterion between Doppler deviation and dispersion.

[2] If I am not under a misapprehension in this intricate subject, it seems to follow that true change of period, such as arises from line-of-sight motion, implies angular displacement of the lines increasing in proportion to the dispersive power of the spectroscope that is employed, while mere anomalous deviation of the rays, without change of period, would give the same angular deviation for all dispersive powers.

is anywhere tangential to a stratum[1], its anomalous constituent is thus thrown off the slit of the spectroscope; but its place is taken there by the same constituent from a source more to the right in the diagram, by a few hundreds of miles at the very most (cf. *supra*), owing to the shallowness of the Solar atmosphere. The light thus caught by the telescope from this source, at one side of the source at which it is pointed, is not quite parallel to the main path of the light; and this difference of direction would come out as a displacement of the spectral line, which might amount to $10^{-2}$ of an angstrom, thus simulating motion of the original source in the line of sight. But the spectroscope could hardly show it.

Effect slight even then.

Fig. 2.

The smallest available breadth of the slit represents hundreds of miles in the region of the Sun whose image is thrown on it. Thus, for these phenomena to be observable without entire loss of sharpness, the parallel rays from a band of sources hundreds of miles wide must be deflected to nearly the same extent where each is tangential to a stratum. This necessary regularity requires that the gradient of density should remain about the same, and of substantial amount along hundreds of miles across the gas; and, in order to account for the observed shifts of lines, bearing in mind the small atmospheric pressure notwithstanding high gravity (cf. the remark on scale, *infra*), this would seem to pile up the density of the gas of the Solar atmosphere to quite impossible values.

*Great homogeneity required to show any dispersive effect; thus debarred.*

Actually in the higher levels the densities are for various reasons probably almost infinitesimally small.

A general grasp of these matters, in their relation to laboratory experience, is perhaps facilitated by use of a principle which follows from the formula for curvature of the rays, viz. that the deviations of rays in the Solar atmosphere are the same as they would be in a model atmosphere, on a scale reduced uniformly in all three dimensions of space, but with the same densities of the gaseous constituents at all corresponding points.

*Deviations by refraction are independent of spatial scale.*

[1] It cannot be anywhere tangential to the strata if its source is in the front half of the mass of gas.

# 88

# PERIODIC DISTURBANCE OF LEVEL ARISING FROM THE LOAD OF NEIGHBOURING OCEANIC TIDES.

[*Proc. Roy. Soc.* Vol. xciv A (1916): abstract of memoir
by K. Terazawa, published in *Phil. Trans.*]

IN Hecker's observations on the lunar deflection of gravity, the force apparently acting on the pendulum at Potsdam is a larger fraction of the Moon's direct attraction when it acts towards east or west than when it acts towards north or south. A similar result has been found by Michelson in his observations of the lunar perturbation of water-level at Chicago. A calculation is here made to ascertain to what extent the tilting of the ground caused by the excess pressure of the tide in the North Atlantic is important for the explanation of this geodynamical discrepancy. Replacing the North Atlantic by a circular basin of radius 2000 kilom., taking the position of Chicago to be 1000 kilom. from the coast, and the rigidity of the Earth to be $6 \times 10^{11}$ C.G.S., it is found that the attraction effect of a uniform tide, per metre of height, is about 0·0024″, while its tilting effect is as much as 0·0069″, the maximum of the direct lunar attraction being 0·017″. If the surface of tide is ellipsoidal, shelving towards the coast, nearly the same result is reached for the same mean tidal height.

## Note by Communicator.

[*July* 16. The following proposition, which is implied in the paper, deserves explicit statement: it is readily proved for a load localized at a point[1], and follows generally by superposition. A load is distributed anyhow on the Earth's surface, over a region not too large to be considered as plane; for each locality it alters the level of the surface by its weight, and the direction of the vertical by its horizontal

Generalized result.

attraction*; for material that is homogeneous elastically, the ratio of these two effects is in all cases the same, being $\dfrac{1-\sigma}{\mu} \dfrac{g}{\gamma} \dfrac{g}{2\pi}$, where $\mu$ is the rigidity of the material and $\sigma$ its ratio of lateral contraction to elongation, $g$ is gravity, and $g/\gamma$ is the mass of the Earth divided

---

[1] Cf. Love's *Theory of Elasticity*, § 135.

* Deviations of the vertical by the direct attraction of the tides were estimated by Cavendish in 1773, see *Scientific Writings*, Vol. II, p. 403.

by the square of its radius[1]. Thus, for steel, taking $\sigma = 0\cdot27$, $\mu = 8 \times 10^{11}$, the deflection of the surface is always and everywhere $2\cdot1$ times the deflection of the vertical, the total effect being their sum. For glass the ratio is $6\cdot7$.

The complete results for various distributions of load can be written down at once from known expressions for cases of attraction: for example, the case of elliptic loading (over any elliptic area), which is considered at the end of the paper.—J. L.]

[1] Since writing the above I have found that this general relation was stated by Dr C. Chree in 1897 (*Phil. Mag.* Vol. XLIII, p. 177), with some particular applications to cases of rectangular loaded areas; like results had previously been given by Sir G. Darwin for the special case of a system of parallel ridges. [For history, see note by Chree next following in *Proc. Roy. Soc.* Aug. 1917; see also G. H. Darwin, *Phil. Mag.* Dec. 1882.]     Historical.

# 89

# PRESIDENTIAL ADDRESS TO THE LONDON MATHEMATICAL SOCIETY.

[*Proc. Lond. Math. Soc.* Ser. 2, Vol. XVI, pp. 1–7 and 8–42 (Nov. 2nd, 1916).]

IN the midst of the universal cataclysm of the War, when all interests are strained towards the national defence, the London Mathematical Society has passed, without notice, its fiftieth year of activity. The

*Early history of Lond. Math. Soc.:* first meeting was held at University College, on January 16th, 1865, and heard an address by Professor de Morgan on the aims and prospects of the Society. The de Morgan medal is a reminder for us of his predominant share in the inauguration of the Society, which he did not survive long to guide. In the early days the publications consisted of a series of pamphlets separately paged, containing single communications; the names of Sylvester, Cayley, Harley, Tucker, occur as authors in the first year. There followed later brief reports of meetings, along with papers by de Morgan, Sylvester, Crofton, Cayley, H. J. S. Smith, Cotterill, and others. These publications now stand as Vol. I of the first series of the *Proceedings*. With Vol. II, which begins with the Annual General Meeting of November 8th, 1866, the *Proceedings* became crystallized into a form which has persisted substantially, except as regards size of page, to the present time. The Society began operations with 27 original members, nearly all of them members of University College, London; at the end of the first year the number of members was 69, rising to 94 in November, 1866; and the Society had already become representative of British Mathematical Science, by having on its roll most of the eminent investigators in our subject belonging to Cambridge and Oxford as well as London.

On January 15th, 1866, it was resolved, "That steps be taken to ascertain on behalf of the Society whether and on what terms rooms can be obtained at Burlington House," and on November 8th a report was made that "by the kindness of the Chemical Society in lending their rooms, the Society had been enabled to hold their meetings at Burlington House, where they now meet for the first time." By 1868, most of the British authorities on Pure and Applied Mathematics of that time, who were resident within reach, including de Morgan, Cayley, Sylvester, Hirst, Crofton, H. J. S. Smith, Archi-

bald Smith, Clerk Maxwell, Spottiswoode, S. Roberts, Clifford, early members.
Stirling, had been taking active share in the work of the Society by
attendance and service on the Council, as well as by the contribution
of papers for discussion at the meetings. We must not omit from this
list Lord Rayleigh, whose memoirs illuminated our *Proceedings* for
many years; who, stimulated by the increasing importance of the
Society, became the donor of our most substantial benefaction, which
has largely increased our resources for publication ever since the early
days. In November 1870, the Society migrated to rooms occupied
also by the British Association, in the house of the Royal Asiatic
Society, 22 Albemarle Street, where accommodation was found for
the library, of which a nucleus had been formed by the books of
Sir J. W. Lubbock, the physical astronomer, presented by his son,
afterwards Lord Avebury; and there by successive forms of tenancy
we have remained until now.

For some years past the library, rendered valuable by accumulation
of scientific journals through exchange, and by donations of books,
has quite outgrown the accommodation available: and weighty
complaints became frequent that, by overcrowding, the books had
become, notwithstanding the zeal of successive honorary librarians,
almost inaccessible to members of the Society. The problem, thus
pressed upon them from many sides, was taken in hand resolutely
by the Council during the last Session: and after various plans had
been proposed and closely considered, a solution was reached.

It came to the knowledge of the Council that the Royal Astronomical Society would probably be willing to extend hospitality to Location.
the Mathematical Society, as regards both place of meeting and
general headquarters, thereby establishing or rather renewing an
alliance between British mathematicians and astronomers, whose
activities have always interpenetrated with the closest mutual benefit.
Following on the confirmation of this plan, subject to the approval
of the Office of Works, arrangements have also been made with great
cordiality by the authorities of the Science Museum at South Kensington, whereby our library will be deposited in their scientific
library under a scheme which will maintain full use of it by the
members of the Society, in surroundings where the cognate scientific
literature, and extensive mechanical applications of mathematical
principles, will be accessible for study.

We have therefore the pleasure now of holding the first of our
meetings under the new conditions, at Burlington House, in very
congenial surroundings.

The necessities of the national emergency have mobilized with
striking success the industrial resources of science, hitherto neglected

too largely in our defensive organizations. A most welcome result is the increased sense that has arisen of the national value of scientific pursuits: but danger is by no means absent that, in the haste to secure the material fruit, the welfare of the tree of knowledge, the pure and fertile source from which it springs, may be neglected or even impaired, and like others of ancient days, as well as recent times, we may succumb to the temptation "propter vitam vivendi perdere causas."

It is our duty here to take into consideration how our own special energies may best be rejuvenated and renewed, so as to become more effective in the enhanced and purified national life which, as we trust, will emerge from our present ordeal. Mathematical knowledge, in all ages the ally of all sustained and exact activities, is now more indispensable than ever, when our material wellbeing depends so much on scientific engineering in its mechanical, electrical, and chemical forms. The highest commendation of any growing department of research is to be able to say that it is approaching the quantitative, the mathematical, form: many sciences, formerly descriptive and classificatory, are even now struggling to assimilate a mathematical method. But if it is just to claim that other sciences, nowadays even the biological, aspire with increasing success to become mathematical, that is, exact, in structure, there is on the other hand a duty enjoined on mathematicians to see to it that the main stream of their discipline is kept accessible—free from specialities and complexities, which, valuable and promising as they may be and usually are, on their own account, to those capable of cultivating them, are yet for the present outside the current of the main advances of human knowledge. The play of human thought knows of no boundaries: it can pursue and clarify itself without limitation into endless mazes. All the more we must be careful, in reclaiming and cultivating our boundless domains of mental evolution, not to lose touch of one another: if a theorist cannot command the attention of his own generation, he is hardly likely to attract the interest or serve the purposes of posterity. The one criterion that is available of the value of an addition to pure knowledge, is the human mental interest it can excite. We have our very being inside a well-ordered cosmos, intellectual and material, which it is our highest mental pleasure to explore in all directions, and learn to comprehend; and we have a not unsafe guide in trained instinct and sense of fitness and symmetry, industriously applied, to appraise aright the value of each new departure. Knowledge thus cultivated, on a broad basis, for its own sake, so far from obstructing industrial applications, is their profound source. The study of curves, especially the conic sections, by the

*Marginal notes:*

Mathematics an essential discipline:

the key to progress in material science.

Greeks, at home and afterwards at Alexandria, is not, as is sometimes asserted, an example of mere useless mental ramifications happening to receive an application in later ages: it was on the direct path of progress, and formed the material, adequate and effective, because not unduly complex or abstract, on which the ideas of the infinitesimal calculus, and may we add the mechanics of Archimedes and Galileo, were gradually matured. And if it became in Newton's hands the weapon for the elucidation of the doctrine of universal gravitation, whereby human science first reached out securely into the illimitable universe, what analyst will deny the preordained fitness of the association?

There was a time, when the annual output of the Mathematical Society was smaller in bulk than it is now, that many of us made a point of taking an interest in all the papers that it published. It would be a great thing if we could get back again towards that state of affairs. At least two of our most distinguished analysts have in my hearing traced the aloofness, and even aridity, of much recent work to the neglect of geometrical ideas, the potent source in the past of mathematical progress and consolidation, and the vehicle for the diffusion of our science. It seems a strange phase of development, when we consider the preponderant graphical, tentative, and practical bent of the national intellect, and remember how much of our most characteristic progress and originality in theoretical physics has been, for the sake of being comprehensively grasped and mastered by the mind, so concisely wrapped up in geometrical imagery, and so freed from analytical technicalities, as to have been even obscure to communities trained in more formal and syllogistic methods. *Increasing abstraction,*

There is always risk in getting too far from the main currents of our times: there is the danger, not always avoided, that in the fog of ignorance and the lack of interest, we may encourage expansion in artificial and unfruitful and even tedious ramifications, while criticizing and suppressing with rigour worthy but immature attempts in the well-explored regions of our science, where improvements are so important and originality is so difficult. The contrast with the difficulty of obtaining publication at all a century ago, except in brief summary, gives ground for reflection. *and volume.*

Of recent years the question must have presented itself to not a few of our authors whether the *Proceedings*, developing in so abstract a direction, are now quite as suitable a place for the publication of mathematical physics as they were in the days when Maxwell and Kelvin, and Rayleigh and Routh, were frequent contributors. Yet the potent source of even the most abstract branches of modern analysis has lain in the seizure and orderly cultivation of the

Fluent physical mathematics, intuitional ideas, largely cast in geometrical mould, that are forged by physical science in the effort to systematize its observations of the uniformities of the rational world around us. To renew our strength for wider flights, we must return frequently to mother Earth. The main feature of the technique of physical mathematics is that we are seldom dealing with a completed, and therefore strictly limited, logical complex: it is of its essence that the specification of the problem is fluent and provisional, always ready to take on new features as the discussion opens out. The student of mathematical physics cannot with safety afford to be a specialist: every department of physics is dovetailed into the other departments and progresses by their aid; knowledge must be as far as possible on an intuitive basis, to prevent it from becoming top-heavy, and all the threads must be in hand. For intuition sees, however imperfectly, all round a problem at a single glance; while analysis afterwards consolidates a permanent structure by fitting brick to brick. Even the most abstract of analysts must work at a disadvantage if he has no informed interest in the problems of external nature for which his analysis might be of assistance; and conversely, even the most recondite constructions of pure analysis would be of interest to a wider audience, if they could be expounded in non-technical manner, without the great detail that is sometimes thought to be essential to the necessary degree of precision. as guide to the rationality of Nature. Nature is never irrational, but our main intellectual aim is the redemption of our views of her operations from that reproach: it is the freshly detected and systematically traced concatenations of her working that enlarge out stock of ideas, become for us a source of new generalizations in abstract procedure, giving fresh points of view to be developed and to react in their turn. It is sufficient to cite the names of Cauchy and Riemann, not to mention the supreme examples of Lagrange and Gauss, to show that the most brilliant originality in abstract analysis, and habitude in the intuitions of physical science, can go together, to great mutual advantage.

Fortunately there are signs, abundant on both sides, that the repulsion which somehow arose with us in the last decades between the tentative, yet essentially progressive, though concise, prospecting of mathematical physics, and the stern but limited rigours associated with undiluted pure analysis, is now beginning to be recognized as cramping and unnatural; it may thus melt away in a better mutual understanding, and may one even say mutual interest, to the great advantage of both disciplines. Our analysts have been turning with success, and with a zest of a kind that seems familiar to their more Empirical methods. physical colleagues, to semi-empirical methods in the Theory of Numbers: speculative interest has again arisen even in divergent series, such as would have rejoiced the soul of de Morgan, logician though he was: and the time-worn problems of partitions and combinations have been yielding their secrets to the powerful leverage of an apparatus of arrays and lattices, that may remind us of crystallography and even of thermodynamics.

Our Society has lost by death not a few of her veteran members during my two years of office. Notices of the work of Morgan W. Crofton, W. H. H. Hudson, Benjamin Williamson, have already appeared in the *Proceedings*. In Sir James Stirling, Senior Wrangler of 1860, lately Lord Justice of Appeal, we have lost another of the survivors of our early days, whose interest in our science never flagged, whose mathematical training and gifts were the foundation of a legal and judicial eminence not often arising in a generation. In William Esson, Savilian Professor, and John Griffiths, we have lost two Oxford mathematicians long connected with us. Rev. M M. U. Wilkinson and Rev. J. White had also been members of long standing, while Professor F. R. Barrell had more recently joined us. Though F. W. Frankland, an early member, had passed out of sight owing to distance of domicile, his combination of mathematical and philosophical interests had not become dormant. The actuarial mathematician, Emory McClintock, a pioneer in the great development of our science in America, was for many years an interested member of our body. I may be permitted to add the name of John Henry Poynting: though his life-work attached him to sister societies, his wide physical outlook, combined with mental exactness and penetration, have made for him an enduring name in mathematical as well as experimental physics. <sup>Obituary.</sup>

*Obituary.*

It is our pride and sad privilege to recall the names of the cultivators of our science who, in response to their country's appeal in time of national peril, have already laid down their lives on her behalf. In E. K. Wakeford, Scholar of Trinity College, Cambridge, not a few of us had recognized a future leader in geometrical science. A colleague more senior and more widely known, S. B. McLaren, Professor of Mathematics at Reading, coming from Australia and taking a high degree at Cambridge, had become a learned and philosophical investigator in the difficult domain of statistical molecular dynamics and the relations of the aether to material systems: the work which formed the basis of the recent award of an Adams Prize may remain, I fear, unpublished\*, in any finally revised form. We are entitled also to recall the name of H. G. J. Moseley, who, though he would not have claimed to be a mathematician, had in a brief and brilliant career at Oxford and Manchester contributed fundamentally to the data of

*War sacrifices.*

---

\* *Scientific Papers of S. B. McLaren*, pp. 112, Camb. Univ. Press, 1925.

the mathematical physics of the future, by revealing the earliest universal and unmistakably quantitative relation in the fascinating domain of the correlations of the chemical elements.

Such heavy sacrifices of colleagues who could so ill be spared we must deeply deplore, but not as if they were made in vain. May we not detect beyond them, and on account of them, the promise of nobler and more disinterested times, when the vast destruction of perishable material resources will be far more than compensated in the remembrance of the heroism of the youth of our generation, and in the gain in moral and intellectual wealth that it will stimulate as an abiding possession?

> The world's great age begins anew,
>     The golden years return,
> The Earth doth like a snake renew
>     Her winter weeds outworn.
>
> \* \* \* \* \*
>
> A brighter Hellas rears its mountains
>     From waves serener far:
> A new Peneus rolls his fountains
>     Against the morning star.

# 90

# THE FOURIER HARMONIC ANALYSIS:
## ITS PRACTICAL SCOPE, WITH OPTICAL ILLUSTRATION*.

[Continuation of previous Address: read November 2nd, 1916.]

THE expression of a periodic function, representable as regards a single period by an arbitrary curve or graph, however complex, in terms of trains of simple undulations with a discrete series of periods, multiples of a fundamental period, has been before the minds of analysts since the days of Euler and Lagrange. To the latter a well-known trigonometrical formula is due, which assigns a train of $n$ simple periodic components so as to pass through $n$ specified equidistant points on the graph. The transition from this finite result to the infinite series of periods representing (under the necessary limitations) the graph in all its detail, is in a way parallel to the transition made by Fourier from his periodic series, with its discrete group of periods, to the Fourier integral, with its continuous range of periods and no limitation to periodicity in the graph which it represents. *Analysis of a periodic graph; as obtained from a discrete set of values: transition to non-periodic graphs.*

It was mainly with the view to rendering the crude numerical sequences of data, supplied by observation of Nature, amenable to analytical discussion, that Fourier became closely occupied with these expansions. Moreover, the graphs representing the initial states in physical problems, such as that of conduction of heat which he formulated and worked on as a typical example, can and often do exhibit what to the mathematician is violent discontinuity †. Fourier *Fourier's practical aim: not deterred by abrupt changes:*

* The practical importance of the Fourier analysis, as a main instrument of physical investigation, may justify this discussion, which abstract mathematicians must regard as semi-empirical. For one thing, contact between the two modes of thought may throw light on both: which the initiation of the abstract theory of integration by Riemann, a mathematician of marked physical proclivities, almost proves. *Riemann as pioneer.*

In questions involving an undefined multitude of active molecules, discussions conducted under the algorithm of integrals may impart only a spurious precision: for the nature of an integral of violent fluctuations is itself a problem amenable to treatment only under severe restrictions.

† One may recall Stokes' practical procedure, in his classical memoir of 1847 (*Papers*, Vol. 1, p. 236), for determination of fields by aid of Fourier analysis, which includes within the solution some selected algebraic expression having the same types of singularities and discontinuities, as known in advance, thus leaving a uniform residue for determination by a Fourier expansion which is rapidly convergent. *Stokes' practical procedure.*

believed, and supported his view by fair reasons, that this was no detriment to the practical application of his method of operation by means of expansion in a series or an integral. But it was not so with great contemporary analysts: Lagrange and Laplace, Poisson even much later, were entirely repugnant to the idea that any degree of graphical or statistical irregularity could be made amenable to analytical representation in this way[1]. And Fourier's early memoir of 1807, in which the content of the *Théorie de la Chaleur* was, as he says (cf. *Discours Préliminaire*, final sections), largely covered, remained at the Institute fourteen years before it was published, in a revised form communicated in 1811, under the care, as it happened, of its author who had become Perpetual Secretary. A large part of the merit of Fourier was in his unwavering insistence for years on the utility and practical advantage of his procedure. Traces of the controversy that surrounded the subject are embedded in the later sections—somewhat miscellaneous, like memoranda (*pensées*) hastily thrown together—in his treatise: and M. Gaston Darboux, in the national edition of 1890, has drawn attention to the fact that Fourier's considerations as to the sufficiency of the expansions at all parts of the range of the graph, however complex the data may be locally, did really go to the root of the matter, though they needed to be systematized, and restricted, into a connected argument by Dirichlet (1827) in order to form a basis for the modern, logically compacted, theory*.

*problem regarded as impossible.*

*So was delayed: has now become a vast abstract theory.*

*Effective practical procedure.*

[1] Cf. the historical account in Riemann, Göttingen *Habilitationsschrift* of 1854, published by Dedekind in 1867; *Werke*, pp. 213–225. Cf. also Arago, *Éloges*, Vol. I, p. 341.

* A brief practical synthesis may be submitted as follows. For any given graph a parabolic curve of the $n$th degree can be drawn through any $n + 1$ assigned points of it: the form of its equation goes back to Newton. To have practical value, the graph would be restricted to a limited range with but slight extrapolation: the representation would be local. On the other hand, for a periodic graph, now of range unlimited, there is a corresponding formula, trigonometrical instead of algebraic, first given by Lagrange. Such formulae are now treated under the head of interpolation: cf. Boole's *Finite Differences*. They may of course depart wildly from the form of the graph even at places between the selected points: the early reluctance of mathematicians to accept the Fourier series was natural. The restrictions that prevent this wandering in value between the assigned points, in the case of the Fourier series for a periodic function, were elicited by Dirichlet, as summed up mainly in a principle of limited variation. Applications in physical science had shown abundantly the practical utility of the Fourier series: indeed in that domain it soon became indispensable, even if only empirically. The question of what happens to the series at an abrupt discontinuity of the graph engaged in early days the attention of Stokes and led him to the general concept of semi-convergence, *i.e.* local loss of convergence in a series, stated also in more abstract terms about the same time by von Seidel. The scrutiny of the Fourier

*Validity local:*

*unless the graph is periodic.*

*Dirichlet's principle:*

*Stokes on semi-convergence.*

In its modern condensed and concise form, nothing could be more cogent and interesting than this argument, as a sheer compulsion of the intellect: perhaps hardly anything could be further removed from an understanding of the physical significance of the terms of the Fourier expansions with which it deals. It affords a striking illustration of the principle that analytical identity consists in its ultimate essence simply of rearrangement of the elements of the expressions under comparison: when the formal collocations correspond to the groupings that present themselves in the physical analysis of Nature, a branch of Mathematical Physics emerges.

A proof may be cogent yet uninforming.

Analysis is intricate rearrangement.

The broad mind of Fourier delighted in general surveys, in preference to minute analysis: the eloquent digressions of the *Théorie de la Chaleur* were probably a greater stimulus and source of inspiration, in their wide sweep, to the masters of Mathematical Physics of the succeeding generations than a more guarded and rigorous analytical exposition of the method could ever have become.

Broad surveys prior to detailed rigour.

*Limitations to Fourier Analysis.* This problem as to the possibility of mathematical representation of an irregular graph, and its limitations where possible, is now far more important for physical science than even Fourier anticipated. Observations, meteorological, statistical, experimental, have been accumulating in vast collections for decades, even for centuries. The problem is insistent, and must have been much in mind with those responsible: What is to be done with them? Can anything substantial be made of them towards the progress of scientific principles, which alone count for progress? To what extent are they worth continuing? Does a century's record of the barometer, after the known diurnal and other periodic parts, of astronomical

Question as to utility of masses of statistics;

series for simple cases of discontinuous graphs had in fact made it familiar to physicists, *e.g.* Kelvin, that in approaching a discontinuity the series fluctuates wildly, but over a range that continually decreases as more terms are included, and on that account of negligible quantitative physical import: for in non-molecular physics what is operative is the element of smoothed-out integral, not a value at a point.

Fluctuations: inessential for physics.

When the conditions for convergence are not satisfied for the graph (a term which need not imply expression by a geometric curve) there is a residue, of rapid fluctuation, superposed on the Fourier series or integral. Physical science has neither perceptions nor instruments to deal with it except in a statistical manner, as one of mean square, *e.g.* of the energy of natural radiation. It is such a residue and its abstract study under various limitations, that has expanded into the vast abstract discipline of the Theory of Functions of a Real Variable. But in physical science the harmonic analysis of Fourier can assert an essential and unlimited importance of another kind, as the mathematical translation of the only feasible practical instrumental analysis of the maze of natural phenomena that can be essayed by us, as Rayleigh in his early time was accustomed to insist. From a different point of view, cf. Rayleigh, *Sci. Papers*, Vol. VI, pp. 131–135, 1912.

Residue:

classed as thermal.

precision of period, have been determined from it and discussed apart, mean anything positive at all, within the range of our powers of analysis, more than a mere congeries of atmospheric instabilities? These and other statistical problems have been too much neglected, to their own practical detriment, by mathematicians. They may

*they present a problem strictly scientific.*

present themselves in an inconvenient manner, hampering the freedom in which the analyst is accustomed to roam; but they are of the essence of the pure science, and do not by any means belong to the rough fringe of technical applications.

The question of the utility of meteorological statistics becomes more

*A partial answer for tidal records:*

emphatic, on account of remarkable success in problems at first sight cognate, achieved concurrently with, even prior to, the activity of Fourier. The theory of gravitation, in Newton's hands, had completely solved the mystery of the tides of the ocean. The practical mathematical problem presented by them, viz. the working out of the subject on dynamical foundations, so as to connect the complex mass of available tidal observations with their thoroughly known main cause, was undertaken in detail with great mastery by Laplace. The hydrodynamical theory of the tides is founded on his work, and has gone on improving on the same lines ever since. Here, as elsewhere, only ideally simplified problems are amenable to complete dynamical treatment: so that the results lie rather in the form of general indications and master principles acquired, than of actual problems resolved. But a different line of procedure was concurrently opened up, which has led, in the hands notably of Lord Kelvin and his colleague Sir George Darwin, to full and most remarkable practical control of the problem. In dynamics, as also in wider fields, periodic

*their periodic components recognized in advance;*

causes produce effects possessing the same periods. Now the gravitational attractions of the heavenly bodies, which originate the tides, are all made up of components having periodicities of astronomical permanence and precision. The periods are not commensurate; but each of them is resolvable into a fundamental period of simple un-

*and extracted:*

dulation and its own harmonics\*. This array of discrete periods ought to stand revealed in unending trains of simple undulations, in the analysis of the tidal graph of each port where a record is maintained. A method of extracting them from the graph, by determining their amplitudes and phases with sufficient accuracy, is all that is required; and their schedule will then express concisely the tidal characteristics

*now by rapid mechanism.*

of that port for all time. An actual tidal record, predicted for the future and verified for the past, can be rapidly run off from them by

---

\* The practical importance of the Fourier harmonics of the pure astronomical periods, for tides as modified by travel into estuaries, has been brought out more recently by the investigation of the Liverpool Tidal Institute.

aid of an appropriate and sufficiently accurate mechanical curve-tracer. But naturally, these permanent periodic components of tide, thus brought out, are not the whole of the fluctuations of level at our port. There is a residue, non-periodic in character unless some periodic influence has been overlooked, due mainly to meteorological causes, wind and rain, thus connected with irregular sporadic in-stabilities in the atmospheric system, that have no element of recurrence, other than transient, involved in them. The contrast here brought into view between true sustained periodicities in a prolonged graph, and a residue, aperiodic and erratic, or containing only transient and imperfect periodic elements such as *seiches* (cf. *infra*), is fundamental. *The unanalyzed residue.*

We have here recalled the essential distinction between graphs which are strictly periodic and those which are not so restricted. The former are decomposable throughout their whole extent, however long, into trains of simple undulations, with their discrete system of frequencies, multiples of a fundamental. Such types may be super-posed to form a compound graph with no apparent periodicity at all, but yet expressible for all its length as a discrete system of periodic simple trains. The wave-trains of higher frequency in the series are required, to include the more abrupt irregularities and fluctuations that may occur along the periodic range of the graph: they tail off to insignificance only when fluctuations, abrupt or of very short range, which would introduce periods of corresponding shortness, are not prominent or not important; in fact a limited series of the components represents the graph with fluctuations of very short range smoothed out. This for practical purposes expresses the now well-known analytical restriction to the validity of the representation of a periodic function by a Fourier series, viz. that the number of its fluctuations within the period must be limited. *Discrete set of periods: the higher ones unimportant and uncertain, under restriction to limited fluctuation.*

The general type of graph, not thus restricted to periodicity, re-quires for its expression a continuous range of frequencies instead of a discrete group: and it is not even expressible in this manner unless it involves only a limited number of fluctuations along the entire graph. The usual case is that of a graph representing a disturbance confined to a finite range along its axis, perhaps tailing away in both directions: when it is reduced to a limited number of fluctuations by smoothing out very minute features, it becomes thus expressible. *Continuous range of periods: the practical case.*

Formally, a single Fourier integral can represent two local groups of disturbance, separated along their common axis by any interval, however great: for the sum of two Fourier integrals is expressible as another such integral. But this transformation holds only under a *Need of limitation in the range of a graph:*

complication which destroys at some stage its practical utility. For example, consider the sum

$$\int f(m) \cos(mx + \alpha)\, dm + \int f(m) \cos(mx - ma + \alpha)\, dm,$$

which combines, with appropriate form of $f(m)$, any two identical distributions local as regards $x$ and separated along $x$ by distance $a$. It is equal to

$$\int F(m) \cos(mx + A)\, dm,$$

where        $F(m) = 2f(m) \cos \tfrac{1}{2}ma, \quad A = -\tfrac{1}{2}ma + \alpha.$

In this resultant Fourier integral* both amplitude $F(m)$ and phase $A$ become more rapidly fluctuating over given range of the period $(2\pi/m)$ as the distance $a$ is taken greater: illustrating the manner in which the representation fails, except for a limited total range of the illustrated data. Subject to such limitation, the portion of a Fourier

*illustrated.*

---

* The expression of these paragraphs has been improved. When there are only two nearly identical groups of disturbance in the radiation that is to be analyzed, the spectrum is widely fluctuating both as regards amplitude and phase when these groups are far apart. It is only the mean square of amplitude, expressing the distribution of energy, that can be the subject of discussion and additive, in the analysis of natural radiation.

*Interference of distant identical groups.*

The main practical phenomena of general harmonic analysis thus appear to be describable briefly as follows. Any graph restricted in length is, under exceptions physical unimportant, expressible as a Fourier integral which can itself be specified in amplitude and phase by a two-fold periodogram. The effect of superposing another identical graph, situated far away along the axis, is that the amplitudes are doubled, while both amplitude and phase become subjected to violent fluctuations over the whole value of each, which become closer without limit as the distance between the two component graphs is increased. The result of superposing a number of such components, identical or nearly so, and ranged irregularly, may be described in similar but of course less definite terms. This is what is implied by the proposition that a graph of unlimited length is not expressible as a Fourier integral, unless it has definite uniform structure: there may however be various types of component graphs contributing to this structure. In integrating to determine the distribution of energy the rapid fluctuations however become smoothed out and so got rid of, while no question remains of phase. The periodogram of energy is thus usually a simple curve, whose ordinates are shown (by the Petzval theorem, p. 549) to be additive as regards combination of component graphs. If the structure of a long graph changes along its length, thus is not homogeneous, its periodograms, as derived from local samples of it, change also: as is exemplified in the discussion on frequency of sun-spots, *infra*, p. 567.

*as regards their Fourier integral.*

*Structure in statistics.*

If component identical graphs are repeated equidistant without limit, the result is expressible exactly by a Fourier series: the periodogram shrinks up into a series of peaks equidistant in frequency, with zero amplitudes or energies between them. Even if these are only two or three or a small number of equidistant identical components, the tendency in this direction, namely towards a line spectrum, should begin to show itself.

*Conditions for a line spectrum.*

integral that is calculated for a given reach of period in any graph belongs in detail to that reach, and makes no contribution to the expression of the parts beyond it. The Fourier integral is, we may say, adopting a concept of fundamental importance as regards fields of energy in Mathematical Physics, a localized representation, not merely as regards the function or graph as a whole, but also in respect of each differential group of periods which it contains: this *aperçu* seems to be the germ of wide corollaries in the dialectic of function theory. The integral is the limit of the Fourier series when the fundamental period of the function analyzed is extended indefinitely, but *only when* the periodic function to which it belongs is not of sensible value except in the middle parts of this long period, remote from the ends: the Fourier series is then, so the say, localized in periodic segments, and the next range where it rises to importance must be far enough away not to produce practical disturbance, rapid crispations in the periodogram being negligible, if we are to neglect all but the local part and so pass from the series to the integral.

The Fourier analysis as localized physically:

as illustrated by transition from series to integral.

In the theory of the tides we deal with unending pure wave-trains: so it is also in the theory of sound, especially for maintained sources such as organ pipes. Mechanical graphs of sound-waves are the ideal illustration of periodic undulations. In the theory of radiation, on the other hand, we have never to do with a single source, but always with a vast congeries of atoms, each radiating temporarily and independently. The resulting stream of radiation may be made up of discrete frequencies forming nearly simple long-sustained trains of waves, as, for example, that emitted by a gas glowing with its line spectrum, and so be periodic in an imperfect sense: or its optical analysis may show a continuous range of frequencies as in natural light. In the latter case, the radiation is not expressible, mathematically, in terms of simple unending trains of waves: still less so, practically, if it is analysed with small resolving power; not so even if the resolving power applied were infinite, unless the train of tumultuous radiation that is analysed is of limited length or duration.

The exact periods of astronomical theory:

contrasted with radiation,

where analysis is possible only for energies.

A true permanent periodicity may be held to be foreign to the Fourier integral mode of representation, for it would be indicated by an ordinate of the periodogram or curve of amplitudes becoming infinite. At first sight one might be tempted even to say that the function of this analysis is to reduce the residue of a graph, given by observation or experiment, to a mathematical form so far as possible amenable to analytical operations, after its truly periodic constituents have been removed by some other process.

Sharp periodicities are outside the representation by an integral.

But there is more than this: the example of a train of radiation has shown that problems presented by sporadic repetition of transient

A welter of sporadic temporary periodicities: periodic elements may be as important as those involving pure periods. One may illustrate by the finite movement of strongly damped light pendulums suspended amid the eddies of a draft of air:

illustrations: or again by the transient *seiches* that are excited in a partially confined bay, such as the Gulf of Lyons where they were shown in the Marseilles tide-gauges, by the irregular disturbance that reaches it from the

question of their existence in meteorology: open sea. Incipient oscillatory motions of such dynamical type might conceivably form part of the general disturbance of the atmosphere with which meteorology deals. If a range long enough to be a fair sample is analysed by the Fourier process, such transient periods, repeating sporadically, will show themselves on the periodogram of amplitude or, in strictness, of energy (cf. p. 548, *infra*). If the analyses

the test. of two adjacent ranges of the weather give different results, we must conclude either that the ranges are too short to be fair samples, or else that the indications of periods which they separately show are only accidental.

The mechanical practice of Fourier analysis: Practical problems such as those just indicated were prominent in the work of the British Meteorological Council about thirty years ago, under the inspiration mainly of Sir George Stokes and Lord Kelvin. Preparations were then made for the mechanical harmonic analysis of long graphs of atmospheric pressure and temperature: but difficulties in running the Kelvin integrating machine seemed to have stopped the project[1]. More recently the basic principles of the subject have been much illuminated, incidentally, by the writings of Lord Rayleigh.

limitations on it suggested by optics. Professor Schuster has greatly stimulated its practical side by clearing the ground for an organized attack, on the lines of the optical resolution of radiation; and he has carried the calculations through for the available statistics of sunspots and other cosmical phenomena, with remarkable and instructive results, some of which will come under

The extraction of a true constituent of known period. notice. The different and much more simple problem, already referred to, of the separation from a mass of statistics of a definite permanent periodic component whose period is known, is illustrated classically by the isolation, and separate discussion as regards origin, of the diurnal oscillation in the data of terrestrial magnetism[2].

Comparison with law of errors: The principles underlying all such discussions, and defining the degree of their validity, are essentially a subject for mathematical development. The writings of Laplace and Gauss on the cognate subject of the distribution of errors and irregularities must ever

[1] Cf. various notes in Stokes' *Math. and Phys. Papers*, Vol. v.

[2] Carried through by A. Schuster, *Phil. Trans.* 1899, pp. 467–518; cf. also the posthumously published revision of the entire Gaussian representation by J. C. Adams, *Collected Papers*, Vol. ii. [What is isolated is of course not an infinitely sharp period, whose amplitude would be infinite, but the aggregate of exceptionally large amplitudes close to the place.]

remain classical, though now they may be not much studied in detail: and modern extensions, by Karl Pearson and other investigators, for application to rougher material, might be more widely known, in abstract at any rate. In de Morgan's and Sylvester's time it was not unusual for mathematicians even to engage professionally, to some extent, in statistical work on the actuarial side. They could perhaps still do much to diffuse economy of arithmetical effort, to guide statistical undertakings, now becoming prominent in many branches of knowledge, to concise and significant expression. For the magnitude of the statistics that present themselves in the meteorological and other domains too readily leads to contentment with laborious and unfruitful routine, inspired possibly in part by exaggerated fear of the consequences of what is known as a break in the continuity of the record[1]. Progress might be assisted in all such departments, with benefits not all on one side, by continuous and sympathetic scrutiny of the nature of the operations, and the scope and the degree of validity of the conclusions, and by interest in their improvement, on the part of modern analysts, to whom minute discussion of the nature and limitations of functional discontinuity in other aspects has become highly congenial. No apology is required for reviewing the practical side of the subject, in an elementary way, before the Mathematical Society, especially as a variation from the purely critical and negative attitude on such matters that is not unfamiliar to us.

*developing into statistical theory:*

*which may degrade into routine:*

*unless stimulated by positive criticism.*

*Analysis of a restricted Range of a tabulated Disturbance: its degree of precision.* If a graph of length $L$ is under consideration, the usual method of extracting the part that repeats itself periodically in equal lengths $\lambda$, is to cut the graph up into segments of this length, and superpose them by numerical addition or else mechanically: if there are enough segments, the resultant divided by the number of segments approximates to that periodic part of the whole which is sought. Analysis by the Fourier series is then applicable to the part, periodic in length $\lambda$, that is thus isolated, and can resolve it into a train of simple undulations and its harmonics. The amplitudes thus obtained are ordinates in the continuous periodogram of the original graph. In the use of an optical grating, which performs precisely this operation of superposition of equal reaches on trains of radiation, the periodic part thrown off in any direction thus includes, as terms in its Fourier series, overlapping elements of the spectra of the various orders; and these can be separated in space by a new optical resolution.

*Isolation of a periodic component:*

*which in optics includes elements of all orders of spectra.*

[1] The question deserves to be put explicitly, in the light of this discussion— What detriment is caused by a break in a meteorological or magnetic record, other than the absence of observations for temporary comparison with those of other stations? [Cf. the case of the sunspot record, *infra*, p. 568.]

*Is a continuous record really necessary?*

It is easy to recognize that this double operation for finding the ordinates in the periodogram of amplitudes is equivalent, except for the repetition[1] of the same periods in the different harmonic groups, to breaking up the Fourier integral process into two stages: by analysing into a series the mean sample of the graph, extracted as regards each particular period of recurrence, it brings out one aspect of the relation of the Fourier series to the Fourier integral.

*Two stages in the analysis.*

This procedure, which is exact for an infinite length of graph of restricted fluctuation, becomes very uncertain except in cases of special simplicity of structure, when the length $L$ that is analysed is not large compared with $\lambda$, the length of the periodicity we are looking for. This point is best elucidated by the converse operation: take a length $L$ of the simple periodic function $A \cos (2\pi x/\lambda)$, the graph abruptly vanishing on both sides of that length, and find what the Fourier analysis will make of it. It will give a continuous range of periods with amplitudes falling off from the central wave-length $\lambda$, on both sides, according to the simple diffraction law of optics. The ordinate in the curve of amplitudes, at the point whose abscissa is $\lambda + \delta\lambda$, is

*Illustration of uncertainties:*

$$A\,\frac{L}{2\pi\lambda}\cdot\sin\left(2\pi\,\frac{L}{\lambda}\,\frac{\delta\lambda}{\lambda}\right)\Big/2\pi\,\frac{L}{\lambda}\,\frac{\delta\lambda}{\lambda}.$$

The sharp wave-length $\lambda$, that would belong to an infinite train, has been blurred into a range of wave-length, whose relative extent $\delta\lambda/\lambda$ is of the order of magnitude $\lambda/L$. By the use of a range $L$ of material, we must not expect to isolate a component of wave-length $\lambda$ more closely than this, unless it stands out clear by itself. The very high peak in the curve of amplitudes that would represent a real permanent periodicity is here flattened down into a band with central ordinate of height $AL/2\pi\lambda$, so that the area concerned remains about equal. In other words, the degree of purity obtainable in the resolution of a limited range $L$ of material, containing true periodic elements of wave-length around $\lambda$, is given by $\delta\lambda/\lambda = \lambda/L$: when pushed within differential ranges narrower than this order of magnitude of $\delta\lambda$, the analysis produces only confused results[2].

*criterion for the range of the record that is necessary*

*to secure purity in its spectrum.*

---

*Successive extractions.*   [1] This repetition, which occurs only when periods are commensurate, can be avoided by taking out a periodicity when found, and analysing the residue for the others.

*Practical test.*   [2] This is only a rough illustrative statement. According to experiments by C. M. Sparrow on spectroscopic resolving power, "On Artificial Infinitely Sharp Doublets," *Astrophys. J.* Sept. 1916, $\tfrac{4}{5}\lambda/L$ would more precisely represent the limit of practical resolution, which corresponds to the vanishing of the subsidiary maximum in the periodogram of intensity after it has assumed its limiting form, a point of inflexion: this value is usually employed in what follows.

Thus if there were a permanent $11\frac{1}{8}$ year sunspot period, analysis of material for eighty years could only bring it out as some kind of band of simple periods extending substantially about $\frac{2}{3}$ of a year on each side of the maximum. If the $11\frac{1}{8}$ year period is not permanent—and the analysis of Schuster has brought out the unexpected result that it does not show itself in the previous eighty years' records—the physical problem is reduced to the question why, among the transient periodic elements involved, those of this particular period have assumed such prominence for the range of time that shows it[1]. But the eleven yearly fluctuations also appear, running parallel as regards phase, in many other physical phenomena, and are therefore probably deep-seated in the Sun: this suggests that they are dynamical, and if so they are probably a permanent phenomenon, composed of true periods. Two true periods, differing by some amount less than a year, of nearly equal amplitudes, and recently in concordant phases, would only imperfectly represent the case: alternatively, the amplitude of the simple undulation might be affected by a factor of other than circular type, which has been prominent in recent times. But more prolonged data are apparently needed for closer discussion on these lines. On a view of mere probability alone, it would seem that close periodic repetition seven or eight times is strong evidence of a permanent rather than an accidental cause[2].

*Application to the sunspot record.*

*Related phenomena tested by phase.*

*Evidence of a unique permanent period in sunspots.*

As already stated, the ideal course as regards true permanent periodic terms of known astronomical periods is to eliminate them from the given graph, necessarily of limited extent, in some other manner, and apply the Fourier treatment only to the residue. Of a given reach of this it will provide a local representation, in a form amenable to mathematical and dynamical operations; but it will be more widely significant only so far as different reaches give concordant results.

*True periods: local residue.*

Here the other question arises: is it of much use to attempt a periodogram analysis when the total material on hand is only, say, a dozen times the length of the component periods that are sought?

*Practicability of an analysis:*

[1] The very remarkable periodogram of Schuster (*Phil. Trans.* 1906, reproduced in Professor Sampson's tract on *The Sun*, p. 90) shows a nearly precise diffraction curve for the range 1825–1900, indicating an approximately pure period of $11\frac{1}{8}$ years and little else: while for the previous 75 years, for which the data are more uncertain, the main feature is two very pronounced summits at about 9 and 14 years, both broader at the base than a single pure period would give. The further details are also suggestive. A discussion "On Permanent Periodicity in Sunspots," by N. Yamaga and the writer, on the lines of the views set out in the text, is in preparation. [*Infra*, p. 567.]

*Sunspot record: discordant features.*

[2] A cognate question—given a range of material, and *assuming* the presence of *one* permanent simple period, to find its most probable value—has recently been treated by J. I. Craig, *M.N.R. Astron. Soc.* 1916.

A true permanent period would reveal itself from such restricted range merely as a band of some width, but its summit would be very near the right place: only when it is high and isolated would its shape give presumption of true rather than transient periodicity. In the case of two adjacent sharp periods the representations, thus widened, would overlap; and the problem of their interpretation would be not unlike that of the separation of two adjacent lines by a spectroscope of insufficient resolving power—the two problems would be the same were we not discussing amplitudes instead of energies. Two adjacent true periods of comparable amplitudes would show as a continuous range of periods, or, if they are very near, mainly as one false period half-way between them. It would seem that nothing but analysis of a greater range of the material could determine whether permanent sharp periods there occur, or merely a diffused and local periodicity which will gradually fade and disappear beyond that range.

*conditions:*

In practice, subject to exceptions where there are only a few prominent summits, as in the sunspot case *supra*, it would usually be of no avail, with a limited range of material, to analyse more closely than a moderate fraction of the range $\delta\lambda$ which determines the purity as estimated above; and it is of no use to take the amplitude $f(\lambda)$ over that limit of range as other than constant at its mean value. We can, of course, analyse closer and separate Fourier elements of shorter range, and so more nearly approximating as regards their own graphs to permanent wave-trains: but we cannot assert that they represent the material in hand any better.

*scale of precision.*

The treatment of known true astronomical periods, where amplitudes and phases alone remain to be determined, is illustrated by the Laplace-Kelvin analysis of the tides at a given port. If there is a permanent periodic component in the graph, say $A \cos\{(m+\kappa)t+\alpha\}$, where $\kappa$ is small and unknown, it will be brought out when the graph multiplied by $\cos mt$ is integrated by one of the mechanical integrating methods. The main term in the graph of the integral which the machine turns out is $A/2\kappa.\cos(\kappa t+\alpha)$, which reveals both the amplitude and phase that were sought: there is the drawback that when $\kappa$ is small a long range is required to bring out its period[1]. But if as in tidal theory the period is exactly known, then $\kappa = 0$, and the integral is $A/2\pi.\cos\alpha$ multiplied by the time from $t = 0$. A second operation with the same multiplier in different phase, viz. $\cos(mt+\beta)$,

*How to track out a suspected period.*

[1] Cf. Stokes (*Proc. Roy. Soc.* 1879; *Math. and Phys. Papers*, Vol. v, pp. 52, 53), who suggested this type of procedure for the case of meteorological graphs. The integrating machine does not seem to have been found efficient. Moreover, in the light of this discussion, our comment would be that the true permanent periods can only be the already known astronomical ones, to be tested in succession.

completes the data; and the amplitude and phase are determined. And the latter course is practically effective when the period is only approximately known, for $\sin(\kappa t)/\kappa t$ is stationary at the origin.

*Natural Radiation*\*. Ordinary light is a very complex conglomerate of transient elements, that may be each more or less periodic: it is compounded of pulses or short trains of waves, which might conceivably be all of a single type but probably vary in type according to some statistical law. The shortest segment of a ray of light that we can appreciate separately is of immense length, hundreds of miles at least: for the most rapid detector that can be used requires a sensible fraction of a second to produce its record. Thus only additive qualities can be recognized in light at all, such as energy; and also polarization when that type of regularity has been impressed all along the ray by reflexion or otherwise. A great addition to the resources of physical analysis of light, one that has made Physical Optics a science, is obtained by splitting a ray into two halves, and sliding one along the other, thus producing effects of "interference": this has regular features in that it occurs between the halves of split constituent pulses, all slid apart through the same distance. The utter and overwhelming impossibility of any sustained and therefore detectable interference between lights that have not a common source is patent.

If the light is made up of component trains of simple waves of length $\lambda$, limited in length and irregularly distributed along the ray, interference will not vanish until the two halves into which each of the trains has been split are slid quite apart. It will reveal itself in a succession of light and darkness as the sliding is increased by each wave-length of the train, making $L/\lambda$ alternations in all. In optical interference arrangements these alternations usually show simultaneously, spread out in a band transverse to the superposed rays[1].

But in thus following out in detail, by way of illustration, light made up of limited sequences of simple waves, we are restricted to amplitudes not energies: for the latter are only additive [other than statistically as *infra*, p. 549] when the pulses do not permanently

\* On the corpuscular or *quantum* idea of radiation, the problem that follows would hardly arise.

[1] But if the superposed rays are passed through a spectroscope, the interference will be assisted to a range far beyond this limit. A channeled spectrum will be produced with alternations that become closer as the one ray is slid further along the other: when they have become so close as to disappear they will be restored by use of a spectroscope of higher dispersive power. The reason is that the instrument resolves the pulses into their Fourier constituent groups, which are each dispersed into shorter range as regards frequency and therefore into greater length of [regular] undulation, as the power is increased. Cf. p. 556, *infra*.

overlap. We might combine two overlapping pulses into a single one of more complex type: but that cannot be carried through either. Recourse must be had to Fourier. We can take a finite reach of the complex radiation, longer than the greatest range of interference that we contemplate employing, but at least some hundreds of the wavelengths with which we are concerned, and analyse it into its Fourier integral. Each element of that integral is one of the set of Fourier pulses, if we may introduce that term, representative of that length of the radiation and of nothing beyond. We can reason about these pulses, which represent disturbances limited in length, on the previous lines. We can then proceed similarly with the next reach of the radiation. If such reaches, long compared with the order of wavelengths, and unpolarized, give the same distribution of energy along the spectrum, they satisfy the only applicable test of identity. A reach of radiation is practically homogeneous when its parts thus give identical distributions of energy along their spectra. A reach could be homogeneous to sense, where hundreds or thousands of miles are the least of it that can be observed, which might be very far from homogeneous on a finer scale. In this way we arrive at an answer to the question as to the nature and limits of the irregularity inherent in natural radiation. It might even be constituted, so far as we can analyse it, of the irregular repetition of a single definite type of pulse: to determine this type (usually of few undulations and fading asymptotically) from the law of distribution of the energy along the spectrum, is a definite problem, which could be readily formulated in an integral equation if the pulses did not overlap[1].

*Limitations of Fourier integral.*

*Criterion of homogeneity of light:*

*practical limitations of test.*

[1] See for an exposition of the general subject on different lines, giving some solved special examples, Rayleigh, *Phil. Mag.*, Vol. xxvii, 1889, pp. 460–469; Vol. v, 1903, pp. 238–243; *Scientific Papers*, Vol. iii, p. 275; Vol. v, p. 98.

On the general discussion, cf. "On the Constitution of Natural Radiation," *Phil. Mag.* Vol. x, p. 574, 1905, and Vol. xi, 1906 [*supra*, p. 306]. Also especially papers by Rayleigh, *Phil. Mag.* Vol. x, p. 401, 1905; Vol. xi, p. 123, 1906; *Scientific Papers*, Vol. v, pp. 272–282: and subsequent papers. Cf. also p. 562, *infra*.

[A valid illustration may be made out of a pendulum hanging free, with a field of aereal disturbance rushing past it. The pendulum will extract from it the energy required to assume a state of vibration, which may be about an inclined position: the amplitude will be constantly changing, and also the phase, on account of the vagaries of the exciting air stream, but its vibration will retain continuity and its own mean period, being fed statistically by the pulses passing over it. Imagine now a medium filled with analogous free vibrators, and a chaotic optical disturbance coming on to it: each of them will be set into vibration and, when its steady state has arrived, will radiate as a secondary source all the energy it absorbs from the field of disturbance. On Rayleigh's synthesis, the wavelets thus radiated in direction nearly along the beam will compound with the incident Fourier component of their own

*Vibrator fed by chaotic disturbance:*

*as in natural radiation:*

*Energy Curve of Natural Radiation: Asymptotic Features.* It is instructive for practical issues to inquire what would be the form of the energy curve of the radiation if the constituent pulses were perfectly irregular, and it were thus devoid of any kind of statistical uniformity for however short a range of its length. Such complete absence of correlation along the ray implies that each linear element of the graph of the total radiation has no relation to any other element. Now for a single element of disturbance situated at the origin, represented by $\phi(x)$, the Fourier expansion, so far as regards wave-lengths much longer than the element itself, is independent of all details of its form, being

$$A\pi^{-1} \int \cos mx\, dm, \quad \text{where } A = \int \phi(x)\, dx,$$

so that $A$ is the total area of the graph of the disturbance. Its law of energy, as given by integrated square of disturbance (cf. p. 549, *infra*) is

$$E = A^2\pi^{-1} \int dm = 2A^2 \int \lambda^{-2}\, d\lambda.$$

The energy is thus distributed according to the law $\lambda^{-2}\delta\lambda$, as regards those wave-lengths which are of a higher order of length than the element; but the main part of it remains undetermined in lower wave-lengths. <sub></sub> The energy spectrum for structureless radiation:

We can infer only that the curve of energy of wholly uncorrelated disturbance tails off as $2A^2\lambda^{-2}\delta\lambda$. But the facts for natural radiation correspond to a law of type $\lambda^{-4}\delta\lambda$. This means that $A$ must vanish, in other words that as regards area of its graph, the constituent pulse is as much negative as positive. *disagrees with natural radiation:*

That property does, in fact, hold good for every mechanically propagated transverse disturbance $\phi(x - ct)$. For, if $\phi(x)$ denote displacement, the disturbance $\psi(x)$ is equal to $d\phi/dt$, which is $-c\,d\phi/dx$, so that $\int \psi(x)\, dx$ is $-c(\phi_2 - \phi_1)$, and so is zero for the whole range from end to end. *as was to be anticipated:*

It is also true for an ordinary electrically propagated pulse for a similar reason, by virtue of the Maxwellian equations: the exception

period, regarded as travelling unchanged, to give the complete beam travelling in the medium, but advancing with smaller velocity on account of their phase difference of a quarter period. The smaller velocity thus determined, differing with the free periods of the vibrators, is the cause, in Huygens' manner, of refraction and dispersion. If the optical vibrators are quite free, steady propagation will promptly be established: if they are matted together, after the manner of the sounding board of a violin, it may be only after many vibrations as Stokes seemed to think (*supra*, p. 311). For the sound analogy of the violin or pianoforte, the time of establishment of the pure tone has doubtless been investigated.] *cause of refraction.*

except for radiation from a projected electron: is an electron suddenly projected, which sends out a positive spherical pulse into the aether, the compensating negative one being sent out far behind when the electron is suddenly stopped, and the steady magnetic field of the moving charge, which is left behind it by the initial pulse, is thus abolished. The energy of the spectrum of the impulsive (non-vibratory) part of the Röntgen radiation, as distinct from the sustained wave-trains in its spectrum characteristic of different sources, should thus tail off according to the law $\lambda^{-2}\delta\lambda$ as $\lambda$ increases.

and perhaps diffuse X-rays.

Alternating pulses: But for ordinary radiation the constituent pulses have a distribution of disturbance which integrated over their length is always zero. Now the Fourier expression for the limited graph $c\phi'(x)$ is

$$c\int X \cdot \sin mx\, dm,$$

where

$$\pi X = \int \phi'(x)\sin mx\, dx = |\phi(x)\sin mx| - m\int \phi(x)\cos mx\, dx.$$

This is equal to $-mA$ for constituents of undulation very long compared with the length of the graph, where

$$A = \int \phi(x)\, dx.$$

Thus the vibration in the pulse is $-c\pi^{-1}A\int m\sin mx\, dm$; and its energy is

formula for natural energy spectrum.

$$E = c^2\pi^{-1}A^2\int m^2 dm = 8\pi^2 c^2 A^2\int \lambda^{-4}d\lambda.$$

If the pulses do not overlap, the radiation of long period that passes per second is expressed by

$$8\pi^2 c^3 B\int \lambda^{-4}d\lambda,$$

where $B = \Sigma A^2$ summed for all the pulses per unit length[1].

If ideas are fixed on transverse pulses travelling along a stretched cord of unlimited length, this formula will give the distribution of energy of very long wave-lengths, when the linear density of the cord is unity and $A$ is the area enclosed between the pulse-form on the cord and the axis. The total energy of the pulse is

Waves on stretched cord:

their energy and momentum.

$$2 \cdot \tfrac{1}{2}\int \left(\frac{dy}{dt}\right)^2 dx = c^2\int \left(\frac{dy}{dx}\right)^2 dx = -c^2\int y\frac{d^2y}{dx^2}\, dx,$$

while

$$A = \int y\, dx.$$

[1] Cf. the cognate argument of E. T. Whittaker, *M.N.R. Astron. Soc.* Nov. 1906, p. 87.

The pulse or wave-group in this case carries momentum equal to *half* the energy divided by c.[1]

If these pulses do overlap permanently, there is no resource but to take the shortest sample length of the radiation as a compound pulse: then $A$ is the (residual) area of the undulatory graph which represents each Fourier element of it, and the same conclusion as regards components of very long period will be maintained.

In illustration, now extended to the complete spectrum instead of its long period constituents, radiation consisting of pulses having displacement of type $\phi(x) = be^{-kx} \cos ax$ on the positive side of the origin, with its image on the negative side, thus of Röntgen [non-alternating] type, may be considered. The velocity in one of them analyses into $\int M \cos mx \, dm$, and its total energy is $E$, where

*(margin: Short pulses of amenable type:)*

$$M = \frac{c}{\pi} bm \left\{ -\frac{2}{m} + \frac{m+a}{(m+a)^2 + k^2} + \frac{m-a}{(m-a)^2 + k^2} \right\},$$

$$E = c^2 b^2 [k^{-1} + k(a^2 + k^2)^{-1}].$$

Thus $\pi M^2/E$ is expressed by a formula of type

$$kF\left(\frac{m}{k}, \frac{a}{k}\right);$$

and the distribution of energy in the spectrum of the radiation is represented exactly by $\int \pi M^2 dm$, which is

*(margin: their law of spectral distribution.)*

$$\int dm \Sigma \left\{ kE_{a,k} F\left(\frac{m}{k}, \frac{a}{k}\right) \right\},$$

where $E_{a,k}$ is the total energy of one of the constituent groups. This may be elucidated graphically for the various simple cases that are contained in it: for the long component periods, $m$ small, $F$ is

$$\frac{4}{\pi}\left(1 + \frac{k^2}{a^2 + k^2}\right)^{-1}.$$

We may recall that Lord Rayleigh, and Mr Jeans, and others following them, have shown that the distribution of energy equipartitioned among the free vibratory modes of an enclosure follows the law of inverse fourth power of the wave-length, with a coefficient which agrees experimentally with the value for natural radiation of low period. The full explanation of this very remarkable agreement, when it comes, will cover the circumstance that it holds good only for the high periods of the enclosure, whereas it is the low periods of natural radiation that are represented by the corresponding formula.

*(margin: Equipartition enclosure must be large.)*

---

[1] See *Proc. International Math. Congress*, Cambridge, 1912, Vol. I, pp. 197–216, "On the Dynamics of Radiation" [as *supra*, p. 434].

*The Fourier Wave-Group.*—We have verified that when a graph along $t$, of homogeneous type, is analyzed into a Fourier integral of form

$$\int dm f(m) \cos(mt + \alpha),$$

Degree of purity attainable: the purity attainable in the resolution depends on the length $L$ of the graph that is employed in the process. Adjacent values of $m$ situated within the limits given roughly by

$$\frac{\Delta\lambda}{\lambda} = \frac{4}{5}\frac{\lambda}{L}, \quad \text{or} \quad \Delta m = \frac{8}{5}\frac{\pi c}{L},$$

will be confused in the result. The effective element of the analysis of the graph will thus not be an unlimited train of oscillations of type

$$\Delta m f(m) \cos(mt + \alpha),$$

but will be a limited Fourier group expressible by

$$\int_{m_0 - \frac{1}{2}\Delta m}^{m_0 + \frac{1}{2}\Delta m} dm f(m) \cos(mt + \alpha),$$

the Fourier effective elements of radiation, which is, when $f(m)$ and $\alpha$ are treated within its range as constants, involving the limitation (cf. *infra*, p. 561) to a smooth distribution of periods,

$$\Delta m f(m) \frac{\sin \frac{1}{2}\Delta m t}{\frac{1}{2}\Delta m t} \cos(mt + \alpha),$$

where $m$ now represents $m_0$, the central frequency for the group.

Writing $t'$, equal to $t - x/c$, for $t$, the practical element of radiation described: is thus expressed as a train of travelling waves of constant period, with amplitude falling off on each side of its centre according to the simple diffraction law of type $\theta^{-1}\sin\theta$, where $\theta = \frac{1}{2}\Delta m t'$, and so becoming negligible after a distance of the order $2\pi c/\Delta m$ on each side. We have taken for $\Delta m$ the value $\frac{8}{5}\pi c/L$, as roughly but sufficiently representing the circumstances; thus this gives a range of extent on the two sides about $\frac{5}{4}L$. The effective lengths of the practical elements of the integral, viz. of the Fourier groups into which it resolves the graph, are thus only slightly greater than the length of the original not an infinite graph itself. No practical question can arise of resolving a given short train, thus length of the radiation into infinitely long trains of waves, a para-avoiding doxical mode of statement of the Fourier theory which seems to have paradox. puzzled even very high authorities[1]. To analyze sodium light within

[1] Thus Poincaré (*C.R.* Vol. cxx, p. 757, 1895), referring to the idea that a temporary disturbance can be represented by harmonic components extending over all time before and after, merely infers (in a critique of the views of Gouy) "la formule est donc fausse." Cf. also A. Landé, "Ueber ein paradoxon der Optik," *Phys. Zeits.* June 1915, p. 204.

[The fact is that such representations cannot combine into an additive statistic and so are nugatory for the optics of natural radiation: that science can draw conclusions only as regards energy, or integral of square, which alone is statistically amenable. See footnote, p. 549, *infra*.]

$10^{-2}$ of an angstrom unit ($10^{-10}$ metres), at least three-quarters of a million wave-lengths are needed (cf. p. 556), extending over about three decimetres of length. Prolonged radiation of white light can be represented in no other way than as an average of some kind of pulses or limited trains; these may be specified most simply as combined into Fourier groups, travelling on along the ray*. This specification by discrete groups is the practical aspect of the limitation, that a graph cannot be expressible as a Fourier integral, unless its length is so short that the number of alternations in it that are important for the purpose in view is finite.  ◁ Homogeneity.

This choice of a Fourier group as the practical element of radiation is convenient because it retains its form unchanged after passage across dispersing media, provided the length of path is not excessive, as will appear immediately.  ◁ Advantage of Fourier type of groups.

*Energy of Travelling Group: Dispersive Medium.*—An expression for the energy of a Fourier wave-group will naturally be required. The displacement in the medium is

$$\phi(x) = \int dm f(m) \cos(mt - nx), \text{ integrated over a range } \pm \tfrac{1}{2}\Delta m.$$

If $\qquad m = m_0 + w, \quad n = n_0 + (dn/dm)_0 \, w + \dots,$

we have, for this medium, supposed for generality to be dispersive so that $n$ is a function of $m$,

$$\phi(x) = \int_{-\frac{1}{2}\Delta m}^{\frac{1}{2}\Delta m} dw \, \{f(m_0) + wf'(m_0) + \dots\}$$
$$\cos\left\{m_0 t - n_0 x + w\left(t - \frac{dn}{dm}x\right) - \tfrac{1}{2}w^2 \frac{d^2 n}{dm^2}x + \dots\right\}$$
$$= \Delta m f(m_0) \frac{\sin \tfrac{1}{2}\Delta m \, (t - x/c')}{\tfrac{1}{2}\Delta m \, (t - x/c')} \cos(m_0 t - n_0 x),$$

◁ It has phase velocity and energy velocity.

approximately, where $\qquad c' = (dm/dn)_0.$

As thus integrated, the individual waves in the Fourier group travel with velocity $c$ equal to $(m/n)_0$; but their amplitudes as distinct from phases, represented by the first factor, travel with the different velocity $c'$, equal to $(dm/dn)_0$. The latter is the velocity of the disturbance itself, and of the energy it conveys. The group, however, begins to change sensibly after travelling for a time $T$ which makes $c\Delta m^2 T d^2 n/dm^2$ a sensible fraction. If the amplitude varies slowly  ◁ The group is temporary.

* They may be extracted from the radiation, and appear in isolated form, by analysis by a grating, or a prism, as discussed *infra* p. 555. The effective length of such a train is another name for the purity of the spectrum that is produced by the analysis, and is conditioned by statistical regularity in the radiation as well as the quality of the prism.

along the group, the total energy in it, kinetic and potential, is practically the value of $2 . \frac{1}{2} k \int (\text{amp.})^2 \, dx$ at any time, say $t = 0$, where $k$ is an elastic modulus of the medium, such that $k/\rho = c^2$. Thus the energy is

$$k \Delta m^2 f \, (m_0)^2 \int_{-\infty}^{\infty} \left( \frac{\sin px}{px} \right)^2 dx,$$

where $p = \frac{1}{2} \Delta m / c'$.

In this expression the integral factor is equal to

$$\left| -\frac{\sin^2 px}{p^2 x} \right| + \int \frac{\sin 2px}{px} \, dx,$$

which is

$$\frac{\pi}{p} \quad \text{or} \quad \frac{2\pi c'}{\Delta m}.$$

Energies of
similar groups
superpose. The total energy of the group, half potential and half kinetic, is therefore

$$2k\pi c' f \, (m_0)^2 \, \Delta m.$$

As this is proportional to the range $\Delta m$ of the group, the energies of groups having a common centre are additive even when they are within the same infinitesimal range of frequency.

Identity
of group
maintained: In so far as groups of finitely different mean periods recover their identity after running through one another [in opposite directions], they, of course, maintain their energies too. But while they are partly superposed, before they have run clear of each other, their energy must also be conserved. This requires that during their encounter, even in a non-dispersive medium, there is slight change of form, the
a consequence. aggregate disturbance not being the sum of the travelling individual groups: for the energy of the sum of two groups is the sum of their energies *plus* a mutual product term, which has to be got rid of.

Interference as regards energy between overlapping Fourier groups becomes observable only when they have the same mean period. The estimation of the mutual part of the energy of two such groups $f(m)$, $F(m)$, with their centres at distance $b$, involves going back to the structure of the group.

*Transformation of Energy Integrals.* This we can do by reverting to the more analytical type of procedure by which the distribution of the energy among the constituent groups was first determined by Lord Rayleigh. We take an arbitrary disturbance $\psi(t')$, which may for convenience be conceived as a travelling one emitted from a
Under
restriction, stationary vibrating source, so that $t' = t - x/c$; it is restricted as regards range and number of fluctuations so as to be of the type expressible in terms of two continuous distributions of Fourier groups,

of the *two standard types*, viz. symmetrical and anti-symmetrical with regard to their central points, in the form

$$\psi(t') = \int f_1(m) \cos mt' dt' + \int f_2(m) \sin mt' dt'.$$

The total energy is proportional to

$$\int \psi(t')^2 dx \quad \text{or} \quad c \int \psi(t')^2 dt',$$

and is given by the formula

$$\int \psi(t')^2 dt' = \pi \int f_1(m)^2 dm + \pi \int f_2(m)^2 dm.$$

energy is analyzable:

This holds whatever be the form of $\psi$; therefore it must represent the distribution of the energy among the Fourier groups as well as its total amount.

To prove this proposition*: first consider the square of the cosine integral in the expression for $\psi(t')$, say $\psi_1(t')$, in the form

Rayleigh on Stokes' demonstration.

$$\psi_1(t')^2 = \int dm' f_1(m') \int dm f_1(m) \cos mt \cos m't.$$

Breaking up the time factor into a sum of cosines, we have, between any assigned limits of $t'$,

$$\int \psi_1(t')^2 dt' = \int dm' f_1(m) \int dm f_1(m) \tfrac{1}{2} \left| \frac{\sin(m'-m)t'}{m'-m} + \frac{\sin(m'+m)t'}{m'+m} \right|.$$

As the fluctuation of the other parts is very rapid, the elements of the integral that contribute sensibly to the result are those for which $m' - m$ is very small. This suggests transformation to a new variable $m' - m$; or slightly better on account of symmetry, to the pair of variables

$$(\mu, \nu) = (m' - m, m' + m).$$

Then

$$d\mu d\nu = 2dm'dm;$$

and the effective part of the integral is a narrow band along the axis of $\mu$, which is the median line $m' = m$ in the original plane of integration[1]

* It is known as Petzval's theorem in the abstract theory of functions of a real variable, in which it demands much deeper treatment, if the functions are to be freed from restrictions.

The definiteness of the energy spectrum while the spectrum of amplitude or of phase has no definite existence is, as in the physical discussion, a main feature.

[1] The procedure of Rayleigh (*Phil. Mag.*, Vol. XXVII, 1899; *Scientific Papers*, Vol. III, p. 273) was, following one of Stokes in a cognate analysis (which he had indeed previously noted, cf. Vol. III, pp. 82, 86, 97, as unnecessary, on grounds akin to the present), to introduce a slight attenuation factor $e^{\mp at'}$ so as to make the integrals thoroughly convergent, and then change the order of integration. It is instructive to verify that a different factor $e^{-a^2 t'^2}$ leads to the same value for the aggregate, but by a very different analysis, indicating different distribution of the elements of energy. The procedure above exhibits how it is that the energy of trains of waves that at first sight purport to be of infinite length, is really finite and local.

with respect to coordinates $(m', m)$. Thus within limits $-t'$ to $+t'$ for $t'$, which can be extended indefinitely,

$$\int \psi_1 (t')^2 \, dt' = \tfrac{1}{2} \int f_1 (\tfrac{1}{2}\nu)^2 \, d\nu \int d\mu \, \frac{\sin \mu t'}{\mu} = \pi \int f_1 (m)^2 \, dm,$$

independent of the range of integration. A similar treatment applies to the sine part $\psi_2 (t')$, of $\psi (t')$.

The part involving products of the two integrands in $\psi (t')$ is

$$\int dt' \int dm' f_1 (m') \int dm f_2 (m) \, (\tfrac{1}{2} \sin \mu t' + \tfrac{1}{2} \sin \nu t');$$

it vanishes, the integrand being of opposite signs on the two sides of the origin.

The theorem thus established is applicable only to the analysis of a limited range of a fluctuating graph, with zero values beyond the ends of that range: for the total number of fluctuations must in general remain finite. The test of the statistical homogeneity of the graph will then be that different ranges of it give energy periodograms with ordinates in the same ratio and proportional to the range that is analysed.

Restriction to homogeneity: test.

The superposition of two groups, identical or of common type, but one with its centre shifted a distance $b$ along the axis from that of the other, produces "interference." The total energy is the sum of the separate energies, for the product term fluctuates along the axis, and represents only the succession of interferences. Two interfering groups of the first Fourier type aggregate, writing $t'$ for $t - x/c$, to

$$\psi (t') = \int f_1 (m) \cos mt' \, dm + \int f_2 (m) \cos \left( mt' - m \frac{b}{c} \right) dm.$$

Thus the product terms in $\int \psi (t')^2 \, dt'$ integrated over the range from $\tau_1$ to $\tau_2$ are

$$\int_{\tau_1}^{\tau_2} dt' \int dm' f_2 (m') \int dm f_1 (m) \cos mt' \cos \left( m't' - m' \frac{b}{c} \right)$$

$$= \int dm f_1 (m) f_2 (m) \int \tfrac{1}{2} \left| \mu^{-1} \cos \left( \mu t' + m' \frac{b}{c} \right) \right|_{\tau_1}^{\tau_2} d\mu$$

on neglecting terms of very rapid fluctuation,

$$= \int dm f_1 (m) f_2 (m) \int \sin \left( \mu \frac{\tau_1 + \tau_2}{2} + m \frac{b}{c} \right) \mu^{-1} \sin \tfrac{1}{2} \mu \, (\tau_2 - \tau_1) \, d\mu$$

$$= 2\pi \int \sin m \frac{b}{c} f_1 (m) f_2 (m) \, dm,$$

when the integration is over the whole range from $-\infty$ to $+\infty$ in $\tau$, as the essential part of the fluctuating integral in $\mu$ is the range, then infinitesimal, for which $\frac{1}{2}\mu\,(\tau_2 - \tau_1)$ is of the order of $\pi^1$. The fluctuation of the local interference, represented by the sine factor under the integral, becomes more rapid as $b$ increases. Fluctuation of energy: its rapidity:

When the two adjacent groups are identical, the range of fluctuation is twice the sum of the energies per unit length of the groups taken separately: in fact the total density of energy fluctuates from that of the sum to that of the difference of the disturbances. its range.

*Energies of Groups must be regarded Atomically*[2]. A train of travelling disturbance has been resolved by analysis limited to a given length of the train, into a system of elementary Fourier wave-groups, of the symmetrical and antisymmetrical types, each of frequency range $\Delta m$, an order of magnitude specified by and representing the degree of purity attainable in the analysis; and the energies of such separate groups have been estimated. Each of these groups is itself bounded in length, not abruptly but fading away on both sides: each of them extends beyond the limits of the original disturbance of which it forms a component. The energy of each constituent group is thus made up of parts, some of which lie entirely beyond the bounds of the aggregate disturbance. In what sense can this be a true conception? As regards amplitude and momentum there is no difficulty, because the superposition of elements can cancel out. But for energy this cannot happen, as each element is essentially positive: the group must in fact as regards energy be treated as an indivisible whole. Summary.Energy paradox:

The energy in the wave-group has been found to be proportional to its range $\Delta m$. We can imagine it divided mathematically into subgroups $\delta m$; each of these, being of shorter range of periods, is longer than the original one, so the same circumstance is involved, that each of them possesses energy beyond the sensible boundaries of the original, of which they form an equivalent. But the essential feature is that the aggregate of the energies is unaltered, as $\Sigma\delta m = \Delta m$. We are fully entitled to introduce into our analysis the aggregate energy of each group; for it is of definite amount and is distributable when the group is subdivided. But in a statistical analysis of fortuitous undulatory disturbance into elementary groups as here defined, it is not allowable to introduce into the discussion the distribution of this energy along the group. The energy must come in atomically, so to speak, by group aggregates alone; like the energy of a vortex ring in hydrodynamics, or of a travelling electron. It is only under this implied limitation that the fundamental relation in optics acquires in what sense specific:energy of group regarded as indivisible,

[1] The expression agrees in form with Schuster's result, *Phil. Mag.* June 1894, p. 533.

[2] Cf. *Proc. Roy. Soc.* 83 A, 1909, pp. 82–95, "On the Statistical and Thermodynamical Relations of Radiant Energy" [*supra*, p. 396].

of necessity. a definite meaning, that the energies or intensities of superposed simple trains of radiation are additive: it is true only in aggregates, not in detail: whereas amplitudes and momenta are additive in minute detail along the rays.

*Groups Disentangled by Dispersion.* We have recognized that a Fourier pulse advancing through a dispersive medium maintains for a long time its form*, which travels with the group velocity $c' = d\tau^{-1}/d\lambda^{-1}$, while its component individual waves run ahead through this travelling configuration, with the greater velocity $c = \lambda/\tau$, where $\lambda$ is wave-length and $\tau$ is period,—increasing in aggregate amplitude as they advance from the rear and then fading away towards the front. A result is that after passing across a prism, if the line of maximum amplitude of wave-crests is traced on the wave-pattern, it has become oblique to the wave-fronts, or lines of equal phase, which specify the individual crests and are thus now undulating in amplitude. The geometrical wave-front or locus of constant phase is in fact determined by constancy of time of passage as regards the individual undulations: the actual disturbance has fallen behind it by a distance $g\,(c - c')/c$, where $g$ is the thickness of the glass or other dispersive medium that has been traversed, measured along the ray considered. If $\Delta g$ is the difference of this thickness for the extreme rays of a parallel pencil which as it emerges is of breadth $b$, the line of maximum of crests makes with the line of phase, or of individual crests, an angle $\alpha$, usually small in optical applications, such that $\tan \alpha = \dfrac{\Delta g}{b}\left(1 - \dfrac{c'}{c}\right)$.

*Picture of pulse in a dispersive medium,*

*skewed by crossing a prism: so that*

*the phase front*

*is not the line of wave-crests.*

If a train of prisms is traversed, instead of only one, the right-hand side will be replaced by a summation over all of them: a Fourier pulse incident on the dispersing system with crests uniform in amplitude along the wave-fronts, comes out with the maxima of the crests in echelon to the wave-fronts at an angle $\alpha$, as above.

Something not very different from what thus holds for a Fourier group may be anticipated to hold for a pulse consisting of an abruptly limited train of simple waves. For its analysis consists of Fourier pulses all ranging around its own period, which will thus all be treated

*Extended to a limited wave-train.*

---

* An *isolated* group of undulations travelling onward, essential to the recent wave mechanics of a free electron after de Broglie and Schrödinger, must thus be founded on a medium dispersion for such waves. That view perhaps has to postulate that a free electron, travelling onward at high speed and so flattened out, carries along *permanently* a group of waves, stimulated from its energy but travelling at their own intrinsic speed (pressural waves are available), somehow as a flat vertical plate pushed through water carries along a group of surface waves also necessarily dispersive: moreover, that electrons can cooperate to produce luminous effects only where the amplitudes of these waves are in phase and so cumulate. (Continued on p. 566, *infra.*)

nearly alike by the dispersive medium; they may be expected to combine again much as before, and thus to emerge as a nearly similar limited train with amplitudes in echelon, of the same depth from front to rear as previously[1]. But there is a consideration which might seem to place a limit on this anticipation. Suppose the incident pulse is restricted to a single undulation, or even a non-undulatory displacement, as it can be for abrupt Röntgen pulses on the electric theory. We have recognized that the function of the prism is, not itself to disperse the incident travelling pulse consisting of a finite series of waves, but rather to prepare for spontaneous dispersion by throwing the lines of maximum displacement oblique to the fronts of the waves: the sorting of the disturbance, thus modified, into a fan of ordinary rays will follow spontaneously in the non-dispersive medium after emergence. But if an incident non-oscillatory sharp pulse could emerge as a pulse of the same sharp type, any distinction between phase front and disturbance front would disappear: such an emergent pulse must travel on unchanged at right angles to its front, so that there could be no dispersion. The paradox is avoided* on reflecting that the analysis of a sharp non-oscillatory pulse involves prolonged Fourier groups spread over a wide range of periods, whereas the previous case of a finite simple train involved such groups over a very short range[2].

> But what happens to a very sharp pulse on refraction?

> Prism makes preparation,

> the medium disperses:

> pulse thus travels single.

These descriptive matters have been put on more precise footing by the answer to the question: What type of pulse or limited train of waves, if any, preserves its form after passage across a dispersive system of prisms or a grating? One answer, the most important but not the only one, is, as already indicated, that the Fourier wave-group, a very small but finite element of the Fourier integral, has this property. For after passage across the prismatic system the train

> Pulses of type permanent in a dispersive medium.

$$\int dm f(m) \cos\left\{m\left(t - \frac{x}{c'}\right) + \alpha\right\} \text{ becomes } \int dm f(m) \cos\left\{m\left(t - \frac{x+\xi}{c'}\right) + \alpha\right\},$$

where $\xi$ is the group-retardation, which is proportional after emergence, for each ray to distance $y$ measured up to it along the wave-front. The central plane of the original Fourier group was determined as that

[1] Cf. Schuster, *Phil. Mag.* Vol. XXXVII, 1894, p. 529. The intuitive form into which he has thrown the explanation of the distinctive features of the Talbot bands in a spectrum is a beautiful illustration of these principles.

> Fox-Talbot bands.

* This implies that they can be ultimately focussed along the direction of phase. The criterion is not yet clarified as to why the echelon envelope of maximum disturbance is not itself the pulse refracted without any dispersion: cf. p. 555.

[2] The dispersive medium might however be such that groups of all periods have the same velocity: then the pulse would emerge as a sharp pulse but of amplitude undulating along it, as Schuster points out: cf. the discussion of these matters in his *Optics*, Ch. XIV.

front for which as origin the phase angle $\alpha$ vanishes: it is a distance $c'\alpha/m$ in front of the origin to which the pulse has actually been referred in the formula. But in the pulse as emergent the centre is at a distance $c'\alpha/m - \xi$ in front of this origin, where $\xi$ varies along the front on account of difference of path in the dispersing medium. Thus it emerges as a new wave-group, of Fourier type in every respect except that it is sheared so as to make its line of centres oblique to the fronts of the waves by an angle $\alpha$ already determined.

*Refraction by shearing.*

If radiation of given mean period is suddenly set free in a dispersive medium, and travels onwards with front nearly abrupt, the front that is practically detectable will advance with the smaller velocity of its constituent Fourier groups. It is in fact the group that is the tangible entity, but its phase, which regulates interference with other groups, is determined by a velocity different from its own. Thus any method of measuring the velocity of light which operates on definite portions of the light and identifies them during their transmission must, as Lord Rayleigh pointed out, determine the velocity of the constituent groups, when the medium is dispersive. This is true equally[1] whether with Fizeau we trace the progress of some kind of interruption in the continuity of the light, or with Foucault we turn the light out of its course by repeated reflexion from a revolving mirror, so that the altered deflexion arising from the rotation can measure the time elapsing between the two incidences. In the latter case the influence of the motion of the mirror—as distinct from its instantaneous position when a constituent group of the radiation meets it—may be shown to be representable by assigning to the mirror a curvature corresponding in practice to a radius hundreds of miles long.

*Double velocity of a group:*

*of propagation and of interaction.*

*Effect of revolving mirror on waves.*

It is on the other hand true that, when one half of a split beam of light is delayed by opposing a film of dispersing substance such as glass or mica in its path, the interval of distance between the resulting bands of interference is determined by the difference of the phases, and these travel with the velocity of the single waves that advance through the group. But even here the prominent feature is the centre of the linear interference pattern, and that is in the position where the centres of the two halves of the Fourier groups, one transmitted through the film of glass and the other travelling alongside it, are coincident. The position of the centre is thus determined by the group velocity in the dispersive film of glass. This argument by

*The centre of an interference pattern:*

*an affair of groups:*

*Fading of groups.*

[1] For the history of this point, cf. Rayleigh, *Scientific Papers*, Vol. I, p. 537 [and especially Vol. VI, p. 565]. It may be noticed that for very long paths there will come a time when $\Delta m\, d^2n/dm^2$ on p. 547, *supra*, cannot be any longer neglected compared with $dn/dm$, and the groups will ultimately cease to be permanent.

groups of waves is equivalent to the principle of Airy (1833) and Stokes[1] (1850), applied more recently by Kelvin (1887) to wave-groups such as ship waves on water, that the centre of the interference system is at the place near which the difference of phases of the interfering trains is stationary. Interference will become obliterated when one half of the split wave-group is so much retarded by the glass that it does not any longer sensibly overlap the other half just after emergence; that is (p. 546) when $g\,(1 - c'/c)$ exceeds the length within which the group, as determined by the smallest ranges $\Delta\lambda$ for which wave-lengths are not confused by the apparatus, is of sensible intensity. *as determined by stationary phase difference.*

*Groups give interference only while crossing.*

Groups at the front of a train cannot possibly advance into a quiescent medium faster than disturbance of their wave-length is transmissible through the medium. But within an established field of undulation, any velocities may occur: in illustration, if two travelling ridges or solitary waves on water cross each other at a small angle, the front will be raised at the place of crossing, and on a flat sea-beach this hump may be observed in shallow water to sweep along the front at surprising speed. *Speed limited at front of train: to which groups may advance and die out.*

*Rapid transverse sweep of a breaking wave.*

We have noted that a wave-group or finite train of waves, advancing in echelon as regards loci of maximum disturbance, and therefore with amplitude undulatory along each front, will disperse itself spontaneously into regular trains of simple waves even in a non-dispersive medium, each complex ray spreading out sideways into a narrow fan of simple rays of wave-lengths extending over the range corresponding to the Fourier analysis of the train. As the rays of the fan, initially overlapping, begin to disentangle themselves at greater distance, the purity and therefore the depth of the wave-group will increase, tending towards a limit equal to the original number of wave-lengths plus an additional number adjoined by the dispersing apparatus. The latter number is of the same order as the number of lines in the equivalent grating, viz. the one which would produce the same angle of echelon $\alpha$ and the same breadth $b$ of emerging front. These additional undulations are not created by the dispersing apparatus, but are a measure of the degree of definiteness (or purity) with which it reflects the constituents ($\Delta m$) of the pulse that are determined by the Fourier mathematical theorem. They are thus evolved from the homogeneity of the disturbance along the ray. *The echelon wave-train disperses spontaneously:*

*into trains prolonged by the apparatus,*

*but not created by it.*

In fact for a grating of $N$ lines $b\tan\alpha$ is equal to $Nn\lambda$ for the $n$th order spectrum, while the angle of the diffraction $r$ is connected with the angle of incidence by the relation $\cos r \tan \alpha = \sin r - \sin \iota$. This *Incident pure train emerges in echelon from a prism,*

[1] Cf. Rayleigh, "Wave Theory," *Ency. Brit.* 1888; *Scientific Papers*, Vol. III, p. 62. It is now usually quoted as the principle of permanence of phase [now familiar in abstract mathematics as the method of steepest descent].

is in exact parallel with emergence from a prismatic system; in each case the emergent pulse, before it has become dispersed into the rays of a spectrum, will have fronts with amplitude undulatory along them. Such an undulating front of periodic disturbance, fixed in space, and sustained for a number of periods by the undulations advancing up to it from behind, has itself the properties of the equidistant spaces as from a grating. of a grating, and this is the root of the correspondence of function between prism and grating[1]. The grating will repeat the group $N$ times, spread along a length $Nn\lambda$ in the $n$th order, involving a number of undulations in the wave-front that varies as $Nn$. Thus the prismatic system will add a number comparable with $Nn$ to the original number of undulations in the group or pulse, where $N$ is given by

$$Nn\lambda = b \tan \alpha, \text{ where } \tan \alpha = \frac{\Delta t}{\lambda}\left(\mathrm{I} - \frac{c'}{c}\right) \text{ as above};$$

so that

$$Nn = \frac{\Delta t}{\lambda}\left(\mathrm{I} - \frac{c'}{c}\right).$$

*Influence of Molecular Structure on Transmission.* This limitation of resolving power does not connote [so far] any optical imperfection in Resolving power of prism, the material of the prism or grating: it is conditioned solely by breadth of the beam of light. A prism of very small angle would, if infinitely broad, admit of infinite magnification, by microscopic aid, without confusion, of the very short spectrum it produces: there not defined by angle: would be no limit to its resolving power. This is illustrated by the objective prisms of small angle used on front of the object-glasses of large telescopes: by Lord Rayleigh's principle, it is the thickness of the Rayleigh's law. prism at its base[2] and not the magnitude of its angle that determines the amount of detail that can be brought out in the spectra of the stars in the field of vision of the telescope to which it is attached.

But a different question arises. Is there any inferior limit to the range $\Delta\lambda$ of an incident Fourier pulse, if it is to pass through the material of the prism without being confused into an aggregate of wider range? It must be the permanence, and constancy of period, of the assisting molecular vibrations in the prism that ensure regular

[1] Cf. Rayleigh, *Phil. Mag.* Vol. x, 1905, p. 401; *Scientific Papers*, Vol. v, p. 276.

Illustration by the echelon grating. [2] Determining as above the total distance by which the plane of disturbance is thrown back, owing to its obliquity, from a plane of phase that coincides with it at one side of the beam: the essential circumstances being in fact directly exemplified on an enlarged scale in the Michelson echelon grating. It is at bottom the same thing whether, following Rayleigh's original procedure (cf. *Sci. Papers*, Vol. III, p. 106), we derive the principle from geometrical optics on the basis of the *mathematically* equivalent aggregate of simple wave-trains each with its own velocity depending upon wave-lengths, or as above evoke the direct correspondence between a prism and a grating on the basis of the actual wave-group with its two relevant velocities.

dispersion, just as it is the constancy of the interval between the rulings on a grating: if such uniformity of repetition is restricted, say to a succession of $N$ vibrations on an average, one cannot expect purity of propagation within narrower limits than a range $\Delta\lambda$ of order given by $\Delta\lambda/\lambda = 4/5N$.[1] In the case of a gaseous medium of propagation, whose molecules are in constant motion with velocity of order $v$, the Fizeau-Doppler effect must come in to limit the definiteness of period of the waves from the excited molecules which take part in the propagation, roughly to an extent $\Delta\lambda/\lambda = v/c$. For a transmitting gas the same limit to the purity of the radiation is thus imposed by the participation of the sympathetically vibrating molecules, as has been recognized for a radiating gas owing to the varying lengths of the waves from the moving molecular sources. The maximum range of interference determined by Fabry, Perot, and Buisson for a complete spectrum line is substantially in accord with the formula of Lord Rayleigh for the Fizeau-Doppler cause alone. Thus there appears to be no room for the other type of cause, limited lengths of the trains; though its influence would probably be greater than the minimum breadth of line that is observed, if we assumed that each encounter between molecules deranged trains of radiation that were being emitted from them. We must infer, apparently, that either the molecules of the gas are not rotated in their mutual encounters, or else rotation does not affect the orientation of their field of radiation. This view is enforced further by Lord Rayleigh's remark that if the molecular energy were distributed equally between translatory and rotatory modes, the latter should introduce a defect of purity much in excess of the translatory effect[2]. The circumstance comes to mind in connection herewith, that the constitution of the aether may in a sense inhibit rotational displacement, if for example the magnetic force were taken to represent in some way a velocity. The sharpness of the Zeeman subdivision of spectral lines seems to demand an explanation of the same kind, that the molecule must somehow be isotropic as regards vibration; anyhow it must persist unchanged when the molecule rotates so to speak beneath it[3].

*Marginal notes:* Purity requires undisturbed molecular vibrators. For a gas, defect of purity by transmission and by radiation the same. Either cause would suffice for the facts. Rotation in molecular encounters: inhibited.

[1] Thus the analysis, in actual practice, of spectra within $10^{-2}$ of an angstrom, pushes $N$, the number in a sequence of the molecular vibrations, up to the order of a million. [In a gas this would imply a free path well over $10^{-3}$ cm., thus density well below one-tenth of an atmosphere.]

[2] Reference should be made to Rayleigh, "On the Widening of Spectrum Lines" (*Phil. Mag.* Vol. XXIX, 1915, pp. 274–284), summing up critically the previous work of himself and others.

[3] Cf. *Phil. Mag.* Dec. 1897; *Aether and Matter*, Appendix F. [The normal Zeeman subdivision is there shown to be necessarily sharp for every molecule of orbital type with a single nucleus.]

*Marginal note:* Intrinsic limit to sharpness of spectra.

*Molecular Scattering of Radiation.* When a Fourier wave-group

Fluorescence usually absent: passes through a material medium, the molecular vibrators receive no sudden shock, for the amplitude of the disturbance rises gradually along the group: their sympathetic vibration is gently established in

because no abruptness. step with the exciting disturbance, and as gently dies away. No free vibrations are excited, such as would arise from abrupt initial circumstances: thus no fluorescence is to be anticipated, unless chemical bonds are snapped, though, when the incident radiation is compound, the elements of it near the molecular periods are selectively scattered. Here

R. W. Wood's structural periodicity. the results obtained by R. W. Wood, on scattering and fluorescence from metallic vapours, should form a clue: incident light adjacent to one of the molecular periods (*i.e.* incident Fourier wave-groups of that period) is found to produce a scattered beam consisting of that period and also a series of others of equidistant frequencies, whose common difference is probably the index of a structural orbital periodicity in the source*.

At first sight the concordance of phase of the secondary molecular wavelets emitted along the front, above described, would appear to cancel them out as regards aggregate scattered amplitude, and so vitiate the Rayleigh formula for the intensity of the blue of the sky, from which inferences so striking have recently been drawn. For a random distribution of molecules very closed packed is very nearly a uniform one: and the phases of the wavelets scattered from them

Statistics of energy not affected by phase interferences. are nearly uniformly spaced. But so long as it is random, the most probable value of the total energy of the scattered wavelets is obtained by summation of the energies that would attach to them separately†: which means that over any appreciable time the actual scattered energy is thus obtained, by simple addition. This fundamental, in fact crucial, point has been brought out by Lord Rayleigh in various connections: cf. *Proc. London Math. Soc.* 1871, in *Sci. Papers*, Vol. I, p. 76; *Ency. Brit.* 1888, "Wave Theory," § 4, in *Sci. Papers*, Vol. III, p. 52; *Theory of Sound*, Vol. II, ed. 2, § 42 *a*. The randomness of phase, which is necessary to the validity of the formula for radiation scattered by a gas, would thus *include* and in fact be derived from the random distribution of the molecules and resulting fluctuation in density—though the opposite has recently been urged, notwithstanding discrepancy from observation. Near the critical point of a gas the randomness fails to be adequate on a large scale,

Opalescence. on account of local sensitiveness to slight change of pressure and consequent sluggish recovery; and this is indicated by the flickering

Raman effect. * Cf. also Raman's recent discovery (1928) that in the light scattered in liquids the spectral lines are accompanied by such sharp satellites, which are at constant distances thus consistent with such an origin.

† The limitations, which *e.g.* make this inapplicable to small dust particles, are now set out in an Appendix at end of this volume.

opalescence that appears. Again in the nearly regular molecular spacing of a crystal the randomness will be only partial, with increased transparency as a result, as has been remarked recently by Lorentz.

Enhanced transparency of crystals.

*Abrupt Pulse in Dispersive Medium.* Even when we contemplate an extremely narrow travelling disturbance, uniform all over its plane, resolution can be made into Fourier groups, each travelling with its own group velocity when dispersive quality is present. After a short interval of time from the origination of the disturbance, the groups will have separated but slightly; thus they will when combined re-form the initial narrow disturbance, which will so far be propagated with their mean velocity without much spreading. But the groups will in time become separated, which means that the aggregate disturbance will broaden out as it travels. After a long time the groups will have separated practically completely; and the disturbance, originally a sharp pulse, will then pass a specified point as a succession of simple waves, with lengths gradually changing*. Strictly indeed, according to this synthesis, a disturbance would be felt at all distances immediately on its origination, though exceeding slightly: but this can only be because the averaged circumstances represented by the differential equation do not faithfully represent such minute detail [*e.g.* pressure assumed to be propagated instantaneously].

A pulse broadens out into regular waves.

This analysis would seem to include the fate of a sharp pulse incident directly on the plane face of the dispersive medium from outside. Part of the pulse is transmitted into the medium, where it gradually spreads out as above, as it travels onward. The only question is whether, if the breadth of the pulse is of the order of molecular coarseness of the medium—if for example it belongs to the impulsive part of the Röntgen radiation—the averaged analysis sufficiently represents the initial abrupt stage: if it does not, part of the pulse is dissipated on incidence by shock among the surface molecules, and the rest travels on as here described[1]. It is not surprising that larger or more massive molecules scatter it more effectively.

Pulse dissipated at abrupt transition:

But the case would seem to be different for a narrow pulse permanently *maintained* in such manner that it is constrained to travel through the dispersive medium at assigned constant speed. The analogy of the standing surface waves on a stream of water flowing over a submerged obstacle is here in point, the relative motion being the same as if it were the obstacle that travelled in stagnant water.

unless maintained.

* Cf. the gradually changing lengths of the waves from a distant storm at sea, as analyzed for a particular case by Stokes, *Scientific Correspondence*, Vol. II, p. 145.

[1] This agrees more closely with Lord Rayleigh's view of the Röntgen pulse radiation (*Phil. Mag.* Vol. XI, 1906, p. 123; *Sci. Papers*, Vol. V, p. 280) than one which I had previously expressed (*Phil. Mag.* Vol. X, 1905, p. 574). [The function of a crystal, in deflecting both rays and electrons, is now fundamental.]

Standing ripples. This regular train of free waves will begin to be established as soon as the obstacle is started: the length of the waves will be that corresponding to a wave-velocity that just keeps pace with the obstacle: but their rear boundary will advance with the slower group velocity[1]. They illustrate forcibly the principle that by virtue of sustained periodic quality in space (grating) or in time (molecular vibrators or elastic pressures) a structure can *resolve* an abrupt disturbance which traverses it into its component Fourier trains separated into different directions in space.

Filtering property of grating or prism.

This discussion on practical harmonic analysis, rather desultory owing to the circumstances of this troubled time, must now terminate. It will have been observed that the original materials as regards most of the questions that have been submitted to review have been drawn Rayleigh: from the writings of Lord Rayleigh: while as regards the recent practice of the Fourier harmonic analysis, as applied to statistical material, we have been chiefly indebted to the papers of Professor Schuster. Schuster. An attempt has been made on revision to indicate these sources by insertion of footnotes: but doubtless there are still omissions.

The tendency of recent abstract analysis on related matters has Molecular assemblages quantitative though unspecified. been to explore the general qualities of the various types of infinite assemblages, rather than to determine the quantitative relations of average or mean which offer themselves for the purposes of physical theory, when the material is too complex for study in detail. The whole subject is so fundamental for modern physical science that a critical survey of what we can expect to achieve in practice, even though it be disputable in parts, may not be unwelcome.

### Added March 3, 1917.

(a) *Effect of Error in Trial Period.* In extracting the periodic part Effect of error in an intrinsic period: $F(x)$ of period $h$ from a graph by the process of reduplication (p. 536), which amounts to wrapping it around a cylinder whose circumference is $h$ and taking the mean of the convolutions, the question arises, what will be the effect of an error in estimating the value of the period $h$? If a trial period $h + a$ is employed instead of the true period $h$, the mean of the $n$ convolutions is as regards this periodic part $F(x)$, thus equal to $F(x + h)$, one $n$th of

$$F(x) + F(x + a) + F(x + 2a) + \ldots + F(x + n - 1 . a).$$

As $a$ is small, this sum is approximately

$$\frac{1}{a} \int_x^{x+na} F(x) \, dx.$$

[1] Cf. Lord Rayleigh, *Phil. Mag.* Vol. x, 1905, p. 406; *Sci. Papers*, Vol. v, p. 274.

Thus the ordinate at $x$ found from the erroneous value for the period is, when $n$ is large, the mean of the ordinates of the true periodic part $F(x)$ taken over a range $na$, on the positive side of $x$ when $a$ is positive. Now this true periodic part may be presumed to be a smooth curve: thus if $na$ is small compared with the period $h$, the small error $a$ in the trial period will not sensibly affect the form of the periodic part that is derived by means of it. The amplitude of the undulations is thus a maximum for the true period, falling away on each side of it. This principle can receive application, for example, in the discussion of the records of sunspots.

slight: for

amplitude is
a maximum
at true value.

(*b*) The travelling disturbance represented by the integral

$$\int f(m) \cos(mt' + a) \, dm, \quad t' = t - x/c,$$

Fourier
unchanging
wave-group.

taken within limits of period of narrow range $\Delta m$, has been here named (p. 544) a Fourier wave-group. Each such group, when expressed in the form

$$\int f(m) \cos a . \cos mt' . dm - \int f(m) \sin a . \sin mt' . dm,$$

is seen to be constituted of two simple centred groups, one symmetrical on the two sides of the origin of $t'$, the other antisymmetrical. Each of them is limited, and therefore so is every Fourier group, fading away on both sides with increasing distance from the centre—provided the amplitude $f(m)$ and the phase $a$ do not change very substantially within the range $\Delta m$. In a dispersive medium, the velocity $c$ depends on $m$, and the group advances through its undulations with a velocity of its own, but subject to ultimate degradation. However great be the range in $m$, the group, though in a dispersive medium it is continually changing its form, is always limited in extent, in so far as $f(m)$ and $a$ change only gradually with $m$: for it is then made up of Fourier unchanging groups for each of which this holds good.

Condition
that a group
does not
dissipate.

(*c*) It is worth while, pursuing further the remark on p. 534, *supra*, to combine two groups, of the same finite range of period, with centres at distance $a$ apart, into a single Fourier integral. Thus

$$\int f_1(m) \cos(mt' + a_1) \, dm + \int f_2(m) \cos(mt' - ma/c + a_2) \, dm$$

is equal to

$$\int F(m) \cos(mt' + a_1 + \beta) \, dm,$$

where the amplitude $F(m)$ and phase $a_1 + \beta$ are determined by

$$F(m) \cos \beta = f_1(m) + f_2(m) \cos(ma/c + a_1 - a_2),$$
$$- F(m) \sin \beta = f_2(m) \sin(ma/c + a_1 - a_2).$$

The "energy" of the total disturbance is expressible by

$$\pi c \int F(m)^2\, dm = \pi c \int f_1(m)^2\, dm + \pi c \int f_2(m)^2\, dm$$

$$+ \pi c \int 2 f_1(m)\, f_2(m)\, \cos\,(ma/c + \alpha_1 - \alpha_2)\, dm.$$

The total energies of the components are thus additive, except for the last term, which expresses the interference between the two groups.

*is local.* That term diminishes, substantially and without limit, owing to increased fluctuation of the integrand, as $a$ increases.

*For crossing groups,* A similar analysis holds for two wave-groups approaching each other, except that $t - x/c$ and $t + x/c$ are both involved. As they get nearer, the total kinetic energy alters; and if this were all, the waves would have to undergo temporary change of form, to be annulled when they recede after crossing. But, in the case of crossing undulations all propagated with constant velocity, *e.g.* on a cord, it is easy to see that there is a compensating change in the potential *energy* energy. Interference of crossing linear waves thus consists, not in gain *redistributed while* or loss of total energy, or in redistribution, but in upsets of the *superposed.* balance at each place between the potential and kinetic energies: in the diverging waves of optics it consists usually of redistribution of the total energy over the region that they occupy.

(*d*) Thus when for example two similar limited wave-groups of complex type, at some distance apart, such as would occur in the Fourier integral expressions for two similar functions of limited range, are combined into one Fourier expression, the *mean* curve of the resulting periodogram of amplitude is obtained by summation, independently of what the distance apart of the constituents may be: but

*Periodogram is only of energy:* this mean curve is affected by large superposed fluctuations which become narrower and closer without limit as the two constituents get further apart. The energies are wholly additive unless the distance apart is small. This progress towards dissolution of the graphical continuity which is essential to physical analysis may be compared with a cognate case: the periodogram representing a finite reach of

*its nature for a long periodic record.* a graph of radiation (repeated however without limit) consists of equidistant abrupt peaks which, around any assigned wave-length, increase in number and so crowd together as the length of the graph increases, finally merging in a sense with the continuous distribution belonging to the Fourier integral. For a thorough discussion of natural radiant disturbance neither of these transitions to a limit carries the requisite degree of validity: it seems to be necessary to proceed by statistical survey, after the manner of gas theory, of the actual limited wave-groups that make up the radiation.

(*e*) A periodogram representing an aggregation of disturbances with interrupted periods is made up of superposed constituents, in relation to which some general statements can be made. As regards amplitude the constituent belonging to a limited simple wave-train has its maximum ordinate proportional to the length of the train, and its dimensions in breadth inversely proportional: its area is independent of the length of the train, but is more flattened out for short trains. As regards periodograms of the energies of the component trains, which are the important ones because they are additive without reference to phase, the maximum ordinate increases as the square of the length of the train (with breadth narrowing as the length itself) and thus much more rapidly than the energy of the train; while the area is proportional to the length. Long component trains of small total energy give more conspicuous peaks than short trains of greater energy: a narrow peak shooting up in the aggregate periodogram thus implies constituent trains of considerable length, but if our analysis is not close it will appear flattened and broadened. Disturbances devoid of periodicity in their constituent parts give rise to flat types of periodogram, without prominent summits.

Dependence on length of the train that is analyzed: as regards amplitude:

as regards energy.

Narrow peaks interpreted.

Flat curve interpreted.

(*f*) A Röntgen pulse, however abrupt and narrow, is expressible as a system of Fourier wave-groups, of limited lengths, travelling together. Each of these groups, since it comes on and leaves gradually, is reflected by a crystal in its own proper direction without initial scattering or sudden disturbance: the sharp pulse is thus spread out by the laminar crystalline reflexion into the now well-known spectrum of regular wave-groups. But this spectrum will be of continuous type, without sharp maxima, if the incident pulse is aperiodic*. The selective spectrum of the periodic rays, that arise probably from the volume vibration of the aether residing in and around the atom of the source, is superposed on this continuous spectrum of the abrupt pulses sent out from the impacts. In the absence of crystalline regularity in the reflector all the waves will be scattered, broken up and lost, inside it.

Sharp pulse gives continuous spectrum:

in regular X-ray trains possibly aether-resonators are effective.

It is the regularity of spacing that turns back the Röntgen rays: it is the periodicity of the forced molecular vibrations that are excited that refracts and spreads out into a spectrum a pulse of ordinary light.

Resolution by periodicity in space, and in time.

In both cases each constituent wave-group comes on gradually: so no initial time of preparation is required for the concomitant molecular vibrations to get into regular swing, nor any final time of relaxation, as I had at one time asserted [p. 312], following Stokes and in opposition to Rayleigh.(*Phil. Mag.* Vol. XI, p. 123, 1906; *Sci. Papers*, Vol. V, p. 279).

Vibrations taken up promptly if molecular encounters are absent.

* It will produce a single row of fragments in echelon, instead of a set of complete rows capable of arrangement into corrugated phase fronts.

Fluorescence. Fluorescence would have to differ in kind from scattering, being of the nature of storage and transformation in a complex aggregate system, with release conditioned perhaps by structural change.

(g) For every finite function that is integrable—and for this type alone the Fourier operations have a meaning—*e.g.* for every function of "limited total fluctuation," it is now formally established (cf. Hobson, *Real Variable*, § 472) that, whether its Fourier constituents form a series convergent in the necessary minute details or not, their "energies" do always form a convergent sum, and so may claim to represent in detail the energy of the distribution specified by the function analyzed. As regards energy then, it would appear that every distribution of disturbance of "integrable" type, without exception, is analyzable by the Fourier procedure: and it is thus not surprising that the energy is the one thing with which the corresponding optical analysis of the complex natural radiation can deal.

The mutual "energy" theorem, *supra*, p. 549, in the form appropriate to two Fourier series instead of integrals, is referred back, as regards its precise enunciation in abstract analysis, to Liapounoff at Kharkoff in 1896 (cf. Hobson, *Real Variable*, p. 716). The range of integration is now *limited* (say from $-\pi$ to $+\pi$) and so the theorem obtains more intuitive assent; viz.

$$\pi^{-1} \int_{-\pi}^{\pi} f(x)\, \phi(x)\, dx = \tfrac{1}{2} a_0 a_0' + \Sigma\, (a_s a' + b_s b_s'),$$

where $a_0, a_1, b_1, \ldots$ and $a_0', a_1', b_1', \ldots$, are the Fourier coefficients for the two functions. The transition to unlimited range and Fourier integrals, under necessary restriction and management[1], leads to the formulae (p. 549) given and utilized by Rayleigh (1889) and Schuster (1894).

(h) The remark originated, apparently, long ago, with Sir John Herschel that the process of finding the coefficients in the representation of a function $f(x)$ between limits $a$ and $b$ by a series $S_n$, of Fourier type with the proper periods, is the same as the process of adjusting to a minimum the aggregate of mean squares of the errors or residues, represented by

$$\int_a^b \{f(x) - S_n\}^2\, dx.$$

It is now formally established, without any limitation on the type of finite "integrable" function that is concerned, that as the number of

---

[1] Mr G. H. Hardy has kindly referred me to M. Plancherel, *Rend. Circ. Mat. Palermo* (1910), for the formal justification of this transition for functions of integrable square. On the general topic, cf. also G. H. Hardy, "On the Summability of Fourier's Series," *Proc. London Math. Soc.* Ser. 2, Vol. XII (1913) p. 365, and M. Plancherel, *Math. Ann.* Vol. LXXVI (1915) p. 315.

*Marginal notes:*
- The abstract theory as regards "energy."
- Historical.
- Fourier series process adjusts mean square of residues of the record to a minimum,

terms $n$ increases, this minimum mean square always approaches zero: that is, though the residue of the series itself may not become insignificant as regards form and fluctuation, as the number of terms is increased, yet the "energy" belonging to the residue does tend to vanish. *which tends to zero with increased number of components.*

Again, if the summation of the "energies" of the separate terms of a Fourier series is convergent, it is established (Fischer-Riesz)[1] that the function represented by the series is "integrable" without limitation, viz. is "completely determined save at an arbitrary set of points of measure zero." Cf. the recent summary by van Vleck, *Bulletin Amer. Math. Soc.* Vol. XXIII, p. 12. The functions representing any kind of natural disturbance, such as radiation, must thus be "integrable," otherwise in the physical analysis energy will run off into the higher periods without limit. For present purposes these abstract theorems, though probably operative in a wider region of their own, do not seem to extend or modify the accepted physical ideas. *Converse.* *An actual radiation graph must be and remain "integrable."*

---

($i$: 1928.) The way in which the Fourier series negotiates an abrupt drop in the value of the function it represents has of course been in evidence since the beginning: it formed a favourite subject of graphical lecture illustration with Lord Kelvin. In passing across the discontinuity the value of the series fluctuates violently, finally again becoming steady at the new level. It was definitely remarked by Willard Gibbs (*Nature*, 1899) that the amplitude of the fluctuations remains limited, as their closeness increases without limit near the discontinuity; though how far it is reasonable to say that the series at the discontinuity really represents a graph jumping up by half its interval then falling symmetrically and then going on at the new level is another affair. It appears that the subject had already been explored adequately by Wilbraham, a young Cambridge graduate, as early as 1848 in *Camb. and Dublin Math. Journal*, at a time when Lord Kelvin was editor of the Journal. *Negotiation of an abrupt transition: Kelvin: Willard Gibbs: historical.*

In extension, may it be possible to justify something like the following? An undulating graph is expressible by a Fourier series. But what a large number of terms of the series represents is this graph with fluctuations superposed, such as would be cancelled by the residue of the series*. Near a steep descent or ascent in the graph *Boundary curve of residue.*

[1] F. Riesz, *Math. Ann.* Vol. LXIX (1910) pp. 449–477.

* The more recent physical *quantum* theories connect the lack of definiteness that has been here discussed simply with inevitable uncertainties of measurement, and so do not ask for improvement of the mathematical theory. Thus even in physics a point of view may be reached, which is more natural in the complexities of biology, that the laws of Nature are relations on the large scale of things, which must not be pressed too closely else their application may be mixed up through entanglement with deeper and possibly undiscoverable principles. *Alternative to convergence.*

the fluctuations are intensified: but everywhere their range is included between two boundary curves, one on each side of the graph.

Physical<br>validity not<br>endangered.

From the physical side, the Fourier expansion is an affair not of values, but of mean values around a locality: thus the unitary quantity involved is an element of an integral in which the fluctuations cancel out. The systematic justification of the theorem was in fact found by Dirichlet just in theorems of mean value, which are another expression of the same range of validity*.

Hamilton.

Cf. Sir W. R. Hamilton's general memoir on "Fluctuating Functions," where the exposition ramifies on one side into special asymptotic expansions, on another into propagation of wave-groups in dispersive media.

($j$: 1928.) In marked contrast to a scheme of permanent free Fourier wave-groups, as adumbrated in p. 552, a leading idea of the original de Broglie scheme (cf. Birtwistle, *New Quantum Mechanics*, ch. 1 and § 114) is to find how the expression of a state of local standing vibration changes when its system is referred to a new frame in which it is convected with speed $v$. If the phenomenon is amenable to the Lorentz transformation, the vibration becomes then expressed as a wave-group, travelling with a speed which is of course $v$, but with phase speed of the enormous value $c^2/v$, so that the group must be very long to make a mental image feasible. Now for any wave-train expressed by $f(mz - nt)$ the wave-speed is $V = n/m$ and the group speed is $V' = dn/dm$. If then this relation $V = c^2/V'$ holds, there results $n^2 = c^2m^2 + b^2$, where $b$ is a constant of integration: as J. J. Thomson has noted, this formula is of the type applicable to the actual physical problem of wireless waves travelling in an ionized atmosphere (cf. *infra*, p. 644). It is this formula, expressive of this special type of dispersive quality required to sustain optically relativist convection of a state of vibration, that was utilized by Schrödinger to adjust the form of his wave-equation, which itself he found adequate to fix the complete set of *quantum* numbers of the system.

Standing<br>vibration<br>referred to a<br>moving<br>frame:

resulting de<br>Broglie type<br>of dispersion,

as utilized by<br>Schrödinger.

In the atomic problem another schematization has been also employed, which offered some promise of connecting up the wave-function with the scalar orbital Action which can formulate, by itself alone, a state of free motion of a dynamical system and its relation to all adjacent states of free motion. Employing the Lagrangian Action

The physical<br>Fourier<br>element.

* The physical conception of an element of a Fourier integral is involved in the question first formulated and discussed by Rayleigh; has a spectral line from a gas any breadth other than what arises from the to-and-fro motion of the radiating molecules? A negative answer would involve infinite intensity, even perhaps on *quantum* theories: but observation seems to show (Fabry) that the intrinsic breadth is relatively small.

$A = \int (T - W)\, dt$, which is the standard and more general form, cf. Vol. I, p. 66, the relation which expresses this property is

$$\delta A = - E\delta t + p_1 \delta q_1 + p_4 \delta q_2 + \ldots,$$

where $p_1$, $p_2$, ... are the momenta conjugate to the coordinates $q_1$, $q_2$, .... The kinetic energy $T$, equal to $E - W$, expressive of the structure of the system, is supposed given as a form

$$T (p_1, p_2, \ldots\, q_1, q_2, \ldots)$$

quadratic in $p_1$, $p_2$, ...; the differential equation characterizing $A$ is thus

$$T \left( \frac{\partial A}{\partial q_1},\ \frac{\partial A}{\partial q_2},\ \ldots q_1, q_2, \ldots \right) + \frac{\partial A}{\partial t} + W = 0.$$

But the Action now to be employed is the original Eulerian form $A' = A + Et$, amenable for a group of motions restricted all to the same constant energy $E$, for which it is therefore $\int 2T dt$; then

$$\delta A' = p_1 \delta q_1 + p_2 \delta q_2 + \ldots + t \delta E,$$

and the characteristic equation takes the form

$$U' \equiv T \left( \frac{\partial A'}{\partial q_1},\ \frac{\partial A'}{\partial q_2},\ \ldots q_1, q_2, \ldots \right) + W - E = 0.$$

Thus the congruence of adjacent free motions, all with the same constant energy, is defined by a general integral for $A'$, annulling $U'$ and involving the sufficient number of arbitrary constants. The Schrödinger equation seems to be derivable (cf. Birtwistle, § 88) by putting $A'$ equal to $K \log \psi$ and finding by the method of variations the condition, as a differential equation in $\psi$ of the second order, that the integral, taken over extent of phase, $\int \psi^2 U' d\tau$ should be everywhere stationary. Here $U'$ is null for the central motion; if the form $U'$ were essentially positive in the neighbourhood, that is if $W$ were minimal along the central motion, so would be every element of this phase integral, and a minimal value for that integral could be presumed: the Schrödinger equation would then be integrable without infinities*. But for a single electronic orbit (hydrogen scheme) $W$ is $- c^2 e^2/r$, and actual solution of the wave-equation has shown that this condition is then secured over a range $E$ positive, implying apparently a hyperbolic orbit, thus hardly here relevant. The quantized solutions for special negative values of $E$, corresponding to the Bohr ellipses, would alone remain pertinent. Thus these explorations, prompted by considerations regarding stability, do not appear to be very pertinent, and the apparent approach of the wave-equation to the Hamiltonian doctrine of Action remains partial and tantalizing.

* This is further developed at end of final Appendix.

Marginal notes:
Dynamics of the general atom:
the Schrödinger equation related to its Eulerian Action.
An explanation indicated:
its failure.

# 91

# ON PERMANENT PERIODICITY IN SUNSPOTS.

[By J. Larmor and N. Yamaga: *Proc. Roy. Soc.* A,
Vol. XCIII, pp. 493–506, 1917.]

Sunspots the key to internal structure.

THE most important index to the dynamical constitution of the Sun lies in its dark spots, which have been known, ever since their discovery by Galileo, to recur periodically. Numerical data for the numbers of spots simultaneously present, and later and more precise data for the extent of spotted area, are in existence for about two centuries; and it is natural that efforts should be made, from all points of view, to extract the knowledge which they contain.

The discussion of the fundamental question, whether there is permanent unbroken periodicity, due either to planetary influence or (as seems much more probable) to a period of dynamical oscillation belonging to the Sun itself, was taken up statistically by Dr Simon Newcomb in a paper "On the Period of the Solar Spots[1]." His criterion was to examine whether the more precise phases were equidistant throughout the record, or on the other hand their deviations

An unbroken periodicity of long range:

increased continually (as $\sqrt{n}$) according to the law of errors. The phases chosen for scrutiny were four, those of maxima and minima and the two more definite intermediate times of mean spottedness*. An analysis by the method of least squares led him to an unbroken period, of $11\cdot13 \pm 0\cdot02$ years, and thus of great definiteness. His conclusion is that "underlying the periodic variations of spot activity there is a uniform cycle unchanging from time to time and determining the general mean of the activity." But to get this very remarkable degree of precision he had to reject the records belonging to the two

but record temporarily weak,

decades around the year 1780, which showed violent irregularity in the phases. "I was at first disposed to think that these perturbations of the period might be real, but on more mature consideration I think they are to be regarded as errors arising from imperfection of the record. The derivation of any exact epoch requires a fairly continuous

---

[1] *Astrophysical Journal*, Vol. XIII, pp. 1–14 (1901).

* A distribution of the form $\phi(t)\,\psi(t)$, where one of the factors is periodic, as on next page, would satisfy this criterion; but the varying amplitude factor is difficult to interpret dynamically, it rather suggests kinematic control.

It has been found by the Greenwich astronomers (*Monthly Notices R.A.S.* May 1918) that the wandering of the Earth's Pole in recent years is rather well expressed by the sum of two simple harmonic terms, but each of varying amplitude, an annual term and one of presumably the free Eulerian period.

series of derivations made on a uniform plan. If we compare and combine the results of observations made in an irregular or sporadic way, it may well be that the actual changes are masked by the apparent changes due only to these imperfections." And, again, "it would seem from what precedes that a revision of the conclusions to be drawn from the observations of sunspots during the interval of 1775–1790 is very desirable."

In the graphical analysis which follows, we shall be driven to reject the record of the two cycles of the spots, here referred to, for much more conspicuous and emphatic reasons; so violent a temporary interruption as they show, in the course of a phenomenon periodic in phase and on so vast a scale, seems to be most anomalous. It is true that the graph (Fig. 1) shows shrinkage in numbers of spots recorded in not a few cycles, involving change of ordinate smaller indeed but yet of the same order of magnitude as the discrepancy in those two cycles; but in no other case is the period deranged. All the rest of the graph is consistent with a law represented by the product *now rejected.*

$$\phi\,(t)\,F\,(t),$$

where $F\,(t)$ is a function strictly periodic in $11\frac{1}{8}$ years, and $\phi\,(t)$ is a slowly changing factor (cf. Fig. 9) which determines the varying amplitude. *Form of periodic undulation, with slowly changing amplitude.*

The next analytical discussion is a memoir by Professor Schuster, "On the Periodicity of Sunspots[1]," which applied the method of quasi-optical analysis elaborated by him, in order to determine the periodogram of the disturbance, viz. the curve connecting amplitude with frequency in the continuous range of simple undulations into which the data for a specified range of time, taken by themselves, can be analyzed. This memoir is the most complete and important analysis that has yet been made, and the data marshalled in it from Wolf and Wolfer, and from the records of Greenwich and of the Solar Physics Observatory, have been largely employed in the present discussion. The very striking, even at first sight startling, result is reached by him, that when the data for the last available 150 years are divided into two equal groups, their periodograms come out totally different; the more recent group showing a nearly pure simple undulation of Newcomb's period $11\frac{1}{8}$ years, when the earlier group hardly shows this period at all, but exhibits prominently two somewhat broad and indefinite periods of about $13\frac{3}{4}$ and $9\frac{1}{4}$ years. Yet Professor Schuster also assigns much weight to Newcomb's feature, that the turning points and mean points on the graph of spot-frequency occur at equidistant times throughout the range, except *Schuster's periodogram analysis* *shows material to be heterogeneous,*

[1] *Phil. Trans.* February 1906.

for the one conspicuous but transient local disturbance after which the phases recover the previous sequence; he considers that this is strong evidence that the $11\frac{1}{8}$-year period thence predicted is a permanent property over the whole range, and of some sort of dynamical type. One of the chief points elucidated in the present more elementary and direct discussion is the cause of the disappearance of this period of $11\frac{1}{8}$ years on the analysis of the record from 1750 to 1820. It is that the two spot-cycles, already referred to, between 1776 and 1798, which are conspicuous and therefore greatly influence the analysis, are entirely abnormal. If this range of the data is left out, the rest **but only on** of the record falls in satisfactorily with a permanent $11\frac{1}{8}$-year period. **account of the** This is brought out especially by the graphs which show, superimposed, **weak part of** the forms of the periodic components having this period, that have **the record.** been extracted from the three ranges separately into which the record has been divided for the purpose of this test.

More recently, Professor Kimura[1] has obtained, by a process of adaptation, a series of trains of unlimited simple undulations which closely represent the sunspot graph over the 164 years of available record, represented here in Fig. 1. A free curve drawn through the ordinates representing these discrete amplitudes, ranged at equidistant frequencies, would trend towards being a periodogram. In **Fourier** two following papers, Professor H. H. Turner has pointed out that **analysis over** this group of components coincides closely with the Fourier series **a limited** representing this record of the 164 years, repeated, however, without **range,** end. This coincidence is not a physically necessary one, for the Fourier components determined from a longer range would, of course, include this range, not however repeated in the same manner. And **valid only** the periodogram curve will not be the same either, unless the original **when the** material is homogeneous, in a statistical sense of which such identity **material is** would be the test. Professor Turner points out that the presence of **homogeneous.** **Interpretation** a periodicity strongly sustained over a considerable length of the **of abrupt** data, but not identical with one of these Fourier harmonics, would **change in a** tend to be revealed by abrupt change of sign in the range of ampli- **periodogram.** tudes; and he shows how the value of this period might be deduced, provided, however, it belongs to an unending simple train*.

**The true** The number of spots is necessarily positive; but, as it fades to zero **dynamical** as well as rises to a maximum, it may be held to be a measure of the **period** **probably the** energy of the Solar disturbance, which is proportional to the square **doubled range.**

[1] *M. N. Roy. Astron. Soc.* May 1913.

* The numerical records have been discussed by the methods of statistics, applied to the residues around a main period, by G. Udny Yule, *Phil. Trans.* 1927, giving results in general agreement with those of his paper, and also rather more exact graphs.

of an amplitude purely alternating from positive to negative, in the simplest case say to $\cos^2 pt$, which is $\frac{1}{2} + \frac{1}{2} \cos 2pt$, representing oscillation around a mean.

At first glance, this view may not be quite in keeping with the antisymmetrical character (*infra*) of the periodic part of the graph, which rather indicates a mean spottedness with this special type of periodic fluctuation superimposed on it; this other special feature, that the total spottedness falls sensibly to zero at the minimum before rising again, must also be kept in view.

The very fundamental character of the Solar phenomena that must be concerned in the spots will justify the application of a direct method, involving perhaps fresh features and principles, to this well-worn problem. A considerable selection from our graphical representations has been here included, as the inferences rest more on visual inspection than on calculation; and, moreover, experience has shown us that it will be convenient for inquirers who in future approach the subject, to have ready at hand the graphs which constitute the material of the problem.

The graph given in Fig. 1* represents the sunspot record, year by year, from 1750 to 1914, the data for the last years from the Greenwich record having now been added to the numbers as tabulated by Dr Wolfer and the areas as recorded by the Solar Physics Committee and at Greenwich up to 1902, which are the data that have been discussed by Newcomb and by Schuster. Each ordinate in the graph is the mean of three consecutive monthly records, so as to give a mean record all round the Sun instead of mainly on one hemisphere. The broken arrows mark off the Wolfer-Newcomb-Schuster cycles of 11·13 years; the full arrows represent an altered trial period of 11·21 years, which perhaps fits more obviously with the phases of the most recent data. In any case it will be of utility to have analyses for two trial periods. The misfit of both sets of arrows between 1776 and 1798 is obvious. A smooth mean curve has been drawn through the jagged peaks of the numerical data in Fig. 1. *[margin: Graphical sketch of the data.]*

The discussions of Wolfer, Newcomb, and Schuster having afforded strong evidence for a very definite secular periodicity unbroken as regards phase, its varying amplitude in the sunspot graph still remains for consideration. The procedure employed here is to multiply the amplitudes throughout each cycle by a common factor, so that all the cycles shall become equal in mean amplitude. The result is represented in Fig. 2; the multipliers required to deduce Fig. 2 from Fig. 1 are represented, smoothed out into a curve $\chi(t)$, in Fig. 9, in which each short line represents the actual constant multiplier for a

* Figs. 1, 2, 7, 8, are not here reproduced.

cycle. If $\psi(t)$ represents Fig. 2, and $f(t)$ represents Fig. 1, which is the actual sunspot graph, then

$$f(t) = \chi(t)^{-1} \cdot \psi(t).$$

The curve of $\chi(t)$ in Fig. 9 shows that this amplitude factor $\chi(t)$ changes gradually throughout the whole range of the data. Thus, if we assume that the $11\frac{1}{8}$-year constituent $F(t)$ of the sunspot record is of constant period but varying amplitude, it will be represented with fair fidelity by

$$\chi(t)^{-1} \Psi'(t),$$

where $\Psi'(t)$ is the periodic constituent of this period that is involved in the modified graph $\psi(t)$ of Fig. 2.

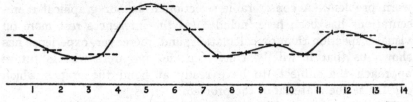

FIG. 9. Curve of amplitude factors.

Separation of a periodic part:    The method of extracting a secular constituent of unbroken period $\tau$ from a graph is (figuratively) to wrap the graph round a cylinder whose circumference is $\tau$, and take the mean ordinates of the various convolutions to form a new curve. All elements of an undulating graph which have not a period at $\tau$, or near it, will then average out; if the range of the graph that is utilized is a large multiple of $\tau$, the result may be presumed to represent with fidelity the periodic con- and Fourier stituent required. This periodic part, of form thus isolated, may then analysis of its be analyzed into a Fourier series representing the fundamental simple form. period $\tau$ and its submultiples.

It will readily appear, on consideration of the nature of this process, that the values of these simple harmonic periods of equidistant frequencies, thus determined, will agree with the ordinates in a con- Contrast with tinuous periodogram[1] which represents the actual disturbance over periodogram the long range that is thus analyzed, taken, however, by itself, with analysis, nothing added outside that range. But we are now dealing, by hypothesis, with the different case of a definite period repeating itself outside that range as well as inside; its components will represent unbroken periods, and the periodogram for them will be, strictly, a series of isolated narrow peaks with vanishing intermediate ampli- tudes; if the rule is retained that the amplitude of a component of range $\delta n$ is the element of area of the periodogram, these isolated

[1] With amplitude plotted against frequency, not length of wave.

ordinates would have to be infinite in length. The continuous periodogram curve thus cannot quite naturally represent sharp permanent periodicities, and a different procedure as here, in fact a much simpler one, is as regards them to be preferred.

On the other hand, the periodogram analysis, as elaborated by Professor Schuster, is well adapted to the discussion of material containing transient periodicities repeated irregularly, but with the same periods, or tendencies to periods, recurring. The standard example is natural radiation. If two reaches of the material that is analyzed, of sufficient lengths, lead to like periodograms, then to that degree the material may be presumed to be statistically homogeneous; otherwise it is fortuitous and sporadic. It seems to be inferable from Professor Schuster's analysis that the sunspot record, apart from the $11\frac{1}{8}$-year period, is in the latter category[1]. Yet it is at first sight startling that the $11\frac{1}{8}$-year period is almost non-existent in the periodogram of the first of the two 75-year ranges; for, although that range contains two anomalous cycles of 22 years, the other cycles are normal, and might be expected to show in the result. But such destructive interference is really not anomalous. For example, if a range of length $L$ contain a train $A \cos (nt + \alpha)$, limited to length $l$ which is a considerable number of periods, this train will, irrespective of its position along the range, add to the periodogram at its own period a term approximately $l/L . A (\cos nt + \alpha)$, exhibiting there a maximum ordinate $lA/L$ of phase $\alpha$. If the range contain a number of such limited trains of uniform amplitude, their contributions can be summed by the vectorial graphical process into a single term $A' \cos (nt + \alpha')$. If there are many of them, and their phases $\alpha$ are sporadically distributed, the resultant of their vector amplitudes will be very small compared with the arithmetic sum, and will often be quite insensible. But if we include a succession of very many such ranges $L$, and the phases $\alpha$ of the component trains vary sporadically all along, then, by Lord Rayleigh's principle[2], the total resultant amplitude has a sharply defined probable value, equal to the most probable value for a single range, multiplied by the square root of the number of ranges, a value, in fact, determined from the total energy along the range, treated as additive. The optical analysis of radiation is an average, then, rendered definite by ranging over an extremely great number of periods; we can, in general, expect nothing so definite, unless under exceptional circumstances, for shorter ranges

which suits transient periodic features repeated sporadically,

*e.g.* radiation.

Deleterious effect of an erroneous range in the data:

like reaches tend to cancel out in the analysis of a long range:

but their energies are additive.

---

[1] On this general subject, see "On the Fourier Harmonic Analysis," in Presidential Address, *Proc. Lond. Math. Soc.* pp. 8–42, Nov. 2, 1916 [*supra*, p. 539].

[2] Cf. *Proc. Lond. Math. Soc.* Nov. 1916, p. 35 [as *supra*, p. 558].

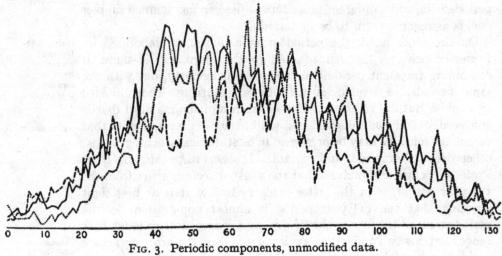

0   10   20   30   40   50   60   70   80   90   100   110   120   130

FIG. 3. Periodic components, unmodified data.

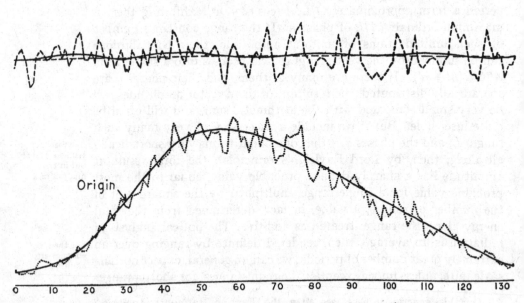

Origin

0   10   20   30   40   50   60   70   80   90   100   110   120   130

FIG. 4. Final mean curves: and doubled departure from antisymmetry.

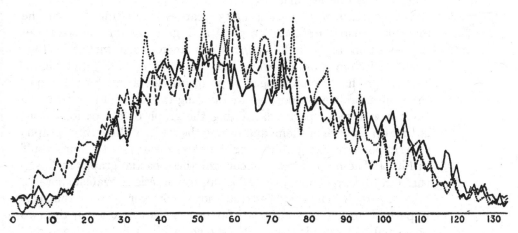

FIG. 5. Periodic components, modified data.

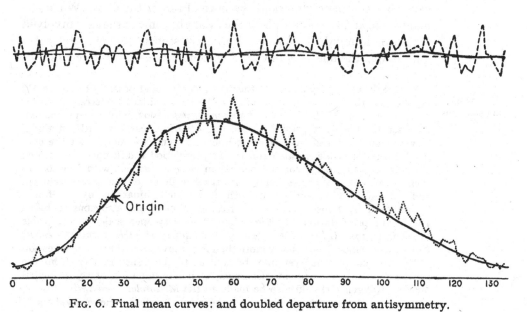

FIG. 6. Final mean curves: and doubled departure from antisymmetry.

of graphs, such as are alone available in observational material or amenable to Fourier analysis.

The graphical isolation of a main sunspot period:

The constituent of period 11·13 years, as determined from the monthly sunspot numbers graphed in Fig. 1, by this procedure of superposition of periodic lengths, conducted arithmetically (but modified by omission as *infra*), is represented in Fig. 3 and Fig. 4. The process has been arranged so as to provide a test for the hypothesis of unbroken periodicity, by dividing the range of the data into three nearly equal parts, extracting the graph of the periodic constituent from each of them, and taking the mean of these three graphs superimposed in Fig. 3, duly weighted if necessary, for the final result in Fig. 4. The more nearly concordant these partial graphs in Fig. 3 turn out to be, the more justified is this hypothesis as against sporadic broken periods. The roughest trial shows, however, at once that this hypothesis, otherwise quite satisfactory, is wholly upset in the two abnormal and conspicuous cycles shown in Fig. 1 between 1776 and

the effect of the doubtful range:

1798. The course very strongly suggested is to omit them from the data to be analyzed, as for some reason wholly foreign to the problem; this will in no way affect the validity of the analyses of the rest of the sunspot graph, but the question will remain over, whether there has been error of record or whether some overmastering transient cause has completely masked the normal run of the data. When the magnitude and the time scale of the solar phenomena that are involved are borne in mind, the former alternative seems to be the preferable one; it was the one adopted by Newcomb for reasons quoted *supra*[1].

Wolfer's investigations.

[1] In reply to inquiry as to whether the records could possibly be so greatly in error for these two cycles, Professor Wolfer has kindly referred us to his classical memoir in the Zürich *Vierteljahrsschrift*, 1902, where in pp. 80 *seq.* he discusses the discrepancies between the various early records; nothing, however, so substantial as is here in question seems to emerge. At the end, p. 95, Professor Wolfer recalculates the 11-year period and its epoch, obtaining a result (11·124 ± 0·030 for the period) in close agreement with Newcomb's then recent one, notwithstanding variations in detail of the data employed, and substantially in agreement with earlier determinations by Wolf and Spörer, and by himself in 1892. Professor Wolfer has also constructed a graphical representation, like Fig. 1 here, on a very open scale, for the whole range from 1750 (*Zürich Astr. Mitteil.* Vol. xcIII; *Monthly Weather Review*, April 1902), which does not any more than Fig. 1 give ground for the suggestion of faulty records, whatever may be said as to the minor matter, from the point of view of our present discussion, of indetermination of the principal phases. Reference should also be made to the *Meteorolog. Zeitschrift*, Vol. v, 1915, where Professor Wolfer arranges the sunspot data for the interval 1901–1914. Earlier knowledge of these references would have facilitated and probably improved our present discussion: but the main issues that arise in it are on a large scale, and would not be sensibly affected by change in the details of presentation. See addition, p. 581, *infra*.

The two sunspot cycles from 1776 to 1798 have been accordingly omitted and the remainder of the data divided into three groups of nearly equal duration:

(- - - - - - -) 1750 to 1822, omitting 1776 to 1798,
(————) 1822 to 1867,
(. . . . . . . .) 1867 to 1912.

The constituents of period 11·13 years extracted from these groups of the original data (Fig. 1) are recorded in Fig. 3, that from the first group being represented by the broken line, that from the second by the continuous line, and that from the third by the dotted line, as indicated above alongside the ranges of time. It will be seen that these three graphs do fairly superpose[1]; the fact that this coincidence is not very good, even when allowance is made for the small number of cycles in each group, arises mainly from the presence of low amplitudes in a range of the cycles (1800–1834) mostly in the first. The (unweighted) mean of the curves for the three groups, which represents the whole of the data, is recorded in Fig. 4, both as a broken graph representing the actual numerical record, and as a smoothed curve such as a physical periodic constituent would naturally be. It will be observed that the smooth curve is here a good representation of the broken graph; and this is in part because the cycles of low amplitudes are now swamped by the greater number of cycles with amplitudes more normal.

*the resulting graphs:*

We have also extracted, for comparison, the periodic part from Fig. 2 in which the amplitudes of the cycles have been equalized. The results are shown in Fig. 5 and Fig. 6. The three constituent ranges in Fig. 5 now agree much more closely; but their mean, Fig. 6, differs little from the previous form, Fig. 4, for a reason already assigned. In the two latter curves (Fig. 5 and Fig. 6), the period has been taken as 11·21 years (somewhat at random, cf. p. 572) instead of 11·13; as was to be anticipated (p. 579, *infra*), this slight change of period does not sensibly affect the form of the periodic constituent, and it is unnecessary to exhibit graphs for both periods.

*result of a different treatment:*

The Fourier analysis of the periodic part of the unmodified sunspot curve thus extracted, is, in arbitrary units,

*harmonic analysis of this main period:*

$$44·5 + 35·4 \sin kt + 6·6 \sin 2kt,$$

and of the same curve modified to equal amplitude is

$$45 + 37·8 \sin k't + 6·5 \sin 2k't + 1·4 \sin 3k't,$$

where the periods chosen are $2\pi/k$, or 11·13 years, and $2\pi/k'$, or 11·21

[1] The steep rise and gradual fall is a well-known feature of each cycle separately.

years. Terms in the series with amplitude less than unity are omitted, as being within the limits of uncertainty. With this limitation, no cosines appear in the expressions. Thus, when the origin is taken on the periodic curve at middle height, that curve is of type satisfying the relation $F(-t) = -F(t)$, the positive side of it being the reversed image of the negative side.

it is anti- symmetrical;

As this antisymmetrical character is remarkable, and probably suggestive for physical theory, we have confirmed it by constructing a graph of each of these Fourier series and fitting these graphs on to the original curves from which they were derived. Nowhere is the discordance more than one-thirteenth of the maximum ordinate, and that amount is very local; the curves are in the main so close that the difference would hardly show on diagrams of the size here given.

The antisymmetrical character of the periodic graph that is thus revealed may also be scrutinized directly, without the analysis into Fourier components. Any graph whatever $G(t)$ can be divided into a symmetrical part $\phi(t)$ and an antisymmetrical part $f(t)$ with respect to any origin, these even and odd functions being determined by the equations

$$G(t) = \phi(t) + f(t), \quad G(-t) = \phi(t) - f(t).$$

In the present case the symmetrical component, equal to

$$\tfrac{1}{2}G(t) + \tfrac{1}{2}G(-t),$$

except for a very slight residue.

is at any rate small. Its value can be plotted by folding the periodic graph over on itself, and then marking the mean (half sum) of the two ordinates, one positive, one negative. The results thus obtained for the rugged graph representing directly the numerical data, and for the smoothed curve, are given by the two flat curves drawn above the periodic graphs in Figs. 4 and 6, when the ordinates in them are reduced by halving. These curves, thus halved, by comparison of their ordinates with the irregularities that were smoothed out in the periodic graph, provide a criterion of the degree of certainty of the property now under discussion. If it is regarded as confirmed, a slightly purified value of the periodic graph itself would be given by $\tfrac{1}{2}G(t) - \tfrac{1}{2}G(-t)$.

If Figs. 3, 5, and 4, 6 are supposed shifted along into contact, each pair will represent roughly two waves of the period, and the lower bend of the curve may be seen by eye to be of the same form and size as the image of the upper bend.

Effect of an error in the adopted period:

We have yet to show that the process of reduplication of the graph, in order to extract permanent periodic components, can be relied on, *i.e.* that the effect, on the periodic graph so obtained, of a small error $\alpha$ in the true period $\tau$ is negligible. If $F(t)$ is the true periodic part of the graph, so that $F(t + \tau) = F(t)$, the use of a trial period $\tau + \alpha$ will produce from the material instead of $F(t)$, a quantity substantially equal to

$$F(t) + F(t + \alpha) + F(t + 2\alpha) + \ldots + F(t + \overline{n - 1}.\alpha)$$

divided by $n$ the number of reduplications or cycles in the data. As $\alpha$ is small this is approximately

$$\frac{1}{n\alpha} \int_t^{t+n\alpha} F(t) \, dt.$$

The ordinate as computed on the trial period $\tau + \alpha$ is thus the mean value of the ordinate of the true periodic part over the range $n\alpha$, on the positive side of $t$ when $\alpha$ is positive. So long therefore as $n\alpha$, the accumulated error along the range, is small compared with the true period $\tau$, an error in the period will lead to but slight error in the form of the periodic component thence derived; for the presumption is that the true periodic part is represented by a smooth curve without violent bends. An illustration is afforded by the result of the trial periods 11·13 and 11·21 years in Fig. 4 and Fig. 6. Had their data been the same, instead of being Fig. 1 and Fig. 2, the criterion of direction of the error would have indicated that the latter may be nearer the correct period. For the amplitude derived from a trial period will be maximum when it coincides with the true period. *illustration from sunspot period. Criterion of direction of error.*

The question arises whether the constituent of period $11\frac{1}{8}$ years thus determined is the only continuously periodic element in the sunspot record. Other likely periods might be tested by the same process, but there is the disadvantage that periods commensurable (or nearly so) have a part in common, made up of their common harmonics[1], which may have to be separated out. It is more convenient and suggestive to take out from the record the periodic part already found and analyze the residue. The residue for the natural graph (Fig. 1) is given in Fig. 7, and that for the modified graph of equal amplitudes (Fig. 2) is given in Fig. 8. In these residues, the two defaulting cycles are, of course, conspicuous. The residue when the fundamental simple constituent alone of the $11\frac{1}{8}$-year periodic part has been removed, is given as a graph in Professor Schuster's memoir (p. 92), and the peaks that it reveals (mainly, however, in its first half, which contains the two abnormal cycles,—they would hardly be noticed in the second half by itself) suggest to him a possible permanent period of 8·36 years. Our residue, when the whole of the $11\frac{1}{8}$-year period is removed, has been tested for this period by the same process as above, dividing up into three groups. The results *Process not strictly repeatable: therefore analysis by residue. No other periods indicated in sunspot record.*

[1] In the optical analogy of spectroscopic analysis by a grating, this corresponds to the overlapping of spectra of different orders.

are shown in Fig. 10: that of the third group containing the most modern and exact data undulates closely along the zero axis, and thus shows no such period; the first and second groups both show rises from the zero axis at the beginning and end, and thus have something in common; but when the two abnormal cycles are omitted, the first group gives a graph lying close along the axis, as did the third; the second result, thus remaining unsupported, is from its form most probably accidental. We thus find no indication of a permanent period near 8·36 years.

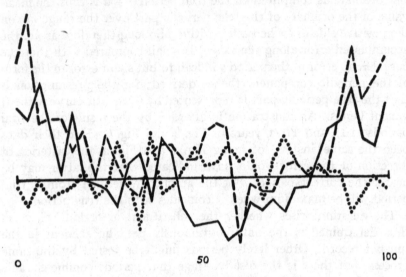

FIG. 10. Components for 8·36 year.

Conclusions.　　Our conclusion is that a secular unbroken solar periodicity of about 11⅛ years, controlling superficial solar phenomena, is established firmly, if we may omit from the record of 160 years the two strongly discrepant cycles from 1776 to 1798. We have found that the graph of this periodic constituent has the remarkable property of being anti-symmetrical with respect to the axis which bisects its undulations, each undulation giving equal areas above and below and the curve repeating as an inverted image; which may contain a clue for a physical theory. We have found no indication that the residue of the sunspot activity, after this periodic part is removed, is of other than fortuitous sporadic character, not amenable to further analysis as a whole. This latter result is supported by the totally different characters of the periodograms found by Professor Schuster for two successive ranges of 75 years, which favour the extended conclusion that the residue involves no sporadically repeated transient periodicities such as might arise from recurrent damped oscillations, as well as no unbroken secular periods.

*Added March* 29. We are permitted to append a letter from Dr Wolfer, giving very important information as to the value of the data on which discussions like the present one depend. The uncertainty which he emphasizes as to the units of measurement along the range of time, seems fully to justify our omission of the very widely discrepant interval, and also enhances the interest of the analysis of the data of Fig. 2, in which the amplitudes of all the cycles have been artificially equalized. The customary mode of discussion, which includes epochs or phases alone, eliminates this uncertainty arising from varying scale; though, of course, individual epochs are liable to non-cumulative displacement by the residue of disturbance, especially when the cyclic amplitude is small. The results of a complete revised discussion of all available data, such as Dr Wolfer promises, will be eagerly awaited.

*Uncertainty of scale along record.*

"Eidgenössische Sternwarte, Zürich, March 19, 1917.

"In reply to your letter of March 14, I take pleasure in giving you some particulars about the sunspot statistics for the critical interval from 1775 to 1795, as far as they are at hand. This interval has always been and will remain a very unsatisfactory one, the spot observations during that time being very scanty, and depending almost exclusively on one single observer, Staudacher, of Nuremberg. On the other hand, just for that reason, they are more homogeneous than many others of remote times, and fortunately they are somewhat more numerous at the time of maximum solar activity than for the minimum epochs, which seemed to be of little interest to the observer. Therefore, the epochs of spot maxima (1779 and 1787) as determined by Wolf are resting upon a tolerably good base, and their uncertainty does not exceed, probably does not even reach, a year. The large departures of the lengths of the periods observed between 1770 and 1805 from the mean value of $11\frac{1}{8}$ years are certainly to be considered as real, and could not be explained by the deficiency of the observations. It is not quite the same with regard to the amplitudes, or the relative heights of the different maxima of that time, especially those of 1778 and 1787. The numbers of spot frequency are, of course, much more influenced by the gaps in the observations, and it is impossible to get reliable monthly and yearly means from perhaps 20 or 30 observations scattered irregularly over the year. Another difficulty lies in the determination of the factors by which the spot numbers of different observers with different instruments are reduced to the same unit, to make them comparable *inter se*. This requires a continuous system of corresponding observations of the successive observers, beginning with Wolf—as normal observer and as unit—and extending backward as far as possible from later observers to former ones. Wolf, in his inaugural investigation, did not employ all the available corresponding observations for the evaluation of these factors, and it is probable

*The doubtful range:*

*the period actually deranged in the graph (but with continuity of phase across).*

that the factors are, in some cases, uncertain, and require corrections. For Staudacher, for instance, I consider it too large, and also for some other contemporary observers whose factors had to be deduced from that of Staudacher, like Oriani, on whose observations the maximum of 1778 is mainly depending. I believe the exceptionally high spot numbers of this maximum are partly due to this fact just mentioned. It will be felt in any investigation in spot periodicity founded upon the whole complex of spot numbers—the "relative numbers"—instead of the epochs of maximum and minimum alone.

"A new reduction of the old observations, back to 1749, based on a new determination of the reduction factors, has been begun here several years ago, but it is a very big task, and has not yet far advanced."

*Added April* 30. The meaning of the term sporadic, as employed above, may be more closely defined. The degree in which the quality described by that term is present varies inversely as the length on the graph within which there is some amount of correlation between the underlying causes; when a graph is highly sporadic, each small range of the curve is an independent entity, practically in no relation to the adjacent ranges except that of statistical averages. In any attempt at harmonic analysis of such a graph we must therefore expand each short element of the graph into its own Fourier integral, and combine the results for all the elements as the laws of chance direct. Now, the Fourier integral for any duly limited graph $\phi(x)$ is given by

*[margin: Sporadic features:]*

$$\phi(x) = \int_0^\infty dn f(n) \cos nx + \int_0^\infty dn F(n) \sin nx,$$

where

$$f(n) = \frac{1}{\pi} \int_{-\infty}^\infty \phi(x) \cos nx\, dx, \quad F(n) = \frac{1}{\pi} \int_{-\infty}^\infty \phi(x) \sin nx\, dx.$$

If $\phi(x)$ represents a physically correlated constituent element of the graph, situated at the origin and of sensible value only very near the origin, of total area $A$, the formulae approximate to

$$f(n) = \pi^{-1}A, \quad F(n) = 0.$$

Thus the local element of the graph, situated at the origin of $x$ and of area $A$, is represented in its harmonic analysis by the integral

$$\frac{A}{\pi} \int_0^\infty dn \cos nx.$$

*[margin: for them all periods are equally present,]*

All the ordinates in its periodogram for frequencies tend to equality, except for wave-lengths that are not long compared with the length of the element. This remark develops into the conclusion of Rayleigh and of Schuster that the periodogram of uncorrelated disturbance, or of any kind of disturbance as regards wave-lengths much beyond its range of correlation, would be a straight line, all periods being equally present. But such a conclusion implies that the area $A$ of the local unless the graph fluctuates constituent element of the graph is not null. If that were so, and the about zero as a mean value: disturbance were thus as much negative as positive, it would be representable by the gradient of some function which vanishes beyond the element, say by $\phi'(x)$; its Fourier integral would thus be obtained from the previous expression by differentiation, in the form

$$- \frac{A}{\pi} \int dn\, n\, \sin nx.$$

The ordinate of the periodogram of amplitudes of an alternating then the amplitude in periodogram disturbance would thus be proportional to the frequency, as regards components whose wave-length is long compared with the length of varies as frequency. the elementary graph or with the range over which correlation holds in the complex disturbance. This agrees with the facts for natural Natural radiation is an illustration: radiation, which dynamically must be constituted of such alternating pulses; the wave-length at which this law that intensity varies as $\lambda^4 T$ begins to assert itself affords a measure of the length on a ray mean lengths of its constituent component trains: over which correlation extends, viz. of the mean effective length of the constituent pulses in ordinary light.

Scrutiny of the variation of the intensity of Rubens' quartz *Reststrahlen* (1901) with temperature seems to indicate correlation extending sensibly over a range of about six wave-lengths of yellow light, for the case of these long waves at ordinary temperatures; but estimates not concordant. the formula of Planck with modern values of the constants, not specially adjusted for long waves, gives as much as eight times this range for natural radiation.

It would, of course, be possible for a graph, like the residual graph of sunspots, to be devoid of correlation over the short range that is available for scrutiny, though it would come in on a much coarser scale over a long range.

*September*, 1926. Each Solar cycle begins with spots in high latitude: they gradually come down in latitude while they increase in number, and the cycle disappears in low latitudes: but by that time a new cycle has been begun in high latitudes. This is expressed by Maunder's "butterfly diagram." An analysis by latitudes would thus be desirable if material were available.

It has been discovered at Mount Wilson Observatory by G. E. Hale and his associates that the magnetic features of the spots reverse at the beginning of each cycle. This confirms the suggestion that $11\frac{1}{8}$ years Doubled period. is a half cycle of the underlying dynamical Solar periodicity, agreeing with the presumption (p. 570, *supra*) that counting the spots gives the measure of the energy of disturbance. The only kind of statistical aggregate in dynamics, over small range of periods, that has continuous existence is the energy distribution. The vector disturbance itself cannot be treated statistically as regards either its amplitudes or phases. Integrals which claim to express it statistically have no convergence. Cf. the Petzval theorem, *supra*, p. 549.

### (1928) *The test of a homogeneous record.*

The practical test that a quantitative record, say of the succession of sunspots, should be homogeneous, lies in its Fourier analysis. It can be expressed in two ways: in each a suitable length of the record is considered by itself. In the one test this length is repeated indefinitely fore and aft, and the Fourier series which then expresses it is determined, the harmonic waves being of this fundamental length and its submultiples. In the other the selected length is analyzed by itself, now in fact completed by null values before and after: this gives a Fourier integral, in which all wave-lengths are represented. If various test-lengths give the same analysis in either form, they belong to a homogeneous range of the record. Physically this quality means that the underlying causes are definite and permanent: analytically it is perhaps expressed by a statement that the record is "integrable." If a periodic cause is suspected, its result can be extracted when the period is known; if it is found to emerge and further to be continuous for different test-lengths, the cause is present and is of permanent type. Analytically, every test-length must be of "limited variation," *e.g.* its gradient may change sign only at a finite number of places: but every physical record is necessarily smoothed out, so this quality must be assumed, residual fine features being transferred to a theory of discrete molecules, such as gas theory would be for the case of an atmospheric record. An example is the question whether the sunspot record is really homogeneous, as discussed in the text: it is fundamental for Solar dynamics.

There seems to be a class of cases, however, in which the test of "limited variation" may decide the legitimacy of a physical theory. The principle is employed with statistical specifications and discussions of natural radiation, that it can be regarded as built up of ray filaments or wavelets: an abstract proposition, especially associated with the names of G. Johnstone Stoney and E. T. Whittaker, is that any distribution of disturbance in a portion of an elastic medium can be analyzed into plane waves. Now consider an isolated region of aether, containing gas molecules, which may be regarded as in steady positions: the radiation enclosed in it is continually being reflected from the walls and dispersed by the molecules. May it not become

*Marginal notes:*
What is homogeneous material?
Fourier series test.
Fourier integral test.
Coterminous with continuous physical theory.

so complicated in the lapse of time that the condition of "limited variation" is no longer applicable? Or is this complication only in inessentials, which beyond a certain limit can be abstracted from a problem of structure of radiation treated as continuous? This seems to be at bottom the same question as the crux of the theory of equipartition of energy: in course of time does the energy become frittered into finer and finer features without limit, or is it permissible to adopt the necessary fundamental postulate of continuous physics, and treat the larger and more continuous features by themselves? The sole justification of this postulate is the actual observed steadiness of the physical system as regards its observable features; if it did not obtain the system could not settle down. In the analogous case of the energy of sound in an enclosed gas, the "limited variation" is imposed overtly and physically by the molecular constitution of the gas: instead of frittering away into infinitely short waves the disturbance becomes distributed ultimately among the translatory energies of the molecules. In the more recent discussions of natural radiation, also, it is a postulate that in some way or other the energy in its transformations always remains concentrated into irresolvable *quanta*. Cf. Appendix IV.

*Frittering away of vibrational energy:*

*unless partitioned in quanta,*

*as of natural radiation.*

# 92

# JOHN MICHELL: HIS CONTRIBUTIONS TO ASTRONOMICAL SCIENCE.

[From a Memoir of John Michell, of Queens' College, Cambridge (Fellow 1749), by Sir Archibald Geikie. Camb. Univ. Press, 1918, pp. 96–105.]

*The torsion balance.* In designing his apparatus to measure the gravitational attraction of a globe of lead, and thence to deduce the mean density of the Earth, Michell was the pioneer in the standard method of determining very small forces by taking advantage of the torsion produced by them in a wire. It was shortly afterwards, as Cavendish remarks, that Coulomb applied the same principle, in a classical series of experiments, to the exact determination of electric and magnetic attractions[*]: and, in various more convenient forms, it is now one of the *Refinements of the Cavendish experiments.* main resources of delicate physical measurement. But Michell's (and Cavendish's) mastery of it, and his just anticipation of its power, went far beyond his age; he designed and constructed appliances with confidence, for a precise estimation of forces so minute that they could hardly even be detected in any other way: even nowadays his application of the principle to gravitation demands the resources of a master.

It is to be expected that a man who could confidently engage in preparations to weigh a ball of lead against one of the celestial bodies would be capable of deep views on other astronomical questions. An examination of his Memoir of 1767 confirms this surmise[1]. As regards *Stellar systems:* general astronomical speculation on stellar systems and their nature, it gives him a place alongside Huygens, Wright and Kant[2]. Further, in more definite fields, it credits him with initiation of the application

[*] It appears that the inverse-square law for magnetic attraction was adumbrated by Michell, as also gravitational deviation of Newtonian corpuscles of radiation: but attractions, electric and gravitational, became an exact modern experimental science in the hands of his friend Cavendish. Cf. the previously unpublished material in *The Scientific Papers of the Hon. Henry Cavendish*, Vol. II (1921) p. 396, containing developments in Geodynamics with suggestions towards Isostatic Distribution, pp. 402–407.

[1] The title of this paper is as follows: "An inquiry into the probable Parallax and Magnitude of the Fixed Stars from the quantity of Light which they afford us, and the particular circumstances of their situation," *Phil. Trans.* Vol. LVII (1767) p. 234.

[2] See R. Grant, *History of Physical Astronomy*, pp. 543, 558, 559.

of mathematical methods, resting on probability and statistics, to the celestial systems. The quantity of material which had then been accumulated was far too small for wide statistical inferences of much certainty; yet Michell amply demonstrated, for the first time[1], the most fundamental fact of stellar cosmogony, the existence of physically-connected stellar groups. In the case of the conspicuous pairs of adjacent stars (the so-called double stars) he anticipated that orbital revolution round each other, owing to their mutual gravitation, would in time be detected,—a prediction afterwards brilliantly realized on a grand scale by Sir William Herschel. He even pointed out that knowledge of the period of their orbital revolution, combined with their distance from the Solar system, would provide means of determining the mass of such a stellar pair in comparison with the mass of the Sun[2],—a problem which is being worked out into exact knowledge by aid of refined determinations of parallax in our own time.

*groups of stars physically connected (1767). Double stars: presumed to be orbital:*

*thus admitting determination of their masses.*

These considerations occur in the course of discussion of a plan for estimating the distances of the stars by comparing their brightness with that of the Sun, on the assumption that they give out an amount of light not greatly different from his. This method had, it seems[1], been first suggested by James Gregory; it was applied roughly by Huygens to Sirius; and it attracted the attention of Lambert and later of Olbers, as well as that of Michell. In Michell's argument the planet Saturn, whose size and distance are known from the Newtonian theory, and whose brightness relative to the Sun could thus be estimated, was used as an intermediary; for it would be impossible to compare directly the dazzling brightness of the Sun with the amount of light received from a star. These astronomers all agree in assigning a parallax less than half a second of arc to the brightest stars; and this is in fact near the values that are now known for the very few nearest stars, which are thus at a distance from our system of about a million times that of the Earth from the Sun, while most stars are very many times more remote.

*Distances of stars as estimated from their absolute brightness,*

*relative to the Sun, with Saturn as intermediary:*

*results.*

Michell was the first[3] to propound, in the same Memoir, just views as to the simple proportionality between the faintness of the stars just visible in a telescope and the area of its aperture, no other circumstance being essentially concerned. He initiated the application of this principle to the estimation of the distribution of the stars at different distances in the depths of space,—a task afterwards carried out so tenaciously and brilliantly in the "star-gauging" of Sir William Herschel. He concluded from a discussion of probabilities that the bright stars were more numerous around our system than

*Telescopic brightness varies as aperture:*

*applied to star-gauging in different regions.*

*Application of probabilities,*

[1] Grant, *op. cit.* p. 547.     [2] *Phil. Trans.* 1784, pp. 36 *et seq.*
[3] Grant, *op. cit.* p. 543.

a uniform distribution in the celestial spaces would permit; and he inferred that most of the bright stars that did not obviously belong to star-groups were our nearer neighbours, and constituted a stellar system of which our own Solar system is a part; while the fainter stars in the depths of space may be grouped in other stellar systems. Thus he thought the nebulae were separated universes of stars, so far away as to defy resolution into their components. Modern astronomical theories are now moving, of course far more definitely, along similar lines, fortified by the immense masses of facts relating to distances, motions and constitutions of the stars and nebulae, which are provided by the photographic plate and the spectroscope in conjunction with large telescopes.

*inferring a star-group containing the Sun.*

*Nebulae as stellar universes.*

In Michell's day the available data were utterly inadequate to guide to safe statistical conclusions on matters of such delicate inference. But the mathematical modes of reasoning in his Memoir of 1767 are still of much interest in the light of modern knowledge, especially as they are illustrated by a discussion of the group of the Pleiades, as it is presented to the naked eye and also in telescopes of various apertures. It may be claimed that these modes of reasoning give Michell a place as the early pioneer in the great modern problem of the configuration and structure of the universe, which first rose to prominence twenty years afterwards, by the labours of Sir William Herschel, founded on similar views.

*The Pleiades group.*

*Forerunner of W. Herschel.*

In regard to optics, Michell was a thoroughgoing Newtonian, as was natural in his time. Light for him consisted of corpuscles projected from the luminous body, rather than waves propagated through an aether. He even thought that, like everything material, they must be subject to gravitation; and he developed a speculation that the velocities of the corpuscles shot out from one of the larger stars must be sensibly diminished by the backward pull of its attraction, and thus be more deviated by a glass prism, a supposition which he proposed to test*. At the end of his Memoir of 1767 (p. 261) he even suggests that the "twinkling of the fixed stars" is due to the small number of luminous corpuscles received by the eye, which might be only a few per second. These corpuscular optical speculations now carry special interest as a curiously definite foreshadowing of the work on electric radioactivity, in which Thomson, Rutherford and

*Light as projected corpuscles,*

*gravitating,*

*so delayed by attraction of source:*

*to be tested.*

*Ideas as to scintillation.*

---

* The Einstein effect on spectra, as recently adapted into the *quantum* theory (cf. Eddington 'Stars and Atoms' p. 123) is just this sort of thing: thus do ideas come round again. The *quantum* of a radiation has energy, $h\nu'$, which is regarded as its natural energy $h\nu$, diminished by the defect of energy due to the position of the emitter in the gravitational field which is its mass $h\nu/c^2$ multiplied by the gravitational potential $V$ at the source. Thus the observed frequency would be $\nu'$, equal to $\nu(1 - V/c^2)$.

others have actually controlled the velocities of the electric corpuscles by the agency of fields of force, and have directly counted the numbers of them that are shot out from active matter.

In his wide outlook over the field of Nature, in the extent of knowledge that was linked together in his active interests, Michell was a true disciple of the British school of physical science, the contemporary members of which were largely his personal friends. They were falling behind in mathematical analysis, owing to too conservative partiality for the geometrical methods of their master Newton. While the great analysts of the continent were closely engaged in the expansion of the infinitesimal calculus, and its improvement by application to the verification and prediction of the motions of the Solar system, the mathematicians of Britain had time for wider, though less intricate, contemplation of the correlations of natural phenomena, not seldom leading into general views which subsequent times were to develop with fuller knowledge.

The British physical school.

# 93

# ON THE NATURE OF HEAT, AS DIRECTLY DEDUCIBLE FROM THE POSTULATE OF CARNOT.

[*Proceedings of the Royal Society*, A, Vol. xciv, Feb. 1, 1918, pp. 326–339; also in *Revue Générale des Sciences*, May 1918.]

Heat engines, invite analysis.

Carnot's one fundamental idea:

1. The contemplation of the actual working of heat engines, and of their great development in England, with which he was well acquainted, suggested, to the mind of Sadi Carnot, the fundamental principle regulating their operation. He postulated that heat can give rise to mechanical work only in the process of carrying through its effort towards an equilibrium[1]. This idea involves immediately the whole of isothermal thermodynamics, including the modern thermodynamic potentials of physical chemistry; for it asserts that, in isothermal circumstances, the heat that is present takes no part in the interchanges of mechanically available energy in the material system, and therefore that the available energy is conserved by itself (or in part dissipated if the operation is irreversible) without any reference to the heat-changes which accompany its transformations.

involves all isothermal science:

its available energy.

His law of reversible thermodynamic operations.

An argument—perhaps the most original in physical science, whether as regards simple abstract power or in respect of grasp of essential practical principles—which was based on combining direct and reversed simplified engines operating in parallel, then led Carnot from this general postulate to a quantitative thermodynamic relation, fundamental for all departments of natural knowledge: that all reversible cyclic thermal operations, involving supply and abstraction of heat at the same two temperatures, have equal mechanical efficiency, which is the maximum possible. But he allowed himself, in his demonstration, somewhat reluctantly, and perhaps hastily in order to fix the ideas, to adopt the view then current that heat is substantial, so

---

[1] "La production d'une puissance motrice est donc due...non à une consommation réelle de calorique, *mais à son transport d'un corps chaud à un corps froid*, c'est-à-dire à son rétablissement d'équilibre, équilibre supposé rompu par quelque cause que ce soit, par une action chimique, telle que la combustion, ou par toute autre.... D'après ce principe, il ne suffit pas, pour donner naissance à la puissance motrice, de produire de la chaleur: il faut encore se procurer du froid: sans lui la chaleur serait inutile...." (*Réflexions*, ed. 1, p. 11 (1824); ed. 2, p. 6.)

Latent heat ineffective.

The unlocking of latent heat into sensible form could not be in itself on these principles a lapse towards equilibrium and so a source of motive power, because it is spontaneously reversible, there being no thermal effort in either direction.

cannot be annulled or created. This point came right later, without trouble, in the corrected expositions by Clausius and W. Thomson, once a net of misconception, arising partly from confusion between total energy and mechanically available energy, had been cleared away. The whole matter ought, however, to be capable of abstract development on broader and more general lines[1]; and the following statement is now advanced to that end. The rough manuscript notes left by Carnot at his death show his own early and very substantial progress towards a more complete doctrine of thermal motive power.

Tacit assumption that heat is indestructible: resulting confusion.

Broader view invited.

2. Suppose that motive power can be gained in a material system, by the carrying out of the effort of some entity, distributed through it, towards an equilibrium. Suppose that the state, as regards fall toward equilibrium, of this entity is determined in each element of mass of the system by only one variable, a potential belonging to it, so that when a path is open it will pass from an element of mass in which this potential is higher to one in which it is lower; also suppose that the effect is in simple proportion to the amount of the entity that enters into this operation, as the idea of a potential implies. Then it can be shown that only two possibilities are logically open. The entity may remain unchanged in amount, but may give rise to motive power in subsiding to a lower potential, just as a stream of water does in falling to a lower level, or a gas in expanding towards a uniform pressure. The other alternative is that, as regards giving rise to motive power by a reversible cyclic process, and therefore also generally, the scale of measurement of the entity may be so chosen that the entity will be used up to an amount equivalent to the motive power gained; and then, with suitable scale of measurement of the potential, the relation between them is defined by the statement that for all reversible paths between the same terminal configurations the amount of the entity that is added to the system at each potential divided by the potential at which it is added makes the same constant sum. Moreover, the first alternative arises in the analysis merely as the limiting form assumed by the second one, when the rate of exchange of the motive power in relation to the entity is indefinitely small.

An entity having levels of potential,

may give power either by subsiding without change,

or more generally by being used up.

But an exact general equation can subsist only for an interlacing plexus of possible operations which are not subject to wearing down or exhaustion: the freedom of the system must then be under

Reversal the test of adequate control.

[1] In an interesting survey of thermodynamic history, Helmholtz expresses the contrary opinion: "...what is still more noteworthy, it is hardly to be supposed that the principle in question could have been deduced from the more correct view, namely, that heat is motion, seeing that we are not yet in a position to establish that view on a completely scientific basis." (*Abhandl.* III, p. 594, in a review of Lord Kelvin's "Papers," from *Nature*, 1885.)

Helmholtz's negative opinion.

adequate control, and the test is that its course is capable of reversal in all respects that are involved in doing work. The equation represents an optimum; when there is defect in the control, so that the system can slip away, the quantity otherwise always constant must increase.

Summary.        Thus, on the basic Carnot idea that motive power arises only from the carrying through of the effort of heat towards equilibrium, and that some of the possible power is dissipated when the operations are not reversible, it follows that there can be *à priori*, only two possible modes of action:

(1) The heat may fall to lower potential unchanged in amount, as in the cases of water power, gravitational or electric attraction, etc.

(2) The heat may be itself consumed in part; in which case the heat can be measured on such a calorimetric scale that there is equivalence between the motive power that is gained and the heat that is lost. This alternative proves to be the actual one for heat; which thus ranks, as regards quantity, but not as regards complete freedom of exchange, with the other forms of natural energies, and is not of the nature of a substance.

There is no third choice open: assuming, of course, that the field of operation partakes of the general characteristic of the order of nature, in that it is rationally explicable, and does not present arbitrary uncorrelated discontinuities.

The fundamental idea of Carnot, in its original form and the most natural one, as above stated in 1824, nearly a century ago, thus involves, in itself alone, either that heat is a substance, or that heat, under the circumstances when it is converted into mechanical energy, always passes in equivalent amount.

3. The argument on which these conclusions can be maintained proceeds as follows, any implication as to the nature of heat being at this stage necessarily avoided. If a reversible cyclic engine, exchanging heat with outside bodies at only two temperatures $\theta_1$ and $\theta_2$, were less efficient than some other type of engine, also taking in heat $(H_1)$ at only one temperature $\theta_1$ and giving out heat $(H_2)$ at another $\theta_2$, then the reversible engine, using the work done by this latter engine to operate it in the reverse direction, would restore heat more than $H_1$ to the source. If, therefore, the reversed engine is operated so as to restore only the same amount of heat $H_1$ to the source, then in the working of the compound engine constituted of the direct and reversed ones thus coupled in parallel, work will, on the whole, be done; while there would be no abstraction of heat from outside bodies at the higher of the two temperatures, and therefore no fall of heat towards equilibrium of temperature; and this contra-

*Engines coupled in parallel, working against each other:*

dicts Carnot's postulate. The work produced would, in fact, have to be done through the mere vanishing of some heat at the lower temperature, if there is any heat-change at all, for no other cause would be assignable; and this process could go on without end. Thus the negation of the proposition that the efficiency, defined by work done $W_{12}$ divided by heat $H_1$ received at the higher temperature, is the maximum possible for a reversible engine, as compared with any other engine working between the same two temperatures of supply and rejection of heat, leads to a result deemed to be impossible. When this is granted, it follows, after the manner of Carnot, that all such simple reversible engines have the same efficiency, which is a function of the two temperatures $\theta_1$ and $\theta_2$ alone. This demonstration seems to be free from any assumption as to the nature of heat.

*Carnot's postulate applied:*

*the far-reaching result: whatever heat may be.*

4. Stated in terms of the simplified reversible thermal engine—introduced by Carnot to render the subject amenable to exact reasoning—which takes into its working system heat $H_1$ at temperature $\theta_1$ alone and rejects $H_2$ at $\theta_2$ alone, thereby doing work $W_{12}$ available for external use in a cyclic manner, the extended mode of argument, on which the further conclusions stated above are based, may be set out briefly as follows. As just shown, for any such reversible engine,

*Idealized simple engine,*

*permits a formula.*

$$W_{12}/H_1 = F(\theta_1, \theta_2).$$

Also, the same formula must apply to the working of the same engine reversed, for there would otherwise be breach in physical continuity; thus

$$W_{21}/H_2 = F(\theta_2, \theta_1)$$

where
$$W_{12} = -W_{21}.$$

Let now, following Carnot, the step of temperature $\theta_1 - \theta_2$ be infinitesimal, say $\delta\theta$. Then we might proceed *tentatively* (in order to exhibit the necessary precautions) to reason as follows:

*Tentative development:*

$$\frac{W_{12}}{H_1} = F(\theta_1, \theta_1) + \frac{\partial F(\theta_1, \theta_1)}{\partial \theta_1}(-\delta\theta)$$

$$\frac{W_{21}}{H_2} = F(\theta_2, \theta_2) + \frac{\partial F(\theta_2, \theta_2)}{\partial \theta_2}\delta\theta$$

where in $\dfrac{\partial F(\theta, \theta)}{\partial \theta}$ the differentiation applies only to the second $\theta$ in the functional bracket. If we represent the function thus denoted by the reciprocal of $-f(\theta)$, we have, since $F(\theta, \theta)$ must itself vanish,

$$W_{12} = \frac{H_1}{f(\theta_1)}\delta\theta = \frac{H_2}{f(\theta_2)}\delta\theta.$$

Thus
$$\frac{H_1}{f(\theta_1)} = \frac{H_2}{f(\theta_2)} = \frac{W_{12}}{\delta\theta}$$

giving for the infinitesimal range with which we are now concerned

$$H_1 - H_2 = W_{12}\frac{f(\theta_1) - f(\theta_2)}{\theta_1 - \theta_2}.$$

If, then, a new scale of temperature $\Theta$ is adopted, such that $\partial f(\Theta)/\partial\Theta$ is unity, *i.e.* if $f(\theta)$ is replaced by $\Theta$, which is W. Thomson's absolute scale, we have

$$\frac{H_1}{\Theta_1} = \frac{H_2}{\Theta_2} \text{ and } H_1 - H_2 = W_{12};$$

the only alternative being an exceptional, or rather limiting, case for which $f(\theta)$ is constant, say $B^{-1}$, and then $H_1 = H_2$ while

$$W = BH(\theta_1 - \theta_2).$$

The argument could now be extended to a finite range of temperature, in Carnot's manner, by coupling engines of infinitesimal range in series, so that the heat from each feeds the next.

must be at fault: 5. Further scrutiny of this mode of reasoning by way of Carnot's function of a single variable is, however, demanded: for at first sight it would seem to be equally open to us to infer, for the infinitesimal range under consideration,

$$\frac{W_{12}}{H_1} = \frac{\delta\theta}{f(\theta_2)}, \quad \frac{W_{21}}{H_2} = \frac{-\delta\theta}{f(\theta_1)}$$

giving

$$\frac{H_1}{f(\theta_2)} = \frac{H_2}{f(\theta_1)} = \frac{W_{12}}{\delta\theta}$$

and so

$$H_1 - H_2 = -W_{12}\frac{d}{d\theta}f(\theta).$$

Thus, if we choose the scale of $\theta$ so as to ensure equivalence of heat and work $W_{12} = H_1 - H_2$, then $-f(\theta)$ becomes $\Theta$ and therefore $H_2\Theta_2 = H_1\Theta_1$, differing from the previous result.

its amendment. But this procedure is ruled out: for the same formula must hold also for infinitesimal ranges shorter than $\theta_1 - \theta_2$, starting with $\theta_1$, which shows that the ratio of the work to the heat supply $H_1$ must be put equal to a function of the temperature $\theta_1$ of that supply, multiplied by the variable infinitesimal range.

6. Indeed, on such an order of ideas as we have just now rejected, this ratio of work to heat ought to be the Carnot function of the mean of the temperatures multiplied by the range; but that would be too narrow a view, as it would necessitate equality of $H_1$ and $H_2$, and therefore a substantial nature for heat. In the earlier papers of W. Thomson's cognate perplexity: W. Thomson, he in fact does compute the value of Carnot's function at the temperature $\frac{1}{2}°$ C. in order to apply it to a range of working from $0°$ C. to $1°$ C.; thus here we have probably a main cause of that

perplexity which led him even to ignore, provisionally, for a year or two, Joule's principle, in order, as he thought, to save that of Carnot. Clausius was more fortunate: his analysis, after Clapeyron, measuring temperature from the first by the expansion of a perfect gas, assumed to have no internal potential energy, and working concretely with the simple gas cycle, went straight to its goal, and did not encounter at all this arresting type of paradox.

evaded by Clausius working with a definite special problem.

7. But even our previous deduction for an infinitesimal range of working requires much further support. For the purpose to which it is applied, the expansion of $F(\theta_1, \theta_1 - \delta\theta)$ ought to proceed up to the term involving $\delta\theta^2$; and then, as it stands, the conclusion could not be drawn.

Closer argument.

A point of exposition and demonstration, cognate to the present one, arises in a closely related case, that of Hamilton's general dynamical equation of variation of the Action, which is sometimes also in this regard faultily developed. There also, it is an affair of passing from one state of motion, between assigned terminal configurations, to another that is variationally near it. In passing from an actual motion to an adjacent motion, frictionlessly constrained in any manner and so possessing constant energy, and having the same terminal configurations, but without further restriction, the variation of the Action, expressed by $\delta\int 2T dt$, always vanishes; so that, provided the path of the dynamical system does not pass beyond the next kinetic focus, the Action is minimum. But notwithstanding, an analytical equation of variation of the Action, in terms of the variations of the terminal configurations alone, and of the energy of the system, subsists only when it is *actual free paths* that are compared. Otherwise, *second* differentials of the Action, derived from this variational expression regarded as implying that it is a function of initial and final configurations and energy alone, could have no definite values, in fact could not exist. The necessity of this distinction, always obvious in special applications of variation of the Action, such (*e.g.* Hamilton's own optical rays) as usually form the guide to wider generalizations, is laid down explicitly in the general statement in Thomson and Tait's *Nat. Phil.* § 329: Jacobi employs a more algebraic order of ideas in which it is latent.

Cognate case in Hamilton's dynamics:

where Varying Action is restricted to actual free paths:

as was indicated by the case of rays.

The thermal procedure above, involving adjacent direct and reversed paths differing by reason of an infinitesimal change of temperature, might be reconstituted on a sound basis after this model. But it is clearer to construct an argument once for all in terms of finite ranges of temperature*.

* Note a family likeness with Einstein's argument (1917) that the laws of natural radiation are implicit in a postulate of exchanges in *quanta*.

8. If in the operation of a Carnot reversible engine, absorption from outside of heat $H_1$ at temperature $\theta_1$ and rejection to outside of $H_2$ at $\theta_2$ leads to the production of motive power $W_{12}$ which can be transmitted mechanically to external uses, then by the argument developed above

$$W_{12}/H_1 = F\,(\theta_1,\theta_2)\colon \text{ where } F\,(\theta_1,\theta_1) = 0.$$

If we may postulate as before that this formula must rest on a rational physical basis, then it must apply also to the reversed working of the engine[1]: thus

$$-\,W_{12}/H_2 = F\,(\theta_2,\theta_1).$$

Therefore
$$\frac{H_1}{H_2} = -\,\frac{F\,(\theta_2,\theta_1)}{F\,(\theta_1,\theta_2)} = f\,(\theta_1,\theta_2),$$

here introducing an abbreviated functional symbol $f\,(\theta_1,\theta_2)$.

But two such engines $\{H_1, H_2\}$ and $\{H_2, H_3\}$ can be coupled to work in series, so that the heat $H_2$ rejected from the first at $\theta_2$ feeds the second at the same temperature. For this compound reversible engine the formula gives

$$\frac{H_1}{H_3} = -\,\frac{F\,(\theta_3,\theta_1)}{F\,(\theta_1,\theta_3)} = f\,(\theta_1,\theta_3).$$

Thus we have the functional relation

$$f\,(\theta_1,\theta_3) = f\,(\theta_1,\theta_2)\,f\,(\theta_2,\theta_3).$$

This expression on the right for $f\,(\theta_1,\theta_3)$ involves an arbitrary parameter $\theta_2$: moreover, it exhibits that function as the product of a function of $\theta_1$ and a function of $\theta_3$. These features can arise only through a relation of the form

$$f\,(\theta_1,\theta_2) = g\,(\theta_1)/g\,(\theta_2);$$

for $f\,(\theta_1,\theta_1)$ is unity and therefore $f\,(\theta_1,\theta_2)\,f\,(\theta_2,\theta_1) = 1$.

Thus $\qquad H_1/g\,(\theta_1) = H_2/g\,(\theta_2) = H_3/g\,(\theta_3), = k$ say.

Moreover $\qquad\qquad W_{12} + W_{23} = W_{13}$

where $\qquad W_{12} = kg\,(\theta_1)\,F\,(\theta_1,\theta_2) = k\phi\,(\theta_1,\theta_2)$ say:

so that we have $\quad \phi\,(\theta_1,\theta_3) = \phi\,(\theta_1,\theta_2) + \phi\,(\theta_2,\theta_3)$

whatever be the value of $\theta_2$: which requires that

$$\phi\,(\theta_1,\theta_2) = \psi\,(\theta_1) - \psi\,(\theta_2).$$

Thus $\qquad W_{12} = k\phi\,(\theta_1,\theta_2) = H_1\,\dfrac{\psi\,(\theta_1) - \psi\,(\theta_2)}{g\,(\theta_1)}.$

---

[1] This can, however, be made a matter of deduction: for even if a different function of $\theta_2$, $\theta_1$ is assumed to apply to the thermal motor when used reversed as a pump for raising heat to higher level of temperature, the same type of argument as in this paragraph will lead to the same final results.

Therefore finally
$$\frac{H_1}{g\,(\theta_1)} = \frac{H_2}{g\,(\theta_2)} = \frac{W_{12}}{\psi\,(\theta_1) - \psi\,(\theta_2)};$$

giving final relation.

so that
$$W_{12} = H_1 \frac{\psi\,(\theta_1)}{g\,(\theta_1)} - H_2 \frac{\psi\,(\theta_2)}{g\,(\theta_2)}.$$

9. The scales of measurement of heat and temperature are in these abstract formulae as yet entirely unspecified. If we now fix the measurement of quantities of heat in terms of a new ideal standard calorimetric substance, whose law of specific heat differs from that of the previous one by the multiplier $\psi\,(\theta)/g\,(\theta)$, the ratio of work gained to heat lost in a cycle of finite range will become unity, $\psi\,(\theta)$ replacing $g\,(\theta)$ in the final formulae. This choice of calorimetric substance will therefore establish quantitative equivalence between work gained and heat lost: and there is no reason why this result should remain restricted to reversible processes. Finally, we are obviously invited to select the one remaining function $\psi\,(\theta)$ as the normal measure of the standard temperature $\Theta$; so that now for any finite range of temperature the formulae for a Carnot engine become

To be simplified by choice of scales,

making heat measurable as work:

$$\frac{H_1}{\Theta_1} = \frac{H_2}{\Theta_2} = \frac{W_{12}}{\Theta_1 - \Theta_2}.$$

The one exception to this deduction is its own limiting case, that $g\,(\theta)$ may be a constant, say $B^{-1}$. Thus $H_1 = H_2$ and
$$W_{12} = BH\,\{\psi\,(\theta_1) - \psi\,(\theta_2)\};$$
so that we would still be invited to selct $\psi\,(\theta)$ as absolute temperature, and the motive power would arise from the fall of the heat, unchanged in amount, through the range of temperature, exactly after the analogy of the fall of a stream of water through a change of level.

and temperature changed to W. Thomson's scale.

The only alternative included as a limiting case.

10. The whole formal theory of heat engines, and of thermodynamic processes in general, is thus developable in a purely abstract manner from Carnot's own initial idea, that heat can give rise to motive power only in the process of carrying through its effort towards equilibrium. Starting from this postulate, it is left to experience merely to decide between two sharply contrasted alternatives, whether (1) heat is virtually a substance doing work merely by falling to a lower level of temperature, or, on the other hand, (2) it is a form of energy, presumably of a fortuitous molecular type, which can be rearranged in part into ordered or mechanical energy by taking advantage of its innate effort to run down towards a dead level of distribution.

Wide scope of the conclusions.

The kinetic energies of the molecules in a small element of mass will, in fact, mix together far more rapidly towards an equilibrium of distribution in that element than will the energies distributed

through a large volume of the substance; thus in the static type of
theory to which thermodynamics is restricted, each element of mass
is assumed to have already a temperature. But the energy existing
at this dead level of temperature in the infinitesimal elements can be
in part recovered or reconstituted into energy arranged in finite
groupings in these elements, thus becoming partly regularized and
therefore of mechanical type, when suitable advantage is taken of
the further effort towards equalization of temperature between the
elements and throughout the whole mass.

*Temperature defined as for elements of mass.*

11. It still remains, on this order of ideas, to identify physically
the scale of temperature for which the formulae become simplified
as above, and also to ascertain that the usual calorimetric substances
do not differ very substantially, within the usual ranges of temperature, from the ideal one which allows heat to be thus measured as
energy.

*Identification with actuality.*

In the ideal perfect gas of Joule and Waterston and Clausius, the
molecules exert mutual forces only in the instant of encounter, their
range being thus so short that only negligible mechanical work is
thereby involved: there is no internal potential energy, and all the
energy that such a substance receives passes into the form of kinetic
energy of the uncoordinated motions of the molecules, which may
be named internal heat. The pressure is, as Joule proved, equal to
$\frac{2}{3}\tau'$, where $\tau'$ is the translatory part of the molecular energy per unit
volume: this is $\frac{2}{3}k\tau$, where $\tau$ is the whole of that energy and $k$ must
be constant. Therefore, the heat received by a mass of such substance
finds a measure in the change of volume produced by it under constant pressure; so that if also temperature is measured by its expansion, this will be the standard calorimetric substance. But what is the
relation of temperature thus measured to the ideal absolute temperature? The working cycle for a gas, expounded and scrutinized[1] in the

*In a perfect gas the energy all kinetic:*

*pressure and energy:*

*it realizes absolute temperature.*

---

[1] The development of heat in the gas by compression he has to accept as a
fact on the basis of the familiar experiments and of Laplace's correction to the
velocity of sound-waves; for by adopting the current view he has debarred
himself from the idea of creation of new heat from work. In the algebraic
analysis for gas cycles (ed. 2, footnote, p. 40), he is thus led to results confused
and in part suspect, and consequent recurring remarks (cf. p. 50) on the doubt-
fulness of the current theory of heat, which he had adopted. He is thus puzzled
by the result of expansion of the gas into a vacuum reservoir, as found by
Gay-Lussac and Welter, the one vessel gaining as much heat as the other loses;
but later, in his posthumous notes (p. 91), he attains to nearly complete
illumination. Thus in the same context (*Notes inédites*, p. 96) we even find a
suggestion of the Joule-Thomson experimental method: "Faire sortir de l'air
d'un vaste réservoir où il est comprimé, *et rompre la vitesse* dans un large tuyau
où se trouvent placés des corps solides, mesurer la température lorsqu'elle est
devenue uniforme. Voir si elle est la même que dans le réservoir. Mêmes

*Carnot's confused results,*

*brought doubts as to heat being a substance:*

*final illumination.*

*He points out the appropriate experimental method:*

necessary close detail by Carnot as a typical precise example of his conception of a reversible cyclic process, shows that the two are identical. Thus a gas satisfying approximately the ideal gaseous laws realizes, practically, as Clausius was the first formally to disentangle, the normal scales of temperature and of calorimetry, which Carnot's idea by itself had already implicitly contained. {.right-note: Carnot to Clausius.}

The experimental checking of the results, to an extent adequate to ensure confidence, is, of course, a necessary supplement to any abstract argument concerning natural processes; and after that there still remains the concrete dynamical realization or reconstitution of the subject, so far as possible, in terms of the molecular constitution of matter. {.right-note: Ultimate explanation.}

12. If then Carnot's original idea is well founded, that heat can give rise to work only in carrying through its innate effort towards an equilibrium, as water does in its effort towards a lower level, or a gas in its effort to expand towards a uniform pressure, it follows necessarily that heat is measurable as regards quantity in the same terms as motive power. The physical alternative, that heat may be virtually a substance, appears in the abstract argument merely as a limiting case, when the ratio of equivalence is indefinitely small and the amount of heat therefore does not sensibly vary. {.right-note: Harvest from the one postulate.}

But although the preliminary idea of Carnot, apart from the semi-practical development that accrued to it in his further logical analysis of heat engines and other thermal processes, was almost co-extensive with the modern doctrine of chemical dynamics, yet in 1824, nearly a century ago, the scientific material did not exist for any systematic search into the ramifications of such a principle of available iso-thermal energy throughout Nature. Carnot was duly impressed with the fact that chemical combination in the furnace is an essential part[1] of the phenomenon which he explores and analyses; but he had to take up the position that for the purposes of his discussion it enters only in a preliminary way, as the means of providing the heat with which the engine works; and elsewhere he notes the production of motive power by electric means as beyond the range of his argument. {.right-note: Available energy concept was premature: now is the universal static principle.}

The principle of available isothermal energy in nature is in fact far wider than its thermal province, though the latter is an inseparable {.right-note: also convective thermal equilibrium in atmosphere.}

expériences avec d'autres gaz et avec la vapeur formée sous diverses pressions...." He points in passing to reduction of temperature with height in the atmosphere as due to the convective equilibrium investigated later by W. Thomson (ed. 2, footnote, p. 16).

[The manuscripts of Carnot have recently (1927) been published in facsimile by the French Academy of Sciences.]

[1] He recognizes that the essential *datum* is the maximum temperature at which the combustion can be sustained. {.right-note: Carnot-Gibbs.}

part of it. Its development must depend on a reasoned survey of the operations of Nature in their wider aspects. Thus it is not so surprising that the principle of Joule should have been recognized in its full scope, as an exact law, only twenty years later. But we find the preliminary notion of conservation and interchange of natural energies already confidently and acutely expanded by Faraday, to whose manifold discoveries in the correlation of the physical agencies it seems, in his own vivid but undefined intuition, to have formed a main guide. No instance of this is historically more instructive than the remonstrance in one of his later researches against the partial interpretation of Volta's original investigations, that the phenomena of the voltaic pile can be founded solely on electric forces of mere contact between different substances. Thus[1]

*Conservation of total energy came later:*

*though Faraday had made it imperative,*

*thinking however of available energy.*

"2071. The contact theory assumes in fact...that—without any change in the acting matter or the consumption of any generating force—a current can be produced which shall go on for ever against a constant resistance, or only be stopped as in the voltaic trough by the ruins which its exertion has heaped up in its own course. This would indeed be a creation of power and is like no other force in nature. We have many processes by which the form of the power may be so changed that an apparent *conversion* of one into another takes place. So we can change chemical force into the electric current, or the current into chemical force. The beautiful experiments of Seebeck and Peltier show the convertibility of heat and electricity; and others by Oersted and myself show the convertibility of electricity and magnetism. But in no case, not even those of the Gymnotus and Torpedo (1790), is there a pure creation of force: a production of power without a corresponding exhaustion of something to supply it.

2073. Were it otherwise than it is, and were the contact theory true, then it appears to me, the equality of cause and effect would be denied (2069). Then would the perpetual motion idea be true: and it would not be at all difficult, upon the first given case of an electric current by contact alone, to produce an electromagnetic arrangement which, as to its principle, would go on producing mechanical effects for ever."*

In a footnote he quotes from Roget (1827) to the effect that physical effort of any kind appears to be unable to carry itself through without drawing upon and partially exhausting its limited reserves: and he might have quoted a footnote of Carnot (*infra*), introducing this very subject of the voltaic pile, had it been known to him.

13. In this special controversy about contact forces, which has persisted down to our own time, a sharp discrimination between two distinct ideas, force and energy, is what had been mainly lacking.

---

[1] *Exp. Res.* Vol. XVII, January, 1840.
* Cf. Appendix VI at end of this volume.

In the eighteenth century a cognate controversy as to whether force should be measured by the *momentum* or by the *vis viva* produced by it, long raged: the final stage in appeasing it was the introduction of the necessary new term *energy* by Young[1] to represent the accumulated *vis viva* of a force, which Leibniz and his school, and many others including the engineer Smeaton, had insisted on taking as its measure, instead of the Newtonian *momentum*. The expansion, rather than correction, of ideas that was thus involved—the distinction between force and energy, between effort and its consummation—may be compared with another classical instance of clarification, W. Thomson's recognition of available energy as distinct from total energy, of which the germ was latent, for development in the fullness of time, in Carnot's term motive power. *Importance of exact definitions.*

Two years after these remarks of Faraday, the idea of energy as conserved and interchangeable in natural processes was grasped firmly and developed in many subtle aspects by the physiologist J. R. Mayer: while Joule was already engaged in the experiments that confirmed his own precise and practical outlook, and constituted it the fundamental generalization of physical science. Finally, the doctrine was crystallized by Helmholtz in his famous essay of 1847, building on the ideas and experiments of Joule, as a quantitative guide through the correlations of natural agencies—stimulated thereto in the main, as was also Young, by study of the mathematical physicists of the previous century, and indeed without much immediate local recognition except from the mathematician Jacobi. But even this formulation of the conservation and interchange of potential motive power in Nature is incomplete until the condition implied in Carnot's fundamental idea is added to it, that the operations must take place at uniform temperature, or else be subject to the other limitations of thermodynamics. *Helmholtz on the wider transformations of energy.*

But the ideas of Carnot on this subject are as definite as those of Helmholtz or of Thomson. On the theoretical abstract side nothing has been added, except by way of further exemplification, to his reasoned footnote on the principle of the *perpetual motion*[2]. The principle has, he states, been demonstrated only for mechanical actions: "Mais peut-on concevoir les phénomènes de la chaleur et de l'électricité comme dues à autre chose qu'à des mouvements quelconques de corps, et comme tels ne doivent-ils pas être soumis aux lois générales de la mécanique?" And he goes on to assert, as an example, the inevitable exhaustion of the power of the pile of Volta, owing to the work that it performs. In the text to which this footnote refers the implication is that heat as well as electricity is a substance, and *Carnot's universal outlook,*

[1] *Lectures*, Vol. I, p. 78 (1807).     [2] *Réflexions*, ed. 1, p. 20; ed. 2, p. 12.

that the phenomena arise from its disturbance. Thus "...tout
extending<br>even towards<br>Kelvin's<br>dissipation. rétablissement d'équilibre [dans le calorique] qui se fera sans production de cette [maximum] puissance devra être considéré comme
une véritable perte;..." The principle of the thermal dissipation of
available energies, in extension of the simpler principle of the complete availability of isothermal energies, is here almost in sight.  His
thoughts recur to the latter subject, now however on a definite
mechanical theory of the nature of heat, but still in a tentative way,
in the posthumous fragments (p. 92) published in 1878 by his brother.

A classic of<br>scientific<br>method. The master thought which presides over all this development of
physical science was enunciated by Sadi Carnot in 1824, in a solitary
essay nearly contemporary with the chief work of his great country-
men Ampère and Fresnel.  As a chapter in scientific method, it seems
desirable even now to bring the full individual potentiality of this
creative idea into view.

(*Added Feb.* 26. The main thesis developed above is that a close
analysis of the postulate of Carnot evolves from it not one but two
equations. These may be interpreted, after suitable simplification of
scales of measurement of heat and temperature, as asserting that
when there is no waste of power two quantities are conserved, viz.
heat *plus* other forms of energy, and entropy: when however irreversible features are present, the latter must increase while the constancy of the former need not be disturbed. Yet, as Helmholtz
remarked, the distinction between the two constituents of this constant sum, the heat in a body and its other forms of internal energy,
has hardly even now become analytically precise.

My attention has now been recalled by Sir Alfred Ewing to a very
interesting and in certain respects cognate discussion by Professor
H. L. Callendar[1], which turns on the idea that the caloric, whose
conservation had been *assumed* by Carnot, is capable of being interpreted as entropy, so far as reversible processes are concerned. He
there recognizes, as above, that the efficiency principle can be established without any assumption as to the nature of heat. Then the
Carnot's<br>caloric as<br>entropy: basic postulate would take the form that motive power can be gained
only in carrying through the effort of this caloric towards its equi-
librium distribution, and would arise from part of the energy, of its
disturbance from equilibrium, being diverted into mechanical work.
The motive power thus gained would be equal to the caloric that is
transferred, multiplied by its fall in potential as measured by the
temperature associated with it: but there would be no necessary
absolute zero. The entity heat would, on this train of ideas, enter as

---

[1] Presidential Address to the Physical Society, February 10, 1911.

mere residual energy, of amount required to satisfy Joule's experimental law of conservation. But wherever the energy of disturbance of the caloric is transformed in a wasteful (because irreversible) manner, there would have to be creation of new caloric, of amount equal to work wasted divided by temperature, thus implying an absolute zero; it would be imaginable, perhaps, as a sort of degenerate residue from lost motive power. The reversible disappearance of free electricity, or even free heat, by becoming latent, is hardly an analogy in point. analogy only partial.

This great fluidity of ideas as to specification of heat, or rather of caloric, has its source in the necessity of postulating a provisional calorimetric substance, after the manner of Black, the unit of heat being proportional to its thermal capacity which may be any function of temperature: thus heat will be an equivalent of vanished energy only when it is measured on the proper scale; while even entropy would assume the *rôle* of caloric provided the thermal capacity of the calorimeter were taken proportional to absolute temperature. Black's calorimetry.

Reference may also here be made to Clausius' long-sustained attempt (which received some favour from Willard Gibbs in his Obituary Notice of Clausius, *Collected Papers*, Vol. II, p. 263, as *supra*, p. 288) to break up the Carnot cycle into two physically distinct parts, the direct transformation of heat $Q'$ into work at temperature $\theta'$, and a compensating transference of heat $Q$ from temperature $\theta_1$ to a lower temperature $\theta_2$.)

### (1928) *On Abstract Entropy and Temperature as indicated by Thermal Continuity.*

A mode of approach *à priori* to an absolute entropy, initiated by Caratheodory[*], has recently become prominent. It rests on the abstract theory of Pfaffian differential relations[†]. The state of a physical system depends on its coordinates of configuration and on some other variable expressive of how it is gaining or losing heat, say in all on a set of variables expressed abstractly by $q_1, q_2, \ldots q_n$. Its mode of acquisition of some quality called heat, from outside, is taken to be amenable to differential expression of standard continuous type, Procedure of Caratheodory.

$$\delta H = \kappa_1 \delta q_1 + \kappa_2 \delta q_2 + \ldots + \kappa_n \delta q_n.$$

Nothing is implied as yet as to what becomes of the heat $\delta H$ thus acquired, whether it is stored up or in part disappears, or even as to how its appearance is recognized: it is only implied that it is some quality that enters in this regular differential manner. As it is not restricted to be conserved in exchanges, $\delta H$ is not an exact differential. Its implications.

[*] *Math. Ann.* 1909; Berlin *Berichte*, 1925, pp. 39–47. Cf. A. Landé, in Geiger and Scheel, *Handbuch der Physik*, Vol. IX, pp. 280–300.

[†] Cf. *supra*, Vol. I, p. 158.

But if we suppose now the system to be isolated by a boundary, impermeable to heat but not to work, its internal transformations are subject to a condition that $\delta H$ must be null: if it is still capable of continuous changes which form an unrestricted group—so that if there is such differential transformation from state $P_1$ to a differentially adjacent state $P_2$ and another from state $P_2$ to an adjacent state $P_3$ then there can be direct return from $P_3$ to $P_1$—then $\delta H$ must be reducible to a form $\vartheta\delta\phi$; and such an adiabatic mutual group of states becomes defined and segregated as represented by points on a hyper-surface $\phi = $ constant in the phase space of the coordinate variables. But the condition involves that there is to be no flux of heat any-where within the system: it is essential that the relation of thermal equilibrium is to remain satisfied throughout it. All possible states of the system could then be classed as belonging in groups to the set of hypersurfaces,—over which however the total heat $\Sigma\delta H$, received during the history of each of the states, has *not* the same constant value.

Helmholtz's<br>monocyclic<br>systems. As Helmholtz noted, this reduced type of differential form $\vartheta\delta\phi$ for $\delta H$ is not unique, for by making $\phi$ equal to $\psi\,(\phi_1)$ it becomes $\vartheta\psi'\,(\phi_1)\,\delta\phi_1$, which is of the same type but with a different $\vartheta$: the effort to attain definiteness in this regard was the source of his discussions on the dynamics of monocyclic systems and on Least Action. If this reduc-tion of $\delta H$ to form $\vartheta\delta\phi$ is not possible, an adiabatic set of states cannot subsist as a continuous sub-group with complete interconnections. From this remark a formal approach towards an Entropy function is held to arise.

The conditions that are here implied, that there is thermal equi-librium throughout the system at each stage of the transformations, but no $\delta H$ supplied to it, make this adiabatic sub-group perhaps too intricate for practical grasp. So the postulate as introduced is turned round into the converse, that if the system admits of *any* states $H$ which cannot be reached at all from a standard state $H_0$ in continuous transition, it can only be because its states are distributable into groups, the states in each group transformable completely among themselves alone but incapable of passing outside the group, just as are the representative points confined to and ranging over the hypersurfaces above; and therefore the differential expression for $\delta H$ must be reducible to the form $\vartheta\delta\phi$. The question whether the purely formal analysis on which this conclusion is based would stand the stress of Reduction<br>to Carnot-<br>Kelvin<br>reversibility. there being only isolated exceptional states which cannot be reached from the others, would side-track the present argument. But how can a correlating conclusion of such vast generality be deducible in any physical sense at all? It is submitted that an implied com-plete restriction to direct reversibility of thermal change is of the essence of the deduction. The differential relation involves that when

$\delta q_1, \delta q_2, \ldots \delta q_n$ all increase gradually in the same ratio, $\delta H$ increases in that ratio: and this growing ratio may become negative, meaning perfect reversal. Of modes of interchange of heat within the system, none of which were even mentioned in the enunciation above, all must be explicitly excluded that are not reversible. The principle has therefore to read as follows: if there are states of the system that cannot be reached spontaneously from a given state by processes admitting of direct thermal reversal, then* the gain of heat $\delta H$ in any part of it must be of the form $\vartheta \delta \phi$. But this implication is just a transcription of the Carnot-Kelvin doctrine of irreversible transformation: though it may perhaps be held to raise the question of the precise amount of irreversibility that is required in order formally to establish temperature and entropy, with assumption however of absolute continuity of change in the sense required by a formula of abstract differentials.

The proposition on this view is that if there are paths of transformation in the phase space that cannot be accomplished in manner that would be reversible, it must, as a matter of abstract interpretation, be because all reversible paths are concatenated into subregions of phase that are sections of the complete phase space. It is as regards the equation for $\delta H$ that the question of reversibility enters: and the reason for the existence of these reversible subgroups is just that for them there is no $\delta H$, it being null as they are adiabatic, so that the limitation to reversibility does not present itself. But whence can arise the idea of such limitation as regards $\delta H$? It can only be because the specification of $\delta H$ by the differential expression is incomplete, that there are processes latent in the system which affect $\delta H$, so that its course cannot be reversed unless they also are reversed, though so far as there is no $\delta H$ at all the system is reversible. But even so, a statical theory such as formal thermodynamics is not wide enough to supply these latent influences: they must be of motional type, and extension into the kinetic formulation by Minimal Action is indicated.

*[margin: Irreversibility implies incomplete specification.]*

*[margin: Action.]*

Anyhow, from a point of view thus presented, one may perhaps venture on the following, which seems to contain the essentials of the scheme. Let there be two such systems, each with imperfect internal adiabatic continuity of states, for which thus

$$\delta H_1 = \vartheta_1 \delta \phi_1, \quad \delta H_2 = \vartheta_2 \delta \phi_2,$$

the systems being at first separated by an impermeable barrier. What is the condition that, when this partition is drawn away, no essential thermal disturbance in the combined system ensues? The two then

*[margin: Thermal equilibrium undisturbed by contacts:]*

---

* As hinted above this is not enough, for it does not exclude a form such as $\vartheta_1 \delta \phi_1 + \vartheta_2 \delta \phi_2$: to restrict to one entropy and one temperature complete hypersurfaces representing interchangeable states must be postulated which are, after Kelvin, the surfaces of available energy.

form one equilibrated system for which again $\delta H$ must, on the same presumption, be of the form $\vartheta\delta\phi$, so that $\delta H = 0$ is an integrable equation. But
$$\delta H = \vartheta_1\delta\phi_1 + \vartheta_2\delta\phi_2:$$
so it can be completely integrable only if $\vartheta_1$, $\vartheta_2$ are functions of the two variables $\phi_1$, $\phi_2$ alone.

Moreover, we can have the system $\delta H_1$ in equilibrium with another system $\delta H_3$, and draw like conclusions. As then $\vartheta_1$ is a function of $\phi_1$, $\phi_2$ only, and of $\phi_1$, $\phi_3$ only, it must be a function of $\phi_1$ alone, affected however by any multiplier which has the same value for the states $\delta H_1$, $\delta H_2$, $\delta H_3$ which are thus in equilibrium of thermal contact (cf. *supra*, p. 596).

*demands an entropy and a uniform temperature.* Thus for any system in static thermal state we derive
$$\delta H = T\delta\eta,$$
where $T$ has the same value for all systems which can be placed in contact without thermal disturbance ensuing between them, so must be constant throughout the parts of any one system itself in internal thermal equilibrium. But this is just the property that defines empirical temperature, though referred to some scale as yet undetermined. Thus for the present combined systems
$$\delta H = \delta H_1 + \delta H_2 = T\delta\eta_1 + T\delta\eta_2$$
where $T$ is the same for both components in thermal equilibrium with each other. In addition to temperature $T$, the existence as well of an entropy $\eta$ which is an additive function of the coordinates and of $T$, and determined as $\Sigma\delta H/T$, has emerged.

*This entropy and temperature purely formal.* This may appear to be a singularly slight foundation from which to get so much: but really the postulate that $\delta H$ is integrable over a set of hypersurfaces in the phase space is in a formal sense far from slight. The real physical problem, which remains over, is to show that the molecular systems of Nature do in fact adjust their transformations within such a standard differential scheme. So abstract indeed are the temperature and the entropy as thus introduced, that nothing has been here said at all about thermodynamics or the work of thermal motors, hardly anything about the nature of the quantity $\delta H$.

*Dynamical and molecular ideas introduced.* The transition would now be made to actual thermodynamics through the principle of conservation of total energy: thus
$$\delta E = \delta H + \delta W, \qquad \delta H = T\delta\eta,$$
while the quantity $E$ is conserved and the quantity $\eta$ cannot spontaneously decrease: and this covers the field.

This is for systems in thermal equilibrium, in which heat is not flowing down a gradient of temperature. When two systems not at the same temperature come into contact, heat is postulated to trend towards the lower temperature; thus the total entropy increases and cannot spontaneously diminish.

The difficulty of forming a conception of the group of transformations $\delta H$ *null* has here been alluded to. Each member of the group of states must be at a uniform temperature throughout, which however changes on transition to other independent members: yet the mode in which the notion of temperature arises is to come out of the result of the discussion. The Carnot-Kelvin systematization may expose new points of view treated in this connection. Instead of $\delta H$ is taken $\delta W$ the work gained from the system in an infinitesimal change: it is there to be implied as before that the changes of the system are capable of analytic differential expression. There are groups of transition for which there is no degradation of energy*, for which therefore $\delta W$ is an exact differential: within each such state there is no spontaneous flux of heat, and no supply from outside, so that in the work of the transformation heat takes no part.

Contrast of cognate procedure from available energy.

This can be satisfied by a form

$$\delta H = \delta h + L\delta M,$$

in which $\delta h$ is a complete differential of a function of state $h$, which is equivalent to

$$\delta H = \delta h' - M\delta L, \quad h' = h - LM:$$

and if $L$ is of form $\psi (M, \vartheta)$ in which $\vartheta$ is a function of the variables of the system, then $\delta H$ is an exact differential as regards the states represented by the points on every hypersurface $\vartheta$ constant. A like result follows from the alternative form involving $h'$. A relation $\chi (L, M, \vartheta) = 0$ of wider form including both may involve complexities of cyclic or other character: avoiding them, we assert that if $L$ is of form $\psi_1 (M, \vartheta)$ or $M$ of form $\phi (M, \vartheta')$, then $\delta H$ is an exact differential over *two different sets* of loci $\vartheta$ constant and $\vartheta'$ constant. These two sets are realized in the two types adiabatic and isothermal transformations, which both transform the energy reversibly. But entropy does not now emerge without further postulations; in fact the previous argument is more special and artificial than this, in that $\delta H$ has to be restricted to null values instead of merely to be an exact differential.

Emergence of adiabatic and isothermal.

* From another point of view there is an indeterminateness, which must have been conspicuous from the first, inherent in the Carnot approach to thermodynamics. The abstract argument deals with operations perfectly reversible: but that is an ideal which is in strictness unattainable, so must be defined relative to some implied scale of efficacy in the minute control of the system. An ideal law would express the limit (for natural radiation on unrestricted or classical lines there is no such limit) that may for some types of structure be approached as the size of the cells, in which the microscopic specification is framed, is diminished. If such a limit as regards state cannot effectively be approached a definite lower limitation of workable extent of cell would have to be involved in order to arrive at any definite result, as Rayleigh noted in a general way, thus introducing a constant, dynamically universal scale or quantification. Cf. also *supra*, p. 411.

Reversibility is relative to a scale of *quanta*.

# 94

# THE PRINCIPLE OF MOLECULAR
# SCATTERING OF RADIATION.

[*Philosophical Magazine*, Vol. XXXVII, pp. 161–163: Jan. 1919.]

A FUNDAMENTAL element in the theory of radiation is the principle first elucidated by Lord Rayleigh[1], that when light is scattered by the particles of a fog or haze, or even by the molecules of the air, they act independently, without sensible mutual interference as regards the distribution of the energy.

Condition for independent energies: The condition necessary for this independence is that the disturbances (such as strain, velocity) must arrive from the scattering particles in phases which are entirely uncorrelated: so that on an average taken over a short interval of time, the square of the sum of the disturbances is equal to the sum of their squares, and thus the total energy would come from addition of energies of independent scattered disturbances.

This condition will be secured if the scattering particles are distributed at random, provided the intervals between adjacent ones are substantial fractions of the wave-length of the radiation that is being scattered: for then the phases of the scattered disturbances coming from adjacent particles will be uncorrelated. It holds good usually case of crowded sources: for particles of dust in the atmosphere. But in the case of a gas there are $10^6$ molecules in a cubic wave-length, and in the case of a liquid or solid $10^9$, giving differences of adjacent phases of the order of only $10^{-2}$ of the period in the former case and $10^{-3}$ in the latter.

Even for a gaseous medium the question thus arises, whether it is wrong to consider the distribution of the scattering molecules as formulation for regular waves in a gas. based upon uniform spacing, but subject to uncorrelated deviations from this regularity which obey the law of statistics of gas theory and amount at most to a few hundredths of the wave-length. This representation would seem to be permissible, at any rate for each group of say about $10^6$ molecules occupying the cubic wave-length; and such groups will be practically independent.

Now if the molecules were spaced with exact uniformity at distances of smaller order than the wave-length, as they are in a crystal, the disturbances scattered from an incident beam, instead of being

---

[1] *Proc. London Math. Soc.* 1870; *Phil. Mag.* Vol. x, 1880; and later papers, including the one under special reference in *Phil. Mag.* Dec. 1918.

additive as regards their energy, would interfere completely; so that there ought to be no radiation scattered in traversing a crystalline medium. This has in fact been remarked by Lord Rayleigh in his recent paper[1], and I think previously by Professor Lorentz. A beautiful experiment by Professor R. J. Strutt, which I had the advantage of seeing some time ago, showed that the actual scattering in a column of quartz crystal was small compared with what occurs in optical glass or even in a liquid such as ether. Quantitative comparison would be of interest on various grounds. Scattering should be nearly absent within crystals:<br>verified for quartz.

This principle that a crystal should scatter no radiation* seems to be unimpeachable, provided the molecules are fixed and do not partake of thermal agitation. And it seems difficult to see why it should not also apply to the molecules of a gas, if they could be regarded as fixed while the radiation is passing, subject to correction for the statistical deviations aforesaid from their mean positions. Why not also for a gas?

If this were so the individual molecules of a gas, and *à fortiori* of a liquid or a solid on account of their closer packing, ought in conjunction to scatter radiation far less than they would do separately, the reason being the vast number contained in a cubic wave-length and the statistical regularity of their distribution. And accordingly Lord Rayleigh's announcement[2] that the blue sky could be due to scattering by the molecules of the air itself came as a surprise, which subsequent quantitative verifications did not wholly resolve. The blue sky.

The suggestion now to be advanced for consideration is that the principle is to be maintained, even to some extent for crystals, but its logical basis is to be shifted.

The molecules of the atmosphere are in thermal motion, with velocities in uncorrelated directions which are at ordinary temperatures of the order of $10^{-6}$ of that of radiation. The wave-length of the radiation scattered from them will thus vary within a range of $10^{-6}$ of itself. If the phases of the scattered radiations are correlated at first, after traversing $10^6$ wave-lengths or 50 cm. they will have become fortuitous, and the energy effects thus additive†. Independent scatterings traced to thermal motions,

---

[1] *Phil. Mag.* Dec. 1918, p. 445, footnote.

[2] *Phil. Mag.* 1899; *Scientific Papers*, Vol. IV, p. 397.

* Except from a residual layer over its surface, representing a broken wave-length. Cf. Appendix v.

† This is confused: the waves of all lengths will travel together, with the same speed, except in so far as dispersion of velocities gradually alters their relative phases; for air this would be sensible only after extremely long paths, but in any case the dispersed wave-lengths are independent, being separated by the spectroscope as in next paragraph. What dispersion does is to make the energy bundles travel rather slower than the waves. For a final decision see Appendix v at end of this volume.

610 Principle of Molecular Scattering of Radiation

This consideration, if justified, would find the source of Lord Rayleigh's principle in the uncoordinated thermal motions of the molecules. And as in a crystal thermal vibrations are contemplated about the mean positions of the molecules in the space lattice, there ought to be some degree of scattering in traversing a crystal*.

not to phase-differences.

This way of envisaging the matter contemplates a scattered beam of radiation of wave-length slightly indefinite, on Doppler principles, in which therefore the phases of the constituent elements [not] become fortuitous after travelling a substantial though not very great distance. Other influences of radiation on the gas which are also adjusted over a considerable range in distance would be implied: the molecular effects connected with pressure of radiation would be expected to belong to this class.

Internal radiation-pressures.

Some confirmation from another point of view seems desirable. The scattered radiation, forming a spectral band of some slight breadth, may be considered as resolved into more homogeneous constituents. The radiation in one of them has been scattered at each instant by molecules whose thermal velocities agree within close limits as regards both direction and magnitude: these molecules constitute a sparsely distributed group whose distances apart can be of the order of the wave-length, so that the condition necessary for fortuitous phases in the scattered disturbances is satisfied.

Molecular scale opened out by dispersion.

The interesting [and fundamental] remark is made by Lord Rayleigh (p. 445) that the radiation scattered nearly in the direction of the primary rays is specially favoured, in that all its components nearly agree in phase. The question arises whether this would not make a clear sky very much brighter within a few degrees of the Sun than at some distance away from it. Here also the slight variety of wave-lengths arising from the thermal motions of the scattering molecules would seem to reduce or perhaps nearly remove such disparity†.

Sky brighter but less blue near the Sun?

* The amount would depend on the order of velocity, and so of amplitude, of the thermal oscillations of the molecules, being slight, and thus perhaps give some measure of it. See Appendix v.

† See end of Appendix v.

# 95

## HOW COULD A ROTATING BODY SUCH AS THE SUN BECOME A MAGNET?

[From *Report of the British Association*, Bournemouth, 1919.]

THE obvious solution by convection of an electric charge, or of electric polarization, is excluded; because electric fields in and near the body would be involved, which would be too enormous. Direct magnetization is also ruled out by the high temperature, notwithstanding the high density. But several feasible possibilities seem to be open. *The problem.*

(1) In the case of the Sun, surface phenomena point to the existence of a residual internal circulation mainly in meridian planes. Such internal motion induces an electric field acting on the moving matter: and if any conducting path around the Solar axis happens to be open, an electric current will flow round it, which may in turn increase the inducing magnetic field. In this way it is possible for the internal *Field possibly* cyclic motion to act after the manner of the cycle of a self-exciting *self-exciting* dynamo, and maintain a permanent magnetic field from insignificant *dynamo.* beginnings, at the expense of some of the energy of the internal circulation. Again, if a sunspot is regarded as a superficial source or sink of radial flow of strongly ionized material, with the familiar *Characteristics* vortical features, its strong magnetic field would, on these lines, be *of sunspot* a natural accompaniment: and if it were an inflow at one level compensated by outflow at another level, the flatness and radial restriction of its magnetic field would be intelligible. [Cf. Appendix VI.]

(2) Theories have been advanced which depend on a hypothesis *Polarization* that the force of gravitation, or centrifugal force, can excite electric *by gravi-* polarization, which, by its rotation, produces a magnetic field. But, *or rotation,* in order to obtain sensible magnetic effect, there would be a very intense internal electric field such as no kind of matter could sustain. That, however, is actually got rid of by a masking distribution of *not excluded,* electric charge, which would accumulate on the surface, and in part *for it can be* in the interior where the polarization is not uniform. The circumstance *electrically,* that the two compensating fields are each enormous is not an objec- *by a free* tion; for it is recognized, and is illustrated by radioactive phenomena, *charge,* that molecular electric fields are, in fact, enormous. But though the electric masking would be complete, the two distributions would not compensate each other as regards the magnetic effects of rotational

39-2

while a magnetic effect remains over.convection: and there would be an outstanding magnetic field comparable with that of either distribution taken separately. Only rotation would count in this way; as the effect of the actual translation, along with the solar system, is masked by relativity.

Natural electric polarization in a crystal must be intense,(3) A crystal possesses permanent intrinsic electric polarization, because its polar molecules are orientated: and if this natural orientation is pronounced, the polarization must be nearly complete, so that if the crystal were of the size of the Earth it would produce an enormous electric field. But, great or small, this field will become

though masked by surface charge;annulled by masking electric charge as above. The explanation of pyro-electric phenomena by Lord Kelvin was that change of temperature alters the polarization, while the masking charge has not had opportunity to adapt itself: and piezo-electric phenomena might have been anticipated on the same lines. Thus, as there is not complete compensation magnetically, an electrically neutralized crystalline

could produce a magnetic field by rotation,body moving with high speed of rotation through the aether would be expected to produce a magnetic field: and a planet whose materials have crystallized out in some rough relation to the direction of gravity, or of its rotation, would possess a magnetic field. But relativity

but not by convection.forbids that a crystalline body translated without rotation, even at astronomical speeds, should exhibit any magnetic field relative to the moving system. [Cf. p. 620.]

The great terrestrial magnetic changes negligible astronomically:The very extraordinary feature of the Earth's magnetic field is its great and rapid changes, comparable with its whole amount. Yet the almost absolute fixity of length of the astronomical day shows extreme stability of the Earth as regards its structure [cf. p. 335]. This consideration would seem to exclude entirely theories of terrestrial

a criterion thus afforded,magnetism of the type of (2) and (3). But the type (1), which appears to be reasonable for the case of the Sun, would account for magnetic change, sudden or gradual, on the Earth merely by change of internal

restricting to one cause, of permanent nature.conducting channels: though, on the other hand, it would require fluidity and residual circulation in deep-seated regions. In any case, in a celestial body residual circulation would be extremely permanent, as the large size would make effects of ordinary viscosity nearly negligible.

## (1927) APPENDIX.—*On the Solar Magnetic Fields.*

The discovery and exploration, by G. E. Hale and his associates, of the intense magnetic fields of sunspots, and of the smaller general field of the Sun, must be significant for general electrical theory: for the Sun is a simple gaseous mass, in contrast with the molecular complexity of solid magnets, and the problem thus restricted should be amenable to molecular theory. *(margin: Problem of Sun's magnetic field:)*

The image that first naturally presents itself is a vortex of whirling electrons or ions: but their numerical density would have to be enormous: and indeed there would have to be as many positive as negative ions in the whirl to avoid impossible electric fields, and then there could be no magnetism.

A modification however invites serious consideration. The counter-vailing positive ions are massive atomic nuclei, as compared with the electrons: in an electric field the two would be urged different ways, yet both producing current the same way, but the electrons would percolate readily through the more sluggish atomic ions: thus the electric current produced by the field would be mainly one of negative electrons, after the manner of an ordinary current in a metallic con-ductor. The problem would be solved if we can recognize some driving field of non-electric type, which on the hypothesis of completed circuital currents need not be of great intensity. *(margin: as due to electric currents not material whirls:)*

It is a feature of magnetic fields that they lie close to their sources: this is on account of the inverse cube law of force. The direct and return halves, joined at their ends, of a flat cyclic current-sheet con-spire in the space between them, but oppose each other outside their circuit. The magnetic field of sunspots is said to be confined to a narrow stratum: if so it could be imagined as due to opposing sheets of current at the two faces of the stratum, or to a row of parallel current vortices, such as might be sustained by radioactive discharge of electrons across the stratum*. *(margin: confined to thin strata.)*

The suggestion here arises, to be tested out, that as regards the Sun as a whole, the electrons may be driven differentially around its axis of rotation by repulsion (p. 448, *supra*) exerted by the outgoing radiation, simulating an electric field, as it would affect the negative electrons far more than the complementary positive ions. For on account of the Sun's rotation this outward stream of radiation is incident obliquely on the electron, after the manner of a shower of rain relative to an observer travelling across it: this Bradley aber- *(margin: Aberration-pressure of outgoing radiation,)*

* The general magnetic field of the Sun is found (G. E. Hale, *Astrophys. J.* 1918) to fall from 50 gauss to nothing in rising 400 km. in the atmosphere.

ration of the rays gives a backward component to the repulsion they exert: the electrons, thus repelled by the rays, would drift through the Solar mass as a negative current opposite to the direction of rotation, the positive ions also but with much smaller speeds; and this gives at any rate the right direction for the general magnetic polarity of the Sun relative to its rotation, being the same as for the Earth.

*could cause electric flux.*

The problem of the intense local magnetic fields of sunspots would be different. Somehow the emerging stream of Solar radiation, instead of passing out across a spot, is in part deflected round it and thus comes out perhaps more intense at its edges, as the surrounding flocculi would indicate. There might here arise a suggestion to connect the magnetic fields with gradients of temperature at the spot: the Kelvin thermo-electric effect in metals carrying currents may have its analogue for the Solar gas. In fact the density of the contained electrons may depend on the temperature: for they are more mobile at higher temperatures on account of greater speed, and so push their way down the gradient of temperature until this tendency is pulled up by their accumulation producing a countervailing electric field. Compare the partial separation of the components of a mixed gas by diffusion along a gradient of temperature, noted and investigated by Chapman. This sifting might, however, lead only to a reacting very slight distribution of electric charge in the medium, of density greater in the region of higher temperatures, which countervails it. But there may conceivably remain over a residual distribution of electromotive force, which could not be compensated by charge because of its cyclic character, yet precisely adapted to sustain currents in complete circuits. It would then be such continuing vortices of current, which may readily be intense, but not involving material whirls, that are responsible for the magnetic fields of the sunspots. The originating electric field, however it be generated, is here regarded as a local vector distribution, made up by Stokes' kinematic principle of a gradient part and a rotational part: it is the former that is compensated and neutralized by a reacting distribution of electric charge; the latter remains over and will produce strong cyclic currents. Such local current cycles associate in Amperean manner into distributions of current in larger circuits, but always without accumulation of any charge such as would choke them.

*A cause of flocculi.*

*Kelvin thermo-electric gradient at sunspots:*

*mainly compensated,*

*but affording a residual cyclic electric field,*

*by Stokes' kinematic principle:*

*only in disturbed regions.*

For a star as a whole a compensating Kelvin thermo-electric force $F$ would on this view be required, radial and outward from the inner higher temperatures, to maintain a steady state: it induces a feeble density of charge $\frac{1}{4\pi c^2}\left(\frac{dF}{dr} + 2\,\frac{F}{r}\right)$ diminishing outward, with which it is in equilibrium, and under these conditions of symmetry there is no cyclic part left to produce a closed current.

Estimates may be adventured which will test the adequacy of Estimate for Sun's general magnetic field, causes such as these, to which indeed in the gaseous Solar medium we appear to be almost restricted. First we consider the Bradley aberration of the stream of issuing radiation. The repulsion of an electron, of charge $\frac{2}{3}.10^{-20}$, by unit incident radiation is $\frac{1}{2}.10^{-24}/c$: as the intensity of the outward radiation at the Sun is $7.10^{10}$, its repulsion of the electron is of the order $10^{-24}$. The mass of the electron is $9.10^{-28}$, gravity at the Sun is $3.10^4$, so its gravitational attraction is $27.10^{-24}$, about 27 times greater than the repulsion. (Cf. p. 448: Vol. I, p. 663.) The angular relative deflection, by aberration, of the rays travelling directly outward is $8.10^{-6}$. Thus the tangential force on an electron at the equator is $10^{-29}/c$, and it persists when the radial repulsion of the electrons by the radiation has become balanced by the electric field of the positive charge left behind and pulling them back. The as due to obliquity of radiation pressure on electrons, tangential acceleration of the electron against the direction of the Sun's rotation would be $\frac{1}{3}10^{-11}$: if the free path is $l$ this gets up a mean velocity $(\frac{1}{3}10^{-11}l)^{\frac{1}{2}}$ or $\frac{1}{3}10^{-5}l^{\frac{1}{2}}$, as compared with velocity $4.10^5$ due to the Sun's rotation. This mean drift involves, for a numerical density $N$ of electrons, as a circuit is open round the Sun, a current round open circuits: at the equator of volume density $\frac{1}{3}10^{-5}l^{\frac{1}{2}}Ne$. As applied to a surface atmosphere this would hardly produce a sensible magnetic field. But for a rotating star composed of gas largely ionized this backward drift of electrons would be established throughout its volume. Inside the star the intensity of the outward stream of radiation varies as $r^{-2}$ and its Bradley aberration, due to the star's rotation on its axis, varies as $r\sin\theta$: therefore the equatoreal intensity of the electronic current has to be multiplied for the interior by $(a/r)\sin\theta$ for the same density $N$ of electrons and the same $l$. For a component ring current $i$ of cross-section $\delta S$ the magnetic moment would be $i\pi r^2 \sin^2\theta\,\delta S$: thus the aggregate moment would be $\Sigma\frac{1}{3}10^{-5}l^{\frac{1}{2}}Ne\,(a/r)\sin\theta\,\pi r^2\sin^2\theta\,\delta S$, of which only the order of magnitude can be estimated: taking $l^{\frac{1}{2}}N$ uniform and dividing by $\frac{4}{3}\pi a^3$ it integrates to a mean intensity of magnetization $I$ for the Sun equal to $10^{-16}l^{\frac{1}{2}}N$, giving a mean magnetic field of order $4.10^{-16}l^{\frac{1}{2}}N$. If this is to be of the order of 20 as observation indicates $l^{\frac{1}{2}}N$ would be of the average order $5.10^{16}$, which if right is far from excessive. Thus the general magnetic field of found to be an adequate *vera causa*, the gaseous globe of the Sun would be ascribed to a recognized cause, the interaction of its rotation and the issuing stream of radiation*. But for the magnetic Earth the stream of radiation is absent. for a star.

* The particles gradually arrest the outward stream of radiation, producing repulsion by dispersing its momentum in all directions. They must renew the stream by fresh temperature radiation of their own, of isotropic type on average: but the kinetic effect of this (pp. 448, 673) is only diminution of their effective inertia, without any reaction of the present type involving their velocities directly.

Field of a sunspot estimated.

A vortex half-ring with its ends on the Solar surface, of sectional radius $b$ and velocity of whirl of the order $v$, would be accompanied by an electric current whirl due to the stream of issuing radiation, producing a magnetic field of about the order $Nebv_0$, where $v_0$ is for the Solar intensity of radiation $2 \cdot 10^{-9} v^{\frac{1}{2}} l^{\frac{1}{2}}$. For this field to be as high as $10^4$ when $b$ is $10^9$, as in some sunspots, $Nv^{\frac{1}{2}} l^{\frac{1}{2}}$ would have to be of the order $4 \cdot 10^{23}$: thus if $v$ were $10^6$ the order of $Nl^{\frac{1}{2}}$ below the surface would have to be as much as $4 \cdot 10^{20}$. One notes that if $l$ were small its square root would be large in comparison.

*Asymmetry of Solar field.*

If there is internal material circulation in the Sun, it would be accompanied by circulating electric current. If it were symmetrical round the axis of rotation, this current would contribute to the magnetic field in the Sun, without displacing its poles: Hale's discovery that the Solar magnetic pole is a few degrees away from the pole of rotation may thus point to internal circulation not symmetrical round the axis, of which there are other prominent indications.

The cyclic electronic drifts producing the magnetic field of a sunspot would have to be sustained somehow, as above, by radiational or other effective thermodynamic energy. We can perhaps recognize at any rate whether the magnetism is essential to the existence of the spot.

*Magnetic stress in a sunspot.*

The opposite sides of a ring current repel each other mechanically: thus the current has expansive force and can balance a defect of pressure inside it, permitting smaller gaseous density there. In short, if we consider Hale's map of the magnetic lines of the local field, those issuing nearly radially from the region of the spot can be regarded as subject to the Maxwellian stress, tension from outside along the radius of intensity $H^2/8\pi$, and equal transverse pressure, which permits an equal defect of internal hydrostatic pressure. If the field is say $5 \cdot 10^3$ this downward tension is $10^6$ dynes, which may hold down the floor of the photosphere below the mean level but only to the order of a km. if its density is comparable with a terrestrial atmosphere, if it does not concentrate the diffuse chromosphere over the spot*.

---

* There is temptation to connect the drift of spots towards the equator and the very precisely opposed polarities even when they meet there (G. E. Hale and S. B. Nicholson, *Astrophys. J.* 62, 1925) with vortical effect of the maximal equatoreal rotation, were it not for the reversal in successive periods. Cf. Lord Rayleigh's last unfinished paper on atmospheric vortices, *Phil. Mag.* 1919; *Scientific Papers*, Vol. VI, pp. 654–7.

# 96

# THE PRESSURE OF WAVES OF SOUND.

[From *Comptes rendus du Congrès International des Mathématiciens*, Strasbourg, 1920.]

THE pressure of sound-waves in air and water against an obstacle has had recent important applications in military science. There has however been difficulty felt on the side of theory as to how trains of waves in a fluid exert pressure at all. The following simple considerations are therefore offered. *Pressure of sound-waves:*

The pressure of light, exactly equal to the density of energy in front of the obstacle, is of course fundamental now in physical theory. The law for waves in a material molecular medium on the other hand could not be exact on account of thermal effects. *contrasted with that of light.*

Let us consider transmission of disturbance of small amplitude in an elastic fluid, constituted so that in the actual circumstances $\partial p/\partial \rho$ is equal to $c^2$, which is the square of the velocity of propagation. Using the hydrodynamic formulae, we are concerned only with velocity and pressure at each point considered, and displacements do not come into the analysis. If there is a velocity potential $\phi$ the pressure is given in terms of it by the usual formula *Compressional waves in fluid:*

$$\int \frac{dp}{\rho} = V - \frac{\partial \phi}{\partial t} - \tfrac{1}{2}\,\text{vel.}^2.$$

For small change of density, if $\varpi$ is the variation of pressure,

$$\rho = \rho_0 + \left(\frac{\partial \rho}{\partial p}\right)_0 \varpi + \dots$$

so that

$$\int \frac{dp}{\rho} = \int \frac{d\varpi}{\rho_0} \left\{ 1 - \frac{1}{\rho_0}\left(\frac{\partial \rho}{\partial p}\right)_0 \varpi + \dots \right\}$$

$$= \frac{\varpi}{\rho_0} \left\{ 1 - \frac{1}{2\rho_0}\left(\frac{\partial \rho}{\partial p}\right)_0 \varpi \right\}.$$

*exact equation of pressure,*

Thus substituting, and transposing a factor, there being no extraneous potential energy $- V$, remembering that $\partial \rho/\partial p$ is equal to $c^{-2}$,

$$\frac{\varpi}{\rho_0} = - \left( 1 + \frac{\varpi}{2\rho_0 c^2} \right)\left( \frac{\partial \phi}{\partial t} + \tfrac{1}{2}\,\text{vel.}^2 \right)$$

$$= - \frac{\partial \phi}{\partial t} + \frac{\varpi^2}{2\rho_0^2 c^2} - \tfrac{1}{2}\,\text{vel.}^2,$$

for in the second term, which is of smaller order, the first-order value

$- \varpi/\rho$ may be substituted for $\partial\phi/\partial t$. Thus the change of pressure is determined up to the second order[1] by

$$\varpi = - \rho_0 \frac{\partial\phi}{\partial t} + \frac{1}{2c^2}\left(\frac{\partial\phi}{\partial t}\right)^2 - \tfrac{1}{2}\rho \text{ vel.}^2$$

$$= - \rho_0 \frac{\partial\phi}{\partial t} + W - T$$

corrected approximate form.

where $W$ is the volume density of potential energy, and $T$ that of kinetic energy.

The effect of changing density.

It is to be noted that the correcting factor above introduced is purely constitutive: thus for statical changes the fall of the potential $V$ of the applied forces must be equal not to $\dfrac{\varpi}{\rho}$ but to $\dfrac{\varpi}{\rho_0}\left(1 - \dfrac{\varpi}{2\rho c_1{}^2}\right)$, where $c_1 = c/\sqrt{\gamma}$, $\gamma$ being the ratio of specific heats of the gas.

If $\phi$ is everywhere purely fluctuating, as it can be taken to be in a disturbance constituted of superposed sound-waves in an enclosure

Steady pressure of the waves,

where their degradation is slow, the first term in $\varpi$ is alternating in each period of the waves, so that any steady pressure arises only from the other part.

its law.

At a nodal surface where there is no velocity, there is thus excess of pressure equal to the mean potential energy per unit volume: at a loop or antinodal surface, where there is no compression, there is defect of pressure equal to the kinetic energy per unit volume.

Maxwell's law holds for direct incidence only.

The first case includes that of plane waves of sound incident *directly* on a reflector, as at the reflector the medium cannot be displaced: then the pressure of the waves is the volume density of the energy at the reflector, in keeping with Maxwell's rule for radiation. But for oblique incidence the relation is more complex. The attraction of a sheet of

An illustration.

paper by a vibrating tuning-fork, described by Guthrie and discussed by Lord Kelvin, is due both to the partial node formed at its surface and to the stronger vibration developed on the nearer side of the sheet.

Repulsion of a resonator.

The repulsion exerted on a resonator has been considered by Lord Rayleigh. When the interior cavity is small compared with the wavelength, there are only changes of pressure inside which are constant throughout but of enhanced amount: and this constant outward pressure on the walls can be replaced by an equal inward pressure over the area of the aperture, of average not null, thus giving a simple rule to estimate the repulsion.

[1] This formula, with indication of its consequences as *infra*, was proposed for proof in the Cambridge Mathematical Tripos Examination, May 30, 1917, afternoon question 12. In the usual discussions initiated by Lord Kelvin, the second term of the formula, which accounts for the pressure on a reflector, seems to have been overlooked.

For a train of progressive waves $\frac{\partial \phi}{\partial t} = -c$ vel.: thus there* is no steady alteration of pressure within the waves, it is all alternating. Thus the waves can exert pressure on being reflected back by an obstacle only because they carry momentum along with them through the medium. And the amount thus convected is sufficient, for it is per unit volume, for a progressive train of waves travelling along $x$,

$$\rho \frac{\partial \phi}{\partial x} = \left\{ \rho_0 + \left(\frac{\partial \rho}{\partial p}\right)_0 \varpi \right\} \frac{\partial \phi}{\partial x}, \text{ where } \varpi = -\rho \frac{\partial \phi}{\partial t};$$

giving a quadratic part equal to $\frac{1}{c}\rho \left(\frac{\partial \phi}{\partial x}\right)^2$, which is the total energy per unit volume divided by $c$. When this momentum is deflected and perhaps also partly absorbed and partly transmitted at a reflecting surface, the appropriate pressure on the surface results.

* For it makes in the previous equation $W$ equal to $T$, so that $\varpi = c\rho_0$ vel.

# 97

# ON ELECTRO-CRYSTALLINE PROPERTIES AS CONDITIONED BY ATOMIC LATTICES.

[*Proceedings of the Royal Society*, A, Vol. xcix, pp. 1–10: Jan. 1921.]

THE method of analysis of crystal lattices by ultra-optical spectra, introduced and applied by W. L. and W. H. Bragg, has led to the view that the ultimate crystalline element is (or may be) the atom[1] and not the chemical molecule[*]: so that each atom, presumably with its ionic electric charge, often stands in identically the same relation to a number of surrounding atoms of the other kind with opposite charge, and therefore cannot be associated physically with any one of them. The alternative view, previously perhaps the more natural on the physical as distinct from the tactical side, was to regard the chemical molecule as the crystalline unit: its associated positive and negative atomic ions might then form an electric diad with definite polar moment orientated in definite direction[†]—and with this the new schemes come into contrast. The following general remarks may

*The ion as crystalline unit.*

[1] Rather the ion, which may be an atom or a group of atoms. The problem as to how stability of the crystalline structures is assured by interaction of the electric forces with the forces concerned in the packing of the molecules is outside the scope of this note, which is concerned with analyzing the results of various types of structure. [It has recently been developed very extensively on the mathematical side by Born, S. Chapman, Lennard-Jones and others, leading to molecular fields other than the electric ones that are extremely local, almost expressive of direct contact.

A general account of the purely topical or configurational side of the problem is given by W. T. Astbury in *Science Progress*, Jan. 1927: also in *Phil. Trans.*]

[*] As crystallography has recently developed, it appears that the elementary cell may be occupied by an ion, or by a complete atom, or by a group formed of a small number of atoms. This applies to the dissection into cells by sets of parallel planes as in X-ray analysis, a more complex scheme being possible in which the cells are occupied by ultimate individuals. For example, quartz would have been classifiable in a more symmetric form until the detection of its secondary chiral faces.

*The ultimate crystalline cell.*

What is however essential here is that the ultimate reflecting stratum for X-ray analysis may be necessarily a stratum several molecules deep, so that the reflection is an affair of periodic structure analyzable into independent Fourier components, rather than of sharp transitions. Cf. the working out of this principle by R. Schlapp from Hill's equation in *Phil. Mag.* Jan. 1925.

[†] There would then be electric polarization the analogue of ferromagnetic magnetization, in which the local electric field would play a prominent part: cf. Mahajani, *Proc. Camb. Phil. Soc.* 1925.

be useful towards comparison of these two fundamentally different conceptions. Other reasons will incidentally emerge, besides the results of analysis by X-rays, favouring the view that facts of Nature point rather towards the atomic mode of crystal structure*.

These considerations deal with fundamental ideas that must be reckoned with, on an ionic view, apart from the detail essential to the science of crystallography. It will however be convenient to fix attention by simple cases; for example, the typical space-lattice for an alkaline compound such as rocksalt NaCl, in which the crystalline unit has been ascertained to be *for each ion* a "face-centred" cube, the corners for one component being the centres of the edges for the other. In the two structural plane diagrams, which represent the atoms projected perpendicular to a diagonal, the letters *a* may represent the negative atomic ions Cl, and the letters *b* the positive ones Na. In the second diagram a diagonal boundary face constituted of Cl atoms is succeeded by parallel planes, alternately of Na and Cl, and finally one of Na, these planes of like atoms being obliquely placed: this diagram invites description at sight as that of a very strongly polarized aggregate, each bipolar element being made up of a Cl combined with an adjacent Na below it. On the other hand, in the first diagram the boundary face is composed of both kinds of ions in equal numbers: and if the aggregate is regarded as sorted into bipolar elements, each consisting of an atom in one face and the adjacent atom in the next parallel plane, the polarities of the consecutive elements in all directions are equal and opposite, so that the mass appears as unpolarized when the effective element of volume is large enough to include a considerable number of such bipolar elements.

*Different types of face.*

| | | | | | | | |
|---|---|---|---|---|---|---|---|
| *a* | *b* | *a* | *b* | | *a* | | |
| *b* | *a* | *b* | *a* | | *b* | *a* | |
| *a* | *b* | *a* | *b* | | *a* | *b* | *a* |
| *b* | *a* | *b* | *a* | | *b* | *a* | *b* | *a* |
| | | | | | | *b* | *a* | *b* |
| | | | | | | *b* | *a* |
| | | | | | | *b* |

These two modes of description of the same lattice can of course be only apparently in contradiction. In the Poisson-Kelvin conception of a polarized medium, the polarity can be replaced, except for locally constitutive purposes[1], by equivalent distributions of simple

---

* Cognate ranges of topics, mainly regarding rotational quality, are explored in theory and experiment by W. G. Burgers, *Roy. Soc. Proc.* 116, A (1927), pp. 553–586: criticized by Friedel, *C.R.* July 1928.

[1] Compare, for example, *Aether and Matter*, 1900, p. 252.

density throughout the volume and over the surface. In a crystal, in which the structure is strictly uniform, the volume density is zero on both these modes of polar specification of NaCl: and to reconcile them it is the surface density that must claim attention.

These regular polarities are special, as being far more precise distributions than the fortuitous aggregates of polar molecules that are the foundation of the usual theory of a polarized medium, as constructed by the process of averaging over an element of volume.

*Essential polarity.* By a mathematical device familiar in the elementary analysis of optical diffraction, it is possible to reduce the polarity overtly associated with the second NaCl diagram. We have only to regard each ionic charge as divided into halves, of which one is associated with half of the next ion in front and the other with half of the next *Compensating half layer of molecules:* ion in rear along a row, thus forming two adjacent bipolar elements, of opposite signs, and therefore practically cancelling each other except as regards local physical quality of the substance. In this way the electric system is reduced to two surface densities of opposite signs, represented by half the distributions of ions on the opposite boundary planes. The factor one-half at first glance may seem to be a discrepancy between this regular discrete distribution of poles and the Poisson-Kelvin averaged theory of continuous polarization: the difference may be traced to the circumstance that the polar pairs in the second diagram occupy only half the intervals along a row of atoms, the rest of the length being vacant.

The equivalent surface density of ions, here contemplated over limited areas, would involve enormous electric fields in the substance between these intensely electrified boundary planes, such as could not possibly subsist. It must be compensated locally on each element of surface. If conducting quality of metallic type is absent so that free electrons are excluded, this can only be by a distribution of ions of the opposite kind added on outside the surface layer, but only of half the number that would make up a complete stratum of the *involves an uneven surface.* crystal. We conclude that a boundary plane, in a crystal, such as contains only one kind of ion, must be a ragged surface instead of a smooth one; for it involves the presence of an additional half stratum of the other kind[1]. When a crystal face grows, the ions must be laid down alternately, and also so that successive pairs constitute opposed polar elements: thus the modes of growth of the two types of face would be different: and the same would apply to modes of dissolution.

The crystalline structure of various metals has been found by

[1] An oblique artificial face should exhibit similar features as regards electric excitation, unless it were made up of minute facets with boundary planes in the principal directions of the crystal.

W. L. Bragg and other investigators to agree with this specially simple face-centred cubical space-lattice of Na, which in the scheme of NaCl is electrically compensated by the similar but displaced lattice for Cl. In case of the pure metal also, electric compensation must be provided, most naturally perhaps[1] by locating negative electrons more or less permanently in the positions occupied in the salt by the Cl ions.

Hitherto there is on this view no provision for the liberation of latent surface electric charge by change of temperature or by stress, as is conspicuous in crystals such as tourmaline. For the number of ions or electrons required for a neutralizing surface distribution depends only on the number of ions in the crystal, and is not affected by change of its form or dimensions.

This additional feature will be introduced if the positive and negative ions are not equidistant along their lattice planes. Thus for the line of ions represented in the diagram, we have to pair into bipoles by resolving each charge, say the one at $B$, into unequal fractions, so that the moments formed by their association with fractions of the adjacent charges at $A$ and $C$ may balance. The uncompensated charge at the terminal ion $A$ is now $e \cdot AB/AC$ instead of $\frac{1}{2}e$; and similarly at the other end of the chain.

$A\ \ B$      $C\ \ D$      $E\ \ F$

$b\ \ \ a$      $b\ \ \ a$      $b\ \ \ a$    ...

*Pyro-electric and piezo-electric:*

These residual charges on the surfaces of the crystal must, as before, be masked by local distribution of additional ions or by electrons. But the masking charge no longer stands in the same simple relation to the number of surface atoms: and whenever, by thermal expansion or by stress, the ratio $AB/AC$ becomes altered in the crystal, the compensation at the surface will become deranged, indications of free surface electrification will appear, and persist until they are wiped out by further accretion of compensating charge. An extremely small change in the ratio $AB/AC$ would suffice to cover the facts of pyro- and piezo-electrification.

*depend on unequal spacing:*

When each crystalline element is itself polarized with regard to a definite axis, the Poisson-Kelvin theory of a medium polarized *in bulk* can be formulated. But in the configuration here under review it seems to require modification; the positive and negative poles are independent ions, in regulated positions, and it has been seen that we can regard them as associated in pairs in various ways; so that there is no definite unique specification in terms of polarity per unit volume of the medium. But the analytical theory of the charges set

[1] Lindemann; Born and Landé; also Borelius, *Phil. Mag.* 1920.

*theory modified to include free ions.* free in crystals by stress or change of temperature seems to have hitherto been developed with regard to an assumed specific electric polarity in the crystal, having reference to some definite axis; with the result that cases were outstanding which would not come under the rules[1], and had to be referred to internal strain set up by cooling. On the present view such cases perhaps cease to be abnormal; for a direction of polarity cannot be ascribed to the compound lattice. The theory thus opens out into a new form of problem in crystalline tactic; the opposing ions within the crystal have to be paired, either as wholes or in fractions, so that the moments of the adjacent pairs shall everywhere cancel out. Then uncompensated ions on the surface remain over unpaired, except so far as they are masked by additional electrons or ions brought there from without; they would give rise to the emergence of free surface charge when the physical state of the crystal is changed.

Such release of latent electricity should not appear on surfaces constituted by equal numbers of the two ions. The specification of its distribution, peculiar as being locally conditioned, should of course be definite and unique, by whatever process of pairing it is reached. *Influence of fissures.* The electric abnormalities caused by cracks and clefts on a crystal surface would now arise naturally, without requiring local stress due to unequal heating, as had to be suggested.

*Dielectric excitation:* Dielectric excitation in a uniform electric field would now be represented by the positive and negative component lattices being drawn towards opposite directions by the field, with resulting electric moment depending on their *relative* displacement. For NaCl in a uniform field of electric force the two similar lattices of the Na atoms and the Cl atoms are thus displaced relative to each other without deformation of either though with very slight change of structure; and this applies generally, for all component lattices are similar.

*Cubic pyro-electric crystals.* [1] It appears that an intrinsic polarity, or regular static collocation of electrons within each molecule, would suffice by interacting with crystalline deformation, to include within the theory the octopolar pyro-electric quality detected long ago in cubic crystals by Haüy: cf. Lord Kelvin's *Baltimore Lectures*, p. 561, or "Aepinus Atomized," *Phil. Mag.* 1891.

The formal theory of Voigt (*Krystalloptik*, §§ 128–9) seems to regard only *induced* polarity, which is expressed as a function of deformation *alone*, whether due to stress or to uniform change of temperature, by means of a sufficient number of moduli. It appears from recent experiments by Röntgen, Hayashi, Ackermann, and Lindman, in which special precautions were taken as to uniformity of temperature, that these moduli are not much, if at all, affected by temperature (cf. Voigt, *Ann. der Physik*, 1915); as would be natural on the present order of ideas. It was in fact held by J. and P. Curie, the discoverers of piezo-electricity, that pyro-electricity is to be regarded as a result of deformation alone and not as essentially thermal.

Here also the electric displacement could be fixed and compensated, by local surface charge masking the polarization—for example, by sweeping a flame over the surfaces while the inducing electric field is in action. In the alternating electric fields of radiation it is, on final analysis, this oscillatory motion relative to each other[1] of the component ionic lattices that influences the speed and mode of propagation of the waves.

*analysis of discharge by flame.*

If the crystal has not sufficient symmetry (in fact, in the lattices that have spiral features) this relative displacement of the two ionic groups in the element of volume, by an applied electric field, may be one of twisting type, involving a rotational part to which is to be ascribed the chiral structural rotation of plane of polarization of light transmitted through the crystal. There are thus the two modes of origin of this chiral optical rotation. One depends on crystalline structure alone, and disappears on fusion of the crystal. In the other the molecule itself, usually a complex aggregate of atoms, undergoes in an electric field this spiral relative deformation of its ionic constituent groups, involving screw quality; and to this extent the optical chirality will remain on fusion or solution, but in altered amount, usually much smaller, as was observed in the case of sucrose by Pocklington. From achiral molecules a crystalline structure of chiral quality can thus be formed. As chiral molecular quality is not of vector type, it could not be annulled by fortuitous orientations as in a fluid, when molecules of opposed chiralities are not present. If the chirality of the molecules is isotropic (cf. Lord Kelvin's isotropic helicoid) there would not be much alteration, by fusion, of that small intrinsic part of the chirality in bulk; if uniaxial, the maximum would be reduced to about one-third.

*Spiral features.*

*Optical chirality,*

*structural (spiral nuance),*

*molecular.*

Double refraction, limited in form[2] in accord with the dynamical equations of MacCullagh and of Maxwell, will arise from any kind of deviation from cubic symmetry in the natural crystalline ionic lattice, supposed of structure very fine compared with the lengths of the waves. But such defect of symmetry can also be produced by mechanical deformation of a regular crystal: and the double refraction, necessarily slight, thereby introduced (in fact by bending of the lattice) even for regular crystals, must hold a clue to the intimate electric characters of the strain. An atomic lattice must be far more amenable to mechanical stress than a set of molecules could be. In

*Double refraction by strain.*

[1] Only partially regular and elastic in the case of metals: the super-conductance of K. Onnes has been ascribed (Lindemann) to complete regularity owing to thermal agitations being suppressed.

[2] The Principle of Minimal Action excludes anything more general, except chirality of the types here reviewed, *provided* the waves of radiation are long in comparison with the lattice structure.

*Long waves simpler.*

amorphous aggregates, such as glass and gelatine, and even in viscous liquids under forced shearing flow, the deformation by stress of some less regular electric structure is indicated[1]: in fluid colloids electric ionic phenomena are of course in other ways conspicuous[2].

The magnetic achiral optical rotation of Faraday, common to all matter however regular in its crystalline form, must arise in quite different way. Here the primary phenomenon is change of free vibrational periods of the individual molecules, or perhaps atoms if the molecule is so loose a structure as is here indicated, by the *Faraday-* extraneous magnetic field, which is the Zeeman effect: and the optical *Zeeman* *relation.* rotation is an affair of resonance due to the change of these free interacting periods[3], after the manner in which ordinary dispersion is produced[4].

In both types of optical rotation these natural explanations concur with the forms for the equations to which the phenomena are restricted by the principle of Conservation of Energy or rather, more generally, the principle of Minimal Action[5]. Thus if we consider the case of chirally isotropic substances, the alternating electric field $(P, Q, R)$ of the travelling radiation would introduce, through the *Scheme for* twisting displacement of the ions, in the molecule or the lattice or *spiral* *structure.* both, a concomitant effective magnetization $k d/dt\,(P, Q, R)$; so that its magnetic induction $(a, b, c)$ would be given by equations of form $a = \alpha + 4\pi A$ where $A = k dP/dt$.

Of the field equations of optical transmission, of type

$$\frac{\partial R}{\partial y} - \frac{\partial Q}{\partial z} = -\frac{da}{dt}, \quad \frac{\partial \gamma}{\partial y} - \frac{\partial \beta}{\partial z} = 4\pi \frac{df}{dt},$$

the second could thus be written, on substitution for $\alpha, \beta, \gamma$,

$$\frac{\partial c}{\partial y} - \frac{\partial b}{\partial z} = \frac{d}{dt} \left\{ 4\pi f + 4\pi k \left( \frac{\partial R}{\partial y} - \frac{\partial Q}{\partial z} \right) \right\}, \quad \text{where } f = \frac{K}{4\pi c^2} P,$$

agreeing in form with the equations derived from the Principle of

[1] The recent optical study of fused quartz by Lord Rayleigh seems to open out a path in this direction.

[2] For example, the particles in a metallic suspension or (so-called) colloidal *Origin of* solution are now known by X-ray analysis to be crystalline [though spherical *electric* in general as regards form (J. C. M. Garnett)]; and a layer of electrons attached *charge on* and locally compensating the intrinsic polarity would account for the electric *colloids.* charges which are the dominating feature of such suspensions. Over the whole surface this charge would be as much positive as negative, but negative electrons would be attached (in appropriate regions) by preference.

[3] Cf. *Aether and Matter*, p. 353 [, on the classical model].

[4] The static weak magnetization of crystals must be an affair of the orientated individual atoms, influenced only slightly by the proximity of the adjacent ones.

[5] *Aether and Matter*, pp. 197, 206.

Extremal Action[1]. When $K$ is uniform, these equations preserve both the electric and the magnetic inductions free from divergence, as they ought to be.

The general outcome is that a set of component similar lattices of electric ions prescribes in a natural manner the existence and character of the electric properties recognized in crystals: double refraction, pyro- and piezo-electricity, chiral polarization, artificial double refraction excited by mechanical stress. It would seem even possible to infer, solely from absence of pyro-electric effect in the more symmetrical forms, that in them the crystalline lattice must be formed of ions equally spaced and not combined into molecular groups. For if the molecule were the unit, expansion by heat would alter the distance between adjacent molecules but not, or at any rate not in the same way, the distance between the two electric poles in each molecule: and it would follow by the argument indicated above, that it must alter the amount of the uncompensated surface charge, thus involving pyro-electric phenomena. The only way to avoid this conclusion would be to imagine that the positive and negative ions are so intimately blended in the molecule that it has no resultant moment, after the manner of the first crystal diagram above, until the positive and negative interlacing groups in it are displaced relatively to each other by an imposed electric field.

*[margin: Inference as to the cubic structure,]*

*[margin: as ionic:]*

*[margin: and induced polarization.]*

Hysteresis in dielectrics, producing the residual charge of condensers, was referred by Maxwell to heterogeneity operating along with transfer of masking charges by conductance. Prompted by this ascription, Rowland found that in crystal dielectrics there was no residual charge. One can imagine dielectric fragments $A, B, \ldots$ separated by a medium of different dielectric and also slight conducting quality. In a field of force, $A, B, \ldots$ are polarized, but the local field is gradually diminished by conductance between $A$ and $B$ conveying masking charges to their surfaces. When the coatings of the condenser are discharged, the whole field throughout the dielectric is suddenly reduced, producing a sudden fall in the polarization of $A, B, \ldots$, but not in the surface charges that masked it. The latter, now uncompensated, gradually make their way back in the reverse direction, involving accumulation of residual charge on the coatings. The actual residual charge would thus come from a small depth, but is permitted to flow back by local compensation in deeper layers. In an alternating field such as that of radiation, conductance would operate by a sort of approximation to a Grotthus chain. But in a very dilute electrolytic solution, the ionic atoms must apparently be

*[margin: Residual electric discharge analyzed:]*

*[margin: null in perfect crystals:]*

*[margin: superficial:]*

---

[1] *Loc. cit.* § 133. The expression there given for the velocity of propagation is misprinted: cf. *infra* at the end of this paper.

wholly isolated from one another, whether the dielectric quality of the solvent be due to incipient atomic lattices or to orientation of polar molecules.

cannot affect capacities across dielectrics.
The internal charges in the dielectric being locally masked, the capacity of a condenser as regards sensible charge on its coatings should not be variable, as R. Kohlrausch in fact verified long ago.

Magnetic hysteresis:
The ferromagnetic hysteresis elucidated by Sir A. Ewing, and so vividly illustrated by his working models, is necessarily of different type, for there are no magnetic ions to compensate the polarization. It requires the molecule, with its own polar magnetic moment, as unit of the structures which acquire instability under increasing extraneous force and finally fall over; yet repeated reversals do not seem to wear down completely such tendency to instabilities, which must be somehow reconstituted.

becomes renewed.

Apart from hysteresis, it has been shown long ago[1] that the law of Curie, paramagnetism inversely as absolute temperature, is the expression of general thermodynamic principles without involving any special theory.

Nature of polarization.
On the same lines it has been argued[2] that, if electric polarization were an affair of orientation of existing polar moments, $K - 1$ should tend to vary inversely as temperature, which it seems to show no sign of doing.

A chiral feature:
The specific optical rotations of the usual active substances are of the same order of magnitude: and ideas advanced to represent their origin must be expected to throw light on that fact.

to be explained.
The equations expressed above lead, on elimination so that the electric force remains, to the type

$$\nabla^2 P = \frac{\partial^2}{\partial t^2}\left\{\frac{K}{c^2}P + 4\pi k\left(\frac{\partial R}{\partial y} - \frac{\partial Q}{\partial z}\right)\right\}.$$

Thus for radiation travelling along $z$, in which therefore both the circuital vectors are on the wave-front $xy$,

$$\frac{\partial^2 P}{\partial z^2} = \frac{K}{c^2}\frac{\partial^2 P}{\partial t^2} - 4\pi k\frac{\partial^3 Q}{\partial z\,\partial t^2},$$

$$\frac{\partial^2 Q}{\partial z^2} = \frac{K}{c^2}\frac{\partial^2 Q}{\partial t^2} + 4\pi k\frac{\partial^3 P}{\partial z^2\,\partial t^2},$$

so that if $U$ represents $P + \iota Q$, $\iota = \sqrt{-1}$,

$$\frac{\partial^2 U}{\partial z^2} = \frac{\partial^2}{\partial t^2}\left(\frac{K}{c^2} + 4\pi k\iota\frac{\partial}{\partial z}\right)U,$$

[1] *Phil. Trans.* 1897, A, p. 287 [as *supra*, p. 116].
[2] *Loc. cit.* p. 283 [: also the recent explorations by Debye].

indicating waves of the two circular types. If they are harmonic wave-trains involving the exponential factor

$$e^{\iota pt - \iota mz}, \quad m = 2\pi/\lambda = p/c',$$

then

$$m^2 = \frac{p^2}{c^2}(K + 4\pi k c^2 m),$$

giving for the velocities of propagation $p/m$ of the two circular radiations of the same period $2\pi/p$ the approximate values

$$c' = cK^{-\frac{1}{2}}(1 \pm 2\pi K^{-\frac{1}{2}}cp).$$

The optical rotation for unit depth of the medium is the difference in $\pi/\lambda$ or in $\frac{1}{2}p/c'$ for these two radiations, and this is $\pi p^2 k$.

Now approximating very coarsely, if there are $N$ molecules or chiral elements per unit volume, in each of which positive and negative ions (say single charges $e$) are drawn asunder a distance $\delta l_0$ by the electric field and screwed through an angle $\theta_0$, then omitting the common factor $e^{\iota pt}$,

$$k\frac{\partial P}{\partial t} = \frac{1}{2}\iota Ner^2\theta_0 p, \quad \frac{K-1}{4\pi c^2}P = Ne\delta l_0,$$

where $r$ represents a mean radius of transverse section of the molecule (or of a spiral filament of the lattice) so that the magnetic moment is say $\frac{1}{2}er^2 d\theta/dt$ for each chiral unit.

On division $N$ disappears and

$$\frac{K-1}{4\pi c^2 k} = 2\frac{\delta l_0}{r^2\theta_0},$$

so that the optical rotation per unit depth is

$$p^2\frac{K-1}{8c^2}\frac{r^2\theta_0}{\delta l_0}, \quad \text{or} \quad \frac{\pi^2}{2\lambda^2}(K-1)r\frac{r\theta_0}{\delta l_0}.$$

If $r$ is of the order of the diameter of a molecule, say $10^{-8}$, this reduces to the order of $10r\theta_0/\delta l_0$ for visible light.

Actually the optical rotation is of the order of two right angles per centimetre for quartz, and seems to range to several times less for pure active liquids. This indicates that the angle of pitch of the relative screw displacement of the positive and negative components of the structures should range around $\frac{1}{10}$, so that the chiral element may twist under an electric field about one-tenth as much as it elongates. This indication is extremely rough; but bearing in mind that powers of the velocity of radiation are involved in the estimate, one may claim that it does not detract from, but rather substantially supports, the type of picture of the phenomena which has been under review.

# (i) ESCAPEMENTS AND QUANTA:
# (ii) ON NON-RADIATING ATOMS.

[*Philosophical Magazine*, Vol. XLII, 1921: communicated
to the British Association, Edinburgh, Sept. 1921.]

(i) Might not an atom be a clock? It is asserted on sufficient
evidence that when an electron is fired into an atom, it sometimes
is retained, but its energy is emitted again in radiation of Röntgen
rays, mostly forming a group of very definite periods but in part
irregular. The former periods are so sharp that a train of many
thousands [almost millions] of undulations, perfectly regular in amplitude and phase, must be sent out, in order to produce such narrow
lines in their Fourier spectrum as appear in crystal analysis. Yet an
unsustained electric vibrating system radiates so intensely that if left
to itself it would be quenched in a few vibrations. How is all this to
be imagined? Such equivalence between the impulse of an absorbed
projectile and a long train of regular waves is unfamiliar and must
therefore be significant.

Consider, however, a common pendulum clock, or better, a clock
with a compound pendular system having a group of sharp periods
in its complex mode of oscillation. If vibrating free the pendulum
will soon slow down to rest; but if it is engaged through an escapement (the great invention of Huygens) with a strong coiled spring or
a suspended weight, it goes on vibrating for weeks, absolutely regular
in amplitude and phase, picking up from the driving spring just the
small *quantum* of energy that is needed to maintain it without disturbing it, each time the escapement is engaged. Now it is Ruther-
fordian isotopic doctrine that an atom is made of an outer shell of
some kind, the seat of all its chemical and radiant properties, linked
up with a core, perhaps sheltered, but anyhow too stiff to reveal itself
sensibly in physical activities, except in the inertia that belongs to
its structure. The outer electron system which emits the lines of a
spectral series (and determines chemical quality) stands for the
pendular system of the clock, which is maintained in absolutely continuous vibration from the energy of the core, imparted gently
through an escapement action, this vibration lasting for many periods,
and renewed every time the core is wound up by its entrapping a
particle with it translatory energy. All which is parabolic, yet to a

*Margin notes:* An electron excites a train of X-rays. Analogy of a clock-escapement, applied to sustained atomic vibration,

vivid imagination may be fertile in analogy. A connection between the feed of energy and the free period ($\epsilon = h\nu$) is, however, wanting. Perhaps also the clock may be reversible, and by absorbing incident radiant energy of proper periods into its pendular system may wind up its spring—pointing ultimately to instability and ejection of a particle from the core, in the manner recognized.

and the mysterious converse ejection.

Such quantized or quantified maintenance of a vibrating system is far from the orthodox smooth vibrator, imported into the usual mathematical theory of optics, but quite ineffective as regards sharp selective spectra. But its case is not uncommon in Nature: a blown organ-pipe, or a bowed string, emits an absolutely regular train of waves maintained as long as is desired. And an escapement may be imagined as modified into a smooth continuous mathematical mechanism of constraint, working without any jerk. Ought one to consent to be driven, even under many-sided compulsion of facts, to introduce such discontinuous gear as a feature in diagrammatic representations of Nature's hidden mechanism of selective radiation?

Contrast with ordinary vibrator. Physical illustrations.

Validity?

But may there be an alternative to such *creation* of natural *quanta* of a sort through mechanism—to the regular breaks of continuity, however slight and gentle, that are perhaps essential to escapement action? One recalls that in the usual periodic vibrating atomic system illustrative of optical theory, each free period is independent, and its own energy fades away by radiation, no transfer or feed of energy from one period to another being possible. Equipartition of the whole energy in the molecule between the periods, so far as it may tend to hold good, can only be assisted during encounters or collisions between molecules, when the periodic character of the internal motions is for a brief time suspended. Here, at any rate, fortuitous discontinuous processes intervene, in order to establish physical law, in a much coarser manner than a regular automatic escapement mechanism would involve.

Equipartition depends on sharp encounters alone:

But let us now contemplate an atom differently—after Rutherford's idea—as an outer physical system having its constant periods of free vibration, linked dynamically—but only to a slight degree so as not much to displace these periods—to an inner system or core, of more complex elasticity or structure so as not to be restricted to constant periods of its own, and so massive or stiff that it can form a receptacle of energy of large capacity, which may be replenished from time to time, from colliding ions or otherwise. There is now nothing to prevent this core from gently feeding out the energy, so acquired at each encounter, to sustain the vibrations and other motions of the outer physical-chemical system linked with it, as the blast feeds energy to the sound vibrations in an organ-pipe? The analogy is rough: yet the action, on Helmholtz's exposition, of the vibrating jet of air which blows the pipe, though alternating, need not be discontinuous.

*unless fed from a feebly linked core.* A new type of vibrating system is thus suggested for general dynamical exploration, possibly fruitful—namely a periodic system linked by slight continuous coupling with another system of perhaps simple type but not of periodic quality, and of large capacity for energy.

It has been implied above that the outer (electric) system of periodic quality is of a type which sinks into a configuration free from further loss by radiation, whenever the feed of energy to it fails: indeed, no other type could subsist, but it is the problem of the note here following how such electric types can be possible.

Finally, a general remark is suggested: that wherever it proves necessary in physical science to treat of discrete *quanta* of energy, it may well be that these are packets separated in the cases concerned *Quanta are separated out within the atom.* by the atomic mechanism—just as period in natural radiation is said in a certain sense to be a creation of the resolving prism or grating—without having to face the difficult assumption that energy is itself necessarily discrete. The *quanta* of practical physics would of course be large multiples of such packets.

*Absence of atomic radiation:* (ii) Is it possible to have orbital systems of electrons which do not radiate unless when disturbed? To the first order, which neglects atomic size compared with lengths of wave, the condition of absence of radiation is (*Aether and Matter*, 1900, § 153) that the vector sum $\Sigma e\dot{v}$, where $\dot{v}$ is acceleration of the electron $e$, must vanish. It might appear that, when pushed to the second order, an infinite number of other conditions would be required in order to annul radiation in all directions. But that seems not to be so. The additional conditions are that for any direction of radiation, say that of $x$, the quantities *to higher orders, yet involving only a finite number of conditions.* $\Sigma e\ddot{y}x$ and $\Sigma e\ddot{z}x$ shall vanish. These are obviously satisfied for every direction if for any one set of coordinates $(x, y, z)$ the nine expressions of this type, including $\Sigma e\ddot{x}x$, $\Sigma e\ddot{y}y$, $\Sigma e\ddot{z}z$, all vanish.

*A fixed magnetic moment involved.* The twelve conditions in all are, in fact, parallel to those of astatic equilibrium (cf. Routh's *Statics*). It may be noted that the latter nine involve, by additions, that the magnetic moment of the atom remain fixed in magnitude and direction: as a consequence, if the atom has *Magnetic interpretation.* a permanent magnetic moment, any rotational motion which changes the direction of that moment will be obstructed and resisted by radiation. This may preclude the existence of a permanent moment: or it may have a bearing on equipartition troubles. Atoms, for instance of permanent symmetry with regard to three planes, can readily be imagined for which the radiation is residual, of the third order, in fact is so small that it ranks with such residual effects as are masked by electrodynamic relativity. An atomic orbital structure could subsist only at or near one of this set of configurations that do not radiate\*: by intense shock it can perhaps be imagined as pushed over from one of them into another, with loss of energy that goes in part into the disturbing system, and in part into radiation.

*Residual radiation may be improbable.*

[A chief significance of the Rutherford-Bohr atomic representation is that it can transfer the quantification of radiant energy solely into operations within the source of radiation, thus avoiding any necessity of contemplating a return from the undulatory theory to a corpuscular theory of free radiation.

*Corpuscular radiation?*

As regards the reversed phenomenon, it is perhaps not inconceivable that transition of the system to a lower "energy level" can occur. For example, absolutely direct encounter of an orbital electron with an equal electron at rest would transfer the whole of its energy of motion. Less direct encounter could transfer enough to reduce the system to the definite lower level: from which a reversed continuous process of absorption of the proper incident radiation would gradually restore it. It is here the free electron that was encountered that plays the part of an ejected electron: the order of events is here changed, the encounter coming first and the absorption later.

*Process of ejection of an electron.*

The older dynamical view of radiation rests on the analogy to sound as originated by Stokes. There the radiator is a material structure, working by continuous material elasticity: and absorption is mainly scattering. It is hardly surprising that in the atomic radiating machine something more than that simple interaction of incidence and scattering should prove to be required.]

*Storage of energy essential.*

\* The slow finite swayings, in form and position, of a Bohr hydrogen orbit precessing in a magnetic field, due to the classical terms of second order, which prove to be stable, have been determined recently (*Phil. Mag.* 1928) by W. M. Hicks by astronomical methods.

# 99

# THE STELLATE APPENDAGES OF TELESCOPIC AND ENTOPTIC DIFFRACTION.

[*Proc. Camb. Phil. Soc.* Vol. XXI, Pt. IV, pp. 410–413: Dec. 1, 1922.]

THE sections which follow set out the proof of statements made in abstract in a paper communicated to the Mathematical Congress at Strasbourg in September 1920. See *Report*, [as follows:

*Ring and ray systems definite.*

If one examines an astronomical photograph of a field of stars, rays projecting from the images are sometimes apparent: and they are precisely the same for all the stars in the field. Such rays are very conspicuous also in other optical point images. The view is here advanced that these rays are due to straight line portions of the boundary of the aperture of the telescope: that then there is the usual central image with its rings, while a ray shoots out from it at right angles to the direction of each such straight edge*.

*J. Herschel's patterns of star-images:*

*theory: readily generalized.*

These ray patterns were figured by Sir John Herschel long ago in the *Encyclopedia Metropolitana*. A striking case is when the aperture of a telescope is limited to an equilateral triangle, which gives three double rays forming a hexagonal system. For this case the mathematical explanation was worked out by Airy in his tract on *Physical Optics*. By a known principle of correspondence for diffraction in an image plane, one can pass from this case to a triangle of any form, the three double rays being then at right angles to the sides through their middle points.

The demonstration of these statements emerges readily on treating the appropriate diffraction integral by integration by parts. The subject seems to merit closer attention in astronomical and other connections.]

*The integral for diffraction in a focal plane.*

Let the coordinates of a point on the aperture plane, which is a wave-front of incident light, be $(x, y)$, and those of a point in the parallel image plane be $(a, b)$, the rays converging as in a telescope to a point image at the origin of coordinates. Let these planes be at a large distance $e$ apart. Then an element of polarized disturbance,

---

* This simple explanation seems to afford an adequate account both of the great length and of the sharpness of the straight rays in a star-image, that are thrown off by every straight edge of the aperture whose length is many times the wave-length. Such rays appear to be well known to astronomers, when the field of the telescope is interrupted by a straight part of the frame. The rays of entoptic images have been ascribed to irregular boundary of the iris in the same way.

of area of front $\delta S$ at $(x, y)$, contributes to the light vector at $(a, b)$, on the Fresnel theory sufficiently valid for small obliquities, an amount varying as

$$\delta S \cos n (ct - R), \text{ where } R^2 = e^2 + (x - a)^2 + (y - b)^2,$$

in which $a$, $b$ are usually small compared with $x$, $y$, and all are small compared with $c$.

Thus

$$R = e + \frac{a^2 + b^2}{2e} - \frac{ax + by}{e} + \frac{x^2 + y^2}{2e} - \dots,$$

so that the light vector at $(a, b)$ in the image plane for the case of a uniform incident beam is

$$\int dS \cos n \left( ct + A - \frac{ax + by}{e} \right),$$

in which $A$ is a constant of phase, and $n$ is $2\pi/\lambda$.

Well-known theorems of correspondence here arise. The distributions on the aperture plane and image-plane correspond in all cases in such manner as maintains $ax + by$ invariant. Thus if the form of aperture is altered so that $x$, $y$ become *[margin: Effect of linear distortion of aperture.]*

$$x' = px, \ y' = qy,$$

then the pattern in the image plane is altered so that $a$, $b$ become

$$a' = p^{-1}a, \ b' = q^{-1}b.$$

Taking the axis of $x$ through the point $(a, b)$ on the focal plane, thereby making $b$ zero, the integral becomes

$$\int dx\, F(x) \cos n \left( ct + A - \frac{a}{e} x \right),$$

where $F(x)$ is the transverse breadth of the aperture at distance $x$ from the projection of the image point on its plane. The diagram [next page] shows it as the length of an infinitesimal slice, for the case of a triangular aperture.

If the aperture is of any form bounded by straight lines, the graph of $F(x)$ the length of the differential element will be a succession of straight lines, with abrupt changes of inclination as it passes each corner of the boundary. This suggests the process of integration by parts, leading to *[margin: Transformation for straight boundary:]*

$$\left| -\frac{e}{na} F(x) \sin + \frac{e^2}{n^2 a^2} F'(x) \cos \right|$$

$$- \int dx\, \frac{e^2}{n^2 a^2} F''(x) \cos n \left( ct + A - \frac{a}{e} x \right).$$

This result may be interpreted in alternative ways. If the corners are regarded as sharp angles, the final integral vanishes over each straight segment: and of the terms at the limits the first vanishes *[margin: Interpretation by effect of corners:]*

except in a crucial special case, while the second provides a contribution from each corner of the aperture. If, however, the corners are regarded as rounded, on account of the continuity thereby introduced the terms at the limits provide nothing, while the integral term gives a contribution spread over each rounding off, on account of the great local magnitude of $F''(x)$. The two modes will agree only when the rounding off is completed in a small fraction of a wave-length. But in either case the diffraction pattern may be regarded as due to a distribution of coherent point sources at the corners, though the relative magnitudes assumed for them will have to alter with the direction of the diffraction.

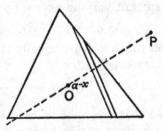

*together with long rays,*  The exceptional case is when the direction of diffraction is exactly perpendicular to one edge of the aperture, say of length $l$; then the first term above at the limits gives, for focal length $e$, a contribution

$$\mp \frac{el}{na} \sin n \left( ct + A - \frac{a}{e} x \right)$$

with choice of sign according as it is a forward or rearward edge. This is to be compared with the contribution from a corner which is of order $\frac{e^2}{n^2 a^2}$. Its ratio to the latter is $\frac{2\pi a}{\lambda} \alpha$, where $\alpha$ is the angle subtended by the edge $l$ of the aperture at the focus. Now the mean radius $a'$ of the diffraction pattern around the focus is readily seen to be of order $\frac{\lambda}{\alpha'}$, where $\alpha'$ is the angle subtended at the focus by the mean diameter of the aperture: thus the ratio is $2\pi \frac{a}{a'} \frac{a}{a'}$, in which the first factor is the ratio of the distance of the point $P$ on the ray to the radius of the central ring and the second is the ratio of the length of the straight edge to the radius of the aperture. The edge thus *pre-ponderant,* strongly preponderates over each corner, while the various corners counteract each other owing to differences of phase. It is only in a *at right angles to boundaries* very definite direction that this preponderance of the edge effect over *nearly straight,* the area effect exists; at an angle $\frac{\lambda e}{la}$ or $\frac{a'}{l}$ with the direction normal to the edge it entirely disappears, while exactly along that direction the diffraction would be sensible to a much greater distance than near it. This would mean that in the diffraction pattern a prolonged sharp *converging to the focus.* ray emerges from the focus perpendicular to each *straight* edge of the aperture. An edge not quite straight will show the effect if its curvature is of smaller order than $\frac{a'}{l^2}$ or $\frac{\theta}{l}$, where $\theta$ is the angle the central diffraction pattern subtends at the aperture.

These rays form a familiar feature in optical images: in Sir John Herschel's drawings (*Encyc. Metropolitana*) of telescopic diffraction with an equilateral triangular aperture, they constitute the hexagonal star diverging from the central rings. The present general mode of argument may be compared with the detailed special integration for this triangular case, as carried out by Airy in his *Tract* on the undulatory theory of light. The present reasoning would predict irregular stars for any triangular or polygonal aperture: in fact it would demand a ray at right angles to each straight part of an aperture elsewhere curved. *Familiar beautiful patterns: explained,*

The diffraction patterns for two apertures whose forms are related after the manner of geometrical projection should correspond linearly according to the relation above stated: it is easily verified that if for one of the related apertures a ray is perpendicular to an edge it will be so for the other, as ought to be the case. For example, the diffraction pattern for a scalene triangular aperture would have the rays shooting out at right angles in both directions from the centres of its sides. The best focus would be their meeting point. *and generalized.*

The very copious illustrations of telescopic diffraction given long ago by Schwerd in his treatise *Beugungserscheinungen* are, it may be observed, in many cases, including the scalene triangle, examples of this simple principle of the ray effect of straight edges. *Examples.*

It has been stated (by de Haas and Lorentz) that Helmholtz ascribed the rays that appear, in normal focused vision of a distant small bright point, to the irregular boundary of the iris diaphragm of the eye. As above, straight segments would be required to produce long sharp rays. *Entoptic stellar rays:*

In short-sighted vision (with one eye) such that the focal plane is in front of the retina, extremely definite ray patterns, it may be radiating from one or two small concentric central rings, are found to develop with years, and could be due, in so far as they are not shadows of the internal structure of the lens, to patches of irregular refraction forming in the lens, a hexagonal star with ringed centre corresponding to a triangular patch, the same in fact as described by Herschel for his telescope. It would be of interest for the sake of comparison to examine the telescopic appearances out of focus*. *also outside the focal plane,*

* These phenomena are treated at length, mainly descriptively, in Helmholtz's *Physiological Optics* under the title "Monochromatic Observations of the Eye," with extensive additions by Gullstrand in the third edition: see Vol. I of Southall's translation for the American Optical Society. (In the writer's own experience they are an initial phase in the degeneration of the lens.)

The effect of locally increased refractive power over a limited region of the lens would be to alter the effective length of path of the ray say by $l$ so that the vibration changes from

$$\cos n\,(ct - x) \text{ to } \cos n\,(ct - x - l),$$

the increase of the vibration vector being thus their difference $2 \sin \frac{1}{2} nl \sin n\,(ct - x - \frac{1}{2}l)$. This added distribution of vibration is imposed on the regular refraction which would produce normal focusing. Thus if $l$ were a constant say $k\lambda$ over an area of the lens, it would be the same as superposing a vibration of relative amplitude $2 \sin k\pi$ integrated over that area but with phase increased by $\frac{1}{2}k$ of a period. If $k$ were $\frac{7}{8}$ it would nearly be equivalent to blocking out that area of the lens: and by Babinet's principle, the bands of diffraction are the same for a blocking screen as for an aperture.

*effects of areas of changed refraction.*

*Beaded rays.*   If there are two parallel edges on the aperture there will be interferences along the ray of diffraction that shoots out normal to them, producing alternations of bright and dark places with homogeneous light merging into the colours of Newton's rings with ordinary light. *Examples.* This appearance is shown, for example, with a square aperture, along the rays of its four-rayed star: and the beaded pattern of the hexagonal rays figured and described by Herschel for a tesselated triangular aperture made up of alternate smaller equilateral triangles would be elucidated on similar lines.

[Compare the (merely descriptive) section on astigmatism in Helmholtz, *Physiologische Optik*: also the treatise of Donders *On accommodation and refraction of the eye*.]

# 100

# WHY WIRELESS ELECTRIC RAYS CAN BEND ROUND THE EARTH.

[*Phil. Mag.* Vol. XLVIII, pp. 1025–1036, Dec. 1924. Read at the Cambridge Philosophical Society, Oct. 27: abstract in *Proceedings*.]

THE earliest announcement that electric signals through free aether Historical. had been successfully detected as far as America by the Marconi operators, gave rise to a prompt query from the late Lord Rayleigh as to how the rays could manage to bend round the protuberance of the curved Earth. In fact, in a single medium propagating waves Argument without absorption or dispersion, all features remain similar when from dimensions: the scales of space and time are altered in the same ratio: thus waves of length $\lambda$ bending round the Earth of radius $a$ will behave similarly to waves of length $k\lambda$ bending round a smaller sphere of like quality of radius $ka$. To change from electric waves $10^4$ cm. long to waves of light $10^{-4}$ cm. long, $k$ would be $10^{-8}$; hence the radius of the sphere corresponding to the Earth would be $10^{-8}.10^9/\frac{1}{2}\pi$ or 6 cm.; and familiar experience indicates that visible light could hardly creep to a sensible degree round one-tenth of the circumference of a sphere of that radius. To elucidate this subject quantitatively in those early days involved difficult and extensive calculations of the problem of diffraction of waves by a boundary, conducting or dielectric, of continuous curvature; and investigations by Lord Rayleigh, H. M. Mac- mathematical donald, A. E. H. Love, J. W. Nicholson, G. N. Watson and others investigations. have widely extended that domain of mathematical optics.

But nowadays, when ray signals are readily received at the an- The actual tipodes, all shadow of uncertainty as to the facts is removed. The problem. rays could not travel free, except in straight lines, in a medium of practically uniform speed of propagation such as ordinary air; either they must be guided by being linked to the surface of the Earth, much as cylindrical Hertzian waves follow a guiding central wire, or else the speed of propagation must for some cause increase notably upwards so as to bend them down. Very early observations by Admiral (then Captain) Jackson, with the help of British cruising squadrons, showed that the high table-land of Spain was not an effective obstacle, which perhaps already pointed to transmission of the rays high in the atmosphere.

The ultra-rarefied upper atmosphere with its long free path must be subject to strong ionization by the incident ultra-violet Solar

radiation, the effect persisting after daytime, and appeal has been generally made to such a conducting layer to provide the necessary
*Conductance cannot help,* increase of velocity of the rays. But ordinary electric conduction operates by introducing a frictional term causing absorption into the equations, and it is a familiar principle that when the absorption is of the first order the change of speed is only of the second order, so that if the latter change is to be adequate, the rays would be damped out immediately instead of being sensibly transmitted. In rough
*on account of the high extinction:* illustration, the usual type (including the optical) of equation of propagation with frictional resistance of modulus $\kappa$

$$\frac{\partial^2\phi}{\partial t^2} + \kappa\frac{\partial\phi}{\partial t} = c^2\frac{\partial^2\phi}{\partial x^2}$$

is satisfied by

$$\phi = Ae^{-imx}e^{int},$$

where $n^2 - \kappa n i = m^2 c^2$, so that, for a real period $2\pi/n$,

$$mc = (n^2 - i\kappa n)^{\frac{1}{2}} = n\left(1 - \tfrac{1}{2}i\frac{\kappa}{n} + \frac{1}{8}\frac{\kappa^2}{n^2} + \ldots\right).$$

Thus, retaining real parts, we arrive at a simple wave-train of type

$$\phi = Ae^{-\frac{1}{2}\frac{\kappa}{c}x}\cos n\left\{t - \left(1 + \frac{1}{8}\frac{\kappa^2}{n^2}\right)\frac{x}{c}\right\}$$

*its law:* with extinction modulus $\tfrac{1}{2}\kappa/c$ and velocity of propagation

$$v = c\left(1 + \frac{1}{8}\frac{\kappa^2}{n^2}\right)^{-1}, \text{ or say } c\left(1 - \frac{1}{8}\frac{\kappa^2}{n^2}\right),$$

thus smaller than $c$, but only to the second order in $\kappa$.

(On the principle of ray curvature as *infra*, a ray of assigned period will adapt itself to the curvature of the Earth only when

$$\tfrac{1}{2}\pi\,10^{-9} = \frac{d}{dh}\log v = -\frac{1}{8n^2}\frac{d}{dh}\kappa^2.$$

The attenuation would vary inversely as the wave-length. If the
*an illustra- tion.* wave-length is one kilometre, $2\pi/n = \lambda/c = \tfrac{1}{3}\,10^{-5}$: so that if the rays are thus adapted over a vertical breadth of only a thousand metres the modulus of attenuation $\tfrac{1}{2}\kappa/c$ must increase downward across it, so that at the lower boundary it mounts to at least $10^{-6}$, which means decay of amplitude in ratio $e^{-1}$ after travelling only ten kilometres.)

To produce sensible bending of the rays without extinction, all action by conductive or other dissipation must thus be excluded. It would not suffice to say that the conductance has to become perfect, for then no waves could travel: the influence must be of dielectric type. A theory satisfying this criterion was hammered out in class-lectures at Cambridge on electric waves last February, and has been, in fact, already expounded in answers in the Mathematical Tripos much as it is proposed here to present it. The attention now excited by long-range free electric transmission, the most wonderful sudden practical evolution since the telephone, may attract the interest of a wider audience.

A sufficient cause for the increase of velocity, without dissipation, for waves travelling horizontally is, in fact, afforded by the free oscillations of ions even sparsely distributed in the very high regions of the atmosphere; though lower down their energy would be dissipated by collisions with the atoms and the travelling waves would be gradually quenched. Current is transmitted without any frictional loss across the vacuum of an electric valve by the jet of free electrons. On the other hand, in conduction in a metal the electrons or light ions get up speed at very rapid rate under the electric field, but at the end of each short free path they render up by collision part of the energy so acquired to the obstructing atoms of the metal, and this dissipation of the energy of the field is the measure of the electric resistance. For electric fields alternating at one-tenth of the rate of the waves of visible light, this process of conduction is, as Rubens showed, fully established at each instant in metal—a fact which leads to some information as to the lengths of the free paths[1]. But for ions travelling in an intensely rarefied upper atmosphere the free path will be long, so that many alternations of the electric force of the field of radiation, even for long electric waves traversing the medium, may occur within the time of one ionic path. This involves that the travelling ions will in the main swing free under the influence of the waves and thus interact without dissipation of their energy, scattering by secondary radiation from the slowly oscillating ions being negligible: there will thus be an influence on the velocity of the waves without absorption. To suit the present explanation the change ought to be a substantial increase, notwithstanding the small number of the ions contributing to it in the ultra-rarefied upper atmosphere. And the reason is simple. The influence of an ion is measured (jointly with the time) by the mean value of the alternating velocity excited in it, and this for a given intensity of field is proportional to the period of the waves, so that for long electric waves the effect may be millions of times as great as for short waves of comparable energy—in fact, it involves the square of the wave-length. Waves of shorter length could travel in a lower layer without excessive absorption, where a greater number of ions may make up for their weaker oscillatory motion[2].

*Side notes:*
Free ions in the rarefied upper air are effective:

are persistent.

Contrast with metallic conduction:

already complete even for infra-red radiation:

in upper air almost absent,

while ions are very potent in increasing speed of long waves:

the reason.

Extension to shorter waves lower down.

[1] Cf. *Phil. Mag.* August 1907.
[2] The only theoretical discussion which the writer has been able to find is by Dr W. H. Eccles (*Proc. Roy. Soc.* p. 86, June 1912). A "Heaviside layer" above, so sharply bounded and so intensely conducting as to reflect the rays without penetration, is assumed, combined with an ionization of the middle

In passing, one may note that the same principle will apply in general terms to the penetration of X-rays of high frequency across metals, especially of low atomic mass; when the time over a free path amounts to several periods of the radiation, the absorption, which must be due mainly to free ions, will be much diminished and the radiation can penetrate deeply—to a limit imposed mainly by simple dispersal or scattering, so proportional to the density of the atomic electrons: the Rubens experiments on reflexion of ultra-red light indicate a like result.

*Law of penetration for X-rays:*

*for ultra-red.*

On general ground the transparency of air for long electric waves would be expected to be great as compared with light; on account of the slow period molecular scattering is practically absent, and the mirage effects of local irregularities are negligible unless their extent is comparable with the wave-length (cf. p. 645, *infra**).

*High transparency for long waves: mirage effects restricted.*

The remark presents itself to contrast the finite free swayings of electrons or light ions in the ultra-rarefied gas of the upper atmosphere, which augment the speed of the exciting waves, with the finite swayings under the alternating attraction of the grid of a stream of electrons across a triode valve, which are thus partially trapped systematically by the grid and so establish its relay action.

*Contrast with valve action of electrons.*

There must thus be a layer high almost beyond the sensible atmosphere, within the auroral domain, in which a sheaf of horizontal electric rays, provided they are long enough, will travel without loss by absorption or scattering, concentration in this stratum being due to the potent influence exerted on the velocity of long waves by the free ions of small inertia interacting with them. But we have to suppose that the energy transmitted in this high stratum is being shed down to sensible degree all along the path, for the signals to be everywhere received. Introducing optical imagery, we can contemplate a special component ray sheaf connecting the transmitter to every receiver situated along the path of the beam; these rays all gather together into a sort of caustic curve or layer up aloft, and so long as

*Perfect propagation aloft:*

*rays shed down from a caustic surface:*

atmosphere which bends them, but "in forming the equations it has been implicitly assumed that the ions are so heavy that they acquire only small velocities and make very small excursions under the action of the waves." A viscous term is inserted in the equation of motion of an ion; if that term is annulled the formulae must agree in form with those justified here for light ions of very long free path. The treatment is mainly by rays, and contains full discussion of the relevant observational material of that date, especially as regards stray radiations.

The diffraction theory of G. N. Watson (*Proc. Roy. Soc.* 1918, more recently studied) seems to make it unlikely that the ground can take any part in the actual [long-distance] transmission.

* Also the general considerations regarding opalescence that are advanced at the end of Appendix v.

they are in it the energy travels onward without loss except by side-way spreading.

.The question arises whether there can be enough energy to be thus showered down to receivers all over the earth. By a known principle a receiver collects energy over a range of order $\lambda^2$; thus the total energy put into the transmitting layer is reduced at the receiver in the order of magnitude $\lambda^2/4\pi a^2$, so that for a wave-length of a kilometre the amplitude of vibration is reduced to the order $10^{-4}$, which is not excessive. *energy sufficient.*

*Propagation of Long Electric Waves in Ionized Highly Rarefied Gas.*
One notes in the first place that the *aurora borealis*, due somehow to free ions, is displayed at a height of the order of fifty miles; this is ten times the height of the homogeneous atmosphere, so putting aside effects of smaller order arising from change of temperature, the mean free path of the atoms would at that height be $e^{10}$ times the value at the Earth's surface, thus of the order of several centimetres. *Atomic free path at* The number of molecules per cubic cm. would be $3 . 10^{19}e^{-10}$, or *auroral* say $10^{15}$. *heights.*

Let us then consider the motion of an ion of charge $e$ and mass $m$, with free oscillation superposed on its translatory motion, due to the alternating electric field of radiation passing over it expressed locally by an electric force

$$F = A \cos pt, \quad 2\pi/p = \lambda/c.$$

For the oscillation thus superposed, all quantities being in electro-dynamic units,

$$m\ddot{x} = eA \cos pt, \quad m\dot{x} = ep^{-1}A \sin pt,$$

so that the amplitude is $ep^{-2}A/m$. There are, say, $N_0$ effective similar ions per unit volume, each supplying its contribution $e\dot{x}$ to the electric flux; all ions, positive and negative, contribute in the same direction, but massive ones may be neglected. An electric current is thus produced of density $C$ expressed in terms of the field $F$ as

$$C = - N_0 \frac{e^2}{mp^2} \frac{dF}{dt}.$$

This current, of precisely the same type, adds, in the electrodynamic *Ionic con-* circuital equations of the exciting wave-train, to the usual Maxwellian *tribution to dielectric* dielectric "total current" of density $\dfrac{K}{4\pi c^2}\dfrac{dF}{dt}$, in which $K$ is the static *current:* dielectric modulus of the medium and is practically unity for ordinary air. On adding them it comes out that the effective $K$ is diminished to *diminishes effective $K$:*

$$K' = K - N\frac{4\pi c^2 e^2}{mp^2}.$$

If $c'$ is the velocity of propagation, $c'^{-2}$ is thereby altered from the

41-2

value $K/c^2$, or say $c^{-2}$ in the absence of ionization, to $K'/c^2$; so that, $\lambda$ being the wave-length in free space,

$$c'^{-2} = c^{-2}\left(1 - N_0\,\frac{e^2\lambda^2}{\pi m}\right).$$

so augments speed, of long waves: This increase of speed* from $c$ to $c'$ arising from free ions mounts with $\lambda$, so for long electric waves it would be sensible, notwithstanding extreme tenuity of the gaseous ions. It would usually be of no account except in metals for the short waves of light.

an estimate. Roughly the ionic charge $e$ is of order $\frac{3}{2}\,10^{-20}$ electromagnetic; taking a wave-length of one kilometre, $\lambda$ is $10^5$; and if the effective ions are electrons $e/m$ is $\frac{1}{4}\,10^7$; the multiplier of $c^{-2}$ on the right then becomes $1 - 10^{-3}N_0$.

The change of speed required: What we are concerned for is that the rays should have a curvature exactly equal to that of the Earth. The index of refraction is

$$\mu = \frac{c}{c'} = 1 - N_0\,\frac{e^2\lambda^2}{2\pi m},$$

or, say here, $1 - \frac{1}{2}\,10^{-3}N_0$†. So the curvature of the rays, which is $-d\mu/dh$ (as *infra*) or $\frac{1}{2}\,10^{-3}\dfrac{dN_0}{dh}$, is not to fall short of $\frac{1}{2}\pi 10^{-9}$. For the present assumed data this condition of curvature requires $\dfrac{dN_0}{dh}$ to be about $3.10^{-6}$, a value which must extend over the effective layer, which may be a considerable range of $h$, say $10^5$ cm.; and this

* Several points remain to be noted.

Colour in mirage. The dispersion in atmospheric optical mirage is negligible, so that effects of colour are slight.

For the wireless rays the velocity $c'$ of propagation of phase exceeds the velocity of radiation. But the velocity of adjustment of the wave at its front cannot exceed $c'$: therefore, as is well known (*Proc. Camb. Phil. Soc.* 1904, as *supra*, p. 301), the individual waves as they run up into the slower travelling front must gradually fade out. When a beam is nearly homogeneous, so variation of $\lambda$ in it is small, the velocity $c''$ of a group of waves is [cf. p. 652]

$$c'' = \frac{d}{dp}\frac{c}{c'} = \frac{c^2}{c'},$$

which may be taken as the velocity of the effective front as regards energy. But the path of the ray is an affair of conspiring phases and is determined as here by $\delta t = \delta\!\int c'^{-1}ds = 0$. We can thus consider phase waves and amplitude waves travelling out from the source on different curved fronts.

Interference is restricted in dispersive media. † Interference when two ray paths, say a horizontal one and one curved downward, meet now requires two conditions: the phases must be concordant, but also groups of waves on both rays must arrive together. On the other hand (*loc. cit.*), the purer the radiation the greater the length of the group, which is of the order of $\lambda/\delta\lambda$ wave-lengths.

Fading of signals. This duplex condition for interference may be relevant to the intermittent phenomena of fading of signals.

mounts up for the terminal value of $N_0$ to only about ·3 electrons or $5 . 10^3$ hydrogen ions per cubic cm., whereas at the height of 50 miles is amply possible. there are still $10^{15}$ molecules of gas per cubic cm.*

*Bending of Optical Rays.* The directions of the wave-fronts are transverse to the rays, and consecutive fronts meet at their centre of curvature: thus, considering as usual two adjacent rays of the coherent beam, at distance $\delta n$, as the time of transit is the same for both from one front to the next, and the element of path of the ray is $v\delta t$, we have $\delta v/v = \delta n/R$, so that the curvature $R^{-1}$ of the ray is Law of bending of rays. $d \log v/dn$, or here simply $- d\mu/dn$. Ordinary atmospheric refraction by the action of the molecules, according to this formula, as applicable to horizontal rays of light or electric rays, would be negligible at the heights here in question of the order of fifty miles; but at sea-level, Density gradient of air is effective only low down: where it is mainly due to temperature gradient, it would give a very substantial though insufficient contribution to the required amount of bending. For the refractive index of air for ordinary light, and also for long waves, is at sea-level 1·0003; and the curvature of a

horizontal ray is $- \dfrac{d\mu}{dh}$, which as $\mu - 1 \propto \rho$ is $- 3 . 10^{-4} \dfrac{1}{p} \dfrac{dp}{dh}$, if there

is no gradient of temperature, or $3 . 10^{-4} \dfrac{g\rho}{p}$; wherein the speed of

sound is $\left(1 \cdot 4 \dfrac{p}{\rho}\right)^{\frac{1}{2}} = 32400$ cm./sec. at $0°$ C., giving $\dfrac{p}{\rho} = \frac{3}{4} (\frac{18}{4})^2 10^8$;

so that $- \dfrac{d\mu}{dh}$ is $3 . 10^{-4}/\frac{3}{4} 10^9 . 10^{-3}$ or $5 . 10^{-10}$, which as the Earth's

radius is $\dfrac{2}{\pi} 10^9$ amounts to as much as one-quarter of the curvature estimates:

of the Earth. But actually the gradient of temperature $d\theta/dh$ affects this result by a factor $1 + 30 d\theta/dh$ if $h$ is measured in metres, and is the chief cause of mirage effects: for instance, the curvature of the ray would agree with that of the Earth if the temperature increased temperature prepotent, upward by $1°$ C. for 9 metres[1]. These temperature variations of the ordinary index of refraction in the lower atmosphere would likewise except for long waves. affect long electric waves, were it not for the saving limitation that

---

* If there are in any region more than the moderate number $10^3$ of electrons per cubic cm., waves of natural length more than one kilometre cannot get through, but will be shed off, while for smaller wave-lengths this limiting number increases as $\lambda^{-2}$. All waves of lengths exceeding $(\pi m/e^2 N_0)^{\frac{1}{2}}$, or roughly $3 . 10^6 N_0^{-\frac{1}{2}}$, must be turned back to the Earth before they reach a stratum of density $N_0$ if such exist, for there the index of refraction would be reduced to zero. When the Earth's magnetic field is taken into account, this limiting condition may be substantially disturbed [p. 653].

[1] Cf. Everett's exposition of James Thomson's theory, *Phil. Mag.* Mar. 1873, p. 169: or Rayleigh, *Theory of Sound*, Vol. II, § 288.

irregularities occupying only a small part of a wave-length would be of no account; while at a height say of 15 kilometres, where the density of the air is reduced seven times, the effect is in any case very slight. As already remarked, the effective stratum is too high for meteorological instabilities to affect it. But it will not, of course, be exactly horizontal: its height may be expected to be some function of the local solar time at the place.

Depending on Solar radiation, not on weather.

*Amplitude of Oscillations of Free Ions.* We have still to verify that the amplitude and time of the oscillation of an ion would not be too great, in comparison with its free path, to allow a large degree of freedom in its vibrations. The value of the amplitude is $\frac{e}{mp^2} A$, or $\frac{e\lambda^2}{4\pi^2mc^2} A$, where $A$ is the amplitude of vibration of the electric force in the waves. If this field of force is $V$ volts per cm. (which compares with ten volts per cm. for the field of full sunlight if it were coherent[1]) so that $A = 10^8 V$, the amplitude for an electron, with wave-length $10^5$ cm. or one kilometre as above, is about $\frac{1}{2} 10^3 V$. It increases as the square of the wave-length. For a hydrogen ion it would be 1700 times smaller. Actually $V$ is so small that even for an electron it fits amply within the gaseous free path at a height of fifty miles, where it is several cm. as above. But sufficient time as well as space must be provided for this postulated regular forced oscillation of the ions. The period of a wave-length of one km. is $\frac{1}{3} 10^{-5}$ sec.: the time for a molecular free path of 3 cm. would be of order rather more than 2 cm. divided by velocity of sound in the gas, which is, say, twenty times longer. But this is for air; for a hydrogen ion the time would be divided by at least 5, for an electron in thermal equilibrium perhaps by 200, unless its path were lengthened by passage through the atoms without much disturbance of its motion. Thus it is only in the ultra-rarefied upper atmosphere that the slow alternations of material ions under the influence of long electric waves would have time or space for many complete oscillations on one free path; and there Ohm's law would be wholly inapplicable.

Actual amplitude of ionic swing not excessive, nor period.

Effect intensified for electrons.

Ohm's law there inapplicable.

[1] Maxwell (*Treatise*, § 793) seems to have miscalculated the results. Solar radiation is 2 cal./min. where 1 cal. $= 4 \cdot 18 \times 10^7$ ergs: for this intensity, if the radiation were coherent, the amplitude of its electric field would be as high as 10 volts/cm., and of its magnetic field 0·034 or $\frac{1}{8}$ of the mean horizontal force in Britain. The pressure of such radiation is 500 grammes weight per square kilometre. But the Solar natural radiation differs from the artificial radiations here discussed (which hardly occur at all in Nature) in being a statistical aggregation from molecular origins and not one coherent undulation: for it, energy is the surviving fundamental *datum*, definite amplitudes of oscillation must be replaced by deviations from a steady average state of disturbance measured by a statistical mean square, and theory would be of entirely different form —that of natural radiation.

If there are $N_0$ ions per cubic cm., each $M$ times the mass of an electron, and the time in the free path is $n$ times the period of the waves, of electric amplitude $F_0$, then in one second energy $\dfrac{F_0{}^2}{8\pi c^2}$ c passes across one sq. cm., while within one cubic cm. energy

$$\frac{N_0}{n}\,\frac{e^2\lambda^2}{8\pi^2 mc^2}\,F_0{}^2\Big/\frac{\lambda}{c}$$

is dissipated in one second: by dividing, the modulus of absorption per cm. comes out $\dfrac{N_0}{n}\,\dfrac{\lambda}{M}\,10^{-13}$. In the circumstances illustrated above $n$ may be 10, $N_0$ is $2.10^3$ hydrogen ions for which $M$ is 1740, the wave-length $\lambda$ is $10^5$ cm., so that plane radiation would be reduced by ionic absorption acting alone in ratio $e^{-1}$ in a path of ten thousand kilometres. This compares with the estimate given earlier of the effect of conductance.

*Dissipation in encounters with atoms (that by radiation negligible)*

*permits great transparency.*

We infer that the present type of explanation by a velocity of propagation increasing with height, which bends down the rays without undue absorption, is sound and reasonable for electric waves of the lengths usually employed, if only a very small proportion of the molecules of the upper air were ionized, so as to provide electrons or light ions, by the intense ultra-violet Solar radiation. Across the region of the dawn the strata of ionization may be deranged: they may be deformed locally so much as to throw the concentrated horizontal sheaf of rays upward into space or else downwards to the Earth so far that it cannot recover the horizontal direction in the new effective stratum. But if that stratum is deep, say a kilometre, such dislocation need not be complete. These considerations retain general validity when the strata of equal ionization are not exactly horizontal.

*Effect on rays crossing the dawn or sunset.*

There seems to be some call here for closer scrutiny of statistics as to the relation of range to wave-length: also as to whether reception is better or worse, or simply irregular, at times of dawn or sunset[1]. It is perhaps significant that "atmospherics" are reported to predominate at sunrise and sunset. It appears also[2] that there is increase of strength but diminished direction-finding at night: that in both respects North-South is superior to East-West: that winter is favourable in the Atlantic, and transmission over sea easier than over land.

*Other features.*

[1] It appears (Eccles, *loc. cit.* p. 97) that in Transatlantic signalling the response in fact does attain a high maximum when either the receiver or the transmitter is in the dawn or the sunset: and other significant features are found.

[2] Cf. Professor E. V. Appleton, Royal Institution Lecture, in *Engineering*, May 30, 1924, p. 709.

Scrutiny of the induced electric moment of a molecule in a strong field is here incidentally suggested. For air at standard density the refractive index is of order $1 + 3.10^{-4}$: thus the induced polarity

$\dfrac{K-1}{4\pi} F$ in electrostatic measure is $\frac{1}{2} 10^{-4} F$ distributed over $\frac{2}{3} 10^{19}$ molecules. If $F$ is $\frac{1}{6} 10^5$ volts per cm. (corresponding to a spark of 5 cm.), which is 60 electrostatic units, this gives a moment $2.10^{-22}$ for each molecule. As $e$ is $5.10^{-10}$ electrostatic, the arm of the moment would be $4.10^{-13}$ cm., while the molecular radius is as much as $10^{-8}$. Thus $K$ would hardly fall away from being constant before the field approached $10^6$ electrostatic, and disruption would have set in long before.

*Undulatory Propagation for a Thin Unchanging Sheaf of Rays.* The electric radiation to great distances thus involves a sheaf of rays whose breadth may be only a fraction of one of its long wave-lengths, which travels along a surface of constant index, while it has to remain concentrated without spreading either by refraction or by diffraction.

The subject thus invites examination dynamically. The equation of propagation, neglecting now the terrestrial curvature of the strata, is of type

$$\frac{\partial^2 \phi}{\partial x^2} + \frac{\partial^2 \phi}{\partial z^2} = c'^{-2}\mu^2 \frac{\partial^2 \phi}{\partial t^2}, \ \mu^2 = 1 + \psi(z),$$

where $c'$ is the speed along the mid-front at which $\psi(z)$ vanishes. The

type of waves here contemplated, with amplitude changing along the front as $f(z)$, is expressed by

$$\phi = f(z) \, F(x - c't);$$

but this must be further specialized for the analysis, into harmonic trains each of type

$$\phi = f(z) \cos m(x - c't), \ m = 2\pi/\lambda.$$

Then
$$\frac{\partial^2 f}{\partial z^2} - m^2 f = -m^2 \{1 + \psi(z)\} f,$$

or
$$\frac{\partial^2 f}{\partial z^2} = -m^2 \psi(z) f, \ \mu^2 = 1 + \psi(z).$$

For a disturbance with its parts keeping together as it travels, $f(z)$ must fall off on both sides of a maximum along $z = 0$: that is, the graph of $f(z)$ must be concave to the axis of $z$, the condition for which is $\partial^2 f/\partial z^2$ negative; so that as the amplitude $f$ is positive, $\psi(z)$ must be positive on both sides of the central plane. This is simply the familiar condition that the stratum is to be one of minimal index,

close to which the rays must cling by virtue of the principle of minimal time of transit.

Over small range of $z$ we may illustrate by taking

$$\mu = 1 + Az^2:$$

the equation to determine the law of transverse intensity for such an unchanging layer of the wave-train is

$$\frac{\partial^2 f}{\partial z^2} + 2m^2 A z^2 f = 0,$$

which is amenable to simple graphical treatment.

The actual effective atmospheric stratum must, however, be curved to the Earth's radius $a$, and the equation of wave-propagation of the curved sheaf of rays travelling along it is of type

$$\frac{\partial^2 \phi}{\partial r^2} + \frac{1}{r}\frac{\partial \phi}{\partial r} + \frac{1}{r^2}\frac{\partial^2 \phi}{\partial \theta^2} = c_0^{-2}\mu^2 \frac{\partial^2 \phi}{\partial t^2}, \quad r = a + z,$$

where $c_0$ is the velocity along $z = 0$ and $\mu$ is the index relative to that layer.

The type of waves travelling along it without spreading may as before be restricted to the harmonic form

$$\phi = f(z)\cos m(a\theta - c_0 t), \quad m = 2\pi/\lambda.$$

As $r = a + z$ this gives

$$\frac{\partial^2 f}{\partial z^2} + \frac{1}{a+z}\frac{\partial f}{\partial z} - \frac{m^2 a^2}{(a+z)^2}f + m^2\mu^2 f = 0.$$

Thus if
$$g = (a+z)^{\frac{1}{2}}f, \quad \mu^2 = 1 + \psi(z),$$

$$\frac{\partial^2 g}{\partial z^2} + \left\{\left(\frac{1}{4a^2 m^2} - 1\right)\left(1 + \frac{z}{a}\right)^{-2} + 1 + \psi(z)\right\}m^2 g = 0.$$

As before, the layer of transmission extends over the values of $z$ which make the coefficient of $g$ in this equation positive; for $z$ small it approximates to

$$m^2\left(\mu^2 - 1 + \frac{2z}{a} - \frac{3z^2}{a^2} + \frac{1}{16\pi^2}\frac{\lambda^2}{a^2}\right).$$

Thus substituting

$$\mu - \mu_0 + \left(\frac{\partial \mu}{\partial z}\right)_0 z + \tfrac{1}{2}\left(\frac{\partial^2 \mu}{\partial z^2}\right)_0 z^2 + \ldots,$$

this equation in $g$ takes the same form as the previous one in $f$, found as the analytical expression of the concentration of the rays for flat strata towards the straight ray of absolutely least time of transit, provided that when $z = 0$ the curvature $a^{-1}$ is equal to $-\left(\dfrac{d\mu}{dz}\right)_0$, as before obtained, and also $\mu_0\left(\dfrac{\partial^2 \mu}{\partial z^2}\right)_0 - \dfrac{2}{a^2}$ is positive[1].

It may be remarked that treatment of the cognate subject of optical mirage involves a like analysis: cf. Everett, *loc. cit.*

* Nevertheless an isolated beam whose vertical breadth is only of the order of one wave-length would disperse rapidly by diffraction were it not that here the efficient part of the curving beam is supported on each side by the rest of the curved front to which it belongs.

## *Appendix* (1928).—ATMOSPHERIC TRANSMISSION IN THE EARTH'S MAGNETIC FIELD.

[From *Proc. Calcutta Math. Soc.*, Jubilee volume, 1928.]

Electric waves can bend round the Earth:

It has appeared above that the presence of free electrons or ions in the auroral region of the upper atmosphere can give an adequate account of the now familiar bending of wireless signals so as to keep parallel to the Earth's surface.

A limited region containing a substantial number of ions of various kinds has a refractive index $\mu$ for the waves, which may be less than unity, given by

$$\mu^2 = \left(\frac{c}{c'}\right)^2 = K - \Sigma \frac{4\pi N_r e_r^2}{p^2 m_r}, \quad \frac{2\pi}{p} = \frac{\lambda_0}{c},$$

where $K$ depends on density being 1·0006 for standard air, and $\lambda_0$ is the wave-length *in vacuo* which expresses the unchanging period: so the rays may be deflected by bending downward which may even become complete gradual reflexion, but yet with only slight absorption when the ionic free paths are long. For wireless waves of the lengths ordinarily in use, the index of the medium would fall even toward zero for quite moderate numerical densities of the electrons, if other causes did not intervene: for a wave-length of one kilometre the required density is only about 9.10², increasing for shorter waves

and above a limiting wave-length they cannot escape into space.

as the inverse square of their length. If there exists a stratum aloft of such electron density, waves longer than a kilometre cannot escape into space: all except the slight proportion travelling nearly vertically must be bent back to the Earth before they get anywhere near it. In the greater freedom of the rarefied high atmosphere ionization would however be dissipated more rapidly, unless it is continually renewed.

Shorter waves liable to local scattering.

Local regions strongly ionized, and comparable in extent with the wave-length, would act as obstacles to transmission, even as total reflectors, turning back and dispersing the waves: in this respect short waves must be at a disadvantage, just as is blue light for penetration through the sky. But nowadays reception is so sensitive that such attenuation is unimportant, unless it is accompanied by change of velocity.

Influence of Earth's magnetic field.

The circumstances of atmospheric electric propagation parallel to the Earth, and also those here noted restricting escape of radiation into space, must be affected sensibly, as E. V. Appleton has remarked, by the Earth's magnetic field, but in complicated ways, that will now be illustrated by estimates for some of the simpler cases.

*A wave-train in an ionized medium travelling in the direction of a magnetic field.*—The equations of motion of an ion transverse to $z$ the direction of propagation of the waves, are, in a magnetic field $H_0$ along $z$,

$$m\ddot{x} = eP + eH_0\dot{y}, \quad m\ddot{y} = eQ - eH_0\dot{x},$$

equivalent together to

$$m\ddot{\xi} = e(P + \iota Q) - \iota e H_0 \dot{\xi}, \quad \xi = x + \iota y.$$

The equations of the electrodynamic field of the radiation, with $N$ electrons per unit volume, $(P, Q, R)$ being the electric and $(\alpha, \beta, \gamma)$ the magnetic field, all functions of $t - z/c'$ alone, become when $R, \gamma$ are taken to be null

$$-\frac{\partial\beta}{\partial z} = Kc^{-2}\dot{P} + 4\pi Ne\dot{x}, \quad -\frac{\partial Q}{\partial z} = -\frac{\partial\alpha}{\partial t},$$

$$\frac{\partial\alpha}{\partial z} = Kc^{-2}\dot{Q} + 4\pi Ne\dot{y}, \quad \frac{\partial P}{\partial z} = -\frac{\partial\beta}{\partial t},$$

$$0 = Kc^{-2}\dot{R} + 4\pi Ne\dot{z}, \quad 0 = -\frac{\partial\gamma}{\partial t},$$

thus requiring also $\dot{z}$ to be null, so that the waves travelling along the direction of $z$ are wholly transverse. Thus

$$\frac{\partial}{\partial z}(\alpha + \iota\beta) = -\iota Kc^{-2}\frac{\partial}{\partial t}(P + \iota Q) - \iota 4\pi Ne\frac{\partial}{\partial t}(x + \iota y),$$

$$\frac{\partial}{\partial z}(P + \iota Q) = \iota\frac{\partial}{\partial t}(\alpha + \iota\beta):$$

so that, eliminating $\alpha + \iota\beta$,

$$\frac{\partial^2}{\partial z^2}(P + \iota Q) = Kc^{-2}\frac{\partial^2}{\partial t^2}(P + \iota Q) + 4\pi Ne\frac{\partial^2}{\partial t^2}(x + \iota y),$$

which is to be combined with the previous equation of movement of free ions

$$m\frac{\partial^2}{\partial t^2}(x + \iota y) = e(P + \iota Q) - \iota e H_0\frac{\partial}{\partial t}(x + \iota y).$$

The system is thus cyclic, as its relations are consolidated into two equations with complex variables, equivalent to four when real and imaginary parts are separated.

Consider a harmonic wave-train expressed by

$$P + \iota Q = F_0 e^{\iota pt} e^{\iota nz}, \quad \alpha + \iota\beta = -\iota\frac{n}{p}F_0 e^{\iota pt} e^{\iota nz}$$

equivalent to

$$P = F_0 \cos(pt + nz), \quad \alpha = \frac{n}{p}F_0 \sin(pt + nz),$$

$$Q = F_0 \sin(pt + nz), \quad \beta = -\frac{n}{p}F_0 \cos(pt + nz).$$

At each instant its vector field $(P, Q)$ has the form of a right-handed spiral staircase with axis along the train, and the field $(\alpha, \beta)$ is a conjugate one at right angles, being a quarter period in advance in phase, the configuration of these fields travelling inward towards the origin with velocity $p/n$.

*the paths of the free ions.* Each ion describes a circle with vector radius along the electric vector, being determined by its equation of vibration involving $x + \iota y$, and with real coefficients, as above.

On substitution, $c'$ representing $p/n$ the velocity of propagation of phase, and $\xi$ representing as before $x + \iota y$,

$$\left(\frac{\mathrm{I}}{c'^2} - \frac{K}{c^2}\right) F_0 = 4\pi N e\xi, \quad eF_0 = -(mp^2 + eH_0 p)\,\xi.$$

Thus*
$$\frac{n^2}{p^2}c^2 = \frac{c^2}{c'^2} = K - \frac{4\pi N e^2 c^2}{mp^2 + eH_0 p}.$$

*The two speeds of travel.* For $c'$ to be real, so that there may be a chance for rays to be transmitted instead of being turned aside, the right-hand side must be positive. Obviously transmission will be precluded if the period $2\pi/p$ is too great, whatever the imposed field $H_0$ may be: all waves of *The limiting periods for escape into space, with given ionization.* length exceeding $(\pi m/e^2 N)^{\frac{1}{2}}$, or roughly $3 . \mathrm{10}^6 . N^{-\frac{1}{2}}$, would thus in any case be turned back to the Earth before they reach a stratum of electrons of numerical density $N$ if such exist: it is only shorter waves that will have a chance to escape upward into space.

Also, the value of the denominator, as affected by the magnetic field, is roughly, when the mass $m$ is that of an electron,

*Limit imposed by the field on dextral waves, for all ionizations,*
$$\mathrm{10}^{-27}p^2 + \mathrm{10}^{-20}H_0 p:$$

thus, whatever $N$ may be, the circumstances become critical, for propagation of right-handed spiral waves in the direction of $H_0$, when $p$ falls towards the value $\mathrm{10}^7 H_0$, or in the Earth's magnetic field when the standard wave-length rises towards a quarter of a kilometre— *thus involving negative circular polarization.* beyond that limit the medium transmits only one component and so circularly polarizes the radiation. This statement is irrespective of the number of electrons the medium contains: the reason is that for the other direction of rotation the motion of each individual electron would increase without limit as this period is approached, until inhibited by other causes.

---

* One may note that a wave-group travels through a nearly homogeneous train of waves with velocity $v$ equal to $(dn/dp)^{-1}$, while the velocity of the individual waves of the train is $c'$ equal to $(n/p)^{-1}$: thus

$$\frac{c^2}{v} = Kc' + \frac{c'}{p}\frac{d}{dp}\frac{2\pi N e^2 c^2 H_0 e/m}{mp^2 + eH_0}.$$

When $H_0$ vanishes this makes $vc'$ equal to $c^2/K$ the standard velocity in air, which as it happens agrees with the form for L. de Broglie's wave-mechanics, as indeed does the Cauchy optical dispersion formula.

When $N$ is given, the precise condition for $c'^2$ to be positive is that
$$mp^2 + eH_0p \text{ exceed } 4\pi K^{-1}Ne^2c^2.$$
Thus, the period $2\pi/p$ being positive, waves cannot penetrate against the direction of an imposed magnetic field $H_0$ if $p$ is less than

$$-\tfrac{1}{2}\frac{e}{m}H_0 + \left(\tfrac{1}{4}\frac{e^2}{m^2}H_0{}^2 + \frac{4\pi}{Km}Ne^2c^2\right)^{\frac{1}{2}}:$$

for the opposite direction of travel the sign of $H_0$ would be reversed in this formula.

Limit for any penetration.

In the absence of a magnetic field the lower limit of wave-length for penetration, in any direction, is a kilometre when $N$ is about $10^3$. The influence of a magnetic field on it may be considerable, and is inversely as $N$, as comparison of the two terms under the radical shows: thus when $N/H_0{}^2$ is $\tfrac{1}{4} 10^5$ the two terms under the radical are of the same order.

A plane-polarized ray travelling along the field $H_0$ would have its plane of polarization rotated. The difference of phase, measured as length, of its two circularly polarized components, would be the difference of the values of $\int \mu \, ds$: and by the Fermat principle that this integral is stationary in value as regards change to all adjacent ray paths, this difference of values can be estimated by integrations made for both along the same path, if the actual paths are not far apart. The values of $\mu^2$ which is $(c/c')^2$ are given above for the two rays differing according to the sign of $H_0$. For periods well below the critical value $10^7H_0$ of $p$, the difference of the indices for the two signs of $H_0$ is approximately

Rotation of plane-polarized waves:

an estimate:

$$4\pi \frac{Ne}{K^{\frac{1}{2}}} \cdot \frac{e^2}{m^2} \frac{H_0}{p^3} c^2:$$

and this multiplied by length of path $l$ and by $pK^{\frac{1}{2}}/c$ will give for the angle of rotation of the plane of polarization in traversing that path the value $Nel\frac{e^2}{m^2}\frac{H_0\lambda^2}{\pi c}$ which is $3.10^{-18}H_0N\lambda^2$ per cm. and increases very rapidly with $\lambda$; taking $H_0$ to be $\tfrac{1}{5}$ it is $2.10^{-11}N$ for waves of 50 metres*. It is to be noted that if this rotation could be observed, it would determine an average value of $N$ along the path of the ray: that of $dN/dh$ is determined by the curvature of the Earth.

may determine the ionization.

Actually it appears (*loc. cit.*) that a vertical atmospheric gradient $dN/dh$ of only a few electrons per kilometre is required to carry the ray round the Earth's curvature. In such a case $N$ may be so small that the square root can be expanded: so that the limits of $p$ for absence of penetration are from $8\pi ec^2N/KH_0$, which may be as low as $10^2$, down to zero, corresponding to impracticable periods of order of more than one-tenth of a second.

* Corrections have been introduced here and on next page.

But though its gradient at the level of the rays is thus fixed, $N$ itself may be large. The limit for transmission is roughly $p$ less than $\mp 10^7 H_0 + (5 \cdot 10^{13} H_0^2 + 2 \cdot 10^8 N)^{\frac{1}{2}}$, $H_0$ being positive and negative for the two directions of rotation. Thus the magnetic field $H_0$ is important if $N/H_0^2$ is much less than $10^6$. For example, if $N$ were of the order of $10^6$, and $H_0$ is taken as $\frac{1}{8}$, which is of the order of the Earth's field, the first term under the square root may be neglected and the limit for $p$ becomes $14 \cdot 10^6 \mp 2 \cdot 10^6$, corresponding to wave-lengths down to 170 metres and to 120 metres, the field being more permeable to dextral waves travelling in its direction. Thus the actual transmission of shorter waves puts a superior limit to the number of electrons that can be present in the strata concerned with it*.

*Imposed magnetic field oblique to wave-fronts.*—For steady propagation of waves travelling oblique to $H_0$ the conditions are much more complex: the planes of the magnetic and electric vibrations† are not along the fronts of the waves, but each has a longitudinal component, the two components combining as will appear into an elliptic vibration. When such waves travel into a region of smaller magnetic field, or one differently directed, the circumstances no longer fit together, and perhaps on this account as well as from change of velocity the ray may tend to turn round horizontally; other causes also, *not* however feasibly a horizontal gradient of $N$, may give rise to horizontal curvature of the rays. On both grounds the usual determinations of the direction of the source may be deceptive. The general problem of propagation in a non-uniform magnetic field is perhaps intractable: but if the velocities of transmission are tabulated for different magnitudes and relative directions of uniform field, the Fermat principle of minimal time may lead by graphical procedure to some notion of the actual ray paths.

These considerations may be further elucidated by working out a simple case. If the imposed constant magnetic field $(0, \beta_0, 0)$ is parallel to the field $(0, \beta_0, 0)$ of the radiation, and thus, as will appear

*Side-note:* Numerical results.

*Side-note:* Non-uniform magnetic field.

---

* An application to the Sun's atmosphere perhaps presents itself. It was found by G. E. Hale and his associates (*Astrophys. J.* 1918) that the general magnetic field of the Sun is rapidly extinguished with height in his atmosphere. Quite a small density of free electrons or ions, provided their free paths are long enough, can prevent any electromagnetic changes, except short-wave radiation, from being transmitted across to the outside. Especially is this so when the atmosphere is in motion: the axial rotation of an atmosphere can shield off all but the meridional components of a star's internal magnetic field: cf. *Phil. Mag.* Jan. 1884; as *supra*, Vol. I, p. 27: also Vol. II, p. 613.

† Goldstein finds that the magnetic elliptic vibration is always in the wave-front.

*Side-note:* Magnetic shielding by Sun's atmosphere.

transverse to the radiation, it may be presumed that the electric field will be transverse to $\beta_0$ and that the plane of polarization will not rotate*. Let us explore therefore the scheme expressed by a field of radiation $(0, \beta, 0)$ and $(P, 0, R)$. All quantities are functions of $pt - nz$ alone. We have therefore, now writing $N_0$ for the density of electrons,

$$-\frac{\partial \beta}{\partial z} = 4\pi u = Kc^{-2}\dot{P} + 4\pi N_0 e\dot{x}, \quad -\frac{\partial \beta}{\partial t} = \frac{\partial P}{\partial z},$$

$$0 = \frac{\partial \beta}{\partial x} = 4\pi w = Kc^{-2}\dot{R} + 4\pi N_0 e\dot{z}.$$

For a free ion at $(x, y, z)$

$$m\ddot{x} = eP - e\beta_0\dot{z}, \quad m\ddot{y} = 0, \quad m\ddot{z} = eR + e\beta_0\dot{x},$$

in which for periodic waves $\partial/\partial t$ is $\iota p$: thus

$$\dot{x} = \frac{-e\dot{P} + \dfrac{e}{m}\beta_0 R}{mp^2 + \dfrac{e^2}{m}\beta_0^2}, \quad \dot{z} = \frac{-e\dot{R} - \dfrac{e}{m}\beta_0 P}{mp^2 + \dfrac{e^2}{m}\beta_0^2}.$$

Hence

$$-\frac{\partial \beta}{\partial z} = L\dot{P} + NR, \quad L = \frac{K}{c^2} - \frac{4\pi N_0 e^2 m}{m^2 p^2 + e^2\beta_0^2},$$

$$0 = \frac{\partial \beta}{\partial x} = L\dot{R} - NP, \quad N = \frac{4\pi N_0 e^2 \beta_0}{m^2 p^2 + e^2\beta_0^2},$$

giving, if $L'$ represent $L\partial/\partial t$,

$$(L'^2 + N^2)P = -L'\frac{\partial \beta}{\partial z}, \quad (L'^2 + N^2)R = -N\frac{\partial \beta}{\partial z}, \quad -\frac{\partial \beta}{\partial t} = \frac{\partial P}{\partial z}.$$

Thus

$$\frac{\partial^2 \beta}{\partial z^2} = \left(L - \frac{N^2}{Lp^2}\right)\frac{\partial^2 \beta}{\partial t^2},$$

giving

$$c'^{-2} = \frac{n^2}{p^2} = L - \frac{N^2}{Lp^2},$$

reducing to the value previously found for $c^2/c'^2$ when $\beta_0$ is null.

To permit penetration $p^{-2}$ must exceed $L^2/N^2$, leading to a discriminating cubic equation in $p$. Reversal of the direction of $\beta_0$ is without influence, for change of sign of $\beta_0$ changes only the sign of $N$ whereas $c'^{-2}$ involves $N^2$.

* In the complementary type of polarization, electric vibration along the imposed magnetic field, the latter will obviously be inoperative on the oscillatory ions. The formulae for the velocities for the two special cases here worked out have been recorded already by E. V. Appleton and M. A. F. Barnett, *Proc. Camb. Phil. Soc.* March 1925, as the writer has been reminded by an early draft of a more general discussion by S. Goldstein. [*Roy. Soc. Proc.* 1928.]

The longitudinal component of the electric field is in quadrantal phase to the transverse, with amplitude in ratio $N/Lp$, so that this field vibrates elliptically.

The general analysis to determine speed of propagation and type of vibration when the magnetic field is inclined to the front involves complex formulae: moreover, in the terrestrial application the inclination changes as the ray progresses. The most convenient expression of results would probably be in tabular arithmetic form. The changing velocity involves horizontal bending of the ray, so that from this cause error may arise in directional findings as regards position of the sources. Perhaps the feasible way of dealing with that problem is, as remarked above, to try to lay off graphically the quickest paths, which are the rays, in the field of propagation with the local velocities marked all over it.

### [Sept. 1928.—*Fading of Signals as due to Interference of Energy Groups*.

When two rays have reached the observer by widely different paths, the relevant criterion appears to be that, if the velocity of phase is $c(1 + k\lambda^2)$ and thus that of the grouped energy $c(1 - k\lambda^2)$, for waves of actual length $\lambda$, then the individual signals will overlap, and there may be fadings by interference of the two paths, only when the difference of the values of $\lambda^2 \int k \, ds$ along them is not a large proper fraction of the path-length of the independent elements of the signal. This perhaps would be an exceptional case in practice, were it not that in waves guided by the ground the delay in the energy groups, due to the different cause, would be very much greater than in the air.]

# 101

# ON THE CONES OF STEADY COMPRESSION FOR A FLYING BULLET.

[Communicated to the International Congress of Mathematicians at Toronto, August 1924.]

WHEN a bullet or shot is flying through the atmosphere, at speed greater than will allow the air to get out of its way by waves carrying off the compression, the natural expectation would perhaps be that it can progress only by smashing its way through. Long ago Thomas Young propounded the correlative idea that when a hammer-blow communicates to the surface of a hard solid a speed greater than that of any possible relieving waves travelling into the interior, the structure of the superficial parts must be shattered. *Significance of crushing.*

But a striking discovery was reached photographically more than a quarter of a century ago by the experimental acumen of C. V. Boys, that in fact no such smashing through is involved. The bullet flies through the air perfectly smoothly, therefore without dissipation of energy, except that a rather narrow wake of fine-grained turbulence is formed in its rear, which contracts with smaller terminal cross-section of the bullet, being, so to say, a sort of misfit in the closing up of the disturbance. This very remarkable type of permanent motion in elastic fluids, as striking in its way (if it be accepted) as Helmholtz's discovery of the permanence of vortex rings in perfect fluid, does not seem to have attracted the full attention that it deserves: yet, if the matter has not been misconceived, the general lines of an explanation may not be remote. *C. V. Boys' revelation of the motion:*

The original photographs by Boys were the subject of correspondence between him and Stokes: see the two volumes of *Scientific Correspondence of Sir George Stokes*. More recent photographs by Professor Dayton C. Miller and J. C. Quayle are in *Journ. Franklin Inst.* 1922; some of them are reproduced by L. Thompson in *Proc. U.S. Nat. Acad.* June 1924, p. 280, and in fact suggested the present note*. *Stokes.*

The concentration of the resistance into a resultant suction located on the rear of the bullet, that they indicate, must have some influence in stabilizing its lengthways direction even when there is no spin. *Stabilizing suction.*

The most significant feature in the photographs is a fine thread-like curve, surrounding the head of the bullet a little way in front and ranging back at the sides so as to take on a hyperbolic form. Another curve precisely parallel strikes off on both sides from the rear of the

* There are extensive reproductions of photographs in a Report, R.D. 63 (1925), of the War Office Research Department, Woolwich.

bullet, or rather converges to a point some distance behind it so that the whole of the bullet is included in the space between, while exhibiting close to it the local misfit above referred to. Such curves are an optical indication of abrupt change of density of the air. It is now

A conical<br>shell of<br>compressed<br>air; suggested as a proposition for scrutiny that the flat cone of changing air between the two blunt conical boundaries, thus shown in section as curves, is maintained at a definite steady compression determined by the velocity of the bullet, air being compressed into it in front and expanding out of it behind as it travels onwards as a motional form across the stationary atmosphere. Thus in each position of the shadow-form, so to say, there is within it a local condensation of the air: if

making room<br>for the bullet, this contraction of volume is just sufficient to make room for the bullet at each stage, there will be no need for the latter to push air aside in mass as it travels onward: so that this conical boundary of a condensed region, which advances steadily in front of the bullet, will shield it and keep the whole motion steady.

In contrast, for velocities of the bullet less than sound, it will

but only for<br>speeds in<br>excess of<br>sound. appear that the same relations hold but give rise to local expansion instead of contraction, so that no room is provided for the bullet and the effects cannot be balanced and merely local. Resistance would then be expected to be produced, mainly, as in the case of liquids, by throwing off vortex rings, in addition to sound-waves, both of which travel away with energy that must have been abstracted from that of the projectile. The features of the motion seem also to be less definite and regular.

Problem of<br>an advancing<br>slab of com-<br>pression: To test this order of ideas one begins naturally with the case of a flat slab-form of air of augmented density, travelling across a still atmosphere, being sustained at the steady excess of density by condensation in front and expansion behind as it flits through the atmosphere. The conditions are simplified by impressing a uniform velocity on the whole system, which by relativity has no dynamical effect, so as to bring the slab of condensation to rest. Then the air moves up to a stationary region of augmented density with velocity $v$, through it with $v'$, and away on the other side with $v$. If the density within this slab* is $\rho'$ and outside on both sides is $\rho$, we have

$$\rho v = \rho' v' = M,$$

where $M$ is the flux across per unit area. Moreover a momentum $Mv$ delivered at the front boundary per unit time has diminished to $Mv'$ on crossing the front boundary: the rest of it must provide surface forces producing change of pressure in crossing the boundary given by

$$p' - p = Mv - Mv' = \rho v^2 - \rho' v'^2.$$

* This assumption of uniform density across the slab restricts to a fluid of slight compression, yet of increased density in the slab due to terminal balanced impulsive forces very great on account of high speed. See modification at end.

This involves change of density inside the slab given by the characteristic equation of air, which is, with sufficient approximation,

$$\delta p = c^2 \delta \rho, \text{ or } p' - p = c^2 \left( \frac{M}{v'} - \frac{M}{v} \right),$$

where $c$ is the velocity of sound, assumed the same throughout as the process can be isothermal.

On equating these values of $p' - p$, a relation emerges in the simple form $vv' = c^2$. *restricted to uniform relative velocity:*

It is to be noted that there is no trouble arising from failure of conservation of energy, for what is imparted to the slab in front is abstracted in the rear [: but there is trouble with the pressure, if the flow remains unbroken round an open edge of the slab of air, for the hydrodynamical term in it for smooth flow is $-\frac{1}{2}\rho v^2$, as on p. 661. See previous footnote].

Impose now a velocity opposite to $v$ on the whole system. The air outside is reduced to rest in front and rear, while the slab-form progresses steadily through it with this new velocity $v$, but the air at each instant within the slab-form also moves forward in mass with a different velocity $V = v - v'$ in the same direction as $v$. The relation which must be satisfied to make this motion possible has been found to be $vv' = c^2$, which now becomes $v(v - V) = c^2$. If $v$, the velocity with which the slab-form advances through the now stagnant air, were less than $c$, the velocity of sound, $V$ would be negative, that is the air in the progressing slab would then move backward: but as above remarked, the density of the air within the slab is now diminished, which is the reason why the solution then fails for the case of a bullet. For the steady density of the air within the slab is given by $\rho' = \rho v/v' = \rho v^2/c^2$. If $v$ were slightly different from the velocity of sound $c$, the change of density and the velocity $V$ of the air would both be small: exact equality would obliterate the slab altogether, which is not really paradoxical. *velocity of slab must exceed sound:* *thus failure for a slow bullet.* *The intermediate case.*

The argument holds also when the compression form advances obliquely to itself, provided the transverse component of its velocity $v \sin \theta$ takes the place of $v$. The air inside would now have a tangential velocity relative to the moving slab, but no such velocity relative to the atmosphere outside, so that there will be no slip.

The same argument is applicable to each part, practically flat, of the bent surface bounding the conical shell of condensed air that accompanies the flying bullet. But now each element of the layer of compression moves outward from the axis with the component of this large transverse velocity $V$ during the time the compression takes to pass over it. The permanent shift thus produced involves a rarefaction that could not proceed to all distances. Hence $V$ must tend to vanish, *Result holds for a cone of compression:*

which actually is prolonged: to its final slope.
and the compression in the layer to disappear, at a distance from the axis. Indeed it is remarkable that the dividing line can be traced so far in the photographs. Anyhow it would be this necessity that fixes the asymptotic value $\alpha$ of the slope of the cone to satisfy $v \sin \alpha = c$.

The experiments, if explained in this way, by sharp transition to a region in compression, point to a definite steady state of movement of the air while it is within the slab-form. That state would be determined at each instant, along with the changes of density which make it possible, by the velocity prescribed along its moving conical boundaries, which is $V$ transverse where at slope $\theta$, as above,

$$V = v \sin \theta - \frac{c^2}{v \sin \theta}, \text{ or } v \sin \theta \,(v \sin \theta - V) = c^2.$$

Problem of internal motion in slab-form.
Also at a distance from the axis of the cone $\theta$ is to tend to the asymptotic value which makes $V$ vanish, and further the total local shrinkage by compression of the air must, for steadiness, be equal to the volume of the bullet. The problem thus furnished—one of a new type in hydrodynamic analysis—is whether these data determine the form of the boundary blunt cone, and, for example, how far their form is sensibly independent of the form of the bullet depending only on its velocity which prescribes the asymptotic angle. The answer to the former question can hardly fail to be in the affirmative, a surface of abrupt transition can exist and is definite: as regards the latter, experiment can probably provide still further crucial indications.

The hydrodynamic problem thus arising as regards the motion of the air in a condensed layer can be elucidated by an analogy derived from electrostatics. The velocity potential of the air relative to the moving form is the analogue of the electric potential of a charged sheet, as it is constant along the boundary cone of the layer. It could, of course, be determined, whatever the form of this sheet subject to the limiting slope $\alpha$: but the density of charge is also specified everywhere in terms of the slope $\theta$, being proportional to $V$, and this restricts the sheet for given $\alpha$ to one standard form. When the steady potential is determined, the compression at each point follows. It would appear, then, if and so far as these ideas are confirmed, that the form of the boundary cone for given velocity is a definite universal one determined intrinsically without reference to the form of the bullet except that its volume would fix the thickness of the condensed layer. Thus the cones belonging to fragments of ejected wadding in the photographs cross each other sharply instead of fusing together. In these respects and others, experiment can still provide further tests and indications for theory.

Intersecting cones do not interfere.
It is to be noted that this compression and adjusting motion within the slab would ensue only where the conical boundary is curved: thus

as the photographs show it could, at high speeds, be localized near the nose of the bullet and presumably alongside it.

If the interpretation here sketched has substance, it would seem that it is largely as regards the rear end of a bullet or shell that further ballistic improvement can come. The smooth motion in front, at speeds greater than that of sound, involves no dissipation of energy, and does not seem to interact much, at any rate overtly, with the local wake of turbulence set up from the rear end. But it is easy to overlook essential features; experiment alone can control, and the existing results are not now accessible to the writer. To what degree does the rumbling sound, other than the sharp ping whose origin seems clear, proceed solely from the wake? To what extent is the length of the bullet really involved in the form of the cones? Where would the rearward conical boundary strike off if the bullet were pointed at both ends? *Limitation of the wake.* *Queries for answer.*

The problem as here formulated is that of the motion of a sheet of air between two boundaries: but near the apex the boundaries could hardly act independently, as suggested, except when their distance apart is large compared with their radii of curvature and the breadth of the bullet.

[1928. The theory as here developed is too simple for the facts. Both the Boys' cones appear in the photographs as surfaces of advancing compression. In the problem treated above, of motion across a resting slab-form, the density was taken uniform all across it, but greater than that outside on account of the balanced forces of impulse on its two faces, which could be intense enough to maintain a greater density if the velocity is very high. But for the actual motion in air the compression may vary across the slab, from density $\rho_2$ to $\rho_1$. Indeed in any case an internal pressure whose variable part is $-\frac{1}{2}\rho v^2$ could not be equilibrated for the slab as a whole by the terminal impulsive forces, which would provide double the reaction required. The exact equations of internal pressure are therefore demanded, namely *The solution not applicable,* *the compression not being uniform.*

$$p = k\rho, \quad \rho v = M, \quad \int dp/\rho = -\tfrac{1}{2}v^2 + \text{constant},$$ *More exact formulation.*

which yield a formula for the values of $p$ in terms of $\rho$ across the slab; and for equilibration this has to satisfy $p_2 - p_1 = M^2/\rho_2 - M^2/\rho_1$, for every stratum in the slab. But these conditions can both be satisfied only when the density within the slab is uniform: the conclusion would have to be that continuity of smooth motion in the air must be broken at the edges of the slab. In the case of a spherical shell there is no free edge and so no difficulty. But the discrepancy as regards the photographs remains. The motion appears surprisingly complex in the shadow photographs in the Woolwich Report, reproduced from Cranz's *Ballistik*.]

# 102

# INSULAR GRAVITY, OCEANIC ISOSTASY,
# AND GEOLOGICAL TECTONIC.

[*Proc. Camb. Phil. Soc.* Vol. XXIII, pp. 130–135: Feb. 8, 1926.]

*Figure of the Earth:* A CENTURY ago geodetic and gravitational universal surveys were mainly concerned with determining the effective (gravitational) ellipticity of the Earth, after due allowance had been made for local anomalies, with especial view to the exact purposes of physical *astronomical interest.* astronomy. One of the chief of these anomalies was exhibited by a remark of Airy, after scrutiny of the available data in his treatise (1830) on "Figure of the Earth" in the *Encyclopedia Metropolitana*, that the observations show gravity to be abnormally in excess on *Gravity: anomaly at island stations:* island stations. It appeared, for instance, that this cause might make the mass of the Moon uncertain even up to 2 per cent. A very refined explanation of this anomaly of island stations (which will be seen presently to be only partially effective) was offered by Sir George Stokes, from whom this last remark is quoted, in the course of a *can lead to no direct knowledge of internal densities.* memoir[1], fundamental for theoretical geodesy, in which he demonstrated that no outside survey could lead to any certain knowledge of the distribution of mass inside the Earth, even in its outer crust, except as a matter of probability when backed up by geological knowledge.

*Oceanic surfaces fall towards their centres:* It is explained there that the form of the sea-level must be locally depressed over a deep ocean, owing to defect of density; and in consequence on insular stations gravity at sea-level is measured abnormally nearer to the centre of the Earth as a whole, so that from this cause its value is greater than that belonging to the mean spheroidal surface. In fact, the form of the ocean is an equipotential surface, including therein the potential of the centrifugal force of rotation in the familiar manner: but the part of the potential at its surface arising from the local water is abnormally small on account of its low density, and this defect must, in absence of local compensation, be made up by a greater potential of the Earth as a whole,

*Historical.* [1] *Cambridge Transactions* (1849): reprinted in *Math. and Phys. Papers*, Vol. II. Some idea of the great debt owed by the Indian and other gravitational surveys to the continuous amateur advice of Sir G. G. Stokes, spread over half a century of their development, may be gleaned from the collection of his *Scientific Correspondence* (Camb. Univ. Press), Vol. II, pp. 253–325.

which demands depression of the local ocean surface towards the Earth's centre.

The opposite result would arise from excess matter of an adjacent mountain or island peak: that would raise the ocean level in its vicinity and thereby indirectly diminish gravity, measured at sea-level, as determined by levelling operations.

For example, at the centre of a circular oceanic basin of radius $b$ and uniform depth $h$, its defect of potential would be with sufficient accuracy $\gamma \rho' h 2\pi r\, dr/r$, where $\rho'$ is the defect of density of the water below that of the average terrestrial crust; thus it is $2\pi\gamma\rho'bh$, where $\gamma$ is the constant of gravitation given by $\gamma E/a^2 = g$. Here $E = \frac{4}{3}\pi a^3\rho$, $\rho$ being $\frac{11}{2}$, is the mass of the Earth of radius $a$. As the potential of the Earth as a whole is $V = \gamma E/r$, this change of local potential, say $\delta V_0$, would be compensated by change of sea-level $\delta h$, where $\delta V_0/V = -\delta h/r$. Thus in the present case the fall of level relative to depth of ocean is given by the expression

$$-\frac{\delta h}{h} = \frac{a 2\pi\rho'b}{E/a} = \frac{3}{2}\frac{\rho'}{\rho}\frac{b}{a} = \frac{9}{22}\frac{b}{a}.$$

If the radius $b$ of the oceanic basin is 50 miles this fall would be the fraction $\frac{9}{22} \cdot \frac{50}{4000}$ or $\frac{1}{200}$ of its depth; if the radius were larger it would increase in direct proportion until it is a considerable fraction of the Earth's radius. A cup-shaped ocean bed could be similarly treated.

The steady sea-level would thus be depressed by $\frac{1}{10}$ of a mile, owing to local causes, at the centre of a basin of 500 miles radius and 2 miles deep, if it is in free communication with the other oceanic waters: and this approach to the Earth's centre would involve increase of $g$ measured at ocean level, given by $\delta g/g = -2\delta h/a$, or here

$$\delta g = \cdot 05 \text{ cm./sec.}^2,$$

where $g$ is about 981, which is over one-third of the order of magnitude of the observed excesses at island stations.

But this explanation fails because there is a predominant offset. The vertical attraction of the local ocean regarded as an extensive flat slab of water is abnormally small by $2\pi\gamma\rho'h$, where $g = \gamma E/a^2$, that is by $g\rho' 2\pi a^2 h/E$ or $\frac{3}{2}\frac{\rho'}{\rho}\frac{h}{a}g$; thus this direct defect in $g$ may be much the greater, being $a/b$ times the indirect excess. There is however some effect in the other direction due to excess density of the local land, which is usually a substantial correction. This preponder-ance destroys and even reverses the Stokes explanation of the oceanic anomaly. Indeed closer examination shows that, as based by him[1], rather confusedly as it seems, it depends on a potential equation used by Laplace which can, in limited manner, apply only to a locally infinitely thin spherical layer\*. The principle of depressed level became familiar, simple examples being worked out, *ab initio* and so correctly, by way of illustration in Chap. IV of Colonel A. R. Clarke's standard treatise on Geodesy (1880), from the point of view however only of levelling operations, not of gravity.

*Pratt's Himalayan isostatic compensation of g.*

But soon the discussion of the data of the Indian geodetic survey, by Archdeacon Pratt in India, revealed new features[2], by showing strong residual defect of gravity on the Himalayas, such as could only be accounted for by a large defect of density underneath the mountains†. Airy's idea that the mountains might be buoyed up by

[1] *Math. and Phys. Papers*, Vol. II, p. 153. Stokes did not make any correction in this reprint in 1883: but Dr Bowie states (*loc. cit. infra*) that there is no generally accepted explanation other than compensating excess of density beneath the ocean.

*Stokes' amended theorem.*

This analysis of Stokes in fact establishes as a general proposition that the effect of *distant* irregularities of surface mass consists of a direct vertical attraction, say $g''$, together with an indirect part due to change of level, equal to $-4g''$, thus countervailing it four times: this influence, of wide range and presumably actually small, is superposed on the *local* effect here considered.

[The local effect has been expressed by Stokes by a formula of integration in his § 31. Dr Vening Meinesz has recently reported a provisional estimate

*Recent oceanic data analyzed.*

(*Geographical J.* 1928), with unexpected results, of the trend of the deviation of the level surface (geoid) from the Clairaut spheroid, obtained by Stokes' formula on the basis of the recorded distribution of intensity of gravity, including his own very remarkable series of oceanic observations across the Pacific in a Dutch submarine. An improved determination, when more extensive data are available, will be fundamental for a practical decision on the basic problem in this domain, then already explored in theory by Stokes in

*The general problem.*

1849, "whether the observed anomalies in the variation of gravity may be attributed wholly or mainly to the irregular distribution of land and sea at the surface of the Earth, or whether they must be referred to more deeply seated causes" (*Math. and Phys. Papers*, Vol. II, pp. 133, 168).]

*Early potential theory.*

\* These principles were in fact not very clear to the founders of potential theory. A formula of Laplace connecting linearly the potential and the attraction at a point on the surface of a spheroid (nearly spherical body) *of uniform density* was challenged by Lagrange and by Ivory (Pontécoulant, *Système du Monde*, II, p. 380). A note by MacCullagh in 1834, justifying it by his characteristic geometrical method, is in his *Collected Papers*, pp. 349–351.

[2] In 1855–9: cf. A. R. Clarke, *Geodesy*, pp. 96–98.

† After a general survey of geodetic irregularities J. H. Pratt had concluded (*Roy. Soc. Proc.* Vol. XIII, 1863, p. 19) in more drastic manner:

*Pratt's view of isostasy.*

"It would seem as if some general cause were at work to increase the density under the ocean, and diminish the density under mountainous tracts of country. The author conceives that, as the Earth cooled down from a state of fusion sufficiently to allow a permanent crust to be formed, those regions where the crust contracted became basins into which the waters ran, while regions where expansion accompanied solidification became elevated without any consequent increase in the total quantity of matter in a vertical column

extensive roots floating in a denser magma, existing beneath a *thin* crust, could not of course now be maintained, at any rate in that form, in view of the high rigidity of the Earth as a whole. But there was much to be said, on various counts, for a thinner and deeper viscid stratum, lying between the crustal material and the solid core, in which in the tendency towards equilibrium the pressure due to the weight of the crust must in course of ages have become equalized laterally, at any rate partially, and the load upon it thus made uniform to that degree everywhere. It is implied that there are no local abnormalities of density in the core, which is reasonable as the core is probably metallic. This is the hypothesis of isostasy, propounded as a universal principle by Dutton and worked out systematically by Hayford and his colleagues of the American Survey, who found that it gave a fair account of the usually slighter anomalies (mainly of levelling) revealed in that great undertaking[1].

*(marginal notes: Airy's earlier idea of flotation, modified to slow viscous adjustment, over a uniform metallic core. The American extension.)*

Circumspection is, however, suggested in applying these ideas to the anomalies at oceanic stations; for the Stokes explanation already claimed to be an effective *vera causa*, without aid from compensation of density underneath. It happens that the subject is amenable in a general way to simple elucidation: and as the essential circumstances for submarine mountains and landscapes can perhaps be more directly estimated, it seems indeed to provide in some respect a closer test. On an ideal very narrow island peak of negligible mass, in a wide ocean of uniform depth, with adjustment as a whole to general isostasy by denser horizontal strata underneath, there would be but slight resultant abnormality of the local part of the attraction. For the totality of the strata could almost be regarded as an extensive thin flat sheet, while local defect of potential on which change of sea-level depends would be still more closely compensated by the extra mass below[2]. Hence, in contrast to the Stokes uncompensated case above, under isostatic conditions gravity and level ought both to be regular over a wide ocean of nearly uniform depth with strata nearly horizontal underneath.

*(marginal notes: Isostasy over a wide ocean would leave no anomaly, in gravity or in level.)*

Thus in considering gravity data, say over the Pacific Ocean, we may on the isostatic hypothesis consider only this excess density, over that of water, of the local land distributions in an ocean with

*(marginal notes: Source of oceanic anomaly when isostatic)*

extending from the surface down to a given surface of equal pressure in the yet viscous mass below. The author considers that the deviations of latitude at the other principal stations of the measured arcs, if not positively confirmatory of, are at least not opposed to this view."

[1] Cf. the chapter in H. Jeffreys' recent treatise *The Earth*.

[2] In the case illustrated above, with radius of ocean about 500 miles and depth of compensation 100 miles, about 10 per cent. of the anomaly both of attraction and of potential would remain after compensation of the ocean.

an ideal conveniently assigned flat bottom: and any gravity anomaly must be due to these excesses alone, together with isostatic compensations of opposite amount underneath to maintain the average total load for the crustal strata. The disturbance of ocean level due to them would be slight if they are merely local peaks. If the station is on a straight ledge even of a narrow island it will be considerable[1].

This note, with its limitations as regards scope that are imposed by imperfect information, has been prompted by an important recent Report on "Isostasy in the Southern Pacific" by Dr W. Bowie[2] in which, from systematic calculation for the five insular stations that were examined, local isostasy is found to account for about three-fourths of the observed local excesses in $g$, which are there of the order of over 0·1 in 981. Since the Stokes effect of change of ocean level is found inadequate, the greater part of the excess of gravity ought to be due solely to direct attraction of the local excess of land density measured from an averaged flat ocean floor with a cancelling compensation underneath, if oceanic compensation is to hold good. A set of gravity determinations over the wide ocean, combined with soundings, would thus throw interesting light*.

*Actual oceanic over-correction.*

For a land survey, as in the Himalayas, heights determined by levelling operations are reckoned from an ideal ocean level which would be affected in the same way as the actual ocean level near island masses, that is only by local excess masses at the surface above a mean compensated distribution when the local compensations extend very deep.

*Influence on a land survey.*

There is a statement near the end of Dr Bowie's Report that, in the cases examined, if the densities of the island "pedestals" were 10 or 15 per cent. (say 12) above the normal value $2\frac{1}{2}$, about one-eighth of the remaining non-isostatic excess of gravity would be accounted for. From this we may perhaps infer that the excess density over that of water, namely $1\frac{1}{2}$, along with a cancelling diminution below on account of the now complete compensation, would account for $\frac{100}{12} \cdot \frac{3}{5} \cdot \frac{1}{8}$, or say $\frac{3}{8}$ of the whole local excess of gravity. This fraction is not discordant with the estimates (around

[1] Cf. A. R. Clarke, *Geodesy*, p. 93; or Thomson and Tait, *Nat. Phil.* (1867); also *infra* for conical forms—and the remark added at the end.

[2] *Proc. Washington Academy of Sciences* (Dec. 4, 1925).

* This approach of the oceanic geoid to the Earth's centre will still act when isostasy has become complete. It appears from V.-Meinesz's gravity observations in submarines that the oceanic defect of density is usually somewhat over-compensated, and this lowering of the geoid has been invoked to account for the excess. See W. Bowie, Presidential Address, in *Proc. Washington Acad.* Mar. 1927.

$\frac{3}{5}$) in the Report, and is in so far confirmation. As the Stokes effect of change of ocean level is now itself obliterated by compensation, this over-compensation by about one-third could be accounted for, as Dr Bowie suggests, by an increase of mean excess density of crustal strata which need only be from $1\frac{1}{2}$ to about 2.  A possible<br>cause of<br>the over-<br>compensation.

In illustration, for a conical island of height $h$ and angle $2\alpha$ the potential at its vertex is $\pi\gamma\rho'h^2(\sec\alpha-1)$ and the component attraction along its axis is $\pi\gamma\rho'h(1-\cos\alpha)$. For a wedge-sector of the cone of angle $\beta$ with its edge along the axis, these are merely affected by a factor $\beta/2\pi$. The extra gravity at the summit of such a sharp promontory is due mainly to its direct attraction, and so differs from that for a flat plate by the factor $\beta/2\pi\,.\,(1-\cos\alpha)$. Due to this uncompensated plate of ocean by itself of depth $h$ and defect of density $\rho'(=\frac{3}{5})$ the direct defect of gravity is (as *supra*) $\frac{3}{5}\dfrac{\rho'}{\rho}\dfrac{h}{a}g$,  Correction<br>for a conical<br>submarine<br>peak:

which is 0·1 in 981 per mile of depth. A conical peak of density $\rho'$ and semi-angle 30° thus would produce an excess half this, diminished however by the compensation below, which is of the order of the fraction $\frac{1}{3}(1+\cos\alpha)\,h/H$ of it, so negligible if $H$, the depth of compensation, is 100 miles. In fact it is only the compensation for sideway excess mass, as it attracts more vertically, that is of sensible direct influence on $g$. Its sideway attraction deflects the plumb line, and so contributes to the oceanic rise of level, but in a way which must be already included in the potential effect, as *supra*. In a sector of a cylinder also, or for a sloping ocean floor, results may readily be calculated in general illustration. It is, of course, the transverse component of local attraction that affects levelling, while pendulums are affected by its vertical component. Theoretically the two are interconnected through the potential, as above indicated in general terms, in a manner illustrated by precise formulae for the case of a set of parallel mountain chains first (1867) in Thomson and Tait's *Natural Philosophy*, or more readily in simpler cases.  for a<br>cylinder.<br><br>Contrast<br>with<br>levelling.<br><br>Simpler<br>problem of<br>parallel<br>ridges.

Generally, the conclusion is that the depression of sea-level at island stations is not adequate by itself to counteract the direct diminution in gravity due to the low density of the attracting water, much less to produce the observed excess. But if the oceanic defect of density is deeply compensated underneath, down to a uniform standard depth of ocean, any local anomaly of gravity should be due simply to the excess mass over that of water up to an ideal level surface of the local land standing in this ocean.  Summary.

An illustration involving the potential is more recently reported[1]. It appears that astronomical observation and direct measurement

[1] *Proc. Washington Acad.* (Jan. 16).

An abrupt change: show a difference of as much as 1 mile in the distance between two stations on the north and south coasts of Porto Rico about 35 miles apart. Thus over the island the radius of a level surface is less than the Earth's radius $R$ by 1 part in 35. Thus the expression for the potential locally is

$$V = - g \left(z - x^2/2 . \tfrac{34}{35}R + pxz + qz^2 + ...\right),$$

Its implications. where $p$ and $q$ are to be determined from observed local gradient of $g$, as to which levelling alone does not give information. Discussions of this kind go back to Bouguer for the Andes, and Cavendish and Maskelyne for Schehallien[1].

## Oceanic Isostasy in Relation to Geological Tectonic.

### [*Nature*, Sept. 11, 1926.]

THE distribution of gravity over an oceanic surface, beneath which local compensations of terrestrial density are taken to be complete, may thus be envisaged, perhaps most simply, by drawing a widely extended arbitrary horizontal boundary beneath the water, and marking out all above it up to the level surface as ocean separately compensated beneath, the law of depth of the compensation being for that hypothetical layer of the density of water unimportant. There Law of will then remain the effect of the surplus of density, over the oceanic sub-oceanic anomaly in g: water, of the solid parts situated above this arbitrary flat boundary; and it is from this reduced submarine mountain landscape alone, together with emergent peaks with density undiminished, and the nature of its compensation, that the amount of the actual local excess of gravity is to be estimated on the hypothesis of isostasy, the cir- simpler than cumstances thus being analogous to those of a range like the Hima- Himalayan. layas, but modified, as all the observations now belong to the same level near the tops of the submarine mountains instead of the bases. The nature of the compensation, in the deep-seated material, of this effective local excess load, would thus permit of being judged by itself; in particular, for steep submarine island peaks it is almost negligible, whatever varying distribution in depth be assigned to it, provided only it extends deep down, say towards the order of $10^2$ kilometres.

The long-recognized excess of gravity at island stations was thus really evidence quite as forcible, and also as direct, as the subsequent Early records of Himalayan surveys, indicating that the defect of density oceanic of the masses of water is actually compensated, even over wide evidence of uniform oceans, at any rate to a very considerable degree, by excess isostasy:

[1] Cf. Cavendish's *Scientific Papers*, Vol. II, pp. 402–407.

of density below[1]. The systematic discussion of the level and gravity surveys of America, primarily by Hayford, has enlarged and forced into prominence the same very striking and surely fundamental type of conclusion, as extended even to the usually smaller and less abrupt anomalies there revealed. *[margin: enlarged by Hayford.]*

The evidence, then, is on all sides remarkably strong, that with increase of depth the terrestrial material gradually becomes softer, so to say, possibly owing mainly to rise of temperature, down to a limit which perhaps at an outside estimate may approach $10^2$ kilometres: that below some such depth the mass of the Earth presents again a perfectly solid, though doubtless elastically deformable, foundation on which the softer strata directly above have flowed gradually in the course of ages towards an equilibrium nearly hydrostatic, depending in detail, however, on the distribution and range in depth of the softness, in a way that is scarcely much amenable to scrutiny. To effect such adaptation, the displacement of deep-seated material need be only over slight distances, unless the yielding layer is thin. An unyielding foundation underneath is essential to any approach to local isostasy; the Earth as a whole must be solid, as it is known to be for dynamical reasons. As regards the relatively shallow upper terrestrial layer which thus becomes viscous with depth, in a way not necessarily uniform nor to the same depth everywhere, the question of rupture or damping of transmission of internal earthquake tremors in crossing these softer layers arises, and is probably ripe for discussion; such a stratum may of course be even completely yielding for slow secular stress while thoroughly elastic for the rapid alternations in seismic oscillations. It is to be remarked, however, that as a result of theory superficial travelling waves, at any rate on uniform elastic material, could scarcely arise from other than a superficial cataclysm, secondary it may be, so that purely superficial seismic undulations would have to come from sources located within their own quite small range of depth. But the velocity would change (dispersively) with wave-length, and this conclusion may be modified, as Professor Love pointed out, if the elastic quality or density, instead of being uniform, changes notably within the depth of a wave-length. *[margin: Physical conclusion: secular flow, over small distances, on an unyielding bed. Relation to seismic tremors. Surface waves must be due to a shallow source: unless they are dispersive.]*

Why distinct settlement of the strata towards isostasy such as is thus variously confirmed should be necessary at all, affords direct scope for fundamental tectonic speculation, of an interest quite apart from geological detail. Is this abnormally small density beneath mountain ranges due to higher temperature or to lighter material? How could such locally varying temperatures have become estab- *[margin: Unsolved problems of terrestrial disturbance.]*

[1] For recent special estimates see a note by W. Bowie, *Proc. Washington Acad.* Dec. 1925.

lished over a consolidating Earth? If the height of the mountains is determined largely by the defect of density beneath, they must to that degree have been pushed up hydrostatically from below rather than elevated by lateral stresses; yet folding of the mountain strata is conspicuous. Subsidence towards isostasy might perhaps induce folding to some degree. If the depression of the Pacific Ocean is thus determined in the main hydrostatically, is there not less room for the cosmic theory that it may represent the cavity from which the Moon was originally shed away?

The literature.

*Postscript.* One observes that these and cognate questions, insistent and fascinating, form the subject-matter of Professor Joly's recent path-breaking book, *The Surface History of the Earth,* which invokes steady evolution of heat by radioactivity of the rocks, interacting with isostatic influences, as the cause of periodic outbursts of surface activity which have fashioned the existing features. There are to be compared the views developed in H. Jeffreys' recent comprehensive treatise, *The Earth.* For a condensed account over an extensive range cf. "A Symposium on Earthquakes," by F. A. Tondorf, N. M. Heck, W. Bowie, A. L. Day in *Journal Washington Academy,* May 4, 1926, pp. 233–254 (also more recently G. R. Putnam). In a less special way, such questions have been prominent since the treatise of E. Suess on the Earth's surface features. There is also the problem of the time scale of development, projecting into vast aeons of the past, yet with clues arising mainly from the fossil traces of the succession of forms of life.

A few special remarks may be significant here.

Time scale.

It appears that the lag in compensation of accumulating great depths of sediment is but small, compared at any rate with the time of accumulation, for the compensation is always well advanced.

Isostatic continents not subject to tidal pull;

Tidal pulls on these adjustable surface sheets would on Newtonian principles be differential, and so extremely slight. Thus even the extreme case of an elastic Earth surrounded by an ocean of molten lava of the order of $10^2$ kilometres in depth, in which continents would be analogous to ice-sheets and mountains to icebergs, is not unthinkable dynamically, however it be thermally; though the existence of actual oceanic tides would demand a rigid and deep crustal layer.

in any case no differential drift.

But even if the lagging tidal pull were large enough, it could only cause a westward drift of the fluid surface material around the Earth as a whole, carrying continents and mountain ranges floating thereon. For the principle of Archimedes asserts itself; as regards the uniform field of force the floating mass can be replaced by the magma which it displaces, up to the level surface; thus it is the same as if the tidal forces acted on a uniform sheet of magma without surface excrescences and no differential drift could arise—except in so far as a uniform drift may be obstructed or deflected locally by the more solid roots of the floating continents that are carried along with it.

The earliest table-lands, of primitive rock, must have been pushed or floated up, and to great heights; it would appear from the literature that their subsequent denudation by aereal influences accumulated stratified deposits along the coasts of the oceanic hollows, which gradually sank into the magma by their own extra weights, perhaps most in the middle so as to curl over by the lateral pressure,—themselves sinking down while the adjacent denuded high land is floated up, until by accumulation combined with sinking, and helped by effusions from below, they attained to considerable slopes and great thicknesses, even five miles or more, that then somehow they were pushed up again bodily, yielding after repetitions of such processes folded mountain ranges of stratified rock such as geologists know, the primitive elevations having passed largely out of sight. At any rate nothing more plausible seems to have been hitherto thought of.

Main headings of physiographic evolution.

# 103

# THE LAW OF INERTIA FOR RADIATING MASSES.

[From *Nature*, Feb. 27, 1926.]

RADIATION of amount $\delta E$, lost by a body, diminishes its effective mass, on Maxwellian principles, by $\delta E/c^2$: and some perplexity is apparent in recent discussion as to how this tells on the dynamical inertia of a planetary mass. According to the Newtonian doctrine, it is the rate of change of momentum $mv$ which must be equated to the extraneous gravitational or electric force acting on the body: and this principle was taken over by Einstein, in extended fourfold form and in concert with relativity, as the key to his brilliant tentative explorations towards a closer view of gravitation. But

$$\frac{d}{dt}(mv) = m\frac{dv}{dt} + v\frac{dm}{dt};$$

so that if the mass is diminishing by radiation, conservation of momentum seems to demand acceleration of velocity of a body isolated and so free from external force. Yet the doctrine of relativity asserts that no standard can exist on which to measure such change of velocity, whether of translation or rotation: such a conclusion therefore would contradict relativity. On this ground it is claimed that the applied force must be equated, following Dr Jeans, to $m\dfrac{dv}{dt}$

**Can an isolated body be said to have velocity?**

and not to $\dfrac{d}{dt}(mv)$. Yet the latter form is based directly on a very keystone of relativity. One way out of the apparent paradox would be to postulate a frame in the aether with reference to which the velocity of an isolated body could be measured: this would institute an exception to accepted doctrine widely verified. Professor E. W. Brown (*Proc. U.S. National Academy*, 1926, p. 2) appears to be troubled, and naturally so, by uncertainty as to which formula to adopt with a view to studies on cumulative long-range effects of radiation in dynamical astronomy.

**Becomes a practical question.**

It seems well worth while, in this connection, to direct attention to a classical memoir by the late Professor Poynting (*Phil. Trans.* 1903) on "Radiation in the Solar System" and to his other investigations, theoretical and experimental, on pressure of radiation, in which astronomical effects are considered. In particular, he revealed (by indirect argument) what amounts to the Bradley aberration effect in

**Poynting's investigations.**

pressure of extraneous radiation on a moving body: and he showed that it produces a retarding force* that would, in times quite short geologically, suck all small bodies such as cosmical dust revolving steadily round the Sun, into that luminary. This is doubtless the explanation, as he remarked, why the celestial spaces are so transparent. Incidentally it may perhaps require that the cosmic dust that reflects the zodiacal light should have some source of replenishment. *[margin: Braking effect of the Bradley aberration of radiation pressure.]*

*[margin: Maintenance of zodiacal light.]*

This, however, is not directly connected with the inertia question. But in the reprint of Poynting's *Collected Papers* (Cambridge University Press, 1920), the question of the effect of diminution of inertia by radiation had to be gone into, in the notes and corrections then appended to the work, as on the face of things it may be quite comparable in importance (for large masses) to the deflection of the radiation pressure by aberration. The explanation proceeds on the principles of Maxwell's great *Treatise* of 1873, without any reference to relativity. It is in effect verified [as *supra*, pp. 444, 448] that the momentum of the whole system, matter *plus* radiation, is conserved, as it ought to be, in the absence of any extraneous force; but a part

*[margin: Conservation of momentum as affected by aberration,]*

* Each ray or filament of radiation may have a long free path in a gas before it is all scattered. The ray carries momentum with it: but in equilibrium of temperature the interchanges of momentum are balanced. The balance is however put out, when the absorbing end of the ray is in motion relative to the emitting end, by the aberrational effect, which is one of relative motion of source and recipient. This unbalanced transfer over long range is effective in mixing up and obliterating the momenta of internal motions in the gas. The analytic methods of Maxwellian gas theory are applicable: and the orders of magnitude of Sir J. H. Jeans' important systematic discussion somewhat later (*M.N.R. Astron. Soc.* May 1926) seem to involve, on the hypotheses adopted, that this radio-viscosity, so to say, may be large enough to suppress any original internal motions in a gaseous Sun, in a time that is astronomically not large—unless they are renewed by irregular thermal or dynamical causes. Once they are gone, the aberration of the resultant stream of radiation travelling outward through the star (which is enormously small compared with the intensity of the flow to and fro of the internal radiation) relative to the rotating layers of atoms on which it is incident, in the play of absorption and emission, would remain to produce some differential backward drift of the outer layers round the Solar axis. If outflow of radiation were prevented by an impervious outer boundary, the final internal state of the star must be rotation as a rigid body with conserved momentum: and if its internal temperatures ($T$) range to millions of degrees the internal natural radiation varying as $T^4$ would appear only to be too rapidly effective as a frictional agent, unless the free path of the radiation is postulated to be correspondingly small. The latter may be the necessary condition for its retention up to such enormous density. This internal natural radiation would be of very high frequency, its maximum varying as temperature by Wien's law: experience of X-rays combined with their recent *quantum* theory gives countenance to such extreme supporting opacity for rays that are soft or of relatively long period, in contrast with the mysterious intensely penetrating cosmical rays: cf. Eddington's discussion in his treatise.

*[margin: Damping of convection currents in stars by internal radiation.]*

*[margin: Retention of very dense internal radiation.]*

part being
sent away in
the radiation.

of its gradient equal to $v \dfrac{dm}{dt}$ where $\delta m = -\,\delta E/c^2$, is momentum carried away by the radiation $\delta E$ issuing from the system, by the mechanism of radiation pressure, while it is the compensating remainder $m \dfrac{dv}{dt}$ that is to be equated to the extraneous force, which acts on the material system itself, not on the radiation that has escaped from it.

This seems to be the adequate practical settlement of this question, and may, it is hoped, encourage Professor Brown to proceed with closer astronomical investigations in improvement of Poynting's pioneer indications: though how far it consorts with an extreme

Cosmogony
of double
stars.

relativist position is a different question. But Poynting's retardation by the aberration influence on pressure of radiation will also have to be taken into account in the cosmical problem of the evolution of the orbits of a double star*, as perhaps of the same degree of importance as change of effective mass by loss of radiant energy.

Clearance
of planetary
spaces.

In Poynting's own special problem there is no change of mass (other than that involved in high velocity in accordance with Least Action), as the loss by radiation is made good by absorption from the Sun.

* In a letter to *Nature* (Mar. 13, 1926) in reply, Jeans reports that he finds the aberrational influence of their mutual radiations on the orbits of actual (presumably long-period) binary stars to be negligible, in comparison with that of extinction of mass and also absolutely.

# 104

# WHAT DETERMINES THE RESISTANCE AND THE TILT OF AN AEROPLANE?

[*Proc. Camb. Phil. Soc.* Vol. XXIII, Pt. VI, pp. 617–630: Feb. 1, 1927.]

IT is recognized that the main hope for advance in a rational theory of aeroplane propulsion lies in extensive use of the general principle of conservation of resultant momentum, even more than on considerations of energy; for energy can be dissipated by friction, while the relation of momentum to force is absolute. *Momentum contrasted with energy:*

For a field of vortices in unlimited fluid, without solid boundaries, remarkable general kinematical expressions for resultant hydrodynamic linear momentum $(P, Q, R)$ and angular momentum $(L, M, N)$ with respect to a selected origin, and for the kinetic energy $(T)$, the former pair involving the distribution of vorticity $(\xi, \eta, \zeta)$ alone, were given long ago by J. J. Thomson in an Adams Prize Essay on *Vortex Motion* (1883, pp. 5–8). They appear, at any rate on first view, to be precisely adaptable to application of the kind above indicated: and they have the advantage moreover of being in three dimensions, dispensing to some degree with the usual limitation of theoretical discussions to plane motions. *expressible in terms of vorticity alone,*

The formulae will now be quoted[1]. They are of types, $\delta\tau$ being an element of volume, and there being either no boundaries or else the suitable vortex sheets spread over the surfaces of the immersed solids,

$$P = \rho \int (y\zeta - z\eta)\, d\tau, \qquad L = -\rho \int (y^2 + z^2)\, \xi\, d\tau,$$

$$T = 2\rho \int \{u\,(y\zeta - z\eta) + v\,(z\xi - x\zeta) + w\,(x\eta - y\xi)\}\, d\tau.$$

It is easier to verify these results than to discover them: in so doing we shall extend them so as to apply to a part of the field of motion instead of the whole. Thus, on substituting[2]

$$(\xi, \eta, \zeta) = \tfrac{1}{2}\left(\frac{\partial w}{\partial y} - \frac{\partial v}{\partial z},\ \frac{\partial u}{\partial z} - \frac{\partial w}{\partial x},\ \frac{\partial v}{\partial x} - \frac{\partial u}{\partial y}\right)$$

and integrating by parts, with some easy reduction, we obtain for

---

[1] The sign of $(L, M, N)$ has been reversed.

[2] The factor $\tfrac{1}{2}$ in this definition identifies vorticity with differential rotation. The formulae imply that the velocity falls off towards infinity owing to all the vortical sources being at finite distance.

these integrals taken over any bounded region of the fluid, itself of any viscous quality, regular or not if only it be of constant uniform density, the values

$$P = \tfrac{1}{2}\rho \int (- yum - zun + yvl + zwl)\, dS + \int \rho u\, d\tau,$$

$$L = - \tfrac{1}{2}\rho \int (wm - vn)\,(y^2 + z^2)\, dS + \int \rho\,(yw - zv)\, d\tau,$$

$$T = \tfrac{1}{2}\rho \int 2\,\{(lu + mv + nw)\,(xu + yv + zw)$$
$$- (lx + my + nz)\,(u^2 + v^2 + w^2)\}\, dS + \int \tfrac{1}{2}\rho\,(u^2 + v^2 + w^2)\, d\tau.$$

The second terms in these values of the integrals $P$, $L$, $T$ are, as anticipated, the actual distribution of the momenta and the energy throughout the region, the first terms being surface integrals over its boundary, which is supposed simply connected unless the motion is acyclic.

For any region of the fluid the content as regards momentum components and energy, which will be denoted for it by accented letters, are thus given in terms of the volume integrals $P$, $L$, $T$, expressed above, and of boundary integrals as transferred across from the other sides of these latter equations. The boundary integrals do not contribute practically towards the forces, as they merely fluctuate *in steady motion.* Indeed if the boundaries are fixed, the surface

*modified into* integral in the rotational momentum round the origin, in the present
*vector form.* form, vanishes. The essential vector character of local momentum may be brought out by a slightly modified formulation of the result. We have, for any region, now in explicit invariant form, $r^2$ being $x^2 + y^2 + z^2$, $U$ denoting velocity, $U_n$ and $U_r$ its components transverse to the boundary and along $r$, and the bracket $|\,...\,|$ representing a scalar product, the momentum and energy of type $P'$, $L'$, $T'$ expressed by

$$P' = \rho \int (y\zeta - z\eta)\, d\tau + \tfrac{1}{2}\rho \int \{u \mid r\, dS \mid - \mid U r \mid l\, dS\},$$

$$L' = - \rho \int (x^2 + y^2 + z^2)\,\xi\, d\tau + \tfrac{1}{2}\rho \int r^2\,(wm - vn)\, dS,$$

$$T' = - \rho \int \begin{vmatrix} x & y & z \\ u & v & w \\ \xi & \eta & \zeta \end{vmatrix} d\tau - \rho \int U_n U_r r\, dS + \rho \int U^2 \mid r\, dS \mid,$$

*Interpreta-* where vorticity is defined as velocity of spin, as above. It thus appears
*tion:* that the resultant translatory momentum is the same as would be derived from endowing each element of fluid with an ideal translatory

momentum equal per unit mass to the moment of its vorticity round the origin; this part for the whole field of motion involves no resultant torque round the origin, but it is supplemented by a moment of momentum with ideal contribution from each element equal per unit mass to its vorticity multiplied by the square of its distance. The significant feature is that only the vortical regions of the field of fluid motion, the current whirls, are involved in these expressions*. The kinetic energy would be derived similarly by taking from each vortical part of the motion an ideal contribution equal per unit mass to the determinantal product of the distance, the velocity, and the vorticity. *[confined to the vortical regions.]*

For instance, a straight endless horizontal vortex of strength $\varpi$ would have a lifting force per unit length $\rho\varpi v$, where $v$ is its horizontal velocity, which is transferred to distance $2v$ from the origin by the accompanying torque, a statement which acquires sense when there are compensating return vortices. For an aerofoil, as will appear, the completion is by a return sheet of vortices above. *[Lifting force of straight vortices.]*

These formulae are moreover of universal application, being merely kinematical transformations of the actual distributions of momentum and energy[1]. The governing dynamical feature that makes them useful now enters in the circumstance that, outside an enclosing boundary far enough away from the local irregular motions near an aeroplane, the movement of the air may be taken to be on the whole differentially irrotational, any rotational character, such as alone enters into the integrals expressing the momentum, being restricted in the main to the substance of vortex rings associated with the wake that it leaves behind[2]. It may be noted that the momenta provide an ideal means *[The formulae valid for any kind of medium if of uniform density.]* *[Momenta always additive.]*

* They can apply to fluid disturbed by the motion of foreign bodies through it, then becoming surface integrals as the vorticities are those of the surface vortex sheets. When the influence of wakes can be neglected, the interaction between smooth-shaped bodies propelled through fluid, at distances apart large compared with their dimensions, depends on the Kelvin coefficients of effective inertia, twenty-one in all for each body unless reduced by symmetry (cf. Kirchhoff's exhaustive memoir of 1869, § 5), which express the reaction of the near fluid, and the components of its vector hydrodynamic moment which express the influence of its translatory motion on distant bodies, that of its rotation being of smaller order. In terms of these essential constants of forms of the bodies an Action function is to be constructed. Cf. the problem, virtually Kelvin's, *Thomson and Tait* (1867), §§ 332–336, of the interaction of the ground with a landing airship. Contrast also the problem of the mutual influence of flexible vortex rings at a distance, treated (1883) by J. J. Thomson. *[Inertia contrasted with hydrodynamic moment of a solid.]*

[1] Applied to a solid of density $\rho_0$ moving in water of density $\rho$, the integrations would naturally be taken over all space, at the constant density $\rho$, leaving over a separate momentum of the solid at reduced (Archimedean) density $\rho_0 - \rho$. [Cf. a cognate case, *iufra*, p. 693.]

([2] or carries with it in renewed circulation, as *infra* at end.)

of descriptive dynamical exploration, as the contributions from different sources are strictly superposable. The confused momentum inside the interface, with which fortunately we are not directly concerned, as it is steady and not left behind, would have to include terms involving the slip-surface or vortex sheet over the boundaries of the solids.

Slip-surfaces:

The genesis of these vortices shed off from the moving body, as traced to the development of unstable surfaces of slip in the fluid, was explored by Helmholtz himself, the great discoverer of the laws of vortex motion, was then developed by Kirchhoff, Kelvin, Rayleigh, and has been improved into a general analytic method mainly in two dimensions by J. H. Michell, Love, Lamb and others. The relation of this slip-theory, so far as it may be accepted, to the phenomena of resistance and support is intimate and fundamental, especially when it is recognized as here that the interchanges of momentum in the field of fluid motion may be associated wholly with the trail of vortex whirls that are shed away from the propelled machine (or carried along with it).

dynamical application.

Let us suppose that a body like an aerofoil, with a blunt front and rather sharply curved trailing edge, is travelling through air, or, more conveniently for the present argument, is anchored as in the diagram in a stream of air. The problem set by Helmholtz was to determine a mode of flow, of type suggested by observation as *infra*, involving a sheet of slip $(v_1, v_2)$ as indicated in the diagram, such as would ease off the infinities at sharp edges that occur in the continuous theoretical solutions for motion of frictionless fluid*. Such a sheet of slip [or more than one] does in actual flow arise, but is soon broken up as it has obviously no stability: it is constantly being renewed close to the edge, but as the fluid moves away carrying it along, with blurring by diffusion, increasing corrugation of the unstable form breaks the sheet of slip into isolated fragments; these, having thus acquired the status of very flat independent vortices, soon shrink each of them into the stable circular form of cross-section. This course of events is presumably what is meant by the phrase used by Helmholtz

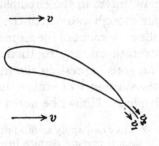

Their mode of curling up,

through corrugation.

\* It has been shown more recently, as *infra*, that this slip-surface may be evaded by modifying the motion by an imposed vortical circulation round the body.

The curves of circulation, when there is only one slip-surface and so no dead water, would presumably be a set of ovals above and another below, with a dividing line meeting the nose of the body.

and his successors, that a surface of slip "curls up" into a row of vortices[1].

It was the practical study of vortices, recognized obviously, in advance of theory, as for some dynamical reason a sub-permanent feature in actual fluid motion, that led Helmholtz into one of the most brilliant chapters in modern theoretical physics. His original problem was the slip-sheets of the air jet, ultimately breaking into vortices, formed at the lip of a blown organ-pipe, which by their rhythmic succession establish and also maintain regular air vibrations with periods that are imposed within the pipe itself: thus providing, outside the acoustic application, an arresting natural dynamical model for general physical rhythm, even possibly quantified, whose potentialities are still far from being exhausted. Helmholtz traced the initiation of vortices in fluid to actual surfaces of slip. But such sharp slip, if it really occurs, being a clean tear in the fluid, stands itself even more in need of elucidation. Lord Kelvin at one time ascribed abrupt slip to an initial cavitation[2] where the fluid flows round a sharp bend with resulting high local speed and thus greatly diminished pressure: such a cavity, drawn out presumably along the flow and continually renewed, would on ultimately closing down provide an adequate cause, initially at any rate, for a sheet of perfect slip. Long after that time, the occurrence of cavitation, leading perhaps by its degradation through such slip-sheets to local whirl of the fluid as a whole along with the screw, imposed conspicuous speed-limits on some of Sir Charles Parsons' practical problems of propulsion by means of turbines.

Circuits made up of the same particles of fluid remain ordinarily [by the laws of continuity] complete: no vortex filaments can cut across them, and the circulations in them remain unchanged whether they traverse vortices or not. The cavitation is what opens a breach in

*Marginal notes:*
- Helmholtz:
- his rhythmic vortical jets.
- Cavitation as cause of vortices:
- its technical importance:
- its mode of operation, by breach of circuits.

[1] More generally, familiar observation appears to indicate that laminar motion, in a layer that would be unstable by the Rayleigh rule, would be stabilized by inserting in it a row of vortices: cf. a theoretical note by Kelvin, *Scientific Papers*, Vol. IV, p. 186. *(margin: Stabilizing vortices.)*

[2] *Scientific Papers*, Vol. IV, Hydrodynamics; a playful epistolary controversy between Stokes and Kelvin on the possibility of slip of the Helmholtz type is there abstracted. The force of the explanation by cavitation is substantially reduced for the case of gases such as air. It is also asserted from observation that vortices break right off from the back of the edge. *(margin: Practical considerations.)*

Incipient cavitations, continually collapsing against the wing itself, might very well maintain an irrotational circulation round it. The loads actually sustained are astonishing. In any case the momentum criterion involves that a free wing has to adjust its tilt so that the double vortex sheet left behind it is at its source of small breadth and so of slight account.

Eddy resistance, in relation to fineness of lines in the rear, was considered by Stokes as early as 1843: cf. *Collected Papers*, Vol. I, pp. 53, 99, 311. *(margin: Stokes on eddies.)*

such circuits, permitting slip which becomes vortex motion to pass across, thus introducing abrupt change into their circulations.

This local discontinuity from velocity $v_1$ to $v_2$ in crossing the slip-surface, which after travelling a short distance thus breaks up into whirls, must [may conceivably] establish circulation in the smooth irrotational flow of the fluid as a whole around the front parts of the initiating body. For the velocity of relative slip is a discontinuity in the value of the tangential gradient of the velocity potential of smooth flow, arising in crossing the sheet: therefore, owing to accumulation of gradient along the sheet, that potential is itself finitely discontinuous in crossing the slip: and the line integral of its gradient in a path around the body, that is, the circulation, as Kelvin named it, instead of being null, as it would be in a complete circuit if the potential were acyclic and so came back to its initial value, represents a cyclic flow determined mainly, as a problem of irrotational motion of a different and discontinuous type as set by Helmholtz, by the general shape of the body, except in the parts close to the slip-surface, where the circulation could be imagined as adjusted for adjacent paths by superposed local disturbance not much affecting the motion at a distance.

The recognition and analysis of this seconday circulation round the body has, as appears, been utilized by Lanchester\*, Prandtl and others, as a practical mode of exploring the character of the supporting pressure on aerofoils, this being greatest underneath, where in the relative steady problem velocity of the fluid as held back by the circulation is least. It presents itself as additional to direct exploration in detail of the effect of the vortex wake: the wake-effect being itself transmitted as in solutions of the Helmholtz slip-problem into a definite modification of the frontal motion with consequent change of pressure along the wing, which may be smooth enough over the greater part of the surface for treatment by standard hydrodynamic methods.

Support by circulation.

In the analytical solutions of the slip-problem, it appears that abruptness of change in the motion near the sharp edge might still remain excessive, notwithstanding easing by the sheet of slip; if that were so it would even prevent smooth formation of any such sheet. The conditions determining the slip-surface have therefore recently, as it appears, been restricted further by Joukowski, by the postulate that near the edge the velocity is to keep within practicable bounds suitable for unbroken flow, or in the abstract mathematical formulation is to remain finite. And this mode of fixing the value of the secondary circulation round the body seems to have had practical application.

Circulation can avoid slip-surfaces.

But from our present point of view the formula for $(L', M', N')$,

* The pioneer in these practical elucidations of fluid support.

which determines the rotational momentum coming within a region, appears to intervene. Referring to the centre of mass of the aeroplane as origin of coordinates, each vortex in crossing our arbitrary boundary separating the turbulent motion near the aeroplane from the more steady flow outside, carries away momentum from the system inside; and the formula shows that the moment of such momentum continually increases as the vortex recedes from the origin by a factor $r^2$, unless a proper balance of vorticity positive and negative is established. Such continued abstraction of moment of momentum, from the inner turbulent system around the aeroplane, would have its reaction on it in the form of a torque increasing as $r$ and tending to rotate the aerofoil. But all the vortices formed along the same slip-trail have the same direction of spin, that of the vortex sheet from which they are developed: they would thus provide torques all conspiring together, so that the aeroplane, if free, would at once turn round by reaction and adjust its presentation to the stream so as to get rid of their torque—or more precisely, so as to obtain equilibrium between the pull of the screw, the weight of the aeroplane and the lift-force and the drag of the wake. *But a double vortex sheet must be thrown off,*

The conclusion appears to be involved, that an aerofoil free to alter its tilt and speed cannot originate permanently a single trail of vortices, for that would produce mainly a torque to turn it round. It must produce a double trail, one sheet starting, as actually appears, from near the edge, the other from the back surface of the foil in its neighbourhood, "dead water" being more or less fully developed between these trails with their two sheets of vortices thus spinning in opposite directions. As the speed of the aeroplane is altered, the pull of this wake would thus have to adjust itself by change of tilt into a single force passing through the intersection of the pull of the screw and the line of weight of the machine, which might then be taken for origin in the formulae. *adjusted to absence of torque.*

The condition that determines the tilt is that the two sheets of vortices should cancel each other as regards moment of momentum about this point of intersection; which requires that the aggregates of their strengths should be everywhere nearly equal and opposed in order to annul $L$. This consideration of $L$ null would be the criterion fixing the value of the secondary circulation of the Lanchester-Prandtl discussions: the slips on the sheets would not be exactly equal and opposite, else it would vanish[1]. It would appear, on the present

---

[1] Recent experiments by Bénard, reported in *Comptes Rendus*, seem to show that regular spacing on the two sheets, which simplifies the Kármán mathematical theory by making the motion steady, is not essential for stability of the vortex trail.

view, in case the opposite vortices in the two sheets were not far from being equal and opposite, that at first at any rate they should be about equidistant in pairs from the centre of mass of the aeroplane, and therefore situated some way above it. If the conjugate straight vortices are thus imagined as bracketed in pairs, in the order of their origination, each pair if of exactly opposite strengths would be analogous to one elongated vortex ring: indeed the vorticity must *Paired into rings.* actually curl round at the ends like a flattened ring. The momentum of a ring is transverse to its plane and acts at its centre of form: as being above the mass centre of the machine it must here be downwards in order that the lifting reaction may be upwards. This is consistent with two sheets of vortices alternately spaced, each edge vortex below, with its determined direction of spin conditioned by the edge trending downward, being regarded as associated with a next following shoulder vortex. Vortices of opposite spin tend to attract *Momentum only dispersed in the wake.* each other. Friction in the wake will waste energy, but cannot alter momentum as a whole.

The relevant considerations may thus be presented on a less abstract basis by formulating the momentum associated with a single isolated thin vortex ring. If $\varpi$ is the total strength integrated over its small cross-section, which is necessarily constant along its circuit, the formula for the momentum $(P, Q, R)$ of its field reduces to a line integration round its circuit, of type

$$P = \rho\varpi \int (y\,dz - z\,dy):$$

*The momentum of a ring vortex.* thus the component of the momentum in any direction is $\rho\varpi$ multiplied by twice the projection of the aperture of the ring on a plane transverse to that direction.

*Lifting of a double trail:* It may be remarked that, on these principles, a Kármán double trail of straight vortices, thus completed into elongated rings, each of strength $\varpi$, spread alternately at intervals $a$ on each of the sheets whose distance apart is $h$, $N$ conjugate pairs being emitted per unit time, ought to exert per unit breadth a lifting force $\rho a\varpi N$, a drag $2\rho h\varpi N$ and a torque $\rho\varpi N \mid r^2 - r'^2 \mid$, where the last factor is the difference between the initial and the ultimate values of the difference *drain of energy,* of squares of their distances from the selected origin. They carry away energy $4\rho h\varpi Nv$ per unit time, which would give a drag twice too great, were it not that their own velocity $v$ is only half of $Na$ on *consistent.* account of the dead water between them.

*Axis of travel of a steady vortex system:* The resultant momentum associated with a vortex ring is in direction such that the projected area $A$ on a plane transverse to it is maximum, and is equal to $2\rho\varpi A$: the ring system must travel along this direction.

The resultant associated momentum for any travelling vortical system however complex must be expressible canonically as a linear momentum $\Lambda$ round a definite central axis and a moment of momentum $\Gamma$, or angular momentum in the terminology introduced by Thomson and Tait, around that axis. a screw-axis.

When a single vortex is isolated, it settles into circular form, and this central axis must be transverse to its plane through its centre. Then $\Lambda$ is $2\rho\varpi A$, where $A$ is the area of the circle, while $\Gamma$ must be by symmetry null. If however the vortex is a compound one, for example, of the type made up of vortex filaments interlaced loosely in spirals, wound round each other like a rope but without contact, after the manner of vortex-atom models long ago investigated by J. J. Thomson, the stranding may be dextral or laeval, and $\Gamma$ can exist. Spin of stranded vortex ring.

Any such complicated interlaced vortex system must have a central screw-axis of momentum. Its system of momentum must be conserved as its form travels through the fluid, either with vacuous core or with a core of fluid convected along with it; therefore it must settle down to travel along a straight line, this central screw-axis.

When a travelling group of vortices becomes obliterated by viscosity, its momentum survives. The expression for angular momentum, as it involves a factor $r^2$ yet cannot increase indefinitely, requires that for an isolated group

$$\rho \int d\tau \, (\xi, \eta, \zeta) = 0,$$

so that the resultant of the component vorticities in any direction is null, as is readily verifiable. It shows also that for a point on the axis of the momentum torque the components of $(\xi, \eta, \zeta)$ in a plane transverse to the axis each multiplied by the square of its distance from the axis must have null vector resultant: thus revealing the locality of the axis of progress by inspection as a sort of central line of the system. Adjustment to steady momentum: location of axis.

These considerations may perhaps be further illustrated by the case of a full-shaped body like a symmetrical airship regarded as throwing off from its rearward shoulders a succession of ring vortices, now situated on a single conical sheet, and also minor secondary vortices from edges: the motion would settle down so that this system of vortices travels along their resultant axis, which is nearly that of the body. From this descriptive point of view a rudder deflects the travelling body by a process which may be regarded as the sending away of a stream of vortices from its edges, abstracting momentum in an asymmetric manner which involves a reacting torque on the body including the steady turbulence around it. When the rudder is Rudder:

released the system settles back into the configuration and mode of progress which preserve unaltered the main regular momentum, unless indeed the body has been deflected so far that it swings away

*function to stabilize lengthwise motion.* towards a new position of stability. Thus any free unsteered body, immersed in water or air, turns flatways to the stream: for bodies such as airships, the further classical theory for steering which operates by arresting the beginnings of such cumulating instability, as constructed by G. H. Bryan in the early days of aviation, has all along been in full practical application.

*Momentum of filament mainly rotational:* The momenta associated immediately with the nearly straight linear elements of a vortex ring are mainly angular momenta around them; but for the complete ring they counteract, and for symmetric

*location.* forms cancel out. The integrations by parts, by which momentum actually belonging to the inertia of the fluid in the field is attached formally to the vortices, have rearranged its location artificially in the general formulae with which we began.

An ideal flattened airship with smooth contours, a long aerofoil however being hardly included even as a limit, may perhaps in suit-

*Pulsating trail.* able conditions be conceived to throw off elliptic vortex rings from its shoulders, which would oscillate through the circular form as they recede and finally rest in that form by virtue of steadying influence of frictional agencies. In no case can a train of rings continue to be shed off obliquely to the motion of the stream, because their axes of momentum, being then inclined to the wake, would acquire greater and greater moment with respect to the source as they recede from it.

*Flat disc eased by an adhering vortex ring.* The motion of a flat body transversely through fluid can also with careful initiation take a different form [Reynolds], sometimes apparently smooth and stable without much wake, by its acquiring a thick vortex ring following close behind it which eases off the motion round the convexities of its surface.

*Banking.* From the present point of view it becomes more obvious that the support of a banked aeroplane is transverse to the wings, and thus oblique and so adapted to counteract the centrifugal reaction to its motion in a curve.

*Interaction of vortex rings on same axis.* The mathematical theory of the interaction of a set of vortex rings travelling along a common straight axis, as developed first by Kirchhoff [the case of a pair was elucidated by Helmholtz], gives another illustration of this descriptive mode of analysis by momentum. The total momentum must be conserved; and now also the energy, as there is in this ideal problem supposed to be no turbulence to accelerate its dissipation. If the vortices are thin rings, the one of strength $\varpi_n$ having radius $r_n$ and velocity $v_n$, these conditions lead by the initial formulae of this note to two relations, sustained as these quantities $r$, $v$ change during the motion, namely

$$2\rho\Sigma\varpi_n\pi r_n{}^2 = M_o, \text{ constant,} \quad 4\rho\Sigma\varpi_n\pi r_n{}^2 v_n = T_o, \text{ constant.}$$

For two vortex rings, continually threading one another, they suffice as is familiar to determine the more prominent features of the motion.

The present order of ideas, to be of utility in practical aeronautics, would have to imply that, as regards aggregate momentum, the turbulent unanalysable wake of the aerofoil can be neglected in comparison with its explorable large-scale vortex motion. The turbulence is itself mainly irregular vorticity on a minute scale, quickly dissipated by internal friction, and perhaps involving negligible resultant momentum. If so, the steady dynamics of the system, propelled in air, could in theory be formulated in terms of estimation of the momentum it thus leaves behind; but how far that plan can be carried into practice, and the principle so become an observational aid to design, must be a matter for the domain of experience*, in which doubtless extensive stores of technical information have already been garnered. *[margin: Practical application.]*

*Postscript.* This note, perhaps too venturesome, has been drawn up by the light of Nature, aided at the end by hints gained from H. Glauert's recent book on *Aerofoil and Airscrew Theory*. Since it was completed the writer has been able to inspect two memoirs in *Phil. Trans.* 1925, tracing the flow round aerofoil models, by L. W. Bryant and D. H. Williams and by A. Fage and L. F. G. Simmons, which contain a mass of interesting experimental plottings and verifications, that only familiarity with the subject could render assimilable in a short time, supported in part by reference to a large and inaccessible technical literature. It is however here relevant to explore the dynamics of the relation of the presumed definite circulation round the machine to the lift, which can be done on lines indicated in an appendix to one of the memoirs by G. I. Taylor. If the axis of $z$ is taken vertically upward, the rate of change of the vertical momentum content of the aeroplane and the region surrounding it is, when the direction $(l, m, n)$ of the normal to the element of the interface $\delta S$ is measured inwards, and $p$ represents the fluid pressure, *[margin: Lift of a travelling circulation:]*

$$Z = \int pn\,dS + \rho\int(lu + mv + nw)\,w\,dS$$

$$= \int(p + \tfrac{1}{2}\rho \text{ vel.}^2)\,n\,dS + \rho\int\{luw + mvw + \tfrac{1}{2}n(w^2 - u^2 - v^2)\}\,dS.$$

* A beautiful series of instantaneous photographs in *Engineering*, May 27, 1927, Plate 33, illustrating a lecture by Prandtl to the Royal Aeronautical Society on "The generation of vortices in fluids of small viscosity", shows how complicated this vortical wake soon becomes.

We have here to consider the relative problem of the aeroplane at rest within this fixed boundary and the air flowing through, in order to get steady conditions in which no momentum is used up in changing the motion of the air within the interface, so that $Z$ may be wholly supporting force. Replacing $u$ by $U + u'$, where $U$ is the reversed speed of the aeroplane and $(u', v, w)$ expresses the actual motion of the air, the second term in $Z$, say of form $f(u, v, w)$, reduces to

$$ - \tfrac{1}{2}\rho U^2 \int n\, dS + \rho U \int (lw - nu')\, dS + f(u', v, w). $$

analysed:   Of this the first term, only however when it has been integrated all over the complete interface, contributes nothing. The second term is $\rho U C$ per unit length along the foil, where $C$ is the circulation round the cross-section of the boundary interface: it applies for the middle part of the motion, where it is practically plane; also more roughly elsewhere except at the ends, because the circulation is nearly equal to its projection on the central plane; while beyond the ends the circulation falls away to a null value.

The third term depends on squares of the actual velocity of the air. Near the aerofoil it is far from negligible, but at a distance of several diameters the velocity begins to vary with distance $r$ as $r^{-2}$, so that the integral varies as $r^{-3}$ and becomes negligible at a moderate distance. Also it is a residual effect between above and below. And the first term in $Z$, which includes the effect of pressure, vanishes everywhere in the irrotational part of the steady flow. Apparently this statement represents the case that can be made for the dependence of the lift on a circulation around the wing without vorticity.

its main   This formula for lift, named as it appears after Kutta and Joukow-
term.   ski, would hold more precisely, as regards its pressure term, for the slow unbroken motion near a fair-shaped body: though on account of the first term, which disappears only by integrating out, the result does not express the distribution of the supporting force but only its total amount. In actuality the argument is upset in the region of the wake; but not much, if the wake is narrow. The formula, applicable apparently only for the value of $C$ round moderately distant circuits, is also incomplete in that it does not involve the line of action of the supporting force, itself be it noted independent of the tilt to this approximation, and so leaves the tilt unregulated.

Vortex   It is essential to note that this circulation $C$ remains constant, in
formation   a vortex field, only when the circuit continues to be made up of the
essential to   same particles of fluid: thus as new fluid comes up to the aerofoil it
wing   has to be continually renewed[1], in fact by creation of new vortices
circulation.

[1] A vortex absolutely steady relative to the wing would by the present formulation have constant momentum in the relative motion, and therefore would exert no force, which is in confirmation of this statement.

from the trailing edge. On the present view the circulation is continu- ally being generated by the slip-process which sets up the Helmholtz type of irrotational motion all round the foil: at a distance of several diameters it tends to a constant value round all circuits and the motion there is on the average a field of cyclic irrotational motion, such as would take place smoothly round an ideal core of some form probably not that of the foil, the part of the circulation in crossing the wake being thus small and alternating.

We may expose the amount of the practical discrepancy in $C$ more definitely, also on lines carried through in detail by G. I. Taylor, by considering the problem of the relative motion for an airship followed by a conical surface of slip having no motion within it. This is per- missible notwithstanding instability: for the slip may for theory be regarded as stabilized, within narrow limits, by very slight internal constraint, such as the presence of an ideal internal friction along it would provide. Continuity in the pressure requires that the velocity ($v_0$) be uniform along the slip-surface, for $\varpi + \frac{1}{2}\rho$ vel.$^2$ must be the same on both sides of the sheet of slip while inside it the velocity is now null. The conversion of an enclosing circuit $ACBA$ crossing the slip cone at $A$ and $B$ into another form $A'C'BA'$ crossing it at $A'$ and $B$ subtracts $v_0AA'$ from the circulation round it. Thus the circulation is not the same for adjacent circuits, but only for those that avoid the wake: for the others it depends on where the circuit strikes and leaves the stagnant relative wake. When the vorticity is now (theoretically) released from the slip-surfaces, and moves away as free vortices into the wake now itself in internal motion, it is the part of the path of the circuit affected by the wake that renders the circulation variable, two such paths making contributions which differ by twice the total vorticity which is enclosed between them. That, by the above, is slight when the paths cross the wake trans- versely. Actually for a Kármán double trail of vortices* the circulation fluctuates, as the crossing place is moved outwards along the wake, with an amplitude equal to twice the vorticity of one of the vortices. This implies an effect on the fluid motion due to the influence of compensating adjacent positive and negative vortices, which may perhaps be presumed to be local and so ineffective except close to the foil. The disturbing effect of this vorticity on the flow, after it has settled down to steady conditions, may thus be presumed to be local: it would be specially intense and turbulent near the ends of the aerofoil, where the direction of the vorticity veers round into the reversed or negative part of the elongated closed vortex, correspond- ing to the rings of the vortex cone following a more rounded body.

---

* An informing discussion by Glauert, *Proc. Roy. Soc.* May 1928, can now be referred to.

Problem with<br>stablized<br>wake.

The result as regards the aerofoil may be nearly the same as if there were no wake, but steady cyclic motion all round the central parts of the foil tending at a considerable distance to the uniform circulation appropriate to irrotational flow, as determined by the considerations here sketched; the supporting force being of this standard value $\rho UC$ per unit transverse span, a value which, of necessity in the present train of ideas, is independent not only of the presentation of the aerofoil to the stream but even largely of the actual inner motion as determined by its size and shape.

Historical.      The theory of lift, drag, and torque, for a body of aeroplane shape, monoplane or biplane, is developed in a very interesting manner, for two-dimensional flow, in terms of strict conformal transformations initiated it seems by v. Mises, in Glauert's book, Ch. VII *seq.* The result includes a condition for constancy of torque, which would determine tilt. At first sight the constancy comes out of the Joukowski condition, which has been here put aside on the ground that this constancy could not subsist for a single vortex trail, and that a double trail with opposite vorticities must be required. But it appears that this condition of single slip-surface with velocity finite is introduced merely to give a value for the circulation round the wing: while calculations made roughly from the slip on the surface of a thin nearly flat wing, with a circulation to express the effect of the wake, lead plausibly without any restriction to relations of the same type. The point is perhaps important, as this latter theory, with which the names of Munk, Glauert, and Birnbaum are associated (Glauert, pp. 87–92), is reported to be in close agreement with experimental determinations.

*Feb.* 1. It was pointed out when the paper was read that the reference (p. 680) to the Joukowski theory is a misunderstanding. That theory aims at determining a vortical whirl round the wing such as will just *abolish all tendency* to a sheet of slip at the edge[1]. The proof that this is theoretically possible is certainly a new departure, which can depose the wake from a controlling position for the case of flat bodies moving nearly edgeways. When this course is taken, as seems however justified only for a wing of unlimited length, this field of circulation would travel on accompanying the wing just as the field of circulation of a vortex ring travels on with its core, without need of renewal. It is now however an additional straight

Lift estimated from momentum rule:      vortex, with its circulation: but it adjusts itself to accompany the wing instead of falling away down the wake, there being no wake at all in the ideal case. The theory of momentum here set out would derive a lifting force $\varpi\rho v$ and no drag, where $v$ is now the velocity of

[1] See L. Bairstow, *Applied Aerodynamics* (1920), p. 360: also L. Prandtl, *Die Naturwissenschaften*, Feb. 1925.

the wing, and $\varpi$ is the total vorticity, reckoned as spin, which is equal to only *half* the circulation $C$ round a circuit *far enough away* to be outside any vorticity in the field[1]. The drag on the wing is on such a view due only to an accidental wake of vortices, now itself really secondary, which also makes a contribution to the lift: and the drag at any rate is reduced in proportion if the breadth of the wake between its compensating positive and negative faces can be adjusted to be small. This application of the momentum formulae replaces, but only for an infinitely long wing, the algebra of this Postscript. It will be observed that there is now no implication that this whirl in which the wing is supposed to become enveloped is a wholly irrotational one sliding by slip rollers at the surface of the ring: what an experimental verification of the relation of lift to circulation can confirm is that a vortex cylinder becomes established around the wing with *some distribution of vorticity* such as can in the actual circumstances, not two-dimensional, travel steadily along with the wing, and that the circulation $C$ of the formula is to be measured round a circuit far enough away to be outside this vorticity. Approximate verification of the theoretical formula for an irrotational flow would imply that the region of this vorticity clings close to the wing; which would permit the distribution of the lift over the surface of the wing to be determined as above indicated. Anyhow this very substantial easing of the conditions of the problem makes it more intelligible that the instability at a sharp edge can really get itself largely surmounted in this way, even when one remembers that the wing is not actually, as the two-dimensional theory requires, very long compared with its breadth. *(margin: Distribution of lift determinable from form of section.)*

(The lesson that is here enforced is that we must deal only with completed vortex rings: thus in two-dimensional analysis each straight vortex must be bracketed with a return vortex, as is familiar for the analogous case of electric currents. [See *infra*.]

There seems to be a hypothesis which does continue the circulation around the wings by vortices trailing from their outer ends and completing a circuit across the wake. The momentum of such a steady completed vortex travelling in fluid is (p. 682) of the order $2\rho\varpi A$, where $A$ is the area of its circuit; the transfer of this momentum seems to involve no supporting force on the wings, but a torque $2\rho\varpi Av \cos \alpha$, where $\alpha$ is the inclination of its area $A$ to the direction *(margin: Imagined hanging vortex without a wake.)*

[1] The discrepancy with the formula for a ring (p. 682) is adjusted by considering the ring made up of two conjugate straight vortices connected by vortical arcs at the ends: the latter make a contribution equal to that of the straight parts. [The end parts, or trailing vortices, are continually being created and left behind, as the aerofoil moves on: and this involves a consumption of momentum equal to that found in the text, thus giving a doubled result. This seems to approach Lanchester's point of view.] *(margin: Half the lift due to trailing end-vortices.)*

of the velocity $v$. Thus if the existence of such a vortex were possible, it must hang transverse to $v$. Contrast p. 688, which implies an infinite straight vortex: the Newtonian dynamical relativity loses its meaning for infinite systems.

The mode in which the experimentally verified body vortex completes itself into a ring, whether by shedding a continual trail of vortices from its ends, and perhaps in any case involving its own continual renewal by slip process, thus opens out a novel and interesting type of pure hydrodynamic problem.)

[Abstract; from *Nature*, May 28, 1927.]

One used to recognize that the exigencies of flight in the tenuous air prescribed a limit to the bulk of a bird, as compared for example with a whale. Yet nowadays everyday loads of twenty tons of stuff or possibly far more are carried over long journeys, owing to the power available, solely on wings. How does the attenuated aereal medium find means of supporting such an astonishing mass? To experts the fact is familiar, and so scarcely demands explanation. Indeed, the source of the support in plain terms is just as wonderful as the fact itself. The load is held up solely by the swirl that it produces and leaves behind; and this vertical support has to be the only dynamical effect of the swirl when the speed is steady: so there remains the question how precisely this result is adjusted. Unfortunately the wakes from screw and wings can scarcely be additive without some mutual interference, though momenta are additive always. For example, the spread of the wings is adapted readily to counteract the rotational grip of the screw. Stability is theoretically (G. H. Bryan) another affair.

*Supported by the swirl of air;*

Whether the supporting medium is air, or water, or even pitch, provided only it is of uniform density everywhere, the momentum, with which the flight is concerned, proves to be expressible at each instant in terms of the distribution of swirl or vorticity alone. Force is experienced by the travelling system equal and opposite to the rate at which momentum is shed away into the wake of its motion. The nature of the swirl passing into the wake thus determines all. In the ideal perfect fluid of abstract hydrodynamics there would be no wake, and therefore no force affecting the translatory motion, though the mass may twirl in permanent precessional spin. If this train of ideas is right it is impossible for a circulation round the wings of an aeroplane to sustain it, except in so far as it has to be associated with a vortical wake*.

*on which alone momentum depends:*

The formula for the momentum associated with each element of whirl in the ambient medium, of whatever kind it be and however

*Analysis of lift on the wing.*

* The computations which decide otherwise are concerned with a wing of infinite span. If its length is finite the circulations round it must complete themselves along return circuits in the fluid. If they all form a vortex ring

complex its internal friction, provided only it is *of uniform density*, turns out to be unexpectedly simple. There is translational momentum according to a simple law, equal, as applied at any chosen origin, to the vector moment of the mass vorticity of the element (mass multiplied by spin), combined with rotational momentum around that origin equal to this mass vorticity multiplied by the square of the distance with sign reversed. Now vorticity has the advantage of being a quality of considerable adapted to exploration. persistence, unless the internal friction is high: if then this field of spin could be sufficiently explored by observation, it would only be necessary, in order to obtain the forces operating, to trace out the rate at which the derived system of momentum thus associated with the travelling machine is changing. Many special illustrations present themselves. For example, a travelling aeroplane adjusts its presenta- Illustration. tion so that this moment of momentum, with regard to the point where the pull of the screw intersects the line of weight, is not subject to loss into the wake: such automatic adjustment would tend to nip together the two boundary sheets of the vortical trail, which thus would open out only at the ends of the wings. Again, a windmill parachute appears to be more effective than the simple umbrella type: if so, the cause doubtless declares itself in a wind-channel by the contrasted types of whirl in the wakes they leave behind. And generally, the performance of any propelling screw wholly submerged would be determinable in terms of the whirl in its wake alone, if only that could be explored.

## Postscript (*April*, 1928).

### ON LIMITATIONS OF MODELS FOR FLOW OF VISCOUS FLUID.

It appears that the Fourier-Stokes-Froude-Rayleigh method of prediction of motions of fluid by aid of models may be subject to limitations, requiring attention in various ways. When there is a

travelling with the wing, part of whose core is the wing itself, they can provide no supporting force on the local system, though there will be such a force on the wing with a countervailing force on the rest of the vortex core. If the rest of the core is virtually unattached to the local system this reacting force will push it away down the wake, as above asserted, and it will have to be continually regenerated as a vortical wake. The side parts of the vortex will assist in sustaining the system to an equal degree with the wing part, thus removing an apparent discrepancy: cf. footnote, p. 689.

The natural mode of description of the flow for a wing of finite length would be a garland of circulation round each section ending on its upper and lower sides in a flat vortex trail thin in the middle, but opening out in dumb-bell manner to breadth and substantial strength at the ends of the wing. It is the natural result of the creation of vorticity by steep slide gradient or actual slip, Origin of the circulation. that part of it is thus completed round the wing itself. In the technical discussions each slice of such vorticity, transverse to the wing, is treated as if it were two-dimensional flow, which it would be if the components of velocity along the length of the wing were negligible.

wake, it is necessary, as is well recognized, that the mode of initiation and development of its instabilities, as well as the steady flow, should be amenable to dimensional correspondence. Experiment in the simpler problems of flow, on the lines initiated long ago by O. Reynolds, seems to have proved there is a substantial empirical foundation for that presumption. Again, it seems to be not always axiomatically certain that the doctrine of relative motion must hold good, so that the flow past a body anchored in a wind would be the same thing as motion of the body in still air. For regular motion of the medium requires, to make it mathematically determinate, that the circumstances both at the boundary of the body and also towards infinity should be specified. There is difficulty about the latter even for smooth motion: where there is a wake, extending theoretically to infinity, it may be far from the same in the two cases. These considerations may perhaps be fixed by a concrete illustration, from the result of a recent investigation by C. W. Oseen. Long ago Stokes found that, when viscosity is predominant, no regular steady motion of fluid can be established past a cylindrical body of unlimited length. It has however been found by Professor Oseen, that a steady motion probably does exist for the infinite cylinder moving transversely through resting fluid. The principle of relative motion is not here applicable though there are no wakes, as the conditions towards infinity differ essentially from those for a compact moving body: even at great distance the infinitude of the length of the cylinder is involved.

It appears that*, though the limiting velocity produced by a given steady force is the same for the case of a sphere in both problems, yet the nature of the relative flow that is obtained is not the same, except, as it happens, close to the sphere; although no wakes are supposed to be set up in either case. The discrepancy that is revealed is very substantial: its main feature is a drift of fluid following the moving sphere with velocity equal to $\frac{3}{2}Ua/r$ at the axis, giving a total inward flow uniform all the way from infinity which is balanced by an outward radial flow from the front of the body.

Contrast is instructive with Rayleigh's paper (*Scientific Papers*, Vol. VI, pp. 229–240) published a few months later, apparently without knowledge of this discussion, which essentially affects the deter-

* The discussion of Professor Oseen's formulae by H. Lamb is here used: *Phil. Mag.* Vol. XXI (1911) p. 112, or *Hydrodynamics*. The motion in the fluid, outside a narrow smooth vortical wake which has been subjected to intense shear in rubbing past the obstacle, is almost irrotational. The steady motion obtained by Stokes is the same for all viscosities and has no wake: his method in terms of a stream function is suitable (cf. *Papers*, Vol. III, p. 57, also Basset, *Hydrodynamics*, Vol. II, ch. 22) for exploring the course of establishment of a steady motion.

minacies or "existence theorems" in the subject of fields of viscous flow if the present interpretation can be maintained.

Objection might possibly be taken that to establish the inward drift from infinity, in the rear of the body moving through stationary fluid, an infinite time would be required, unless the speed of transmission of fluid pressure, which is the velocity of sound in the fluid, were infinite. *establishment of the motion of the fluid.* But what actually becomes established is a circulation, outward in front changing to inward in the rear, whose range spreads outward with the velocity of sound as the motion of the body continues.

### Postscript (*August*, 1928).

#### ON THE KELVIN THEORY OF THE IMPULSE.

A remarkable result has been established by G. I. Taylor (*Proc. Roy. Soc.* 1928) which invites consideration from the point of view of the footnote, p. 677. The theorem amounts essentially to the following statement. For a solid body moving in fluid of uniform incompressible *G. I. Taylor's* density the impulse (after Kelvin) is made up of the reversed gradients *theorem.* (rates of increase) of the energy with respect to the components of velocity translatory and rotatory: and the equations of motion of the solid itself are expressible in terms of these component momenta. Where the solid is constrained to move without rotation, and without any wake such as a rudder of an air-ship would produce, in infinite fluid it appears that its hydrodynamic moment, *i.e.* the moment of a doublet to which the distant motion of the fluid would be due, is identical with the translational part of the impulse divided by $4\pi$: this is on the supposition that the energy of motion of the solid is included in the expression for the energy, reckoned however for a solid of the same density as the fluid, the rest being an excess to be dealt with in Archimedean manner. It is this direct correlation of the remote motion with the local effective momentum of the solid, irrespective of its shape, that invites attention.

The significance of the impulse of the motion may be put in evidence. *The impulse* When the solid is moving through the fluid, it produces velocity *in fluid motion:* and momentum in its parts which integrate over the whole to a finite value, the impulse, with its six components: yet the forward momentum, in a conical region in the front or rear part of the unbounded fluid, is infinite, and it is only by balance with that of backward flow at the sides that a finite value is arrived at. This is a paradox that *paradox:* ought not to be permitted to remain. Incompressible quality implies infinite speed of readjustment of pressures: whereas for actual fluid this speed is only very great, being that of sound: the adjustment is thus made by wave-motion, of amplitude never sensible because the speed is great, spreading out from the source, so that at any instant the momentum has become established throughout a sphere with the source as centre. Integration throughout this sphere gives the finite impulse: yet it still remains remarkable that it can be a difference of two actual parts forward and sideways, each far greater than itself: in each region it is acquired from the pressure that is impulsively established around it.

*its solution.*

We now touch on the general theory of the impulse in kinetics (Thomson and Tait, 1867, *passim*). When the accelerations are very great the terms, usually quadratic, in the kinetic reaction that involve the velocities can be neglected in comparison, and the equations simplify into a formula that change of momentum represents the imposed impulse. The general equations of motion can then be constructed in terms of gradient of momentum in the Eulerian manner, as impressively represented by the Kelvin forms for a solid moving in fluid. For a fluid of uniform density $\rho$ subjected to impulse from a solid set in motion within it, the dynamical result is a distribution of impulsive pressure $-\rho\phi$ where $\phi$ is the velocity potential of the motion thus produced. As an illustration, suppose a solid of the same density $\rho$, moving through the fluid, instantly itself to become fluid: the velocity potential will be suddenly altered within it, becoming discontinuous at its boundary: the distribution of momentum, of density $-\rho \operatorname{grad} \phi$, will not integrate to zero over the mass as it would for a continuous form of $\phi$, there is an impulsive resultant over the boundary, transverse to it and equal to $-\rho\Delta\phi$ per unit surface where $\Delta\phi$ is the discontinuity in $\phi$. In other words an impulse of intensity $\rho\Delta\phi$ distributed over the boundary is required in order to prevent the motion of the boundary and of the exterior fluid from changing. There is only one case in which no such impulse is required, that in which the motion of the solid, of equal density, was wholly translational and so does not alter on liquefaction.

*The general Kelvin theory.*

*Impulsive fluid pressure:*

*impulses on interfaces.*

Now consider a very distant ideal interface in the fluid, and suppose its outward velocity annulled by an impulse spread over it: that will annul all motion outside it while it will alter the motion inside it in a way that is determinate. The result of the original impulse on the solid that started the motion and this impulse over the interface is that there is motion only inside it, but such as does not displace the centre of mass. Thus those two impulses equilibrate each other, in magnitude but not line of action. The impulse over the distant interface is readily calculated: in fact the resultant impulse required to annul a velocity potential $Ax/r^3$ over a spherical boundary comes out to be $4\pi A$: and this verifies Taylor's result. If the solid has also rotational motion, a cognate result, involving now the motion of fluid as liquefied from the solid and confined by the rotating boundary, may be formulated.

*Distant motion determined by impulse on solid.*

# APPENDIX I.

## ABSTRACTS: FROM REPORTS OF THE BRITISH ASSOCIATION.

### *Molecular Distances in Galvanic Polarization*[1].

[Aberdeen, 1885.]

IT has been shown, principally by Helmholtz (*Wissen. Abhand.* Vol. I, Galvanismus, and Faraday Lecture, *Journal of the Chemical Society*, 1882) that polarization involves a condensing action on the surface of the metallic electrodes. The particles, say of the cation, which exist in the electrolyte in a state of temporary dissociation, are drawn towards the cathode by the electromotive force till further approach is prevented by chemical forces. Thus along the surface of the cathode we obtain a sheet of positively charged cation molecules, which are in an equidistant arrangement on account of their mutual repulsion; and opposite to each on the material of the electrode there is the equal and opposite charge drawn there by electrostatic induction. This double sheet is equivalent to a condenser.

Electric double layer:

Kohlrausch and Helmholtz measured the charge required to produce a given polarization difference of potential, and therefrom estimated the thickness of this double layer.

Lippmann (*Comptes Rendus*, 1882) determined the surface energy of the charge per unit area from the variation of the capillary constant of a mercury electrode; and thereby gained an estimate in agreement with the above.

We should expect that the capacity of the double layer per unit surface would remain constant until the distance between contiguous cation particles became of the same order as the thickness of the double layer, *i.e.* until the contiguous particles began to feel one anothers' chemical forces, and be influenced thereby. An examination of Lippmann's results shows this constancy of capacity to a very close degree for a range of about a volt. We may, accordingly, conclude that with his acidulated water polarized to a volt the particles of the polarization layer have just become so numerous as to be in chemical contact with one another. This leads to a third estimate of a molecular distance which we find to agree with the two former.

mono-molecular up to a limit:

The first estimate is based on the absolute electrostatic measurement of the charge; the second on the measurement of a surface tension; the third on the absolute electrochemical equivalent of the electrolyte. The complete agreement of three estimates founded on

giving dimensions of molecules.

---

[1] See *Phil. Mag.* Nov. 1885 [: *supra*, Vol. I, pp. 133–145. Cf. also experiments over a wide range by Bowden and Rideal, *Roy. Soc. Proc.* 1928].

physical constants so various and of so different orders of magnitude, is strong evidence of the validity of this method of interpreting the phenomena of polarization.

They are all in satisfactory agreement with the estimates of Sir W. Thomson and others from different considerations (*Nature*, 1870, Thomson and Tait's *Natural Philosophy*, Part II, Appendix F), and give an average result of about one $10^{-10}$ metre, more or less.

### On the Physical Character of Caustic Surfaces.

[Leeds, 1890.]

The diffraction produced at a caustic is peculiar, in that it is not conditioned by a beam of light limited by an aperture or otherwise. The theoretical explanation why undulations, propagated by unlimited wave-fronts, cannot penetrate beyond a certain geometrical surface, was slightly indicated by Thomas Young, and fully worked out on Fresnel's principles by Sir G. B. Airy, for the special case of the rainbow.

*Caustic as formed of parallel sheets:* It is worth while to formulate the general law of the thicknesses of the bright bands which, for any kind of homogeneous light, lie parallel to the principal physical caustic. As the beam of light is determined solely by the geometrical caustic, or ray-envelope, it is clear that the law must be expressed in terms of this surface. It comes out that

*its structure.* The bands form, along with the geometrical caustic, a system whose relative distances apart are in every case the same, and whose absolute distances at any point are proportional to the two-thirds power of the radius of curvature of the caustic surface at the point, measured in the direction of the rays. [Cf. *supra*, Vol. I, pp. 214–219.]

### The Action of Electrical Radiators, with a Mechanical Analogy.

[Cardiff, 1891.]

In an electrical vibrator of rapid period the currents in the metallic parts are confined to the surface; the periodic times are therefore *Vibrations as located in the dielectric:* independent of the metals of which the vibrators are made, being determined only by their forms, and there is no considerable loss due to degradation into heat in these conductors. The question occurs, what are the surface conditions that must be imposed under these circumstances at the boundaries of the dielectrics, in order that the vibrations may be discussed with reference only to the dielectric in which they exist and are propagated?

It appears that the vibrations are analogous to those of an elastic solid, when elastic displacement is made the analogue of the electric displacement in the dielectric. It is demonstrable[1] that if the velocity an elastic parallel: of propagation is the inverse square root of the specific inductive capacity, this auxiliary solid must be considered as incompressible, and the scheme of electrodynamics must be that of Maxwell. The surface condition will then be absolute stiffness in the surface layer for all tangential displacement, and freedom for normal displacement.

The mathematical examination of a typical case shows that this way of presenting the phenomena is practically exact for all wave- except for slow periods. lengths greater than a centimetre for copper or other highly conducting metal. For very minute waves the circumstances are not independent of the material of the conductor, but are similar to those which actually exist in the case of the metallic reflexion of light waves.

By aid of this representation a qualitative view of the possible modes of vibration is rendered feasible in cases where the mathematical analysis would be difficult or impossible[2].

## On a Familiar Formation of Plane Caustic Curves.

### [Nottingham, 1893.]

The illustration of the formation of caustics by the reflexion of the light of the Sun or other point-source from a band of polished metal is in everyday use. But it seems worth while to call attention to the fact that obliquity of the incidence on a cylindrical reflector does not vitiate the experiment as an exact representation of the geometrical caustic. The bright caustic surface formed by reflexion Cylindrical caustic surfaces; from a cylinder is, in fact, always itself cylindrical; and the caustic curve depicted on any screen placed across it is merely the section of this cylinder formed by the screen [, and is the caustic curve by reflexion of the projection of the incident system of rays]. Although, once this proposition is propounded, its reason is plain, yet it does not their sections as plane caustics. seem to have occurred to any of the writers on optics or plane geometry.

It may be shown, also, that when the reflector is a piece of a conical surface the caustic surface is always a conical surface with the same vertex—thus suggesting an extension of the theory of actual caustics Conical caustics: into spherical geometry, or rather realizing in actuality the analogous theory in spherical geometry. More generally, when the reflector is such as may be bent flat without stretching, *i.e.* when it is a piece developable caustics. of a developable surface, the caustic surface is one of the same kind, and a geometrical correlation may be established between them.

[1] *Proc. Roy. Soc.* May 1891 [: *supra*, Vol. 1, pp. 232–247].
[2] *Proc. Camb. Phil. Soc.* May 1891 [: *supra*, Vol. 1, pp. 221–231].

## The Influence of Pressure on Spectral Lines.

[Toronto, 1897.]

Orbital type of radiating molecule: A definite picture of the relations of the aether and matter is obtained by assuming the material molecule to be made up of electrons or intrinsic strain-centres in the aether[1]. A system of electrons describing steady orbits round each other, after the manner of the bodies of a solar or stellar system, would represent a molecule; any disturbance of this steady motion would induce radiation across the aether, which would last until it had reduced the motion again to a state of steadiness. The natural configuration of a molecule would, however, be the unique one of minimum energy corresponding to its non-radiating states must exist. intrinsic constant rotational momenta, for the influence of radiation would set towards this configuration, and would not allow much departure from it.

Gaussian theory of secular oscillations applicable. The wave-lengths of luminous radiation are about $10^3$ times the linear dimensions of the molecules; thus the intrinsic luminous periods are those of rather slow periodic inequalities (in the sense of physical astronomy) in the orbital motions. This circumstance allows us roughly to appreciate the order of magnitude of the influence of the surrounding medium on these free periods. On account of their slowness the aethereal oscillations which are governed by the inequalities of the orbits of the electrons are sensible over the space occupied by some thousands of molecules each way, and this number is so great as to tempt us to form an idea of the influence of these imbedded molecules by considering them to form a continuous medium. If now the molecules were vibrating in a homogenous medium, say, surrounded by simple aether, the free periods would vary inversely as the square root of the elasticity of this ambient medium, provided we could assume that change of the medium did not involve change of type of the steady intramolecular orbits. This latter circumstance, however, will also operate to alter the periods, and will be of the same Influence of electric elasticity; order of importance as the other. Now the effective elasticity of the gaseous medium surrounding the vibrating molecule, when thus treated as continuous, varies inversely as its dielectric constant. We should thus expect on the above hypothesis that increase of pressure would lower the free periods roughly in the same ratio as it raises the square root of the dielectric constant. To reduce to figures: a shift of $\frac{1}{40}$ of the distance between the $D$ lines would correspond to $\delta\lambda/\lambda = \frac{1}{4} . 10^{-4}$, while the dielectric constant of air at $0°$ C. and atmospheric pressure is $1\cdot0006$. Thus this shift towards the less refrangible

---

[1] Cf. *Phil. Trans.* 1895, pp. 695–743 [: also *supra*, Vol. I, p. 524, Vol. II, p. 342].

end would indicate a change of density of the surrounding air of the a numerical<br>estimate of<br>effect of order of that due to a pressure of $\frac{1}{10}$ of an atmosphere at $0°$ C.

This would make the effect about $10^2$ times too large for the obser- density, vations: thus the main seat of the aether strain maintaining the is too large: vibrations of the molecule is the free aether immediately surrounding it, and the loss of stiffness due to the other molecules which are some way off diminishes the free periods only about $10^{-2}$ times as much as so field is if it were averaged right up to the vibrator. With similarly constituted closely concentrated. lines, it is the relative shift $\delta\lambda/\lambda$ that is proportional to the change of density of the medium.

## On the Application of the Method of Entropy to Radiant Energy.

[Belfast, 1902.]

The entropy of a material system has been defined by Boltzmann as proportional to log. probability of its molecular configuration; and this definition has recently been applied by Planck to the radiation in an enclosure, thereby obtaining a law for the constitution of natural radiation at a given temperature, which is in close accord with the facts. His argument involves simple vibrators in the region, and it is their fortuitous arrangement that enters. It was explained that Statistical theory of various difficulties attending this procedure are evaded, and the same radiation: result attained, by discarding the vibrators and considering the random distribution of the permanent element of the radiation itself, quantified phase space: among the differential elements of volume of the enclosure, somewhat on the analogy of the Newtonian corpuscular theory of optics[1].

## On the Relation of Voltaic Potential Differences to Temperature.

[Belfast, 1902.]

It was shown, by means of Carnot's principle, that if material A Carnot voltaic substances have no special affinity for electricity, voltaic potential condenser differences should be proportional to the absolute temperature[*]. The cycle: experiments of Majorana with liquid air as a cooling agent show that they actually fall at a more rapid rate than this law would give, from which it is inferred that imparting an electric charge to a metal involves absorption of heat owing to direct affinity between its molecules affinity of a metal for and the charge. electricity.

[1] Cf. *Brit. Assoc. Report*, 1900 [: *supra*, p. 222: also Appendix v].

[*] The junctions in a thermoelectric circuit of two metals at $T$ and $T_0$ can be replaced by identical voltaic cells at those temperatures, into which the metals Thermo-voltaic are led as electrodes. If $H$ is the reversible absorption of heat in a cell per circuit: unit electric flux, on the lines of the Peltier effect at a metallic junction, the

## On the Dynamical Significance of Kundt's Law
## of Anomalous Dispersion.

[Cambridge, 1904.]

It was pointed out that the energy of a train of approximately homogeneous radiation must be propagated forwards; on the principles of O. Reynolds and Rayleigh this requires that the group velocity must be positive, provided there is no absorption. Thus, outside an absorption band the index of refraction must always increase with increasing frequency of the wave-train. Inside the absorption band the argument does not apply, and the curve of dispersion there bends round so as to connect the two arcs, both with upward trend, on the two sides of the band. These are the features of actual dispersion curves to which Kundt drew attention. [Cf. *supra*, pp. 54, 299.]

*Explanation of trend of dispersion.*

## On the Relation of the Röntgen Radiation
## to Ordinary Light.

[Cambridge, 1904.]

Arguments were offered in support of the view advanced by Sir George Stokes ("Wilde Lecture," *Lit. and Phil. Soc.*, Manchester, 1897) that a single radiant pulse, incident on a molecular medium, would not suffer regular refraction. As [If] natural radiation consists

*worked out after Kelvin:* Kelvin equation $\Sigma H/T$ null for a cycle becomes $\dfrac{H}{T} - \dfrac{H_0}{T_0} - \displaystyle\int_{T_0}^{T} \dfrac{\sigma_2 - \sigma_1}{T}\, dT = 0$ when his specific heat $\sigma$ of electricity is introduced, which is now heat actually abstracted from the metal by the electrons that constitute the flow, the flux being taken as from metal 1 to metal 2 through the hot junction $T$ in the direction in which the cell works. This neglects the unavoidable irreversible effects of conduction; they can, after Kelvin, be made very small by reducing the intensity of flow, but at the expense of prolonged time of transit. The *justified.* argument on the Carnot principle, based on finite range of temperature as on p. 586, *supra*, appears to make this a matter merely of approximation. This equation in differential form is $\dfrac{d}{dT}\dfrac{H}{T} = \dfrac{\sigma_2 - \sigma_1}{T}$.

*Theory of the reversible cell:* As regards the cell at the junction, the Kelvin equation of available energy (*supra*, p. 103) may be applied, after Gibbs and Helmholtz. The available energy abstracted from it per unit electric flux is its electromotive force $E$; the heat that has left it is the total energy abstracted less the available part, it is thus given by $H = T\dfrac{dE}{dT}$.

*the effect of temperature deduced,* These are the two equations of the circuit. They combine with $\dfrac{d^2E}{dT^2} = \dfrac{\sigma_2 - \sigma_1}{T}$.

*uniform except for an electrode part.* They may perhaps admit of experimental scrutiny: but that involves cells closely reversible. When $\sigma_2 - \sigma_1$ is negligible compared with the effects in the junction cells, $dE/dT$ is constant, which is the result above stated. It is to be noted that in any case the rate of change of this gradient $dE/dT$ would depend only on the metallic electrodes, and is extremely small compared with voltaic effects.

of a succession of pulses, propagated from molecular shocks occurring at the surface of the incandescent solid or liquid radiator, it becomes necessary on this view to specify some kind of regularity in the shocks, in order to explain refraction and dispersion. It is held that a statistical regularity, such as is connected in molecular theory (*e.g.* gas theory) with the existence at each point of a definite temperature, is adequate for this purpose. On the other hand, if the Röntgen rays consist of absolutely independent pulses, devoid of all regularity in their succession, thus comparable to the traffic along a street or to the fire of a skirmishing company of soldiers, they would undergo no regular refraction*.

*Are narrow pulses of radiation refracted?*

### On the Range of Freedom of Electrons in Metals[1].

[Leicester, 1907.]

It was remarked that perhaps the most obscure present problem in abstract physics is the mechanism of the transfer of electricity (the electron) from molecule to molecule. A hopeful plan is to study it in its time relations; the optical phenomena of metals introduce times, the periodic times of the vibrations, that are small enough for this purpose.

*Metallic conduction in its optical aspect:*

The experiments of Hagen and Rubens show that the behaviour of metals to long infra-red radiation depends on their steady ohmic resistance alone. Thus the time required to establish conduction completely is a small fraction of the period of such waves. If the same free electrons to which conduction is due have velocity of mean square determined by the gas laws, this restricts their range of freedom almost to the interspace between the molecules. On the other hand, the fact that the square of the *quasi*-index of refraction of light for the nobler metals is not far removed from being a real negative quantity, indicates that the number of such free electrons is of about the same order of magnitude as the number of the molecules. This again recalls the electrochemical principle that the number of *transferable* electrons in an atom represents its valency.

*velocities of electrons and their freedom:*

*their number.*

### [*From* "A discussion on Radiation," with contributions by Jeans, Lorentz, Love and others.]

[Birmingham, 1913.]

Sir J. Larmor derived the impression from the trend of the discussion that it would turn out that in the new low-temperature determinations there was nothing in direct conflict with the classical

* But cf. paper "On the Constitution of Natural Radiation," *Phil. Mag.* 1905 [: *supra*, p. 306, also Appendix v].

[1] For details see *Phil. Mag.* August 1907 [: *supra*, p. 336].

dynamical principles. The essential argument for equipartition among
vibrational types of energy is, briefly, that these types enter similarly
into the total energy, and thus, other things being indifferent, there
is no reason that can be assigned to the contrary. They enter similarly
merely because the energy is a sum of squares of their "momentoids."
But other things may not be indifferent; for example, in the kinetics
of a rotating atmosphere the distribution of energy must be modified
so as to maintain constancy of the angular momentum as well as of
the energy, giving as the result equipartition relative to the rotation
instead of absolutely*. Moreover, in an isolated region of aether there
is no way open for any interchange of energy at all between one type
of vibration and another; here also other things are *not* indifferent.
The exchange must be effected through the mediation of material
molecules. It is true that a single electron, moving erratically between
complete reflexions from the ideal impervious walls of the chamber,
would suffice. But the structure of an electron, including the mechanisms by which it exchanges energy with the aether, is totally unknown.
Such a fundamental fact as the pressure of radiation is involved in
that structure; we can only establish it theoretically as pressure on
systems of electrons†; it must be transmitted by the aether in some
way, but we do not know how, except by speculating, for it is a second-
order phenomenon not involved in the Maxwellian linear scheme of
equations. In the very intense kinetic phenomena in the mechanism
of the electrons or molecules, by which they serve to transfer energy
from one type of aethereal vibration to other types, the energy must
be expressible as a sum of squares of definite momentoids if the
transfer is to lead ultimately to equipartition. This restriction in its
form is unlikely; the transfer may even be of a discontinuous character,
involving release of electrons into freedom. The Planck formula for
the constitution of natural free radiation may be obtained by statistical
reasoning, strictly on the lines of Boltzmann's entropy theory for
gases, in which atoms or vibrators are not considered at all, but the
pressure of radiation is introduced instead, as I have tried to show
(*Proc. Roy. Soc.* 1909 [: *supra*, p. 402]); the only implication is that
the process of interchange of aethereal energy between different
vibrational types, by the mediation of matter, though unknown,
must be such as to provide a pressure of radiation. If, then, there is
no reason to press equipartition as regards free natural radiation, the
atomic vibrations, which are set up by its agency and must be in
equilibrium with it, are also absolved therefrom.

The new knowledge relating to specific heats at very low temperatures has already suggested most interesting speculations and tentative adjustments, and will certainly lead to definite expansion of our

Conditions
for equi-
partition
of energy,

are not
actually
satisfied.

Pressure of
radiation,
how trans-
mitted?

Radiation
from atoms
in *quanta.*

Statistics
of free
radiation,

as set up
through the
atoms,

in unknown
ways.

* Cf. Appendix III, *infra*, p. 742.          † Cf. however Appendix VIII, *infra.*

theoretical schemes; but it can be held [p. 744] that there is nothing in it that is destructive to the principles of physics which have led to so rich a harvest of discovery and synthesis in the past.

## On Lightning and Protection from it.

[Birmingham, 1913.]

The *rationale* of electric discharge in a gas is now understood. When a small region becomes conducting through ionization by collisions in the electric field, it should spread in the direction in which the field is most intense, which is along the lines of force. Thus the electric rupture is not a tear along a surface but a perforation along a line. This is roughly the line of force of the field: the electrokinetic force induced by the discharge, being parallel to the current, does not modify this conclusion. A zigzag discharge* would thus consist of independent flashes, the first one upsetting adjacent equilibria by transference of charge. Successive discharges between the same masses would tend to follow the same ionized path, which may meantime be displaced by air currents. *(Nature of electric rupture in air: consequences.)*

If the line of discharge is thus determined by the previous electric field, the influence of a lightning conductor in drawing the discharge must be determined by the modification of this electric field which its presence produces. For a field of vertical force, such as an overhead cloud would produce, it may be shown that the disturbance caused by a thin vertical rod is confined to its own immediate neighbourhood. Thus while it provides a strong silent discharge from earth into the air, it does not assist in drawing a disruptive discharge from above —except in so far as the stream of electrified air rising from it may provide a path. It is the broader building, to which the rod is attached, that draws the lightning: the rod affords the means of safely carrying it away, and thus should be well connected with all metallic channels on the building as well as with earth. It is the branching top of an isolated tree that attracts the discharge; a wire pole could not do so to a sensible degree. Separate rods projecting upwards from the corner of a building do not much affect the field above it, but if they are connected at their summits by horizontal wires, the latter, being thus earthed, lift up the electric field from the top of the building itself to the region above them, and thus take the discharge which they help in attracting, instead of the building below them. Similarly, when the lines of force are oblique to a vertical rod, its presence does somewhat modify the field and protect the lee side; but generally the presence of a rod should not ever be a source of danger, unless the ionized air rising from it provides an actual path for discharge. [Cf. *supra*, p. 457.] *(Influence of a lightning rod, slight, contrasted with a wire frame.)*

* Except when only apparent: cf. Faraday, *Exp. Res.* Vol. II (1841) p. 277.

## The Limitations of Relativity.

[Bournemouth, 1919.]

The following propositions are believed to be valid, on the basis of a concise symbolic calculus, subject, of course, to critical verification.

*The converse of optical relativity.* (1) If a field of physical activity possesses the two characteristic properties (i) that the quantities which define it are propagated through the aether with a single constant velocity, and (ii) that translatory uniform convection as a whole through the aether produces no modification in the field, then it is necessarily restricted to the special type of the electrodynamic field as formulated by Maxwell.

*Connection with gravitation.* (2) A field of gravitation is included as the limiting form of such a type when the velocity of propagation becomes very great. As like source is now to attract like, the energy of the field must be kinetic and not elastic. The question of interaction between a field of gravitation and electric fields or rays of light is, of course, a separate and fundamental one, independent of theories of relativity, and is now being put to refined test by astronomical observation.

*Material sources must be enduring.* (3) If time were linked with space after the manner of a fourth dimension, relativity in electrodynamic fields would be secured as above, but the sources of the field could not be permanent particles or electrons. If physical science is to evolve on the basis of relations of permanent matter and its motions, time must be maintained distinct from space, and the effect of convection must continue to be thrown on to the material observing system in the form of slight modification of its structure. [Cf. Appendix VIII, *infra.*]

### A Kinematic Representation of Jacobi's Theory of the Last Multiplier.

[Toronto, 1897.]

[*Condensed Abstract.*—A system of linear differential equations, when reduced to a standard form $dx_1/u_1 = dx_2/u_2 = \ldots = dx_n/u_n$, may be visualized as representing streaming flow of the particles of a material fluid in $n$ dimensions, with velocity $(u_1, u_2, \ldots u_n)$ steady at each point and thus specified as a function of position. The density $\rho$ of the fluid is therefore restricted to satisfy the equation of continuity of flow, of form $\dfrac{\partial \rho u_1}{\partial x_1} + \dfrac{\partial \rho u_2}{\partial x_2} + \ldots + \dfrac{\partial \rho u_n}{\partial x_n} = 0$ for Cartesian coordinates. If all the integrals of this system except one are known, it reduces to $dX_1/U_1 = dX_2/U_2$; the known integrals restrict the lines of flow, so that it is confined to thin layers or sheets of flow on which $X_1$, $X_2$ are the coordinates. In each sheet there is a stream function, readily constructed in terms of $\rho$ and the law of thickness as involved in the configuration, so that its increment from $P_1$ to $P_2$ is the flow across any arc whatever connecting these points. This stream function is the final integral of the system; for it also is constant along every line of flow. Cf. Boltzmann, *Abhandl.* III (1893) p. 497: E. T. Whittaker, *Analytical Dynamics*, § 119.]

# APPENDIX II.

## ON GENERALIZED FORMAL THERMODYNAMICS.

### (i) Professor J. Clerk Maxwell (1876) *On Gibbs' Thermo-dynamic Formulation for Coexistent Phases.*

[From South Kensington Conferences (1876): reprinted in
*Philosophical Magazine*, November 1908, pp. 818–824.]

(THE paper here reprinted is the report of an Address delivered by
Professor Clerk Maxwell on May 24, 1876, at the South Kensington
Conferences in connection with the Special Loan Collection of Scientific
Apparatus. It is contained (pp. 144–150) in the official volume of
reports of the Conferences, which has long ago dropped out of notice.
An earlier and less complete version of this summary of Professor *[Maxwell's contemporary exposition of Willard Gibbs' thermodynamics.]* Willard Gibbs' developments of the doctrine of available energy, on
the basis of his new concept of the chemical potentials of the con-
stituent substances, was communicated to the Cambridge Philosophical Society on March 8th, 1876, and appeared in abstract in
Vol. II of their *Proceedings*, pp. 427–430; it was reprinted in Maxwell's
*Collected Papers*, Vol. II, pp. 498–500[: also *infra*, p. 713]. The fact that
the energy associated with any constituent substance is proportional
jointly to the mass of that constituent and to another factor representing energy per unit mass, is of course involved in the very notion
of conservation of energy; but the general factorization into magni-
tudes (or quantities) and intensities (such as the potentials of Gibbs) *[Quantities and intensities.]*
is, it would seem, here set forth formally and explicitly for the first
time, though it is thoroughly implied throughout Gibbs' work. This
paper reviews only the first portion of that work, *Proc. Connecticut
Academy*, Oct. 1875–May 1876; the second part of it, mainly special *[Willard Gibbs: dates:]*
applications, did not appear in the same *Proceedings* until May,
1877–July 1878. The principle, as restated in less guarded terms by
G. Helm (1887), that each type of (available) energy strives to pass *[G. Helm's energetics:]*
from positions of higher to positions of lower intensity, has been
supposed sometimes to mark a new departure in physical ideas: cf.
Professor W. Ostwald, *Die Energie* (1908), p. 103, whose regret, *[W. Ostwald's:]*
expressed in this connection, that he has not had access to Maxwell's
paper, has revived the project of reprinting it. Some years before
this time, in a paper "On the Mathematical Classification of Physical *[previously W. J. M. Rankine's.]*
Quantities," *Proc. Lond. Math. Soc.* Vol. III, Maxwell had pointed
out, in connection with Rankine's idea of factors of energy, that this
conception loses most of its definiteness and efficacy when applied to

kinetic phenomena. It is only for static or steady material transformations that it is effective; and it became so, in development of
the general Kelvin doctrine of available energy, only by virtue of

Gibbs' static the fundamental step involved in Gibbs' recognition of the existence
potentials: of quantitative chemical potentials for the independent constituent
substances of a mixture or solution—involving the formulation of
the general criterion of chemical coexistence of complex substances
in contact, that the potential of each constituent should be the same
in both of them*, and of the trend of chemical change, as towards
positions of lower aggregate total energy in an adiabatic system, or
of available energy (Kelvin) in isothermal change.

Most interesting also as regards this evolution of ideas is the postscript of a letter from Maxwell to Stokes of an earlier date, August 3,
1875, in which, in connection with Andrews' experiments on the

Maxwell's condensation of mixed gases, Maxwell sends a tentative sketch of
early the whole theory, using a provisional term *reaction*, printed between
anticipations. inverted commas, for the quantity which his friend Gibbs soon afterwards named chemical *potential*: see *Memoir and Scientific Correspondence of Sir G. G. Stokes*, Vol. II, pp. 33–35 (Cambridge, 1907),
also reprinted *infra*, p. 712.—J.L.)

The warning which Comte addressed to his disciples, not to apply
dynamical or physical ideas to chemical phenomena, may be taken,
like several other warnings of his, as an indication of the direction in
which science was threatening to advance.

Dynamics of We can already distinguish two lines along which dynamical
chemistry: science is working its way to undermine at least the outworks of
Chemistry; and the chemists of the present day, instead of upholding
the mystery of their craft, are doing all they can to open their gates
to the enemy.

Of these two lines of advance one is conducted by the help of the
molecular, hypothesis that bodies consist of molecules in motion; and it seeks
to determine the structure of the molecules and the nature of their
motion from the phenomena of portions of matter of sensible size.

The other line of advance, that of Thermodynamics, makes no
and thermo hypothesis about the ultimate structure of bodies, but deduces
dynamic. relations among observed phenomena by means of two general
principles—the conservation of energy and its tendency towards
diffusion. The thermodynamical problem of the equilibrium of
heterogeneous substances was attacked by Kirchhoff in 1855, when
the science was yet in its infancy, and his method has been lately
followed by C. Neumann. But the methods introduced by Professor

* The formula for the potential in each phase is to be determined from
experiments conducted within that phase.

J. Willard Gibbs, of Yale College, Connecticut[1], seem to me to be more likely than any others to enable us, without any lengthy calculations, to comprehend the relations between the different physical and chemical states of bodies; and it is to these that I now wish to direct your attention.

In studying the properties of a homogeneous mass of fluid, consisting of $n$ component substances, Professor Gibbs takes as his principal function the energy of the fluid, as depending on its volume and entropy together with the masses, $m_1$, $m_2$, ... $m_n$ of its $n$ components, these $n + 2$ variables being regarded as independent. Each of these variables is such that its value for any material system is the sum of its values for the different parts of the system. *(margin: Energy and entropy.)*

By differentiating the energy with respect to each of these variables we obtain $n + 2$ other quantities, each of which has a physical significance which is related to that of the variable to which it corresponds.

Thus, by differentiating with respect to the volume, we obtain the pressure of the fluid with its sign reversed; by differentiating with respect to the entropy, we obtain the temperature on the thermodynamic scale; and by differentiating with respect to the mass of any one of the component substances, we obtain what Professor Gibbs calls the potential of that substance in the mass considered. *(margin: Characteristic equation.)*

As this conception of the potential of a substance in a given homogeneous mass is a new one, and likely to become very important in the theory of chemistry, I shall give Professor Gibbs' definition of it. *(margin: Potentials of components.)*

If to any homogeneous mass we suppose an infinitesimal quantity of any substance added, the mass remaining homogeneous and its entropy and volume remaining unchanged, the increase of the energy of the mass, divided by the mass of the substance added, is the potential of that substance in the mass considered.

These $n + 2$ new quantities, the pressure, the temperature, and the $n$ potentials of the component substances, form a class differing in kind from the first set of variables. They are not quantities capable of combination by addition, but denote the intensity of certain physical properties of the substance. Thus the pressure is the intensity of the tendency of the body to expand, the temperature is the intensity of its tendency to part with heat; and the potential of any component substance is the intensity with which it tends to expel that substance from its mass. *(margin: Intensity factors:)*

We may therefore distinguish between these two classes of variables by calling the volume, the entropy, and the component masses the *(margin: contrasted with magnitudes.)*

[1] *Transactions of the Academy of Sciences of Connecticut*, Vol. III [: Gibbs' *Scientific Papers*, Vol. I].

*magnitudes*, and the pressure, the temperature, and the potentials the *intensities* of the system.

The problem before us may be stated thus: Given a homogeneous

**Stability of a chemical phase.** mass in a certain phase, will it remain in that phase, or will the whole or part of it pass into some other phase?

The criterion of stability may be expressed thus in Professor Gibbs' words:

For the equilibrium of any isolated system it is necessary and sufficient that in all possible variations of the state of the system which do not alter its energy, the variation of its entropy shall either vanish or be negative.

The condition may also be expressed by saying that for all possible variations of the state of the system which do not alter its entropy, the variation of its energy shall either vanish or be negative.

Professor Gibbs has made a most important contribution to science by giving us a mathematical expression for the stability of any given phase (*A*) of matter with respect to any other phase (*B*).

If this expression for the stability (which we may denote by the letter *K*) is positive, the phase *A* will not of itself pass into the phase

**Passive resistances may delay.** *B*; but if it is negative the phase *A* will of itself pass into the phase *B*, unless prevented by passive resistances.

**Measure of stability:** The stability (*K*) of any given phase (*A*) with respect to any other phase (*B*), is expressed in the following forms*:

$$K = \epsilon + vp - \eta t - m_1\mu_1 - \ldots - m_n\mu_n,$$

where $\epsilon$ is the energy, $v$ the volume, $\eta$ the entropy, and $m_1$, $m_2$, etc. the components corresponding to the second phase (*B*), while $p$ is the pressure, $t$ the temperature, and $\mu_1$, $\mu_2$, etc. the potentials corresponding to the given phase (*A*). The intensities therefore are those belonging to the given phase (*A*), while the magnitudes are those corresponding to the other phase (*B*).

**its meaning.** We may interpret this expression for the stability by saying that it is measured by the excess of the energy in the phase (*B*), above what it would have been if the magnitudes had increased from zero to the values corresponding to the phase *B*, while the values of the intensities were those belonging to the phase (*A*)†.

If the phase (*B*) is in all respects except that of absolute quantity of matter the same as the phase (*A*), *K* is zero; but when the phase (*B*) differs from the phase (*A*), a portion of the matter in the phase (*A*) will tend to pass into the phase (*B*) if *K* is negative, but not if it is zero or positive.

* The signs of the second and third terms in *K* have here been reversed.

† That is, if any rearrangement, with potentials supposed unchanged, would require more energy for the same entropy. The argument is clearer in Gibbs.

If the given phase (A) of the mass is such that the value of K is Protected stability: positive or zero with respect to every other phase (B), then the phase (A) is absolutely stable, and will not of itself pass into any other phase.

If, however, K is positive with respect to all phases which differ from the phase (A) only by infinitesimal variations of the magnitudes, while for a certain other phase, B, in which the magnitudes differ by finite quantities from those of the phase (A), K is negative, then the question whether the mass will pass from the phase (A) to the phase (B) will depend on whether it can do so without any transportation broken by contacts: of matter through a finite distance, or, in other words, on whether matter in the phase B is or is not in contact with the mass.

In this case the phase (A) is stable in itself, but is liable to have its stability destroyed by contact with the smallest portion of matter in producing explosive change, certain other phases.

Finally, if K can be made negative by any infinitesimal variations of the magnitudes of the system (A), the mass will be in unstable equilibrium, and will of itself pass into some other phase.

As no such unstable phase can continue in any finite mass for any or catalytic lapse. finite time, it can never become the subject of experiment; but it is of great importance in the theory of chemistry to know how these unstable phases are related to those which are relatively or absolutely stable.

The absolutely stable phases are divided from the relatively stable phases by a series of pairs of coexistent phases, for which the intensi- Coexistent pairs of phases separate stable ones. ties $p$, $t$, $\mu$, etc. are equal and K is zero. Thus water and steam at the same temperature and pressure are coexistent phases.

As one of the two coexistent phases is made to vary in a continuous manner, the other may approach it and ultimately coincide with it. The phase in which this coincidence takes place is called the Critical Phase*.

The region of absolutely unstable phases is in contact with that of Thomson-Andrews diagram: absolutely stable phases at the critical point. Hence, though it may be possible by preventing the body from coming in contact with certain substances to bring it into a phase far beyond the limits of absolute stability, this process cannot be indefinitely continued, for before the substance can enter a new region of stability it must pass continuity must ultimately break, out of the region of relative stability into one of absolute instability, when it will at once break up into a system of stable phases.

Thus in water for any given pressure there is a corresponding temperature at which it is in equilibrium with its vapour, and beyond

---

* See the reasoned description and analysis of the Gibbs' earlier thermodynamic surface, for a pure substance such as carbonic acid or water, inserted in later editions of Maxwell's *Theory of Heat*.

yet per-
manence of
super-heated
liquid may
be ensured,
which it cannot be raised when in contact with any gas. But if, as in the experiment of Dufour, a drop of water is carefully freed from air and entirely surrounded by liquid which has a high boiling point, it may remain in the liquid state at a temperature far above the boiling point corresponding to the pressure, though if it comes in contact with the smallest portion of any gas it instantly explodes.

within limits:
But it is certain that if the temperature were raised high enough the water would enter a phase of absolutely unstable equilibrium, and that it would then explode without requiring the contact of any other substance.

also of
super-cooled.
Water may also be cooled below the temperature at which it generally freezes, and if the water is surrounded by another liquid of the same density the pressure may also be reduced below that of the vapour of water at that temperature. If the water when in this phase is brought in contact with ice it will freeze, but if brought in contact with a gas it will evaporate*.

Early
examples of
multiple
coexistent
phases.
Professor Guthrie has recently discovered a very remarkable case of equilibrium of a liquid which may be solidified in three different ways by contact with three different substances. This is a solution of chloride of calcium in water containing 37 per cent. of the salt. This solution is capable of solidification at − 37° C., when it forms the solid cryohydrate having the same composition as itself. But it may be cooled somewhat below this temperature; and then if it is touched with a bit of ice, it throws up ice, if it is touched with the anhydrous salt it throws down anhydrous salt, and if it is touched with the cryohydrate it solidifies into cryohydrate.

## (ii) *Historical Note* (1927): Andrews, Maxwell, Willard Gibbs.

In *Proc. Roy. Soc.* 1875 there is a "Preliminary Note" by Andrews giving results of a long continued investigation on mixed gases, conducted with his usual extreme experimental precision. On the final page (*Scientific Papers*, p. 392) there occur the following sentences.

Critical points
of mixed
gases:
"The most important of these results is *the lowering of the critical point by admixture with a non-condensable gas.* Thus in the mixture mentioned above of carbonic acid and nitrogen, no liquid was formed at any pressure till the temperature was reduced below − 20° C. Even the addition of only $\frac{1}{10}$ of its volume of air or nitrogen to carbonic acid gas will lower the critical point several degrees." Thus,

---

* On the degree of definiteness of the critical phenomena, as explored by Andrews, cf. *supra*, p. 512.

[That subject has been re-opened fundamentally by the measurements of H. L. Callendar (*Roy. Soc. Proc.* Aug. 1928) on the influence of traces of dissolved air on steam-pressures. Cf. *infra*, p. 740.]

as he concludes, laws like those of Boyle and Gay-Lussac, and the law of Dalton which asserts that gases are immune to each other in mixture, "are interfered with, and in certain conditions the interfering causes become so powerful as practically to efface them." The subject was developed into most significant detail in the Bakerian Lecture for the following year, *Phil. Trans.* 1876.

These experiments naturally excited the close attention of Clerk Maxwell, and are the subject of the letter to Stokes, of date 3 August, 1875, here reprinted, with footnote, from *Scientific Correspondence of Sir George Gabriel Stokes*, Vol. II (1907) p. 34. This letter in effect sketches the general theory of coexistent complex phases of matter, as exemplified by Andrews' concomitant liquid and gaseous states of mixed substances, the circumstances of their coexistence and of their merging into one homogeneous phase. When Maxwell a few months later came into possession of the earlier part of Willard Gibbs' memoir[*], he became the expositor of Gibbs' general methods: his own independent synthesis, which had naturally stimulated his interest, came to light only thirty years later in sorting out the scientific manuscripts left by Stokes. suggested Maxwell's theory of coexistent states:

It may be noted that the final equation of equilibrium between two phases in contact, as here reprinted, is the equivalent of Willard Gibbs' integral equation of state

$$e = -pv + \theta\phi + r_1q_1 + r_2q_2 + \dots + e_0,$$

where $e_0$, now added, is the intrinsic atomic energy of the unit mass to which the equation belongs, which does not change when the system changes its physical and chemical state.

"I have been thinking about Dr Andrews' experiments on mixtures of N and $CO_2$.

I find the conditions of equilibrium between two mixtures in different states to be[1]:

[*] The dates of printing of the sheets of Gibbs' great memoir are Oct. 1875 onwards, the phase rule appearing Jan. 1876.

[1] These are precisely the general analytical conditions for the equilibrium of coexistent heterogeneous substances that were formulated by Willard Gibbs in his epoch-making memoir, of which the first part was communicated to the Connecticut Academy in October 1875. Professor Gibbs employed the term "potential" of a constituent substance for what is here provisionally named "reaction." The theory was probably elaborated by both Gibbs and Maxwell from Gibbs' Thermodynamic Surface, of which an account was inserted by Maxwell in the fourth edition (1875) of his *Theory of Heat* (pp. 195–208), giving the method he employed in constructing the surface to scale for water substance. On March 8, 1876, Maxwell, then President of the Cambridge Philosophical Society, communicated to the Society an account of Gibbs' methods "which seem to me to throw a new light upon Thermodynamics" (*Proceedings*, Vol. II, pp. 427–430). The special problem which here suggested to Maxwell this general investigation, has been developed with much success by van der Waals and his pupils on the same lines as those indicated above. Historical.

Let $e$ = energy of unit of mass of mixture ($e = f(v, \phi, q_1, \dots q_{n-1})$),

$v$ = volume of do.

$\phi$ = entropy of do.,

$q_1, q_2, \dots q_{n-1}$ the masses of the substances $_1, _2, \dots _{n-1}$ in unit of mass of the mixture (there being $n$ constituents). Then

his potentials of components,

$$\text{pressure} = p = -\frac{de}{dv}, \quad \text{temperature} = \theta = \frac{de}{d\phi},$$

$$\text{``reaction,''} = r_1 = \frac{de}{dq_1}, \quad r_2 = \frac{de}{dq_2}, \quad \text{etc.,}$$

and the conditions of equilibrium between two mixtures, one of which is distinguished by an accent, are

$$p = p', \quad r_1 = r_1',$$
$$\theta = \theta', \quad r_2 = r_2', \quad \text{etc.,}$$

and

$$e + pv - \theta\phi - r_1 q_1 - r_2 q_2 - \dots = e' + pv' - \theta\phi' - r_1 q_1' - r_2 q_2' - \dots.$$

and conditions for co-existence:

These conditions seem to me to be all right. The difficulty is to conceive clearly the energy as a function of volume, entropy, and composition, say in liquid $CO_2$ saturated with absorbed N in contact with a gaseous mixture of $CO_2$ and N.

assisted by the Gibbs' thermo-dynamic surface.

But it seems plain that if the thermodynamic surface* (coordinates, volume entropy and energy) has a valley or hollow place which works itself out at the critical point, after which the surface is convexo-convex, and if the surface for N is everywhere convex, then the effect of mixing N with $CO_2$ will be to make the head of the valley more convex, that is it will work itself out sooner, or the critical point will be lowered.

But the difficulty is to see how to form the surface for a mixture when those of the constituents are given.

However, I must interpret my conditions in this case, and in that of two liquid mixtures, as of benzol and (alcohol and water), and in solutions of salts (supersaturated, etc.), and in solutions of gases (supersaturated, etc.)."

The writer now ventures to round off this history by reprinting also the very brief and weighty communication of Professor Maxwell, then President, to the Cambridge Philosophical Society, *Proceedings*, Vol. II, March 8, 1876, from his *Scientific Papers*, Vol. II, pp. 498–500. The lecture at South Kensington, reprinted above from the report, was of date May 24, 1876.

* See Maxwell's *Theory of Heat*, ed. 2.

### (iii) Professor J. Clerk Maxwell (1876) *On the Equilibrium of Heterogeneous Substances.*

[From *Proc. Camb. Phil. Soc.* Vol. 11, March 8, 1876.]

THE thermodynamical problem of the equilibrium of heterogeneous substances was first attacked by Kirchhoff in 1855, who studied the properties of mixtures of sulphuric acid with water, and the density of the vapour in equilibrium with the mixture. His method has recently been adopted by C. Neumann in his *Vorlesungen über die mechanische Theorie der Wärme* (Leipzig, 1875). Neither of these writers, however, makes use of two of the most valuable concepts in thermodynamics, namely, the intrinsic energy and the entropy of the substance. *Kirchhoff.*

It is probably for this reason that their methods do not readily give an explanation of those states of equilibrium which are stable in themselves, but which the contact of certain substances may render unstable.

I therefore wish to point out to the Society the methods adopted by Professor J. Willard Gibbs of Yale College, published in the *Transactions of the Academy of Sciences of Connecticut*, which seem to me to throw a new light on thermodynamics.

He considers the intrinsic energy ($\epsilon$) of a homogeneous mass consisting of $n$ kinds of component matter to be a function of $n + 2$ variables, namely, the volume of the mass $v$, its entropy $\eta$, and the $n$ masses, $m_1, m_2, \ldots m_n$, of its component substances. *The general method of Gibbs:*

Each of these variables represents a physical quantity, the value of which, for a material system, is the sum of its values for the parts of the system.

By differentiating the energy with respect to each of these variables (considered as independent), we obtain a set of $n + 2$ differential coefficients which represent the intensity of various properties of the substance. Thus,

$\frac{d\epsilon}{dv} = -p$, where $p$ is the pressure of the substance;

$\frac{d\epsilon}{d\eta} = \theta$, where $\theta$ is the temperature on the thermodynamic scale;

$\frac{d\epsilon}{dm_1} = \mu_1$, where $\mu_1$ is the potential of the component ($m_1$) with respect to the compound mass.

Each of the component substances has therefore a potential with respect to the whole mass.

The idea of the potential of a substance is, I believe, due to Professor Gibbs. [Cf. footnote *supra*, p. 711.] His definition is as follows: *his potentials:*

If to any homogeneous mass we suppose an infinitesimal quantity of any substance to be added, the mass remaining homogeneous, and its entropy and volume remaining unchanged, the increase of the energy of the mass, divided by the mass of the substance added, is the _potential_ of that substance in the mass considered.

The condition of the stable equilibrium of the mass is expressed by Professor Gibbs in either of the two following ways:

I. _For the equilibrium of any isolated system it is necessary and sufficient that in all possible variations of the state of the system which do not alter its energy, the variation of its entropy shall either vanish or be negative._

II. _For the equilibrium of any isolated system it is necessary and sufficient that in all possible variations of the state of the system which do not alter its entropy, the variation of the energy shall either vanish or be positive._

contact criterion.

The variations here spoken of must not involve the transportation of any matter through any finite distance.

Conditions for coexistence of phases:

It follows from this that the quantities $\theta, p, \mu_1, \dots \mu_n$ must have the same values in all parts of the mass. For if not, heat will flow from places of higher to places of lower temperature, the mass as a whole will move from places of higher to places of lower pressure, and each of the several component substances will pass from places where its potential is higher to places where it is lower, if it can do so continuously.

Hence Professor Gibbs shows that if $\Theta, P, M_1, \dots M_n$ are the values of $\theta, p, \mu_1, \dots \mu_n$ for a given phase of the compound, and if the quantity

$$K = \epsilon - \Theta\eta + Pv - M_1 m_1 - \dots - M_n m_n$$

for stability of a phase.

is zero for the given fluid, and is positive for every other phase of the same components, the condition of the given fluid will be stable.

If this condition holds for all variations of the variables the fluid will be absolutely stable; but if it holds only for _small_ variations but not for certain finite variations, then the fluid will be stable when not in contact with matter in any of those phases for which $K$ is positive, but if matter in any one of these phases is in contact with it, its equilibrium will be destroyed, and a portion will pass into the phase of the substance with which it is in contact.

Discussion of a special case.

Thus in Professor F. Guthrie's experiments, a solution of chloride of calcium of 37 per cent. was cooled to a temperature somewhat below $-37°$ C. without solidification.

In this state, however, the contact of three different solids determines three different kinds of solidification. A piece of ice causes ice to separate from the fluid. A piece of the cryohydrate of chloride of calcium determines the formation of cryohydrate from the fluid, and the anhydrous salt causes a precipitation of anhydrous salt.

The phase of the fluid is such that $K$ is positive for all phases differing slightly from its own phase, and its equilibrium is therefore stable, but for certain widely different phases, namely, ice, cryohydrate and anhydrous salt, $K$ is negative.

If none of these substances are in contact with the fluid, the fluid cannot alter in phase without a transport of matter through a finite distance, and is therefore stable; but if any one of them is in contact with the fluid, part of the fluid is enabled to pass into a phase in which $K$ is negative. The conditions of coexistent phases are that the values of $\theta$, $p$, $\mu_1, \dots \mu_n$, and $K$ are equal for all phases which can coexist in equilibrium, the surface of contact being plane.

This was illustrated by Mr [P. T.] Main's experiments on coexistent phases of mixtures of chloroform, alcohol and water.

### (iv: 1928) *The Types of Characteristic Equation determining a Thermodynamic System.*

THE following systematic synopsis is now submitted. The parallel graphical scheme involved in Maxwell's discussion of the Gibbs Thermodynamic Surface (1872), with entropy, energy and volume as coordinates, contained in the later editions of the *Theory of Heat* (pp. 195–208), can illuminate the scope and the limitations of the principles that are involved. *(margin: Maxwell on the Gibbs' thermodynamic surface.)*

The characteristic equation for any mass of homogeneous substance in a steady state is *(margin: Equation of internal energy:)*

$$\delta E = \theta \delta \eta + \mu_1 \delta m_1 + \mu_2 \delta m_2 + \dots - Q_1 \delta q_1 - Q_2 \delta q_2 - \dots.$$

Herein it is implied that the energy of the system is a definite function of its quantity of entropy $\eta$, the quantities $m_1, m_2, \dots$ of its independent component substances, and its geometric coordinates $q_1, q_2$, on which forces $Q_1, Q_2$, elastic or other, of origin *internal* to the system do work: there is only one thermal quality, temperature, necessitating one characteristic equation, of whatever form.

Also forces of extraneous origin $P_1, P_2, \dots$ may act on coordinates $p_1, p_2, \dots$ of the system, including say an external pressure $p$ acting against its volume $v$. When the processes are slow so that no energy of motion is involved, there must be mechanical equilibrium at each stage: therefore by Lagrangian virtual work *(margin: equation of mechanical equilibrium.)*

$$Q_1 \delta q_1 + Q_2 \delta q_2 + \dots + P_1 \delta p_1 + P_2 \delta p_2 + \dots - p \delta v = 0.$$

The forces here involved may be in part frictional: then this quantity, which vanishes in equilibrium, will not be the differential of a potential energy function.

Thus free of all restriction except to homogeneity of the phases and absence of motional energy in bulk,

$$\delta E = \theta \delta \eta + \mu_1 \delta m_1 + \mu_2 \delta m_2 + \dots + P_1 \delta p_1 + P_2 \delta p_2 + \dots - p \delta v.$$

For the usual chemical reactions the system is in the fluid state, so the only extraneous force is atmospheric or other pressure $p$, and

$$\delta E = \theta \delta \eta - p \delta v + \mu_1 \delta m_1 + \mu_2 \delta m_2 + \dots.$$

The extraneous forces such as $p$, $P$ do not depend elastically on the coordinates, and so on the form of the energy of the system, as the internal forces $Q_1$, $Q_2$, ..., now eliminated by the condition of mechanical equilibrium, would do. They can be adjusted arbitrarily. Even if the system reacting to them is a gas, it can be imagined as enclosed in a flexible envelope, and so subject to the pressure $p$ of an atmosphere outside; this can for instance be maintained at a constant value, to which the course of steady reaction inside the system must adjust itself.

Under circumstances now to be determined, an integrated characteristic equation between the variables of the system can be formed. Each potential $\mu$ is constant throughout the whole of the homogeneous component mass $m$, to which it belongs: the only exception is close to an interface of transition between two phases coexisting in contact, where the correction takes the form, intimately discussed by Gibbs for the first time, of an interfacial distribution of energy. If also the external pressure is the same for two regions of the system, both under identical internal conditions, they can be united into one equation: thus by successive additions an integral form is obtained

$$E = \theta \eta - p v + \mu_1 m_1 + \mu_2 m_2 + \dots.$$

Formally we may reason thus: if addition of substance is made to the system so that $\delta m_1$ is $k m_1$, $\delta m_2$ is $k m_2$, ... where $k$ is infinitesimal, then $\delta \eta$ is $k \eta$ and $\delta v$ is $k v$, while the variables of state $p$, $\theta$, $\mu_1$, $\mu_2$ are unaltered (cf. Bryan, *loc. cit.*): the significant feature is that this addition can be made anyhow, for example the system may be stirred up and it will settle down again to the same *state* of internal equilibrium. Substitution of these values for the differentials and division across by $k$ produces this integral equation.

But now it is proposed to take the complete differential of this equation, with all the quantities including $p$, $\theta$, $\mu_1$, $\mu_2$, ... subject to variation. This addition to the variables means that we now compare the system in the original state with the same system in an altered state, and subject to a different uniform pressure. The transition as regards $p$ can be made so as to be traceable in detail, by imagining the system to be pushed across a partition separating an atmosphere at $p_1$ from one at $p_2$, which involves work done on it of amount

$p_2v_2 - p_1v_1$: an illustration is afforded in the thermometric explorations of Joule-Kelvin by pushing a gas through a porous plug into a region at lower pressure. The transition as regards potentials arises, without need of explanation, from gradual change of the densities of the components in the system.

Combining the unrestricted total differential of the expression for $E$ as thus justified, with the previous equilibrium form for $\delta E$, there arises, with Gibbs, another universal equation of total differentials, *combined with equilibrium equation of energy,*

$$0 = \eta\delta\theta - v\delta p + m_1\delta\mu_1 + m_2\delta\mu_2 + \dots.$$

It expresses that $p$ cannot change so long as $\theta, \mu_1, \mu_2, \dots$ remain unchanged: therefore that $p$ is a function of the temperature and these potentials of the components, and of nothing else, say *leads to variational equation of pressure,*

$$p = f(\theta, \mu_1, \mu_2, \dots).$$

This is an integral equation of *state* of the system, contrasted with an equation of *quantities*, or *magnitudes* in Maxwell's phrase, for example as *infra*,

$$p = F(\sigma, \rho_1, \rho_2, \dots),$$

where the density $\rho_1$ is $m_1/v$, amd $\sigma$ is density of entropy.

This fundamental equation, expressive of state irrespective of quantity, when it is known determines the densities immediately by the form of its differential *involving densities in terms of potentials:*

$$\delta p = \sigma\delta\theta + \rho_1\delta\mu_1 + \rho_2\delta\mu_2 + \dots.$$

We can modify it to the form

$$\delta(\sigma\theta + \mu_1\rho_1 + \mu_2\rho_2 + \dots - p) = \theta\delta\sigma + \mu_1\delta\rho_1 + \mu_2\delta\rho_2 + \dots.$$

Thus $\sigma\theta + \mu_1\rho_1 + \mu_2\rho_2 + \dots - p$ must be a function of $\sigma, \rho_1, \rho_2, \dots$ alone, and the partial gradients of this function are the temperature and the potentials: it is in fact none other than the density of energy in the system. *or of energy, involving potentials in terms of densities.*

Other modified characteristic functions, expressible in other sets of variables, may be formed similarly, after the manner initiated by Hamilton (Vol. I, *supra*, Appendix I) in optics and dynamics: and each of them gives rise to a scheme of conjugate relations between cross gradients of the coefficients in its complete differential, with physical interpretations in the manner of Hamilton, Thomson and Tait, Helmholtz, Rayleigh, and in thermodynamics of Maxwell. *Mixed forms of characteristic: reciprocal relations involved.*

We thus have completely determined a set of fundamental equations, in Gibbs' terminology, two of them of predominant interest— one purely of quantities, another purely of physical states irrespective of quantity[1]—and a variety of equations, not so fundamental, of mixed

It seems strange that, so far as the writer can make out, the standard expositions, in treatises and encyclopaedias, nearly all omit any reference to the more remarkable half of Gibbs' formulation, named here the characteristic quantities and states. Two of them are independent, as mechanical equilibrium and thermal equilibrium are both presumed.

This procedure has been set out minutely because, for example, in the case where the system is a mass of gas, the first impulse may be to take $p$ as the pressure of the gas determined internally by its volume and temperature, instead of an independent arbitrary pressure imposed slowly from an outside atmosphere to which its internal state has to conform. The former procedure, being too narrow, would lead straight to contradictions.

To illustrate this distinction we have, following Gibbs, for a quantity $m$ of perfect gas

$$E = m\kappa\theta + E_0, \quad dH = dE + p\,dv, \quad pv = mR'\theta, \quad R' = R/M,$$

so that

$$\eta = \int \frac{dH}{\theta} = m\kappa \log \theta + mR' \log v.$$

The integral form of equation

$$E = \theta\eta - pv + m\mu$$

can thus give the value of the potential $\mu$; which turns out to be such as to make the equation

$$0 = \eta\,d\theta - v\,dp + m\,d\mu$$

an exact differential, thus verifying the validity of the procedure.

In the theoretical procedure adopted by Gibbs for a mixture, the necessary postulate defining perfect gases, without recourse to molecular theory, is taken to be the principle of Dalton, that each constituent gaseous substance exerts its own independent pressure as if the others were not there. Then for the mixture the characteristic equation of state gives $p$ is a sum of independent expressions, one for each gas: and the properties of the mixture are evolved from the differential of the characteristic equation of state so expressed, its form as a simple summation indicating independence of properties of the component gases.

It may be recalled that the original procedure of Rayleigh (1875) took the entropy $\eta$ to be additive, in the same way as $p$ here, on the

equation of state of the given phase of the substance: in that familiarly necessary equation, connecting, say for a fluid, pressure with temperature and constitution, by itself alone, all the properties of the phase to which it belongs are implicitly involved. This may possibly in some cases arise from confusion of imposed external forces with natural forces belonging to the energy of the system, from which the writer has in some degree suffered, and may be a justification for setting out the subject in more explicit exposition. Exceptions are G. H. Bryan's *Thermodynamics* (Teubner, 1907) and an exposition by the writer in the Supplement (1902) to the *Encyclopaedia Britannica*. The general procedure of Bryan is on the lines of the Kelvin available energy, *supra* (cf. p. 605), as handed on in Maxwell's *Theory of Heat*.

basis of a postulate that reversible absorption of a component by contrasted with Rayleigh's earlier principles. a liquid phase in contact with the gas may be contemplated among thermodynamic processes: thence were deduced the other properties of mixtures of perfect gases—including Henry's law of independent solubilities each proportional to pressure, which is subject to limitation as regards densities, and can itself be made an empirical foundation for the theory.

The problem of the thermodynamics of concomitant gaseous and liquid phases of a mixture of imperfect gases, originated by the searching experimental investigations of Andrews on the critical Critical points. points of mixtures of carbonic acid and nitrogen, led Maxwell (*supra*, p. 712) independently into the field of thermodynamic potentials.

A strictly cognate theory arises for the problem of solutions of Osmotics. mixed substances, where it is osmotic pressures instead of gas pressures that are involved.

Finally the theory has asserted itself on the lead of Saha (cf. Vol. I, Stellar atmospheres. *supra*, p. 666) in the study of the stratification of stellar atmospheres. The appearance of each spectral line is associated with an energy *quantum*, identified with the energy of an impinging exciting electron, on the hypothesis of an electron gas as a constituent of the mixed atmosphere in equilibrium of dissociation: the earliest appearance of a radiation becomes correlated, through this minimal energy required for the exciting electrons, with the pressure and temperature of the region in which it appears.

Let us now revert to the underlying principles. The recognition is The entropy of Clausius: due to Clausius that a definite thermodynamic quantity entropy, whose value is a function of the state of the system, subsists alongside the other universal quantity energy, of which the existence had for at least half a century been presumed. (Cf. *supra*, p. 603.)

This function has physical properties, which mathematically are of original and peculiar type. In an isolated system it cannot spontaneously diminish: thus in a change that is reversible in the Carnot sense it is conserved, but otherwise it must increase. A map or model prescribes trend of change: expressing, by the points in its space, a range of possible states of an isolated system, can envisage the natural trend of entropy as always uphill so to speak. This feature of course by no means fixes a path: the information thus afforded is slight. But it becomes definite when the steady states of the system, its equilibria, are sought: for they are the peaks of maximum entropy on the representative curve or determines equilibria. surface. Each of them represents a permanent equilibrium of the system—until by a suitable modification some path permitting a further continued ascent of entropy into higher values is discovered

Catalysis. and opened up, that being the opening out of a catalytic reaction in the system, such as was previously excluded by absence of the necessary contacts.

If a system reacts internally so slowly that uniform temperature is maintained, the constitutive processes are reversible and therefore its entropy remains constant. But this result is modified if the system gains heat from outside: a gain $\delta H$ increases its entropy by $\delta H/\theta$, so the internal constitution has to change in order to correspond with this new value for its entropy as a function of its structure.

The energy of the system is another function of its internal configuration: if it gains thermal energy $\delta H$, and work $\delta'W_e$ from the operation of external forces, including say $-p\delta v$ as that of a uniform external pressure—the accent on $\delta$ implying that the work need not be cyclic, thus is not a perfect differential of a function of state—then the internal constitution of the system must adapt itself also to this new value of the energy.

Thus $E$ and $\eta$ being the energy and entropy of the system, both determined intrinsically in each stage by its constitution alone, the principles of equilibrium, mechanical and thermodynamic, give rise as above to

$$\delta E = \delta'H + \delta'W, \quad \delta\eta = \delta'H/\theta.$$

For our system, supposed to be maintained at uniform temperature throughout, we can now get rid of $\delta'H$, which, not being connected with any function of constitution, is entirely unconditioned—retaining however the work $\delta'W$ of the external forces, also unconditioned, but essential as being the subject matter of all physical experiment. The resulting universal form is

Fundamental equation.

$$\delta E = \theta\delta\eta + \delta'W.$$

The system may be made up of two or more distinct states in free contact, coexistent phases, with matter transferable from one to the others. The part of the energy that is in each phase then depends on the amounts of each independent kind of matter, say of each component, that are present in it.

Free reversible interchange of matter between the phases in the system at uniform temperature cannot alter its total entropy: but forced interchange could do so, the entropy in each phase being therefore a function of $m_1$, $m_2$, ... as well as of the configuration.

The value of the change of entropy $\delta\eta$ for any such change, if infinitesimal, is in each phase given by the equation

$$\delta E = \theta\delta\eta + \mu_1\delta m_1 + \mu_2\delta m_2 + \dots + \delta'W,$$

where, in the standard chemical case of interaction of fluid phases, $\delta'W$ is simply $-p\delta v$.

The principle of Clausius requires that for the whole system the natural trend of the entropy, as aggregated by summation over all its phases, must be towards increase, provided the system is isolated and so receives no accession of energy of any kind from outside. If we are to extend the application to systems which are not isolated, for which $\delta E$ is thus equal to accession of heat $\delta h$ plus that of work $\delta'W$, this principle must be enlarged. The essential thing is that the trend of internal entropy $\eta_i$ of the system must be upward owing to inevitable irreversible internal interactions operating with diffusion of heat. But the variation of the total entropy $\eta$, which is a function of the constitution of the system, is $\delta\eta$ equal to $\delta\eta_i + \delta h/\theta$. As $\delta E$ is equal to $\delta h + \delta'W$, this criterion that the change of $\delta\eta_i$ must be positive absolutely is the same as that the trend expressed by $\delta\eta - \delta h/\theta$ must be positive, which is that $\delta\eta - (\delta E - \delta'W)/\theta$ must be positive or that $\delta E - \delta'W - \theta\delta\eta$ must be negative[1].

This criterion is unconditioned, and applies to any system whatever receiving heat and work from outside. But its application is to be regulated by the intrinsic characteristic equation defining the system by a restriction on the independence of its variables of quantity, or of state, or any mixed set of variables of equal number. The equation of state merely expresses the familiar restriction that the temperature, as recognized empirically, is a function of the pressure and the constitution of the fluid substances. As here evolved, there must be a conjugate fundamental equation of quantities: in it the variable conjugate to $\theta$ is the entropy $\eta$, or rather the density of entropy $\sigma$. There exists in Nature only one such thermal variable $\theta$ and its one conjugate $\eta$. Finally in this ideal evolution of the system from its equation of state the density of energy can arise as above as the function reciprocal to $p$. All this may be formulated from various aspects according to choice of variables, aspects which consolidate into one essential or invariant form in a spatial model, after the manner of Gibbs. *Universal criterion for trend of the system, subject to characteristic equation.*

Various special cases of this rather intangible general criterion, $\delta E - \delta'W - \theta\delta\eta$ negative, acquire practical physical significance. *Sub-cases physically significant:*

(i) If the system is isolated, so that $\delta E$ and $\delta'W$ are null, it can change spontaneously only in directions which make $\delta\eta$ positive, that is of increasing internal entropy, as above. It trends towards maximal values of $\eta$, the higher maxima being the ultimate secular states.

(ii) If heat is supplied so as to maintain constant the uniform temperature $\theta$ of the system during transformation, and there is no

---

[1] This argument agrees, in substance, with the procedure of Planck, *Thermodynamics*, § 140, as *supra*, p. 104.

work imparted to the system from outside, it can change freely only in directions of diminution of the constitutive function $E - \theta\eta$, a quantity which is for this reason named the available energy, or free energy, at that temperature.

The change of free energy can by a natural extension be defined so as to include the work imparted by the applied outside forces and absorbed at each stage into change of the internal mechanical energy function of the system; unless they do no work, as for example in a fluid system when there is constraint to constant volume. As a sub-case,

(iii) If the extraneous pressure is maintained constant instead of the volume, under maintenance of a constant temperature as before, it is the function $E - \theta\eta + pv$ that must diminish in all spontaneous change: while

(iv) At constant entropy instead of maintained temperature it is $E + pv$ that must diminish, but this restricts to reversible internal operations.

in relation to Gibbs' thermo-dynamic surface.    All these results refer to mere trends of change: they are best visualized diagrammatically, in a hyperspace in which each point represents in Cartesian manner a state of the system, in which therefore paths of transition are expressed as curves. Their ultimate intuitive source is in the thermodynamic surfaces of Gibbs (1873), immediately grasped and illustrated by Maxwell's construction at the Cavendish Laboratory of a model to scale of the surface for a substance such as water or carbonic acid, as described in the second edition of his *Theory of Heat*, where he includes an interesting attempt to introduce the formal geometers to this fundamental application of the ideas of their abstract science.

Actual model.

These special corollaries to the general principle may be widely extended, keeping within this mode of formulation. For instance in a solution of a salt, the dissolved substance, say $m_1$, may be prevented by the nature of the interfacial transition from spontaneous passage into the vapour phase, which is pure steam mixed it may be with air. One condition for a steady state, that the potential $\mu_1$ is the same in the two phases, liquid and vapour, thus disappears, being replaced by this restriction of excluded transfer. Yet the density of $m_1$ in the liquid phase affects the value of $\mu_2$, the potential of the other substance, the solvent water, that is present in that phase, therefore also its equal potential $\mu_2$ in the steam phase into which it is permeable, which in turn involves an influence on the density, or pressure, of the steam. This is the simplest example of an interface of so-called semi-permeable type, between two phases. Generally an osmotic partition

Limited inter-change:

introduces modification:

solution and its vapour:

semi-permeable osmotic partition:

—first scientifically imagined as permissible in the present connection by Gibbs in advance of experiments, botanical or other—can prevent passage of certain constituents while the others are free to pass: but the densities of the permeable constituents on the two sides of it nevertheless affect the potentials of the other impermeable constituents there, and therefore the densities of these other components in a final state of osmotic equilibrium—a far-reaching conception. It is thus that Andrews' investigation on the critical points of mixed gases immediately suggested, as *supra*, the principle of equated potentials, or as he called them provisionally "reactions," of the components to Maxwell. Or again, in ionized solutions, a partition impermeable to one ion will affect the densities of the other ions on the two sides and so, in conditions when a steady state proves to be possible, produce an osmotic difference of electric potentials on the two sides, after the manner of G. Wiedemann and more precisely of Nernst, *supra*, p. 105.

*types of significant inferences:*

*material,*

*voltaic.*

The existence of a fundamental equation of state alone, exclusive of quantities, as above, for each of the coexistent phases, rendered in Gibbs' hands a limitation to the possible number of coexistent phases intuitive. For there are $n$ potentials of the $n$ independent constituents, which are the same in the steady state in all the $r$ phases; but there is a fundamental equation in each phase connecting them with two new universal variables of state, say $p$ and $\theta$, there being $r$ of these equations in all: thus there are $n + 2 - r$ degrees of freedom left. Therefore the maximum possible numbers of phases is $n + 2$. If these all actually coexist, it can only be at determined values of the pressure and temperature: if only $n + 1$ phases are present, there will be one necessary relation between $p$ and $\theta$: when only $n$ phases or fewer coexist, the temperature and pressure are both variable at choice: and so on.

*The Phase Rule an immediate result.*

The usual modes of deduction of the Phase Rule, which ignore Gibbs' fundamental equation of state, are much more intricate, though they too must reduce ultimately to a mere count of degrees of freedom.

# APPENDIX III (1927).

# THERMODYNAMIC CYCLES IN RELATION TO STATISTICS.

## (i) *Formal Synthesis of Available Energy and Entropy*\*.

THE main perplexity of formal thermodynamics relates to the dividing line between mechanical energy and thermal energy. In which divi-

*Are Brownian movements thermal?* sion are the visible Brownian movements of large particles within a gas to be classed? Might they even not be amenable to Carnot's principle, work being got out of them individually as a gliding aeroplane extracts available energy from confused air currents? Energy

*Diffusion as reversible but with loss of availability.* molecularly available runs down in the process of mixing of two gases, though no thermal exchanges are involved: for, as Rayleigh pointed out (1875), the mixed gases can be separated by an ideal absorption of one of them into a solid or liquid substance with which it has definite affinity, so that it can be emitted later by itself into a space of equal volume, the gases, though intimately mixed originally, thus being separable, and all in a purely reversible manner.

The availability would thus be unaltered by the process of separation†, and the same as for the unmixed gases each by itself at the actual volume of the mixture; which may be regarded as an exact

*Dalton's postulate,* statement (not a demonstration) of Dalton's principle that in a mixture in the ultimate steady state each component gas disregards the others. For this reversibility in the process of absorption implies small density, and therefore perfect gases. As at constant temperature the value of available energy is $E - \theta\eta$, and $E$ is not altered by separation, there being no internal energy, it follows that the entropy $\eta$

*as a law of entropy.* is additive for the constituent gases when reckoned as if each occupied by itself the entire volume of the system.

But the idea of entropy, and of available energy also, originally arose in connection with transfer of sensible heat in the cycles of practical thermal engines, machines which existed in effective forms long before thermodynamics. May we now extend the idea and assert

*Entropy as a universal function,* that there is always an entropy, expressible as a function of constitution, even when transfers of heat have become infinitesimal? That would be a natural enlargement of the Carnot chain of ideas: and formal thermodynamics would thus be extended (Thomson, Clausius)

---

\* Cf. *supra*, pp. 101, 603: also the treatment in the later editions of Maxwell's *Theory of Heat* or in G. H. Bryan's *Thermodynamics* (Teubner, 1907).

† Just as the electromotive power of a voltaic cell is unaltered by spontaneous solidification of its materials.

from the theory of heat engines to a general theory of molecular energy and of the limitations imposed on its relations to mechanical phenomena on a large scale.

In Carnot's time the perplexity as to the nature of heat was fundamental; it was then not merely a question of a line of demarcation of heat from other types of energy. The relevant problem would present itself at that period in the form that there exists an entity named heat, measurable calorimetrically after the manner of Black: that there is temperature, with the fundamental tendencies that are revealed directly, as so happens, by the perceptions from our organs of sense: and then there stands Carnot's postulate defining the conditions under which cyclic subsidence of heat to lower temperature could be utilized to obtain mechanical work, in the manner already achieved very conspicuously and suggestively by the engineers. What can be inferred about the nature of heat from these limited foundations alone? The formal answer as developed above (p. 596) is that logically there must be, on the basis of the Carnot postulate, a standard (though ideal) calorimetric substance on which heat can be measured, and a standard (ideal) thermometric arrangement on which temperature can be registered, so that for every simplified Carnot cycle a standard formula then holds good,

$$\frac{H_1}{\theta_1} = \frac{H_2}{\theta_2} = \frac{W_{12}}{\theta_1 - \theta_2},$$

where $W_{12}$ is the work done on outside bodies in such reversible cycle in which the operating system takes in heat $H_1$ at temperature $\theta_1$ and rejects heat $H_2$ at $\theta_2$. This formula involves that $H_1 - H_2$ is equal to $W_{12}$, thus that heat is convertible in the course of running down, but only in part and under limitations of availability, into an equivalent of mechanical work.

There can however possibly arise within this formulation a limiting case, when the zero of temperature is supposed to be infinitely remote, while $\theta_1 - \theta_2$ is finite: in such cases $W_{12}$ is still $H_1 - H_2$ but is infinitely small compared with either $H_1$ or $H_2$. If heat is measured by calorimeter, $H_2$ and $H_1$ are now practically equal, and the relative value of the Joule mechanical equivalent becomes infinite: so that under these limiting conditions heat passes virtually unchanged in amount —as Carnot assumed in accordance with the ideas of his time, which soon ceased to be his own—while the work is done owing to mere fall in temperature, just as the work of a stream of water on a turbine wheel is due to mere fall of level.

When thermodynamics is expanded from the special theory of heat engines, the overt fundamental principle is the Kelvin inevitable

The principle
of dissipation:

recovery pos-
sible but
improbable:

entropy as
the con-
venient
measure of
probability.

Limitations.

Reversible
paths always
presumed,

as required
to explore
the entropy
as a function
of state of
the system,

constant for
internal
change, when
reversible,
otherwise
increasing.

Contrast with
energy of the
system.

dissipation of energy, which postulates that, owing to the molecular organization of energy in matter, its availability for operations in bulk tends continually to run down. In an early illustration, Kelvin figured out the infinitesimal value of the chance there would be of any proportion of abnormal molecular velocities becoming restored again in a gas which had subsided into equilibrium. Soon came Boltzmann's illuminating remark that the logarithm of this probability of occurrence as regards the microcosmic state of any molecular system, in the simple cases where it could be analytically formulated, would possess all the specific qualities of the entropy which Clausius found to be inherent in the Carnot postulate: it would be an additive function of constitution, and it must tend naturally to increase.

The question whether there is for every system a definite entropy, or whether that function can be formulated definitely in terms of constitution only for the simpler and physically manageable systems, is one that in the nature of the case does not admit of an answer. Fundamental issues relative to vital activities here emerge. Cf. *supra*, p. 603.

We can imagine, for clearer intuition, the various states of a material system as suitably mapped out, for example as expressed by points in a hyperspace with one dimension for each coordinate of state of the system and another for the temperature. Each pair of states may then be connected by a multiplicity of possible paths that are reversible, at any rate this is presumed, and by a much greater multiplicity of paths that cannot be spontaneously reversed. Along every reversible path $\Sigma \delta H/\theta$, where $\delta H$ is heat gained by a part of the system that is at temperature $\theta$, is a complete differential, because round a reversible cycle the summation vanishes: its integral along such path is thus $\eta_2 - \eta_1$, where $\eta_2$ and $\eta_1$ are the values of a definite function of the variables defining its constitution at the two ends of the path. The function $\eta$, so determined definitely by reversible transfer, is the entropy of the system in that state. But if the path is not reversible the value of the integral must exceed $\eta_2 - \eta_1$. This inequality contemplates that $\delta H$ will be in part heat communicated in the course of the path from outside the system. If the system is self-contained, so that it is wholly heat transferred from one part of it to another without change of temperature that is concerned, the one part of it will gain as much entropy as the other loses; therefore the entropy of the isolated system is not altered in the reversible change from one phase to another: for irreversible change it inevitably increases. There is also presumed to be another possession of the system, long previously surmised, its total energy $E$; this also is expressed by a definite function of state, now as regards all transformations whether reversible or not: while in contrast the trend of spontaneous change in a self-contained system must make the change of $\eta$ positive when $E$ is conserved. Cf. *supra*, p. 603.

In each cross-section of the hyperspace defined by $\theta$ constant, in which paths thus represent the course of change at constant absolute temperature $\theta$, there is by Carnot's postulate, as emphasized and opened out by Kelvin, a work-function $A$, expressing the available energy that is operative in all possible modes of transformation restricted to that temperature. Moreover two closely adjacent finite paths in such isothermal consecutive cross-sections can be connected into a flat Carnot cycle consisting of a finite path along the section of the hyperspace at $\theta$ and a return path along the section at $\theta - d\theta$, the cycle being completed by infinitesimal adiabatic paths connecting their ends*. The work done on external systems around this elongated cycle is made up of an amount arising from the finite change $- \delta A$ along the path at $\theta$ and another $+ (\delta A - \partial A/\partial\theta \, . \, d\theta)$ along the return part at $\theta - d\theta$, the work involved along the terminal transitions being negligible: thus it amounts for the cycle to $- \dfrac{\partial}{\partial\theta} \delta A \, . \, d\theta$, say $\delta W$.

Available energy at each constant temperature:

extended so as to be a function of temperature,

If $\delta h$ is the heat taken in along the isothermal path at $\theta$, then the formula for this reversible cycle is

$$\frac{\delta h}{\theta} = \frac{\delta W}{d\theta}, \text{ or } \delta h = - \theta \frac{\partial}{\partial\theta} \delta A.$$

The gain of total energy along the isothermal path at $\theta$ is $\delta E$, equal to $\delta h + \delta A$; for $\delta A$ can change into work but not at all into heat in a reversible path. From these two relations we can eliminate $\delta h$, which is merely an accidental quantity, not the differential of a definite function of state except in calorimetry where mechanical work is excluded. Thus along a reversible isothermal path $\delta E = \delta A + \delta h$, where as above $\delta h = - \theta \dfrac{\partial}{\partial\theta} \delta A$, yielding a relation, now between definite functions of state and so unrestricted,

thus replacing the unconditioned transfers of heat,

$$\delta E = \delta A - \theta \frac{\partial}{\partial\theta} \delta A.$$

If we measure from assigned origins of $A$ and $E$, the difference symbol $\delta$ may be omitted: thus, as on p. 102, *supra*, and fundamentally,

by a relation connecting Available with total Energy,

$$\frac{\partial A}{\partial\theta} - \frac{1}{\theta} A = - \frac{E}{\theta}, \text{ or } E = - \theta^2 \frac{\partial}{\partial\theta} \frac{A}{\theta}.$$

Thus *e.g.* if $E$ varied as $\theta^n$, $A$ would be $E/(1 - n)$ not applicable however to natural radiation.

* This arrangement may be compared with Kelvin's own procedure, as expounded in the first number of the *Quart. Journ. of Math.* April 1855: see obituary notice, *Roy. Soc. Proc.* Vol. LXXXI (1908) p. xlv, or *Math. and Phys. Papers*, Vol. I, pp. 291–300.

From the relation utilized above $\delta E = \delta h + \delta A$, it might appear at first glance that as $\delta E$ is an exact differential, but not $\delta h$, so neither can be $\delta A$: but the equation is there restricted to isothermal change, and such inference does not lie.

Entropy
analytically
more con-
venient than
Available
Energy. The fundamental advance made by Willard Gibbs was the consolidation (cf. *supra*, p. 711) of all formal thermodynamics into one characteristic equation of variation of state or as he called it, of phase, made possible by its involving these two dynamical quantities $E$ and $\eta$ that are intrinsic to each state. It appears that our present pair $E$ and $A$ would not serve directly for such purpose: this is because the equation connecting them, as above expressed, involves a gradient of one of them. In this respect entropy can play a more direct *rôle* than available energy, though the whole theory may be formulated in terms of either concept.

We may also recall the original derivation by Kelvin (and Rankine) of a "thermodynamic function," identifiable with entropy, by adjoining to the system a reservoir of heat of infinite capacity at $\theta_0$. Relative to this reservoir a gain of heat $\delta h$ at $\theta$ in the system has mechanical availability $\delta h - \theta_0 \delta h/\theta$: but in this and every other reversible process, when cyclic and so recovering the initial state, no heat can have disappeared in the aggregate during the cycle: thus $\Sigma \delta h/\theta = 0$ for the cycle, involving $\int_1^2 dh/\theta = \eta_2 - \eta_1$ as above.

## (ii) *On the Statistical Method in Thermodynamics: with Illustrations from Magnetism.*

### *The Statistics of Magnetic Orientations in Gases.*

IN the statistics of orientation of magnetic polar elements, for a gas in which the molecules have certainly the requisite freedom, we are concerned only with the rotational motions of the elements, each say of magnetic moment $\mu$ and all in the same field $H$; the distribution may be treated as a local one, regarded as specified per unit volume, and subject to the usual adjustment of the field to include a local part $\frac{4}{3}\pi I$. The rotational inertial momenta of the molecules provide the reacting influence that attains to a statistical balance with magnetic orientation. For molecules having material symmetry around their magnetic axes, the orientation of this axis being referred to a directional frame in co-latitude and longitude $\theta$, $\phi$, the number of them within the range $\delta\theta\delta\phi\delta\dot{\theta}\delta\dot{\phi}$ is, after the Maxwell-Boltzmann principle,

$$A e^{-h\,(\frac{1}{2}i\dot{\theta}^2 + \frac{1}{2}i\dot{\phi}^2 \sin^2\theta - H\mu\cos\theta)}\,\delta\dot{\theta}\delta\dot{\phi}\delta\theta \sin^2\theta\delta\phi,$$

where $i$, $i$, $j$ are of the nature of moments of inertia of the molecule transverse to its axis and around it. For by the Liouville dynamical theorem it is $\delta\theta\delta\phi\,\delta\Theta\,\delta\Phi$ that is invariant, where $\Theta$, $\Phi$ are the generalized momenta pertaining to $\theta$, $\phi$, thus being $\partial T/\partial\dot\theta$, $\partial T/\partial\dot\phi$: while the exponent of $e$ is the energy, of amount not altered by encounters, thus leading to a law of distribution for a steady state by the usual Maxwellian argument.

The complete kinetic energy of rotation of the molecule is

$$T = \tfrac{1}{2}i\dot\theta^2 + \tfrac{1}{2}i\dot\phi^2 \sin^2\theta + \tfrac{1}{2}j\,(\dot\psi + \dot\phi\cos\theta)^2$$

where $\psi$ is a third Eulerian coordinate expressing orientation round the axis: thus the momenta $\Theta$, $\Phi$, $\Psi$ pertaining to $\theta$, $\phi$, $\psi$ are

$$i\dot\theta,\quad i\dot\phi\sin^2\theta + j\cos\theta\,(\dot\psi + \dot\phi\cos\theta),\quad j\,(\dot\psi + \dot\phi\cos\theta).$$

Each of the two latter is conserved along the course of a free path between encounters of the molecule. But it is only $\Psi$ that, on account of the complete axial symmetry, is unaffected by the encounters: it is for this reason that the statistics of the unchanging $\Psi$ and the inessential $\psi$ are in the formula omitted from the reckoning. *omitting axial spin,*

The number $N$ of magnetic molecules per unit volume, thus statistically distributed, and their resultant magnetic moment $M$, are therefore given, on completing the integration with respect to $\dot\theta$, by

$$N = A' \iint e^{-\tfrac{1}{2}hi\dot\phi^2\sin^2\theta}\, e^{hH\mu\cos\theta}\sin^2\theta\, d\theta\, d\dot\phi,$$

$$M = A' \iint e^{-\tfrac{1}{2}hi\dot\phi^2\sin^2\theta}\, e^{hH\mu\cos\theta}\,\mu\cos\theta\sin^2\theta\, d\theta\, d\dot\phi.$$

The integrations can be completed with regard to $\dot\phi$ between its limits $-\infty$ to $+\infty$, by aid of the formula

$$\int_0^\infty e^{-ax^2}\, dx = \tfrac{1}{2}\sqrt{\frac{\pi}{a}},$$

yielding a factor $\left(\dfrac{2\pi}{i}\right)^{\tfrac{1}{2}}\dfrac{1}{\sin\theta}$, and so leading, if $A'' = A'\left(\dfrac{2\pi}{i}\right)^{\tfrac{1}{2}}$, to

$$N = A''\int e^{hH\mu\cos\theta}\sin\theta\, d\theta, \quad = \frac{2A''}{hH\mu}\sinh hH\mu,$$

$$M = \int \mu\cos\theta\, dN = \frac{2A''}{hH}\left(\cosh hH\mu - \frac{1}{hH\mu}\sinh hH\mu\right),$$

so that $\dfrac{M}{N} = \mu\left(\coth hH\mu - \dfrac{1}{hH\mu}\right)$, of the form $\mu f\,(\mu H/RT)$ reducing to $\tfrac{1}{3}\mu^2 H/RT$ in weak fields, as $h$ is $(RT)^{-1}$. It involves $\mu^2$, if this is right: so that for paramagnetic gases the susceptibility $\kappa$ in weak fields would be $\tfrac{1}{3}\dfrac{N}{R}\dfrac{\mu^2}{T}$, varying as the square of the molecular moment $\mu$, and inversely as temperature for given density $N$ of molecules: $\sqrt{\kappa}$ for different gases should thus afford indication of magnetons, cf. Stoner, Ch. VI.

The integration with respect to $\dot\phi$ thus introduces a factor just

reducing
finally to
the Langevin
spatial
formula. cancelling the extra $\sin \theta$ here involved in the multiple differential of the momentum domain; so that the final result agrees with the law of spatial distribution advanced by Langevin (1905), who initiated this beautiful application of the generalized gas theory that has formed the basis of much recent discussion of the extensive magnetic experimental data.

The feature brought out by the Langevin statistical formulation Modification
of the law
of Curie
involved. is that, even for a gas, though only in extremely high fields, $I$ is no longer proportional to $H$ at constant temperatures, though there is no hysteresis, but tends to an obvious limit expressive of complete orientation; thus destroying for such fields the linear relation which formed part (*supra*, p. 116) of the thermodynamic foundation for the Curie law. In the Carnot cycle (p. 116) as thus modified the mechanical work becomes $\int I dH$ and the heat absorbed $- \int H dI$, so that the more general formula is

$$\frac{M}{T} = \frac{dH}{dT} = \frac{dI}{dT} \Big/ \frac{dI}{dH};$$

while the heat developed in entering the field is

$$\int dI \cdot T \frac{dI}{dT} \Big/ \frac{dI}{dH}$$

instead of $\frac{1}{2}HI$.

Here, as in the theory of natural radiation and indeed also in gas theory, the statistics, dealing with intimate structure, and apparently only with binary encounters, can transcend the limitations of formal thermodynamics.

The small influence of temperature on diamagnetic excitation was accounted for (p. 116), following Curie, by the implication that the negative moment thereby induced electrically, internal to the molecule, is parallel and proportional to the inducing field, and thus involves no mechanical torque. This is natural for the cognate dielectric (and optical) polarization in symmetric molecules*: but as regards diamagnetism affairs might conceivably go differently. It is

Influence of
temperature
on dielectrics. * Thus in certain cases inference has been made, from modern views of ionic molecular structure, that owing to symmetry the molecule can have no permanent electric moment, *e.g.* for the gases $CH_4$ and $CCl_4$ there is tetrahedral symmetry, and for these substances it is found (Debye, *Brit. Assoc.* 1927) that the dielectric modulus does not vary sensibly with temperature; but for the intermediate compounds $CH_3Cl$, $CH_2Cl_2$, $CHCl_3$, where symmetry is absent, there is presumably an intrinsic electric moment in the molecule, and as one would expect the dielectric modulus is found to be much greater and to rise with fall of temperature. See p. 117, *supra*. A systematic summary of the cognate modern literature, largely experimental, on electric and magnetic polarizations of the molecules, is provided by Debye, in Marx's *Handbuch der Radiologie*, Vol. VI (1925) pp. 597–786: cf. also E. C. Stoner's book, *Magnetism and Atomic Structure* (1926).

there at first glance plausible that the moment may be induced in direction transverse to the resultant of the electronic orbits in the molecule, instead of along the field, and perhaps proportional to the component of the field in that transverse direction. But for any orbital atom with a single positive nucleus the effect of the magnetic field is to change it into an atom of the same type with a Zeeman precession superposed around the direction of the field (*supra*, p. 142): this involves an induced moment along the direction of the field, and therefore gives no torque and so null thermodynamic influence, in verification of Curie's idea. A magnetization of the kind above entertained, if it existed, say $- \epsilon H \cos \theta$, would be subject to a torque tending to rotate its molecule and thus have potential energy of orientation of amount $- \int H \cos \theta \, d \, (\epsilon H \cos \theta)$ or $- \frac{1}{2} \epsilon H^2 \cos^2 \theta$. If this were so the law of distribution would be changed to

$$\delta N = A e^{h H \mu \cos \theta - \frac{1}{2} h \epsilon H^2 \cos^2 \theta} \sin \theta \, \delta \theta,$$

and the total induced magnetic moment would be, writing $w$ for $\cos \theta$,

$$M = A \int_{-1}^{1} e^{h H \mu w - \frac{1}{2} h \epsilon H^2 w^2} \left( \mu w - \frac{1}{2} \epsilon H w^2 \right) dw,$$

where $h^{-1}$ is $RT$: which would lead to a different type of formula, in which naturally $M$, or $I$, is no longer exactly a function of $H/T$ even in weak fields.

### Generalized Formulae for Spatial Distribution.

We can penetrate to a general reason why a formula of this type should, over a large range of cases, be valid for the distribution relative to coordinates alone. Reference is pertinent here to Rayleigh's weighty remarks, interpreting Maxwell (as *infra*), on the degree and conditions of validity of the Maxwell-Boltzmann dynamical statistic, and in defence of that procedure[*]. The kinetic energy is a quadratic function of the component velocities, whose coefficients involve the coordinates. In treating each *local* statistic of distribution, because it does not involve their gradients these coordinates can be regarded as constants. We have then a quadratic function with constant coefficients, which is reducible to a sum of squares of new local velocity components, corresponding to new coordinates which might have been chosen in advance with this end in view. With these local coordinates $q_1, q_2, \ldots$ and the correlated momenta $p_1, p_2, \ldots$ which being $\partial T / \partial \dot{q}_1$, $\partial T / \partial \dot{q}_2$, $\ldots$ are now $a_1 \dot{q}_1$, $a_2 \dot{q}_2$, $\ldots$, the statistical formula integrates with respect to $dp_1, dp_2, \ldots$ to a result

$$dN' = C \left( \frac{\pi a_1}{h} \frac{\pi a_2}{h} \ldots \right)^{\frac{1}{2}} e^{-hW} dq_1 dq_2 \ldots.$$

*Margin notes:*
Relation of diamagnetism to temperature:

Curie's view justified.

Local momentoids.

[*] *Phil. Mag.* Vol. XXXIII (1892): *Scientific Papers*, Vol. III, pp. 534–535.

This distribution has now nothing peculiar to the locality: though it expresses a universal analytic result only when the coordinates $q_1, q_2, \ldots$ are such that the kinetic energy $T$ is a universal quadratic function of the relevant velocities such as involves no products. The relations $p_1 = \partial T/\partial \dot{q}_1, \ldots$, holding locally, suffice to assure, after Liouville, the invariance of the multiple differential factor in $dN'$.

The expression for $\delta N'$ in the Langevin magnetic statistics is covered by this result. There one of the coordinate differentials in the local statistic is $\sin\theta\,\delta\phi$ and the corresponding local momentum $i\sin\theta\,\delta\phi$, for $\sin\theta$ can be regarded as constant throughout the local domain: indeed the integration with respect to $\phi$ has been made above just by changing to $\phi\sin\theta$ as a new *local* variable.

Similar considerations are implied in Willard Gibbs' rather intricately expressed analysis of about the same date. The results that

Willard Gibbs' treatment: follow are transferred* into the present notation: their invariant form amounts to proof of their validity. In an *ensemble* of similar systems referred to coordinates $q_1, q_2, \ldots$ and their correlated momenta $p_1, p_2, \ldots$, a distribution depending on the energy $E$ alone (and so in his terminology *canonical*) will be steady: its expression is

$$dN = Ce^{-hE}\,dp_1 \ldots dp_n dq_1 \ldots dq_n,$$

where, in Hamiltonian form, with $T$ expressed as a quadratic function of $p_1, p_2, \ldots$,

$$T = \tfrac{1}{2}p_1\frac{\partial T}{\partial p_1} + \tfrac{1}{2}p_2\frac{\partial T}{\partial p_2} + \ldots, \quad E = T + W.$$

The distribution of the kinetic energy is expressed by $T\delta N$. Integrating $T$ locally by parts with respect to each component $p_1$ of momentum from $-\infty$ to $\infty$ gives (Gibbs, *e.g.* formula 130)

$$\int \tfrac{1}{2}p_1\frac{\partial T}{\partial p_1}\,Ce^{-hT}dp_1 = \int \frac{1}{2h}p_1\frac{\partial}{\partial p_1}\left(Ce^{-hT}\right)dp_1 = \frac{1}{2h}\int Ce^{-hT}dp_1:$$

therefore if the component $\tfrac{1}{2}p_1\partial T/\partial p_1$ of the kinetic energy is expressed by $T_1$, we have *in the aggregate* $T_1 = (2h)^{-1}N$, which expresses an

leading to a modified equipartition of energy. equipartition of the energy into equal local packets. But it is only when $T$ is expressible analytically in terms of $p_1, \ldots p_n$ as a sum of squares, without products, that these components† can be said to belong to degrees of freedom of the system expressed by coordinates $q_1, \ldots q_n$: though universally the total energy $E$ is $n \cdot N/2h$. This mode of statement avoids a spurious invariance of the theorem of equipartition, as usually expressed, which had been challenged by Kelvin and by Bryan and, as Rayleigh remarked, invited further elucidation.

* *Statistical Mechanics* (1902), pp. 49–56.

† If the energy corresponding to a coordinate $q_r$ is defined to be $\tfrac{1}{2}\dot{q}_r p_r$, then there is universal equipartition. Cf. as here relevant, Vol. I, *supra*, pp. 48, 67, on reduced Action.

Alternatively, we may employ directly the universal form
$$T = \tfrac{1}{2}a_{11}\dot{q}_1{}^2 + \tfrac{1}{2}a_{22}\dot{q}_2{}^2 + \dots + a_{12}\dot{q}_1\dot{q}_2 + \dots,$$
where the $a_{rs}$ are functions of the coordinates $q_1, q_2, \dots.$ The law of distribution is
$$dN = Ce^{-hT}e^{-hW}dp_1 dp_2 \dots dq_1 dq_2 \dots.$$
The multiple differential factor is invariant: thus the form of $dN$ remains the same when the coordinates are transformed linearly so that $T$ becomes a sum of squares. When that is done integrations over the infinite ranges of $p_1, p_2, \dots$ are at once effected separately, with the result
$$dN = C'\Delta^{\frac{1}{2}}e^{-hW}dq_1 dq_2 \dots,$$
where $\Delta$ is the determinant (Hessian) of the $a_{rs}$ which in this special case is their product. As the Hessian is invariant for all such transformations, this formula expresses the result of the integration over the range of $p_1, p_2, \dots$ whatever be the quadratic form of $T$.

Gibbs' general formula.

In the symmetrical magnetic case the relevant coordinates are $\theta, \phi$, and
$$T = \tfrac{1}{2}i\dot{\theta}^2 + \tfrac{1}{2}i\sin^2\theta\,\dot{\phi}^2:$$
thus $\Delta$ is $i^2\sin^2\theta$ and the free distribution of axes of the permanently magnetic molecules is obtained in the form

Result for axial case verified:

$$dN = C'i\sin\theta\,e^{-hW}d\theta, \quad I = \int dN\mu\cos\theta.$$

The result for the more general case, when there is no axis of symmetry in the molecule such as could have rotational momentum assumed to be immune from change by encounters, may be noted. It will suffice for illustration to restrict the moment of the permanent molecular magnets to be along a principal axis of inertia: then
$$T = \tfrac{1}{2}i\dot{\theta}^2 + \tfrac{1}{2}j\sin^2\theta\,\dot{\phi}^2 + \tfrac{1}{2}k(\dot{\psi} + \cos\theta\,\dot{\phi})^2$$
giving
$$\Delta = \begin{vmatrix} i & . & . \\ . & j\sin^2\theta + k\cos^2\theta & k\cos\theta \\ . & k\cos\theta & k \end{vmatrix} = ijk\sin^2\theta$$
$$N = 2\pi C'(ijk)^{\frac{1}{2}}\int e^{hH\mu\cos\theta}\sin\theta\,d\theta, \quad I = \int dN\mu\cos\theta:$$
so that, writing $w$ for $\cos\theta$,
$$N = C''\int_{-1}^{1} e^{hH\mu w}\mu\,dw, \quad I = C''\int_{-1}^{1} e^{hH\mu w}\mu w\,dw,$$
reducing again to the Langevin function, as was to be anticipated from the idea of momentoids, p. 731. As in all such cases, the statistics transcend formal thermodynamics, thus here making the magnetization $I$ a more complex function of $hH$ or of $H/T$, in place of the Curie simple proportion. See p. 745, *infra*.

applies also for a more general case.

The distribution problem for molecules of helical inertial quality (Vol. I, Appendix IX) must give rise to a different scheme.

### Statistical Equilibria, temporary and secular.

Opportunity seems to arise here for extended discussion. In a case like the present one there are three angular coordinates, say $q_1$, $q_2$, $q_3$, and their three rotational momenta $p_1$, $p_2$, $p_3$: and the law of final steady distribution is

$$\delta N = C e^{-\lambda(\frac{1}{2}a_1 p_1{}^2 + \frac{1}{2}a_2 p_2{}^2 + \frac{1}{2}a_3 p_3{}^2) - \lambda W}\, dp_1\, dp_2\, dp_3\, dq_1\, dq_2\, dq_3.$$

For the case of symmetry, geometrical as well as inertial, around the magnetic axis of the molecule, $p_3$ is constant: therefore the statistics of $p_3$ may be separated into a self-contained integral, so independent of those of $p_1$ and $p_2$, being merely a multiplying factor. This leads as above to the usual Langevin formula. But

*Secular growth of magnetization.* if while inertial symmetry is preserved, geometric symmetry is slightly violated, the value of $p_3$ will be disturbed at each encounter of molecules; the general formula involving $\Delta^{\frac{1}{2}}$ becomes applicable through resulting variation of $p_3$, though it takes a long time to establish the statistic as regards that slow variable. The case may be envisaged as a trend of slow secular change of the steady magnetization.

The case of a molecule inertially isotropic, so that $i$, $j$, $k$ are equal, is noteworthy, but involves no analytical simplification.

For the case of an ideal atom consisting of one electron in orbital motion round a nucleus, in addition to the Zeeman precession of the orbit there is a very slow oscillation of its plane, as W. M. Hicks has worked out by astronomical methods (*Phil. Mag.* 1928, as *infra*, p. 633), negligible however when the orbit is circular. A question arises how far this involves a displacement of the mean plane of the orbit, and so links up with paramagnetism: it shows features not unlike the fission of a ray of magnetic atoms discovered by Gerland and Stern.

However slight be the defect from geometrical axial symmetry, this kind of trend of change towards a secular final state will go on, but the more slowly as the defect is slighter. The only way to avoid it is to imagine the axial rotational momenta to be inaccessible to encounters, after the manner of the Kelvin original realization of latent momenta by freely spinning flywheels buried within the model of the molecule.

When a number of conserved momenta are thus treated as constants of the system, we may envisage the application of the general Gibbs formula above to this temporary statistic of the remaining ones. As regards them, the energy is reducible to a form

*Formula for secular equilibrium.* $E = \frac{1}{2}a_{11}(p_1 - \alpha_1)^2 + \frac{1}{2}a_{22}(p_2 - \alpha_2)^2 + \ldots + W - \frac{1}{2}a_{11}\alpha_1{}^2 - \frac{1}{2}a_{22}\alpha_2{}^2 - \ldots.$ The multiplier $\Delta$ is now the determinant of the new $a_{11}$, $a_{12}$, ..., while the final terms function as an addition to the potential energy $W$: in such cases the varying momenta $p_1, p_2, \ldots$ thus prove to be distributed around mean values $a_1, a_2, \ldots$ instead of symmetrically around null values. But there is gradual lapse towards a secular equilibrium as before*.

This illustrates how the statistic, as well as formal thermodynamic, has to deal only with ultimate steady states, affording but little knowledge of the modes or rapidities, as distinct from the mere trend, of transition towards them.

Weakness of the statistical method.

There may also be groups of momenta forming temporarily steady sub-systems, with gradual lapse towards a secular equilibrium. Cf. the statistic of *ensembles* worked out by Maxwell† in 1879, of all the molecules in a gas, when each molecule possesses its own such group of internal momenta, isolated and so conserved except to the degree that it is gradually worn down in the encounters with other molecules. Throughout all this type of theory, the ideal, even diagrammatic, precision and simplicity of the Maxwell-Boltzmann exponential statistical formula in terms of the energy, permitting simple processes of integration throughout, as developed *e.g.* in the *Treatise* of Gibbs, is the arresting feature.

Boltzmann: Maxwell: Willard Gibbs.

## (iii) *The Magnetic Carnot Cycle Scrutinized in Detail.*

ALTHOUGH the historical Carnot Cycle has been worn down in the course of time to a very simple standard formulation, that does not diminish the essential profundity of the ideas that are involved in it, and appearing in the more intricate special applications. In further general illustration of the cycle as applied to latent physical processes, and for its own sake, it seems desirable to probe more closely the thermodynamics of magnetization or dielectric polarization, as briefly formulated *supra*, p. 116.

Ideas may be fixed on the magnetic process as representing a physical type. The thermal system is there a paramagnetic body, in the field $H_e$ of an extraneous permanent magnet supposed itself protected from all change by its own rigid internal constraints. When the material system is transferred statically into this field through a range $\delta H_e$, and an addition $\delta I$ to its magnetism is thus induced, this change has produced a fall in the energy of position of its magnetic

A type of physical Carnot cycle:

its closer analysis:

* We may recall that the secular stabilities of cyclic dynamical systems, with their conserved cyclic momenta, are determined from the modified Lagrangian function of Routh and Kelvin treated as a potential energy: to this function the Poincaré graphical exploration of the trend of stabilities, evolved in relation to the problem of a rotating liquid mass, naturally belongs.

General theory of lapse to secular stability.

† The idea, as Gibbs remarks, is at the foundation of Boltzmann's memoir of 1871, as the title of Maxwell's memoir indicates.

poles whose total amount is in each position $- HI$ per unit volume, where $H = H_e + H'$, the part $H'$ being the field due to the magnetized body itself. As energy is supposed to be never destroyed, whatever one may think about matter in that regard, we have to trace what becomes of this loss. A portion of it, namely $\int I dH$, has been abstracted as work done in resisting external force as the system is brought up with slow acceleration into the field. The other part, namely $\int H dI$, must have been emitted as heat from the system to outside bodies if the temperature is maintained constant; but if the approach is adiabatic this heat is retained and raises the temperature of the system. Thus when the susceptibility $\kappa$ is a constant as regards change of intensity of field, only half the loss represents abstraction of palpable heat from the system as it changes along the isothermal arm of a Carnot cycle, in a manner in analogy with the typical scheme of the air-engine cycle. This supposes that the polarization arises wholly from orientation of polarized molecules by simple overt rotation, such as can come to an equilibrium of equipartition with the other energies that are present, which are the rotational and translational energies of the molecules in the case of a gas. Absorption of energy into the intimate constitution of the molecule so as to produce another kind of polarization by change of internal organization would constitute diamagnetism, after the original Weberian model, also not excluding effort, but of reversed type, towards orientation, were it not that, as *supra*, p. 730, this induced polarization is presumed to be exactly along the field.

represented on a diagram: The Carnot cycle can be visualized, in the usual manner, on a Watt-Clapeyron indicator diagram, here a diagram of $H$ plotted against $I$, of the form of a curvilinear parallelo-gram with two opposite sides along isothermal paths of constant temperature $T$, and the other two along adiabatic paths of constant entropy. This cycle is made up of an introduction of the material system further into the field iso-thermally at $T$, then a withdrawal adiabatically, in accordance with the explanation above, until the temperature has fallen to $T - \delta T$, then further withdrawal iso-

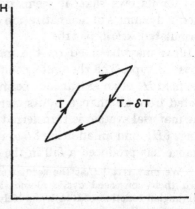

thermally until the stage when an introduction again adiabatically into the field can exactly complete the cycle. In this cycle the supply of heat along the arm for which the temperature is kept at

$T$ is thus negative, and equal to $- H\delta I$. Note that for small fields for which $\kappa$ is constant when $T$ is constant, this is $- \delta\left(\tfrac{1}{2}\kappa H^2\right)$ which is also equal to $- I\delta H$ as above. The energy lost from the system as work done on external bodies is $I\delta H$ under all circumstances. The two losses, of heat and of mechanical energy, for any open path necessarily always make up $- \delta(HI)$, the change in the total positional energy of polarization in the system. The mechanical work done by the system in the Carnot cycle is $- \int I dH$ taken around it, in the direction of its progress as represented by arrows in the diagram; it is thus expressed on the diagram by its area with negative sign, which is for a small cycle equal to that of a rectangle $\delta H\delta I$, where $\delta H$ is measured across the area parallel to the ordinates, and $\delta I$ parallel to abscissae: thus the standard formula for the Carnot cycle, temperatures being absolute, gives

$$- \frac{H\delta I}{T} = - \frac{\delta H\delta I}{\delta T},$$

in which $\delta H$ is measured along the abscissa on which $I$ is constant, so that when $\kappa$ is taken to be a constant at constant temperature $\delta H$ for constant $I$ is $I d\kappa^{-1}/dT \cdot \delta T$, while $H$ is $\kappa^{-1}I$. Hence

the resulting relation.

$$- \frac{\kappa^{-1}}{T} = - \frac{d\kappa^{-1}}{dT}, \text{ yielding } \kappa = \frac{C}{T},$$

which is the law discovered by Curie: compare footnote, p. 116.

The work expended by the magnetic system on external bodies, along any path on the $(I, H)$ diagram, is $\int I dH$: if $I$ is not a definite function of $H$ its amount will depend on the path. In a repeated cyclic process, such as on the average restores the initial state of things after each cycle, this work equal to $\int I dH$ integrated round the cycle must have been dissipated, which is in agreement with the standard formula for hysteretic loss in electrotechnics. If the body is not isotropic $I dH$ will be replaced by a scalar product of the two vectors $I$ and $dH$, as *supra*, p. 237.

The loss of energy by hysteresis.

When diamagnetism is regarded as thermodynamically neutral (cf. p. 730) the induced moment $I$ per unit volume, conforming to the molecular statistics, *supra*, p. 729, for the case of gases, would alter with temperature as some function of $H/T$, following Langevin: so that in fields not very intense, for which at constant temperature $I$ would be proportional to $H$, the law of Curie, that the magnetic susceptibility is $C/T$ less a constant diamagnetic part, would hold good.

Results from a special hypothesis:

The thermodynamic results of applying a hypothesis of this kind, $I = f(H/T)$ thus indicated by the statistic for free molecules, for very intense fields still reversible, may be followed out. Taking now $H$ and $T$ as independent variables, instead of $I$ and $T$ as on p. 116, *supra*, as

analytic formulation.

we may do since on the restriction to no hysteresis $I$ is a function of these two variables, we have, by a similar argument, $\delta h$ being heat imparted to the system, and $H$ the magnetic field as before,

$$\delta h = \mu \delta H + \nu \delta T, \quad \delta E = \delta h - I \delta H;$$

and formal thermodynamics requires that $\delta E$, and also $\delta h/T$ or $\delta \eta$, must be exact differentials, so that

$$\mu = T \frac{\partial I}{\partial T}, \quad \frac{\partial \mu}{\partial H} - \frac{\partial \nu}{\partial T} = \frac{\partial I}{\partial H}.$$

If now $I$ is equal to $\kappa H$, where for fields small enough $\kappa$ is a function of $T$ only, these relations become

$$\mu = HT \frac{\partial \kappa}{\partial T}, \quad \frac{\partial \nu}{\partial T} = T^2 \frac{\partial}{\partial T} \left( \frac{\kappa}{T} \right),$$

*Influence on specific heat.* thus also expressing the magnetic influence on thermal capacity $\nu$. The total heat absorbed at constant temperature is $\mu \delta H$, which is in excess of its directly magnetic part $- H \delta I$ by an amount

$$HT \frac{\partial \kappa}{\partial T} \delta H + H \delta (\kappa H), \text{ which is } \left( HT \frac{\partial \kappa}{\partial T} + \kappa H \right) \delta H,$$

thus vanishing when the law of Curie holds good.

*Result of less restricted hypothesis.* But when the less special hypothesis suitable to extremely strong fields is made as above, after Langevin, which is suggested by the application of molecular statistics in the ideally simple exponential form, that $I$ is a function of $H/T$, so that

$$\frac{\partial I}{\partial T} = - \frac{H}{I} \frac{\partial I}{\partial H},$$

all that can be inferred thermodynamically regarding these moduli $\mu$, $\nu$ is that

$$\frac{\partial \mu}{\partial H} - \frac{\partial \nu}{\partial T} + \frac{H}{I} \frac{\mu}{T} = 0,$$

the heat absorbed at constant temperature being $- (H \partial I / \partial H + \mu) \delta H$ as before*.

On interaction of mechanical stress and magnetization see *supra*, p. 130.

### (iv) *The Carnot Pressure-Cycle developed for an Imperfect Gas.*

THE contrast, and the parallel, with the circumstances of the gas *Gas-engine cycle.* cycle for an imperfect gas seem worth setting out†. On the dynamical theory, if all energy is kinetic there being negligible work of attraction

[* Magneton values for many elements, nearly integers when expressed in terms of one-fifth of the Bohr theoretical magneton, but based on the modified Curie law, are collected by P. Weiss and Foëx, *C.R.* Oct. 1928: the diamagnetic correction comes out to be slight.]

† On the general ramifications of this subject see Boltzmann's *Lectures*, or their French translation, *Théorie des Gaz*.

between the molecules of a perfect gas, expansion along an isothermal ought, by the principle of Joule, not to alter its internal energy per unit mass: but it contributes work $p\delta v$ to external bodies, therefore to keep its temperature thus unchanged the gas must receive heat of this amount from outside. The work done in the cycle is represented by the area on the indicator diagram reckoned around it in the direction of the cycle, which is $\delta p\delta v$ interpreted as above. Thus for the gas-engine cycle

$$\frac{p\delta v}{T} = \frac{\delta p\delta v}{\delta T};$$

so that at constant volume $p$ varies as $T$, where $T$ is the Kelvin absolute temperature.

Thus if a substance has no potential energy of expansion, and therefore its specific heat at constant volume is a function of the temperature only, then for it $p = T/f(v)$, so that at constant pressure $dT/dv$ is $pf'(v)$. But if Boyle's law also holds, then $f'(v)$ is constant, so that thermodynamic temperature is as measured on the scale of a thermometer constituted of a perfect gas defined by $pv = R'T$. {Substance devoid of internal energy: and also obeying Boyle's law: is an absolute thermometer.}

The practical problem presented to Kelvin and Joule in 1851, at that stage of thermodynamic development, was to explore the deviation from absolute temperature, of a thermometer constituted of an actual gas for which the internal potential energy of mutual molecular influence though small is not quite null. {Deviation of an actual gas thermometer.}

We may illustrate by exploring the discrepancy from the ideal law, for a thermometric substance obeying the characteristic equation of van der Waals,

$$p = \frac{R'}{v-b}T - \frac{a}{v^2},$$

in which $R'$ is the gas constant $R/m$, where $R$ is absolute and $m$ is molecular weight. Or more generally, we may work with the formula tested extensively by Ramsay, {Approximate formulae explored.}

$$p = \beta T + \alpha,$$

where $\beta$ and $\alpha$ are functions of the volume $v$ per unit mass.

Writing the differential relations in the form, $\delta H$ being heat imparted,

$$\delta H = M\delta v + \kappa\delta T, \quad \delta E = \delta H - p\delta v,$$

the thermodynamic principles, namely $\delta E$ and $\delta H/T$ exact differentials, give $M = p - \beta$, and $\kappa$ a function of $T$ only, as *supra*, p. 25. All that is involved as regards the specific heat at constant volume is that it is not altered by expansion: there is change of internal energy of the molecules $\delta E = \kappa\delta T - \beta\delta v$, integrating immediately, of which the latter term is potential energy.

The illustration may be extended* to a form of characteristic per unit mass, found by Clausius to suit Andrews' early measurement for carbonic acid more closely, which is exhibited in graphic form in Planck's *Thermodynamics*, § 26,

$$p = \frac{R'T}{v - a} - \frac{c}{T(v + \beta)^2}.$$

This, as Kirchhoff remarks, agrees over its range when $a$ and $\beta$ are neglected compared with $v$, with the equation of Joule and Thomson, —derived from observation of the heat generated in pushing the gas through an obstructing partition to a lower pressure, which first determined the zero of absolute temperature—namely

$$v = \frac{R'T}{p} - \frac{c}{R'T^2}.$$

In fact, considering unit mass of the gas, the work $p_1 v_1$ put into it in this operation exceeds the work $p_2 v_2$ taken out of it, by the heat developed in the process, as signified by change of temperature of the gas in passing across the obstructing porous partition. For a perfect gas this condition would be satisfied, as Joule had recognized and verified to a first approximation, by temperature entirely unchanged: actually however $\delta(pv)$ was found to be $R'\delta T$ as for a perfect gas together with a correction which proved to vary as $\delta(pT^{-2})$.

It may be noted that this form may be linked up to some extent with the virial theory of Clausius. The usual expression of the virial equation is, for unit mass,

$$\tfrac{3}{2}pv = \tfrac{3}{2}R'T + i,$$

where $p$ is transmitted pressure, and $i$ is the virial of intrinsic pressure of molecular attractions together with any other forces operating mutually in encounters of the molecules, all expressed as mean values over time. The latter should be proportional to the number of the encounters if they are binary, and depends on their closeness in a way representable by a function of their speeds, say in the mean $f(T)$. Each molecule may be regarded roughly as occupying an effective space proportional to its velocity of mean square $V$, and the number of encounters an assigned molecule makes is proportional directly to its $V$ and inversely to its free path which varies as the volume $v$: thus the total number would vary as $V^2/v$, therefore as $T/v$. Hence the virial formula would become

$$p = \frac{R'T}{v} + \frac{g}{v^2} + \frac{cTf(T)}{v^2}.$$

[* All this domain of transition to the liquid state requires fundamental revision in the light of H. L. Callendar's experimental determinations for steam, and resulting theory based on partial aggregation of molecules, perhaps unsettling the doctrine of definite phases. *Roy. Soc. Proc.* Aug. 1928: on the precise history of equations of state cf. his lecture reported in *Engineering*, Oct. 9.]

To reconcile with the formula of Clausius, the aggregated virial of an encounter would have to vary as $T^{-2}$, thus as the inverse fourth power of the mean speeds of the colliding molecules.

### (v) *Molecular Statistics and Temperatures for Mixed Fluid Substances in Rapid Flow.*

By the Maxwell-Boltzmann fundamental statistic a distribution of gas molecules given by

$$\delta N = Ce^{-\lambda E}\delta\tau\,\delta\mu$$

is permanent, notwithstanding mutual encounters, where $E$ is the energy of a molecule, in part kinetic $T$, and in part potential, due whether to imposed or to internal fields of force of total potential $W$, also $\delta\tau$ is the element of the multiplex geometrical phase space and $\delta\mu$ an element of the correlative multiplex momentum space, or more generally $\delta\tau\,\delta\mu$ represents an element fusing both into one. *[The Maxwell statistical principle: as regards energy:]*

The same Maxwellian argument permits the adjoining of another factor of type $e^{-\lambda'M}$, where $M$ is any component of the momentum, translational or rotational, of the molecule: and any other quantity may be similarly adjoined whose aggregate is not altered by an encounter. *[extended to include momentum also:]*

A distribution expressed by

$$\delta N = Ce^{-\frac{1}{2}hm\,(\dot{x}^2+\dot{y}^2+\dot{z}^2)+hm\,(u\dot{x}+v\dot{y}+w\dot{z})-hmW}\delta\tau\,\delta\mu,$$

where $u, v, w$ enter as any parameters that are functions of position, thus remains steady notwithstanding internal mutual encounters.

The exponent of $e$ is here

$$-\tfrac{1}{2}hm\left[(\dot{x}-u)^2+(\dot{y}-v)^2+(\dot{z}-w)^2\right]+\tfrac{1}{2}hm\,(u^2+v^2+w^2)-hmW.$$

Of this the first term represents the normal statistics of internal or *relative* translatory motions when the local mass centre has a velocity $(u, v, w)$. The other two terms, not involving internal motions, belong to the motion of the medium in bulk: according to them the local density is modified, as if the potential energy per unit volume $\rho W$ were diminished by $\tfrac{1}{2}\rho$ vel.$^2$ *[thus giving the statistics for gases in motion,]*

This latter relation ought to be involved in the hydrodynamics of steady motion of the gas in bulk. In fact its equations of motion, of type *[result agreeing with formal dynamics of flow,]*

$$-\frac{\partial p}{\partial x}-\rho\frac{\partial W}{\partial x}=\rho\left(u\frac{\partial u}{\partial x}+v\frac{\partial u}{\partial y}+w\frac{\partial u}{\partial z}\right),\quad p\rho=\frac{R}{m}T,$$

lead to the result along any stream line

$$\frac{R}{m}T\log\rho = \text{const.} - W + \tfrac{1}{2}\text{vel.}^2,$$

that is,

$$\rho = Ce^{-hm\,(W-\frac{1}{2}\text{vel.}^2)},\quad h = \frac{1}{RT},$$

which is the equivalent of the distribution law aforesaid, and also indicates the value of $h$ appropriate to gas theory in terms of the universal gas constant $R$.

This restriction of the thermodynamic formula to a stream line can be withdrawn only when the velocity of flow is irrotational, that is, when the streaming is devoid of local cyclic quality: otherwise the statistic must be invalid for a rapidly flowing medium—as might have been anticipated. In fact no flow of cyclic quality could arise naturally from local mutual encounters of point molecules, provided there is no viscosity: while the present statistical formulation ignores viscosity, which would have to be expressed by interaction between the phenomena in adjacent phase regions here taken as independent.

*viscosity being outside the scheme.*

The standard illustration of this molecular statistic, for gases in motion, is the case of the atmosphere rotating with the Earth: the result is as if the atmosphere were at rest and gravity modified by the addition to it of the usual centrifugal force.

*Centrifugal force of the rotating atmosphere.*

In this and all other cases of this statistic, the gas, treated as perfect, may be a mixture of different types of molecules. Encountering molecules may be of different kinds, involving different masses $m$ and even different fields of force $W$: yet, provided the parameter $h$ has managed to settle down to be the same for all kinds of them and everywhere, the encounters will by the Maxwellian argument maintain this statistical condition unaltered. This involves, by virtue of the exponential character of the formula, that the distributions of the various kinds of molecules will have each its own independent statistic; and the gas constant $R$ being the same for them all, $T$ must be the same for them all and everywhere, in the steady state of which such uniformity of temperature is thus expressive. It involves that in the fields of forces each constituent of the gas will then have its own independent law of density: as expressed in Maxwell's early result that in an atmosphere ideally steady the density of each constituent falls away with altitude according to its own law, so that at high altitudes only the lighter constituents remain to any sensible degree[*].

*Mixed gases:*

*the constituents distributed independently:*

*a uniform temperature involved when steady.*

*Historical.*

This mode of argument is not however applicable to liquids and solids, where the encounters are not binary—unless so far as it is modifiable after Boltzmann and Maxwell as *infra*, p. 744. But for the more general deduction of the existence of temperature by way of a statistical entropy[†] see *supra*, p. 402: in the statistic of vast assemblages, the state that shows no tendency to change is the one that

---

[*] When the gases are not perfect the mutual influence of the components comes into the characteristic equation as small terms, and the thermodynamic analysis of Maxwell-Andrews and Gibbs (p. 711) applies.

[†] The connection between the two procedures has been explored by very general analytical methods by C. G. Darwin and R. H. Fowler. Cf. also a *Faraday Soc. Report* on "Strong Electrolytes," 1927.

occurs oftenest and almost exclusively. And for the case of gases we may even, after Boltzmann (and Maxwell, *loc. cit.* (1879) p. 724) for purposes of theory of temperature, suppose the various constituents, of molecular masses $m_1, m_2, ...,$ to have their own different fields of force $W_1, W_2, ...$ so intense that the species of molecules will ultimately be separated into different regions*. In that ultimate steady state, the temperature as determined by $h$ will thus be the same throughout all of the separated gases. And this law thus established for gases can then be extended indirectly to liquids and solids: for we can imagine them as existing surrounded by a gaseous medium— indeed it is only for states known in advance to exist that thermo- dynamic reasoning can apply: there can then be no steady state until the temperature of this gas, as affected by the adjacent bodies, is uniform everywhere. The temperature of a liquid or solid is in fact to be defined, for all practical purposes, as that of the gas in steady contact with it, just as the thermodynamic scale for all temperatures can be defined as that provided by a practically perfect gas.

*Temperature the same for separated gases:*

*extended to other substances in thermal contact.*

## (vi) *On the Maxwellian General Doctrine of Statistical Dynamics.*

CLOSER reference seems to be required to the ordered train of ideas in the path-breaking memoir† of Maxwell, constructed near the end of his life, "On Boltzmann's Theorem...," which surely, in connection with Boltzmann's papers‡, marks the emergence of Statistical Dynamics into the rank of a special exact science.

*Historical.*

It appears to have been Maxwell who expanded this subject into generalized Hamiltonian dynamics§. But the fundamental character of the general science became in time temporarily eclipsed, through its close connection with the difficulties inherent in equipartition of energy as applied to the degrees of freedom of natural molecular systems.

* There is a practical case in point, that of the complementary ions of an ionized medium in an electric field, though somewhat vitiated by the circumstances that the law of the mutual electric forces is not a molecular law restricted to the immediate locality.

† *Camb. Phil. Trans.* Vol. XII (1879): *Scientific Papers*, Vol. II, pp. 713–741.

‡ Cf. especially the *referat* of Maxwell's memoir of 1879 by Boltzmann (*Ann. der Phys.*) in *Abhandlungen*, Vol. I, pp. 582–595; translated in *Phil. Mag.* 1882.

§ This statement is on Boltzmann's authority (*referat, loc. cit.*). Generalized coordinates as distinct from Hamiltonian analysis had been introduced by Boltzmann himself in 1871, as Rayleigh has remarked, *infra*. Cf. also the historical appreciation in the preface to Willard Gibbs' *Statistical Mechanics* (1902), the first formal treatise in this domain, where the introduction of the method of *ensembles* of systems is ascribed to this memoir of Boltzmann: also and especially Rayleigh's critical summary of the whole subject, *loc. cit. infra* (1900).

The procedure, as modified from Maxwell's original distribution law with factor $e^{-\hbar E}$, treats, in this memoir of 1879, the complete physical system as the subject of one vast interconnected dynamics, instead of confining the dynamics to one of its molecules: thus evading, at the great cost however of the enormous number of concomitant freedoms thus introduced, his earlier restriction, on the gas theory, to molecular encounters purely binary*. Cf. the alternative procedure, narrower but more manageable and suggesting his minimal $H$ analysis, for such avoidance by Boltzmann's interpretation of entropy by the logarithm of statistical probability.

*Restriction to gas theory removed.*

In Maxwell's method in this memoir, the special exponential law of canonical distribution as among various energies does not arise, for the statistic is restricted to systems all of the same total energy by use of a modified form of Action, namely the original Eulerian form $\int 2T\,dt$ with energy unvaried along the path, a form which in more general problems is somewhat artificial, as Rayleigh remarks, on account of its introducing $E$ as a variable in place of $t$. It has also to be assumed that a system with this energy can in course of time pass, practically at any rate, through all possible configurations in the domain of phases that is under consideration.

*Distributions in relation to energy.*

*Maxwell's ergodic postulate:*

Finally, by refined argument he arrives, only however for the case when $W$ is small compared with $E$, but evading as above the restriction to binary encounters, at the law of distribution in space for a mixture of gases, each constituent having identical molecules, in a field of force, as involving $e^{-\hbar W}$ where $W$ is the potential of the field which may be in part that of the system itself. Incidentally the memoir develops finally into determinations relating to the time required for partial separation of gases by introduction of a centrifugal or other field of force, and relevant also to separation by temperature gradient after the manner recently discussed by Chapman.

*his general gas theory for small mutual energy:*

*time required for settlement of a mixture of gases.*

The later weighty paper of Rayleigh†, passing provisional judgment on a prolonged controversy relating to the conditions required to ensure equipartition of energy in physical systems, is a necessary

---

* The distribution, according to this law, is not altered by multiple encounters: the criticism is that each encounter must be self-contained, which requires the interactions to be very intimate and the distribution to be sparse, so that multiple encounters do not occur. Cf. *supra*, p. 374. The debated question, whether in a thermostatic aggregate the different types of interchange are separately equilibrated, thus excluding all cyclic processes, appears to be intimately related.

† *Phil. Mag.* Vol. XLIX (1900): *Scientific Papers*, Vol. IV, pp. 433–451: cf. his use of the Hamiltonian principal function with time as variable in place of energy, p. 433, and his essential introduction of what were later named momentoids, as *supra*, pp. 444, 731.

supplement to the fundamental memoir of Maxwell of 1879. It had doubtless also an influence on Gibbs' systematization in the Treatise of 1902: while the Gibbs partial integration of the components of the energy (as *supra*, p. 732) is calculated to reflect back an essential light on the whole.

This subject, as a fascinating domain in pure dynamics, is far from obsolete. The formal conclusion of Poincaré, and of Jeans, that the actual law of intensity of natural radiation demands *quanta* of energy and admits of no other conceivable solution, has naturally excited surprise, though it is doubtless inevitable in the necessarily procrustean framework imposed in the analytic argument.

*Possibilities wider than analysis.*

### (1928).—*Limitations of Exponential Statistical Dynamics.*

But not only may a statistical outlook be wider than the thermodynamic, conceivably the two might even be discrepant (cf. p. 376). The Langevin procedure for paramagnetism in gases, by statistics of the exponential type, serves by its simplicity as an effective illustration: its final formula is wider than the Curie law, which has itself appeared as the result of a Carnot cycle (pp. 115, 737), the additional diamagnetic effect being slight, but only when $I$ varies as $H$.

*Relation to Carnot cycle.*

Generally, it appears to be implied in the proof of the Maxwellian exponential statistic involving the distribution factor $e^{-h\epsilon}$ that there is a definite energy $\epsilon$ of its own for each molecule or other ultimate unit, that its energy is not affected by its neighbours except when they are in encounter, which is postulated at any rate tacitly to be for a very brief part of the whole time. The argument is simply that an encounter does not alter the joint distribution factor $e^{-h\epsilon_1} e^{-h\epsilon_2}$, because it does not alter the total energy $\epsilon_1 + \epsilon_2$ of the pair (unless a *quantum* of radiation intervenes, when the encounter is not binary), and therefore that this law of distribution represents a steady state (cf. p. 411). Even for a gas, in states of high density, there does not appear to be strictly any such intrinsic energy $\epsilon$: when Boyle's law is departed from, the value of this energy depends sensibly on the positions of the interacting neighbouring molecules, even the internal motions of the molecule being thereby sensibly affected,—at all events it is a postulate derived from experience and not a necessity of the case, if they are not. If this principle fails, it becomes a question whether there is any such thing as equipartition at high densities, and the paradoxical discrepancies between theory and measured specific heats of gases may lose at any rate part of their force*. In the early

*Restriction of energy statistic to dilute systems.*

---

[1] This is in fact equivalent to the restriction to binary encounters, which has been asserting itself (p. 376) even for chemical interactions, where it was found to involve separate equilibration of each possible type of reaction.

fundamental application to steady stratification of an atmosphere under gravity (p. 718) such troubles do not arise.

*Procedure in generalized statistical dynamics.* Maxwell and Boltzmann were perhaps aware of this consideration: it may have led to their final far more complex formulations in which the whole material system is the effective unit, being a sort of molecule with its number of freedoms tending to infinity, the statistics now being those of comparison with all other equally likely independent states of the whole system. A point in the pure mathematics may be passed over, whether an integral of multiplicity tending towards infinity retains a sufficiently definite meaning: at any rate the result now is that the ideally simple exponential law survives under restrictions (Maxwell, *Scientific Papers*, Vol. II, p. 728) which seem to exclude dense distributions. The general method of statistical distribution among cells, which can extend also to natural radiation (*supra*, p. 401), is a sort of combination of these two procedures.

*The paramagnetic paradox resolved.* In the illustrative case of paramagnetism for gases, the mutual energy of the bipolar molecules varies as $r^{-3}$, thus with increasing density the energy for a molecule derived from position of its neighbours mounts rapidly (cf. p. 342). But in the statistic (*supra*, p. 733) it is only the energy due to the imposed field that has been taken into account; that suffices for densities small enough, and there it is consistent with the law of Curie.

*Electric polarity and ions also excluded.* If this general consideration has weight, it applies with increased force (cf. p. 102) to the wide class of theory in which the exponential energy statistic is assumed to subsist for media constituted in part of electric polar molecules, and *à fortiori* of ions or electrons: it would become a tentative exploring procedure which may be expected to converge to actuality for great dilution.

### *Limitation of Equipartition for Radiation.*

*Unlimited equipartition merely presumed:* Unlimited equipartition as regards radiation cannot pretend to be justified by the Rayleigh mode of argument: restrictions are required which would hardly have been uncongenial to its originator. The rectangular block of aether (p. 409) has to contain a fragment of matter, in order that interchange of energy between its free periods may be possible. The Stewart-Kirchhoff principle correlating emission and absorption is here implicated: but its proof involves rays of radiant energy as definite entities, which requires that the dimensions of the ideal bodies that are concerned in their reflexion should be large compared with the wave-lengths. If not, the rays are nebulous, cf. *its physical specifications inexact.* p. 406: which is just another mode of expression within its domain of the recently emphasized doctrine that physical measurement is essentially indeterminate, or as we would rather say, physical specification in bulk is necessarily inexact in an atomic system. Now this self-contained block of aether is not comparable with infinite extension: but it can be expanded by including the sets of images of all the contained sources as reflected in its boundary planes. It has then become an unbounded periodic structure, a kind of crystalline medium: and any laws of radiation deduced for it are restricted to that type of medium.

But this periodic structure may be smoothed away as regards the large-scale phenomena, including here radiation of long wave-length, just as the discrete atomic structure of matter is legitimately smoothed away into continuous schemes of differential equations. Indeed this is none other than the Bohr idea of correspondence between the actual discrete atomic system and the continuous medium which is correlated with it. *Continuous analysis and atomic structure.*

If there is a distribution of sources, or even only one moving source continually reflected by the walls, this system smooths out into an averaged crystalline medium, but with exceptional conditions of image symmetry close to the planes of its lattice: it is only for waves long compared with the scale of these peculiarities that they may be smoothed away. The paradox thus disappears: for we have no right to expect on this foundation any general result that can be applicable to short periods without limit. *Essential implied limitation of the Rayleigh law of radiation.*

### *Ionic Statistic with Viscous Control.*

The preceding statistic, which concerns itself only with conserved energy, and so must involve an equipartition, whether it be free or quantified, may present contrasts (cf. p. 742) with the other extreme case of an imposed viscous control.

The problem of this type now prominent, after Debye (p. 106), is conduction across a completely ionized electrolyte. If the negative ions, in viscous flow past a positive, tend to accumulate on the side of it in advance, they will be pulled back by its attraction thus producing a decrease in their effective mobility. As both types of ions are mobile, and are of compensating numbers, any close analysis must be very complex. But if both kinds were identical except as regards sign, so of equal mobility, the positives flowing past each negative and the negatives flowing past each positive, ought perhaps by symmetry to lead to a cancelling result. If this be so, a main element in an actual effect would be unequal mobilities of the two ions. *Current in an ionized electrolyte:*

In illustration, one type, say positive, may be taken so massive or resistant, that they are virtually fixed in the medium. There is then a problem of negatives flowing past it. If we imagine further, very artificially, that each positive is a multiple ion, their number can be *limiting case.*

small in comparison with the negative, and the problem, now remote from the facts, even for mere illustration, would be reduced to flow past a single multiple ion. But a formulation can now be attempted for a steady flow.

In a cylindrical frame $(z, \rho)$ if $w$, $v$ are the component drifts of a negative ionic distribution of density $D$ along the axis $z$ and along $\rho$, in a uniform field $F$, and $\psi$ is the ionic stream function,

$$w\rho D = -\frac{\partial \psi}{\partial \rho} = k^{-1}\left(F - \frac{\partial V}{\partial z}\right)D,$$

$$v\rho D = -\frac{\partial \psi}{\partial z} = k^{-1}\left(-\frac{\partial V}{\partial \rho}\right)D,$$

$$\nabla^2 V \equiv \frac{\partial^2 V}{\partial z^2} + \frac{1}{\rho}\frac{\partial}{\partial \rho}(\rho V) = -4\pi D.$$

Thus

$$\frac{\partial}{\partial z}\left\{\left(F - \frac{\partial V}{\partial z}\right)\nabla^2 V\right\} = \frac{\partial}{\partial \rho}\left\{\frac{\partial V}{\partial \rho}\nabla^2 V\right\},$$

Drifting atmosphere around an ion. an equation determining the distribution $D$, and thence the mode of drift, in which the modulus $k$ of viscosity has disappeared. It is altered by change of sign of $F$, so involves the effect under discussion. It will become amenable by the substitution $V = A/r + U$, at any rate when $U$ is small compared with $A/r$ and with $F$: but the illustration of possible orders of magnitude that is involved, if interesting, is very artificial.

No steady thermionic atmosphere. Application of cognate ideas to the electron distributions around metals suggests itself. If the potential $V$ is taken to be constant within the metal, and determined by an electron-density $\rho$ outside, we would have

$$\nabla^2 V = -4\pi\rho, \quad \rho = \rho_0 e^{heV}, \quad h^{-e} = RT,$$

leading to
$$\nabla^2 V + 4\pi\rho_0 e^{heV} = 0.$$

This is on the assumption, in default of a better, that the electrons constitute an atmosphere, distributed in the field of force according to the Maxwell-Boltzmann exponential statistics. But it does not appear that this characteristic equation, when discussed, *e.g.* for the symmetrical case of a spherical mass of metal, is consistent with any finite distribution of $V$ and therefore of $\rho$, unless there is an outer metallic containing boundary.

# APPENDIX IV.

## THE DOCTRINE OF MOLECULAR SCATTERING OF RADIATION.

THE principle (cf. Vol. I, *supra*, Appendix VII) that atomic sources are to be expected to radiate energy independently, except to the extent Historical. that their phases of vibration may be definitely found to be correlated, originates with Rayleigh; it was disentangled, and expounded in relation to the diffused light of the sky, in some of his very earliest papers (1871), also later as illuminating general statistical theory (1880, *Scientific Papers*, Vol. I, pp. 491–496). In 1899, resuming the subject of an early private letter in reply to a provocative query from Maxwell in 1873, he shows that if the molecules of the air are themselves the scatterers of light, instead of the particles of dust which the atmosphere contains, the result that must follow, on the basis of the accepted value of the Avogadro number of the molecules for gases, is in good agreement with the observed transparency of the high atmosphere: cf. also incidental remarks of about the same date in Maxwell's lecture (of 1875) on molecules, at the Chemical Society, reprinted in *Scientific Papers*, Vol. II, p. 435. In Rayleigh's estimates, subsequently confirmed with astonishing closeness by observations in Italy and America, the molecules are treated as electrically polarizable each as a whole, to moments indicated by the index of optical refraction of the gas: this implies that the $n$ electrons which are the scatter- The internal electrons scatter coherently: ing agents in each molecule vibrate in step, thus deflecting about $n$ times as much energy as they would do if they oscillated incoherently. But in contrast it is assumed, in accordance with the general presumption, that the complete molecules, however numerous within but not the molecules. a wave-length, scatter independently or incoherently. Moreover, as Barkla and Thomson found much later, the electrons in the same molecule do scatter the energy of the short X-rays independently*. Perhaps not much confidence could have then been placed from theory alone on these varied results, were it not for the striking observational verifications†.

---

\* Thus a modified index of refraction arises, after the manner of the effect of ions on atmospheric electric waves (*supra*, p. 644). When they may be regarded as distributed sporadically, the effect on light would be negligible: for a crystal it is an affair of the distribution of the molecules in the lattice, as arising in terms of the Fourier analysis: cf. R. Schlapp, *Phil. Mag.* Dec. 1925, as *supra*, p. 503.

† The verification of the Rayleigh formula may be held to show that the molecule vibrates as one physical system, like a cord or bell, as regards the relatively very great wave-lengths of the emitted light, the periods belonging to the whole system as *supra*, p. 50.

This recital gives rise to the question, what are the conditions that discriminate between the types of scattering of energy of radiation *Rayleigh:* here described? For the light of the sky the Rayleigh formula has been verified very closely. Its originator returned again to the general principles of the subject in 1906 (*Scientific Papers*, Vol. v, pp. 279–282) in remarks, partly critical, called forth by a discussion here reprinted *supra*, p. 306. Since that time the theory has been examined *Raman:* by various writers, notably in a recent memoir by C. V. Raman and K. R. Ramanathan, *Phil. Mag.* Vol. xLV (1923) p. 113, as a preliminary to very fruitful experimental work mainly devoted to scattering in dense media. These papers, as well as a further detailed discussion by Rayleigh in 1918 (*Scientific Papers*, Vol. vi, pp. 565–582), are usually concerned with emitting or scattering sources which are regarded as at rest. In the paper by Prof. Raman and his colleagues they adopt the view, in a preliminary discussion, that the motions as distinct from the collocation of the molecules in a gas, which had been appealed to (as *supra*, p. 609) as a main cause of their independent scattering, cannot have influence: while Rayleigh (*loc. cit.* p. 579 at foot) expressly puts it on the redistributions of phase due to their mutual encounters. The consequent Doppler changes in period can indeed make the relative phases in which the lights scattered in any direction by adjacent molecules pass any fixed point rapidly change: but a group of waves scattered locally will not change its form as it travels away, unless the medium is dispersive so that phase velocity changes with period. The molecules of a gas are held to scatter incident light in detail practically as if they were at rest at each instant, but this result is supposed to be complete only for a gas which closely obeys Boyle's law. When it does not, then on a view that within the sphere of influence of each molecule there is a deficiency of other molecules which if they were present would not, on account of their proximity, scatter in random phases but all in phases close to that of the central one, it is found on the exponential statistical hypothesis that the law becomes modified into a very re-*Einstein:* markable form given by Einstein (*Ann. der Phys.* 1910), namely for unpolarized scattering $\frac{8\pi^3 RT\beta}{27N\lambda^4}(\mu^2-1)^2(\mu^2+2)^2$, with a modification after Rayleigh as applied by Cabannes for the effect of polarization, where $\beta$ is the compressibility of the substance which thus curiously appears and $N$ is Avogadro's number, reducing to the Rayleigh form as the limit for an ideal gas.

*using discordant principles.* The method of Einstein, as explained (*loc. cit.*) by these investigators, appears to differ from their own, even in principle, as they remark. Formulae may be of wider range than the model: here the unifying feature common to both schemes is perhaps the assumption of the Maxwell-Boltzmann exponential partition law. The work of Smulochowski and of Einstein treats of fluctuations of density in an opalescent medium, such as a gas close to its critical point, regarding each fluctuation (cf. p. 512) as a local group of coherently vibrating molecules which deflects light just as the molecules in a particle of dust or the electrons in an atom prove to do, the scattering being thus as the square of their number rather than the first power; and the formula above quoted, for opalescent scattering, is thence derived. It becomes in fact an affair of irregular density of a medium in bulk; reducing to the Rayleigh formula on passing to a limit, which stretches even to contributions of individual molecules as independent scatterers.

But just as we have been certain of the Rayleigh formula for gases, perhaps mainly because it has been closely verified by observation, so for this extension it seems difficult, as is recognized, to admit security except so far as it also may be confirmed by the authority of experiment. The problem of a medium consisting of a perfect gas permeated by very small and ideally thin balloons containing gas slightly denser, can illustrate the nature of the question. *The test of fact.*

This slight sketch discloses an interesting variety of procedures which, as Prof. Raman remarks, ought to be significant, as affording a fundamental means of test for statistical and thermodynamic constructs as applied to radiation in dense media. In the earliest stage of optical science the presumption was to take a pure substance like the atmosphere, with millions of molecules per cubic wave-length, as virtually smoothed-out optically into a uniform refracting medium, thus transmitting light without any deflection due to its granular structure. A closer justification is now required.

The subject is a delicate one: in molecular science an algebraic analysis, unless it pursues abstract openings of its own creation, can only develop what is once for all presented to it, and that of necessity has to be smoothed out somehow into manageable simplicity. The following attempt towards disentangling the *criteria* is now submitted. *The problem resumed.*

We are in face of a train of radiation, passing across a distribution of secondary sources each of atomic type*. Each of these, when active, sends off a scattered train consisting of a succession perhaps large of regular harmonic waves, thus representable as regards each direction by an undulatory line of finite but limited length, travelling

* The subsidiary rays, equally spaced in frequency, produced by diffusion of homogeneous rays, and recognized by Wood (p. 559), and Raman, and with special definiteness by Landsberg and Mandelstam (*C.R.* July 1928) in traversing crystals of quartz and calcite, are a different subject in close connection with the Mathieu-Hill equation (*supra*, p. 503).

away from the source. If there is no dispersion of speeds all these lines or ray-segments travel away exactly in company, thus with unchanging relative phases: and the compounded wave, however complex in resultant profile, of which they are the components, preserves its form as it recedes, without any change. Propagation in groups or bunches of waves, with some persistence of form, is of course a feature that can subsist for all wave disturbance. What is peculiar to a dispersive medium is that the bunches travel through the substratum, by no means unsubstantial, of regular waves in which they subsist: though (p. 552) if the form of the group is properly chosen it can travel for a time isolated. They carry an excess of energy which, after Reynolds, is made up for by slower speed.

*Combinations into groups:*

*unchanging,*

*shifting.*

But in a gas the atomic secondary sources have their own velocities, whereby the secondary wave-lengths that issue from them are altered slightly in the Doppler manner: if the phases concur at one end of two of the coinciding ray-segments above, they may thus be discordant at the other. If this effect is superposed for all the ray-segments, there may be a temptation (cf. *supra*, p. 309) to regard the phases at any point as erratic, with the energies therefore additive and not the amplitudes. But this is not permitted by the analyzing spectroscope: it separates out the ray-segments up to its degree of resolving power, into the groups of identical wave-length which have travelled in company without change, thereby giving breadth and fine structure to the spectral line. Yet the result remains, in different form, with the Doppler effect thus eliminated: the radiation coming to a given position in the breadth of the spectral line is emitted by a sub-group of the sources, so sparsely distributed that their phases at the spectral line do range widely over the whole period, so that it is their energies that are there additive.

*Doppler effect can be analyzed out,*

*yet effective towards law of added energies.*

If we inquire further what would be the influence of the dispersion of the atmosphere in mixing up the phases of ray-segments, the answer is, after O. Reynolds and Rayleigh as above, that the energy travels, concentrated more or less in packets, lagging through the component ray-segments with a group speed of their own.

*Its modification by dispersion.*

Why is the same argument not applicable to the secondary atomic vibrators in a scattering dust particle? It appears to be an affair of dimensions. The ray-segments from these secondary sources travel away in company, and because the atomic excursions of thermal type are alternating and with small extent, the relative phases on coincident ray-segments, while changing irregularly along each segment, never get far apart. As the phases thus nowhere differ except slightly the amplitudes are additive up to the second order, and the energy scattered varies as the square of the number of sources: the fragment

*A dust particle an exception to law of energies:*

of matter, whether crystalline or amorphous, thus scatters by its polarization in the manner of a coherent vibrator. This however requires the fragment to be small compared with the wave-length, else delays of propagation would come in. It however establishes the validity of the usual continuous theory of propagation in a medium, which is based just on this type of behaviour of differential elements of its volume each small compared with the wave-length: it is thus that the scattering of a globe or ellipsoid of matter, of dimensions comparable with a wave-length, is rightly calculated from field equations and boundary conditions by the usual harmonic analysis*. {how this confirms a continuous analysis, even for scattering by small bodies.}

Even a volume of gas imagined as confined in an infinitely thin ideal shell of dimensions small compared with the wave-length, should scatter incident light proportionally to the square of the total number of molecules: for their excursions are restricted by the boundary so as to be small compared with the wave-length, which does not allow phase differences in the ray-segments travelling together to accumulate up to disturbing values. But even so, why will not the analyzer evade this conclusion just as before by resolving the radiation into independent rays converging to the parts of a broadened spectral line? The answer must be, as it seems, either that the analyzer, being itself a structure composed of molecules, requires time to get vibrating in step and so cannot be effective for very short ray-segments, or else that there are not enough sources in the sub-group for their ray-segments to overlap. But we have hardly reached the bottom of the matter: in this connection energy means energy thermodynamically available, and that depends on the faculties of the sorting or resolving apparatus in face of the numerical density of units involved: if that apparatus were imagined to be more finely constituted dynamically than are actual atomic structures, availability relative to it might be augmented. Cf. the remarks on quantification in relation to the more refined control of ideal Maxwellian demons, *supra*, p. 411. {Gases not an exception, when Doppler dispersion is excluded. Contacts with thermodynamics.}

The outstanding fact that guides and compels this synthesis is the verification of the Rayleigh formula for atmospheric scattering, show-

* The repulsion of cosmical dust by radiation, as applied to comets' tails, after FitzGerald, Arrhenius, and Schwarzschild, and to the clearance of celestial spaces by Poynting, would thus be expected far to exceed that of clouds consisting of gaseous molecules. The question of loss of energy by radiation left open in Vol. 1, p. 667, is not yet settled: cf. p. 759.

[Rather the position seems to stand, with application to dust nebulae, that although a particle of condensed matter interacts with an incident beam of radiation so as to set up an intense field of vibration of its own, yet only a small part of the energy of this field is propagated away without limit and lost, so that in the main what is produced is a kind of extended field of standing waves. But this field would require renewal, and so exert an aberrational pressure, when the scattering body is in motion. See p. 755, *infra*.]

Relation to radiation by *quanta.*

ing that each molecule of the air scatters as a whole and not by its electrons independently. Yet for the short waves forming the X-rays (excluding the anomalous Compton phenomena of forward scattering for extremely short ones) this is no longer a fact: each ray-segment there contains enough wave-lengths (perhaps $10^6$) to render relative phases variable through the range of a whole period and therefore separate energies additive. But what are we to make of intermediate cases between these two limiting classes of periods? Are we to conclude that it is the small number of scattering electrons in the atom that affords not more than one ray-segment operative at a time in producing each physical effective element of breadth of the spectral line, here as it would seem much widened by their high internal speeds?

Material viscosity arising from internal radiation,

by aberration of its mutual repulsions.

The principle introduced and applied by Poynting, to explain the clearance of celestial spaces with respect to diffuse matter, more especially of the nature of dust, by the retarding influence of radiation, may be further considered. It has been reduced (*supra*, pp. 448, 573) to its ultimate terms of unadapted fact, by referring the viscous diffusion of momentum thus arising as between an exciting source and an absorbing source to the Bradley mutual aberration of the momentum carried across by the radiation between them, which makes the momentum transmitted by the rays oblique so as to have a component opposing any transverse relative motion of the particles, each supposed to remain steady in state. It is the more potent as a viscous agent the longer the free paths of the rays between emission and absorption. It may be presumed to assert a prominent part as regards recent views on the origin and illumination of nebulae, one held

Why do nebulae become spiral?

to be far less potent (but see p. 769) for gaseous masses than for clouds of particles. Thus on a current view that a nebula, so transparent that we can see through it, is so because it is full of its own radiation, of wave-length gradually degrading in time but originally small compared even to an atom, perhaps to an electron, so that the transparency may even be comparable (somehow) with the dimensions of the nebula, such internal radiation must be a potent mixer of velocities of remotely distant parts, conceivably reducing the motion of the whole system in a time not astronomically unreasonable towards a state of spiral rotation*. Some means for evolution of such regular

Can cosmical evolution be accelerated?

structures, that would not be infinitely slow in its action for these vast tenuous masses, is doubtless a *desideratum*†: cf. the appearances in

* But in the Compton type of scattering, conspicuous for very penetrating rays, the deviation of momentum is small, which is a factor in the opposite direction.

† "Abîmé dans l'infinie immensité des espaces que j'ignore...les effroyables espaces de l'univers qui m'enferment."

different lights of some of the ring nebulae. As this idea, ascribed by Hale to Jeans and Millikan, can be plausibly sustained, it may serve as an indication that perhaps in other respects also the field is not absolutely closed against modes of hurrying-up of evolutional processes, such as would allow the universe to function with reasonable duration without eating up its own ultimate substance to keep it going. Yet such considerations, involving very great and changing transparency to excessively short waves, seem to be far out of touch with the result that extinction by scattering by bipolar molecules, at any rate on classical lines, becomes ultimately independent of the frequency.

We may follow up the application of the same principle to the slowing of the rotation of a gaseous star by the stream of emerging radiation which traverses it, after the recent initiative of Sir J. H. Jeans. The molecules are not isolated as are Poynting's particles of dust, which are kept from falling directly into the Sun by their orbital motion which the aberration of the incident Solar radiation converts into very slow spiral approach; thus the backward push exerted by the stream of emergent radiation, on their rotation as part of the star's atmosphere, must here be braked by its viscosity, of which the viscous interaction set up by the general internal flux of radiations may form an important part. In fact for plane strata of molecular matter of any kind gliding over each other with velocity $v$ having a shearing gradient $dv/dz$, this braking by aberration per unit volume

for radiation of density $\epsilon$ and free path $l$ is $-\tfrac{1}{3}\epsilon c . c . \dfrac{v}{c} . l \dfrac{\partial v}{\partial z}$, the

*Application to a rotating gaseous star.*

factor $\tfrac{1}{3}$ arising as usual from averaging various directions of rays: which expresses an induced viscosity $\tfrac{1}{3}\epsilon cl$, in agreement with Jeans' formula transferred from gas theory. This adds to the direct viscosity $v$ due to molecular motions, itself great if it is presumed to come largely from free electrons with their high speeds due to averaged energies and their long free paths. It appears that, on current special formulae for atomic radiation applied to gaseous stars, near the centre of the Sun this radiational viscosity might be three times the molecular, the ratio increasing very rapidly for larger stars (Eddington's treatise, § 197, quoting Jeans' memoir). According to Jeans' final verdict in his later treatise, § 249, the general retardation in a star is slight, as was perhaps to be anticipated for all viscous effects owing to largeness of the linear scale, except in the outer rarefied layers which may be braked down to quite a small fraction of the angular velocity of the central mass. Yet it would appear as *infra* that just below the surface, where the natural field of temperature radiation has become established, the rotation ought not to change sensibly with depth: for

*The viscosity induced by internal radiation.*

though the internal viscous forces are balanced, there would be nothing to counteract the drag on the surface proportional to rate of change of the rotation. But here we are verging on the problems of the overlying chromosphere, as now tackled successfully in detail on the initiatives of Saha and Milne.

It is worth while to examine how far we can proceed here without special hypothesis regarding atomic radiation. Let $\pi\epsilon_0$ be the energy density at the surface, so that the flux of the emerging stream is $c\epsilon_0$: the outward stream at distance $r$ from the centre is $c\epsilon_0 a^2/r^2$ for a sphere, or for cylindrical analysis as *infra*, $\epsilon_0 a/r$. The aggregate viscosity is $\nu$, equal to $\frac{1}{3}\epsilon cl + \frac{1}{3}\rho\,(kT)^{\frac{1}{2}}\,l'$, where $l'$ is the mean molecular free path:

*Radiational viscosity preponderant.* $l$ and $l'$ vary in about the same manner, inversely as $\rho$. As $\epsilon$ varies as $T^4$, the radiational viscosity largely preponderates at the temperatures to be expected inside a star. If the angular velocity of the stratum at distance $r$ relative to the outward rays is $\omega$, the viscous shearing traction is $\nu r\,d\omega/dr$, and this has to balance the transverse aberrational push of the emergent directed beam, which is per unit volume $\epsilon_0\dfrac{a^2}{r^2}\,l^{-1}\,\dfrac{\omega r}{c}$

for spherical analysis. For the equatoreal parts a cylindrical scheme will suffice for preliminary illustration, with its factor $a/r$: thus

$$\frac{1}{r}\frac{d}{dr}\,r\nu\,\frac{r\,d\omega}{dr} = -\,\epsilon_0\frac{a}{r}\frac{\omega}{cl}\,r,\quad \nu = \tfrac{1}{3}\epsilon cl + \tfrac{1}{3}\rho\,(kT)^{\frac{1}{2}}\,l',$$

*Law of damping of rotation.* which is an equation for $\omega$ when $\nu$, $l$, $\epsilon$ and $T$ are expressed in terms of $r$ from some special stellar theory. The density does not enter except through $l$ and in the second term of $\nu$, as a steady state is assumed. Then an integration can be made giving

$$\log\frac{\omega}{\omega_0} = \tfrac{1}{2}\epsilon_0 a\int_r^a (cl\nu)^{-1}\,(1 - a^2/r^2)\,dr,$$

while deep in the star $\nu$ is $\frac{1}{3}\epsilon cl$. This permits a rough estimate, at any rate in the outer layers where $\rho$, $l$, $\nu$ can be guessed at *.

The enormous density of internal radiation is kept from bursting outward by repeated dispersal by scattering at the atoms, requiring free path enormously small, yet its bombarding pressure being more than overcome by their inward gravitation. A stage may arrive when *Eddington's criterion for stability.* the preponderance of gravitation is precarious: and Eddington finds in the ensuing instability the cause which limits the masses of the stars. A scientific imagination could perhaps conceive if necessary an explosion which would turn the star inside out, and then when the

---

* Cf. also p. 613, *supra*, on the Solar magnetic field as due to this aberration of radiation pressure in an electronized medium. Wherever, as possibly in close double stars, streams of intense radiation traverse electronized gases in rapid motion, electric currents with strong magnetic fields are perhaps to be expected, with Zeeman disturbance of the spectra.

resulting smoke-screen had subsided it might appear as a double star, the two components having been pushed apart by their mutual radiations, practically still of the great internal intensity. This trapping of the intense internal radiation, depending on frequent scatterings and therefore on extremely short free path, would diminish $v$. And the "mutilation" of the atoms by the high internal temperature, shrinking them so as to permit in the companion to Sirius a mass density of $10^5$, would involve that they all became ions, with increased effective radius for encounters, were it not that the high speeds counteract.

The potent stimulus to all this recent theory has been the remarkable empirical discoveries of surface clues to internal structure: thus surface brightness appears to be not very far from proportionality to surface gravity, except for diffuse giant stars which hardly have a surface: while features of surface spectra (after W. S. Adams) present other rather precise indications. *Superficial criteria for stars.*

It must have been remarked that the shape of the sunspot diagram (cf. p. 577) recalls that of the light curve of a Cepheid variable star, if the long period of $11\frac{1}{3}$ years may be taken to involve much reduced amplitude of the pulsation or other oscillation. The very remarkable estimate (Eddington, *Stars and Atoms*, p. 88: cf. however Sanford, Struve, *Ap. J.* 1928) that the period of pulsation of δ Cephei has shown no change of the order of $10^{-7}$ of itself, in the century that it has been under observation, seems to carry the result there enforced that neither the state of the star nor even the amplitude of its periodic change can have altered to this extent, a degree of permanence surely very remarkable notwithstanding its size when the very large amplitude of the very rapid pulsation, relative to its radius, and the fluctuating output of radiation, are considered. Rotating ellipsoidal or pear-shaped masses would be periodic variable stars; but the variety of form as their evolution proceeds appears to be in strong contrast to the uniqueness of constitution claimed on statistical grounds for the Cepheid type. *The permanence of pulsating stars.*

A remark is invited relating to the forward scattering of X-rays mentioned above. By the principle of Rayleigh, the secondary waves from the molecules, however they be distributed, always conspire in phase exactly in the direction of propagation without any mutual interference, at a half-period from the main wave, causing the reduced velocity in the medium—with index of refraction thus independent of scattering unless it is great—and ensuring propagation mainly towards the forward direction*. For $x$-radiation of mean frequency with its heavy scattering this ought still to hold: for high frequency the *Forward scattering.*

---

* But this concentration of scattering in forward directions soon falls away with obliquity, yet not instantly. Thus the resultant amplitude at inclination

Compton sideway degradation of frequency enters in addition to this predominant scattering of energy in the forward direction, entailing the familiar theory of *quanta*.

As regards index of refraction, provided the wave-length is so short that the electrons in the atom are affected independently, and therefore only for X-rays not long relative to $10^{-8}$ cm., the effect is the same as for the same number of electrons existing free, and the formula for wireless propagation in an ionized atmosphere (p. 644) applies, unless where the Compton effect becomes important. The total Maxwellian displacement is there $D$, equal to $F/4\pi c^2 + Ne\ddot{x}$, where $m\ddot{x} = eF$: and the index $\mu$ is given by

*Refraction of X-rays very slight.*

$$\mu^2 = 4\pi c^2 \frac{D}{F} = 1 - 4\pi N \frac{e^2 c^2}{mp^2} = 1 - \frac{Ne^2\lambda^2}{\pi m},$$

which for waves of $10^{-9}$ cm. could hardly differ from unity by as much as $10^{-3}$. Their extinction by scattering would also be slight; and it appears to be independent of $\lambda$.

*Scattering:*

The circumstances of scattering of radiation appear however to submit to analysis maintaining the original position set out on p. 608. When the molecules of a gas are closely packed relative to the wave-length, then if they were at rest they ought to scatter practically in coherence, and so internally not at all (p. 609); for the aberrations of phase owing to deviations from mean positions equally spaced would be small fractions of a period. But as they have the translatory velocities of a gas, changes of period come in, and the molecules which contribute to divert a train homogeneous within a very small range of periods would be sparsely distributed in the cubic wave-length, so that they would be wholly erratic in phase, being around no mean spacing, and they would scatter energy independently. These component waves, of the various differential intervals constituting a broad spectral line, are themselves independent, although they travel away in step, for they can be separated by a grating: in fact their energy for any considerable length of the train is the sum of the separate energies. More definitely, for two trains such as

*the original position re-asserted.*

$$\cos(pt - mx) \quad \text{and} \quad \cos(p't - m'x),$$

in passing any given point the energies are additive in time, on the

*Main influence is on propagation.*

$\theta$ is as $\Sigma \cos m \{x(1 - \cos\theta) + y\sin\theta\cos\phi + z\sin\theta\sin\phi\}\,\delta\tau$, where $x, y, z$ is position of the scattering atom, $m$ is $2\pi/\lambda$, and $\delta\tau$ is $\delta x\,\delta y\,\delta z$. This is

$$\Sigma\cos\{2m\sin\tfrac{1}{2}\theta\,(x\sin\tfrac{1}{2}\theta + y\cos\tfrac{1}{2}\theta\cos\phi + z\cos\tfrac{1}{2}\theta\sin\phi)\},$$

in which the defect of each term from its maximum value unity, appropriate to the forward direction, is at first of the second order.

The principle that in free space radiation disentangles itself (p. 146) and travels out in compact shells (cf. p. 261) without leaving any trail behind, is essential in all this discussion: in actuality the component shells may be uniformly undulatory over a range of $10^6$ wave-lengths or less.

average by change of period, whatever be $m$ and $m'$, while along the ray they are additive by change of phase, whatever be $p$ and $p'$: when both kinds of averages are taken into account the scattered energies are cumulatively additive as if the trains were incoherent. In a condensed atomic mass such as a particle of dust, small compared with the wave-length, the scattering will be much greater, the waves being nearly coherent except in so far as thermal atomic motions are effective to prevent it. In a gas, though the distribution in space is almost uniform in relation to the wave-length, scattering would remain, being due to the random distribution of velocities, as originally asserted.

In fact if, following Rayleigh's remark, we represent each scattered component by a vector radius whose length is amplitude and inclination phase, these vectors as regards positions of the scattering atoms will be smoothly distributed around a circle. Their resultant will therefore be null, unless there is excess of density over considerable parts of the circle which would mean concentration of atoms into denser local groups, the result being, with Smulochowski, opalescence*. This would conform to the position of Raman (p. 750) that the molecules of a gas such as air obeying Boyle's law ought not to scatter sensibly on account of random positions: yet they do scatter, and that independently, on account of their random motions, to an amount given precisely by the Rayleigh formula, and variations of density should not be invoked. Expressed analytically, the resultant amplitude as regards positions is $\Sigma a_r \cos (\alpha_r + \epsilon_r)$, where $a_r$ is a component amplitude, $\alpha_r$ is phase uniformly distributed and $\epsilon_r$ is the small actual deviation from this uniform spacing: this is $\Sigma a_r \cos \alpha_r - \Sigma a_r \epsilon_r \sin \alpha_r$: the first term vanishes when integrated round the circle, the second is negligible to the second order for cumulating reasons, as $\Sigma a_r \sin \alpha_r$ vanishes and the mean value of $\Sigma \epsilon_r$ vanishes compared with $\epsilon_r$. Thus the random positions alone give negligible resultant amplitude, except over ranges expressed by broken residues at the surface.

*[marginal note: Scattering can arise only from random velocities, or from fluctuations of density.]*

Thus in the case of a dense medium, which scatters coherently, in contrast with a gas, for an obstacle large compared with the wavelength, the total amplitude of scattering determined by integration cancels between successive half wave-lengths in the source measured in the direction of the ray, so that the result is an integral effective only at its ends, that is over a surface layer—as Rayleigh has already remarked: it is thus the same in type as would arise by the usual continuous analysis involving a beam reflected from the surface of a finite obstacle. Yet there ought to be powerful internal scattering, by the atmosphere, of $x$-radiation of long period: why that does not

*[marginal note: Relation to continuous analysis.]*

* But in opalescence the aggregations satisfy the condition, here held to be necessary, of being of diameters comparable with the wave-length.

extend to hard rays, whose wave-lengths are small compared to the atomic diameters, must open up other considerations.

If the sources of scattering, supposed stationary, are dense ($\rho$) per unit cross-section, for lengths comparable with the wave-length, we can be more precise. The amplitude is then $\int C\rho dx \cos(mx - pt)$, where $x$ is distance and $m$ is $2\pi/\lambda$, which integrates by parts to

$$\left| Cm^{-1}\rho \sin(mx - pt) \right| - \int Cm^{-1}\partial\rho/\partial\kappa \sin(mx - pt)\, dx.$$

*Opalescence.* If the graph of the density $\rho$ of sources has sporadic humps at intervals comparable with $\lambda$, over which alone $\partial\rho/\partial\kappa$ is of sensible amount, each of them makes its own contribution to the scattered amplitude, depending on its form and length: out of them the *internal* scattering of the medium is made up. On the other hand, if the gradient $\partial\rho/\partial\kappa$ is nearly uniform along the medium, another integration by parts will remove its effect also to the surface. Intermediate cases are less simple to describe.

But there is perhaps more to be said on this topic. Combining two trains along $x$ the energy is proportional to $(a \cos mx + a' \cos m'x)^2$: *Absence of* on expanding, its mean value is $\frac{1}{2}a^2 + \frac{1}{2}a'^2$ together with means of *approach to* periodic terms one of which is $aa' \cos(m - m')\,x$. These additions *a limit.* average to null values; but when $m$ and $m'$ are nearly equal, the range necessary to secure an average of the latter term is very long. If the train is limited in length, a factor $e^{-kx}$, where $k$ finally vanishes, will ensure convergence: then a limit is attained, after Stokes and Rayleigh (p. 549), by integrating for all the trains of waves over a range $\delta m$ including zero, and this would be adequate as a basis for the present result. But the question left unsettled in Vol. I, p. 668, is hardly cleared up: the unlimited length of the rays is the obstacle, which with *quanta* may not arise.

As regards the doctrine of equipartition, things are different in another respect. The procedure of Rayleigh was to reason from sources of sound isolated within a rectangular cavity with perfectly reflecting *The paradox* walls; and Jeans adapted it to electric sources. The Rayleigh formula *of Rayleigh:* deduced from equipartition of energy among all possible modes of freestanding waves in the enclosure was found, perhaps with surprise, to agree with the actual distribution but only for long periods: and he anticipated that there is in actuality some mode of prohibition that prevents the radiation from running into the shorter periods without limit. There is, however, in the actual problem no limitation of the field of radiation by reflecting walls. They can be removed if each source is replaced by the source itself together with an infinite lattice of images in the reflectors: thus the circumstances are not those of free radiation in unlimited space, which indeed is a problem of steady flux, not of steady stationary energy. In other words, this solution applies only to a system of sources repeated periodically in space—a sort of crystalline structure: and the feature of scale comes in. Yet natural radiation is, on a large scale, held to be of constitution un- influenced by atomic configuration; but the proof is perhaps restricted to matter capable of averaging, thus without the resonance appropriate to crystalline grouping*. Cf. p. 747.

*a possible solution.*

What has been advanced, *supra*, p. 444, as a proof that loss of energy $\delta\epsilon$ by radiation diminishes inertia by $\delta\epsilon/c^2$ is of the same type as the transformation of L. de Broglie (p. 805) with its wave-complications. It is there shown that if the force on the system is calculated from the Maxwellian stress around it as referred to its own "proper" frame, and the point of view is then transferred to an external frame in which the system is in uniform motion, a term $vd/dt - (c^{-2}\delta\epsilon)$ appears in the force. This may be held to account for the discrepancy of pp. 444, 447 with the result $(\delta\epsilon - 2\delta W)/c^2$ in Vol. I, p. 673, the formula $\delta\epsilon/c^2$ belonging to the world of the artificial fourfold, not to personal experience in space and time: when it is an affair only of translatory and not internal energy the two agree.

*In what sense has the energy inertia?*

The relativity stress-tensor argument (p. 444) gave a loss of momentum due to a radiation $\delta\epsilon$ from an isolated body of amount $v . \delta\epsilon/c^2$: if the factor of $v$ is the loss of inertia, total momentum for an isolated radiator is conserved without requiring any change of its velocity, a result which is consistent with the relativity from which it came. But if the change of mass is not $\delta\epsilon/c^2$, a change of velocity $v$ is required in order to conserve momentum which would be proportional to $v$, so that there would be negative acceleration of the radiator, of different values in different uniformly convected frames, —which would involve the need of a standard frame, here as *supra*, that of the Solar system as a whole: cf. p. 770. This is naturally so because change of translatory energy is relative to its frame, in contrast to the invariance of the Action.

* The discrepancy can also exhibit itself in another way. Compare two Rayleigh boxes, one of double the dimensions of the other. All the free modes of the latter occur in modes of the former but repeated eight times in each, for it can be divided into eight of the former boxes by partitions. But the total associated energy is to be the same for each complete mode of either box, being that of one degree of freedom of a gas-molecule. Thus, now comparing equal volumes, each octant of the larger box has only one-eighth of that energy in each period. This defect is made up for it by the greater number of free frequencies within assigned range $\delta\nu$, so that the energy density is the same in both boxes, and varies as the temperature. Then the principles of Boltzmann and Wien complete the specification. Thus in neither case does opening a small hole in the box disturb the energy equilibrium with the natural radiation outside. If we attempt to quantify by taking the energy in each mode to be $h\nu$, further discrepancies supervene. Contrast the procedure of p. 406. This topic of correlation can be pursued, for in comparing different Rayleigh boxes there is a one to one correspondence between their modes of vibration: these very various ultimate quantifications however approach, after Rayleigh, to the same macroscopic limit for the higher frequencies.

*Varieties of quantification.*

# APPENDIX V.

## FARADAY AS CONTEMPORARY WITH AMPÈRE: MOTORS, DYNAMOS, STRESSES, AND THRESHOLDS.

IT may be startling to recognize that Faraday's essential electro-dynamic activity goes away back to 1821, the year after the science was founded by Oersted and Ampère, when he was thirty years of age. Yet it was the preparation of a connected account (anonymous) of the great activity (*e.g.* electromagnetic screening had been ob-served\*) including that of Fresnel, in electromagnetism, for Phillips' *Annals of Philosophy* but originally intended as a synopsis for his own use†, that led straight to the invention of his appliances for electro-dynamic rotations, in Sept. 1821, which were pregnant at the very beginning of the subject with the foundations from which both the electromotor and its converse the dynamo would evolve. The story of how this brilliant discovery, which was no accident ‡, got him into trouble and even endangered his election into the Royal Society, is recounted, with other matters of interest for scientific history, at the end of the first volume (cf. p. 299) of Bence Jones' *Life*. It appears that W. H. Wollaston had entertained general notions of rotation, of similar type, as was indeed natural, of which Faraday, then merely

*Faraday's early study:*

*led to his electro-magnetic rotations.*

*Principle of magnetic wave-detector.*

\* A very sensitive test used much at this time for transient currents was magnetization of a steel needle inside a helix. Even an electrostatic discharge passed through the helix could do that: an extreme form later was Rutherford's early magnetic detector for wireless signals, further developed under Marconi's auspices. In fact if the period of oscillatory discharge in the circuit is $10^{-6}$ seconds, the whole charge passes initially to and fro in that time: while a sparking distance of one cm. involves a potential 120 electrostatic units, and so on a globe only 10 cm. in radius a charge 1200: thus the mean intensity of current produced by this feeble discharge initially would be about $2 . 1200 . 10^6/c$, which is of the order of an ampere. The steel is gripped by the transient peak of the current, and manages to retain part of the impression: but the energy of the current is very small, and it can grip only the outer skin of the needle. Cf. the phenomena of disruption by a lightning flash, and of the puncture of glass by a spark.

† In Maxwell's words, copied from Faraday's own emphatic assertion of 1823 in revealing his authorship (*Exp. Res.* II, p. 161), "he repeated almost all the experiments he described." It is of interest also to note that Maxwell in the same brief sketch (*Encyc. Brit.*) makes special mention of Faraday's last piece of experiment, a premature attempt to detect the Zeeman effect, revealed in 1897 long after Maxwell's death.

‡ In the repetition of the Amperean attractions Faraday's experimental resource was already conspicuous: his wires were suspended from the ceiling by silk threads: when the circuit was completed by their dipping into mercury complication arose which he ultimately traced to capillary deformation of the surface by the current at the place of immersion.

a chemical assistant at the Royal Institution, was at first suspected (not by Wollaston himself) of having overheard more than he acknowledged: cf. Maxwell's pungent remarks in a notice of his life written for the *Encyclopaedia Britannica*, as in *Scientific Papers*, Vol. II, p. 788. To neither of them was Ampère's powerful mathematical analysis of 1820 congenial or perhaps even intelligible. Faraday could get no grasp of the Amperean theory, abstruse and mathematically exact when rightly interpreted, expressing all the forces between complete currents as made up of direct mutual attractions between elements. The necessity for this limitation to direct attraction arose in Ampère's mind from his regarding the current, or rather Ampère's electric conflict as it was called, as made up of linear elements all guiding idea. physically independent: for the only way in which such self-subsisting elements could act directly on one another across a distance must, as he thought, be by a force in the line joining them, as otherwise the essential Newtonian conditions of balanced action and reaction between them could not be satisfied*. Maxwell extended the Amperean theory as a mathematical exercise on the possibilities of forcives of intrinsic character, in his *Treatise*, by adding a mutual torque, which also satisfied that condition but in a more unlikely manner. Instead of this systematic but artificial synthesis Wollaston appears to have simply conceived the excited wire as endowed in a general way with an electric power acting along it and an electromagnetic power (now to be interpreted as magnetic) acting circumferentially around it: and Faraday's guiding idea was of the same kind. Cf. *Life*, p. 315: letter of Sept. 12, 1821, in answer to remonstrance from his friend De la Rive†, on his lack of appreciation of Ampère's procedure.

The Faraday rotations, so-called, were soon refined by him to an astonishingly simple appliance, and in this compact form he distributed copies of the apparatus to friends such as De la Rive. The Simplified current passed from a point $A$, into a ring $B$ connected to the rest apparatus for of the circuit, through a stiff wire $AB$ which was free to revolve round rotations: $A$, while keeping contact with the ring. Between $A$ and $B$ in the axis

* Cf. on this subject generally, the systematic historical collection *Mémoires sur l'Electrodynamique*, edited by Joubert (1885) for the French Physical Society.

† The main paper on electromagnetic rotations, dated Sept. 11, 1821, was reprinted from the *Quart. Journ. of Science*, along with subsequent relative notes, by Faraday, and also with some much later history from which Dr Bence Jones' biographical account is mainly taken, in Vol. II. (1844), pp. 127–163, 230, of the *Experimental Researches*. The general notion of regions of power around Early origin magnets and currents there proved to be adequate to guide Faraday's very of Faraday's refined experimenting: and as involving the earliest practical indication of the fields of field stress which was afterwards set up in mathematical form by Maxwell, Maxwell's power and they seem to be still worthy of close attention for the history of scientific stress. thought.

of the ring was a bar of soft iron which could be magnetized, in either
direction at will, by a magnet applied to its end. This adjacent
magnetism set the wire $AB$ carrying the current into revolution
conically with increasing speed, until the
brake due to increasing resistances produced a steady balance. Conversely, if the
conductor carrying the current were fixed
he could make a magnet revolve. Original
specimens of this, the primeval motor and
dynamo, doubtless are still in existence*.

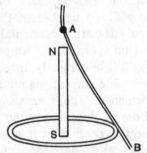

The revolving wire was competent to
push onward extraneous bodies to which
it could be attached, and so to abstract
mechanical work into other systems. Whence could it arise? It must
electro- draw upon the energies of the current: thus either the electric battery
magnetic must do more work or the strength of the current must be diminished.
induction an
easy inference. But the time was not ripe for this fundamental inference, now so obvious. A period of enforced rest, for nearly four years, intervened—
not the only one in his career: in 1831 (*aetat.* 40), after seven years of
renewed occupation with the subject, with full cognizance of the
efforts of others as his early digest shows, there came the crowning
systematic discovery of the production of currents from magnetism,
and electric science advanced with a rush.

Let us now simplify the conditions by removing the soft iron
Refined cylinder inside the cone of the revolving wire, replacing it by a coil of
further into the wire circuit that carries the current, wound into an Amperean
a self-starting
dynamo: helix. If a current is flowing in this single circuit it has, on modern
ideas, electrokinetic energy $\frac{1}{2}L\iota^2$, mostly in the helix, it dissipates
energy according to linear laws of friction and therefore proportionally
to the square of the current so at the time rate $R\iota^2$, and it receives
energy from work done on the system in driving the movable part
round its conical path with angular velocity $\Omega$, at a rate $\mu\iota\Omega.\iota$ where
$\mu\iota\Omega$ is the number of magnetic tubes of the current itself which this
revolving wire cuts across per unit time. Thus if the circuit contains
an electromotive source $E$ providing energy at rate $E\iota$, the equation
of transfer of energy in the system per unit time is

$$\frac{d}{dt}\left(\tfrac{1}{2}L\iota^2\right) - R\iota^2 + \mu\Omega\iota^2 + E\iota = 0,$$

* It was twelve years later, in 1833, the discovery of induction of currents
intervening in 1831, that Faraday discovered, in the sweep of his explorations
based as usual with him on intense critical knowledge of previous work, and
almost against his will, the atomic character of electricity: for he only half
believed in atoms.

which is, when the configuration remains the same so that $L$ is constant,

$$L \frac{d\iota}{dt} - (R - \mu\Omega) \iota = - E,$$

giving

$$\iota = \frac{E}{\rho} (1 - e^{-\rho t/L}), \quad \rho = R - \mu\Omega.$$

The current settles down to a steady value $E/\rho$ belonging to the source by itself, so long as the angular velocity $\Omega$ permits this effective resistance $\rho$ to be positive. But for values of $\Omega$ greater than $R/\mu$ the resistance is more than annulled, and the current, now reversed, increases very rapidly to a limit, which may be much greater, imposed by the new resistances that its magnitude has made important.

But the arresting consideration is that there need be no electromotive source $E$ provided in the circuit. If this argument be right, as the wire begins to be rotated round the cone nothing happens till its velocity is raised to this threshold value $R/\mu$: then a current should *at a threshold* suddenly appear, and promptly rise to its full limiting value for the *speed.* actual supply of mechanical power, the direction of the current being such as by its magnetic field to use up this supply of energy by opposing the rotation of the wire.

This device affords in fact the model, reduced to simplest terms, for the essentials of a modern self-starting dynamo. The wire-system, in steady motion involving a sliding contact, becomes unstable in its relation to the ambient aether on reaching a threshold speed,—really a very startling phenomenon which its enormous industrial consequences can never make common or familiar. It can only arise from *Source of* the electrons in the wire having an intrinsic twisting grip on the *such electro-* aether such as can unfold itself by actions originated at the com- *instabilities,* mutating slider and guided by an incipient surrounding electrodynamic field, though itself remaining infinitesimal (cf. p. 42), into drift uniform all along each wire constituting an electric current. When this principle of self-excitation was evolved in practical construction long after by H. Wilde and Werner Siemens, the start was *is not residual* ascribed to residual magnetism in the iron cores: but here there is a *magnetism.* non-magnetic ionic circuit guided into electric instability, and nothing more.

An instructive alternative experimental arrangement (evolved by *Alternative* joint discussion in class work many years ago) may now be mentioned. *model.* It is a circuit completed by a downward jet of mercury, with an adjacent bar horizontally mounted so that when it spins round an axis not at its centre, cutting across the mercury jet periodically, one of its magnetic poles threads the electric circuit in each revolution while the other describes a path not so linked with it. If the magnet is driven the system is a dynamo: if the circuit carries a current it drives the magnet and is a motor. Here also the iron magnet is inessential, and may be replaced by a coil pivoted in the wire circuit; also the arrangement may be self-starting like a dynamo.

The complete principles of the operation of dynamos and electromotors were thus on the point of emerging, with Faraday, in the year of Oersted and Ampère: so simple are the essentials of fundamental processes.

Where there are, as in modern practice, two circuits, stator and rotor, in relative motion but now without direct connection through sliding contacts, alternating currents each uniform throughout its circuit at each instant may be excited after a like threshold speed has been attained.

One recalls the electric instability excited, also after a threshold
*General* speed, in an electrostatic inductor apparatus after the manner of the
*threshold* Varley or Holtz or Wimshurst machines: cf. Maxwell, *Treatise*, Vol. I,
*principle.* §§ 209–213, where however the unavoidable resistances and the threshold to which they probably give rise do not come into the discussion.

In like manner it is the introduction, into the oscillating circuit of a vacuum valve, of an electromotive force changing sign with the current in it, and produced from a mutual inductance in a linked circuit, that starts the spontaneous electric oscillations in it as soon as that inductance has been increased up to an initial effective value.

Thus as a general principle, in self-starting dynamical apparatus, forces of the type of resistance may be expected to involve inhibition up to a minimum limit of speed. Viscosity arising from loose bearings may thus postpone instability due to spin. Gyrostatic influence provides
*Radiation of* another type of effective force linear in the velocities (cf. Vol. I, *supra*,
*an atom.* p. 49), the whirling of a rotating shaft is a case of such instability: even perhaps suggesting that cyclic activity in an atom, without any dissipation of energy, may delay radiative response to external excitation until a threshold value is reached. Cf. Thomson and Tait, *Nat. Phil.* ed. 2.

*Graphical* These rudimentary general ideas may be extended further. If the
*scheme.* equation of increase of $\iota$ is $d\iota/dt - a\iota + \psi\,(\iota) = 0$, the current, once it gets going, rapidly establishes itself up to a maximum value given by $d\iota/dt$ null, therefore given by $\psi\,(\iota) = a\iota$. In such circumstances a graph of $\psi\,(\iota)$ would express the course of phenomena: for example, an undulatory form of $\psi\,(\iota)$ would involve a succession of thresholds, of type obvious to inspection. Thus simple, analogically, may also be the essentials

in other ordered natural agencies, however complex, not excluding physiological types such as the mechanics of the animal circulation.

Faraday was specially interesting when he allowed himself to think aloud, which was mainly in his brief letters to Dr Phillips, republished from the *Phil. Mag.* by Faraday himself in *Exp. Res.* Vol. II. Cf. especially the penetrating note on the production of the extra current, with its spark and shock, on coiling the wire, which influenced Maxwell so much: also on inserting a bar of iron in the helix, and the nature of the efficiency of soft iron compared with hard. This extra impulsive current he ascribes to a kind of momentum, but not one of flow in the wire, rather momentum located in the magnetic field around it which may alter on coiling while the current remains the same.

It appears that Faraday himself drew a current from sliding contacts between a pole and the equator of a spinning magnet: referred to in *Exp. Res.* Vol. II (1834) p. 209.

He also had the idea at an early stage that all metals have a temperature of transition to a ferromagnetic state, but cooling experiments did not support it. He noted however how sharp the transition was in the case of iron; and he found that the coercive property disappeared at a lower temperature than the ferromagnetic quality.

# APPENDIX VI.

## THE TIME AND SPACE OF ASTRONOMICAL OBSERVERS.

In recent papers (Berlin *Berichte*, 1927–8) Prof. Einstein appears to approach his problem of the laws of motion of a particle under gravitation, whether with electric charge or not, by locating it in the extraneous fourfold field specified by its vectors and tensors, and constructing the additional mechanical stress-tensor arising from the presence of the particle regarded as an isotropic singularity of the simplest type. The self-equilibration of the combination of these two quadratic tensors of stress would impose differential equations, involving for the particle-source relations between its four coordinates, thus prescribing formulae defining a curve in the fourfold which ought, when translated into a frame of space and time, to give

*Geodesic orbital postulate now verified approximately from field-stress theory.* equations of its orbital motion. He concludes that the analysis does verify, to a first approximation at any rate, that each fourfold path is everywhere of minimal length, by itself alone, provided there is absence of electrodynamic influence\*. It would follow from this that the assumed reduction of the distribution of Action as equilibrated in the fourfold to an orbital form $\Sigma\!\int\! m_r ds_r$, which is the foundation of a discussion of orbits by the writer in *Phil. Mag.* Jan. 1923, and had in fact been widely customary, must be given up. Yet complete abandonment would imply that the interaction between two bodies

*The medium cannot be eliminated from law of gravitation.* is not expressible in any sense, even for the slowest motions, as an affair of the two bodies alone, but cannot avoid the intervening region appearing as an overt contributor after the manner of an aether: that in fact atoms are tense knots in the active medium whose relations are not reducible, even in the simplest cases, to a direct interaction between them of the Newtonian-Amperean type. A further scrutiny of the underlying ideas is therefore necessary.

The observer, who is himself in motion in the orbit of his own planet, would carry along with him his personal frame, local and uniform, of space and time, in which he mentally sets and actually measures

*Knowledge purely local were it not for rays.* the phenomena that come to his cognizance. He could, in any case, be aware of nothing physical except what arrives into the locality of that frame: it is from this alone that he and his associates extrapolate, from the purely local impressions thus gained, into their science of an extraneous astronomy, a feat which would be a sheer impossi-

---

\* The misfit with the type of actually recognized electrodynamic Action is further remarked on *infra*, p. 794.

bility in the absence of organic memory and exact developed records, and even then unless the phenomena were recurrent with convincing regularity.

The negation of any law of reduced material Action of such type as $\delta\Sigma\int m_r ds_r = 0$ would involve that it would be impossible for the group of astronomers, isolated in their own environment in space and time on one planet, to extrapolate so as to come into touch with the records of astronomers on another moving planet, without passing outside space and time altogether—except approximately for slow orbital speeds. This may be indeed the basic conclusion of gravitational relativity, when reduced to its ultimate terms; and, however startling, it could not be rejected as paradoxical. Even then, however, it seems strange that the problem of the orbital motion of two stars around each other, so brilliantly and easily achieved by Newton, should on the tensorial scheme be held to be so hopelessly, even in rough approximation, beyond the range of practical discussion. Comparison of frames transcends space and time.

A mode of reconciliation of the eclipse values of the gravitational deflection of light, as now generally accepted, yet saving a fourfold orbital form of Action of the type described above, was however advanced in the course of a supplemental Appendix in *Nature* (April 9, 1927) on "Newtonian Time Essential to Astronomy," which, after a period of hesitation, the writer has been again inclined to maintain as valid, in modified form. Briefly, as the $dt_r$ was taken in the formula for an orbital Action to be the same $dt$ for all the moving systems $m_r$, it had to be identified with the universal (Newtonian) time which is the same for the local frames of all groups of observers domiciled on the various planetary bodies. It is this in fact that permits and prescribes a Riemannian fourfold, as the proper graphic formulation, such absolute local time being the direct significance of the absolute $ds^2$ of that type of geometry. But if this identification is made, the coordinates such as $x_r, y_r, z_r$ of the formula would also, as well as $dt$, have to belong, for each moving observer, to his frame attached to his observatory; whereas in astronomy all the local systems have of necessity to be referred to one standard spatial frame, most conveniently to the frame attached to the central Sun, whereby alone indeed can any idea of motion and its velocity be introduced. The time in the Action formulation must therefore be transformed to the special time $T$ belonging to the central Solar frame, within which as basic frame the frames of all the systems are to be regarded as immersed and in motion with astronomical velocities $v_r$. The Lorentz transformation to Solar time $T$ proves to be adequately expressed for this purpose as $dt_r = (1 - v_r^2/c^2)^{\frac{1}{2}} dT$, because the changes of epoch for local events in each system's own local frame which enter Not absolute times in astronomical Action, but times belonging to the standard Solar frame:

into the transformation can be neglected on account of the slow

velocities therein\*: while the spatial shrinkage of the correlation merely alters things to the slight order of $v^2/c^2$. The transfer from local to Solar frame in space and time is thus practically only change of measure of time from $dt_r$ to $dT$ by this formula: though a reversed transfer from Solar to local would for the local observer be complicated (cf. *infra*, p. 790) because the local variation of epoch of time with position now involves displacements with the full velocity $v_r$ in the Solar frame and could not be ignored†. Thus when gravitation is negligible, the orbital equation of Action changes from

$$\delta\Sigma \int - \tfrac{1}{2}cm_r\,(c^2dt_r{}^2 - dx_r{}^2 - dy_r{}^2 - dz_r{}^2)^{\frac{1}{2}} + \delta A_e = 0,$$

where $A_e$ is electric action, to approximately

$$\delta\Sigma \int - \tfrac{1}{2}cm_r \left\{ c^2\left(1 - \frac{v_r{}^2}{c^2}\right) dT^2 - dX_r{}^2 - dY_r{}^2 - dZ_r{}^2 \right\}^{\frac{1}{2}} + \delta A_e = 0:$$

in this the new term $- v_r{}^2 dT^2$, arising from the transformation, is equal to $- dX_r{}^2 - dY_r{}^2 - dZ_r{}^2$, so that the effect of the transformation is just to double this latter term under the square root. With this factor 2, the original form can be modified to express gravitation, as before. But as thus altered the Action form is invariant only relative to the central Sun and only approximately so. When interaction of the bodies can be neglected, the path curves in the fourfold, then uniform, remain straight.

More explicitly, on this modification, due to velocity of the observer's system in the Solar frame, there may be grafted the effect of

a gravitational field of potential $V$, found originally by Einstein to be sufficiently expressed within his generalized optical relativity by affecting $dt^2$ with a factor $(1 - 2V/c^2)$. The two influences, being both small, may be added together. The factor 2 introduced (*loc. cit. Phil. Mag.* Jan. 1923), when a single Action formula is used instead of geodesic postulates for each orbit, would compensate the factor 2 here

arising from the velocity: and reconciliation of the Action formula with the eclipse results would be effected.

\* The perihelion effect for Mercury goes deeper.

† The ultimate origins of these theories were conceivably along lines somewhat similar to the above. The changes of speed due to gravitation, combined

with the absoluteness of the free optical periods of the atoms, confused however to the observers by Doppler effects, demanded a continual adaptation of their measures of time: this adaptation of time, most easily discussed in connection with the Minkowskian fourfold conglomerated representation of optical relativity, required also continual adaptation of their frames of space: this would affect the expression of gravitation, and might even, with Einstein, be its essence—not however as an optical spatial illusion, but as conditioned by the nature of the atoms on which it is concentrated.

The natural unit of time is absolute, and must remain unchanged everywhere, while in the Solar frame more time must be regarded as pressed in, so to say, owing to adjacent matter. Gravitation would make no alteration in local phenomena, with their absolute scale of time, but it would imply an excess of astronomical time near larger masses according to the factor $(1 - 2V/c^2)^{-\frac{1}{2}}$. The appropriate type of analogy here implied is a flat sheet of elastic metal, where all lengths in the metal have to be increased on account of local rise of temperature, thus requiring a local bulge on the sheet that would otherwise be plane. The geometry of this bulge, if it is slight, would naturally be discussed relative to the geography of the standard plane sheet on which it has appeared: so the motions in the Solar system are symbolized in the basic uniform or flat fourfold into which the bodies introduce slight local warps.

> Analogy with thermal bulging of a plate.

The natural universal form of $ds^2$ in the observer's own frame, expressed by $c^2 dt^2 - dx^2 - dy^2 - dz^2$, corresponds in the Solar frame to a form $c^2 (1 - g) dt^2 - dx^2 - dy^2 - dz^2$, any modifications of $dx^2, \ldots$ being negligible in effect compared with that of $dt$ because in it $dt^2$ enters affected by the factor $c^2$. The geometrical theory would have to limit $g$ to satisfy an approximate differential equation: this would be expected to be intrinsic in terms of the basic uniform fourfold on which the small warp is superposed: the one available linear operator other than a constant factor $\kappa^2$ is $c^{-2} d_t^2 - d_x^2 - d_y^2 - d_z^2$ or $-\square^2$, thus the equation might be presumed to be of type $(\square^2 - \kappa^2) g = 0$ or, as there is no cause as yet recognized for the presence of $\kappa^2$, of type $\square^2 g = 0$: and so the geometric theory does make it out to be. When this is interpreted into Solar time and space, it means that a local gravitational disturbance would be smoothed out by being shed off with the velocity of radiation, until only a steady part remains, which is keyed on to the masses and satisfies the Laplacian $\nabla^2 g = 0$. The process is however merely one of possible modes of adaptation of gravitation into optical relativity, to be tested by the results which it compels.

> The approximate gravitational result not unexpected.

The extension of considerations such as these into variation of inertia with speed seems to be a different affair, involving a hybrid frame of reference, terrestrial space measurements and the astronomical Solar time: cf. Vol. 1, *supra*, p. 674.

> Hybrid actual astronomical frame.

It is mind that perceives the universe: its organization into a working system of knowledge rests with mind: it is the same universe that must present itself to a mind wherever its bodily vehicle is situated and however it be moving, if all minds operate identically.

The fundamental criterion in the discussion above is the ubiquitous potential observer: it is postulated that minds may exist anywhere, and that there is a unique material universe running exactly parallel in consistent manner to the impressions conveyed by light to them all. Time and space belong to the observer, not to the material cosmos; they are part of the mental frame in which the observer sets the impressions that come to him from the world outside: his place and time are thus local to him, but are instinctively prolonged, as a uniform Cartesian frame, so as to cover the whole universe. If all observers were relatively at rest these universal frames or nets would be the same for all, indistinguishable where they overlap. For we have made the far-ramifying postulate that all observers are mentally identical, just as all atoms of matter are physically identical. But if the observers are in motion relative to each other, the circumstance that the causes of their impressions travel to their localities with the finite speed of light becomes an essential feature for exact theory, and a process of light-ranging is necessary after the manner of the sound-ranging in modern artillery science. Where the natural frame common to the astronomers on the Earth, as thus mentally prolonged, over-laps the frame of potential astronomers say on Mercury travelling with different speed, they do not exactly fit together, as many recent ways of partial illustration have enforced. But all such frames prove to be very simply related as one interlacing group, as was noted earliest by Poincaré on the basis of the Lorentz electrodynamic corre-lation between every pair of them, and by Einstein in a more ultimate train of thought. This group relation was formally systematized, even made in some degree intuitive, by Minkowski, who imbedded the whole group of frames in one continuous extension of higher dimen-sions.

The fourfold is thus made up of flat frames as its elements, not of points: just as, to take a simple but illuminating case, a surface in geometry can be regarded as made up of its flat tangent planes, but the elements of the planes are not elements of the surface. Yet in each case there is a potent mathematical correlation between the group of frames or tangent planes, as the case may be, and the group of points of contact.

Our potential observers on Mercury and everywhere else, as thus introduced, are not merely a transcendental subordination of the cosmos to mind. The source of our setting of all practical knowledge in the astronomical domain is the sweep of the actual terrestrial observers round the Sun, with high velocity which will be different in direction in a quarter or half a year from what it is now. Their own frames, all of them measuring absolute space and time, are identical when each is referred to its owner, but the same frame may not belong to two observers unless they have the same velocity. Thus the

*Marginal notes (left column):*

Space and time sub-jective,

but universal.
The observers that have the same frame of reference,

are those in relative rest.

The delay of light-signals the sole com-plicating cause:

requiring a group of frames more complex than the Newtonian,

consolidated into a four-fold map,

by a tan-gential geometry.

Observers' frames change relative to Solar:

frames of the two observers, prolonged uniformly, will not coincide where they overlap, but will be related by the Lorentz correspondence involving their relative velocity. Yet it is the same absolute material universe that has to be capable of being framed, presumably in manageable terms, in each of them. This is what is secured to a certain extent by the nature of the correspondence between the frames: it is from such a postulate of invariance of the material world that the group relation of the frames to each other has to be derived. These statements constitute the basic principle underlying the original electrodynamic and optical relativity: their content can be set out symbolically with some graphical intuition as well as much algebraic control within the Minkowskian fourfold continuum. *in manner determined by uniqueness of the cosmos:*

Now it can be imagined, with Einstein, that this fourfold, while remaining flat or uniform on the whole, is essentially distorted around each material atom. It is precisely the problems of the Riemannian *quasi*-geometric continuum that thus arise: the absolute character of his local frame for each observer is expressed completely and concisely by the invariance of the differential element of *quasi*-distance which defines such a continuum*. What type of warp in the fourfold extension may exist around an isotropic centre without vitiating such Riemannian continuity? This is the mathematical problem opened out by Einstein, who found that a mode of representation of what in space and time exists as gravitation is imbedded in its solution. Electrodynamics had already been reduced to minimal Action in a uniform aether: its invariance for the frames of all potential observers merely involved that this Action was an intrinsic form when transferred into the Minkowskian uniform fourfold. The problem as now modified proved to be equivalent (Lorentz, Hilbert, Einstein) to generalizing suitably into a form of Action intrinsic to the warped fourfold. When the fourfold is nearly flat, the warping being slight except near the masses, the astronomical process ought perhaps to be capable of being carried through (cf. p. 792) by referring the changes to the flat original as a reference system: but instead, an additional *widened by absorbing a potential gravitation into the frames.*

*A fourfold field Action:*

---

* It has been noted by Eddington that in the affine geometry of vectors, developed by Levi-Civita, one quadratic scalar can be constructed with the invariance, though only *local*, that is characteristic of distance and is thus expressive of a continuum locally Euclidean. It is the characteristic of Riemann's geometry, which adapts it for the expression of an aether for molecular matter, that there is such a scalar, but one transferable with invariance to all localities. On the other hand in the Action formula for electron orbits in a field of potential *Helmholtz-Lie-Eddington quadratic geometry:*

$(FGHV)$, namely $\delta\Sigma \int \frac{1}{2} m ds \left\{ -c + \frac{e}{m} (F\dot{x} + G\dot{y} + H\dot{z} - c^2 V\dot{t}) \right\} = 0$, in which $\dot{x}$

represents $dx/ds$, the original proposal of Weyl seems to amount to regarding the multiplier of $ds$ as a factor of scale whereby the interval is no longer of quadratic form. *a contrast.*

with geodesic postulate superposed.

general postulate, of extraneous yet intrinsic character, was imported that every orbit purely gravitational is geodesic everywhere locally, without any direct reference to the bodies elsewhere in the field, though things were not so ordered in any sense in the electric field. On the other hand the procedure sketched above simply inserts the orbital gravitation into the previously formulated electrodynamic orbital Action: when the electric influence is negligible a gravitational orbital Action naturally has to survive*.

Inertia changes with frame:

As already remarked (cf. *supra*, Vol. I, p. 674) the correction on effective inertia due to velocity is involved here, for velocity has no *locus standi* except relative to an extraneous frame. As regards the ultimate electronic projectiles, with their great speeds, and also as regards spectral applications, it does not matter practically whether

its astronomical value.

it is the terrestrial or the Solar frame that is adopted. But in astronomy, based throughout as *supra* on the astronomical time belonging to the Solar frame, the term expressing inertia, which however now mixes up with the gravitation, would have to be $m_r \left(1 - 2v_r^2/c^2\right)^{\frac{1}{2}}$ instead of $m_r \left(1 - v_r^2/c^2\right)^{\frac{1}{2}}$.

Form of intrinsic Action for system of moving electrons.

* The general orbital formulation by Action (if such can exist) which includes everything orbital within itself alone, would be, for a system of masses and electrons, of type

$$\delta \int \Sigma - \tfrac{1}{2}c\left[ m_r \left\{ c^2 . dt^2 \left(1 - \frac{v_r^2}{c^2}\right).\left(1 - 2\frac{V_r}{c^2}\right) - v_r^2\, dt^2\right\}^{\frac{1}{2}} \right.$$

$$\left. - \left(\tfrac{1}{2}e\dot{x}_r F_r + \tfrac{1}{2}e\dot{y}G_r + \tfrac{1}{2}e\dot{z}H_r - \tfrac{1}{2}ec^2 V_r\right) dt \right] = 0,$$

in an electric field, of potential $(F, G, H, V)$ equal to $(\Sigma e\dot{x}, \Sigma e\dot{y}, \Sigma e\dot{z}, \Sigma ec^{-1})\,|r|^{-1}$. The unusual factors $\tfrac{1}{2}$ are required by each product such as $\dot{x}_1\dot{x}_3$ occurring twice in the summation. In the first term $v_r^2$ represents $\dot{x}_r^2 + \dot{y}_r^2 + \dot{z}_r^2$ and the time factors are grouped to indicate their origins. But see *infra*, p. 794.

# APPENDIX VII.

## MIND, NATURE AND ATOMISM*.

THE fundamental doctrine that underlies all modern scientific speculation is the atomic theory. The natural tendency, judging from one's early memories, is to regard a substance apparently uniform, for example water, as divisible into smaller and smaller parts without limit. Yet even in early Greek times prominent schools of philosophy felt compelled to decide that it is not so. What were the grounds that induced Leucippus and Democritus, even Newton at the beginning of the modern age, to hold that the universe is made up of small indivisible particles? The Greek idea seems to have been materialist and naïve. All change arises from motion: if matter were continuous there would be no room for its parts to move about. The universe is constituted therefore of atoms and the void, as proclaimed in imperishable verse by Lucretius along the lines of his master Epicurus. The early atomists claimed to be the discoverers of the void. For them there is no aether, only the universal vacuum, which makes room for motion: the movements of the atoms are thus free of all mutual control except their occasional collisions with one another, combined with a vague innate tendency to fall in some definite direction.

Origins of atomism.

In the eighteenth century a more precise idea of control arose, along what were then called Newtonian lines, as effected by imagined forces of attraction at a distance which the atoms exert on each other across the void, but according to laws of a mathematical type potent only while the atoms are near together: so that, following Clerk Maxwell in his systematic dynamical development of the Theory of Gases, it became appropriate to speak of soft encounters between atoms swinging round each other replacing more shattering collisions. But these forces of attraction and repulsion had to be themselves accounted for: in theory they still survive as provisional modes of expression of intimate interactions deeper but as yet unexplored.

Laws of molecular force.

To Dalton belongs the definite formulation of a practical atomic theory, as regards its main overt realm that of chemical change, at the beginning of last century. Yet not long ago a school of chemists still flourished who rejected, as fantastic and artificial in the light of ideas of gradual evolution in Nature, then very prominent, the notion

Energetics as substitute for atomism.

---

* This is essentially an abstract of a lecture entitled "The Grasp of Mind on Nature," in *Proceedings of the Royal Society of Edinburgh*, July 1927, pp. 307–325.

that matter could be made of atoms of a few unalterable types, in each type all essentially identical; who preferred to refer all natural phenomena to interchanges of one abstract entity called energy, only recently recognized, subsisting as continuous substance, in fact more ultimate than even matter. Though their groundwork of energy as transformable but indestructible came from dynamics, the school never included any mathematical physicists, at any rate in those days. The usual account is that it succumbed to the revelations of radioactivity, which presented to direct inspection the activities of definite sets of individual atoms, in a manner that could no longer be gainsaid.

We have it then that the universe is made up of self-constituted units, of an astonishingly limited number of types, all individuals of each type essentially alike, with no slight variations such as those exhibited in animals or plants of the same species or even variety. This underworld is revealed to us, its spectators and interpreters, only through impressions made on the limited number of our senses, marvellously sensitive, yet themselves also entirely material and atomic in structure. The messages that pass through them to the mind must surely bear traces of their atomic origins: though the mental image that is evolved cannot in any case be other than a mere partial shadow of its extraneous cause. How limited it is one may realize, with Berkeley, from the case of the predominant sense, that of vision, which can provide nothing more than a somewhat rough photographic impression on a flat plane, the retina of the eye, as the sole material for the mind to work upon, all else being interpretation and inference.

*Reaction of atomism on perception.*

These physical limitations to our actual means of knowledge may impose like restrictions on our mental modes of comprehending the world, may have reactions in psychology and even in metaphysic. This is the field that it is proposed to explore: the aim is thus to try to approach more closely, from both sides, the chasm that separates mind from matter.

Abstract material science puts aside the observer and treats only of knowledge. It is now proposed to bring him back, to consider that we have to deal with a mental interpreter, as well as a universe. We take it moreover, for purposes of discussion, that it is not merely a case of one self-conscious mental observer, but of numbers of potential observers distributed throughout the universe and capable of comparing results. It is in the spirit of the atomic theory to regard them all, interacting with the world only through their own material senses —at any rate to regard all human observers,—as identical in the normal structure of their reasoning faculties without any trace of variation. This absolute supremacy of reason is indeed the fundamental postulate of all: for it provides the criterion of truth and error.

*The mental observers of the universe:*

We have to regard this plurality of minds, potential observers as we have called them, distributed throughout the universe, as capable theoretically, of communicating with one another. Then the question arises: is it the same universe that they all explore? Or will they find out on close comparison of their mental acquisitions and records that each of them has his personal universe, which may or may not be completely intelligible to himself, but is not necessarily capable of being placed in congruent relation with the universes seen by his correspondents situated on other whirling planets or on remote stars?  *their means of intercommunication:*

The answer of naïvely material philosophy would perhaps be that these suppositions are impracticable, and therefore the question cannot arise. Nor would it, except for the essential extra-physical fact that the minds of these observers are endowed with the faculty of memory*. An astronomer on the Earth cannot communicate directly with one on Mars or on a dark star. But he does what is an equivalent. *virtually instantaneous,* He can compare his own exact observations, taken and recorded now, with his future observations a quarter of a year hence when he will be far away along the Earth's orbit, and travelling through the universe with his orbital speed of thirty kilometres per second towards very different stars. The records of his measurements and those of his colleagues under such changing conditions constitute an astronomical memory, of most imposing power and exactness, extending backward now for hundreds of years. The practical form of our question is thus:—Is it possible for our terrestrial astronomer to *astronomy being constructed on this basis* elaborate a picture of an external world into which all his wealth of recorded observations, spread over different times and locations and speeds, shall fit precisely? Or will the notion of a unique external world, evolving consistently within the mental spaces and times of all potential astronomical observers, the counterpart of their physically measured records, be found to be an illusion? One may perhaps hold that this issue focuses the questions which, in more abstract forms, are in recent discussions grouped under the name of relativity. On the practical side the subject calls up the great constructive astronomical achievement of Bradley.

We have had to mention the spaces and times belonging to the various potential observers. They too may be an illusion: we may be *The problem:* driven by the logic of comparison of the records to the conclusion that in strictness there is really nothing corresponding to an observer's

* The immense power of organic memory of various kinds as a direct guiding agent in the phenomena presented by animals and especially insects, as contrasted with man, may be recalled. An active memory is a chief ingredient in intellectual power.

simple personal notion of a frame of space enshrining a cosmic history evolving in his personal time.

It is here that the modern atomic doctrine can have a claim, even transcendental, to intervene. The construction that now follows is perhaps still common form in physical science, admitted from all the *as controlled* points of view. The atom of each chemical element, say hydrogen or *by atomism.* helium or lithium, is presumed to be a self-centred entity, identically the same wherever it be in the cosmos—in mathematical language is absolutely invariant: it is relative to nothing except its own structure. The modern science of spectroscopy, by intimate analysis of the very definite and complex radiations coming from the vibrations of the individual atoms, has compelled that point of view. Our astronomer, in his travels around the Earth's orbit, carries his standard atomic vibrators along with him: it is recognized that each type of vibration of the hydrogen atoms travelling with him, when measured *The observer's* upon his own spectroscope, gives him primarily an absolute measure *personal* of length, the wave-length; and that it gives him, by inference, an *frame is* *absolute.* absolute measure of time, the period of the vibration, in his scheme of space and time, through intervention of a universal constant the same for all observers yet to be identified with the velocity of radiation. Indeed the historic bars of metal, located in Paris, which define the standard metre, are now largely obsolete for ultimate purposes, having been replaced by the wave-length of a specified line in the spectrum of cadmium: a unit which, as Clerk Maxwell remarked in the early days, probably little suspecting that his idea would ever be realized, is absolute and does not need to be transferred precariously, like a standard bar, from one place to another for comparison, as it can be directly recovered by angular measurement, which is absolute, anywhere in the universe at any time.

Varying our imagery to the lines of the most recent mode of pre-*The ultimate* cision-measurement of time, we may describe the atom of cadmium as *standard of* the cosmical astronomer's free pendulum which, itself immune from *time.* disturbance, checks and controls any slight aberrations of the complex astronomical time-keepers that are his recording standards for everyday use. The actual standard clock for our practical astronomy is the rotating Earth, as it appears that we cannot for some reason refer to the revolutions of the inner satellites of Jupiter as a court of closer appeal. The constancy of the Earth's period, the sidereal day, was brought under suspicion long ago by the scientific imagination of Lord Kelvin; and the increasing refinement of the Lunar Theory has given actuality to his challenge. In humorous semi-practical vein, he insisted that the time would come for a clock to be instituted to check the accuracy of the Earth as a time-keeper, by aid perhaps of a pendulum swinging *in vacuo*. That indeed would still be very far from the ideal limit, in which the vibrations of our absolute atom might be invoked to control both the Earth and the checking installation and so assert themselves as the ultimate standard of time, as their wave-length has already become the practical unit of length. Anyhow, of these two projects relating to measurement of absolute space and time, at first regarded as ideal humorous flights of genius, that of Clerk Maxwell has been literally realized, and we may not deny that the idea of Lord Kelvin is possibly on its way towards some actuality.

Now let us make direct contact with the main question. Every one of our potential astronomers, wherever situated in the universe, however great the motion of his environment may be, provided its acceleration is not large enough to be a disturbing cause, has absolute time and absolute length at his command, by means of his own spectroscopic equipment: are the universes which are constructed and measured by each group of them in their space and time identically the same universe for all*? The only possible test is:—Do all relations between bodies within the universe show exact parallelism for these groups of observers? If a discrepancy showed itself, there would indeed remain open a conceivable path of escape from it; so hard is it to escape the relative. One has to be certain that the time-keeping of the observer's free atomic clock is not affected by the changes in the speed and direction of motion, the accelerations, of its frame as it travels with him round the Earth's orbit. But no philosopher, in his pursuit of the strictest relativism, has yet stressed that point: rather they have, in order to make some practical developments possible, adhered to the opposite. It has been assumed with reason, in discussing the amount of possible influence of a field of gravitation on radiation from atoms, that such effect is in actuality practically negligible: just as other fundamental possibilities are not absolutely excluded, such as influence of acceleration of their relative motion on the law of correspondence between two optically permitted frames of reference for physical phenomena. It comes therefore on various grounds practically to this: that our test of the reality of the external world is feasible, not absolutely, but up to the second power of the ratio of the actual astronomical speeds to the speed of light: while beyond that order all is speculative and abstract. Yet this practical test, in the physical domain as now widened into astronomy, will be very convincing if, as is likely, the minute astronomical inter-

*The objective absolute as the invariant in relation to the group of permitted frames.*

*The closest physical and astronomical tests:*

* It seems very remarkable that the simple original scheme of a rotationally elastic fluid aether with its essential electrons should satisfy automatically (cf. p. 809) the whole of this psychological criterion of personal knowledge.

actions of gravitation with light, foreshadowed as necessary to the consistency of the records by the genius of Einstein, become (cf. p. 789) definitely cleared up in theory and verification.

It is to be observed that the complications involved in the comparison of the records of our groups of potential astronomers arise solely from the circumstance that their observations are delayed in transmission, by minutes or hours or even years, on account of the slowness of the light which alone carries the information across the celestial spaces. If some kind of instantaneous signalling were available, the astronomers could recognize very simply, by direct survey, whether their spatial universes were identical or not: there would be no need for a Mathematical Theory of Relativity. Absolute space and time would stand justified: there is no ground for an absolute exclusion of them as a scheme of reference.

*made complex solely by the slowness of light.*

The absolute scales of space and time, as belonging to each astronomer, may however actually be restricted to his own region in which relations can be explored by direct spatial superposition: this region includes his observatory, and his landscape so far as to cover his fixed marks for astronomical collimation. It is not essential that his uniform space should extend to regions that are inaccessible to him except indirectly by the messages of light: the wider spatial frame, convenient to enshrine optical knowledge, may prove to be one that curves round like the surface of a globe. As regards time, the argument from the atoms confirms, or rather explains from the material structure of his organs of sense, the presumption that for every observer local physical time is in agreement with his mental time regarded as a mode to which universal thought conforms: the adequate test being that there is for him and his fellows one ultimate order of succession of the recurring events of experience, the same in physical nature as in mental representation.

*Space and time need not be uniformly measurable:*

*but must be identical everywhere locally.*

This consistency of the universes of different observers has been known and verified, in the minutest scientific detail, as regards all local optical and electrical phenomena, as far in fact as practical test could go (Michelson, Rayleigh and Brace, FitzGerald and Trouton), ever since the beginning of this century. But to astronomers concerned with celestial distances the orbital motions due to gravitation take priority over the equally sensitive local tests, optical and electric. This sets an ideal abstract problem, whether such correspondence could be exact without any approximation, which has been subjected to impressive analysis (Einstein, Minkowski) by a very refined analytical method.

*Physical and astronomical tests equally sensitive.*

This mathematical problem is to establish, so far as it may be possible, complete parallelism between relations subsisting in the

worlds seen and recorded by two distant observers travelling with
different speeds each comparable with light. In all such circumstances
precise correspondence, as regards all internal relations, is the sole
test of identity. The frame of space and time of one observer may not
perhaps conveniently be extended across, to the distant locality of
the other, without great complication: and the misfit, being amenable
to law, may even constitute, with Einstein, the appropriate expression
of gravitation within this representative scheme. There is no reason
indeed why uniform continuation of space and time should give the
most convenient enlarged frame. What may be actually achieved is
that an auxiliary model of some kind, of enlarged scope, covering a
limited range of properties of the universe, is constructed mathematically, to which the local worlds of both observers can be referred
as two local surveys are referred to a map of the world, on which a
connection across, sufficiently wide for many purposes but by no means
complete for all purposes, can be traced.

The scope
and depth of
the external
world

There is thus no reason why this auxiliary map of the history of
the cosmos, whose exploration constitutes the *quasi*-geometric analysis now known as the Mathematical Theory of Relativity, should be
a complete representation: any more than why a map of the Earth,
either on a globe or on the flat, on which connection can be traced
between our local region and, say, an Australian landscape, need be
a representation of anything except geographical position, all physical
activity omitted. In fact this auxiliary fourfold model for reference
can represent directly neither spaces nor times, but only an algebraic
blend of them both, in Minkowski's phrase, in which however their
identities, as we must expect, are not completely fused: it cannot
represent motion at all, so it cannot express even a ray of light except
as a line, because being purely geometrical it cannot mark its direction. It is admitted that it cannot distinguish between past and
future. Yet within its own scope it can be of great utility, though of
the historic record of the cosmos it is only a partial expression. Its
effectiveness indeed arises from its being expressed in terms that are
beyond space and time: for it must be possible to project the model
into the uniform space and time belonging to any one astronomer,
when they are regarded as prolonged over the cosmos, though in such
application it would, except in the very simplest cases which suffice
for actual astronomy, become enormously complicated as well as
particularized. Our mental world is taken to be a shadow of the outside physical reality: therefore it is as rational as this reality, though
only an incomplete expression thereof. There may be other kinds of
shadows of the physical reality, also rational but incomplete in other
respects. Two types of shadow combined may afford a clearer repre-

The fourfold
map of uni-
versal history:

its composite
character:

its limitations.

The Platonic
shadows.

sentation. In this way our scheme in space and time, and the algebraic fourfold auxiliary construct may supplement each other.

The classic illustration of the other side of the shield, that, subject
*Frames chosen only for convenience.* to complications that may be impracticable, any world content can be expressed in any frame of reference whatever, is F. Klein's actual map, essentially revolutionary as it had at first seemed, of the elliptic or hyperbolic type of uniform space, constructed upon an ordinary Euclidean framework.

To resume from another point of view. Every atom carries its own individual measure of time, which is continuous and, where its motion is not accelerated, is the unique intrinsic or absolute time. Where
*Local gravitational frames as shading into one another in differential fashion,* atoms are collocated by the laws of Nature into the structure of a corporeal observer with his necessary material environment, a mean local frame of reference would naturally be elaborated, involving this local standard scale of time combined with a convenient Euclidean space: and the details of behaviour of the material environment can be mapped as its motions within that frame, conforming to local dynamical law. Such a frame of uniform space and time would function universally, may be prolonged without modification up to all distances, were it not for gravitation which may complicate relations within parts of it and so make it inconvenient. This complexity
*the fourfold,* will not involve any contradiction, or exclusion of such localities, on comparison with other observers, if its features can be expressed for all of them in mutual relation on one comprehensive auxiliary map of history, just as the geography on a curved surface may be sufficiently but not completely expressed by a flat representation of it as a plane map.

This incomplete auxiliary cosmos, as developed in the modern Theory of Relativity, is a *quasi*-geometrical world of fourfold tensors: it involves a geometry of lines rather than of points. There is no room in it for anything that is not capable of expression as a component
*which is a construct only of geometric tensors,* of such a tensor: a point is relative to some other point. Thus time regarded by itself could not exist in it: it must arise as interlaced with length and direction to form a fourfold tensor. There appears to be no room in the fourfold for the thinking astronomer himself: and that is the essential point. But locally this hiatus can disappear:
*not of space and time and movement.* each observer has his intrinsic time and his independent landscape and his local world of light and movement. A model consolidating an astronomical history which cannot include astronomers everywhere and their local records in space and time, in fact has no room for mind as universal, can be effective, a shadow of the reality, incomplete, but rational as all shadows must be, so far as it carries.

Any way that may be available to express a stellar system of orbits so as to fit in with the astronomical records must be in terms of our universal time, atomically given. There seems to be no feasible alternative of sufficient precision: which in itself may be an indirect indication of an essential intellectual validity of the atomic theory*. This intrinsic time must be combined with a suitable approximate spatial frame, to some degree at our choice: the most convenient one, which proves to be the Euclidean space subject to slight local modifications, is the best. The test of reality is that it frames a world whose relations are congruent when compared with the framed worlds of the astronomers in other remote regions. Every partial dynamical formulation in space and time, say by Least Action, must therefore be of a type which, when translated into the language of the auxiliary fourfold, shows itself to be intrinsic therein, and so fits consistently into the whole cosmic history of which the fourfold is a consolidated partial model. This, it may be held, is the way in which the auxiliary but incomplete relativist fourfold map of complete history assists and guides a consistent formulation of a dynamical Action in space and time for each observer, definite because it has to satisfy the criterion of a unique external world the same for all observers however travelling and everywhere. *[The physical world perhaps adequately formulated by Action:]*

All this consistent complexity of the permitted frames of our knowledge has been found to be involved, or rather ought to be involved if knowledge is to be coherent, in the modern concept of a definite material universe of atomic constitution built on the simple lines of actual physical science, taking on consistent aspects as surveyed by a plurality of potential astronomers, everywhere of identical mental structure, whose observations are complicated by the delays inherent in the slowness of light but yet are conserved as a permanent exact record. *[on an atomic basis.]*

The world ought to be rational to human observers everywhere, working with human intellect in space and time. Each group of observers, however they be rushing through the cosmos, have their own absolute space and time for all local purposes that concern them —this the atomic theory demands: for relations at a distance they depend on vision, and its messages are in fact delayed. How are these groups of observers to get into consistent touch with one another? Is the problem even possible? The Minkowski formal procedure, after Poincaré, is by aid of a fourfold which transcends and can include all *[Scope of any theory of relativity.]*

---

* This absolute time is every observer's time in his own frame: the analysis of astronomy must however concentrate on one spatial frame, thus must abandon this attractive simplicity of absolute time. The natural frame for the Planetary Theory is the Sun's frame, and his time which is the astronomical time is related to each observer's local absolute time by the optical transformation as on pp. 770, 790. *[The frame of dynamical astronomy.]*

their frames of absolute space and time: and Einstein went a stage further, utilizing the hypergeometrical analysis originated by Riemann, by removing the restriction that the fourfold is to be uniform. Thus the theory of relativity would resolve itself into the invention of a formal procedure for coordinating the modes of knowledge of the world as presented to identical absolute observers everywhere and convected anyhow. As the senses of perception are material and so atomic, it suffices if all the ultimate atoms correspond in the various observers' frames: the laws regulating the intervening fields need not do so, which may be an incitement to including them as part of the frame. The velocity of light and the intensity of gravitation are prescribed *à priori* at their actual values, which are not explained: thus neither is the fourfold auxiliary construct expressing the course of history explained: but formally it subsists, and thereby knowledge in the domain of physics is assured of rationality. A scheme of wider scope than the electrodynamic could equally conform to all this. This retarded light travelling practically as rays, being the sole source of knowledge at a distance, the modern inverse abstract procedure starts with a geometry, of whatever type, of a *plenum* of the possible retarded ray-paths, and aims at finding out what sort of universe they could be concordant with: but in the course of this exploration the rays themselves come up for re-consideration, and in the new aspect they become very complex messengers, with periods and polarizations which have revealed so much, simulating very closely the familiar elastic medium of transmission which can provide the most effective and coherent imagery.

Provinces of knowledge:

Finally and repeating, may not various theories of natural phenomena be possible, for example the atomic physical and the biological, each of them valid as far as it can go, no essential contradiction between them being involved as they may never come together, for the simple reason that they are partial yet precise shadows from different angles, in Platonic phrase, of the same reality that transcends them all?

powerful but independent.

But what are we to think of the practical control of mind over Nature, as exercised in modern times by man, physically one of the feeblest of animal creatures? That can only remain a subject of astonishment, relieved by study of the historical record of its evolution.

# APPENDIX VIII.

## SYNOPTIC VIEW OF A PHYSICAL UNIVERSE AS OPTICALLY APPREHENDED.

THE subject matter is the analysis of the aspects of the world as viewed by intellectual observers, situated and in action all over the universe, and in a position to compare their records and impressions —a subject rendered complex only by the slowness of the rays of light on which these astronomers depend for all their outside knowledge. Actually these observers are restricted to the working astronomers situated on and moving along the Earth's orbit; but the treatment of their records on this presumption of universal consistency is the foundation idea on which the exact science of practical astronomy has been constructed. The problem is to explore the ways in which these observers are compelled to frame the world so that it may be the same world for all—which means that its internal relations referred to frames appropriately chosen shall be in definite correspondence for every pair of them.

Problem of a universe consistent for all observers:

The science of pure dynamics has found a universal synthesis, within its own scope, in the Principle of Minimal Action. When it is applied, in extension of the initiative by Lagrange and Green and MacCullagh, to a universal aether in which the atoms of matter are the controlling singularities after the manner of the permanent vortex rings in perfect fluid, the Action takes on the form of a fourfold integral extended over space and throughout time, and based on the atoms. It has to be invariant, that is to retain the same algebraic form, at any rate so far as concerns the atoms, in all the coordinate frames of reference that are appropriate for these astronomers, if all are to be viewing the same world: that implies that the two factors in the integral of Action, one the fourfold differential of space in time, the other describable as Action density, shall both of them be invariant except for local multipliers which cancel in their product. The restriction to celestial messages transmitted with the slowness of light implies that a quadratic, reducible locally to a standard form $c^2 \delta t^2 - \delta x^2 - \delta y^2 - \delta z^2$, shall remain the same except for a factor when transferred from one of the appropriate frames to any other: for its vanishing expresses the relation between interval of length and interval of time along a ray, and if the rays on which knowledge depends altered intrinsically from frame to frame the world could not be consistent as viewed in those frames. Here an absolute

treated from invariance of Action:

quantity c, the link connecting length with time, and in the actual world expressive of the velocity of light, enters unavoidably.

Both these conditions can subsist together if the frames are such that $c^2\delta t^2 - \delta x^2 - \delta y^2 - \delta z^2$ is itself invariant, without a factor. Such an invariant defines a pseudo-space, after Riemann, in which the permitted frames of reference for space and time all inhere: in electrodynamic theory this spatial extension would be uniform everywhere, as are the frames; while when it is taken to be warped the new features have proved to be the mode of adaptation of the frames, no longer uniform when continued beyond the locality of the observers, to include also an optically invariant gravitation.

When this integral expressing the Action, whose possible forms are closely restricted by the necessary quality of invariance in the fourfold, without which it could not exist, is minimized, it leads to equations of the electrodynamic field, and also in the wider case of the gravitational field, in which the sources, the atoms of matter, would be local intrinsic structures, expressible each mathematically as a complex of singularities, largely unknown.

When the fourfold has thus been minimized as regards Action, so is equilibrated internally to the form in which it can subsist, conforming to the resulting equations of the field, the distribution of Action so particularized can be expressed by integration by parts in terms of the surfaces that limit the field, here the boundaries of the filamentary paths of the atomic sources: no residual fourfold integral will remain over, provided the Action density was originally
a quadratic function of the first gradients of the quantities expressing potentials of the medium, and provided there is no free radiation whose sources would have to be located on an outer boundary. Further minimizing of this Hamiltonian Action, now of paths, regarded at first as unrestricted, will then determine the actual free paths by resulting orbital equations: and, what is the foundation of general
systematic analysis of a grouped complex of possible free paths, it will establish the existence of Hamiltonian functions of Varying Action each defining such a complex and expressing the mutual relations of its component paths.

This reduced Action of the material sources, thus eliminating the intervening medium, can readily be seen to subsist (within its appropriate limitations) under any quadratic form for the fourfold Action density. We may therefore begin afresh without the medium, by formulating a reduced Action as made up of line integrations along the paths, of type restricted so as to be invariant for the group of frames that are optically permitted in keeping with our postulate of an actual world, for the companies of potential astronomers throughout the universe. In the familiar field of electrodynamics, this develops (Vol. 1, p. 235) by the analysis of Neumann-Maxwell-Helmholtz into forces of Amperean type between ideal current elements, or in the modern form between the actual electrons in motion. For $x, y, z$ and $t$ of the fourfold may be interpreted as the variables in the uniform frame, gravitation locally negligible, of any group of observers having a common speed of convection. Abstractly however in the fourfold, it is really a geometry of lines that arises, developed in the Mathematical Theory of Relativity into an algebra of vectors and other tensors, for a point enters only relative to another point: it is even restricted largely to a theory of one special complex of lines in the fourfold, those appropriate to represent rays.

Geometry of the ray-complex.

But a *hiatus* reveals itself here. We find (p. 795) that the permitted line-integral forms, those which are invariant, though they are static in the fourfold yet do lead to propagated potentials in space and time such as are appropriate to radiation; but the radiation must be inward to the sources as well as outward. Which is perhaps really not un-natural, for it has been implied that the disturbance in the medium all comes from the electronic sources: this involves that there is no free radiation in the field, or rather that it be condensed into its own distant sinks which when included in the system would make great complication. Also in the reduction from spatial to orbital integrals an integral over the boundary at infinity has been neglected, which implies that all emitted radiation is absorbed regularly by other sources, none of it escaping to infinity. We ought not then to expect that this reduction, by elimination of the medium, will apply to radiating sources as we know them: they here have to again absorb all the radiation that is emitted by them. In actuality it appears to be restricted to electrons that are slow so that their radiation is negligible, or else grouped into atoms restrained from radiating; for approximately for slow actions the incoming and outgoing radiated potentials are found to combine into stationary potentials fashioned after the manner of the primitive Amperean theory, correct up to the order of the first power of $v/c$ which neglects radiation. In astronomy also gravitation may have to be completely absorbed: but the speeds are always there small enough for gravitational free waves to be negligible.

But radiation must be negligible,

excluding second-order phenomena from the scheme.

*Influence of gravitation on light.* The symbolic fourfold is constructed from the absoluteness of $ds^2$. We can imagine a short measuring rod, or interval, which can slide through the fourfold as a tape along a Gaussian surface: but how may such a process be interpreted into the actual world? Close to any position in the fourfold, $ds^2$ can be reduced to a standard form $c^2 dT^2 - dX^2 - dY^2 - dZ^2$ by change, practically linear, of local coordinates: in this form $dT$ is absolute and so is $dX^2 + dY^2 + dZ^2$ around that position, expressing absolute scale-relations for intervals of time and of length, which it is the function of the mobile fourfold rod to extend throughout the medium. In cases where we may take the fourfold to be an astronomical space, relative say to the Sun, *i.e.* in which the Sun is at rest, or even any space nearly uniform except for slight local warps at the Sun and planets, and a distinct Solar time with slight local warp which is multiplied into importance by occurring as $c\,dT$, the mobile measuring body may be taken to be a transferable vibrating atom. Wherever it is, it vibrates equal intervals of $dT$ at that place, relative to its own frame which travels in company with it, but while it is being transferred from one place to another with acceleration relative to its own frame its absolute period is thereby thrown out*;

*Atomic time absolute but without an epoch.* this involves that what is given in $dT$ is an absolute scale of intervals of time, but with origin of measurement (epoch) of $T$ not recoverable, or more precisely that $dT$ is not integrable along a path—in fact just because $ds$ is so. According to Einstein's early determination $ds^2$ near the Sun is approximately $c^2 (\mathrm{I} + 2V/c^2)\, dt^2 - dx^2 - dy^2 - dz^2$, which reduces to the standard form above if $dT = dt\,(\mathrm{I} + V/c^2)$. Moreover the waves of radiation carry away with them, as registered in their periods of vibration, the values of intervals $dt$ unchanged through space, say from the Sun to the Earth, where $t$ is the time-like coordi-

*Relation between two measures of duration.* nate of the universal $ds^2$. Thus intervals $dT$ are conserved in atoms, but intervals $dt$ along rays. This transmitted value of $dt$ corresponds to a smaller value of $dT$ at the Earth than at the Sun, on account of the smaller $V$: therefore an atom of hydrogen vibrating at the Sun appears in the spectrometer to go slow compared with one vibrating at the Earth to this degree, which is the accepted result.

The reduced Action in the fourfold for a planetary system $m_1, m_2, \ldots,$ including now gravitation, may be surmised to be of type

$$A = \int - \tfrac{1}{2} c m_1\, ds_1 + \int - \tfrac{1}{2} c m_2\, ds_2 + \ldots,$$

where $- \tfrac{1}{2} c$ is an adjusting factor, this being an invariant form. To sufficient first approximation in the locality of $(x_1, y_1, z_1, t_1)$

$$ds_1{}^2 = c^2 (\mathrm{I} + 2V_1/c^2)\, dt_1{}^2 - dx_1{}^2 - dy_1{}^2 - dz_1{}^2,$$

where $- m_1 V_1$ is the gravitational potential energy of the intrinsic

---

* The mode of statement that follows may obviate the early demur of Bergson, in his eloquent book *Durée et Simultanéité*, against the validity of accidentally changing standards of intrinsic time. It may also mitigate the criticism that his own personal interpretation of activity has to be at the expense of a rational order of external nature.

mass $m_1$. In the local system of the planet $m_1$ the integrated interval of absolute time is not $\delta T$ equal to $\delta t_1 (1 + V_1/c^2)$, because that system is moving, with its orbital acceleration, in the coordinate scheme of $ds_1{}^2$. But we can express $dt_1$ at each position in terms of $dt_s$ the interval of time in the Solar frame, the Sun being practically fixed and so having no acceleration in the general planetary astronomical frame, by a Lorentz transformation that is here sufficiently expressed by $dt_1 = dt_s (1 - v_1{}^2/c^2)^{\frac{1}{2}}$. Thus in terms of the Solar astronomical time, which has a definite epoch in the uniformly moving Solar frame of the field and so is integrable, Effect of gravitation on time scale, superposed on effect of velocity,

$$A = \Sigma \int - \tfrac{1}{2} c^2 m_1 \left(1 + \frac{2V_1}{c^2} - \frac{v_1{}^2}{c^2}\right)^{\frac{1}{2}} \left(1 - \frac{v_1{}^2}{c^2}\right)^{\frac{1}{2}} dt_s + \ldots;$$

and this is to sufficient approximation

$$A = \text{constant} + \int (T - W)\, dt_s,$$

where $T$ is $\Sigma \tfrac{1}{2} m_r v_r{}^2$ and $W$ is $\tfrac{1}{2}\Sigma m_r V_r$. Thus $A$ agrees with the dynamical Action of the Newtonian planetary problem, $T$ and $W$ being there the expressions for the energies, kinetic and potential, of the orbital system. It also happens to agree, to sufficient approximation, with making each orbit a geodesic by itself as determined by $\delta \int ds = 0$, an original hypothesis which Einstein has recently verified to this order from the necessary equilibration of the stress tensor of the field. But without the factor here introduced reducing from the unsettled times $dt_r$ to a common time $dt_s$ in the Solar frame this agreement with the planetary dynamical orbits would have failed; unless $\tfrac{1}{2} V_r$ were substituted for $V_r$ in the Einstein formula for $ds^2$, which would reduce both the gravitational optical deviations, of rays and of free periods, to one-half their observed amounts, as had been previously argued (*Phil. Mag.* Jan. 1923). Corrections for relativity within an ideal atomic orbital system could hardly at first sight come within the scope of this argument: cf. however footnote, p. 792. recovers the Newtonian orbital Action, now without discrepancy in gravitational influences on light.

Let us proceed to closer exposition of these ideas, covering the domain of dynamical astronomy. We consider a system of bodies of which the one of mass $m_r$ has its configuration expressed in its own internal Cartesian coordinates $(X_r, Y_r, Z_r)$, and the intervals of time $\delta T$, which are absolute time, the same for them all. If there is to be a function expressing physical Action, depending only on the bodies mutually when their velocities are small enough, the internally equilibrated aether having been eliminated, after the manner of the Action (cf. Vol. I, *supra*, p. 524) of a current system composed of slow electrons in electrodynamics, it may be expected as regards the terms involving masses and their inertias to have the form $\Sigma \int k m_r\, ds_r$, where Action form for a field of orbits:

$k$ has to be identified as *supra* with $-\frac{1}{2}c$. For this form is intrinsic in the Minkowskian symbolic fourfold into which the Lorentz group of transformations between the permitted frames in space and time has condensed itself. Here $ds_r{}^2$ is, after Einstein, a quadratic function which can everywhere be reduced locally into the standard form

$$c^2 dT^2 - dX_r{}^2 - dY_r{}^2 - dZ_r{}^2$$

with a universal c. We have now to change from the coordinate frames attached to the separate bodies, and partaking in their motions, to a common astronomical frame for all, which in the Planetary Theory will be the frame of reference, represented by $(x, y, z)$ and $t$, attached to the Sun. For the domain of the body $m_r$ thus in motion relative to the Sun, with velocity $v_r$ in the direction chosen for $X_r$, this transformation is linear, so of type

*transferred to the Solar frame:*

$$\delta X_r = a\,(\delta x - v\delta t), \quad \delta T = b\delta x + b'\delta t;$$

and it has to make $c^2 \delta t^2 - \delta x^2$ invariant, which restricts it to the form

$$\epsilon^{-\frac{1}{2}} \delta X_r = \delta x - \frac{v_r}{c}\,\delta ct, \quad \epsilon^{-\frac{1}{2}} \delta cT = \delta ct - \frac{v_r}{c}\,\delta x, \quad \epsilon = \left(1 - \frac{v_r{}^2}{c^2}\right)^{-1}.$$

What we can deal with much more simply is however the reversed transformation*, which is, changing sign of $v_r$, and transferring the factor $\epsilon^{-\frac{1}{2}}$,

$$\delta x = \epsilon^{\frac{1}{2}}\,(\delta X_r + v_r \delta T), \quad \delta t = \epsilon^{\frac{1}{2}}\left(\delta T + \frac{v_r}{c}\frac{\delta X_r}{c}\right);$$

for in all motions concerning the observer's local material system $m_r$, of which alone he can be directly cognizant, the $\delta X_r$ now relate to the effects of internal velocity, which is very small compared with $v_r$ the velocity of his system as a whole relative to the Sun. This permits us practically to neglect the change of epoch of $T$, depending on $X_r$, in the second equation, and so warrants a first approximation

$$\delta x = \epsilon^{\frac{1}{2}} \delta x_r', \quad \delta t = \epsilon^{\frac{1}{2}} \delta T, \quad \text{where } x_r' = X_r + v_r T,$$

thus involving simply change in the same ratio $\epsilon^{\frac{1}{2}}$ as regards time and also components of length in the direction of $v$. Thus an appropriate, as being optically invariant, orbital Action form, so far as such may prove to exist by itself without reference to the medium, is

$$A = \Sigma \int km_r\,(c^2 dT^2 - dX_r{}^2 - dY_r{}^2 - dZ_r{}^2)^{\frac{1}{2}}$$

expressible in the Solar frame as

$$A = \Sigma \int km_r\,(c^2 \epsilon_r{}^{-1} dt^2 - dx_r{}^2 - dy_r{}^2 - dz_r{}^2)^{\frac{1}{2}}$$

approximately, a term $(1 - \epsilon_r{}^{-1})\,dX_r{}^2$, or $v^2/c^2\,.\,dX_r{}^2$, being perforce

---

* The previous form is expressive of a real rotation in an imaginary correlated flat frame in which $\iota ct'$ is put for $ct$ and $\iota v'$ for $v$; this restricts artificially the extent of the group in an imaginary *quasi*-geometry in such way as involves adhering directly to the algebraic method of correspondence, as in p. 794, *infra*.

neglected, but not the corresponding term in $dt^2$ because as $dt^2$ enters it is always affected by the very great factor $c^2$.

We have hitherto taken $c^2$ constant, thus neglecting the Einsteinian expression of gravitation: for the orbital motions of the system of bodies we then arrive at

$$\delta A = 0, \text{ where } A = \Sigma \int km_r \, (c^2 - 2v_r^2)^{\frac{1}{2}} \, dt,$$

in which

$$v_r^2 = \left(\frac{\partial x_r}{\partial t}\right)^2 + \left(\frac{\partial y_r}{\partial t}\right)^2 + \left(\frac{\partial z_r}{\partial t}\right)^2.$$

The presence of the factor 2 in the application to dynamical astronomy, which has here come out of the multiplier $\epsilon_r^{-1}$, cannot be regarded as unexpected: for this expression determines the orbits relative to the particular frame of space and time in which the central Sun is at rest, and that can be no direct affair of separate geodesics in a fourfold symbolic consolidation of all possible permitted frames. Though on either hypothesis the orbits are straight lines until mutual influences are introduced. *(a modified result:)*

Now let us introduce the field of gravitation in the guise of a warp of this fourfold. To the first approximation its simplest possible type is represented, after Einstein's discovery, by changing $c^2$ in the Action formula to $c^2 (1 + 2V_r/c^2)$ where $V_r$ is sufficiently expressed as $\Sigma\gamma \, m_s/R_{rs}$, $\gamma$ being a constant as yet arbitrary, and $R$ distance in the flat mean space of the Solar frame supposed prolonged so as to extend over the whole system of planets. The effect of the new factor 2 affecting $v^2$ that is here introduced is to bring the optical astronomical effects derived from the Action form into agreement with observed values as now generally accepted, removing the discord of the previous formulation* by Action, now recognized as invalid at the outset *(the Einstein gravitation introduced,)*

---

* *Philosophical Magazine*, Jan. 1923.

The consideration here advanced, the crucial one, may be expressed differently. In the fourfold $\int ds$ relates solely to transition from one position to another. To each position there belongs a set of frames convected with various relative speeds, yet all counting as one unit as regards $\int ds$. For the system constituting an atom, including its environment, absolute time is the time given by its vibration as reckoned in that special frame among those of this set in which its system is nearly at rest. Solar time would be that measured in the frame in which it has velocity $v$ that of the Earth in its orbit, in which therefore $\delta x$, $\delta y$, $\delta z$, being the components of $v\delta t$, are no longer negligible. This would hold, as stated, were it not for gravitation: the effect of gravitation is to make the Solar time no longer uniform all over space. This double operation of transfer from the system's own frame to the Solar frame affects $dt$ by the factor $(1 - v^2/c^2)^{\frac{1}{2}}$, by the reasoning in the text *supra*, and $c^2$ by the factor $(1 + 2V/c^2)$, where $v^2 \delta t^2$ is for the orbit equal to $\delta x^2 + \delta y^2 + \delta z^2$. *(Frames not separately represented in the fourfold.)*

In absence or neglect of gravitation *(The relativity spectral fine structure.)*

$$\int - \tfrac{1}{2}mc \, ds \text{ would be } \int - \tfrac{1}{2}mc^2 \, (1 - 2v^2/c^2)^{\frac{1}{2}} \, dt,$$

where $v$ is the velocity of the system relative to the frame of the Sun or other

giving an orbital Action twice modified.

because it failed to avail itself of the Solar frame for all the planets, but with which the present discussion may still be profitably compared.

Transfer from the set of personal absolute frames.

The fundamental weakness in this connection of the Lorentz transformation is that it applies only to uniform convection: it fails even in face of uniformly rotating frames. The way it is proposed to get round this, so far as may be, is to recognize the absolute character of the relations in the local frames, and transfer them directly, but only approximately because acceleration is ignored, to one standard frame which for astronomy is that of the Sun.

Orbits approximately geodesic.

The procedure by an orbital Action, in either case, when followed out, does make each orbit correspond, but only approximately, to a geodesic in the fourfold: so that it does not conflict with Einstein's recent approximate verification* of his original geodesic assumption, on the foundation that it has naturally to be consistent with the implications of the balanced fourfold stress tensor which had absorbed all the dynamical features in his scheme.

The transformation is on the time variable:

This approximate procedure, for passing from the fourfold symbolism into the space and time of actual astronomical observers, thus works by manipulation of the time variable, the curved fourfold being referred to a flat one with which it nearly coincides except closely adjacent to the disturbing masses†. Yet spectroscopic atomic theory is held to postulate that there is a measure of true time, everywhere absolute. The situation here may perhaps be illuminated by an analogy.

analogy of warped plate.

Consider a flat plate of elastic metal, and suppose the thermal temperature is raised over a limited region of it: the measure of length remains absolute and Euclidean, in analogy with our atomic measure of time, so that the observed effect is to be expressed by asserting that more length is thereby put into the material of that part of the plate, which thus expands, so that to be accommodated thereto this material has to bulge locally out of the flat form. So here the standard of time is absolute and universal; and the effect of Einsteinian gravitation to a first approximation is expressed figuratively by a statement that more of this absolute time, in a ratio $1 - V_r/c^2$, has to be

central mass: this would upset the scale of the relativity fine spectral structure, as reckoned geodesically by Sommerfeld, but not when regarded as an orbital problem under electric attraction treated by the principle of Action. The uncertainty about the effect of radiation on the potential energy *infra*, p. 795, is not of sensible influence here.

* Berlin *Berichte*, 1928.

† If this mode of approach is right, the orbital problem for a double star presents complication on the Einstein scheme only when the distance between its components is comparable with their radii.

pressed into the flat fourfold in the neighbourhood of each mass $m_r$; that time expands locally, just as length does for the case of the plate, which involves warping of the fourfold frame.

A mode of interpretation of the algebraic formulae, in some degree alternative, contemplates an influence of velocity on inertia, or, perhaps under limitation, a connection between mass and energy[*]. As energy is relative to the frame of reference, such effective mass would have to be relative also. The atomic theory requires an intrinsic mass, that measured in the system's own frame: and an intrinsic energy of the same type, therefore describable as the internal energy of the system. The further question thus arises: Are these two related for a dynamical atomic system, as wider than a point-mass, so that the internal Action of the local system as well as the translatory is concerned in its translatory inertia? A decision must take us deeper into invariant dynamics, and will come from the Principle of Action: cf. p. 769, in relation to Vol. I, Appendix VIII.

May we then sum up: that to a first approximation and practically, the effect of gravitation is, on the analogy of linear expansion of a plate by temperature as above, to press more of absolute time, according to a factor $(1 + 2V/c^2)^{\frac{1}{2}}$, into the uniform basic fourfold symbolic frame that condenses within itself the group of permitted frames of space and time: also that the effect of velocity of the observer, due to transfer into a standard Solar frame not his own, of space and time, is to compress more time according to a factor $(1 - v^2/c^2)^{\frac{1}{2}}$. But beyond this first approximation, rendered possible with disregard of spatial change from frame to frame by the variable being $ct$ and not $t$, that is by the smallness of astronomical speeds, our resources of ordinary description fail, and the problem of establishing connection between groups of observers on different planets has to struggle into precise mathematical expression, after the manner of Einstein, in terms of a fourfold Riemannian symbolic calculus.

*Transfer into frame of Solar system, requires fourfold algebra.*

This mode of statement identifies figuratively the influences at work with extra supply of time in both cases, whether gravitation or motion, instead of extra time in one case and extra mass in the other:

---

[*] It may not be too irrelevant to quote an earlier synthesis of the *Renaissance*. Cf. p. 601. "Weight [inertia] by its own nature perishes when it reaches the desired position....Weight is material and force is spiritual....If force yearns continually for flight and death weight yearns for perpetuity....Propulsion results from the death of motion and motion from the death of force....Force is born of restraint and dies of freedom—and the greater it is the more rapidly it is consumed....Whatever resists it, it expels with violence, wishing to destroy the very conditions of its existence, and in victory causing its own death" (Leonardo da Vinci, about 1510: from his MSS. as quoted by K. Ludwig from Duhem, *Études*, Vol. II, p. 227).

addition to density of time supplants addition to intensity of inertia, the method of variation of Action supplying the fundamental process, free from ambiguity whichever mode of expression be adopted.

Path of flying electron.
The expression for the inertial part of the Action for a free travelling electron would be given by the integral on p. 789 but with $v_0$ the Earth's orbital velocity instead of $v_1$ in the second factor: the difference thus introduced from the observer's local frame is usually negligible for electronic projectiles, which constitute the chief application of the formula, on account of the magnitude of $v_1$ compared with the Earth's velocity.

The permitted types of fourfold Action for a system of electrons moving slowly, so that the aether is only locally disturbed without sensible emission of radiation, and can presumably be eliminated from the statement, may now be considered.

Orbital Action for system of electrons,
First simplify to the cases in which the complication of gravitation is negligible. The fourfold frame is then uniform over all the cosmos: and the translation of its symbolic data into a comparison of the presentations of the world to two observers travelling with different velocities is independent of their positions in it. The flat threefold spatial section of the fourfold can thus pass over with its relativity shrinkage into the universal space of the observers. The differing aspects of the world, to the observers in their respective frames of space and time, are all related back into correspondence with the content of the one auxiliary symbolic fourfold. An Action formula for an electronic system, of the type now usual,

$$A = \Sigma \left[ \int km_r \, (\mathrm{c}^2 dt_r{}^2 - dx_r{}^2 - dy_r{}^2 - dz_r{}^2)^{\frac{1}{2}} \right.$$
$$\left. - \int \tfrac{1}{2} \, (e_r F_r \, dx_r + e_r G_r \, dy_r + e_r H_r \, dz_r - e_r \mathrm{c}^2 V_r \, dt_r) \right],$$

in the fourfold must be invariant,
where $k$ proves as above to be $-\tfrac{1}{2}\mathrm{c}$, is, or ought to be adaptable so as to be, intrinsic to the fourfold. The second part can be the equivalent of

$$- \Sigma\Sigma \iint \frac{e_r dx_r e_s dx_s + e_r dy_r e_s dy_s + e_r dz_r e_s dz_s - \mathrm{c}^2 e_r dt_r e_s dt_s}{- \mathrm{c}^2 (t_r - t_s)^2 + (x_r - x_s)^2 + (y_r - y_s)^2 + (z_r - z_s)^2}$$

as a double integral form,
for which both numerator and denominator are invariant in the uniform fourfold, while the expression satisfies the $\square$ characteristic equation (that of propagation in space and time without any trail) for fourfold potentials, which leads to an inverse square law of potential in the four dimensions. If time is identical for both the sources $e_r$ and $e_s$, so that $t_r$ and $t_s$ are the same, the denominator has thus to be $R_{rs}{}^2$ the square of the fourfold distance, which contrasts with the first power of $R$ of the spatial electrodynamic formula. If the sources are in motion we have to transform into the time $t$ of the frame of

space and time to which the motion of both is referred. It is this form of Action, in which optical relativity is ensured, that we must transfer into the spaces and times of the observers.

There are, however, other intrinsic constituents that might enter into its formulation. The most elementary are two of type

$$c^2 (t_r - t_s) \delta t_r - (x_r - x_s) \delta x_r - (y_r - y_s) \delta y_r - (z_r - z_s) \delta z_r.$$

But they do not lend themselves at this stage to any double integral form of reasonable simplicity: though, as will appear immediately, they are fundamental in another way.

To obtain the value of the potential $(F, G, H, V)_s$ at $e_s$ the integration must be carried out with respect to $d\, (x, y, z, t)_r$. The integrands of the line integrals become infinite at two places, namely where $t_r - t_s \pm R_{rs}/c$ vanishes. This means, not that the integral really loses itself in the infinite, but that the elements of the integral thereabouts vastly predominate, and to estimate their effect the infinitely thin line of integration must be replaced by the finitely thin filament which it represents, so that locally it has to be treated as a volume integral, which converges. To separate these local singularities we decompose the integrand into two terms by the formula *[reducible to retarded potentials,]*

$$\frac{1}{t^2 - r^2/c^2} = \frac{1}{2t} \left( \frac{1}{t - r/c} + \frac{1}{t + r/c} \right),$$

showing that the result of integration as regards $dx_s$ is logarithmic, and leads to a form

$$- \tfrac{1}{2} \int \frac{e_r\, dx_r}{- 2c\, (t_r - t_s)} \{| \eta_s |_- + | \eta_s |_+\},$$

where $| \eta_s |_-$ is as regards $dx_r$ a delayed value corresponding to $(t_r - t_s) - R_{rs}/c$ null*, and similarly $| \eta_s |_+$ is a hastened value.

Here new potentials have emerged, now inversely as distance instead of its square, but propagated both outwards and inwards as regards the source, and both necessary to invariance in the fourfold. *[but involving both incoming and outgoing rays.]* Can it be held that the analysis is applicable to express actual transmission in space and time in face of this dual character?

* This procedure, if it stands the test of scrutiny, and especially if it is sufficiently wide in scope, may indicate a path, of fundamental import, for transition from the compacted fourfold static history into progress in space and time with the spatial potentials travelling as rays and delayed in transit. A theory of line integrals having infinities of the present type has recently been constructed on an abstract basis by Hadamard: here the infinities are evaded on physical grounds. *[Relation of observer to fourfold paths.]*

The observers' framings of the universe are cognizant only of rays of light: the far wider complex of fourfold straight lines is outside interpretation. This construction of an Action scheme for its atoms and electrons while maintaining material invariance has now managed to some degree to eliminate such paths and their intervals, only optical rays remaining, so that there can be a claim that it is built on suitable lines.

In the first place it is rather an overstatement to assert that the fourfold cannot take cognizance of rays, on the ground that owing to $ds^2$ involving only squares it has no criterion of direction of travel: more certainly however the fourfold has no direct means of expressing motion of a body. The equation $ds^2 = 0$ expressive of a cone of rays, being quadratic, in fact comprehends both classes of rays, outgoing and incoming: for instance, its cross-section $c^2 dt^2 - dx^2 = 0$ factorizes into two such rays, one of each type.

It is to be noted that the integrated local moduli in the values of the line integrals expressed by $|\eta_s|_+$ and $|\eta_s|_-$ are taken to be all equal, thus reducible to unity by choice of unit of c: this is in keeping with the atomic theory which makes all electrons alike.

We have arrived at potentials both outgoing and incoming: but as between two sources $e_r$ and $e_s$ the mutual potentials may be distributed so that only incoming parts are ascribed to each source. May then the original equation expressing the Action $A$ as a sum of single line integrals be valid in terms of only incoming potentials? Even if this separation were legitimate, entanglement would enter afresh in the process of its variation, when the effects are regarded as propagated*.

The pro-
cedure thus
defective.
The separation aforesaid into incoming and outgoing rays would spoil the invariance of the variational procedure, which could only hold when the velocities are so small that $v/c$, and therefore the effect of delay, is negligible. The reason is analogous to the elementary fact that though $t^2 - r^2/c^2$ is optically invariant, its retarded and hastened factors $t - r/c$ and $t + r/c$ are not so. When this misfit is probed to the bottom, we seem to be required to fall back on the formulation of last century (cf. *supra*, p. 39) that in the transformation from one optical frame of space and time to another, all relations between electrons persist unchanged in their mode of expression, but the aethereal fields do not so persist. The invariance for the group of frames of space and time was partial, but adequate for purposes of correlation of electrons and atoms. Cf. the fundamental distinction *infra*, p. 808, between relative invariance, intrinsic to the fourfold, and personal invariance as regards the observer. The symbolic fourfold is a consolidation of the frames that are permitted so far as concerns the mere delay of light signals. The essential feature is that the field equations which regulate the structure of radiation, and in fact all electrodynamics, are also inherent intrinsically in it, in the sense that they can be extracted from it, though they are not invariant for all

---

* It does not follow, because the differential equations of the fourfold are invariant, or intrinsic to it, that every integral of them is also so: thus the equation $\Box\phi = 0$ is invariant, but the form $r^{-1}f(t - r/c)$, which is consistent with that equation, is not invariant.

observers as are the electrons and their positions. They inhere in terms of a sixfold vector intensity of field, the gradient in the Clifford symbolic geometry (after W. J. Johnston, as on p. 804) of a fourfold vector potential. This implies that the equations of the field are involved in the frame of reference: that it has physical as well as geometric qualities, that in fact it is the aether or in Faraday's presentation the field of activity. If the fourfold potential is interpreted as a displacement in the fourfold, after the manner of the strain in the early rotational aether model, the whole consolidates into a kinematic formulation obtaining expression in the fourfold as a widened scheme of the type of geometry that is based on displacements. The aether transformed to fit the fourfold.

If there is to be an Action in terms of sources, subject as here to optical invariance, both the inward and the outward potentials along their rays must be kept: but for small velocities their sum is very approximately an instantaneous potential in the space, giving rise in fact to the original vector potential of Amperean analysis. But that instantaneous vector potential, like the gravitational one, would have now only to do with interactions between the slowly moving charges, not with their mode of transmission. A separated Action holds, for slow speeds,

In fact we ought to have anticipated this result. The double line integral form is expressive of mutual relations between pairs of linear elements alone: when radiation is originated integration over the infinite boundary across which it escapes ought also to have come in. The line integral remains suitable for determining mutual forces between the moving electrons due to the overlapping of their intrinsic fields, after the Neumann-Helmholtz manner. The general formulation would have to be in terms of atoms which interact with each other and also free radiation which interacts with them, the latter perhaps treated independently*. else free radiation must come in.

So too the procedure by Action for gravitational orbits neglects gravitational radiation from the moving masses, presumably of order $(v/c)^2$ which is outside actual planetary astronomy.

The equations of motion of each electron, in the field referred to the local frame, provided all their motions in it are small, would be derived as usual (*supra*, p. 163: *Aether and Matter*, Ch. VI) by variation of the double integral form of $A$, but now in the fourfold: its form as a summation may afterwards possibly be reduced to a mode of expression in terms of potentials.

* The universe which we know is subject to degradation, its activities running off by radiation into the infinite in space. The auxiliary fourfold cosmos with its completely achieved Minimal Action must avoid this. It can be self-contained only if it returns on itself like the surface of a sphere: then there is no place for the radiation to run off into: it can only be concentrated back into its sources, the fate to which the analysis in the text has pointed.

These considerations attempt an answer to the question: What is the practical astronomer to make of the theory of optical invariants as sublimated into a four-dimensional symbolic cosmos? The materials with which he can work are the rays of light alone, that come into his locality. The orbits that he arrives at from his records are mental constructs, which may run parallel in some degree to tracks in the symbolic, and also absolute, timeless cosmic fourfold. What the potential observers throughout the universe could in any case be in touch with is not a fourfold defined by a formula for $ds^2$, but only one feature of that fourfold expressed by the content of the ray-equation $ds^2 = 0$: the data presented to the astronomer are restricted into the geometry of this special Plückerian—or Hamiltonian—ray-complex within the fourfold. The fourfold intervals, other than rays, would be purely ideal auxiliaries, admitting in general of no interpretation into actuality as it is presented to the observers.

*Concern only with the ray-congruence in the fourfold,*

A locality in the fourfold includes not one astronomer but a colony with their various individual velocities of convection therein: the course of each of them is involved in the local element in his world-track $(\delta x, \delta y, \delta z, \delta ct)$, say $\delta s$, expressing his velocity and its direction. The fourfold is thus largely one constituted of Hamiltonian conjugates in space, of direction vectors rather than points: so far as it can be regarded as approximately uniform the same directions at all its points will correspond to identical frames of space and time.

The interpretation of the variable $t$, or rather of its interval $dt$, near the position $(x, y, z, t)$ in the fourfold is that it is intrinsic time for an observer there situated and identified with a direction $(0, 0, 0, \delta ct)$; and for him it is privileged, when his acceleration is neglected, by reason of the atomic postulate, so that this time flows equably at the same Newtonian rate for all observers everywhere. The only means by which this observer can get into touch with a body in another locality in the fourfold is by the ray, endowed with direction, that connects them, but only when such a ray exists. Along it $t_1 - t_2$ is determined as $s_{12}/c$ where $s_{12}$ is spatial length, which always subsists even when a connecting ray does not. The analysis *supra* (p. 795), from the form of orbital Action that was chosen on grounds of fourfold intrinsic invariance alone, arrived in its own way at the conclusion that effective connection in the fourfold between any two bodies can only be along retarded rays: which constitutes a test of the consistency of this fourfold ray-doctrine. As regards relations between bodies we can eliminate the $t_2 - t_1$ of the fourfold, replacing it by $s_{21}/c$ measured along the connecting ray, which is thus the way to attain to interpretation of $t_2 - t_1$ in space and time*.

*and the common space.*

*The translation into space and time.*

* Astronomical space and time would be the uniform space and time of

But this refers to bodies at rest in the fourfold frame, thus expressed by elements such as $(0, 0, 0, \delta ct)$: for a body not at rest $\delta t_1$ would be expanded into the closest equivalent invariant scalar product: now $c\,(t_2 - t_1)\,\delta t$ expands into

$$c\,(t_2 - t_1)\,\delta t_1 - (x_2 - x_1)\,\delta x_1 - (y_2 - y_1)\,\delta y_1 - (z_2 - z_1)\,\delta z_1$$

which is such a form, while the term $c\,(t_2 - t_1)$ translates into $s_{21}$ along the ray. As above noted, the ray in this expression includes its direction.

It is to be noted incidentally that the expression

$$\left(\frac{\cos \epsilon}{r} + \tfrac{1}{2}\frac{d^2 r}{ds_1 ds_2}\right) \iota_1 \delta s_1 \iota_2 \delta s_2, \quad \text{or} \quad \tfrac{1}{2}\,(2u_1 u_2 + v_1 v_2 + w_1 w_2)\,\delta s_1 \delta s_2$$

for the mutual energy of two "rational" current elements, with components along $r$ here represented by $u$, in Vol. I, *supra*, p. 523[*], is applicable to two electrons in motion slow compared with radiation therefore not radiating sensibly, for they constitute current elements $e_1 v_1$, $e_2 v_2$ each with its circuit completed by aethereal displacement: and so their complete mutual energy is $\quad$ *The Action for two travelling electrons:*

$$\frac{e_1 v_1\, e_2 v_2}{r_{12}} \cos\,(v_1 v_2) - \frac{e_1 \dot{r}_1\, e_2 \dot{r}_2}{r_{12}},$$

where $\dot{r}_1$ is $\qquad \dfrac{x_2 - x_1}{r}\,\dot{x}_1 + \dfrac{y_2 - y_1}{r}\,\dot{y}_1 + \dfrac{z_2 - z_1}{r}\,\dot{z}_1,$

$v$ being $(\dot{x}, \dot{y}, \dot{z})$. From this expression the mutual forces between two electrons are to be deduced by variation (cf. Vol. I, p. 524), being definite when they are moving so slow that their losses of energy by radiation can be neglected.

But we have found direct reason to doubt, as was in fact natural, *restricted to* whether any such formula, not taking independent account of the *slow speeds.* radiation in the medium, constructed for higher speeds by choice of expressions conforming to optical relativity, could be adequate[†].

We recall here that the Einstein spectral effect has lately been subsumed directly under the principles of *quanta* of radiation, just as if relativity were an affection of quantification.

There is also a question how far this synthesis, constructed on the foundation of optical invariance, is in line with the classical Liénard form for the retarded electrodynamic potential, Vol. I, *supra*, Appendix IV, p. 653.

Newtonian astronomy for the Solar system as a whole: near the Sun the intrinsic local frames do not fit into this astronomical frame without a slight warping. This appears to be the meaning attachable to current phrases such as "the slowing of the atomic clock in the Sun."

[*] As given by Lamb, *Proc. Math. Soc.* (1883), and Heaviside (1888), *Electrical Papers*, Vol. II, p. 501.

[†] In brief, results that are of relativist type would have to be established in the fourfold, and then translated where possible into space and time with a view to ascertain whether they are there relevant to actuality as we know it.

The direct deduction of the Maxwellian quadratic stress tensor from the Action (cf. p. 136), which is significant of the searching completeness of that formulation, is rather difficult to visualize in direct dynamical terms.

Electrostatic stress tensor.

In very simplest illustration, consider the field of an ordinary steady attraction, as specified by its potential function $\phi$ itself determined by its simple poles. The equation of internal equilibrium of $\phi$ is known to be the Laplacian $\nabla^2\phi = 0$: this characteristic equation can arise from, and so suggests, internal equalization of a field energy $W$ arising from latent local structure, annulling change of $W$ for all variation with regard to $\phi$ alone, so that

$$4\pi \delta W = \delta \int \tfrac{1}{2} \left\{ \left(\frac{\partial\phi}{\partial x}\right)^2 + \left(\frac{\partial\phi}{\partial y}\right)^2 + \left(\frac{\partial\phi}{\partial z}\right)^2 \right\} d\tau = 0.$$

Let the poles or masses $m_r$ of the field be referred to a frame in which the position of $m_r$ is expressed as $(x_r, y_r, z_r)$, and consider a Lagrangian virtual displacement expressed by $(\xi_r, \eta_r, \zeta_r)$ which are functions of position. The variational equation holds for all possible local distributions of $\delta\phi$ in its frame, therefore certainly for the values arising from virtual displacement by strain of the position of the frame alone, carrying along $\phi$ with this straining frame but leaving behind the masses: this special variation which changes $\phi$ locally, only on account of its transfer as attached to this frame, is given by

$$\delta\phi + \frac{\partial\phi}{\partial x}\xi + \frac{\partial\phi}{\partial y}\eta + \frac{\partial\phi}{\partial z}\zeta = 0.$$

Thus
$$-\delta\frac{\partial\phi}{\partial x} = -\frac{\partial}{\partial x}\delta_0\phi + \frac{\partial\phi}{\partial x}\delta\frac{\partial\xi}{\partial x} + \frac{\partial\phi}{\partial y}\delta\frac{\partial\eta}{\partial x} + \frac{\partial\phi}{\partial z}\delta\frac{\partial\zeta}{\partial x}.$$

Also there is variation of the volume factor $d\tau$ of integration in $W$ arising from this variation of the frame, expressed by

$$\delta d\tau = \left(\frac{\partial\xi}{\partial x} + \frac{\partial\eta}{\partial y} + \frac{\partial\zeta}{\partial z}\right) d\tau.$$

The first term in this expression for $-\delta\partial\phi/\partial x$ arises from a positional variation $\delta_0\phi$ of $\phi$ by itself, namely one without any alteration of the frame: thus it is already minimized by itself by virtue of $\nabla^2\phi = 0$ and can be omitted.

Obtained by variation of field energy.

On carrying through the variation as regards both factors of $W$, subject to the internal equilibrium as regards variation of $\phi$ itself without strain of frame, already secured by $\nabla^2\phi = 0$, the result comes out

$$4\pi \delta W = \int \begin{vmatrix} \{\phi_x{}^2 - \tfrac{1}{2}(\phi_x{}^2 + \phi_y{}^2 + \phi_z{}^2)\}\,\xi l & \phi_x\phi_y\eta l & \phi_x\phi_z\zeta l \\ \phi_y\phi_x\xi m & \{\phi_y{}^2 - \tfrac{1}{2}(\phi_x{}^2 + \phi_y{}^2 + \phi_z{}^2)\}\,\eta m & \phi_y\phi_z\zeta m \\ \phi_z\phi_x\xi n & \phi_z\phi_y\eta n & \{\phi_z{}^2 - \tfrac{1}{2}(\phi_x{}^2 + \phi_y{}^2 + \phi_z{}^2)\}\,\zeta n \end{vmatrix} dS - \int (\ldots)\,d\tau,$$

subscripts here denoting gradients.

The first term represents the Maxwell stress working on virtual displacement of the boundary of the region, $(l, m, n)$ being the direction vector of the element $dS$ of that boundary: the second term must vanish everywhere to ensure $\delta W = 0$, and is the expression of the internal equilibrium of this stress, as already secured.

A similar procedure applies to a static magnetic stress. In each case the stress may be no more than a merely formal synthesis of forces acting directly between the poles.

But for a joint field, static and magnetic, two such stress systems Electro-<br>magnetic<br>stress do not superpose by addition, unless the field is steady: there is an outstanding momentum distribution in the field. The equations of deduced from<br>Action: equilibrium of the field, which are employed in the reduction of the variation, are in fact of mixed type involving both sets of variables together. The stress around a source expresses the force it experiences.

We must, therefore, then proceed dynamically, by variation of Action* instead of energy, thus operating over a fourfold field of integration $dx\,dy\,dz\,dt$. The specification of this Action must conform to optical relativity, if possible without redundant variables. This its specifica-<br>tion *à priori*, demands invariance of the fourfold differential element $dx\,dy\,dz\,dt$, at any rate with a local factor: it points directly, in advance of all special theory, towards the Minkowskian fourfold as the simplest formulation, with or without a local factor expressive of divergence or dilatation of volume. In this fourfold an Action density would now be chosen so as to be of invariant type, which ties it down to the electrodynamic field form when dilatation, in the form of a divergence of the potential, is excluded: cf. p. 809. The most condensed expression containing just sufficient variables is that arising from the fourfold potential $U$, equal to $(F, G, H, c^{-1}V)$: then the Action in the by way of<br>the vector<br>potential: field is

$$A = \frac{1}{8\pi} \int\!\!\int (\nabla U)^2 \, d\tau\, dt,$$

where

$$(\nabla U)^2 = \left(\frac{\partial H}{\partial y} - \frac{\partial G}{\partial z}\right)^2 + \dots + \dots - c^{-2}\left(\frac{\partial F}{\partial t} + \frac{\partial V}{\partial x}\right)^2 - \dots - \dots,$$

when a divergence term of type $(\operatorname{div} U)^2$ in the Action density is excluded: the variables with regard to which it is intrinsic, in the Clifford geometry (p. 804), being $(x, y, z, \iota ct)$ and $(F, G, H, \iota c^{-1}V)$. When invariance is thus secured for the Action expressed in terms of a fourfold electric potential, the stress tensor derived from it is an intrinsic system of fourfold stress, but translated into space and time it will change with the frame.

The point to be emphasized is that, after the manner of Lagrange

* On this mode of development the limitations of the Maxwellian field stress in Preface (p. vii) to Volume I can be regarded as unfounded.

and Green, the stress in the medium is involved directly in the variation of its Action, when the boundary on which it acts is taken to be an interface within the region of stress, instead of, as has been usual in this type of analysis regarded as merely formal, taken so far away that surface terms can be ignored. By the latter procedure half *its essential feature,* the significance of the dynamical formulation by Action, especially in its Hamiltonian aspects, goes out of sight and is lost.

This process of deducing the stress tensor, when an invariant of curvature is included in the Action density, after Einstein with a view to express gravitation, has been discussed from various aspects *generalized by Klein.* to some degree cognate to the above by F. Klein and Frl. E. Noether*.

It is to be observed that as this curvature invariant introduces gradients of the tensor potential higher than the first into the Action, the stress, so far as a gravitational term in the Action density is concerned, is not expressible as a distribution of simple traction on an interface. But it can, if worth while, be expressed as a distribution of force *Stress with torque elements.* and torque thereon, or as its equivalent a double sheet of force—with the accompanying bodily distribution of momentum now in part rotational—and this appears to be the proper form. But in the usual procedure after Einstein, the term involving second gradients, being *Why the gravitational stress tensor cannot be invariant.* linear therein, has been removed from the Action in advance by integration by parts: then the resulting Action form is not invariant by itself, and therefore gives as is recognized a spurious stress of the ordinary traction type, not merely changing with changed convection of the coordinate system as is permitted, but not even intrinsic to the fourfold.

The null convergence of the stress in free space involves an alternative mode of expression of the dynamical equations for the material sources. For when sources are included within the boundary, the convergence of stress integrated over that boundary is not null, but is the same for all boundaries including the same sources: the resultant *Convergent stress tensor with contributions from the sources.* over a boundary closely surrounding any source is thus intrinsic to that source and is the reaction to its own dynamical contribution to the field of stress. By including this contribution with changed sign,

* *Goettinger Nachr.* 1918, following after Hilbert: reprinted in part in Klein, *Abhandlungen,* at end of Vol. 1. The discussion, by general tensor analysis, is as usual difficult to visualize even algebraically, and much more in respect of physical interpretation. The convergence of the stress tensor on to the atoms, the material sources of the field, does not enter, perhaps is foreign to their synthesis.

The point of view of the local equilibrium of the free field, as settled by local minimizing of the Action with respect to the variables of the field, and then the kinetic balance of its sources as settled by further variation of this field *Sources are necessarily singularities.* equilibrium with regard to displacement of the sources, ought perhaps to be capable of direct analytical expression in the tensor analysis of the general formulations. But it is entirely lost when the atomic sources are replaced by volume densities.

a composite stress tensor may be formed, after the manner of Einstein for his continuous density of material sources, which is everywhere convergent.

This field procedure by stress ought to give the same result as the previous reduction of the Action, already equilibrated for the field, to expression in terms of the sources (for slow motions whose radiation is negligible) followed by further variation with regard to their positions*.

The direct correlation between observers, here presenting itself as inevitable, amounts to no other than a reversion from the fourfold back to the direct correspondence between pairs, as *supra*, p. 39. The frames of the observers in space and time are not the same, nor are their electrodynamic fields in their frames the same; but all the electrons, adjusted so as to be of identical values in the two systems, continue to occupy corresponding positions in the frames. The electrons need not be all alike: but so long as they are of purely electric (aethereal) constitution, and within the approximation that they are so far apart in the atoms relative to their sizes that they may be regarded in their mutual relations as points, the correspondence will subsist. The impressions coming to the observers in the correlated systems are effects on those electrons of which their organs of sense are constituted, therefore these impressions must wholly correspond: the material correlation is thus complete, if it be postulated that even the nervous system is amenable to the same relations that hold for dead matter, though the ambient aethereal fields, as located in space and time, are not the same in the two frames†. Such precisely was the original scheme as *supra*, p. 39, out of which generalized relativities have developed. It was the imposition of this condition, that all electrons are to retain identical values in the correlated systems, that made the correlation of the fields

*Marginal notes:*
- Reversion to spaces and times.
- The correspondence for convection covers all material phenomena.

---

* Cf. F. Klein, *Abhandlungen*, I, p. 553. The present form of Action thus may claim to be unique in leading up to, and consolidating, the Maxwellian scheme with its latent electronic pressure (p. xi). May we assert also that it is the only one that can formulate simple personal spaces and times for all observers, consistently with knowledge derived from delayed optical signals, which may be made (p. 777) our ultimate psychological foundation for the whole affair?

† The frames must however consolidate into an invariant complex in the fourfold, else there would be no fourfold. A procedure which is not unusual in relativist determinations is to set up a scheme of local equations (for example an equation of Action) in a frame travelling locally with the system just as if it were at rest, then to change to another frame whose velocity thus appears in the scheme, then to alter the result into a form that is optically invariant, and to take this to be the necessary universal expression of things. There appears to be no assurance, other than relativist prepossessions, that different processes of this unsystematic type will be consistent with one another.

*Marginal note:* Unrestricted relativity merely a presumption.

definite—except as regards a possible superposed general expansion of scale uniform and the same in both space and time. But, as Lorentz was the first to note (1904), such expansion of scale could not be arranged so that the change from system $A$ to system $B$ would be reversible, and the original representation thus recoverable when the convection imposed on the observer was annulled. It was further demonstrated, immediately afterwards, by Poincaré that the aggregate of all such transformations constitutes mathematically a closed or self-contained group, in that if system $A$ could thus transform to $B$, and also $A$ transform to $C$, then $B$ could change directly to $C$ by a transformation within the group*. This latter result is the essence of the new principle of relativity, valid within its scope, that no one of the permitted optical frames is privileged over the others: in contrast with the indications of an intrinsic universal frame that are presented by the small velocities of the stars when compared with light.

It may naturally be held that the reason for this fundamental property was first exposed to suitably direct inspection by Minkowski's remark fitting the aggregate of permitted frames into one uniform fourfold extension. But that mode of argument seems to require careful scrutiny. A new explicit correlation appears to be involved tacitly, between the fourfold $- \delta s^2 = \delta x^2 + \delta y^2 + \delta z^2 - c^2 \delta t^2$ and another one $- \delta s^2 = \delta x^2 + \delta y^2 + \delta z^2 + \delta T^2$, by the imaginary transformation $\delta T = \iota c \delta t$, involving however also as *infra* change of $v$ to a pure imaginary but no change of c. The latter fourfold is of spatial type, and in it the correlative to a Lorentz transformation in the former (for a convection however which is a pure imaginary) is simply the intuitive transfer of the correlated fourfold system to another Cartesian geometric lattice. This remark appears however to establish the correspondence and its group character only as regards those properties whose correlations do not get entangled into imaginary expression. The Lorentz transformation of electrodynamic vectors has to be correlated in the real fourfold without introducing imaginaries: for this to be so, when $t$ is made imaginary so must the convection $v$, but not the absolute constant c. Under this restriction alone the essential feature is preserved that in the correlated real geometric fourfold the relations remain real. They are there expressed immediately in the Clifford symbolic geometry as derived from a fourfold potential, as W. J. Johnston pointed out (*Roy. Soc. Proc.* 1919): it is an affair of the relations of the relevant intrinsic vectors which are a fourfold potential and its derived sixfold field of force—and possibly also its convergence. And only after this stage does the question naturally arise, to what

*Marginal notes:*
Group property involves correlation with a spatial fourfold.

Radiation is outside the direct correlation,

---

* These conditions for relativity are all satisfied automatically by the original scheme of a rotationally elastic aether and its electrons, held together by an absolute pressure: cf. pp. 809, xiii.

degree is this field potential expressible in terms of the sources, themselves also vectors inasmuch as they are electrons in movement.

The essential function of the Minkowskian fourfold* would thus be to exhibit the group relations between the frames and fields belonging to the various pairs of observers, both of them finding the world to consist of identical collocations of electrons. Yet *qua* isotropic geometry there is no room in it for rays, nor for progress in time, because $\delta s^2$ is not endowed with direction. When the fourfold is stretched so as to be non-uniform, which makes room for a possible gravitation, in the groups of frames now localized instead of universal, $\delta s^2$ becoming the general local quadratic form, a like subsidiary correlation would establish geometric visualization, which cannot in general subsist unless where only $\delta t^2$ and $v\delta t$ enter in its expression. The alternative to such geometrical direct view is the tensor theory, in which all results are real algebraic forms, usually of great complication were they not symmetric, and hard to interpret into specific experiences of the actual world.

Some further remarks relating to the wave-mechanics (cf. p. 566) may find a place here†. Consider any permanent material system, not merely a wave-system, executing steady internal vibrations relative to its own frame of reference which accompanies it, and not subject to decay by radiation: this activity is expressible by functions of type $f(x, y, z)\, \phi(t)$, the spatial factor indicating the general steadiness of form of the system. Suppose that this system is transferred to observers relative to whom it is moving onwards with velocity $v$ along $z$: they get information by rays of light alone, and in their frame of space and time it is therefore expressible, after L. de Broglie, by functions transformed to the type

$$f\{x, y,\ \epsilon^{\frac{1}{2}}(z - vt)\}\, \phi\{\epsilon^{\frac{1}{2}}(t - vz/c^2)\}, \quad \epsilon^{-1} = 1 - v^2/c^2,$$

where c is the cosmical constant. The interpretation of its aspect, as thus presented, to the new observers, is that while the system changes

perhaps gravitation also.

Limitation of the quasi-geometry evaded by theory of real tensors.

Relativity scheme of L. de Broglie as applicable to any steady system.

---

* The remark of F. Klein, following on E. Cunningham and Bateman, that an electrodynamic field in this fourfold, therefore also its sources, which however become unequal point-electrons, is amenable to the general conformal transformation, thus passing into a non-uniform fourfold yet without gravitation, offers an enlarged field of interesting even if artificial interpretations into space and time.

† The parallelism between travelling electrons and wave-groups has recently become very precise in various ways through the experiments of Davisson and Germer, and of G. P. Thomson on rings of diffraction, and lately of Szczeniewski (*C. R.* July 1928) on the spectrum of a stream of electrons produced from a single crystal and closely registered by the method of the Bragg spectroscope—just as if each incident electron excited its own specific radiation which was then diffracted in crossing the crystal as it travelled away. (Cf. p. 807).

Electron waves.

as regards $z$ in manner corresponding to its relative velocity of convection $v$ along $z$ (combined with shrinkage $\epsilon^{\frac{1}{2}}$ of the system itself along $z$), it exhibits also changes of amplitude of vibration which indicate a drift of energy with velocity $dz/dt$ which is $c^2/v$, relative to these observers. In planetary astronomy, this speed is so immense that the energy relative to the new standard frame can be regarded as adjusted over the field instantaneously at each moment of time: for the high values of $v$ within the atoms this is not allowable. The factor $\epsilon^{\frac{1}{2}}$ affects, through the Action theory, the energy and the convective inertia of the system: for any component vibration for which $\phi$ is a sine function the energy varies as the frequency for all values of velocity $v$ of the system, which concurs, in a welcome manner but one to be expected, with the original *quantum* formulation of Planck. Any self-contained vibrating steady system when referred to a frame not its own, thus appears unsteady, to the extent that it moves across the frame with apparent pulsatory analogy to a group of waves in a dispersive medium, a circumstance that seems fitted (like other such influences within the atom, cf. that applied in the magneto-optics by L. H. Thomas) to offer some clue towards an essential interpretation of the scheme of analytical relativity*. On the other hand, if the steady state of internal vibration expressed by $\phi(t)$ is replaced by an internal wave-train $\phi(pt - lx - my - nz)$, the exact laws of aberration of light become exhibited in the transformation, just as it was carried out in *Aether and Matter* (1900), § 113. The essential feature there, as here, is that relations involving radiation, or other field activities, change their form, are not invariant, on change of frame.

May be a clue to closer interpretation of the relativity consolidation.

Some further considerations may be pertinent, relative to diffraction of streams of electrons. If a beam of electrons of translatory energy $\epsilon$

* The most acute and lucid probing of the points of view, and the discrepancies they present, is the exposition by L. de Broglie in his thesis (1924), which has only now attracted the writer's closer attention. In particular, the position that the fourfold, as an independent construct on its own foundation, has to be correlated, but only just so far as it can agree, with the spaces and times of the observers, is in evidence.

Time not of spatial type.

The conclusion compelled in the present argument is that one of the limits that impose themselves of the utility of the fourfold is that it cannot afford a representation of outgoing radiation. The tendency of the new wave-mechanics in this regard is to adhere to unlimited optical relativity and reconstruct if possible the idea of radiation so as to suit it. Thus it is suggested (L. de Broglie, *Ondes et Mouvements*, 1926, ch. VII) that time ought to be reversible, therefore incoming radiation is as natural as outgoing: that the expectation (cf. p. 795) ought to be an equable combination of both, expressing the standing periodic waves within the source which need not radiate but are a starting point of the new formulations.

is contemplated as incident on a crystal lattice, of ionic type, in which there is thus a stratified electrical potential $V$, their free paths, and therefore their caustic curves which are the spectral lines, are determined by the Action relation $\delta\int(\epsilon + eV)^{\frac{1}{2}}ds = 0$. This contrasts with the determining relation $\delta\int K^{\frac{1}{2}}ds = 0$ for a beam of X-rays (cf. *supra*, Vol. I, p. 39; Vol. II, p. 503). The intervals of reinforced reflection in each case depend on the grating interval alone, but the maxima do not coincide in the two cases; in agreement with the experimental graphs of Davisson and Germer, *Proc. U.S. Nat. Acad.*, Aug. 1928, p. 619. The crystal would be expected also to function for electrons after the manner of a surface grating. But the rings of diffraction (G. P. Thomson) imply undulatory processes.

*[margin: Electron paths and ray paths in crystals: not the same.]*

Such undulation may however be adequately provided by spiral progress of the electron itself, after the manner of the periodic feature of Newtonian light corpuscles. This spirality has been already contemplated (Vol. I, *supra*, Appendix VIII) in order to obtain for the electron a sensible effective magnetic moment and related angular momentum, without introducing impossible amount of stress. The electric twist around the electron had in fact suggested helical constitution, which need not interfere with the effectiveness of the model (cf. Preface, p. xi): for one can imagine a screw structure superposed on the vacuous core. These moments depend on the speed of the electron in the atom or in free space, by relation there set out: how far they may be effective there is not space here to examine.

*[margin: Helicoidal adjustments for electron orbits]*

In this kind of model, which is at bottom nothing more than a vivid aid to consistent schematization, huge numbers are involved. Thus (p. 809) there is an intrinsic aether-pressure $(2\pi\sigma^2c^2)$ of order $10^{30}$. If $v_0$ is the velocity of an electron nucleus of radius $a$ the fluid velocity of the aether at its equator is $\frac{1}{2}v_0$ and the energy of hydrodynamic flow by itself would be $\frac{1}{2}\rho\int v^2 d\tau$ which is $\frac{1}{6}\pi a^3\rho v_0^2$; while if $v$ is $kH$, the electrokinetic energy is $(8\pi)^{-1}\int H^2 d\tau$ or $\frac{1}{6}k^{-2}a^3\rho v_0^2$, which is $\frac{1}{4}e^2a^{-1}v_0^2$, so that $k^{-2}\rho$ is $3e^2a^{-4}$, of order $10^{10}$. According to the determination of Lodge (Vol. I, p. 484) $k$ could not exceed $\frac{1}{3}$, thus $\rho$, the density of the aether, could not exceed $10^{12}$, which is not too small to permit of permanent cyclic magnets. The hydrodynamic part of the energy would be a small fraction of the whole of order $4k^2$, so that even with this limit of $k$ the nucleus pushing its way through the fluid medium would not much disturb the electrodynamic scheme, and permeability to electrons is secured. But the present helicoidal effect and its spiral path need not be negligible.

*[margin: The electron model: permissible magnetic and helicoidal features.]*

Here again it will be noted that there has been reversion from the composite fourfold to one special transformation. Neither of these formulae of p. 805 embodies a relation that is absolute, in the sense that it is common to all observers, though it is intrinsic to the fourfold.

Personally absolute contrasted with relatively absolute: But the first form is absolute (p. 792) in a different sense, that it is intrinsic to its own personal (or in the equivalent epithet "proper") frame of reference; and the second form is the expression of how it appears, in a world explored by the rays of light, to another observer relative to whom its own system is in uniform motion. The distant external world is thus confined to a world of relations, which are common to the various observers, as distinct from the personal local worlds possessed by the observers individually: and the relevant problem is to explore it with a view to determine its extent and depth.

the latter inherent in the actual astronomical constructs. The science of astronomy has had *of necessity* to be constructed as a scheme in which no observer, wherever he be, in his motion along the Earth's orbit, is specially privileged. Therefore, so far as it is consistent with itself, it is compelled to be relatively invariant, the student of the science and his own personal world counting for nothing: and that is what various investigators, and Einstein with striking successes, have essayed to exhibit it to be. There is involved that the formulation of the phenomena of light, the sole astronomical messenger, and therefore of all electric transmission, should also be subject to a co-variance relatively as regards all possible orbital velocities of the observers, but not personally to each of them. All this inheres in the postulate of rationality of our pictures of the universe. The mathematical tensor theory constructs a fourfold world-history that is relatively invariant: some features of it are translatable into space and time, and others are not, while of the former some may be too narrow for experience. The personal presentation, which made so strong an appeal to the synthetic intuition of Bergson, is thus an independent affair with its own basis.

The arresting feature of the Schrödinger process is that it presents a coherent problem, isolating the possible states of finite vibration of some kind, thus occurring around definite configurations of stability in the system, in contrast with the accidental features of the previous orbital quantifications. The special orbits of Bohr are replaced by the types of free vibration of a system, mainly in groups of cyclic oscillations such as may be formally associated with the orbits, with nodes regarded, after L. de Broglie, as in a way spread around them.

From the other point of view*, covering far wider problems, a form of relation (p. 567) of the Schrödinger scalar variable $\psi$ to the Eulerian scalar Action $A'$, itself universal and sufficing for formal dynamics, which will bring the characteristic differential equations

---

* The mode of formulation of the general Action theory, indicated from Liouville (1856) in *Thomson and Tait*, § 368, may have relevance here.

of the two into identity, is found to be $A' = K \log \psi$. Thus if a solution for $A'$ with the adequate number of arbitrary constants can be obtained broken up into a *sum* of special types of solution, some of which may involve cyclic *quanta* after the original Wilson-Sommerfeld manner, there corresponds a solution for $\psi$ expressed after the manner of the usual harmonic analysis, as a *product* of special functions for none of which however is there to be any cyclosis, but which all are restricted to finiteness throughout the field. This might suggest interchange of relations as regards momenta and coordinates: yet if for this purpose transformations were made from $A'$ to a modified Action $A''$ equal to $p_1 q_1 + p_2 q_2 + \dots - A'$, the characteristic equation for $A''$ would take on forms impracticably complex, being no longer quadratic.

*Equation of Schrödinger correlated alternatively with varying Action.*

Finally, a prominent feature of the new spectroscopic theories at first sight is the facility with which apparently very different formulations converge to the same essential results, reminiscent almost of the *dictum* in fourfold relativity that it is only the order of the intersections of the world-lines of particles that is essential to the cosmos, all else being filled in as one pleases. The subject of this internal atomic dynamic appears now however to have been dissected in some degree towards its primitive symbolic elements by various analysts, most coherently perhaps in the hyperphenomenal procedure of Dirac —elements in which the forms of Hamiltonian dynamics persist while restriction of the sources, whether of electron or atom type, as expressed in terms of radiation the fundamental element of perception, is not required to the earlier illustrative forms.

### The Absolute Hydrostatic Pressure of the Aether

For an ellipsoid with free surface charge, convected steadily through space, the mutual repulsion of the electrons of the charge has to be counteracted by a balancing force of compression, of non-electric type. It has long been recognized, and was noted especially by Poincaré, that the necessary force, called extraneous, is in this type of model of an electron simply a hydrostatic pressure uniform over the outer surface. In the original rotationally elastic scheme of aether and electrons (*Phil. Trans.* 1894 as *supra*, Vol. I, p. 499) the aether has fluid quality, and such a pressure must be expected to belong to it.

*Equilibration of electron structure,*

We may here probe further. We can imagine a model electron as represented by a cavity in the aether, on to which the lines of the elastic twist which expressed the electric field converge: its equilibrium under this electric stress is sustained by this pressure in the aether. Suppose that the speed of convection through the fluid aether

*by a constant pressure,*

is altered: the electron assumes a new form relative to the same frame, still ellipsoidal, on account of the relativity shrinkage. Equilibrium *intrinsic to the aether:* can again be assured by a hydrostatic pressure. The question now essential is whether this pressure remains the same as before. If it is the same however the electron be convected, it is a uniform universal pressure intrinsic to the aether: then the types of electrons that are possible in the rotational aether conform to optical relativity, already known to be inherent in the ambient Maxwellian field. Such electrons can make up a definite external world that presents consistent aspects to all observers, however convected, in space and time: and the original model aether from which electrons were evolved does not crack under this searching test: though of extreme simplicity, it is not on wrong lines within its proper and very limited range. The problem of the existence of positive electrons remains however inscrutable.

It is involved logically, that optically no frame is privileged, up to the order $(v/c)^2$ to which our procedure is allowable as regards in-
*compels a limited optical relativity.* teriors of actual atoms of orbital type. But the arresting fact that astronomical speeds are all very small compared with the velocity of radiation, which alone is available for comparison, remains strongly to support, in the light of this limitation, the simplifying hypothesis of an underlying aether.

This property of unaltering aether-pressure pertains to any steady *Generalized proof of invariance of pressure.* system, keyed on to hollow regions in the aether, with internal stress none other than aethereal and including such intrinsic pressure. For the electric force in the field of a steady convected system has a potential, else the cyclic path of an electron subject to it could be a perpetual source of work removable from the system: indeed the existence of this potential is involved in the Faraday circuital relation. When the system is referred to a frame not its own, this electric force includes a part due to the convection in the magnetic field. In the system's own frame there is an electric potential $V_0$, and in the other frame the potential is $V$: and it follows as on p. 40 that as the charges are the same in the two fields $V$ is equal to $\epsilon^{-\frac{1}{2}} V_0$ for corresponding positions, where $\epsilon^{-1}$ is $1 - v^2/c^2$. The reason for this factor $\epsilon^{-\frac{1}{2}}$ was that the value of the electron is the convergence towards it of the electric field, but measured in its own frame—that is of what has been named the aethereal force. The equipotential surfaces for an isolated convected electron are thus ellipsoids similar (not here confocal) to its own surface. For any general static system the charges on corresponding elements of surface $\delta S_0$ and $\delta S$, in the two frames, one shrunk relative to the other, are the same: the intensities of force acting on them are $\delta V/\delta n$ and $\delta V_0/\delta n_0$, where $\delta n$, $\delta n_0$ are linear elements normal to $\delta S$ and $\delta S_0$ and $V$ is $\epsilon^{-\frac{1}{2}}V_0$. Thus the forces per unit surface, to be balanced by pressure, are as $\delta V/\delta n\delta S$ and $\delta V_0/\delta n_0\delta S_0$. But by the optical shrinkage, corresponding volumes such as $\delta n\delta S$ and $\delta n_0\delta S_0$ are in the ratio $\epsilon^{-\frac{1}{2}}$, while as above $\delta V$ is to $\delta V_0$ in the same ratio $\epsilon^{-\frac{1}{2}}$. On dividing, the intensity of a balancing pressure comes out the same in the two cases; thus it is independent of the convection, as required above.

The pressure, when regarded as instantaneously propagated, arises from the field Action, after the Lagrangian manner, through a term providing for incompressibility, thus proportional to the square of the divergence of the aethereal displacement. It conforms to the Laplacian equation, after the manner of a static potential.

This aether-pressure, being $2\pi c^2\sigma^2$ or $c^2e^2/8\pi a^4$ at the electron of radius $a$, would be vastly greater at the surface of a proton regarded as a construct on the pattern of an electron of smaller radius. The excess would be a local self-balancing pressure, varying for an incompressible aether as does the potential of a free charge on the surface. Thus this type of proton would possess a field of diminishing aether-pressure extending from the minute nucleus to a radius of the order of that of the whole atom. When an electron travels into this field of excess pressure it is compressed to smaller radius, with inertia increased in proportion. Thus the aether-pressure around a proton based on the hollow electron model would be an essential feature in the dynamics of an orbital atom, leading it far away from the Bohr successful scheme. But on the other hand if there were nothing but negative and positive electrons the aether-pressure would be uniform everywhere and so latent.

*[margin note:] Orbital atom as amended for aether-pressure around the proton.*

# INDEX TO VOLUME I

# INDEX TO VOLUME II

ited States

Printed in the United States
By Bookmasters